Philosophy of Information

Handbook of the Philosophy of Science

General Editors

Dov Gabbay
Paul Thagard
John Woods

ELSEVIER

AMSTERDAM • BOSTON • HEIDELBERG • LONDON • NEW YORK • OXFORD
PARIS • SAN DIEGO • SAN FRANCISCO • SINGAPORE • SYDNEY • TOKYO
North-Holland is an imprint of Elsevier

N·H

Philosophy of Information

Volume 8

Edited by

Pieter Adriaans and Johan van Benthem

ELSEVIER

AMSTERDAM • BOSTON • HEIDELBERG • LONDON • NEW YORK • OXFORD
PARIS • SAN DIEGO • SAN FRANCISCO • SINGAPORE • SYDNEY • TOKYO
North-Holland is an imprint of Elsevier

N·H

North-Holland is an imprint of Elsevier
Radarweg 29, PO Box 211, 1000 AE Amsterdam, The Netherlands
The Boulevard, Langford Lane, Kidlington, Oxford OX5 1GB, UK

First edition 2008

Library of Congress Cataloging-in-Publication Data
A catalog record for this book is available from the Library of Congress

British Library Cataloguing in Publication Data
A catalogue record for this book is available from the British Library

ISBN: 978-0-444-51726-5

Printed and bound in the United Kingdom

08 09 10 11 10 9 8 7 6 5 4 3 2 1

GENERAL PREFACE

Dov Gabbay, Paul Thagard, and John Woods

Whenever science operates at the cutting edge of what is known, it invariably runs into philosophical issues about the nature of knowledge and reality. Scientific controversies raise such questions as the relation of theory and experiment, the nature of explanation, and the extent to which science can approximate to the truth. Within particular sciences, special concerns arise about what exists and how it can be known, for example in physics about the nature of space and time, and in psychology about the nature of consciousness. Hence the philosophy of science is an essential part of the scientific investigation of the world.

In recent decades, philosophy of science has become an increasingly central part of philosophy in general. Although there are still philosophers who think that theories of knowledge and reality can be developed by pure reflection, much current philosophical work finds it necessary and valuable to take into account relevant scientific findings. For example, the philosophy of mind is now closely tied to empirical psychology, and political theory often intersects with economics. Thus philosophy of science provides a valuable bridge between philosophical and scientific inquiry.

More and more, the philosophy of science concerns itself not just with general issues about the nature and validity of science, but especially with particular issues that arise in specific sciences. Accordingly, we have organized this Handbook into many volumes reflecting the full range of current research in the philosophy of science. We invited volume editors who are fully involved in the specific sciences, and are delighted that they have solicited contributions by scientifically-informed philosophers and (in a few cases) philosophically-informed scientists. The result is the most comprehensive review ever provided of the philosophy of science.

Here are the volumes in the Handbook:

Philosophy of Science: Focal Issues, edited by Theo Kuipers.

Philosophy of Physics, edited by Jeremy Butterfield and John Earman.

Philosophy of Biology, edited by Mohan Matthen and Christopher Stephens.

Philosophy of Mathematics, edited by Andrew Irvine.

Philosophy of Logic, edited by Dale Jacquette.

Philosophy of Chemistry and Pharmacology, edited by Andrea Woody and Robin Hendry.

Philosophy of Statistics, edited by Prasanta S. Bandyopadhyay and Malcolm Forster.

Philosophy of Information, edited by Pieter Adriaans and Johan van Benthem.

Philosophy of Technological Sciences, edited by Anthonie Meijers.

Philosophy of Complex Systems, edited by Cliff Hooker and John Collier.

Philosophy of Earth Systems Science, edited by Bryson Brown and Kent Peacock.

Philosophy of Psychology and Cognitive Science, edited by Paul Thagard.

Philosophy of Economics, edited by Uskali Mäki.

Philosophy of Linguistics, edited by Martin Stokhof and Jeroen Groenendijk.

Philosophy of Anthropology and Sociology, edited by Stephen Turner and Mark Risjord.

Philosophy of Medicine, edited by Fred Gifford.

Details about the contents and publishing schedule of the volumes can be found at http://www.johnwoods.ca/HPS/.

As general editors, we are extremely grateful to the volume editors for arranging such a distinguished array of contributors and for managing their contributions. Production of these volumes has been a huge enterprise, and our warmest thanks go to Jane Spurr and Carol Woods for putting them together. Thanks also to Andy Deelen and Arjen Sevenster at Elsevier for their support and direction.

CONTENTS

CONTRIBUTORS

Samson Abramsky
Oxford University, UK.
Samson.Abramsky@comlab.ox.ac.uk

Pieter Adriaans
University of Amsterdam, The Netherlands.
pietera@science.uva.nl

F. Alexander Bais
University of Amsterdam, The Netherlands.
bais@science.uva.nl

Alexandru Baltag
Oxford University, UK.
Alexandru.Baltag@comlab.ox.ac.uk

Johan van Benthem
University of Amsterdam, The Netherlands, and Stanford University, USA.
johan@science.uva.nl

Margaret A. Boden
University of Sussex, UK.
m.a.boden@sussex.ac.uk

John Collier
University of Natal, South Africa.
collierj@ukzn.ac.za

Keith Devlin
CSLI, Stanford University, USA.
devlin@csli.stanford.edu

Hans van Ditmarsch
University of Otago, New Zealand, and IRIT, France.
hans@cs.otago.ac.nz

Fred Dretske
Duke University, USA.
fred.dretske@mindspring.com

J. Michael Dunn
Indiana University, USA.
dunn@indiana.edu

J. Doyne Farmer
Santa Fe Institute, USA.
jdf@santafe.edu

Luciano Floridi
University of Hertfordshire, and Oxford University, UK
l.floridi@herts.ac.uk

Peter Grünwald
CWI, Amsterdam, The Netherlands.
Peter.Grunwald@cwi.nl

Peter Harremoës
CWI, Amsterdam, The Netherlands.
P.Harremoes@cwi.nl

Hans Kamp
University of Stuttgart, Germany.
Hans.Kamp@ims.uni-stuttgart.de

Kevin T. Kelly
Carnegie Mellon University, USA.
kk3n@andrew.cmu.edu

John McCarthy
Stanford University, USA.
jmc@cs.stanford.edu

Maricarmen Martinez
Universidad de los Andes, Colombia.
m.martinez97@uniandes.edu.co

Lawrence S. Moss
Indiana University, USA.
lsm@cs.indiana.edu

Duska Rosenberg
Royal Holloway University of London, UK.
D.Rosenberg@rhul.ac.uk

Hans Rott
University of Regensburg, Germany.
hans.rott@psk.uni-regensburg.de

Martin Stokhof
University of Amsterdam, The Netherlands.
M.J.B.Stokhof@uva.nl

Flemming Topsøe
University of Copenhagen, Denmark.
topsoe@math.ku.dk

Paul Vitányi
CWI, Amsterdam, and University of Amsterdam, The Netherlands.
paul.vitanyi@cwi.nl

Bernard Walliser
Ecole des Hautes Etudes en Sciences Sociales, France.
walliser@mail.enpc.fr

LIST OF COMMENTATORS

Epistemology and Information
Fred Dretske
Commentator: John-Jules Meijer, Utrecht

Information in Natural Language
Hans Kamp and Martin Stokhof
Commentator: Menno Lievers, Utrecht

Modern Trends in Philosophy of Information
Luciano Floridi
Commentator: Jan Heering, CWI

Learning and the Cooperative Computational Universe
Pieter Adriaans
Commentator: Peter Harremoës, Amsterdam

The Quantitative Theory of Information
Peter Harremoës and Flemming Topsøe
Commentator: Robert van Rooij, Amsterdam

The Stories of Logic and Information
Johan van Benthem and Maricarmen Martinez
Commentators: John Perry, Stanford; David Israel, SRI

Algorithmic Complexity
Peter Grünwald and Paul Vitányi
Commentator: Peter Harremoës, Amsterdam

Ockham's Razor, Truth and Information
Kevin T. Kelly
Commentator: Wolfgang Spohn, Konstanz

Epistemic Logic and Information Update
Alexandru Baltag, Hans P. van Ditmarsch and Lawrence S. Moss
Commentator: Donald Gillies, Michigan

Information Structures in Belief Revision
Hans Rott
Commentators: Horacio Arló-Costa, CMU Pittsburgh; Cristiano Castelfranchi, Siena/Rome

Information, Processes and Games
Samson Abramsky
Commentator: Jan van Eijck, CWI Amsterdam

Information and Beliefs in Game Theory
Bernard Walliser
Commentator: Boudewijn de Bruin, Groningen

Information in Computer Science
J. Michael Dunn
Commentator: Luciano Floridi, Hertfordshire and Oxford; Katalin Bimbó, Indiana

The Physics of Information
Alexander Bais and Doyne Farmer
Commentators: Seth Lloyd, MIT, MA; Peter Harremoës, Copenhagen

Information in the Study of Social Interaction
Keith Devlin and Duska Rosenberg
Commentator: Ruth Kempson, London

Information in Artificial Intelligence
John McCarthy
Commentator: Rich Thomason, Pittsburgh

Information and Cognitive Science
Margaret A. Boden
Commentators: Herman Philipse, Utrecht; Harald Baayen, Nijmegen

Information in Biological Systems
John Collier
Commentator: Patrick Forber, Tufts

ACKNOWLEDGEMENTS

This Handbook was produced with the help of many people who made it a rewarding experience far beyond generating just words and pages. We thank Mark Theunissen and Olivier Roy for their help with various technical aspects, including an Amsterdam workshop plus a public event in 2005 which turned a mere set of authors into a group with shared agency. We especially thank Jane Spurr for her indispensable help in the crucial stages of production, showing all her proverbial qualities at their best. And finally, we thank our authors and commentators. This Handbook reflects the efforts of a galaxy of prominent colleagues from many fields, and it is fair to say that the editors have experienced nothing but enthusiastic and generous cooperation, learning a lot in this enjoyable process.

Part A

Introduction and Scene Setting

INTRODUCTION: INFORMATION IS WHAT INFORMATION DOES

Pieter Adriaans and Johan van Benthem

1 INTRODUCTION: WHY THIS HANDBOOK?

Information is a high-frequency and low-content phrase that permeates our ordinary language without attracting much attention, since its meaning has long eroded. Even so, is there more to the notion, and in particular, is there philosophy to it? The editors of the series of 'Handbook of the Philosophy of Science" thought so, when they invited us to contribute a volume, more years ago than we care to remember. But right at the start, a distinction must be made concerning the aim of this text, which comes from the philosophy of language. A Handbook for an established field has a descriptive function in terms of 'what there is', serving as a record of insights and issues. But other, activist Handbooks have a *performative* use, trying to create a new field by a 'let it be'. The present volume is definitely of the second category.

Clearly, one cannot just create an academic discipline by fiat when there is no material to go on. But as it happens, information is a unifying notion across the sciences and humanities, with a backbone of serious mathematical theory. Moreover, there is even a whole discipline of 'informatics' ('computer science', in the unfortunate terminology used in some countries) which studies the structure of representation and transformation of information by machines, but gradually also by humans, and various hybrids of the two. Indeed, universities in several countries have created schools of Informatics or Information Sciences, highlighting the central role of information and its associated themes of computation and cognition in the modern academic landscape.

But this observation again calls for a distinction, this time concerning our purpose. 'Philosophy of information' might mean philosophy of the information sciences, just as there is philosophy of the natural sciences, the life sciences, or humanities. Such methodological reflection on specific fields is absolutely necessary given the explosion of relevant technical research. It will be found in abundance in the pages of this Handbook, with authors engaging in foundational analysis of disciplines such as computer science, economics, linguistics, or physics. But there is also the parallel, and in some ways more ambitious aim of information as a major category of thought within philosophy itself, which might have the potential of transforming that whole field. Indeed, major philosophers like Fred

Handbook of the Philosophy of Science. Volume 8: Philosophy of Information
Volume editors: Pieter Adriaans and Johan van Benthem. General editors: Dov M. Gabbay, Paul Thagard and John Woods.

Dretske or John Perry have argued that perennial questions of epistemology and other core areas of their field can be solved, or at least taken much further, from an information-oriented stance. Beyond that largely analytical tradition, in recent years, Luciano Floridi has been arguing forcefully that a well-conceived philosophy of information might affect the field as a whole, making distinctions like 'analytical' vs. 'continental' irrelevant.

We are sympathetic to both purposes: foundations of the information sciences and transformation of core philosophy, even though the second seems more programmatic than the first right now. In what follows we will discuss some more concrete themes in this Handbook, and then return to these broad purposes.

2 A VERY BRIEF HISTORY OF INFORMATION

Philosophy

The term information is of Latin origin, and authors like Cicero and Augustine used it in the context of Plato's theory of ideas (or forms) and its successors. In particular, Cicero uses 'in-formare' to render the Epicurean notion of 'prolepsis', i.e., a representation implanted in the mind [Capurro and Hjørland, 2003]. In the Middle Ages, a significant shift occurred. In the 15th century, the French word 'information' emerges in colloquial language with a cluster of meanings: 'investigation', 'education', 'the act of informing or communicating knowledge' and 'intelligence'. The technical term 'information' then vanishes from philosophical discourse as though it had lost its appeal. Instead, when the English empiricists went back to the original Platonic inspiration, they coined the term 'idea' (derived from Platonic 'eidos'): "whatsoever is the object of understanding when a man thinks ... whatever is meant by phantasm, notion, species, or whatever it is which the mind can be employed about when thinking" [Locke, 1961, Essay I,i,8]. The philosophical adventures of this notion of 'idea' run from Hume, Kant, and the German idealists up to Husserl and beyond. But like famous Cats through history, 'information' has had many more lives than just one — and to these, we now turn.

Coding

Information has long been associated with language and *coding*. Like theoretical philosophy, the practical ambition to hide information in messages and to then decode these messages with, or without a key dates back to Antiquity [Kahn, 1967]. Cicero's contemporary Julius Caesar used code systems to communicate with his generals, and so did his Hellenistic and Chinese predecessors — and code breaking must be equally old. Reflection on this practice soon followed. The efficiency of assigning shortest codes to most frequent signals has long been known, witness the 10th century Arabic texts on cyphers and decoding via frequencies mentioned in Singh [1999]. With the invention of book-printing in the 15th century, typesetters soon discovered that they needed more *e*s than *z*s in a font. Characteristic frequencies of letters in languages were used to decode simple replacement ciphers.

The 18th century saw the emergence of 'black-rooms' in Europe with the task of encoding and decoding messages for political purposes. With the development of the first electronic communication media, efficient coding systems became of wider use. In 1838, Samuel Morse designed his telegraph code on the basis of a statistical analysis of a Philadelphia newspaper.

Physics

Another step toward the modern concept of information occurred in 19th century physics. When explaining macroscopic events in terms of large quantities of discontinuous microscopic ones, Rudolf Clausius [1850] introduced the statistical notion of *entropy*. Entropy measures the number of different microscopic states a macroscopic system can be in. The entropy in a container is higher if the particles are evenly distributed over the space in the container. With this concept, Clausius formulated what we now call the Second Law of Thermodynamics: a closed system either remains the same or becomes more disordered over time, i.e., its entropy can only increase. The philosopher Henri Bergson once called this "the most metaphysical law of nature" [Bergson, 1998]. Clausius' famous paper ends with a disturbing observation from an informational point of view: "The energy of the universe is constant — the entropy of the universe tends toward a maximum."

Mathematics

In the 20th century, 'information' became a subject for mathematical theory, with the pioneering work of Ronald Fisher on the foundations of statistics [Fisher, 1925]. Indeed all of probability theory might be seen with some justice as a form of information theory, with objective probability closer to physical perspectives, and subjective probability closer to information as used by rational human agents. While this is true, we have decided to concentrate on more specific 'information theories' as such. The pioneering example is the work of Claude Shannon on channel transmission [Shannon, 1948], which may well be most people's association with 'information theory'. Shannon defined the amount of information in a message as the negative base-2 logarithm of the probability of its occurrence from a given source over a given channel — thus measuring in 'bits', which has become a household term.

Actually, this notion fits with the physics tradition via one transformation. The total entropy of two independent systems is the sum of their individual entropies, while the total probability is the product of the individual probabilities. Already Ludwig Boltzmann proposed to make the entropy of a system proportional to the logarithm of the number of microstates it can be in. Shannon's quantitative approach is a momentous shift away from the common-sense conception of meaningful information, but it has been spectacularly successful, witness its use in many chapters of this Handbook.

Computer science

Even so, Shannon's is not the only quantitative version of information to appear in the 20th century. In the 1960s, Kolmogorov, Solomonoff and Chaitin [Solomonoff, 1997; Chaitin, 1987; Li and Vitányi, 1997] developed a new information measure in terms of optimal coding by a computational device. The information in a string X is now an absolute number, viz. the length of the shortest code of a program that would lead a universal Turing Machine to output string X. It can be shown that this definition makes sense independently from accidental features of code language and computing device. Now, highly regular strings will have low complexity, while highly random strings have high complexity. Thus the information content of a string 'reverses' in an obvious way. Kolmogorov complexity is a major tool in computer science (the most authoritative source is Li and Vitányi [1997]), with foundational uses in complexity theory and learning theory.

Again, there are strong links here with the earlier traditions. For instance, strings with low Kolmogorov complexity have low entropy, random strings have high entropy. As we shall see in several chapters of this Handbook, the kinship between thermodynamics and mathematical and computational information theories ensures an almost seamless translation of concepts and applications.[1]

Logic and linguistics

So far, our historical tour of information has taken us from abstract philosophy to hardcore quantitative science and computation. But the 20th century also produced another strand of technical information theories, which will be very much in evidence in this Handbook. For a start, our human information is most obviously expressed in natural language, and indeed, analyzing even the simplest episode of language use quickly reveals a host of subtle informational phenomena. What is a speaker trying to convey, on the basis of what knowledge about the hearer's information? Figuring out this communication-oriented sense of information — which Shannon acknowledged explicitly as significant, but then ignored — involves a study of semantic meaning, knowledge, and other notions that form the domain of linguistics, philosophy, and logic. Modern logical modeling of information dates back to the 1930s with Alfred Tarski's fundamental work on the concept of truth (cf. [Tarski, 1944]). Of course, traditionally, logic already studied informational processes like inference, which work largely on linguistic code, without an explicit model of reality attached. Logical accounts of information tend to be qualitative, in terms of sets and orderings rather than numbers, but they are just as rigorous as quantitative accounts. The chapter by van Benthem & Martinez in this Handbook is a broad survey of sources and varieties. Finally, logic-based accounts of information, too, have strong connections with the foundations of mathematics

[1] In a slogan, information theory is the thermodynamics of code strings, while thermodynamics is the information theory of particles in space. Some authors take this analogy to extremes, viewing black holes and even the universe as a computational system [Lloyd and Ng, 2004].

and computer science, and so we have another major kind of 'information theories' that goes into the total picture of this Handbook.

Broader uses in society

A history of the emergence of 'information' as a staple of public discourse in the 20th century is yet to be written. It appears to be connected with modern intelligence services and communication technologies like the telegraph, and later, the computer. At the end of the 19th century, several countries started systematic collection of military information. The US Office of Naval Intelligence was established in 1882, followed by a Military Information Division — with one clerk and one officer — in 1885. Its task was to collect "military data on our own and foreign services which would be available for the use of the War Department and the Army at large." A modern use of the term information in this context can be found in the *'World Fact Book'*, an annual publication of the CIA:

> *Information is raw data from any source, data that may be fragmentary, contradictory, unreliable, ambiguous, deceptive, or wrong. Intelligence is information that has been collected, integrated, evaluated, analyzed, and interpreted.*[2]

In this compact passage, various broad themes running across this whole Handbook occur in a nutshell, viz. 'information as the act of informing', 'information as the result of the act of informing', and 'information as something that is contained in the message used to inform'. In addition to the impact of this military usage, much broader reflection on information has been generated by recent technologies like the Internet, again related to issues in this Handbook in interesting ways. Just as in 17th century physics, what we see is an intriguing parallelism, and indeed a lively stream of interaction, between scientific, technological and social developments [Castells, 1996; Kahn, 1967; Capurro and Hjørland, 2003].

Philosophy once more

While scientific and social developments made information a crucial notion, little of this penetrated into modern philosophy. Although Gödel's incompleteness results, the Church-Turing thesis, and Turing's ideas on machine intelligence generated much philosophical debate, this did not lead to widespread philosophical reflection on the notion of 'information' itself. To be sure, there were some serious philosophical responses to Shannon's theory around 1950, witness Bar-Hillel and Carnap [1953], which took a closer look at the interplay of what they saw as equally viable quantitative and logical notions of information, starting off a tradition in 'confirmation theory' continued by Jaakko Hintikka, and many others.[3]

[2] `https://www.cia.gov/library/publications/the-world-factbook/docs/history.html`

[3] Cf. [Hintikka, 1973; Kuipers, 2000]. Our companion publication "Handbook of the General Philosophy of Science" presents the current state of the art in confirmation theories.

Solomonoff, who is one of the founding fathers of algorithmic information theory, and whose work was partly motivated by philosophical questions concerning the nature of probability and the induction problem, studied with Carnap in the fifties. Until now this work never percolated to mainstream philosophy. 'Information' is not mentioned, for instance, in the well-known history of logic [Kneale and Kneale, 1962], nor does it have a lemma in Paul Edwards *"Encyclopedia of Philosophy"* of 1967. Things started changing around 1980. Fred Dretske gave information theory its due in epistemology [Dretske, 1981], and the same is true for the work of Jon Barwise and John Perry in the philosophy of language [Barwise and Perry, 1983]. On the latter view, triggered by ideas from cognitive 'ecological psychology', logic should study the information flow in rich distributed environments with physical and human components. All these philosophers use the notion of information to throw new light on classical issues of knowledge, objectivity, representation and 'aboutness', thus facilitating 'second opinions' and new solutions. Finally, we already mentioned Luciano Floridi's seminal work on a new 'Philosophy of Information' at the start of the 21st century [Floridi, 2003A; 2003B].

Modern interdisciplinary trends

This historical sketch provides the background for the main themes that the reader will find in this Handbook. But maybe we should also explain our cast of authors, which mixes philosophers with practitioners of other disciplines. This combination is well in line with what has happened over the last two decades in foundational studies of information, with topics moving in and out of philosophy. Indeed, Barwise and Perry already started the interdisciplinary *'Center for the Study of Language and Information'* (*CSLI*) at Stanford, a hot-bed of encounters between philosophers, linguists, computer scientists, mathematicians, and psychologists. Its current director Keith Devlin is one of our Handbook authors.

At the same time, in Europe, natural language semantics took an informational turn. Jeroen Groenendijk and Martin Stokhof[4] introduced information of language users in defining meanings of key linguistic constructions, including speech acts like questions. With Peter van Emde Boas, a pioneer in the study of parallels between natural and programming languages, and Frank Veltman, who had developed an update semantics for conditional expressions, they redefined meaning as 'potential for information update' based on abstract computation in appropriate state spaces. Similar ideas underlie the influential discourse representation theory of Irene Heim and Hans Kamp. Details on this linguistic paradigm shift may be found in the chapter by Kamp and Stokhof in this volume. By 1986, this led to the foundation of the 'Institute for Language, Logic and Information' in Amsterdam, better known today as the *ILLC*, the *Institute for Logic, Language, and Computation.* Similar initiatives include the European Association for Logic, Language and Information, and its annual *ESSLLI* Summer Schools, as well as its international off-spring in other continents.

[4]Editors of the companion volume *Handbook of the Philosophy of Language* in our series.

One more major interdisciplinary strand in the 1980s was the rise of *epistemic logic* describing agents' knowledge 'to the best of their information'. Epistemic logic was first proposed by Jaakko Hintikka [Hintikka, 1962] as a tool for philosophers, and taken further by David Lewis [Lewis, 1969] and Robert Stalnaker [Stalnaker, 1984]. Epistemic logic was invented independently by Robert Aumann in economics in the 1970s, in his eventually Nobel-Prize winning analysis of the foundations of Nash equilibrium through common knowledge of rationality. Since the 1980s, when Joe Halpern and colleagues at *IBM* San Jose started the still-thriving *TARK* conferences on *'Reasoning about Knowledge and Rationality'*, while themselves making major contributions to the study of information and communication, the field has lived at the interface of computer science, philosophy, and economics.[5]

In the 1990s, a further notable new force was the rise of 'Informatics': a new academic conglomerate of disciplines sharing a natural interest in information and computation as themes cutting through old boundaries between humanities, social, and natural sciences. By now, there are Informatics schools and institutes in Bloomington, Edinburgh, Philadelphia (IRCS), and Kanazawa (JAIST), to name a few, and the founding dean of such a School at Indiana University, Mike Dunn, is one of our Handbook authors.[6]

While all this organizational and social information may grate on ears of traditional philosophers (how far away can the Mammon be?) — to us, it seems highly relevant if Philosophy of Information is to have a significant future as a vibrant endeavour with many sources.

3 INFORMATION THEORIES, THREE MAJOR STRANDS

We have sketched a rich history of information studies ranging through the whole academic spectrum into society. The reverse side of this wealth is the diversity. What do all these themes and fields, worth-while as they may be *per se*, have in common, except at best a metaphor? This impression of diversity may even be reinforced when the reader gets to our actual chapters. Before sketching their content, then, let us first draw a few lines confronting some doubts and worries.

Just a metaphor?

'Information' may be a ubiquitous phrase, and even a real phenomenon, and yet it might be just a metaphor leading to vague philosophy, like 'system' or 'game' have done in the past. The real situation seems less bleak, however. As with terms like 'energy' or 'money', there is indeed a general usage of information where little can be said beyond generalities. Energy is what drives inanimate processes and

[5]Epistemic logic as information theory is a new view, proposed in [van Benthem, 2006], and the chapter by van Benthem and Martinez on 'Logic and Information' in this Handbook.

[6]Dunn's chapter in this Handbook provides much additional detail beyond our historical sketch, while also mapping out connections to major approaches to information in the foundations of logic and computer science.

animate activities, and what allows us to relate the effort involved. Money is what makes transactions possible without undue real transportation of goods. In both cases, general usage is backed up by pockets of precise use in expert circles, grounded in mathematical theory: thermodynamics, or economics. This interplay causes no real problems: we understand the broad usage, and we specialize and make it more precise as needed. These lessons transfer to information.[7] Indeed, when Keith Devlin says tongue-in-cheek to broad audiences that "information is the tennis ball of communication", he actually formulates a very similar role for information as for money, viz. as the abstract currency that gets transferred when people say or observe things. And he also gets the idea right that information usually arises in complex multi-agent settings, where interaction is of the essence. But on that topic, we will have more to say below.

Go for a larger family of notions?

Can information stand on its own in conceptual analysis? Compare the case of *knowledge*. Most standard philosophical analyses, mainstream like Plato's, or more avant-garde like Dretske [1981] or Nozick [1978], make it part of a larger cluster of notions, involving also *truth, belief, information (...)*, and perhaps even *counterfactuals*. We are usually not after single concepts in philosophical analysis: we are also charting their closest relatives and friends. This is an issue on which we have not arrived at a final position. Natural candidates for a clan of related concepts — not identical, but naturally intertwined — in our case would be: *information, probability, complexity, meaning, coding,* and *computation*. Our Handbook does not really take a stand here. While using information as its running theme, it does give extensive coverage to many of these related notions.

Three major concepts of information

One might assume a priori that there is just one notion of information. But one striking feature, even in our brief history, is the existence of respectable, but very different mathematical views of what makes it tick! We have seen approaches, roughly, from logic, physics, and computer science. Should we first assure ourselves that these all amount to the same thing? Perhaps not. The plurality of mathematical theories of information may reflect a genuine diversity in the concept itself, which needs to be frankly acknowledged.

Compare the case of probability, another crucial foundational notion across the sciences whose precise nature has been under debate ever since its rise in the 17th century. Carnap 1950 proposed a famous conceptual dichotomy between two irreducible, complementary notions: *Probability-1* for objective frequency, and *Probability-2* for subjective chance, and this is still widely seen as a major duality

[7]That 'money' leads the way need not be a bad thing, if we recall Karl Marx' famous saying that 'Logic is the Currency of the Mind'. A mere slogan perhaps: but, how rich and suggestive!

between two different legitimate concepts in both mathematics and philosophy.[8] And legitimate stances on this concept do not even stop here. One can think of Ludwig von Mises' views on randomness as a *Probability-3*, explaining statistically random sequences of outcomes via algorithmic notions of recursive place selection.

Whatever one's final verdict, it seems uncontroversial that there are three main stances in the technical literature on information theories, which we dub

Information-A Knowledge, logic, what is conveyed in informative answers

Information-B Probabilistic, information-theoretic, measured quantitatively

Information-C Algorithmic, code compression, measured quantitatively

Over-simplifying a bit, A is the world of epistemic logic and linguistic semantics, B that of Shannon information theory, linked to entropy in physics, and C that of Kolmogorov complexity, linked to the foundations of computation. We do not feel that these are opposing camps, but rather natural clusters of themes and research styles. Thus, we felt that all of these need to be represented in our Handbook, since only their encounter gives us the proper canvas for philosophical enquiry.

A first comparison

What are the paradigmatic informational scenarios described by these approaches? We start with a first pass, and draw a few comparisons.

(A) The typical logic-based setting lets an agent acquire new information about what the real world is like, through acts of observation, linguistic communication, or deduction. A simple example would be an agent asking a question, and learning what things are like from an answer. Thus, three features are crucial: *agents* which represent and use the information, *dynamic events* of information change, and *'aboutness'*: the information is always about some relevant described situation or world. Here, we measure quality of information qualitatively in terms of new things agents can truly say: a quantitative measure may be handy, but it is not required. Finally, the formal paradigm for the theory is mathematical or computational logic.

(B) By contrast, the typical Shannon scenario is about a source emitting signals with certain frequencies, say a 'language' viewed as a global text producer, and the information which a receiver picks up from this is measured in terms of expected reduction of uncertainty. This is the sense in which seeing a particular roll of a fair die gives me 3 bits of information. No specific agency seems involved here, but the scenario does analyze major features of communication which are absent on the logical approach, such as *probability* of signals (i.e.,

[8]Carnap tried a similar move with 'information' in the early 1950s, juxtaposing Shannon's quantitative notion with his own qualitative logical information spaces. (Cf. [Kohler, 2001].)

the long-term behaviour of a source, maybe as viewed by the receiver), *optimal coding*, and *channel capacity*. Finally, mathematical paradigms for the theory are probability theory and physics.

Clearly, scenarios A and B are not mutually contradictory. They are about different aspects of sometimes even one and the same scenario of information flow, omitting some and high-lighting others. Still, the two stances meet at various points. For instance, coding systems relate to the efficiency of natural language (or lack thereof), signal probability relates to reliability of sources (also relevant to logicians), and Shannon theorists often use question-answer scenarios to motivate their notion, in terms of minimal numbers of questions to pin down the truth.

(C) Next, take the basic Kolmogorov scenario. We receive a code string, and ask for its informational value. The answer is the algorithmic complexity of the string, defined as the length of the shortest program that computes it on some fixed universal Turing machine. While this looks like a totally different setting from the preceding two, there is a direct link to Scenario B. Working with the enumerable set of all 'prefix-free programs', we can easily find an associated probability distribution.[9] In this way, the shortest program for a string becomes an optimal code in Shannon's sense. Thus the following 'traffic' arises: Information-B starts with the notion of probability as fundamental and derives an optimal code. Information-C starts with the notion of shortest code as fundamental and derives an a priori probability from it. Further details may be found in the chapters of Grünwald & Vitányi, Topsøe and Harremoës, and Adriaans in this volume.

Stating technical transformations between notions of information is one thing, understanding their philosophical consequences another. For instance, consider the following intriguing questions. What is the status of a computational device like a Turing machine in grasping the available information in Nature [Wolfram, 2002]? Does algorithmic complexity still apply if we go from computer code to datasets of observations? Is Nature a computing agent sending us encoded messages? To some computer scientists [Schmidhuber, 1997], Information-C is indeed the basis for a general theory of induction that commits us to 'metaphysical computationalism'.

Relations between Information-C and Information-A are even more delicate. The latter seems closer to information flow in human settings and purposeful activities. But here, too, some researchers see algorithmic data compression as a universal principle governing human-level information flow, leading to what may be called 'cognitive computationalism': the idea that the human brain is a universal computational device [Pinker, 1997; Chater and Vitányi, 2003; Wolff, 2006]. If an agent has background knowledge, in the form of optimal descriptions of a set of objects (e.g., animals), then identifying such an object (e.g., a cow) via a picture amounts to finding a shortest algorithmic description of the picture conditional on

[9] By Kraft's Inequality, for any finite or infinite sequence $l_1, l_2, \ldots$ of natural numbers, there is a prefix code with this sequence as the lengths of its binary words iff $\sum_n 2^{-l_n} \leq 1$.

that background knowledge. While not uncontroversial, this philosophical view, too, has interesting consequences, and even some degree of empirical support.[10]

This brief discussion may suffice to show that Information-A, Information-B, and Information-C make sense on their own, while engendering many intriguing interactions. As editors, we do not have a final view on the relation between these approaches, and whether a Grand Unification is possible. We do feel that they need to be compared in an open fashion, questioning even the usual labels 'qualitative' vs. 'quantitative'.[11] Our own sense, developed partly thanks to insights from our authors in this Handbook, is that B and C are close, while the relation to A-approaches is much less settled. Even so, the B scenario clearly shares some features with A-type views of information update, and thus one might view Shannon's theory as go-between for A and C. But still, we may have to 'do a Carnap' in the end, putting the three side-by-side, just as we saw with probability.[12]

4 THE CHAPTERS OF THIS HANDBOOK

This is a good point to interrupt the editors' story, and let another voice speak for itself, viz. the list of chapters of this Handbook. The idea behind its composition has been to put two things at the reader's disposal. One is a Grandstand View of serious studies of information in the various sciences, and the styles of work as done by leading practitioners. The other item offered are a number of major leads toward a philosophy of information, written by distinguished philosophers. The latter include both senses that we have described earlier: philosophical foundations of the information sciences, and also informational turns inside philosophy itself. We give some cameo descriptions, while also briefly 'presenting' the authors.

After this editorial Introduction, the Handbook starts with a first Part on *Philosophy and Information*. The opening chapter by Fred Dretske, a pioneer in bringing information theory to philosophy, discusses how the notion of information plays in epistemology, and merges well with current debates. Next, Hans Kamp and Martin Stokhof examine the role of information in the philosophy of language and the theory of meaning, drawing upon their long experience in philosophical logic and formal semantics at the interface of philosophy and linguistics. Pieter Adriaans, a classical philosopher turned machine learning expert (amongst other things), continues with major issues in the philosophy of learning, exploring in particular the knowability of the physical universe from a computational standpoint. Finally, Luciano Floridi, mentioned several times already, maps out

[10]The most efficient current program recognizing musical styles uses algorithmic information theory [Cilibrasi and Vitányi, 2005]. Adriaans [2008] even proposes an algorithmic esthetics.

[11]Indeed, all three types can have more qualitative or quantitative versions, witness Carnap's Inductive Logic on the A-side, or the basic 'representation theorems' of Shannon information theory on the B-side.

[12]Indeed, von Mises third probability intuition in terms of randomness and computable 'place selection' does look a bit like an algorithmic Type C approach to information, through its links with recursion theory in the work of Per Martin-Löf, Michiel van Lambalgen, and others.

the broader agenda for a philosophy of information as he has been advocating it over the recent years.

Next comes a foundational part on *Major Technical Approaches.* Mathematicians Fleming Topsøe and Peter Harremoës give a lucid exposition of Shannon's quantitative theory of information and its embedding in general mathematics. Next, Peter Grünwald and Paul Vitanyi, leading theorists in the foundations of Kolmogorov complexity, statistics, and recently also quantum information, follow up with a state-of-the-art account of algorithmic complexity theory, including its connections with probability and Shannon information. Finally, logicians Johan van Benthem and Maricarmen Martinez, representing the different traditions of epistemic logic and situation theory, investigate the role of information in logic, and describe what this discipline has to offer by way of general theory.

Our third part, *Major Themes in Using Information,* zooms in on some key themes in the foundations of 'informatics'. Kevin Kelly, who has been instrumental in bringing topology and recursion theory to the philosophy of science, writes about learning, simplicity, and belief revision, with Occam's Razor as a running theme. Logicians Alexandru Baltag, Hans van Ditmarsch, and Lawrence Moss describe knowledge and information update as studied in recent 'dynamic epistemic logics', showing how informational themes are creating new logics right now. Hans Rott, one of the architects of belief revision theory, follows up on this with a formal account of how agents change their beliefs when triggered by new information, and discusses optimal cognitive architectures for this. Moving to other information-producing activities, Samson Abramsky, a leader in the current interest in 'information dynamics' in computer science, discusses the information flow in computation, drawing upon recent game-based models of interactive processes, with surprising connections to quantum information flow in physics. Information in games and rational agency per se is then discussed in depth by Bernard Walliser, an economist who has published extensively on the conceptual foundations of game theory.

The final part of the Handbook collects a number of representative case studies of *Information in the Sciences & Humanities.* Mike Dunn, logician, philosopher, computer scientist, and prime mover in the formation of Indiana University's School of Informatics, surveys the various uses of information in computer science, from Scott 'information systems' to algebraic theories of data structures and informational actions. Well-known physicists Sander Bais and Farmer then present a masterful treatment of the notion of information in physics, opening up to connections with Shannon information and Kolmogorov complexity. Information in the social sciences is represented by the chapter of Keith Devlin and Duska Rosenberg, who give an in-depth transaction model for linguistic communication using tools from situation theory. Next, John McCarthy, one of the founders of AI, surveys the uses of information in artificial intelligence, stressing the role of representation, context, and common sense reasoning, and throwing out a list of challenges to philosophers. The final two chapters move to the natural world of the life sciences. Margaret Boden discusses the role of information in cognitive psychology, including recent neuro-science perspectives. And the last chapter in our tour of Academia is John Collier's critical study of current uses of information and coding in biology, whose repercussions are all around us in bio-technology and its hybrids with computer science.

In addition to the authors, we should also mention the official commentators, who have played an important role in this Handbook. Each chapter has been read by its assigned commentator, and their extensive responses and the ensuing discussions have kept authors alert and fair to what has been achieved in their fields. The commentators behind this Handbook are as distinguished and diverse a group as our authors, including prominent philosophers, computer scientists, linguists, and psychologists, and their names will be found in the separate chapters.

Of course, no system is fool-proof, and as with every Handbook, the editors might have made some choices of chapters differently, while there are also bound to be strands in the field that remain under-represented. One can look only so far. Even so, we feel that the present collection provides ample material for substantial reflections, and in the rest of this Introduction, we present a few of our own.

5 INTEGRATIVE THEMES AND NEW QUESTIONS

When collecting the material for this Handbook we have toyed for a moment with the ambition of providing one unified account of information that would satisfy all our authors, and even a more general audience. While this has proved somewhat illusory at our current state of enlightenment, we do feel that we are now in a much better position to draw some main lines. Here are a few themes that we see running through many of our chapters, found not by looking top-down at what information should be, but bottom-up, looking at stable patterns in existing research. We start by re-analyzing the three streams we identified earlier, 'unpacking' these paradigms into a number of general themes that seem relevant to information generally. In this manner, we hope to find a unity through themes instead of 'all-in' packages.

Logical range and reduction of uncertainty

One simple, yet powerful theme in many of our chapters is this — and it may even be the common sense view. Information may be encoded in a *range of possibilities*: the different ways the real situation might be. For instance, at the start of a card game, the range consists of the different possible deals of the cards. Numerically, this view reflects in the standard representation of information in bits being the (weighted) base-two logarithm of the size of the range. More dynamically, on this view, new information is that which reduces my current range — that is: more information leads to a smaller range. This is the standard logical sense of information in which a proposition P *updates* the current set of worlds W to $\{w$ in $W|w$ makes P true$\}$. This notion is relative to a 'logical space' describing the options. It is also relative to agents, since the update happens to what they know about the world. In our reading, this is the main notion of information used in our Handbook chapters by Baltag, van Ditmarsch and Moss, van Benthem and Martinez, Dretske, Kamp and Stokhof, McCarthy, Rott, and Walliser. It is an A-type account in our earlier sense, which revolves around agents' logical spaces of alternative options, set up for some purpose (information is *"for"* something), zooming in on some yet unknown actual situation (the latter is what the information is *"about"*), and new information typically has to do with dynamic events of observation, communication or inference updating the current state.

Yet there are also links with B and C types of information. If a range of n messages has maximum Shannon entropy, the optimal code for each message takes $log_2 n$ bits. And as for update, if I know that John lives in Europe, I need some 30 bits to identify him, but after new information that he lives in Amsterdam this effort is reduced to 20 bits. And as to Information-C, the shortest program p for a string x in the sense of Kolmogorov complexity can also be interpreted as a measure for the smallest set of $2^{|p|}$ possible worlds that we need to describe x. Thus, 'range' truly seems an integrating feature across information theories.

Correlation and channel transmission

The next pervasive notion in our Handbook emphasizes another key aspect of information flow, viz. the correlation between different systems that drives it. One situation carries information about another if there is a stable correlation between the two. This is the sense in which dots on a radar screen carry information about airplanes out there. Note that this information may be there, even when there is no agent to pick it up.[13] In philosophy, this sense of information is central to the earlier-mentioned work of Dretske and Barwise and Perry, who were inspired by Shannon's paradigm, and who stress the essential 'situatedness' and 'aboutness' of information. Indeed, correlation seems of the essence there, and the view of information transmitted across less or more reliable channels is dominant in our chapters by Bais and Farmer, Boden, Collier, Devlin, Dretske, Kelly, Topsøe and Harremoës. One of its key features is that information is crucially *about something*, and thus a relation between a receiving situation and a described, or sending situation. In this scenario, the 'quality' of the information depends essentially on the reliability of the correlation. But it is also possible to find these same concerns implicit in our more 'A-type chapters'.

The two themes identified so far play in various fields. For instance, our chapter on logical theories of information finds range and correlation right inside logic, and shows how they are highly compatible there, combining into a single mathematical model. But also, Shannon's information theory contains aspects of both range and correlation. It is definitely about reducing ranges of uncertainty — in a quantitative manner asking for the *average* reduction of uncertainty, summarizing many possible update actions. But is also crucially about correlation between

[13]Thus, unlike in the classic Procol Harum song 'Homburg', `http://www.lyricsdomain.com/16/procol_harum/homburg.html`, in situation theory, "signposts" do not "cease to sign" when there are no human beings left on our planet.

a source and a receiver across a channel. In algorithmic information theory the notion of correlation seems less pregnant at fist sight, as Kolmogorov complexity is a priori and universal, being a measure of 'self information' of a data set. But even there, in principle, it is always correlated with an abstract computational device, its source.[14] More technically, correlation between data sets and what they describe has been studied in terms of 'conditional Kolmogorov complexity', with the reference universal Turing machine providing the 'channel' in the above-discussed correlational sense.

Temporal dynamics and informational events

But there are further general themes in the A, B, and C stances that seem of general significance for information. In particular, the Shannon scenario and correlation generally, seems to presuppose a *temporal dynamics.* Information is not a one-shot relation between single events: it presupposes an objective pattern of matched events over time, and this frequency information is one essential function of the probabilities employed.[15] This temporal perspective is also in evidence on the logical side, and it even plays there in two different ways. Locally, the flow of information is driven by specific informational events that produce it, such as an observation, or an answer to a question.[16] But there is also a global long-term process of repeated observations, which establishes reliability and information flow in some higher sense. In computer science terms, the local dynamics calls for an account of stepwise informational actions, while the global dynamics calls for a *temporal logic*, or a statistical dynamical systems model, of long-term program behaviour over time. We have nothing to add to the latter feature here, but the local dynamics bears some separate discussion, since it seems intimately related to our very understanding of information. We start with the basic information-handling process, and discuss some generalizations later.

---

[14] Again, this at once raises philosophical questions. Kolmogorov complexity claims to be a priori and objective. But the price is high: the notion is asymptotic and non-computable. Three key results from Turing govern this setting: (a) Enumerability: there is a countable number of Turing machines, (b) Universality: there is an unlimited number of universal Turing machines that can emulate any other Turing machine, (c) Undecidability: there is no program that can predict, for all combinations of input X and Turing machines M, whether M will stop on X. A universal Turing machine can be defined in less than 100 bits. Given all this, we can select a small universal Turing machine U on which any digital object O will have a shortest program. On the C-view, the length of this program will be the 'objective' amount of information in O. This program cannot be found by any effective computational process, because of point (b), but the work of Solomonoff, Kolmogorov and Levin shows that under certain constraints we may still use all this as an adequate information measure.

[15] Of course, these probabilities also have a subjective aspect, since they may be seen as describing agents' views of the situation.

[16] Note that performing an experiment is asking a question to Nature, cf. [Hintikka, 1973].

Information and computation

One can teach a course on information theory without mentioning computers, and conversely, one can treat computation theory without reference to information. Yet the interplay of information with computation as a way of producing or extracting it is subtle and challenging. Here is one issue which plays in several chapters of this Handbook. Due to the 'data processing inequality' (see [Cover and Thomas, 2006]) deterministic computational processes do not create information: though they may discard it. Thus, the amount of information in a computational system can never grow on B- or C-type views! Indeed, the only processes in our world that generate maximal information-rich sets are pure random processes like quantum random number generators. A string generated by such a device will with high probability have maximal Kolmogorov complexity. And yet, our world seems a very information-rich place, and clearly not all information is random. Many natural processes generate new information by a non-deterministic device under deterministic constraints. Thus, evolution and growth seem to create complexity 'for free', and though we can simulate them on a computer, the merit of these simulations in terms of the creation or annihilation of information is not clear. The chapters by Abramsky, Bais and Farmer, Topsøe and Harremoës, Floridi, and Adriaans contain a wealth of material shedding light on the general interplay of information and computation, but key issues like the one mentioned here are far from settled. It may call for a deeper understanding of connections between B- and C-type accounts with A-type accounts.

The process stance: information in action

Next, generalizing from computation in a narrower sense to cognitive activities of agents, let us develop a methodological idea from computer science — and philosophy — in its appropriate generality. In a computational perspective, it makes little sense to talk about static data structures in isolation from the *dynamic processes* that manipulate them, and the tasks which these are supposed to perform. The same point was made in philosophy, e.g., by David Lewis, who famously said that 'Meaning Is What Meaning Does'. We can only give good representations of meanings for linguistic expressions when we state at the same time how they are *used* in communication, disambiguation, inference, and so on. In a slogan: *structure should always be studied in tandem with a process!* The same duality between structure and process seems valid for information, and indeed, all of our stances, and al of our chapters, have specific processes in mind. *No information without transformation*! The logical A-stance was about information update, the Shannon B-view stressed transmission events, and the Kolmogorov $C-$view is all about computational activities of encoding and decoding. And these process scenarios are not just 'background stories' to an essentially static notion of information, they are right at the heart of the matter.

But then, *which processes* would be paradigmatic for the notion of information? The chapters of this Handbook show a great variety: from questions and answers

(Kamp and Stokhof), observations (Baltag, van Ditmarsch and Moss), communication (Devlin and Rozenberg), learning (Adriaans, Kelly), belief revision (Rott), computation (Abramsky), and inference (van Benthem and Martinez) to game-theoretic interaction (Walliser). And this list generates many questions of its own. What does information *do* for each process, and can we find one abstract level of representation for them that stays away from details of implementation? Also, some of these processes concern single agents, while others are intrinsically multi-agent 'social' events. Is the basic informational process a multi-agent one, with single-agent activities their 'one-dimensional projections'?[17] We will not attempt to answer these questions here, but we do think they are central to a philosophy of information that bases itself on the best available information-theoretic practices.

Information as code and representation

While the preceding tandem view seems to high-light the dynamic processes, it equally well forces us to think more about the details of representation of information. Here is where the linguistic study of natural language has much to offer (see our chapter by Kamp and Stokhof), in particular in connection with A-type views of information. In another setting, the chapter by Devlin and Rozenberg highlights subtleties of linguistic formulation in informational transactions in social settings. But other abstraction levels, even when far removed from 'meaningful discourse', carry insights of their own. Recall the mathematical fine-structure of our C-stance. The Kolmogorov complexity of a data set was the length of the shortest program that generates this data on a computer.[18] Now consider an apparently strange feature here, viz. the definition of *randomness*. A string X is random if it cannot be compressed, i.e., no program shorter than the length of X produces X on our universal Turing machine. Thus, random strings have the highest amount of information possible: say, a radio transmission that only contains noise! This runs head-long into the idea of information as 'meaningful'. But it does reveal an intriguing connection elsewhere, with thermodynamics as in the chapter of Bais and Farmer. Kolmogorov complexity can be viewed as a theory of string entropy, with random strings as systems in thermodynamic equilibrium. This suggest intriguing equivalence relations for translating between complexity theory and physics, for whose details we refer to Adriaans [2008].[19]

---

[17] For instance, is 'learning' as in formal learning theories just a one-agent projection of a shared activity of a two-agent system {Learner, Teacher}? Likewise, is a logician's 'proof' as a formal string of symbols the zero-agent projection of a multi-agent interactive activity of argumentation?

[18] Here is one more common sense way to understand the different stances here. You are at an information booth at the airport, trying to book a hotel. The information in statements like "There is a room free in the Ritz", is probably best analyzed in A- or B-terms, but when the official shows you a city map that tells you how to get to the Ritz, something else is going on. The map contains information which can be measured: a detailed map contains more information then a sketch. The computer file that the printer uses to produce a detailed map contains more bits than the file for a large scale one. This is the structure measured by Kolmogorov information.

[19] Here is a summary. Consider these 'identities': (a) Length $|x|$ of a string $x \approx$ the internal energy U of a system, (b) Kolmogorov Complexity $C(x) \approx$ Entropy S of a system, (c) Ran-

This concludes our list of general themes, showing how systematic reflection on the various stances in information theory raises questions of interest to all.

6 CONCLUSION, AND THE PURPOSE OF THIS HANDBOOK ONCE MORE

The main scientific ingredients

This Handbook presents a panorama of approaches to information, drawing for its methods on at least three major scientific disciplines: *logic*, *computer science*, and *physics*. It might be thought that all of these strands have already been integrated in current broad academic 'informatics' environments, but this seems more of a hope than a reality so far. In particular, while it is true that, over the 20th century, computer science has yielded a host of fundamental insights into the representation and processing of information,[20] its foundations remain an exciting open field. It may even be true eventually that the complete scientific background for the foundations of information should include *cognitive science*, but we have not chosen this as major focus in our scheme yet — though we do have chapters by Boden on information in cognitive science, and Collier on biology.

From unification to co-existence

What we have not achieved in this Handbook is a Grand Unification of all major technical approaches to information. We do not know if one is possible, and we sometimes even wonder whether it would be desirable. What does happen here is that different bona fide traditions meet, and what we hope will happen is that they find a common language, and a research agenda including new shared concerns. We think this is possible because our analysis in the preceding sections, largely based on the contents of this Handbook, has not revealed incompatibility, but rather a *complementarity* of perspectives.

---

domness deficiency $|x| - C(x) \approx$ the Helmholz free energy $U - TS$ of a system (T = absolute temperature), (d) Random string $\approx$ system in equilibrium. Here the randomness deficiency of a string is its length minus its Kolmogorov complexity, just as the free energy of a system is the internal energy minus its entropy by equal temperature. Free energy is linked with meaningful information. A system in equilibrium cannot do any work, just as a random string does not contain any meaningful information. Thus the meaningful information in a string may be defined as follows. The *facticity* $F(x)$ of a string x is the product of the normalized entropy $C(x)/|x|$ and the normalized randomness deficiency $1 - (C(x)/|x|)$. The term is motivated by Heidegger's notion of 'die unbegrundbare und unableitbare Faktizität des Daseins, die Existenz..." [Gadamer, p. 240]. If p is the shortest program that generates x on U, then p is by definition a random string. Nothing can be said about it or derived from it other than that $U(p) = x$. The string p is completely meaningless outside the context of U. Kolmogorov complexity maps all meaningful strings on to meaningless random strings.

[20] Just think of automata theory, complexity theory, process theories, AI: the list is impressive, and it immediately belies the modest 'handmaiden' role that some want to relegate the field to.

Successful merges

Concrete examples of the potential for merging will be clear to any serious reader of our chapters — if only, because many ingredients of one paradigm make immediate sense in another. For instance, one might, and probably should, introduce correlationist information *channels* in a more realistic logical range view, and several proposals to this effect were made recently. Or, our chapter on Shannon theory involves questions and answers at crucial stages, and introducing explicit *dynamic multi-agent* perspectives in B- and C-type accounts of information might be worth-while. This would reflect a recent general move toward studying 'interaction' as a basic phenomenon in the foundations of logic and computer science. But many further desiderata emerge from the material collected here. For instance, various chapters make surprising new moves towards *physical models* of information, including those by Abramsky and Adriaans. This connection seems important, and it might lead to possible new academic alignments. Finally, even the austere code-based view of information really occurs throughout this book, witness the chapters on natural language, on computation, and on logic. Indeed, the latter discusses the related 'scandals' of computation and deduction: which reflect long-standing philosophical discussions. How can a code-based process of valid computational or inferential steps generate information? How can we harmonize algorithmic and semantic views? The reader will find some answers in the relevant chapters, including links to the foundations of logic, Hilbert's proof theory, and Gödel's completeness theorem — but again, the issue is far from settled.

Indeed, fruitful combinations of the different perspectives in this Handbook already exist. Useful combinations of logical range spaces and Shannon-style correlation measures co-exist in modern semantics for natural language: cf. [van Rooij, 2004] on questions and answers, or [Parikh and Ramanujam, 2003] on general messaging. Indeed, a recent special issue of the *Journal of Logic, Language and Information* [van Benthem and van Rooij, 2003] brought paradigms together in the following simple manner. Just consider one basic informational scenario like a question followed by an answer. Now ask a logician, an information theorist, and an algorithmics expert to analyze the very same scenario. It was highly instructive to see what features they picked up on as important, but also that, despite their differences in concerns and methodology, no deep contradictions arose.[21]

Creative tensions

Indeed, fostering some residual differences can be creative. Consider the editors themselves. Their 'gut views' on information are different. Adriaans is on the quantitative side, van Benthem on the qualitative one. At first sight, this seems a sharp divide. Scientists and engineers love computation, since we can now 'compute with information'. Philosophers and logicians feel that all the content and

[21] See also [Kooi, 2003] for a case study of strategies for question answering combining ideas from logic, probability theory, and information theory in a practical manner.

drama of an informational event is 'flattened' into a one-dimensional number. Messages with totally different content can become equivalent in this way.

But this difference in direction can easily become a productive force. Even from a logical point of view, adding numerical measures seems relevant and natural, and many hybrids exist of logical and probabilistic systems for various cognitive tasks. Thus, there are already many areas of fruitful confrontation between logical and quantitative, often probabilistic methods. Consider evolutionary game theory or current methodological debates in ethics, where the role of norms and moral behaviour can be analyzed either in traditional logical terms, based on conscious reasoning from moral principles,[22] or as inevitable statistical equilibrium behaviour in large-scale long-term populations. Indeed, from a more practical viewpoint, Adriaans [2007] points out that in most realistic scenarios involving informational events, logical micro-descriptions are either unavailable, or the cost of computing them becomes prohibitive. In that case, the statistical approach is *the only way we have* of finding essential macro-features of the relevant process. The same might be true for information on a large scale and in the long run — and here, despite the, perhaps, one-dimensionality of the numerical bit measure, it has amply shown the same 'unreasonable effectiveness' that mathematics has for Nature in general.[23]

Philosophy of information once more: two levels of ambition

Let us now take all this back to the title theme of this Handbook. The same difference in perspective that we discussed just now may be seen in the different scenarios discussed throughout this Introduction. And here is one way in which the editors have come to see it. Information plays at quite different levels in our human and natural world. One focus for many of the scenarios discussed here are episodes from our daily cognitive practice: language use, observation, communication, or other interaction between agents. Logical and linguistic models of information used by agents in small situations, acting on their private intentions, are meant for this fine-structure of informational transactions. But around all these private episodes, there is the global physical universe that we live in. And another highly significant question is the amount of information that we can hope to extract from that in our theories. At this level, single agents with their private purposes are totally irrelevant, and we are interested only in the large-scale structure of learnability. And the latter question seems to fit much better with the abstraction level provided by Kolmogorov complexity, where we can think of the universe as the output of a single Turing machine producing all data that we see.

In line with this distinction, we also see a distinction between philosophical themes connected to this Handbook. Agent-oriented episodes of meaningful *A*-type information flow seem closer to the concerns of epistemology today, and what people may be said to know about specific issues, perhaps kept from slum-

[22]Cf. Kant's Categorical Imperative, or Rawls' initial scenario in "A Theory of Justice".

[23]This discussion of aggregation levels does show the importance of probability to our Handbook, and we might give the logic/probability interface even more attention in future editions.

bering by skeptics. Several chapters of our Handbook show what clarification arises from making information a major concern here, tying in to fundamental questions about the nature of knowledge, language, and logic. In contrast to this, global knowability of the universe in terms of its information content comes closer to the Grand Questions of the classical philosophical tradition, and asks what we could achieve in principle through observation and theory formation. Taking the mathematical perspectives in this Handbook seriously raises fundamental issues as well, this time, involving the nature and reach of the computationalism implicit in both B-type and C-type views. Is it more than just a convenient methodology? We have briefly discussed some positions in our earlier list of general themes, from metaphysical computationalism about nature to cognitive computationalism about human agents, though of course much more could be said.[24]

While all this may sound like a new-fangled 'technological' view, we see the roots of computationalism in the history of philosophy, going back at least to Descartes' mechanistic analysis of the 'res extensa'. Indeed, it still shares some of the weaknesses of that tradition — but there is also one obvious gain: the precision and clarity provided by the sophisticated mathematical models now at our disposal. Both strengths and weaknesses of philosophical claims can now be stated and investigated in ways that were simply unavailable before.[25] For instance, even if the whole universe can be simulated on a simple Turing machine, given enough time, this does not yet imply a *simple* model. The 'Turing Machine of Nature' could still be a universal computational device of any finite complexity.[26]

Now our point with these final thoughts should not be misunderstood. We are not saying that somewhere above the local level of informational episodes in daily life, and even beyond the whole history of science, there lies some Platonic reality of learnability that we can grasp a priori, making detailed studies redundant. What we do want to say is that the tools in this Handbook allow us to think about both the 'small questions' of philosophy, concerning language use, knowledge, belief, and reasoning of single agents, and the 'big questions', about the intelligibility of the universe, and what we can hope to achieve by collective enquiry.

[24]Many pioneers of computer science have implicitly endorsed metaphysical computationalism. 'The entire universe is being computed on a computer, possibly a cellular automaton' according to Konrad Zuse (cf. [Zuse, 1969]). Similar views have been considered by John Archibald Wheeler, Seth Lloyd, Stephen Wolfram, Nick Bostrum, and many other serious thinkers.

[25]For instance, identifying computability with recursiveness, we can assign an objective, though inevitably non-computable information measure to all objects/messages in this universe. This is precise computational metaphysics. Of course, this, too, has its presuppositions, which might be questioned. How harmless is the choice of a Universal Turing machine, defined up to a 'constant factor'? Could even a leeway of 100 bits prevent us from using Kolmogorov complexity for the analysis of human intelligence? (Our brain has roughly 10^{15} neurons.)

[26]Moreover, the point at which Kolmogorov complexity asymptotically approaches the actual complexity of objects in our world might lie well beyond a horizon that is useful and practical.

Philosophy of information: some major issues

To summarize, we list the broad research issues emerging in this Handbook that we see as central for the development of the field:

1. *Information per se.* What is information? Is there one general notion that encompasses all others, or do we merely have a family of loosely related concepts, or perhaps 'complementary stances' in practical settings, making the peaceful co-existence of approaches as described in this editorial the best that can be achieved?

2. *Information and process.* What is the relation between information structure and computation, deduction, observation, learning, game playing, or evolution? These processes seem to create information for free. How to understand this? Can we unify the theory of information, computation, dynamic logics of epistemic update and belief revision, and the thermodynamics of non-equilibrium processes?

3. *Information and philosophy.* The chapters in this Handbook tie the notion of information to fundamental issues in classical philosophy, 'analytical' but equally well 'continental'. Can we 'deconstruct' classical philosophy with modern information-theoretic tools, and bridge the culture gap between the two traditions? The tools of logic and mathematics at least have no bias for one over the other.[27]

Thus, though this Handbook is full of answers to anyone interested in a serious study of information, we end with open questions, as true philosophers should.

BIBLIOGRAPHY

[Adriaans, 2007] P. W. Adriaans. Learning as Data Compression. In S. B. Cooper, B. Löwe and A. Sorbi, eds., *Computation and Logic in the Real World,* Springer, Lecture Notes in Computer Science, Vol. 449, 11-24, 2007.

[Adriaans, 2009] P. W. Adriaans. Between Order and Chaos: the Quest for Meaningful Information, submitted by invitation to *Theory of Computing Systems*, CiE special issue.

[Bar-Hillel and Carnap, 1953] Y. Bar-Hillel and R. Carnap. Semantic Information, *The British Journal for the Philosophy of Science* 4:14, 147–157, 1953.

[Barwise and Perry, 1983] J. Barwise and J. Perry. *Situations and Attitudes*, The MIT Press, Cambridge (Mass.), 1983.

[van Benthem, 2006] J. van Benthem. Epistemic Logic and Epistemology: the state of their affairs, *Philosophical Studies* 128, 49–76, 2006.

[van Benthem and van Rooij, 2003] J. van Benthem and R. van Rooij, eds. Connecting the Different Faces of Information, *Journal of Logic, Language and Information* 12, 2003.

[Bergson, 1998] H. Bergson. *Creative Evolution,* Dover Publications, New York, 1998.

[Capurro and Hjørland, 2003] R. Capurro and B. Hjørland. The Concept of Information, *Annual Review of Information Science and Technology (ARIST),* Ed. Blaise Cronin, Vol. 37, Chapter 8, 343-411, 2003.

[27] Incidentally, Adriaans comes from the continental tradition, van Benthem from the analytical one, though their paths have crossed repeatedly in logic and mathematics.

[Carnap, 1950] R. Carnap. *Logical Foundations of Probability*, The University of Chicago Press, Chicago, 1950.

[Castells, 1996] M. Castells. *The Rise of the Network Society, The Information Age: Economy, Society and Culture,* Vol. I. Blackwell, Oxford, 1996.

[Chaitin, 1987] G. J. Chaitin. *Algorithmic Information Theory*, Cambridge University Press, New York, NY, 1987.

[Chater and Vitányi, 2003] N. Chater and P. M. B. Vitányi. Simplicity: a Unifying Principle in Cognitive Science, *Trends in Cognitive Science*, 7:1, 19–22, 2003.

[Cilibrasi and Vitányi, 2005] R. Cilibrasi and P. M. B. Vitányi. Clustering by compression, *IEEE Transactions on Information Theory*, 51(4), 1523-1545, 2005.

[Clausius, 1850] R. Clausius. Über die bewegende Kraft der Wärme und die Gesetze welche sich daraus für die Wärmelehre selbst ableiten lassen, *Poggendorffs Annalen der Physik und Chemie*, Vol. 79, 368–97, 1850.

[Cover and Thomas, 2006] T. M. Cover and J. A. Thomas. *Elements of Information theory*, 2nd edition, Wiley and Sons, New York, 2006.

[Dretske, 1981] F. Dretske. *Knowledge and the Flow of Information*, The MIT Press, Cambridge (Mass.), 1981.

[Fisher, 1925] R. A. Fisher. Theory of Statistical Estimation, *Proceedings Cambridge Philosophical Society* 22, 700–725, 1925.

[Floridi, 2003a] L. Floridi. Information. In L. Floridi, ed., 2003, 40-61, 2003.

[Floridi, 2003b] L. Floridi, ed. *The Blackwell Guide to the Philosophy of Computing and Information*, Blackwell, Oxford, 2003.

[Hintikka, 1962] J. Hintikka. *Knowledge and Belief,* Cornell University Press, Ithaca, 1962.

[Hintikka, 1973] J. Hintikka. *Logic, Language Games, and Information*, Clarendon, Oxford, 1973.

[Kahn, 1967] D. Kahn. *The Code-Breakers, The Comprehensive History of Secret Communication from Ancient Times to the Internet,* Scribner, New York, 1967.

[Kneale and Kneale, 1962] W. and M. Kneale. *The Development of Logic*, Clarendon Press, Oxford, 1962.

[Kohler, 2001] E. Kohler. Why von Neumann rejected Rudolf Carnap's Dualism of Information Concepts. In M. Redei and M. Stoeltzner, eds., 2001.

[Kooi, 2003] B. Kooi. *Knowledge, Chance, and Change*, Philosophical Institute, University of Groningen and ILLC Dissertation Series 2003-01, University of Amsterdam, 2003.

[Kuipers, 2000] Th. Kuipers. *From Instrumentalism To Constructive Realism*, Kluwer, Boston, 2000.

[Lewis, 1969] D. Lewis. *Convention, A Philosophical Study*, Harvard University Press, Cambridge (Mass.), 1969.

[Lloyd and Ng, 2004] S. Lloyd and J. Ng. Black Hole Computers, *Scientific American,* Vol. 291, Nr. 5, 30–39, 2004.

[Locke, 1961] J. Locke. *An Essay Concerning Human Understanding*, J. W. Yolton, ed., Dent, London and Dutton, New York, 1961.

[Li and Vitányi, 1997] M. Li and P. Vitányi. *An Introduction to Kolmogorov Complexity and its Applications*, Springer-Verlag, second edition, 1997.

[Nozick, 1978] R. Nozick. *Philosophical Explanations*, Oxford University Press, Oxford, 1978.

[Parikh and Ramanujam, 2003] R. Parikh and R. Ramanujam. A Knowledge Based Semantics of Messages, *Journal of Logic, Language and Information* 12, 453–467, 2003.

[Pinker, 1997] S. Pinker. *How the Mind Works*, Allen Lane, New York, 1997.

[Redei and Stoeltzner, 2001] M. Redei and M. Stoeltzner, eds. *John von Neumann and the Foundations of Quantum Physics, Kluwer Academic Publishers,* Dordrecht, 2001.

[van Rooij, 2004] R. van Rooij. Signalling Games select Horn Strategies, *Linguistics and Philosophy* 27, 493–527, 2004.

[Schmidhuber, 1997] J. Schmidhuber. A Computer Scientist's View of Life, the Universe, and Everything, *Lecture Notes in Computer Science*, Vol. 1337, 201–208, 1997.

[Shannon, 1948] C. Shannon. A Mathematical Theory of Communication, *Bell System Technical Journal*, 27, 379–423, 623–656, 1948.

[Singh, 1999] S. Singh. *The Code Book: The Science of Secrecy from Ancient Egypt to Quantum Cryptography,* Anchor Books, New York, 1999.

[Solomonoff, 1997] R. J. Solomonoff. The Discovery of Algorithmic Probability, *Journal of Computer and System Sciences*, vol. 55, nr. 1, 73–88, 1997.

[Stalnaker, 1984] R. Stalnaker. *Inquiry*, The MIT Press, Cambridge, Mass. 1984.

[Tarski, 1944] A. Tarski. The Semantic Conception of Truth, *Philosophy and Phenomenological Research* 4, 13–47, 1944.

[Wolff, 2006] J.G. Wolff. *Unifying Computing and Cognition. The SP Theory and its Applications.* CognitionResearch.org.uk., 2006

[Wolfram, 2002] S. Wolfram. *A New Kind of Science*, Wolfram Media Inc., Champaign, Ill., 2002.

[Zuse, 1969] K. Zuse. *Rechnender Raum*, Friedrich Vieweg and Sohn, Braunschweig, 1969.

Part B

History of Ideas: Information Concepts

EPISTEMOLOGY AND INFORMATION

Fred Dretske

Epistemology is the study of knowledge — its nature, sources, limits, and forms. Since perception is an important source of knowledge, memory a common way of storing and retrieving knowledge, and reasoning and inference effective methods for extending knowledge, epistemology embraces many of the topics comprised in cognitive science. It is, in fact, a philosopher's way of doing cognitive science.

Information, as commonly understood, as the layperson understands it, is an epistemologically important commodity. It is important because it is necessary for knowledge. Without it one remains ignorant. It is the sort of thing we associate with instruction, news, intelligence, and learning. It is what teachers dispense, what we (hope to) find in books and documents, what measuring instruments provide, what airline and train schedules contain, what spies are used to ferret out, what (in time of war) people are tortured to divulge, and what (we hope) to get by tuning in to the evening news.

It is this connection between knowledge and information, as both are commonly understood, that has encouraged philosophers to use mathematically precise codifications of information to formulate more refined theories of knowledge. If information is really what it takes to know, then it seems reasonable to expect that a more precise account of information will yield a scientifically more creditable theory of knowledge. Maybe — or so we may hope — communication engineers can help philosophers with questions raised by Descartes and Kant. That is one of the motives behind information-based theories of knowledge.

1 NECESSARY CLARIFICATIONS: MEANING, TRUTH, AND INFORMATION.

As the name suggests, information booths are supposed to dispense information. The ones in airports and train stations are supposed to provide answers to questions about when planes and trains arrive and depart. But not just *any* answers. *True* answers. They are not there to entertain patrons with meaningful sentences on the general topic of trains, planes, and time. Meaning is fine. You can't have truth without it. False statements, though, are as meaningful as true statements. They are not, however, what information booths have the function of providing. Their purpose is to dispense truths, and that is because information, unlike meaning, has to be true. If nothing you are told about the trains is true, you haven't been given information about the trains. At best, you have been given misinformation, and misinformation is not a kind of information anymore than decoy ducks

Handbook of the Philosophy of Science. Volume 8: Philosophy of Information
Volume editors: Pieter Adriaans and Johan van Benthem. General editors: Dov M. Gabbay, Paul Thagard and John Woods.

are a kind of duck. If nothing you are told is true, you may leave an information booth with a lot of false beliefs, but you won't leave with knowledge. You won't leave with knowledge because you haven't been given what you need to know: information.

So if in formulating a theory of information we respect ordinary intuitions about what information is — and why else would one call it a theory *of information*? — we must carefully distinguish meaning, something that need not be true, from information which must be true. There are, to be sure, special uses of the term "information" — computer science is a case in point — in which truth seems to be irrelevant. Almost anything that can be put into the memory of a computer, anything that can be entered into a "data" base, is counted as information. If it isn't correct, then it is misinformation or false information. But, according to this usage, it is still information. Computers, after all, can't distinguish between "Paris is the capital of France" and "Paris is the capital of Italy." Both "facts", if fed into a computer, will be stored, retrieved, and used in exactly the same way. So if true sentences count as information, so should false ones. For computational purposes they are indistinguishable.

This approach to information — an approach that is, I believe, widespread in the information sciences — blithely skates over absolutely fundamental distinctions between truth and falsity, between meaning and information. Perhaps, for some purposes, these distinctions can be ignored. Perhaps, for some purposes, they *should* be ignored. You cannot, however, build a science of knowledge, a cognitive science, and ignore them. For knowledge is knowledge of the truth. That is why, no matter how fervently you might believe it, you cannot know that Paris is the capital of Italy, that pigs can fly or that there is a Santa Claus. You can, to be sure, put these "facts", these false sentences, into a computer's data base (or a person's head for that matter), but that doesn't make them true. It doesn't make them information. It just makes them sentences that, given the machine's limitations (or the person's ignorance), the machine (or person) treats as information. But you can't make something true by thinking it is true, and you can't make something into information by regarding it as information.

So something — e.g., the sentence "Pigs can fly" — can mean pigs can fly without carrying that information. Indeed, given the fact that pigs can't fly, nothing can carry the information that pigs can fly. This is why, as commonly understood, information is such an important, such a useful, commodity. It gives you what you need to know — the truth. Meaning doesn't.

Information (once again, as it is commonly conceived) is something closely related to what natural signs and indicators provide. We say that the twenty rings in the tree stump indicate, they signify, that the tree is twenty years old. That is the information (about the age of the tree) the rings carry. We can come to know how old the tree is by counting the rings. Likewise, the rising mercury in a glass tube, a thermometer, indicates that the temperature is rising. That is what the increasing volume of the mercury is a sign of. That is the information the expanding mercury carries and, hence, what we can come to know by using this instrument.

We sometimes use the word "meaning" to express this sentential content (what we can come to know) but this sense of the word, a sense of the word in which smoke means (indicates, is a sign of) fire, must be carefully distinguished from a linguistic sense of meaning in which the word "fire" (not the word "smoke" nor smoke itself) means fire. In a deservedly famous article, Paul Grice [1957] dubbed this informational kind of meaning, the kind of meaning in which smoke means (indicates, is a sign of) fire, *natural meaning*. With this kind of meaning, natural meaning, if an event, e, means (indicates, is a sign) that so-and-so exists, then so-and-so must exist. The red spots on her face can't mean, not in the natural sense of meaning, that she has the measles if she doesn't have the measles. If she doesn't have the measles, then perhaps all the spots mean in this natural sense is that she has been eating too many sweets. This contrasts with a language related (Grice called it "non-natural) meaning in which something (e.g., the sentence "She has the measles") can mean she has the measles even when she doesn't have them. If she doesn't have the measles, the sentence is false but that doesn't prevent it from meaning that she has the measles. If e (some event) means, in the natural sense, that s is F, however, then s has to be F. Natural meaning is what indicators indicate. It is what natural signs are signs of. Natural meaning is information. It has to be true.

This isn't to say that we must know what things indicate, what information they carry. We may not know. We may have to find this out by patient investigation. But what we find out by patient investigation — that the tracks in the snow mean so-and-so or shadows on the film indicate such-and-such — is something that was true before we found it out. In this (natural) sense of meaning, we discover what things mean. We don't, as we do with linguistic or non-natural meaning, assign, create or invent it. By a collective change of mind we could change what the words "lightning" and "smoke" mean, but we cannot, by a similar change of mind, change what smoke and lightning mean (indicate). Maybe God can, but we can't. What things mean, what they indicate, what information they provide, is in this way objective. It is independent of what we think or believe. It is independent of what we know. We may seek information in order to obtain knowledge, but the information we seek doesn't depend for its existence on anyone coming to know. It is, so to speak, out there in the world awaiting our use (or abuse) of it. Information is, in this way, different from knowledge. Information doesn't need conscious beings to exist, but knowledge does. Without life there is no knowledge (because there is nobody to know anything), but there is still information. There still exists that which, if knowers existed, they would need to know.

2 INFORMATION AND COMMUNICATION

If this is, even roughly, the target we are aiming at, the idea of information we want a theory of, then a theory of information should provide some systematic, more precise, perhaps more analytical, way of thinking about this epistemologically important commodity. If possible, we want a framework, a set of principles, that will illuminate the nature and structure of information and, at the same time, reveal the source of its power to confer knowledge on those who possess it.

In Dretske [1981; 1983] I found it useful to use Claude Shannon's Mathematical Theory of Communication [1948] for these purposes (see also [Cherry, 1957] for a useful overview and [Sayre, 1965] for an early effort in this direction). Shannon's theory does not deal with the semantic aspects of information. It has nothing to say about the news, message, or content of a signal, the information (that the enemy is coming by sea, for instance) expressed in propositional form that a condition (a lantern in a tower) conveys. It does, however, focus on what is, for epistemological purposes, the absolutely critical relation between a source of information (the whereabouts of the enemy) and a signal (a lantern in the tower) that carries information about that source. Shannon's theory doesn't concern itself with what news, message or information is communicated from s (source) to r (receiver) or, indeed, whether anything intelligible is communicated at all. As far as Shannon's theory is concerned, it could all be gibberish (e.g., "By are they sea coming."). What the theory does focus on in its theory of mutual information (a measure of amount of information at the receiver about a source) is the question of the amount of statistical dependency existing between events occurring at these two places. Do events occurring at the receiver alter in any way the probability of what occurred at the source? Given the totality of things that occur, or that might occur, at these two places, is there, given what happens at the receiver, a reduction in (what is suggestively called) the uncertainty of what happened at the source?

This topic, the communication channel between source and receiver, is a critically important topic for epistemology because "receiver" and "source" are just information-theoretic labels for knower and known. Unless a knower (at a receiver) is connected to the facts (at a source) in an appropriate way, unless there is a suitably reliable channel of communication between them, the facts cannot be known. With the possible exception of the mind's awareness of itself (introspection) — there is always, even in proprioception, a channel between knower and known, a set of conditions on which the communication of information — and therefore the possibility of knowledge — depends. What we can hope to learn from communication theory is what this channel must look like, what conditions must actually exist, for the transmission of the information, needed to know

At one level, all this sounds perfectly familiar and commonplace. If someone cuts the phone lines between you and me, we can no longer communicate. I can no longer get from you the information I need in order to know when you are planning to arrive. Even if the phone lines are repaired, a faulty connection can generate so much "noise" (another important concept in communication theory) that not enough information gets through to be of much use. I hear you, yes, but not well enough to understand you. If we don't find a better, a clearer, channel over which to communicate, I will never find out, never come to know, when you plan to arrive.

That, as I say, is a familiar, almost banal, example of the way the communication of information is deemed essential for knowledge. What we hope to obtain from a theory of communication, if we can get it, is a systematic and illuminating generalization of the intuitions at work in such examples. What we seek, in its most general possible form, whether the communication occurs by phone, gesture, speech, writing, smoke signals, or mental telepathy, is what kind of communication channel must exist between you and me for me to learn what your plans are? Even more generally, for any A and B, what must the channel, the connection, between A and B be like for someone at A to learn something about B?

The Mathematical Theory of Communication doesn't answer this question, but it does supply a set of ideas, and a mathematical formalism, from which an answer can be constructed. The theory itself deals in amounts of information, how much (on average) information is generated at source s and how much (on average) information there is at receiver r about this source. It does not try to tell us *what* information is communicated from s to r or even, if some information is communicated, how much is enough to know what is happening at s. It might tell us that there are 8 bits of information generated at s about, say, the location of a chess piece on a chessboard (the piece is on KB-3) and that there are 7 bits of information at r about the location of this piece, but it does not tell us what information this 7 bits is the measure of nor whether 7 bits of information is enough to know where the chess piece is. About that it is silent.

3 USING COMMUNICATION THEORY

We can, however, piece together the answers to these questions out of the elements and structure provided by communication theory. To understand the way this might work consider the following toy example (adapted from [Dretske, 1981]) and the way it is handled by communication theory. There are eight employees and one of them must perform some unpleasant task. Their employer has left the job of selecting the unfortunate individual up to the group itself, asking only to be informed of the outcome once the decision is made. The group devises some random procedure that it deems fair (drawing straws, flipping a coin), and Herman is selected. A memo is dispatched to the employer with the sentence, "Herman was chosen" written on it.

Communication theory identifies the amount of information associated with, or generated by, the occurrence of an event with the reduction in uncertainty, the elimination of possibilities, represented by that event. Initially there were eight eligible candidates for the task. These eight possibilities, all (let us assume) equally likely, were then reduced to one by the selection of Herman. In a certain intuitive sense of "uncertainty", there is no longer any uncertainty about who will do the job. The choice has been made. When an ensemble of possibilities is reduced in this way (by the occurrence of one of them), the amount of information associated with the result is a function of how many possibilities there were (8 in this case) and their respective probabilities (.125 for each in this case). If all are equally likely, then the amount of information (measured in *bits*) generated by the occurrence of one of these n possibilities, I_g, is the logarithm to the base 2 of n (the power to which 2 must be raised to equal n):

(1) $I_g = \log_2 n$

Since we started with eight possibilities all of which were all equally likely, I_g is $\log_2 8 = 3$ bits. Had there been 16 instead of 8 employees, Herman's selection would have generated 4 bits of information — more information since there is a reduction of more uncertainty.[1]

The quantity of interest to epistemology, though, is not the information *generated* by an event, but the amount of information *transmitted* to some potential knower, in this case the employer, about the occurrence of that event. It doesn't make much difference how much information an event generates: 1 bit or 100 gigabytes. The epistemologically important question is: how much of this information is transmitted to, and subsequently ends up in the head of, a person at r seeking to know what happened at s. Think, therefore, about the note with the name "Herman" on it lying on the employer's desk. How much information does this piece of paper carry about what occurred in the other room? Does it carry the information that Herman was selected? Would the employer, upon reading (and understanding) the message, know who was selected? The sentence written on the memo does, of course, mean in that non-natural or linguistic sense described above that Herman was selected. It certainly would cause the employer to *believe* that Herman was selected. But these aren't the questions being asked. What is being asked is whether the message indicates, whether it means in the natural sense, whether it carries the information, that Herman was selected. Would it enable the employer to *know* that Herman was selected? Not every sentence written on a piece of paper carries information corresponding to its (non-natural) meaning. "Pigs can fly" as it appears on this (or, indeed, any other) page doesn't carry the information that pigs can fly. Does the sentence "Herman was selected" on the employees' memo carry the information that Herman was selected? If so, why?

Our example involves the use of an information-carrying signal — the memo to the employer — that has linguistic (non-natural) meaning, but this is quite irrelevant to the way the situation is analyzed in communication theory. To understand why, think about an analogous situation in which non-natural (linguistic) meaning is absent. There are eight mischievous boys and a missing cookie. Who took it? Inspection reveals cookie crumbs on Junior's lips. How much information about the identity of the thief do the crumbs on Junior's lips carry? For informational purposes, this question is exactly the same as our question about how

[1]If the probabilities of selection are not equal (e.g., probability of Herman = 1/6, probability of Barbara = 1/12, etc.), then I_g (average amount of information generated by the selection of an employee) is a weighted average of the information generated by the selection of each. I pass over these complications here since they aren't relevant to the use of communication theory in epistemology. What is relevant to epistemology is not how much information is generated by the occurrence of an event, or how much (on average) is generated by the occurrence of an ensemble of events, but how much of that information is transmitted to a potential knower at some receiver.

much information about which employee was selected the memo to the employer carries. In the case of Junior, the crumbs on his lips do not have linguistic meaning. They have a natural meaning, yes. They mean (indicate) he took the cookie. But they don't have the kind of conventional meaning associated with a sentence like, "Junior took the cookie."

Communication theory has a formula for computing amounts of transmitted (it is sometimes called *mutual*) information. Once again, the theory is concerned not with the conditional probabilities that exist between particular events at the source (Herman being selected) and the receiver (Herman's name appearing on the memo) but with the average amount of information, a measure of the *general* reliability of the communication channel connecting source and receiver. There are eight different conditions that might exist at s: Barbara is selected, Herman is selected, etc. There are eight different results at r: a memo with the name "Herman" on it, a memo with the name "Barbara" on it, and so on. There are, then, sixty four conditional probabilities between these events: the probability that Herman was selected given that his name appears on the memo:

Pr[Herman was selected/the name "Herman" appears on the memo];

the probability that Barbara was selected given that the name "Herman" appears on the memo:

Pr[Barbara was selected/the name "Herman" appears on the memo];

and so on for each of the eight employees and each of the eight possible memos. The transmitted information, I_t, is identified with a certain function of these 64 conditional probabilities. One way to express this function is to say that the amount of information transmitted, I_t, is the amount of information generated at s, I_g, minus a quantity called equivocation, E, a measure of the statistical independence between events occurring at s and r. [2]

(2) $I_t = I_g - E$

The mathematical details are not really important. A few examples will illustrate the main ideas. Suppose the employees and messenger are completely scrupulous. Memos always indicate exactly who was selected, and memos always arrive on the employer's desk exactly as they were sent. Given this kind of reliability, the conditional probabilities are all either 0 or 1.

Pr[Herman was selected/the name "Herman" appears on the memo] = 1
Pr[Barbara was selected/the name "Herman" appears on the memo] = 0

[2]Equivocation, E, is the weighted (according to its probability of occurrence) sum of individual contributions, $E(r_1), E(r_2), \ldots$ to equivocation of each of the possible events (eight possible memos) at r : $E = pr(r_1)E(r_1) + pr(r_2)E(r_2) + \ldots pr(r_8)E(r_8)$ where $E(r_i) = -?pr(s_i/r_i)\log_2[pr(s_i/r_i)]$. If events at s and r are statistically independent then E is at a maximum ($E = I_g$) and I_t is zero.

Pr[Nancy was selected/the name "Herman" appears on the memo] = 0

⋮

Pr[Barbara was selected/the name "Barbara" appears on the memo] = 1
Pr[Herman was selected/the name "Barbara" appears on the memo] = 0
Pr[Nancy was selected/the name "Barbara" appears on the memo] = 0

⋮

And so on for all employees and possible memos. Given this reliable connection, this trustworthy communication channel, between what happens among the employees and what appears on the memo to their employer, the *equivocation*, E turns out to be zero.[3]

$I_t = I_g$: the memo on which is written an employee's name carries 3 bits of information about who was selected. All of the information generated by an employee's selection, 3 bits, reaches its destination.

Suppose, on the other hand, we have a faulty, a broken, channel of communication. On his way to the employer's office the messenger loses the memo. He knows it contained the name of one of the employees, but he doesn't remember which one. Too lazy to return for a new message, he selects a name of one of the employees at random, scribbles it on a sheet of paper, and delivers it. The name he selects happens, by chance, to be "Herman." Things turn out as before. Herman is assigned the job, and no one (but the messenger) is the wiser. In this case, though, the set of conditional probabilities defining equivocation (and, thus, amount of transmitted or mutual information) are quite different. Given that the messenger plucked a name at random, the probabilities look like this:

Pr[Herman was selected/the name "Herman" appears on the memo] = 1/8
Pr[Barbara was selected/the name "Herman" appears on the memo] = 1/8

⋮

Pr[Herman was selected/the name "Barbara" appears on the memo] = 1/8
Pr[Barbara was selected/the name "Barbara" appears on the memo] = 1/8

⋮

The statistical function defining equivocation (see footnote 2) now yields a maximum value of 3 bits. The amount of transmitted information, formula (2), is therefore zero.

[3]Either $pr(s/r) = 0$ or $\log_2[pr(s/r)] = 0$ in the individual contributions to equivocation (see footnote 2). Note: $\log_2 1 = 0$.

These two examples represent the extreme cases: maximum communication and zero communication. One final example of an intermediate case and we will be ready to explore the possibility of applying these results in an information-theoretic account of knowledge.

Imagine the employees solicitous about Barbara's delicate health. They agree to name Herman on their note if, by chance, Barbara should be the nominee according to their random selection process. In this case I_g, the amount of information generated by Herman's selection would still be 3 bits: 8 possibilities, all equally likely, reduced to 1. Given their intention to protect Barbara, though, the probabilities defining transmitted information change. In particular

Pr[Herman was selected/the name "Herman" appears on the memo] = 1/2
Pr[Barbara was selected/the name "Herman" appears on the memo] = 1/2

The remaining conditional probabilities stay the same. This small change means that E, the average equivocation on the channel, is no longer 0. It rises to .25. Hence, according to (2), I_t drops from 3 to 2.75. Some information is transmitted, but not as much as in the first case. Not as much information is transmitted as is generated by the selection of an employee (3 bits)

This result seems to be in perfect accord with ordinary intuitions about what it takes to know. For it seems right to say that, in these circumstances, anyone reading the memo naming Herman as the one selected could not learn, could not come to know, on the basis of the memo alone, that Herman actually was selected. Given the circumstances, the person selected might have been Barbara. So it would seem that communication theory gives us the right answer about when someone could know. One could know that it was Herman in the first case, when the message contained 3 bits of information — exactly the amount generated by Herman's selection — and one couldn't know in the second and third case, when the memo contains 0 bits and 2.75 bits of information, something less than the amount generated by Herman's selection. So if information is what it takes to know, then it seem correct to conclude that in the first case the information that Herman was selected was transmitted and in the second and third case it was not. By focusing on the *amount* of information carried by a signal, communication theory manages to tell us something about the informational *content* of the signal — something about the news or message the signal actually carries — and, hence, something about what (in propositional form) can be known.

4 THE COMMUNICATION CHANNEL

Let us, however, ask a slightly different question. We keep conditions the same as in the third example (Herman will be named on the memo if Barbara is selected), but ask whether communication theory gives the right result if someone else is selected. Suppose Nancy is selected, and a memo sent bearing her name. Since the *general* reliability of the communication channel remains exactly the same, the amount of transmitted information (a quantity that, by averaging over all possible messages, is intended to reflect this general reliability) also stays the same: 2.75 bits. This is, as it were, a 2.75 bit channel, and this measure doesn't change no matter which particular message we happen to send over this channel. If we use this as a measure of how much information is carried by a memo with Nancy's name on it, though, we seem to get the wrong result. The message doesn't carry as much information, 3 bits, as Nancy's selection generates. So the message doesn't carry the information that Nancy was selected. Yet, a message bearing the name "Nancy" (or, indeed, a memo bearing the name of any employee except "Herman") is a perfectly reliable sign of who was selected. The name "Nancy" indicates, it means (in the natural sense) that Nancy was selected even though a memo bearing the name "Herman" doesn't mean that Herman was selected. The same is true of the other employees. The only time the memo is equivocal (in the ordinary sense of "equivocal") is when it bears the name "Herman." *Then* it can't be trusted. Then the nominee could be either Herman or Barbara. But as long as the message doesn't carry the name "Herman" it is an absolutely reliable indicator of who was selected. So when it bears the name "Nancy" ("Tom" etc.) why doesn't the memo, contrary to communication theory, carry the information that Nancy (Tom, etc.) was selected? A 2.75 bit channel is a reliable enough channel — at least sometimes, when the message bears the name "Nancy" or "Tom," for instance — to carry a 3 bit message.

Philosophical opinions diverge at this point. Some are inclined to say that Communication Theory's concentration on averages disqualifies it for rendering a useful analysis of when a signal carries information in the ordinary sense of information. For, according to this view, a message to the employer bearing the name "Nancy" does carry information about who was selected. It enables the employer to know who was selected even though he might have been misled had a message arrived bearing a different name. The fact that the average amount of transmitted information (2.75 bits) is less than the average amount of generated information (3 bits) doesn't mean that a particular signal (e.g., a memo with the name "Nancy" on it) can't carry all the information needed to know that Nancy was selected. As long as the signal indicates, as long as it means in the natural sense, that Nancy was selected, it is a secure enough connection (channel) to the facts to know that Nancy was selected even if other signals (a memo with the name "Herman" on it) fail to be equally informative. Communication Theory, in so far as it concentrates on averages, then, is irrelevant to the ordinary, the epistemologically important, sense of information. [4]

Others will disagree. Disagreement arises as a result of different judgments about what it takes to know and, therefore, about which events can be said to carry information in the ordinary sense of information. The thought is something

[4]This is the view I took in Dretske [1981] and why I argued that the statistical functions of epistemological importance were not those defining average amounts of information (equivocation, etc.), but the amount of information associated with particular signals. It was not, I argued, average equivocation that we needed to be concerned with, but the equivocation associated with particular signals (see [Dretske, 1981, 25–26]).

like this: a communication channel that is *sometimes* unreliable is not good enough to know even when it happens to be right. A channel of the sort described here, a channel that (unknown to the receiver) sometimes transmits misleading messages, is a channel that should never be trusted. If it is trusted, the resulting belief, even it happens to be true, does not possess the "certainty" characteristic of knowledge. If messages are trusted, if the receiver actually believes that Nancy was selected on the basis of a message bearing the name "Nancy," the resulting belief does not, therefore, add up to knowledge. To think otherwise is like supposing that one could come to know by taking the word of a chronic liar just because he happened, on this particular occasion, and quite unintentionally, to be speaking the truth.

Imagine a Q meter designed to measure values of Q. Unknown to its users, it is perfectly reliable for values below 100, but unpredictably erratic for values above 100. Is such an instrument one that a person, ignorant of the instrument's eccentric disposition[5], could use to learn values of Q below 100? Would a person who took a reading of "84" at face value, a person who was caused to believe that Q was 84 by a reading of "84" on this instrument, *know* that Q was 84? Does the instrument deliver information about values of Q below 100 to trusting users? If your answer to these questions is "No," you are using something like communication theory to guide your judgments about what is needed to know and, hence, about when information is communicated. This instrument doesn't deliver what it takes to know (i.e., information in the ordinary sense) because although the particular reading ("84") one ends up trusting is within the instrument's reliable range (the instrument wouldn't read "84" unless Q was 84) you don't know this. You would have trusted it even if it had registered "104". The method being used to "track" the truth (the value of Q) doesn't track the truth throughout the range in which that method is being used.[6]

Externalism is the name for an epistemological view that maintains that some of the conditions required to know that P may be, and often are, completely beyond the ken of the knower. You can, in normal illumination, see (hence, know) what color the walls are even if you don't know (because you haven't checked) that the illumination is normal. Contrary to Descartes, in normal circumstances you can know you are sitting in front of the fireplace even if you don't know (and can't show) the circumstances are normal, even if you don't know (and can't show) you are not dreaming or being deceived by some deceptive demon. According to externalism, what is important for knowledge is not that you *know* perceptual

[5] If users were aware of the instrument's limited reliability, of course, they could compensate by ignoring readings above 100 and, in effect, make the instrument completely accurate in the ranges it was used (i.e., trusted). Practically speaking, this represent a change in the communication channel since certain readings (those above 100) would no longer be regarded as information-bearing signals.

[6] This way of putting the point is meant to recall Robert Nozick's [1981] discussion of similar issues. If the method being used to "track" (Nozick's term) the truth is insensitive to ranges of unreliability, then the method is not such as to satisfy the counterfactual conditions Nozick uses to define tracking. One would (using that method) have believed P even when P was false. See, also, Goldman's [1976] insightful discussion of the importance of distinguishing the ways we come to know.

conditions are normal (the way they are when things are the way they appear to be), but that conditions actually *be* normal. If they are, if illumination (perspective, eyesight, etc.) are as you (in ordinary life) routinely take them to be, then you can see — and, hence, know — that the walls are blue, that you are sitting in front of the fireplace, and that you have two hands. You can know these things even if, for skeptical reasons, you cannot verify (without arguing in a circle) that circumstances are propitious. Information-theoretic accounts of knowledge are typically advanced as forms of externalism. The idea is that the information required to know can be obtained from a signal without having to know that the signal from which you obtain this information actually carries it. What matters in finding out that Nancy was selected (or in coming to know any other empirical matter of fact) is not that equivocation on the channel (connecting knower and known) be known to be zero. What is crucial is that it actually — whether known or not — *be* zero. This dispute about whether a memo bearing the name "Nancy" carries the information that Nancy was selected is really a dispute among externalists not about what has to be known about a communication channel for it to carry information. Externalists will typically agree that nothing has to be known. It is, instead, a dispute about exactly what (independently of whether or not it is known) constitutes the communication channel. In calculating equivocation between source and receiver — and, therefore, the amount of information a signal at the receiver carries about a source, should we count every signal that would produce the same resulting belief — the belief (to use our example again) that Nancy was selected? In this case we don't count memos carrying the name "Herman" since although these memos will produce false belief, they will not produce a false belief about *Nancy's* selection. If we do this, we get an equivocation-free channel. Information transmission is optimal. Or should we count every signal that would produce a belief about who was selected — whether or not it is Nancy? Then we count memos carrying the name Herman, and the communication channel, as so defined, starts to get noisy. The amount of mutual information, a measure of the amount of information transmitted, about who was selected is no longer equal to the amount of information generated. Memos — even when they carry the name "Nancy" — do not carry as much information as is generated the choice of Nancy because equivocal messages bearing the name "Herman" are used to reckon the channel's reliability even when it carries the message "Nancy." Or — a third possible option — in reckoning the equivocation on a communication channel, should we (as skeptics would urge) count *any* belief that would be produced by any memo (or, worse, any signal) whatsoever? If we start reckoning equivocation on communication channels in that way, then, given the mere possibility of misperception, no communication channel is *ever* entirely free of equivocation. The required information is *never* communicated. Nothing is known.

I do not — not *here* at least — take sides in this dispute. I merely describe a choice point for those interested in pursuing an information-theoretic epistemology. The choice one makes here — a choice about what collection of events and conditions are to determine the channel of communication between knower and known — is an important one. In the end, it determines what conclusions one will reach about such traditional epistemological problems as skepticism and the limits of human knowledge. I refer to this as a "choice" point to register my own belief that communication theory, and the concept of information it yields, does not solve philosophical problems. It is, at best, a tool one can use to express solutions — choices — reached by other means.

5 RESIDUAL PROBLEMS AND CHOICES

What follows are three more problems or, as I prefer to put it, three more choices confronting anyone developing an information-theoretic epistemology that is based, even if only roughly, on an interpretation of information supplied by communication theory. I have my own ideas about which choices should be made and I will so indicate, but I will not here argue for these choices. That would require a depth of epistemological argument that goes beyond the scope of this paper.

A. **Probability**

In speaking of mutual information within the framework of communication theory, we imply that there is a set of conditional probabilities relating events at source and receiver. If these conditional probabilities are objective, then the resulting idea of information is objective. If they are subjective, somehow dependent on what we happen to believe, on our willingness to bet, on our level of confidence, then the resulting notion of information is subjective. If information is objective, then to the extent that knowledge depends on information, knowledge will also be objective. Whether a person who believes that P *knows* that P will depend on how, objectively speaking, that person is connected to the world. It will depend on whether the person's belief (assuming it is true) has appropriate informational credentials — whether, that is, it (or the evidence on which it is based) stands in suitable probabilistic relations to events at the source. That will be an objective matter, a matter to be decided by objective facts defining information. It will not depend on the person's (or anyone else's) opinion about these facts, their level of confidence, or their willingness to bet. If, on the other hand, probability is a reflection of subjective attitudes, if the probability of e (some event at a source) given r (an event at a receiver) depends on the judgments of people assigning the probability, then knowledge, in so far as it depends on information, will depend on these judgments. Whether S knows that P will depend on who is saying S knows that P.

I have said nothing here about the concept of probability that figures so centrally in communication theory. I have said nothing because, as far as I can see, an information-theoretic epistemology is compatible with different interpretations of probability.[7] One can interpret it as degree of rational expectation (subjective), or (objectively) as limiting frequency or propensity. In developing my own

[7]But see Loewer [1983] for arguments that there is *no* extant theory of probability that will do the job.

information-based account of knowledge in [Dretske, 1981] I assumed (without arguing for) an objective interpretation. There are, I think, strong reasons for preferring this approach, but strictly speaking, this is optional. The probabilities can be given a subjective interpretation with little or no change in the formal machinery. What changes (for the worse, I would argue) are the epistemological consequences.

If probability is understood objectively, an informational account of knowledge takes on some of the characteristics of a causal theory of knowledge.[8] According to a causal theory of knowledge, a belief qualifies as knowledge only if the belief stands in an appropriate causal relation to the facts. I know Judy left the party early, for instance, only if her early departure causes me to believe it (either by my seeing her leave or by someone else — who saw her leave early — telling me she left). Whether my belief that she left early is caused in the right way is presumably an objective matter. It doesn't depend on whether I or anyone else know it was caused in the right way. For this reason everyone (including me) may be wrong in thinking that I (who believes Judy left early) know she left early. Or everyone (including me) may be wrong in thinking I don't know she left early. Whether or not I know depends on facts, possibly unknown, about the causal etiology of my belief. If probability is (like causality) an objective relation between events, then an information-theoretic account of knowledge has the same result. Whether or not someone knows is a matter about which everyone (including the knower) may be ignorant. To know whether S knows something — that Judy left early, say — requires knowing whether S's belief that Judy left early meets appropriate informational (i.e., probabilistic) conditions, and this is a piece of knowledge that people (including *S* herself) may well not have.

If, on the other hand, probability is given a subjective interpretation, information — and therefore the knowledge that depends on it — takes on a more relativistic character.

Whether or not *S* knows now depends on who is attributing the knowledge. It will depend on (and thus vary with) the attributor of knowledge because, presumably, the person who is attributing the knowledge will be doing the interpreting on which the probabilities and, therefore, the information and, therefore, the knowledge depends. As a result, it will turn out that you and I can both speak truly when you assert and I deny that *S* knows Judy left early. Contextualism (see [Cohen, 1986; 1988; 1999; DeRose, 1995; Feldman, 1999; Heller, 1999; Lewis, 1996]) in the theory of knowledge is a view that embraces this result.

B. Necessary Truths

Communication theory defines the amount of transmitted information between source and receiver in terms of the conditional probabilities between events that occur, or might have occurred, at these two places. As long as what occurs at the source generates information — as long, that is, as the condition existing at a source is a *contingent* state of affairs (a state of affairs for which there are possible

[8]Goldman [1967] gives a classic statement of this theory.

alternatives) there will always be a set of events (the totality of events that might have occurred there) over which these probabilities are defined. But if the targeted condition is one for which there are no possible alternatives, a *necessary* state of affairs, no information is generated. Since a necessary state of affairs generates zero information, every other state (no matter how informationally impoverished it might be) carries an amount of information (i.e., ≥ 0 bits) needed to know about its existence. According to communication theory, then, it would seem that nothing (in the way of information) is needed to know that 3 is the cube root of 27. Or, to put the same point differently, informationally speaking anything whatsoever is good enough to know a necessary truth. Bubba's assurances are good enough to know that 3 is the cube root of 27 because his assurances carry all the information generated by that fact. Mathematical knowledge appears to be cheap indeed.

One way to deal with this problem is to accept a subjective account of probability. The village idiot's assurances that 3 is the cube root of 27 need not carry the information that 3 is the cube root of 27 if probability is a measure of, say, one's willingness to bet or one's level of confidence. On this interpretation, the probability that 3 is the cube root of 27, given (only) Bubba's assurances, may be anything between 0 and 1. Whether or not I know, on the basis of Bubba's assurances, that 3 is the cube root of 27, will then depend on how willing I am to trust Bubba. *That* will determine whether Bubba is a suitable informant about mathematics, a suitable channel for getting information about the cube root of 27.

Another way to deal with this problem is to retain an objective interpretation of probability but insist that the equivocation on the channel connecting you to the facts, the channel involving (in this case) Bubba's pronouncements, is to be computed by the entire set of things Bubba might say (on all manner of topics), not just what he happened to say about the cube root of 27. If equivocation (and, thus, amount of transmitted information) is computed in this way, then whether or not one receives information about the cube root of 27 from Bubba depends on how generally reliable Bubba is. Generally speaking, on all kinds of topics, is Bubba a reliable informant? If not, then whether or not he is telling the truth about the cube root of 27, whether or not he *could* be wrong about that, he is not a purveyor of information. One cannot learn, cannot come to know, that 3 is the cube root of 27 *from him*. If Bubba is a generally reliable informant, on the other hand, then he is someone from whom one can learn mathematics as well as any other subject about which he is generally reliable.

A third way to deal with the problem, the way I took in Dretske [1981], is to restrict one's theory of knowledge to perceptual knowledge or (more generally) to knowledge of contingent (empirical) fact. Since a contingent fact is a fact for which there are possible alternatives, a fact that might not have been a fact, a fact that (because it has a probability less than one) generates information, one will always have a channel of communication between knower and known that is possibly equivocal, a channel that might mislead. If a theory of knowledge is a theory about this limited domain of facts, a theory (merely) of *empirical* knowledge, then communication theory is prepared to say something about an essential ingredient in such knowledge. It tells you what the channel between source and receiver must be like for someone at the receiver to learn, come to know, empirical facts about the source.

C. How Much Information is Enough?

I have been assuming that information is necessary for knowledge. The employer can't know who was selected — that it was Herman — unless he receives the required information. Following a natural line of thought, I have also been assuming that if information is understood in a communication-theoretic sense, then the amount of information received about who was selected has to be equal to (or greater) than the amount of information generated by the selection. So if Herman's selection generates 3 bits of information (there are eight employees, each of which has an equal chance of being selected), then to know who was selected you have to receive some communication (e.g., a message with the name "Herman" on it) that carries at least that much information about who was selected. If it carries only 2.75 bits of information, as it did in the hypothetical case where employees were determined to protect (i.e., not name) Barbara, then the message, although enough (if it carries the name "Herman") to produce true belief, could not produce knowledge. In order to know what happened at s you have to receive as much information — in this case 3 bits — about s as is generated by the event you believe to have occurred there.

My examples were deliberately chosen to support this judgment. But there are other examples, or other ways of framing the same example, that suggest otherwise. So, for instance, suppose the employees' messages are not so rigidly determined. Messages bearing the name "Herman" make it 99% probable that Herman was selected, messages bearing the name "Barbara" make it 98% probable that Barbara was chosen, and so on (with correspondingly high probabilities) for the remaining six employees. As long as these probabilities are neither 0 nor 1, the individual contributions to equivocation (see footnote 2) will be positive. The equivocation, E, on the channel will, therefore, be greater than 0 and the amount of transmitted information will be less than the amount of information generated. Messages about an employee's selection will *never* carry as much information as is generated by that employee's selection. Full and complete information about who was selected, the kind of information (I have been arguing) required to know who was selected, will never be transmitted by these messages. Can this be right? Is it clear that messages sent on this channel do not carry the requisite information? Why can't the employer know Herman was selected if he receives a memo with the name "Herman" on it? The probability is, after all, .99.

If a probability of .99 is not high enough, we can make the equivocation even less and the amount of information transmitted even greater by increasing probabilities. We can make the probability that X was selected, given that his or her name appears on the memo, .999 or .9999. As long as this probability is less than 1, equivocation is positive and the amount of transmitted information less than information generated. Should we conclude, though, that however high the probabilities become, as long as $E > 0$ and, therefore, $I_t < I_g$), not enough information is transmitted to yield knowledge? If we say this, doesn't this make the informational price of knowledge unacceptably high? Isn't this an open embrace of skepticism?

If, on the other hand, we relax standards and say that enough information about conditions at a source is communicated to know that what condition exists there even when there is a permissibly small amount of equivocation, what is permissibly small? If, in order to know that Herman was selected, we don't need all the information generated by his selection, how much information is enough?

Non-skeptics are tugged in two directions here. In order to avoid skepticism, they want conditions for knowledge that can, at least in clear cases of knowledge, be satisfied. On the other hand, they do not want conditions that are too easily satisfied else clear and widely shared intuitions about what it takes to know are violated. Reasonable beliefs, beliefs that are very probably true, are clearly not good enough. Most people would say, for instance, that if S is drawing balls at random from a collection of balls (100, say) only one of which is white, all the rest being black, you can't, before you see the color of the ball, know that S selected a black ball even though you know the probability of its being black is 99%. S might, for all you know, have picked the white ball. Things like that happen. Not often, but often enough to discredit a claim that (before you peek) you know it didn't happen on this occasion. Examples like this suggest that knowledge requires eliminating *all* (reasonable? relevant?) chances of being wrong, and elimination of these is simply another way of requiring that the amount of information received about the state known to exist be (at least) as much as the amount of information generated by that state.

There are different strategies for dealing with this problem. One can adopt a relativistic picture of knowledge attributions wherein the amount of information needed to know depends on contextual factors. In some contexts, reasonably high probabilities are enough. In other contexts, perhaps they are not enough. How high the probabilities must be, how much equivocation is tolerated, will depend on such things as how important it is to be right about what is occurring at the source (do lives depend on your being right or is it just a matter of getting a higher score on an inconsequential examination?), how salient the possibilities are of being wrong, and so on.

A second possible way of dealing with the problem, one that retains an absolute (i.e., non-relativistic) picture of knowledge, is to adopt a more flexible (I would say more realistic) way of thinking about the conditional probabilities defining equivocation and, therefore, amount of transmitted information. Probabilities, in so far as they are relevant to practical affairs, are always computed against a set of circumstances that are assumed to be fixed or stable. The conditional probability of s, an event at a source, given r, the condition at the receiver is really the probability of s, given r within a background of stable or fixed circumstances B. To say that these circumstances are fixed or stable is *not* to say that they cannot change. It is only to say that for purposes of reckoning conditional probabilities, such changes are set aside as irrelevant. They are ignored. If the batteries in a measuring instrument are brand new, then even if it is possible, even if there is a non-zero probability, that new batteries are defective, that possibility is ignored in calculating the amount of information the instrument is delivering about the quantity it is being used to measure. The non-zero probability that B fails, that the batteries are defective, does not contribute to the equivocation of instruments for which B holds, instruments whose batteries are functioning well. The same is true of all communication channels. The fact — if it is a fact — that there is a non-zero probability that there were hallucinatory drugs in my morning coffee, does not make my current (perfectly veridical) experience of bananas in the local grocery store equivocal. It doesn't prevent my perception of bananas from delivering the information needed to know that they (what I see) are bananas. It doesn't because the equivocation of the information delivery system, my perceptual system, is computed taking as given the de facto condition (no hallucinatory drugs) of the channel. Possible (non-actual) conditions of this channel are ignored even if there is a non-zero probability that they actually exist. The communication of information depends on their being, in fact, a reliable channel between a source and a receiver. It doesn't require that this reliability itself be necessary.

BIBLIOGRAPHY

[Cherry, 1957] C. Cherry. On Human Communication, Cambridge, MA: MIT Press, 1957.

[Cohen, 1986] S. Cohen. Knowledge and Context. *Journal of Philosophy* 83:10, 1986.

[Cohen, 1988] S. Cohen. How to Be a Fallibilist. *Philosophical Perspectives*, Vol. 2, J. Tomberlin, ed. Atascadero, CA; Ridgeview Publishing, 1988.

[Cohen, 1991] S. Cohen. Scepticism, Relevance, and Relativity. In *Dretske and His Critics*. Cambridge, MA; Blackwell: 17-37, 1988.

[Cohen, 1999] S. Cohen. Contextualism, Skepticism, and the Structure of Reasons. In [Tomberlin, 1999, pp. 57-90].

[DeRose, 1995] K. DeRose. Solving the Skeptical Problem. *Philosophical Review*, 104.1: 1-52, 1995.

[Dretske, 1981] F. Dretske. Knowledge and the Flow of Information, Cambridge, MA: MIT Press, 1981.

[Dretske, 1983] F. Dretske. Multiple book review of Knowledge and the Flow of Information, Behavioral and Brain Sciences 6 (1): 55-89, 1983.

[Feldman, 1999] R. Feldman. Contextualism and Skepticism. In [Tomberlin, 1999, pp. 91–114].

[Goldman, 1967] A. Goldman. A Causal Theory of Knowing. *Journal of Philosophy*, 64: 357-72, 1967.

[Goldman, 1976] A. Goldman. Discrimination and Perceptual Knowledge. *Journal of Philosophy*, 73: 771-791, 1976.

[Heller, 1999] M. Heller. The proper role for contextualism in an anti-luck epistemology. In [Tomberlin, 1999, pp. 115–130] .

[Lewis, 1996] D. Lewis. Elusive Knowledge. *Australian Journal of Philosophy*, 74.4: 549-567, 1996.

[Loewer, 1983] B. Loewer. Information and Belief. *Behavioral and Brain Sciences* 6: 75-76, 1983.

[Nozick, 1981] R. Nozick. *Philosophical Explanations*. Cambridge, MA; Harvard University Press, 1981.

[Sayre, 1965] K. Sayre. Recognition: A Study in the Philosophy of Artificial Intelligence, South Bend, IN: University of Notre Dame Press, 1965.

[Shannon, 1948] C. Shannon. The mathematical theory of communication. *Bell Systems Technical Journal* 27: 379-423, 623-56, 1948; reprinted with introduction by W. Weaver, Urbana, IL: The University of Illinois Press, 1949.

INFORMATION IN NATURAL LANGUAGE

Hans Kamp and Martin Stokhof

1 INTRODUCTION

Natural languages are vehicles of information, arguably the most important, certainly the most ubiquitous that humans possess. Our everyday interactions with the world, with each other and with ourselves depend on them. And even where in the specialised contexts of science we use dedicated formalisms to convey information, their use is embedded in natural language.

This omnipresence of natural language is due in large part to its flexibility, which is almost always a virtue, sometimes a vice. Natural languages are able to carry information in a wide variety of ways, about a seemingly unlimited range of topics, which makes them both efficient and versatile, and hence useful in almost every circumstance. But sometimes, when pinpoint precision is what counts, this versatility can get in the way, and we make use of formal languages, such as those of mathematics.

The variety of ways in which the use of natural language involves information, reveals itself immediately if we look at the various functions that utterances of natural language expressions may have. First, many of the utterances we produce serve to directly impart information to our readers or listeners — usually information which we take to be new and of interest to them. We describe situations, stating what we take to be facts ('Mary is in Paris'), or contemplating what we regard as possibilities ('John might be accompanying her'). This declarative use of language is perhaps the most obvious way in which natural languages are used to convey information.

But, of course, this doesn't hold for all utterances. We also ask questions ('What time does the meeting start?'), in order to elicit information rather than to impart it; we give directives ('Open a window', 'Stay away from her'), in order to get the other to do certain things, or to keep him from doing them; we issue warnings and threats ('Look out, a bus!', 'If you do that, I'll tell the boss'), we express regret and joy ('I apologise for the belated reply, ...', 'Congratulations!'), and so on. But in these cases, too, our utterances carry information, and that they do so is essential: a question must convey what information is requested; a directive must specify information about what is to be done, or to be refrained from; a warning or threat must identify a particular situation or event; and if we don't convey what it is that we regret, or what we are happy about, the point of our speech is lost.

Handbook of the Philosophy of Science. Volume 8: Philosophy of Information
Volume editors: Pieter Adriaans and Johan van Benthem. General editors: Dov M. Gabbay, Paul Thagard and John Woods.

These humdrum observations also illustrate a further point. Not only do natural language utterances involve information about a variety of types of situations: factual and possible, past, present and future; they also convey information about the specific attitudes that natural language users have concerning these situations: that the speaker takes them to be factual, or merely possible; that they are to be avoided, or to be realised, by the hearer; that the speaker regards them with regret, or with joy. Thus the information carrying capacity of a natural language encompasses not just what its expressions are about, but also the various attitudes that its users may have towards that.

Another way in which information might be conveyed is more indirect than in the examples above, where it is coded in the syntactic form or indicated by a particular expression or turn of phrase. Making use of the context in which an utterance is produced, for example by relying on the presence of certain expectations on the part of the hearer, we may also indirectly convey information about a certain situation. For example, when answering a question about the whereabouts of Jane by means of a disjunction ('She's either in Paris, or in London, with Mary'), we indicate that we do not know exactly where she is. This is information that is not explicitly stated, but only suggested. However, an addressee who expects the speaker to be as co-operative as he can will pick up this information without hesitation.

And it doesn't stop there. Besides utterances of the above kinds there are those which serve various social purposes: greeting someone, acknowledging a gesture or utterance she is making, expressing concern or empathy. Many utterances we use to such ends — 'Hi, how are you?', 'I am sorry to hear that', and so on — are formulaic. They carry information, not in virtue of being about something and expressing an attitude to that, but by being fixed through special conventions, which bypass the general mechanisms by which information is associated with linguistic form. But these utterances do carry information nonetheless, as is indicated by the fact that the purposes they serve can as a rule also be accomplished by means of other, non-formulaic utterances.

Yet another way in which information is conveyed by natural language is through mechanisms that relate a specific utterance to its linguistic context. After all, an utterance hardly ever occurs on its own, out of the blue; usually it is part of a larger whole, a text or a conversation, that serves a specific purpose and accordingly may have a specific form. When it is part of such a larger textual or conversational complex an individual utterance may contribute to the meaning of that complex as a whole through mechanisms that relate it to other parts of the complex. The use of pronouns to refer to an entity mentioned previously in a conversation (A: 'I met John the other day.' B: 'How's he doing?') is a simple example; the specific form of a question — answer dialogue, in which answers more often than not are fragments of complete sentences, yet do express complete propositions (A: 'Who's chairing the meeting?' B: 'Bill.'), provides another.

As they stand, all these observations, with their repeated references to 'information', are, we take it, hardly controversial. But to say precisely what the information is of which they speak is not so easy. For one thing, it is not at all clear that we are dealing with a uniform concept: when trying to explain in what sense natural languages are information carriers, we may well find that it is necessary to distinguish various 'kinds' of information. And should that need arise, there will be the further task of saying exactly how these different notions are related to each other and how natural languages are able to handle various notions of information in such elegant and efficient ways. To outline some aspects of the current state of thinking about these issues is the goal of the present chapter.

But first we should make clear what the information concept that we talk about in this chapter is not. We are not concerned with information based on mere likelihood, according to which the information carried by a symbol or symbol string is some inverse function of the probability of its occurrence. Common to such a concept of information and the one that will be relevant in this chapter is the conception of individual events that are classified as each being of some particular event type. In our case the event types are the types of symbols and symbol strings and the individual events are particular occurrences ('utterances') of symbols and strings of them. The probability-based notion of information presupposes in addition to such a space of classifiable occurrences a probability distribution over possible occurrences, which assigns each occurrence of an individual event an a priori probability in terms of the classification-related properties it has. On the other hand, what is essential to the concept of information that will be discussed here is that symbols and symbol complexes have denotations, i.e., that they stand for, or represent, entities and situations, and that the information they carry is about those denotations.

On a simple-minded, purely causal conception of how symbols denote the two conceptions of information would be compatible. On such a view, the occurrence of symbols (both simple and complex) is prompted by the occurrence of their denotations. So the space of symbol occurrences maps onto a corresponding space of denotations, and the probability of a symbol occurrence is the direct reflection of the occurrence probability of the denotation that is its cause. In that case the information represented by the occurrence of a given symbol would be the occurrence of its denotation and the quantity of that information could be meaningfully assessed in terms of the probability that the denotation should have occurred. We will see, however, that in connection with natural languages such a conception of denotation is untenable. The simple causal nature of the denotation relation which it presupposes is belied, both at the level of the simple symbols of the language (its 'words') and at that of its complex symbols (its phrases, sentences, texts and conversations), by the way in which natural languages actually work.

The first, and most widely acknowledged difficulty with a purely causal conception of denotation concerns the denotation of complex symbols. The denotations of phrases and sentences are determined by their syntactic form and by the denotations of the words from which they are made up. In principle the recursive process by which the denotations of complex symbols are built from the denotations of their constituents might have a purely causal grounding. But any serious exploration of the way in which natural language expressions denote soon reveals the extreme implausibility of this. Complex expressions denote what they do because of the denotation building rules that are part of the language as a conventional system; and speakers can use these expressions to refer to their denotations because they know those rules and thus know that a given complex phrase or sentence does have the denotation to which they want to refer. In other words, information as conveyed by natural language utterances depends on a conceptualisation of what the information is about that, at least to a large extent, is shared between the users of the language.[1]

The existence of a set of conventional rules for building complex expressions to denote complex entities or situations is something that natural languages share with the formal languages of logic, mathematics, and computer science. But, as we will argue in some detail below, there are also important differences between natural and formal languages. One of these is that in natural languages the principles which govern the building of expressions to denote complex things or situations are far more complex than the comparatively straightforward recursive principles that define formal languages (like those of the predicate calculus or the lambda-calculus). This greater complexity is connected with the remarkable flexibility and adaptability of natural languages, which makes it possible to use them for the purpose of conveying information about a vast and open-ended range of different subjects.

This is connected with another feature of natural languages, viz., that they can be used to speak about non-existent objects, unrealised situations. In some cases expressions and linguistic constructions are even meant to do just that, e.g., when we use a counterfactual sentence ('If I had left home earlier, I wouldn't have missed the five o' clock train'), in other cases the possibility is left open, e.g., when we utter a conditional sentence ('If John comes to the party too, Mary will be upset', where the actual appearance of John is neither affirmed nor denied). And then again we may be convinced that what we say is true, whereas in fact things are not as we assert them to be. 'Information', then, is used here in such a way that it can also be false (just as it can be misleading, or partial): the information provided by an utterance, i.e., what anybody who understands its linguistic meaning might assume to be the case, need not actually hold. By no means should this be considered as a defect of natural languages. In fact, it is an unavoidable consequence of the partial and fallible nature of human knowledge and our ability to imagine what we know or have reason to think is not the case

[1]More on this below, in section 3. It should be noted that this observation is not meant to rule out the possibility of 'non-conceptual content': it pertains to the information expressed by means of utterances of linguistic expressions only and remains neutral with respect to the question whether objects, events, situations —including linguistic expressions and their use — may also convey information of a different nature. Also note that we take utterances (i.e., the production of 'tokens') to be primary to expressions (conceived of as 'types') when it comes to what are the entities that carry information. But in as much as there are systematic relations between the two, we sometimes also talk about expressions in that vain. We assume that no confusion will arise from this.

on the one hand, and on the other the fact that natural languages are means of expressing not only what we think is the case, but also what we suspect may be the case, what we hope, fear, would wish to be the case, and so on.[2]

There is an obvious connection between denotation and meaning: the meaning of a linguistic expression is given by what it denotes, in actual situations or in non-actual ones. Since the notion of linguistic information we are after is also closely tied to denotation, there is an intimate connection between linguistic information and linguistic meaning. The fact that both linguistic meaning and linguistic information are connected with denotation entails an important moral for either. Both linguistic meaning and linguistic information are inherently relational concepts, both involve the form-governed relation between linguistic expressions and their denotations. This is a moral that some would consider too obvious to merit stating. But the relational nature of meaning is not something that has always been self-evident to everyone. In fact, the moderately clear picture of the ways in which the meanings of linguistic expressions are relational that we now possess is the outcome of a long process of philosophical analysis. Because the two notions are so closely intertwined the history of the concept of linguistic meaning is at the same time also the history of linguistic information. Therefore we will devote the first part of this chapter to tracing what from the perspective of one particular tradition, viz., that of formal semantics and its immediate predecessors, are seen as some of the salient stations in the historical development that has led up to the current state of thinking on these issues. Probably, from other angles different pictures would emerge, but it is beyond the scope of this chapter to sketch those as well. So the emphasis is on history as perceived by the discipline, not as it actually occurred (if there is such a thing). For it is the former, and not the latter, that best explains its development.

This concise historical overview also shows how the formal semantics tradition has struggled to come to grips with the variety of ways in which natural language utterances carry information that we briefly touched upon above. That process is one of both contraction and expansion, as is so often the case in the development of a scientific discipline. At one stage there is a focus on one specific aspect of a phenomenon, which often allows the use of formal tools and leads to precise accounts. At another stage the resulting notions are extended to deal with other

[2]It is this very potential of natural languages to be about about non-actual objects and situations that according to Frege liberates the human mind and sets us apart from other animals. In his 'Über die wissenschaftliche Berechtigung einer Begriffsschrift' ([Frege, 1882]; the translation is Bartlett's [Frege, 1964]) he writes:

> Nonetheless, our imagery [...] would be limited to that which our hand could form, our voice intone, if it were not for the grand discovery of the symbol which calls to our mind that which is absent, out of sight or perhaps even unseeable.

And George Steiner regards the very possibility it gives us to talk about the non-actual as fundamental for human language [Steiner, 1975, page 215]:

> Hypotheticals, 'imaginaries', conditionals, the syntax of counterfactuality and contingency may very well be the generative centres of human speech.

aspects, or complemented by other notions, as the situation may require.

After the historical sketch we turn to an overview of various elements that would be needed in an adequate theory of natural language information. Such a theory must do several things. First of all, it should give an account of how expressions of natural language come to have meaning, and of the ways in which meaning depends on the context of utterance. This involves, among others things, coming up with a suitable set of formal concepts that can be used to define adequate representations of natural language meanings, to model the relevant features of context and to characterise the way in which these interact. Second, the scope of an adequate theory should encompass the fact that natural languages are used in interactive situations: natural languages are used to convey meaning for a reason, and that reason lies in information exchange, broadly conceived. Thus, the information conveying capabilities of natural languages are tailored to their use in a discourse context, be it dialogical, textual, or of some other form. As a matter of fact, many features of these capabilities depend on structural features of such discourses, which, hence, need to be modelled. Third, it is the language users that need to be taken into account: what information the utterance of a natural language expression conveys, and how it does that, obviously depends also on the language users involved, both as speakers and as hearers. Modelling them, then, is yet another essential ingredient of an adequate theory of natural language information. The overview we will be giving follows the broad outlines of an approach that has become well-established in empirical work in natural language semantics over the last couple of decades. But we should emphasise that it is not this particular approach we want to propagate, but rather the underlying ideas that together form a theme that allows for many variations: some of these we will refer to when appropriate.

2 A TALE OF TWO DEVELOPMENTS

As is so often the case, a systematic discipline has a somewhat distorted picture of its own history, one that usually takes 'a bird's eye view' and focuses on those aspects that retain a certain present relevance. For us, taking such a bird's eye view of what we see as the important stages in philosophical and linguistic thinking about the concepts of information and meaning, one development that stands out is that from 'thick' and fairly concrete notions of meaning, closely tied to (perceptual) experience, judgement and application, to rather 'thin' and abstract conceptions and (ultimately) to a view of natural languages as purely information coding and information transferring devices.[3] This line of development is complemented, however, by another one that extends the restricted, 'descriptive' conception of linguistic meaning that is the outcome of the former, and tries to

[3]This is not unlike Gøran Sundholm's [to appear] view on development of logic: from a theory about judgements and reasoning as psychological acts to (ultimately) formal symbol manipulation. The distinction between 'thick' and 'thin' concepts is taken from Bernard Williams, who developed it with regard to ethical concepts.

enrich it by encompassing a wider variety of aspects and by reinstating connections with other components of human cognition. The first development is assumed to have its starting point in traditional philosophical thinking,[4] it gains momentum with the development of formal logic at the end of the nineteenth century, comes to fruition in the 1970s and 1980s, and still remains strong until the present day. The latter development is partly a reaction to the former and mainly dates from the last two or three decades, after systematic thinking about linguistic meaning developed into a distinct discipline.

2.1 Uncovering structure

An, admittedly very rough, sketch of the first line of development distinguishes the following stages. At the first stage, which is assumed to start in classical philosophy and to extend right up to the rise of modern philosophy in the sixteenth and seventeenth century, thinking about the concept of meaning usually is intimately related with metaphysical and epistemological concerns. The latter obviously take precedence, and meaning, and language more generally, as such are by and large not distinct and independent topics of concern. Language and linguistic meaning are viewed and analysed primarily as means to express judgements, and it is the origin, content and justification of judgements that most philosophers are interested in.

For example, Plato's discussion of the possibility of false statements in the Sophist is motivated by a metaphysical concern about the possibility of knowledge and the relation between thought and reality, not by any autonomous interest in natural language meaning.[5] Similarly, the main motivation behind the work of the scholastics on language and logic is metaphysical (and some of it theological). And the 'idea theories of meaning' of the classical empiricism and rationalism of the sixteenth and seventeenth centuries are mainly motivated by questions and problems in epistemology.[6] From a modern, systematic perspective there seems

---

[4]In this short sketch we limit ourselves to developments in Western philosophy. That is not to deny that very interesting theories and views, that are highly relevant from a systematic point of view, have been developed in other traditions. Especially in India there is a rich tradition of sophisticated thinking about language, as is witnessed by the great works of Pāṇini and other Indian grammarians (cf., [Cardona, 1988. 2nd ed 1997]). However, historically these have not played a major role in shaping present day theories in semantics, and it is for that reason that we feel it is justified to leave them out.

[5]Thus the discussion of word and sentence meaning and of truth and falsity, in the Sophist, 261c6–264b3 [Plato, 1921], ends as follows:

> Then because speech, we saw, is true and false, and thinking is a dialogue of the mind with itself, and opinion is that completion of thought, and what we say by "it seems" is a combination of perception and opinion, it must be that because all of these are like speech, some thinking and opinion must also be false.

Evidently, the linguistic analysis is subservient to the metaphysical point that Plato wants to make.

[6]Hence Ian Hacking [1975] called idea theories 'nobody's theory of meaning': since meaning as such is not a separate concern, nobody had a *theory* about it, or even felt the need to come up with one.

to be no such thing as a separate philosophy of language in this period, nor is there a distinct and substantial empirical discipline of linguistics that is concerned with the analysis of natural language meaning for its own sake.[7] It would take a century or so for linguistics to really come into its own, with the seminal work of Humboldt and others, and almost another one for language to become a separate and central topic in philosophy.

Nevertheless, a case can be made that this is the stage that most closely resembles a 'common sense theory of meaning'. From a common sense perspective it seems plausible that the meaning of a declarative sentence and the judgement that it serves to express are the same.[8] No strict separation between 'propositional content' and 'illocutionary force' seems to be called for. Also, what the sentence means, the judgement it expresses, and what in reality justifies that judgement seem to be not really distinguished: how language relates to reality, the question that haunts much of the later philosophical thinking, thus never comes into proper focus. The way in which we form judgements about reality — be it either in empiristic fashion, by receiving impressions through the senses and manipulating them, or more rationalistically, with more of the content being innate to the mind — is the way in which language 'hooks up' with it, except that it needs no hooks, since the relation is immediate.

The second stage in the development towards a more independent and more abstract conception of linguistic information is characterised by the rise of 'meaning proper' in the wake of the development of modern logic, mainly through the work of Frege, Russell, and early Wittgenstein. One of the hallmarks of Frege's philosophy of logic is his anti-psychologism: in order to give logic its proper due, he claims, we need to separate it from 'psychology', i.e., we need to distinguish the subject of logic, viz., the systematic explication of the validity of inference, from the empirical study of actual judgements and actual reasoning. In his logical theorising Frege developed his position gradually. In the *Begriffsschrift* [Frege, 1879] he distinguishes between judgement and content, noting that the content of, e.g., an hypothetical judgement cannot be expressed in terms of the judgement of the antecedent and that of the consequent, but has to be defined in terms of their respective contents. However, he does still formulate his logic using a separate sign, the 'judgement stroke', for the judgement as such. Later on, he states that only the content of a judgement, but not the actual act of judging that content as

[7]Which is not to say that no work was being done that we could call 'linguistic' or 'semantic', for there certainly was. There is a whole tradition of thinking about grammar that goes back to Hellenistic times, at least, and within logic, there are penetrating analyses of the functions of expressions in, e.g., the Stoic school and in medieval scholastic thinking. The point is that in many cases the analyses developed are subservient to different goals, and that both the results and the ways in which these are argued for, are rather different from how the issues are approached in modern times. But, of course, that does not mean that no interesting insights were developed along the way. Cf., [Robins, 1990] for an overview. [Seuren, 1998] is an example of an approach that is not purely historical, but attempts to connect the development of linguistics with modern systematic theories.

[8]Which is not to say that it can not be, and has not been, challenged. Cf., e.g., [Dummett, 2004, page 1] for a dissenting opinion from an anti-realistic point of view.

true (or false, or plausible, or ...), plays a role in the normative theory of valid deductive reasoning.[9] As Frege states in 'Der Gedanke', which dates from 1918: logic is concerned with the 'laws of truth', and these laws can be regarded also as the 'laws of thought', but *not* in the sense of laws covering 'general features of thinking as a mental occurrence'.[10] His argument, characteristically concise, is that 'error and superstition have causes just as much as correct cognition' and a study of actual thought would need to treat them on a par with correct judgement and valid reasoning, which contravenes the true task of logic. Rather than being a description of how we actually think and reason, logic is a normative theory that states how we should. Similarly, where in the early *Begriffsschrift* Frege uses the term 'Vorstellungsinhalt' (lit., 'content of imagination') to refer to contents of judgements, he later acknowledges that this term may lead to confusion since it is (also) used in a psychological sense, and instead settles on the term 'Gedanke' ('thought'), which is supposed not to carry such connotations.[11] No doubt also inspired by Frege, Wittgenstein claimed in the *Tractatus* that 'psychology is no more closely related to philosophy than any other natural science', immediately following up with the claim that epistemology is 'the philosophy of psychology'.[12] Thus the idea that it is possible to treat language and meaning separately from questions regarding judgement and justification is gaining ground, and with that, the contours of modern philosophy of language become visible.

The separation from epistemology did not carry with it a similar move away from metaphysical concerns: the analysis of language and meaning remained strongly related to ontology. In part this is due to the particular philosophical aims to which people at the time made the analysis of meaning subservient. Stimulated by the success of the use of formal languages in the 'new logic', the age-old quest for a philosophically transparent ('ideal') language gained new momentum. This time it would be a strictly formal one, and it would provide philosophers with an analytic tool that could be used with scientific precision and mathematical rigour. This 'linguistic turn' put language centre stage in philosophy, and consequently turned philosophy of language into a distinct and central discipline.[13] This is not the place to trace what happened to the idea of linguistic analysis as a philosophical tool

---

[9] A point concisely expressed by Wittgenstein in the *Tractatus*, where, referring to the *Begriffsschrift* he remarks parenthetically in 4.412: '(Frege's "judgement stroke" "⊢" is logically quite meaningless)'.

[10] Cf., [Frege, 1918–19]; quotations are taken from the English translation by Peter Geach in [Frege, 1977].

[11] Cf., Frege's comment from 1910 to Jourdain, who had written a summary of the *Begriffsschrift* in a paper on the history of logic: 'For this word I now simply say 'Gedanke'. The word 'Vorstellungsinhalt' is used now in a psychological, now in a logical sense. Since this creates obscurities, I think it is best not to use this word at all in logic.' [Frege, 1879, page 11].

[12] [Wittgenstein, 1960, 4.1121]. Cf., also Wittgenstein's attempt to give a extensional analysis of so-called 'propositional attitude' statements in 5.541 ff.

[13] It is interesting to note that in the history of phenomenology, associated with the work of Husserl, Heidegger, Ricoeur, Merleau-Ponty and others, a similar development took place, but without the strict separation from epistemology that is characteristic for analytic philosophy. Cf., [Dummett, 1996] for more details.

employed outside the analysis of meaning proper.[14] What is relevant here is how it influenced subsequent theories about natural language information and natural language meaning. And from that perspective it is important to briefly point out three general characteristics that have been very influential, viz., 'universalism', 'intensional referentialism', and 'compositionality'.

'Universalism' refers to the nature of the task that philosophical analysis sets itself, viz., to give an account of 'how language operates' in general, with no reference to any specific features of any specific language in particular. What is of interest is not the way a certain language works, but what underlies the possibility of any language to express meaning. A straightforward feature, perhaps, of any philosophical analysis worth its salt, but one that will turn out to have repercussions for the form and the application of theories that are subsequently based on this idea. For in the application to concrete, empirical cases the universalistic and a prioristic features of these theories do not simply disappear. In many cases they become consolidated in the use of certain formal tools and in the adherence to particular basic methodological principles that are applied 'across the board' and that are even taken for granted as defining characteristics of the enterprise.

'Intensional referentialism' indicates the central role of the notions of reference and truth in the analysis of meaning, combined with the use of an intensional ontology consisting of possible situations and the entities of which such situations consist. Together these two assumptions, or requirements, tend to favour a fairly abstract notion of meaning, one that is grounded in the possibility of words having referential relations to objects, properties and relations in the world, where the relation of reference is understood as a truly intensional concept, not in any way restricted to reality as we know it: 'the world' can be any one from a set of logically possible ones.

Meanings of complex expressions, including sentences, are then assumed to be somehow constructed from these basic referential relations, which means that compositionality is assigned a key role.[15] The result is an approach to meaning that is detached from actual reality and actual language use, one that works in a bottom up fashion, constructing complex meanings from basic ones, and that assigns the basic meanings a fairly independent status: they are 'self-sufficient' in

[14]There are a large number of studies dealing with this topic; cf., [Biletzki and Matar, 1998; Soames, 2003].

[15]Compositionality extends the expressive power of a language — the range of different meanings it is able to express — beyond that of its simplest expressions (its 'words'). How far it does, depends on the kind of compositionality that the language allows. It is commonly assumed that most (and presumably all) human languages display a kind of compositionality that is genuinely recursive and that permits the construction of infinitely many expressions of unbounded complexity from a finite vocabulary. This sets human languages, as well as many formal languages, such as that of the predicate calculus, apart from simple signalling systems, in which each of a certain finite set of signs corresponds to one state of the system's fixed application domain (like, say, the set of traffic signs of the traffic code of a given country), and also from language-like systems with limited forms of compositionality, such as the 'language' of the bee-dance or the languages used by chimpanzees who have acquired the ability to produce the sign for 'green banana' on the basis of having separately learnt the sign for 'banana' and that for 'green'.

so far as they have determinate and thoroughly non-contextual identity conditions.

The third stage that we need to distinguish in this brief historic sketch is that in which semantics arises as a separate discipline. This happened in the late 1960s, early 1970s, when developments in philosophy, logic and linguistics came together and gave rise to the idea that a formal theory of meaning can be developed and applied in the description of actual natural languages. This is the time in which people like Donald Davidson, Richard Montague, David Lewis, Max Cresswell, and a great many others did their seminal work.[16] This time is the heyday of 'Montague grammar' (and its various variations and rivals) as a grand unifying framework, in which the conception of meaning that was developed mainly from a philosophical perspective at an earlier stage, was formalised using various logical techniques (borrowed from model theory, modal logic, type theory, tense logic, etc.), and applied in the description of natural languages. This approach to natural language semantics, aptly dubbed 'formal semantics', proved very successful and was the dominant one for quite some time, in particular in philosophy, less so in linguistics at large.[17]

One thing that is important from the perspective of this chapter is that through the extensive use of formal languages as tools for modelling natural language meaning yet another shift in that concept occurs: meanings now are first and foremost formal constructs, and theories of meaning are primarily differentiated in terms of the formal machinery one deems necessary for the description of semantic features of natural languages:[18] concerns with epistemology or ontology become less and less important as semantics becomes more and more autonomous, and the nature of the concept of meaning reflects this. Montague's claim, in 'Universal Grammar' [Montague, 1970b], that 'there is in my opinion no important theoretical difference between natural languages and the formal languages of logicians' and that therefore 'it [is] possible to comprehend the syntax and semantics of both kinds of languages within a single natural and mathematically precise theory' testifies to this shift. The consequences are far-reaching. For one thing, although Montague seems to think of logic and semantics as some kind of 'equal partners', the practice is less symmetrical: it is formal languages that are used as models for natural languages, and this implies a sharpened focus on those aspects of meaning that can indeed be dealt with using existing logical techniques, and a proportionate

[16]Cf., [Davidson, 1967; Montague, 1973; Lewis, 1970; Cresswell, 1973]. Other seminal work was done by Barbara Partee [1973]. Though less directed to natural language the work done by Jaakko Hintikka, David Kaplan and Saul Kripke in that period was also of fundamental importance.

[17]In particular within the Chomskyan tradition people tended to reject the use of model-theoretic techniques, and pursued a different approach, that is more in line with Chomsky's idea that linguistics is a branch of cognitive psychology, and, ultimately, of biology. Cf., further below.

[18]One could say that as a result of this shift semantics deals with an altogether different type of phenomena. Although this may seem exaggerated — and it probably is — it does point to a curious and slightly worrisome fact, viz., that there seems to be no theory-independent agreement about what exactly the domain of semantics consists of. This is reinforced when one takes a closer look, e.g., at the kind of arguments that formal semanticists and Chomskyan semanticists bring to bear on their dispute. Cf., [Stokhof, 2002] for some more discussion.

neglect of those that can't. The distinction between 'structural semantics' and 'lexical semantics', arguably one that is not in any sense inherent in meaning itself but rather an artifact of the kind of instruments one wants to use, is maximally exploited and the resulting concept of meaning becomes both more formal and more 'thin'.

At this third stage the three features identified above are very much present, although not always explicitly so. Looking at the abundance of descriptions of various semantic phenomena in a wide variety of languages produced in the 1970s and 1980s, one might think that 'universalism', the idea that a proper semantic theory deals with natural language semantics as such, isn't something that people subscribed to. And indeed, the very fact that semanticists deal with actual phenomena, some of which are specific to a particular language, indicates that their concern is not that of the philosophers at an earlier stage. Nevertheless, the use of a unified framework has universalistic consequences, whether intended or not. The point is that the framework itself embodies assumptions about what meanings are, how they are related to each other, how they are expressed, and so on. So right in the framework itself there is a conceptual structure, explicated by means of the formal properties of the concepts and languages that are used, that shapes a concept of natural language meaning that is independent of any concrete manifestation in any concrete natural language.[19]

The other two features, pertaining to the central role of reference and truth and the use of an intensional framework, and to compositionality as the basic principle for dealing with semantic complexity and creativity, are less hidden and more explicitly adhered to. Despite discussion about the kinds and the number of intensional concepts that one needs to employ, the common denominator is the use of a formal framework that models 'the world' — i.e., that to which the expressions of the language bear a referential relation and in terms of which the concept of truth for the language is defined — in an abstract, and, one might almost be tempted to say, 'detached' way. 'The world' is reduced to the bare minimum of components and structure that is needed to define what kinds of things the referents of various types of basic expressions are, compositionality being understood to take care of the rest. It is important to note that it is not actual reference that is defined or explicated, it is only the formal type of relationship involved that is being accounted for.

The resulting picture, which for a long time served as the classical model for semantics of natural language and which we will refer to as such in what follows, in many ways comes close to that of a natural language as a formal language — significantly, a formal language without a concrete application. It portrays natural

[19]This is particularly clear in the work of Donald Davidson, who actually uses the logical structure of a semantic theory, which according to him takes the shape of a Tarski-style theory of truth, in a transcendental argument against 'the very idea of a conceptual scheme', arguing that because the semantics of any language can only be described by means of such a theory and because the very framework of that theory implicates substantial properties of the meanings expressed in the language, all languages are essentially translatable into each other [Davidson, 1974].

languages as information carrying devices in the fairly abstract sense in which the same characterisation can be given of many other information carrying systems, ranging from signalling systems to mathematical notation. But, as was already indicated in the introductory section, if we look more closely at the various ways in which natural languages convey information, at what kind of information that is and what it is about, we encounter a much richer structure, and one that is tied more closely to the actual world that we live and use our language in than is accounted for in this approach. Small wonder, then, that after its initial success and broad acceptance the classical model became gradually discredited. At first one tried to augment it with additions that put more semantic flesh on its formal bones; later it was supplanted altogether by approaches in which the flesh is taken as seriously as the bones.

2.2 Reinstating content

The 'counter current' that contributed to a much more balanced picture of the specific characteristics of how natural languages act as information carrying devices does not represent one, homogeneous conception of meaning, rather it springs from a number of sources. These do have one thing in common, though, which is a profound dissatisfaction with the conception of linguistic meaning that informs the formal semantics of the 1970s. Different people addressed different aspects of that dissatisfaction; together they effected a shift in the orientation of natural language semantics that is still taking place today. Again, we should note that this development primarily is a reaction to a 'self-styled' history, which only partly covers what actually occurred in philosophical thinking about language and meaning. Obviously, there is the work of a number of authors who already early on explored different directions that implicitly challenged some of the basic assumptions of the classical model, e.g., the 'linguistic phenomenology of J. L. Austin, H. P. Grice's work on meaning and intention, and the work on speech acts of John Searle,[20] much of which was inspired by Wittgenstein's later work.[21] But this work only became influential after formal semantics had gone through an autonomous development, and even then it was taken up not in semantics proper, but mainly in a theory of pragmatics, which was supposed to complement it.

The conception of meaning that people reacted against can be dubbed 'classical descriptivism'. Central to this conception is the essentially Fregean principle that the meaning of an expression determines its reference by providing a specification of the conditions that something needs to satisfy in order to count as being the referent. The Fregean concept of 'Sinn' is explicated formally by reconstructing it as a function that takes a possible world (or other such intensional construct) as its argument and delivers an entity (an individual, set of individuals, or set of n-tuples of individuals, as the case may be) that acts as the referent in that world. In line with the above-mentioned distinction between structural and lexical

[20]Cf., e.g., [Austin, 1962; Grice, 1957; Searle, 1969].

[21]Primarily his *Philosophical Investigations* [Wittgenstein, 1958].

semantics, the actual specification of these functions for concrete expressions was by and large considered not to belong to the subject matter of semantics. Instead one focused on the various ways in which these largely unspecified functions can be combined to form appropriate meanings for larger expressions, in particular sentences, yet another illustration of the pivotal role of compositionality.

By thus reducing both the 'world' (that which natural languages are about) and the meanings of particular words and phrases to formal structures many questions were bracketed out that both linguists and philosophers would consider it their task to answer: questions as to how concrete expressions actually refer to concrete objects or properties, how such referential relations arise, what role contingent features of the way the world is have to play in that process, how considerations regarding the communicative functions of natural language utterances might interfere, how the use of language interacts with other cognitive functions, how utterances employ features of the linguistic and non-linguistic context in which they are produced, and a host of others. It is to the neglect of such questions that people reacted and which motivated them to develop alternative approaches.

Consequently, we can, admittedly somewhat arbitrarily, identify four separate sources of this counter current, one that is concerned with the role of the world, another that focuses on the variety of communicative uses, a third that insists on taking indexicality and the linguistic context seriously, and a fourth that investigates the cognitive status of language and its relations to other cognitive structures and functions. Of course, these divisions are to some extent artificial, but they serve to indicate major trends.

The first source of dissatisfaction with classical descriptivism relates to the minimal role that it assigns to the world and our interactions with it. One central question here is how linguistic meaning comes about, a question that actually reinstates the connection with traditional, basic epistemological concerns. And, as in the tradition, there are two main points of view, an internalistic and an externalistic one.[22] The central claim of semantic externalism is that meaning derives from the world, at least substantially.[23] It is from our environment and our interactions with it that natural language expressions get their meanings, and to a large extent the processes involved are of a causal nature. Hence this view is often also referred to as a 'causal theory of reference'.

According to this externalistic view natural language meanings can, to a large extent at least, be naturalised: the contents of many natural language expressions

[22]What is called 'internalism' and 'externalism' here, in the context of semantics, should not be confused with the 'internalism — externalism' debate in the philosophy of mind and in epistemology, although there are connections, of course. In the philosophy of mind internalism and externalism are rival views on the nature of mental content, centring around the question whether mental content can be completely understood in terms of internal mental representations, and, ultimately, perhaps entirely in terms of brain states..

[23]This is a crude generalisation, of course. There are many, often subtle variations on this theme that are lumped together here under the one heading 'externalism'. Cf., [McGinn, 1989] for an overview. The locus classicus of semantic externalism is Putnam's 'The meaning of "meaning" ' [Putnam, 1975], which is also one of the classic sources of the theory of direct reference to which Kripke, Donnellan and others contributed.

can be identified with real situations, events, objects, properties and relations — entities belonging to the external world but with which the language user can interact through perception and action. The most explicit realisation of this viewpoint within formal semantic theorising, and the most systematic formal attempt to restore the relationship between meanings and their naturalistic determinants,[24]are situation semantics and its logical-philosophical foundation, situation theory.[25]

Taken to its extreme, radical externalism involves naturalising all aspects of meaning. One reason why someone might think that such a radically externalistic account of linguistic meaning ought to be possible is that, arguably, all our concepts are ultimately the product of our interactions with the world in which we live, and thus are, in some fashion, reflections of the ways in which that world imposes itself upon us in the course of those interactions. But this consideration overlooks the fact that even where experience of the world is causally involved in the construction of the kind of information that linguistic expressions convey, this information cannot be equated with that experience.[26] There is no a priori reason to suppose that the world, our experience of it, and how we conceptualise and express it in natural language have the same fine structure. Rather, there are a number of good reasons to doubt that this is the case. For one thing, there is the moulding role that our cognitive system may exert on the form and structure of the experience. And many of our expressions refer to things in the world that exist at least partly because of our shared language and the way we use it. In fact, for all we know, linguistic representation makes its own contributions to the ontology of natural languages, which includes entities the reality of which is confined to aspects of the 'world' that our language projects, and which have no right of being in any language-independent sense. So it seems that although experience allows information — in the sense of enabling it by anchoring the terms of our language in an external world, thereby creating the possibility of objective reference and truth — it does not determine it completely: in general, the information that is conveyed by means of natural language is the product of more factors than experience alone. And that entails that a complete naturalising of meaning is not possible.

The next question then is what else might be needed for meaning. Several answers are possible, one of which is provided by internalism. Like externalism, internalism aims to enrich the meaning content of expressions. But it does so via a different route, viz., through an appeal to substantial contents and structures that are supposed to be resident in the human mind. From the internalistic perspective the mind contains a rich repertoire of basic contents, in the form of innate concepts and features and of structural operations, that together allow for the formation of the huge variety of actual meaning contents that we find expressed in natural languages. As such, internalism naturally allies with the equally rational-

[24]And thereby also traditional connections between philosophy of language, epistemology, and psychology of a particular bend, viz., naturalistic and empiricist psychology,

[25]Cf., [Barwise and Perry, 1983; Barwise and Seligman, 1997].

[26]Cf., also Dretske's analysis of the concept of information in epistemology, in this volume.

istic conception of grammar that is so characteristic for the Chomskyan paradigm in linguistics. Nevertheless, internalism, too, faces some difficult questions, some of which are conceptual: 'What explains the origins of all these mind contents?', 'How can we account for the application of mental content to reality?', others empirical: 'What is the explanation of semantic variety across languages?'.[27]

Also, it should be noted that externalism (usually) and internalism (by definition) are individualistic: they take the individual human being as their point of reference when discussing linguistic and mental content and its relation to the world. This is in accordance with much of the main stream thinking in philosophy of language, philosophy of mind, semantics, and linguistics. Again, a reflection of this is the central role that is played by compositionality. From an individualistic perspective what is often called the 'creativity' of language, viz., the potential infinity of structures and meanings that together make up a language, poses a serious problem. How can individuals, being finite creatures with finite memory and finite computational resources, be considered competent users of their language? Compositionality comes to the rescue: it not only characterises languages conceived as formal objects, but is also posited as an inherent feature of human linguistic competence.[28]

Nevertheless, there remains the interesting question whether the individualism that characterises both externalism and internalism makes these accounts too restrictive. Internalism seems to have a hard time accounting for the availability and contents of concepts that rely on the existence of social institutions, and faces serious problems when dealing with phenomena such as distributed information and reliance on expert knowledge.[29] That we could locate the concepts involved in such phenomena exclusively 'in the mind' seems improbable. For the externalistic perspective individualism becomes problematic when it is robustly physicalistic. A lot of mental content and linguistic meaning seems to defy a straightforward reduction to physicalistic causes. Note that the problem here is not one for physicalism as a doctrine concerning the nature of scientific explanation. Whether or not that is a tenable position does not depend on the possibility of giving a physicalistic account of all of linguistic meaning, for one could argue that some such meanings simply have no role to play in an ultimate scientific account of the world. But from a semantic point of view this is different, since we obviously want a semantic theory to account for all linguistic meaning, including the meanings of those parts of the language that certain views on scientific explanation would consider irrelevant to their concerns. This does not rule out externalism per se, but it does indicate that an externalistic account of natural language meaning needs

[27]Cf., [Farkas, 2006] for a recent overview of externalistic and internalistic perspectives in the philosophy of language.

[28]Cf. [Groenendijk and Stokhof, 2005] for some discussion about how these various elements are usually linked up, and for some discussion of possible alternative ways of accounting for competence.

[29]Which is one of the central points in Putnam's original 1975 paper (cf., footnote 23). For attempts to account for such issues in terms of the distinction between 'broad', externally determined content and 'narrow', internal and individual content, cf., [Fodor, 1987].

to take into account that whatever causal relations are involved in producing it, they are not monostratal and uniform, but rather play at different levels and are of different types; that they involve radically different kinds of entities, including various sorts of social entities; and that they work in both directions, from the world to meaning and vice versa. The structure of meaning is partly due to the structure of the world, but the structure of our world is also partly a linguistic one.

Such an approach transcends the structure of radical externalism as we characterised it above. In particular, the causal processes which it would take into account are not simply just the ones that govern the perceptions of individual speakers. This applies specifically to those processes that are needed to account for the meanings of social terms, among them those that pertain to the interactions between verbally communicating speakers of a given language.[30] One effect of the impact of these additional causal relations, which connect the members of a given (speech) community rather than any one of them to a particular content, is that these linguistic meanings aren't the private property of individual speakers, but rather a shared possession of the language community as a whole. For such expressions the ultimate linguistic competence rests with the community, and the competence of any particular member of that community is determined by the degree to which he partakes in that common good. Such a move away from mainstream individualism could also account for the real diversity of experience and the diversity of information, not necessarily parallel to the first, that we find across individual language users. Viewed from the perspective of a community, experience is heterogeneous, but connected, and the same holds for information. It is precisely this diversity that is one of the main reasons why humans use such complicated, expressive languages as they do.

The last observation is connected with the second source of the counter current to the classical model, which is a concern for the complexity and the variety of the communicative uses that are made of natural languages. In the introduction we hinted at this by giving some simple examples of other uses than the straightforwardly declarative use. Quite in line with its ancestry in logic and philosophical analysis the classical model focuses on declarative utterances. Actually, just as the 'judging' element from the traditional notion of a judgement was first isolated and then dropped by Frege, leaving only the contents of judgements as the material to which logic was supposed to apply, analogously looking just at declarative utterances made it easy to first isolate the 'use' part of an utterance and then focus exclusively on the resulting content, turning formal semantics into a theory of pure contents, radically dissociated from the various ways in which these can be used. Such a separation between what is often called 'mood' (or 'illocutionary force') and 'radical' (i.e., propositional content) goes back to Frege and was taken up later in various forms by people like Austin, Stenius, and Searle.[31] The result-

[30]This could also be called a form of externalism, viz., 'social externalism'. Cf., e.g., [Burge, 1990].

[31]Cf., [Austin, 1962; Stenius, 1967; Searle, 1969].

ing speech act theory made this distinction into one of its basic principles. Thus a division of labour arose between formal semantics as an account of propositional content and speech act theory, or pragmatics in a wider sense, as a theory of the use that is made of these contents.

However, some have questioned whether this strategy will work. For one thing the variety of uses we make of natural language expressions does not seem to be one-to-one related to the mood-radical distinctions we can make within these expressions, be it on the basis of syntactic form (interrogative, indicative, ...), the presence of lexical items, or a combination thereof. And then there are aspects of meaning, i.e., information conveyed through a speaker's utterance to other interlocutors, that are not in any obvious way coded into the expressions uttered, but that arise from the interplay between the context in which the utterance occurs, the intentions and expectations of the various speech participants, and other meaning elements. These 'implicatures', as they are called, have been studied extensively; and they have given rise to serious questions about the tenability of the classical model. Like in the case of speech acts, the initial approach towards an account of such indirectly conveyed meaning depended on a division of labour, in this case between semantics as conceived in the classical model and a pragmatic theory called the 'logic of conversation', developed by H. P. Grice.[32] Grice's central idea was that language use is a cooperative task and that therefore language users can be expected to obey certain rational principles of communication, such as telling the truth (as they see it), giving sufficient but no superfluous information, and so on.

One problem with Grice's original approach concerns one of its starting points: one of Grice's main motivations was to show that certain aspects of the meaning of natural language connectives that are not captured by their extensional two-valued logical counterparts (for example, the order sensitivity of natural language conjunction) can be accounted for by an appeal to cooperative principles. A closer look at the apparent meanings of 'and' co-ordinations in English (the same also applies to other languages) reveals that their meaning depends on factors that go beyond the mere truth table of classical logical conjunction and are also different from the conversational principles Grice invokes. In particular, the order-sensitivity of 'and' co-ordinations is largely the effect of the mechanisms of interclausal temporal anaphora, mechanisms that are operative also where no 'and' is in sight, and that any theory of natural language meaning and information will have to account for in any case.

What goes for 'and' goes for most applications to which Gricean conversation theory has been put: The principles of the theory are important and indispensable, but so are other principles, which also transcend the restricted conception of meaning that is part of the classical model. And again and again it has been found that deciding which of these principles should be counted as semantic and which as pragmatic is possible only on theory-internal grounds.[33] This has led

[32]Cf., [Grice, 1975]; [Levinson, 1983] is an excellent introduction to this and related subjects.

[33]Cf., the discussions in [Recanati, 2004; van Rooij, 2004b; Stanley, 2005], and the contributions

to the view that the demarcation between semantic and extra-semantic (= pragmatic) aspects of meaning is to a considerable extent arbitrary, and has thereby undermined another fundamental assumption of the classical model.

What has thus emerged in lieu of the classical model is a far more complex account in which a great variety of principles and mechanisms collaborate in the construction of utterance meanings out of the meanings of the words contained in them. Some have taken the reasoning that has led to this line of development one step further and argued that even the concept of 'literal meaning' that the classical model, speech act theory and Gricean pragmatics all rely on is a myth. In a theory of literal and non-literal meaning the words of the language have literal meanings, which are encoded in the lexicon. These serve as a starting point for the derivation, via inferential processes that take various pragmatic factors into account, of other, non-literal meanings, and, on the basis of these, of the specifications of individual utterance contents. But here too, it is argued, we are dealing with a distinction — that between the literal meanings of words and their non-literal meanings — which proves to be slippery and hard to draw except on theory-internal grounds. One major empirical problem is the existence of (productive) polysemy. The assumption of literal meaning forces one to try to account for the various meanings of, e.g., 'running' as it occurs in 'The tap is running', 'John is running', 'The program is running', 'My nose is running', etc., by picking one meaning as the core, or 'literal' one and then accounting for the others on the basis of some contextual derivational process. A more plausible alternative is to forgo the choice and account for this type of variability by making lexical meanings themselves contextual and flexible, in effect viewing linguistic meaning as something that is the result of interaction between a language user and his environment.[34]

Emphasis on interaction with the environment, especially the communicative environment, consisting of other speech participants, conversational goals, information about the world (individual and shared), and so on, is characteristic for the third source of the counter current, the one that focuses on context in this broad sense. An important shift in the way meaning is viewed that is characteristic for this development is the result of a turn away from the exclusively descriptive orientation, with its emphasis on the language – world relation, that is a central feature of the classical model, to a perspective on language and language use that analyses them primarily in terms of information and information exchange.[35] The resulting view is one in which the primary focus is on the 'horizontal' relation between language users engaged in an information exchange discourse, with the 'vertical' relation of language to world entering only indirectly, and no longer playing the lead role. The information exchanged in a discourse can be quite diverse: usually, part of it will be information about the world, but at least as important

in [Szabó, 2005].

[34]Cf., [Bartsch, 1996] for more discussion and a concrete model along these lines.

[35]Stalnaker's work on presupposition and assertion is an early representative of this conceptual turn. Cf., the two seminal papers [Stalnaker, 1974] and [Stalnaker, 1979].

is information of speech participants about each other, and information about the discourse itself. When engaging in a conversation, but also when reading a text or listening to a speech, what the participants know, or think they know, about each other plays a crucial role in interpretation, and, derivatively, also in production. (A speaker will choose the expression she utters so that it will lead, to the best of her knowledge, her audience to assign to it the interpretation she intends, given the total package of information, about world, antecedent discourse and her own state of mind, that she assumes is available to that audience.) Stalnaker's notion of 'common ground', i.e., the information that the speech participants assume they share, is an important element in this, since it provides them with common resources for picking out individuals, properties and situations, solving (co)referential relationships, and so on. But the common ground will normally also include information of all the different kinds we have mentioned, not just assumptions that directly concern the topic of conversation.

In addition to what is being said by whom to whom, i.e., content in the narrow sense, it is also form that matters for determining what information gets exchanged. Among the natural language devices that serve this purpose we find: anaphoric expressions of various kinds, among them pronouns, tenses, and certain temporal and spatial adverbs, which permit resumption of entities previously introduced into the discourse; presupposition-inducing expressions, that enrich and structure the common ground; the order in which the events that make up a narrated episode are described, which usually indicates the temporal ordering of those events; and so on. These and other devices help the hearer to relate the information conveyed by an utterance to the information he already has, and thus to identify exactly what the new information is. As such they are an integral part of what linguistic meaning is and how linguistic expressions convey information. At yet another level, not so much concerned with linguistic form or narrow content, there is information about the aims with which speech participants have entered the conversation, their rhetorical strategies, and other features of their linguistic personae. This type of information is crucial for the detection of irony or sarcasm, the appreciation of a verbal sleight of hand or a clever play on words, and for the recognition of an implicit reproach or a concealed request. These aspects of discourse, too, are factors that enter into the way in which natural language utterances play their information conveying role.

These considerations have given rise to a variety of alternatives to the classical model. In as much as all these models share the shift from the descriptive to the information exchange perspective, along with a shift from the sentential to the discourse level, they can be captured under a common denominator, that of 'dynamic theories of meaning'.[36] These theories take the development outlined

[36]Thus the original model of discourse representation theory developed by Kamp in [Kamp, 1981] (cf., also [Kamp and Reyle, 1993]), explicitly aims to combine a declarative and a procedural view on natural language meaning. Other models of dynamic semantics include Heim's file change semantics [Heim, 1983], Veltman's update semantics [Veltman, 1996], and Groenendijk and Stokhof's dynamic semantics [Groenendijk and Stokhof, 1990; Groenendijk and Stokhof, 1991], cf., also [Groenendijk *et al.*, 1996].

above one step further and change the notion of meaning itself: the descriptive and referential and hence truth oriented perspective of the classical model is replaced by a dynamic one that views the meaning of expressions in terms of what is called their 'context change potential', with information being one, central aspect of the context. This further shifts, or rather blurs the original distinction between semantics and pragmatics (i.e., the distinction between what is supposed to be a matter of meaning proper and what belongs to the realm of use). Accordingly the focus of research in these theories is no longer on the referential and logical features of linguistic meaning but on issues involving information structure (topic – focus, presupposition, anaphoric relations, intonation and prosody) as linguistic devices that can be used to link a new sentence in a text or a new utterance in a conversation to what went before, or to prepare the ground for what comes next. This increasing focus on information exchange and information structure also weakens the link with ontology that in the classical model was secured through the central role of reference and truth. In a dynamic perspective truth becomes a mere limit concept of the more general notion of acceptance by the speech participants of information that is being exchanged.[37]

Integral to the dynamic view on meaning as context change potential is a renewed interest in the cognitive function of meaning. This ties in with the fourth source of the counter current that we discerned above, viz., a renewed interest in the cognitive aspects of language and its relations to other cognitive systems. The development of the classical model in the 1970s brought along a new and somewhat problematic relationship with psychology. On the one hand its proponents, sometimes explicitly, more often implicitly, took over Frege's anti-psychologism, that made a principled distinction between logic as a normative science and the empirical study of actual reasoning, and they applied it to the study of natural language meaning, separating formal description of semantic structure from the study of the way in which language is produced and interpreted. But unlike logic, semantics never really was conceived as a purely formal discipline; after all, its aim is to describe and explain empirical facts, and it is therefore considered to be as much a branch of empirical linguistics as phonology or syntax.[38]

From that perspective the classical model should have been quite compatible with the Chomskyan approach to grammar. But in fact the relationship turned out to be more complicated. For one thing, the Chomskyan model involved a close alliance with rationalistic thought and with the computational approach in cognitive psychology that developed from the 1960s onwards. But not everybody felt comfortable with these particular philosophical presuppositions, and many semanticists working within the classical model preferred to keep their distance. In turn, many Chomskyans, including Chomsky himself,[39] kept formal semantics

[37]This does not necessarily imply that the concept of truth is a completely epistemic notion. That depends on how states of complete information relate to states of the world. In fact, in the theories mentioned in footnote 36 the notion of truth is as objective as it is in formal theories that implement the classical conception.

[38]Cf., [Stokhof, 2002] for some more discussion of this tension.

[39]Thus early on, replying to a suggestion from Bar-Hillel that formal logic might contribute

at bay, arguing that the use of the concepts of truth and reference as central tools in the explication of meaning disqualified the classical model as far too externalistic to be compatible with the internalistic approach that they favoured. Semantics in the Chomskyan framework accordingly concentrated primarily on the way in which conceptual structure is expressed, mainly in lexical semantics.[40] In this connection the approach of 'cognitive semantics'[41] should be mentioned as well. Though in many ways adverse to the generative framework as developed by Chomsky in his later work, it shares with that approach a focus on lexical semantics and an unwillingness to account for meaning in terms of reference, using the tools of logic in the way exemplified by the formal implementations of the classical conception, in particular in model-theoretic semantics. Characteristic for cognitive semantics is the emphasis on the fluidity of the distinction between semantic knowledge and encyclopedic knowledge and on the embodied nature of meaning.

With its focus on formal properties of natural language meanings, the classical model initially succeeded in maintaining something of a 'splendid isolation' from empirical work in psychology and biology. But as the counter current grew stronger, as more aspects of use were taken into account, context became more and more important, users and their conversational goals and strategies were incorporated as significant aspects of the context and as the emphasis accordingly shifted from formal structure to actual content and its use, these barriers began to crumble. For many it has become increasingly obvious that one of the tasks of semantics is a realistic modelling of language users and their interactions, for in the end natural language meaning can be properly understood only if we understand how it functions in real information exchanges and other linguistic interactions.

This has brought about a certain rapprochement between semantics and psychology, and to some extent also between formal semanticists and people working in the Chomskyan tradition. This rapprochement has also been helped by a growing interest in lexical semantics on the part of formal semanticists, who at long last have begun to respond to the charge that if all meanings are derived from lexical meanings, then explaining how they are derived is not good enough if one has nothing to say about what they are derived from. Nevertheless there remain substantial differences between the internalistic and the externalistic perspective (the former being preferred by those who take the Chomskyan approach). But as the computational model in cognitive psychology began to loose its grip, it became clear that the study of how language functions as one of the human cognitive faculties does not necessarily commit one to an internalistic view. There is room for a variety of perspectives, some working with a model that is individualistic and

to linguistics, Chomsky stated that 'the relevance of logical syntax and semantics [to the study of natural language] is very dubious' [Chomsky, 1955]. And throughout the years Chomsky expressed similar sentiments on a number of occasions. For example, in [Chomsky, 2005] he states, in keeping with his internalistic perspective, that 'even the most elementary concepts of human language do not relate to mind-independent objects by means of some reference-like relation between symbols and identifiable physical features of the external world'.

[40]Cf., [Jackendoff, 1990; Pustejovsky, 1995].

[41]Cf., [Lakoff, 1987; Talmy, 2000].

internalistic,[42] others favouring a more externalistic set up that emphasises the role of the linguistic community.[43]

The rapid development of new techniques for studying brain processes and the consequent rise of cognitive neuroscience during the last decade also has greatly contributed to a renewed interest in the underlying mechanisms of meaning. Language being one of the core cognitive functions of humans, it has always been an important object of study in cognitive psychology, as witnessed by a long tradition of studies in language acquisition, language pathologies, and language processing. For a long time such studies were generally based on computational, internalistic models of language, although some more empiristic and community oriented studies were undertaken as well.[44] The prospect of being able to study the brain almost 'in vivo' as it processes language, holds much promise. Particularly enticing is the possibility of experimentally testing different theoretical models that account for more or less the same linguistic data. The advent of more performance oriented models, such as dynamic semantics, optimality theory and game theoretical semantics have greatly facilitated this reorientation.[45] However, as our earlier discussions concerning externalism, internalism and individualism illustrate, we should be careful in our assessment of what exactly can be achieved in this fashion. The idea that research in cognitive neuroscience will be able to arbitrate between rival semantic frameworks all by itself is certainly not unproblematic: for one thing, the relationship between neurological correlates of semantic concepts and these concepts themselves cannot simply be taken for granted, and it seems that the relation between the two is much more symmetric than a reductionist approach would predict.[46] And the contributions of physical and social reality need to be taken into account as well.

Finally, it should be noted that the shift towards information exchange and other aspects of use that is embodied in these new approaches also has spurred a renewed interest in the biological and cultural origins of language, both phenotypically and genotypically.[47] Using techniques from evolutionary game theory and learning theory, semanticists have begun to study the way in which expressive systems can arise within a population of interacting agents, trying to isolate which factors are responsible for the characteristic features of human languages, notably recursive structure and semantic compositionality.[48]

---

[42] Cf., the references given above.

[43] Cf., [Tomasello, 2003].

[44] Cf., work by Bruner and others [Bruner, 1983; Garton, 1992], and early work in the connectionistic paradigm.

[45] Cf., footnote 36 for references to work on dynamic semantics in relation to natural language; cf., [van Eijck and Stokhof, 2006] for a more general overview of various concepts from dynamic logic. For optimality theoretic semantics, cf., [Hendriks and de Hoop, 2001], for game theoretical approaches, cf., [Hintikka, 1983].

[46] Cf., [Baggio *et al.*, to appear] for an in-depth discussion.

[47] Cf., [Tomasello, 1999], for an early, influential study.

[48] Cf., [Christiansen and Kirby, 2003] for a collection of papers that gives an overview of current thinking about language evolution; for recursive structure and compositionality, cf., [Kirby, 2000].

In conclusion, it seems fair to say that the current state of thinking in philosophy of language and natural language semantics about meaning is one of diversity. There seems to be no one, dominant conception of natural language meaning, and many, sometimes quite divergent approaches to its analysis are being pursued concurrently. The resulting situation might strike some as somewhat paradoxical: on the one hand all these abstractions have led to success, yet what it is they purport to study, viz., natural language meaning, seems to fade from view, at least as one coherent, unifying concept.[49]

Indeed, in some cases we appear to be dealing with incompatible underlying conceptions, as for example in the case of internalism and externalism. But more often it seems that differences arise because people focus on different aspects, and that, although it might not always look that way, the results could be unified in a single, more encompassing theory. The contours of such a theory are beginning to emerge, although no generally accepted format has been established as yet. It treats natural language meaning as a 'thick', i.e., substantial concept that gets its content and structure from a variety of sources (conversational goals, with a pivotal role for information exchange, the world, reflexive models of language users) and that ties in closely with other cognitive functions (perception, the emotional repertoire, everyday skills). Thus it reinstates the close relationship between meaning, information, judgement and the world that was characteristic for many of the earlier views on linguistic meaning that predate the classical model. But it does so based on a firm grasp of the underlying formal structure of the concepts involved, thus allowing for descriptions that have extended empirical scope and greater explanatory power.[50]

In the following sections we will illustrate a few important aspects of the present state of thinking about meaning and information in natural language by outlining in somewhat more detail the main elements of one particular way of describing and analysing how natural language expressions perform their information conveying roles. In section 3 we will discuss how the relational nature of linguistic meaning can be captured by means of representational techniques that are derived from model theory, allowing us to define the linguistic meaning of an expression in terms of the information carried by an utterance of it in various circumstances. The starting point of our exposition will be something akin to the classical model, which we will then subsequently modify and refine to capture more aspects of content and context. Next, section 4 will be devoted to an illustration of the way in which this particular conception can be used to capture how information is

[49]Which gives rise to difficult methodological questions as to what the nature of the success is: What is it that current semantics and philosophy of language are successful at? What are the measures of success here? Are these measures (relatively) theory independent? What do they apply to? And so on.

[50]It should be noted that there is also a continuing tendency toward the use of notions of meaning and information that are at a greater distance from what we could call the qualitative, common sense notion, as witnessed by the rise of purely quantitative, statistical notions of information in combination with the use of 'shallow', non-rule based techniques in certain approaches in natural language processing, information retrieval, semantic web, and so on.

conveyed in larger units of linguistic material, such as texts and conversations. We will illustrate the way in which anaphoric relations and presuppositions establish discourse connections that enter into the specification of the informational content of utterances, and we will show how the model set up in section 3 can be enriched so that it accounts for this. In section 5 we analyse how the state of the recipient enters into the picture, again indicating how the model can be adapted to account for this as well. In section 6 we come back to the question of what is characteristic of linguistic information. We conclude this chapter with a short overview of current further work in this area.

3 MODELLING MEANING IN CONTEXT

In the introduction to this chapter we observed that the notion of linguistic information is inseparable from that of linguistic meaning, that both are relational and that the richness of linguistic meaning is due in large part to the fact that the syntax and semantics of human languages involve recursion. In this section we discuss these issues in more detail.

First a few words on syntax. One of the oldest insights into language is that sentences have grammatical structure. For instance, the observation that the typical sentence of a language such as Latin, French, or English, contains a verb and that this verb has a subject can be found in the earliest grammars; and it is something that speakers of those languages will accept without demur when it is pointed out them, and that they might find out without much trouble for themselves. It is also plain, and no doubt always was, that simple sentences can be used as building blocks for larger sentences, e.g., as conjuncts, or as relative clauses, or as subordinate clauses beginning with subordinate conjunctions such as 'when', 'although', or 'because'. Speaking more generally, it was from the beginning a central aim of the 'Art of Grammar' to describe how grammatically correct sentences can be analysed into their grammatical constituents, as a way of proving that they are in accordance with what Grammar demands.

Modern generative grammar starts from a superficially different point of view, according to which sentences and other complex linguistic expressions are built from basic constituents (the words and morphemes of the language) according to rules that guarantee their grammaticality (or 'syntactic well-formedness', as terminology has it). And the way in which a grammatical expression is built from the words and morphemes occurring in it according to the rules of syntax shows its grammatical structure and is thus, once again, a demonstration of its grammatical correctness. What makes a generative grammar recursive is that some of its rules can be used repeatedly in the construction of a single sentence. More explicitly: the grammar is recursive if it has recursive rules — where a rule R is a recursive rule of a given grammar G if and only if for any number n there are sentences generated by G in which R is used at least n times. (In the generative grammars that have thus far been proposed for natural languages all or nearly all rules are recursive in this sense.)

In the end there is not much to choose between the generative and the analytical approach to grammar. In fact, on the basis of a generative grammar it is generally possible to construct parsers which compute syntactic analyses for those strings of words and morphemes that the grammar generates, by tracking how the string can be built using the grammar's rules. So, when we proceed, as we do, from the assumption that grammaticality is defined in terms of generative grammars we do so without loss of generality.

The first formally explicit accounts of natural language meaning made use of generative grammars that fit our characterisation of such grammars perfectly in that they consisted exclusively of generative rules, which serve to build complex expressions out of simpler ones. The accounts assumed that for each such rule R that tells us how expressions $e_1, \ldots, e_n$ can be combined into a complex expression e there is a corresponding semantic rule R' which states how the denotations $d_1, \ldots, d_n$ of $e_1, \ldots, e_n$ must be combined to obtain the denotation d of e.[51] As a matter of fact, natural languages do not take well to the comparatively rigid regime that is imposed by generative grammars of this strict and simple generative form, and more complex rule systems are needed if their syntax is to be captured in intuitively plausible and theoretically convincing terms. But for our present purposes the way in which these more complex systems determine the meanings of complex expressions is the same as it is for the simpler generative grammars described above and the extra complications can safely be set aside.

We will therefore assume that the syntax of natural languages can be given as consisting of (i) a set of rules, determining how complex expressions can be built from smaller ones, and (ii) a lexicon, specifying words and morphemes.[52]

All grammars make use of grammatical categories. This is true in particular of generative grammars: like other grammars they classify well-formed expressions into different categories. These categories are essential to generative grammars in as much as the rules refer to them. The perhaps best known illustration of this is the rule $S \rightarrow NP\ VP$, which, in some form or other, is part of most generative grammars that have been proposed for English. This rule says that an expression of the category 'S(entence)' can be formed by concatenating an expression of the category 'N(oun) P(hrase)' with an expression of the category 'V(erb) P(hrase)'. The members of a grammatical category can be either lexical items or complex expressions. Lexical categories are those which contain at least some words. (It is possible for a lexical category to consist of lexical items only, but in general this

[51]Cf, [Montague, 1970a].

[52]Among the building rules for a language like English there are those which state how full words can be built out of their stems by addition of certain morphemes. For instance, the past tense form 'called' of the verb 'to call' is formed by concatenating the stem 'call' with the past tense morpheme '–ed'. In what follows we will ignore the distinction between words, stems and morphemes. For our purposes morphemes and stems can both be thought of as 'lexical items', i.e., as elements of the vocabulary of the language, and full forms like 'called' can be thought of as complex expressions. (The interaction between syntax and morphology is one of the aspects of natural languages that make it awkward to press natural language grammars into the strict format of sets of construction rules.)

won't be so.) Examples of familiar lexical categories, which will be found in any grammar for a language such as English, are 'Noun', 'Verb', 'Adjective', 'Adverb' and 'Preposition'. In addition to lexical categories many grammars also postulate certain non-lexical categories, which contain no lexical items but only complex expressions.

For the theory of meaning grammatical categories are important in that expressions of the same category will have denotations of the same logical type. For instance, the denotations of expressions of category Noun generally are properties — or, in another formulation which we will favour in what follows, the extensions of those properties, i.e., the sets of entities of which a given property is true.[53] Another example: the denotations of elements of the category S, i.e., of well-formed sentences, are always propositions — the denotation of a sentence s is the proposition expressed by s — or, in the formulation favoured, the truth values of those propositions. It should be noted that being of the same category is a sufficient but not in general a necessary condition for having denotations of the same type. For instance, the denotations of verb phrases (i.e., members of the category VP) in many semantic theories are properties (or, alternatively, property extensions) just like the denotations of nouns.

So much for syntax. The standard method to account in a formally precise way for denotation and meaning is that of model theory. The method consists in (i) defining structures — the so-called 'models' — in which expressions of the different grammatical categories can be assigned suitable denotations; (ii) a specification in each model M of denotations for each of the lexical items of the language or language fragment L under consideration; and (ii), in order to account for the denotations of complex expressions, a general definition of how the denotation of any complex expression is determined by the denotations of its constituents. (Cf., the remarks made earlier about the semantics of generative rules.) Together (ii) and (iii) will assign in each model M a denotation to each well-formed expression. In particular we obtain a denotation in M for each of the sentences of L (which, as noted, will, depending on how the theory is set up, either be a proposition or a truth value ('true' in case the sentence is true on the interpretation provided by M or 'false' in case the sentence is false on that interpretation).[54]

The model-theoretic concept of meaning is relational in that it connects expressions and models. This can be seen most clearly for the case of sentences, assuming that their denotations are construed as truth values. On this assumption a given sentence s is positively related to those models in which its denotation is 'true' and negatively to those in which its denotation is 'false'. For expressions of other categories the matter is somewhat different insofar as their denotations aren't simply truth values. But here too the denotation is the product of the interaction

[53]'Generally' because, e.g., so-called 'natural kind terms' (nouns such as 'water' and 'gold') may be taken to denote, not properties, but abstract essences.

[54]In what follows we will, as indicated above, assume that the denotations of sentences are truth values and the denotations of nouns and other property expressions extensions; but we will briefly return to the other option, according to which sentences denote propositions and nouns properties, in section 3.2, footnote 59.

between expression and model, and can be seen as the manifestation of the way in which the two are related. We can isolate the contribution that the expressions make to these manifestations by associating with each expression e the function which maps each model M to the denotation of e in M. It has been suggested that the meaning of an expression can be identified with this function, in particular, that the meanings of sentences can be regarded as functions from models to truth values. We will see in sections 3.1 and 3.2, however, that such an identification isn't possible in general.

In order that a model-theoretic account for a language L does justice to our intuitions about what words and complex expressions mean, great care must be taken with the definition of its models. Of particular importance is that only such models be admitted in which the denotations of words represent realistic possibilities. To give just one example, assume that our account identifies denotations of nouns as sets. Let $n_1, \ldots, n_k$ be nouns. Then as a rule not any combination $S_1, \ldots, S_k$ of sets of entities will constitute a conceptually possible combination of denotations for these words. Suppose for instance that n_1 is the noun 'woman' and n_2 the noun 'man'. Then in any model M the denotations of n_1 and n_2 should be disjoint. This is a rather simple case of a semantic connection between two words that imposes a restriction on the models that should count as admissible in a satisfactory account of meaning. In general the connections are much more complex and more difficult to identify. And at the present time semantics is nowhere near a comprehensive inventory of the constraints that such connections impose.[55]

The reason why this issue is relevant for the specific concerns of this chapter is that what information a sentence carries depends on the models it excludes — i.e., those models which are incompatible with what the sentence says, and which we are entitled to ignore when we take the sentence as giving us true information. But evidently, what the set of those models is, and how large a proportion it represents of the totality of all admissible models, depends on which models are admissible to start with.

In view of the importance that the question of constraints on models has for the central topic of this chapter, it is appropriate to dwell on it a little more. First something that has been implicit in the assumption that the denotations of nouns are sets. On that assumption the elements of those sets must be in some way part of M. The way in which this requirement is met is to assume that each model M comes with a domain of entities, or 'individuals', which includes all denotations of nouns as subsets. The domain of individuals can be used as a foundation on which a hierarchy of further domains of other, higher logical types can be built, using certain elementary set-theoretical operations. Some of these higher type domains correspond to the logical types that are associated with grammatical categories of L, with the understanding that the denotations in M of expressions of a category C will be members of the domain of the logical type associated with C. For example,

[55] It should be noted that such conceptual restrictions derive from language, and one may well argue, as Quine has done in his attack on the analytic–synthetic distinction [Quine, 1953b], that they may not hold as such.

the denotations of nouns are members of the domain of the logical type associated with the category Noun. So, if noun denotations are sets of individuals of M, then this domain must be the power set of the domain of individuals, i.e., the set of all subsets of that domain.

A model M must thus have at a minimum the structure of a range of domains of different logical types, held together by the relations which are entailed by the way in which higher type domains are constructed from the individual domain. But this is only one, comparatively simple part of the structure that is presupposed by the constraints that single out the conceptually admissible models. For one thing, we need more structure within the different domains. This is true in particular of the domain of individuals itself. We already noted that the denotations of 'woman' and 'man' should always be disjoint. The same constraint applies to the nouns 'wife' and 'husband'. But connected with these last two nouns there is a further constraint. Their denotations are delimited by the restriction that they can be felicitously applied only to human beings; or — to put this in the current terminology used by many linguists — both 'wife' and 'husband' come with what is called a 'selection restriction' to the set of human beings. When we look more closely at the use of nouns (in English or other languages), we see that pretty much all of them come with selection restrictions of some kind. Furthermore, the sorts that form the selection restrictions of the different nouns of the language form a complex hierarchical structure, with some sorts being proper sub-sorts of others. A simple example: the noun 'bachelor' is, in its most prominent use, restricted to men of a certain age and social position (excluding for instance those who have made a formal vow of celibacy). So its selection restriction is a sub-sort of the selection restriction of 'husband' and 'wife'. A more thorough exploration of this phenomenon also makes clear that part of what is needed is an articulation of a sort hierarchy that provides, among other things, the selection restrictions of the different nouns of the language.

Another source of complexity is that most denotations change over time. In fact, this is so for two reasons, as can be seen plainly for nouns such as 'wife' and 'husband'. First, the set of human beings changes over time, as new people are born and other people die. So death affects the denotations of these nouns directly in that they lose members because these disappear from the scene altogether. But people also enter and leave the denotations of 'wife' and 'husband' while alive — viz., by getting married or divorced.[56] To do justice to this temporal dependence of the denotations of nouns and expressions of other categories, models must include a time structure. Since this is an aspect of model-theoretic meaning accounts that is especially important in connection with what will be said below, we do well to be a little more explicit about it. We keep things as simple as possible, assuming that each model M includes a time structure $\langle T, < \rangle$, where T is a set of temporal instants and $<$, the 'earlier-later' relation, is a linear ordering of T. Furthermore, the domain of individuals of M may now vary as a function of time — that is,

[56] As the tabloids keep reminding us, weaving your way in and out of these denotations can become a form of life in its own right.

as a function of T — and the same goes for the higher type domains that are constructed from domains of individuals, for the sortal hierarchies that subdivide these various domains, and for the denotations in M of the words of L. Thus each expression e of L no longer has a single denotation in M, but a possibly different denotation for each $t \in T$.

These are just some of the complications that model-theoretic accounts of meaning must address. This is not the place to do more than indicate that these issues need to be dealt with, and give a rough idea of what they are. But one further remark, of a more general tenor, is in order. Both the structure of sort hierarchies and the nature and structure of time are matters of ontology, the science of 'what there is'[57] This endeavour, of determining the kinds of entities that must be assumed to exist and their logical properties and relations, was for centuries the exclusive province of philosophy. In more recent times it has become a major concern in artificial intelligence and cognitive psychology and this is where now much of the kind of work on ontology is being done that is relevant to the theory of meaning. That is indicative of an important aspect of the meanings of linguistic expressions and the information they carry: ontology is not just a part of a theory of the meanings of words (although, as we have argued, it is an indispensable part of such a theory too), but rather a general theory of the structure of the world that presents itself to, and is projected by our cognitive faculties — of the different kinds of entities of which that structure is composed and of the principles that hold this multiplicity of kinds together. Up to a point the languages we speak presuppose and mimic this structure, it would be there even if we didn't speak a language, or didn't speak the particular languages that we do speak. But as pointed out earlier, languages also contribute to this ontology by projecting certain kinds of entities and structures on it. Thus, the 'ontology of language' is a complex affair, the result of external, causal influences from reality, the structuring principles underlying general cognitive abilities, such as perception, and the contributions made by linguistic structure.

Assuming that this assessment of the nature of ontology is correct, the claim that an account of natural language meaning must include parts of it amounts to the acknowledgement that the meanings of words (and, by implication, also those of larger expressions) are not 'autonomous', but are also constrained by general conditions that relate to the ways in which we perceive the world and structure our expectations about its regularities. A similar conclusion follows for the information that is carried by linguistic utterances: it too depends on the structure that cognition imposes on what it receives as input. Note that this implication is two-sided. On the one hand utterances could be said to succeed in carrying as much information as they do because their meaning implicitly relies on, and thus implicitly incorporates, so much of the cognitively based, though not specifically linguistic structures that our languages presuppose and exploit. On the other hand, the new information that an utterance conveys is limited by the fact that one must already be in possession of much of this implicit information

[57]Quine's happy phrase, cf., [Quine, 1948].

in order to be able to interpret the utterance in the first place.

3.1 Utterance dependence of content

The next aspect we must consider of the way in which the denotations of natural language expressions are determined is very different from the one we have just discussed and it requires a shift of perspective, from expressions as such to their uses on particular occasions — that is, to utterances. For an illustration of the point at issue consider the following sentences:

1. (a) That is a man.

   (b) He is a widower.

First (1a). One of the things one must understand in order to understand an utterance of (1a) is what is denoted by the word 'that'. What is its denotation? Well, that depends on whom the speaker of the utterance intends to denote by her use of 'that'. An interpreter will be able to determine what that entity is only insofar as the speaker provides him with some clue, for instance by pointing at the individual that she intends as denotation, or by gazing pointedly in its direction. This is a general property of 'that' and other so-called 'demonstrative' expressions: they can be used to denote pretty much anything, and what they denote on a particular occasion is determined by what the speaker wants them to denote, as long as she conveys this to her audience by providing the right clues. Much the same goes for personal pronouns like the 'he' occurring in (1b). The denotations of 'he' are more restricted in that they always must be male (and usually human). But which male is again a matter of the speaker's current intentions and her ability to get her intention across.

Because the denotations of the words 'that' and 'he' may vary from utterance to utterance, this is also true for the denotations of the sentences (1a) and (1b) themselves, since these depend on the denotations of the words they contain. There is however also another reason why the denotations of (1a) and (1b) vary, and this is a very general one. Time determines which denotations of the nouns occurring in (1a) and (1b), 'man' and 'widower', are to be combined with the denotations of 'that' and 'he', respectively. For instance, an utterance of (1b) at time t is a statement to the effect that the denotation of 'he' belongs to the denotation of 'widower' at t (rather than to the denotation of 'widower' at some other time). It is clear that this temporal dependence of (1a) and (1b) has to do with their tense, viz., that it is the present, rather than a past or future tense. For example, had the tense of (1b) been the simple past, as in (1c) below, then an utterance made at t would not have expressed that the denotation of 'he' belongs to the denotation of 'widower' at t, but to its denotation at some time before t:

1. (c) He was a widower.

Note well, however, that the utterance time is as indispensable to the interpretation of (1c) as it is to that of (1b). For although in the case of (1c) it is not the denotation of 'widower' at the utterance time t itself that is involved, the denotation times that are relevant are those which stand to t in a certain relation. They are times that precede t, and not t itself or times following it.

The temporal dependence exemplified in (1c), with its one verb in the simple past, is relatively simple. But a closer look at the full range of tenses, as well as at other expressions with which the tenses interact reveals a very complex field of temporal relations in which the events described may be linked to the utterance time complicated ways.[58] Here, however, it is not the complexity of these relations that matters, but the mere fact that what is needed to determine the denotations of sentences containing past or future tenses are not just the denotations of words and morphemes at the utterance time t, but also their denotations at other times. This is important because it entails that the denotations in a model M of sentences uttered at time t will in general require not just the denotations of their lexical constituents in M at t, but the entire 'temporal history' of M, providing denotations for all instants of its temporal structure.

Echoes of what we have just observed in connection with time can be found in the realm of modality, that part of the theory of meaning that has to do with the difference between the actual and the possible, the difference between what is true and what isn't but could have been. In fact, in the early days of the model-theoretic approach to meaning time and modality were treated as two dimensions of a simple ontological structure. Since then the general perspective has changed. According to more recent views the differences between time and modality outweigh the similarities, and most current formal treatments reflect this. But there is one similarity between the temporal and the modal that is as prominent in recent treatments as it is in older ones. This similarity can perhaps be brought home most forcefully by a look at subjunctive conditionals. Consider for instance an utterance of the conditional in (1d):

1. (d) If he had been a widower, she would have married him.

The sentence in (1d) relates two constituent sentences, the 'if'-clause and the main clause, and it is this relation which determines whether the conditional claim as a whole is to be counted as true. Moreover, whether the conditional is true does not just depend on what is the case in the world as it is. The conditional implies that both 'if'-clause and main clause are false in the actual world. But that is not enough; what is required in addition concerns other worlds than the actual one. Roughly, the additional requirement is that in any relevant possible world in which the 'if'-clause is true, the main clause should be true as well.

---

[58]Examples of such expressions can be found among adjectives (e.g., 'former', 'repeated'), conjunctions (e.g., 'while', 'after', 'before') and prepositions (e.g., 'after', 'before', 'during', 'ago'). Within the class of temporal adverbials we find representatives of a whole spectrum of distinct functions, as the following examples illustrate: 'Monday', 'the twentieth of March', 'last week', 'often', 'every other Sunday', 'still', 'again', 'the second time', 'for the second time'. The semantics of tenses and other temporal expressions in English and a few other languages is one of the most assiduously researched areas of the theory of meaning, and much in this area is by now quite well understood. For a recent study, cf., [Rothstein, 2004].

There is a large literature on conditionals (different in spirit from that on temporal reference and tense, but comparable in size). A large part of this literature is concerned with the difficult question how to define the concept of a 'relevant' possible world, as it occurs in the truth requirement we just stated for (1d). But once again, it is not such details that matter here, but only the general fact that to determine the denotations of utterances of conditionals (and other sentences containing modal terms or constructions) in one world we must have recourse to denotations in other worlds.

If we want our model-theoretic approach to deal with modality along the same lines that we have outlined for dealing with tense, then we must extend our models with yet another layer of complexity. What we need are not simply models that provide the development of denotations through time, but whole bundles of such models which cover not only the actual world but also other worlds that are relevant to modality-involving sentences of L. We will call such bundles 'intensional models' and refer to the models considered up to this point as 'extensional models'. (In other words, an intensional model is a bundle of extensional models.) Since intensional models will play an important part in all that follows, it will be useful to stipulate a specific form for them. The following definition is simple, but suits our needs.

By an intensional model $\mathcal{M}$ for a given language or language fragment L we understand a pair $\langle W, M\rangle$, where W is some non-empty set (of 'possible worlds') and M is a function which maps each $w \in W$ to an extensional model for L, i.e., to a model for L of the kind considered so far. (We write M_w (rather than $M(w)$) for the extensional model that M associates with each $w \in W$.)

This definition gives what you might call a 'bare bones' characterisation of intensional models. For many purposes the models it specifies won't be enough. For instance, in order that a model yield a satisfactory analysis of various kinds of conditionals, it must provide, apart from what is specified by our definition, also certain relations between worlds (which tell us which worlds are relevant to the denotations of various modal sentences in which other worlds). But as we have said, the exact analysis of the denotations of particular sentences is not what concerns us here. And as we will see, for the purposes of this chapter our present definition gives us just what we want.

There is one aspect of intensional models, however, that does require our attention. This is the structure of time. So far we assumed that each extensional model has its own time structure. But what can we say about the time structures of the different extensional models that make up a single intensional model? Are all these time structures the same, or may we expect them to vary from one extensional model to the next?

Behind this question lurks an age-old debate about the nature of time and the formal properties that follow from it. It is a debate that started out within philosophy, but that spread to several other disciplines once these had taken on their own topical and methodological identity, most notably to physics and psychology. The various positions that have been argued in the course of this debate can be divided into two main groups. On the one side there have been those who see time as a given absolute, either along the lines of Newton's *Principia*, or, more in the spirit of psychology or cognitive science, as some sort of Kantian category. On the other hand time has been seen as an immanent feature of an unfolding world of successive events, as an abstraction from its flow of events. (Well-known representatives of this second view are Leibniz and Russell.) On this view it cannot be excluded a priori that time structures inherit also some of the contingent properties of the event flows from which they are derived and thus that they vary from one world to the next. For proponents of a view of time of the first type it will go without saying that all extensional models come with the same time structure; for proponents of a view of the second kind this will not be self-evident, and some at least will want to deny it.

This is not the place to take sides in this debate. In general we should allow for intensional models that are consistent with either position, thus including those in which time structures may vary between their component extensional models. Variability of time structure within intensional models, however, leads to certain conceptual and technical complications that it is better to side-step here. We will therefore, in the interest of presentational perspicuity, restrict our attention to intensional models that each have a single time structure.

3.2 Content and meaning

In section 3.1 we drew attention to two complications that model-theoretic accounts of meaning must deal with: (i) the dependence of denotations on utterance features other than the linguistic form of expression uttered, and (ii) the power of an utterance to make a statement not just about its here-and-now, but also about what lies beyond — in the past, in the future or even in other possible worlds. These are by no means the only complications that theories of meaning have to deal with. But we have singled them out because it is they which affect the general form of a theory of meaning most deeply and therefore it is they also which have the greatest impact on answers to the questions that are the principal business of this chapter.

The main questions that will occupy us in the remainder of the chapter are: What is the propositional content of an utterance? What, if anything, are the contents and meanings of sentences? And what is the information carried by a natural language utterance? We will deal with the first two of these in the remainder of this section. The last question — which is the central question of this chapter — will be discussed in section 5.

As a preamble to answering the first question recall that earlier in this section, when we first spoke of denotations, we mentioned that the denotations of sentences could be either construed as propositions or as truth values, and that meaning theories vary on this point. There is a close relation between those two concepts of sentence denotation, just as there exists a close relationship between properties and property extensions: a proposition can be either true or false, depending on whether the situation to which it is applied is compatible with what the proposition says or not. It has been argued that these manifestations of the proposition, its being true in some situations and false in others, are all there is to its identity, i.e., that a proposition is nothing but the truth values it takes in different possible situations. On this view a proposition can be identified with the function which returns for each situation the truth value that it has in that situation. We use this as our leading idea in formulating our answer to the first question.[59]

We have seen in section 3.1 that in general it is only utterances of natural language sentences that can be said to have definite denotations, but not those sentences by themselves. By the same token it is only to utterances that we can attribute definite content, and not to sentences per se: it is utterances, not sentences, that express propositions. So, if we want to stick to the spirit of our leading idea, it is to utterances, and not to sentences as linguistic expressions, that we should apply it. That is, utterance content should be defined as 'propositional content' — viz., as the range of truth values that an utterance determines in different possible situations. Or, stated in terms of intensional models: the content of an utterance relative to an intensional model $\mathcal{M} = \langle W, M \rangle$ should be defined as the range of truth values that the utterance determines in the extensional models M_w associated with the different worlds $w \in W$.

At first sight it may look as if defining utterance content along these lines runs into a snag. Consider once more a sentential utterance u, for instance one of sentence (1e) (which is like (1b), except that it doesn't have the pronoun 'he' so that the only utterance feature that its interpretation depends on is the utterance time):

1. (e) Helmut is a widower.

[59]This is the point to return to the question how one might choose between model-theoretic accounts which construe sentence and noun denotations as truth values and sets, respectively, and those which construe them as propositions and properties. At the level at which the answer to this question is at all relevant to the issues of this chapter, it is quite simple, and also quite uninteresting. In model-theoretic accounts which exclusively make use of extensional models only the former denotations (truth values and sets) are well-defined, so it is only in that way that sentence and noun denotations can be understood. In accounts that use intensional models, both construals are possible, but there is little to choose between them. First, everything that can be done with truth values and sets as denotations can also be done when the denotations of sentences and nouns are taken to be propositions and properties, since we can always pass from propositions to the truth values they have in particular worlds or models, and from properties to their various extensions. Conversely, when propositions are defined, as suggested above, as functions from possible worlds to truth values, then in an intensional model $\mathcal{M}$ it is in principle possible to recover propositions from the corresponding truth values in the different extensional models that are part of $\mathcal{M}$ (and a similar reconstruction is possible when properties are construed as functions from worlds to extensions). So in theories that make use of intensional models the two ways of construing denotations are equivalent so long as the technical machinery is in place for going from propositions and properties to truth values and sets, and back. In all model-theoretic accounts of which we are aware, however, this machinery is available. As noted earlier, there appears to be a preference for theories in which denotations are construed, like we have been doing here, as truth values and sets. But as far as we can see, there are no compelling reasons for this preference.

According to what we have just suggested, the content of u relative to $\mathcal{M}$ should be identifiable as the function which maps each world $w \in W$ to the truth value of u in the model M_w. But what is this truth value of u in models associated with worlds in which u has not actually been made? Since the truth value determined by u depends not only on the sentence uttered, but also on some further properties of u, notably the utterance time, it cannot, one might think, be taken for granted that truth evaluation is possible also in relation to other worlds.

Fortunately this worry can easily be put to rest. Intuitively it seems clear that the world in which the utterance u is made could have been different from what it is, but that this would not have made any difference to the possibility of enquiring whether or not it makes the statement that u expresses true. The only difference might have been that the enquiry might have led to a different outcome. The intuitive reason why this should be so is that once the utterance time t of our utterance u has been fixed, as the time at which the utterance act is performed in its world w, the denotation of u at that time t can be computed just as easily in models $M_{w'}$ that are associated with worlds w' different from w as it can be in the model M_w associated with w itself. All we need to assume for this is that t can be identified as a time of those other worlds too.[60]

Now that we have resolved the apparent snag, nothing stands in the way to the intended characterisation of utterance content:

> The *propositional content* of an utterance u of a sentence s, relative to an intensional model $\mathcal{M} = \langle W, M \rangle$, made at a time t (of the time structure of $\mathcal{M}$), is the function which maps each world $w \in W$ to the truth value of s at t in M_w.

We now turn to the second question: What, if anything, is the content or meaning of a sentence? As regards sentence content we can be brief and simply repeat what we have noted already: given that there can be no definite sentence content without definite sentence denotations — that is, definite truth values — there can't be a definite content for any sentence of which the interpretation depends on additional features of its utterances. It is still possible, however, to make sense of the notion of sentence meaning, viz., as that which enables the different possible utterances of a sentence to express their respective propositional contents. Understood in this way the meaning of a sentence s can be identified with the function that maps each possible utterance of s to its propositional content. This brings us to the following formal characterisation:

[60]It is here that our assumption that all extensional models belonging to a given intensional model have the same time structure is being used. Without this assumption arguing for the present conclusion becomes more complicated, since it will involve the question how the possibility of identifying t in other worlds than w correlates with the relevance of those worlds for determining the denotation of the utterance in w. Other complications arise when additional utterance features besides the utterance time play a part in the content of u. All in all there are many non-trivial details that an elaboration of the argument we have sketched here must deal with. We refer the reader in particular to the locus classicus for these issues [Kaplan, 1989]. Further discussion can be found, e.g., in [Almog *et al.*, 1989].

> The *meaning* of a sentence s relative to an intensional model $\mathcal{M}$ is that function which maps each possible utterance u of s in some world w from $\mathcal{M}$ at some time t of its time structure to the propositional content of u.

We have already pointed out several reasons why the information state of a recipient who is in a position to interpret an utterance, and thereby profit from the information it contains, cannot be a tabula rasa. But there also is a further reason, which is connected with the fact that natural language utterances show a strong tendency to build upon those that precede it in discourse. In fact, human languages are rich in devices that serve this very purpose — devices for linking the sentences in which they occur to the sentences that precede them in the texts or dialogues of which they are part. These devices enable the recipient to interpret the sentences that contain them in the way the speaker intends — viz., as integral pieces of a larger discourse. But of course this can work only if the recipient has already interpreted those preceding sentences and has thereby acquired the information which they carry. In this sense too the information he will get from the new sentence takes the form of an increment to the information he already had. In the next section we will have a closer look at this incremental dimension of interpretation, and of the acquisition of linguistic information that goes with it.

4 MODELLING DISCOURSE CONNECTIONS

Much of what we want to say we say in several sentences. Single sentence utterances suit only the simplest of messages, as soon as the message becomes a little more complex, a single sentence won't do. Strictly speaking, of course, conveying a complex message in a single sentence isn't impossible in principle. But the sentence that one would have to use would be so long and convoluted that others would have the greatest difficulty in unscrambling the message; and even the speaker himself would be likely to get tied up in knots and lose track of what he was saying. This humdrum fact about the use of language points at an aspect of our language handling capacities which is also quite obvious. Our ability to parse sentences, i.e., to ascertain their syntactic form, is not commensurate with our capacity for grasping and retaining content. For whatever reason parsing is, apparently, something that we humans find hard as soon as we are confronted with strings that exceed a certain length or structural complexity. Such sentences should therefore be avoided, and instead the story one has to tell must be broken up into a sequence of sentences that are each of manageable size.

But breaking up a message into a sequence of sentences each of which covers some part of it comes at a price. It requires that each sentence can be recognised as making a particular contribution to the larger content. That is, the recipient must be able to see how and where the contribution of each new sentence fits within the part of the message that he has already reconstructed from preceding sentences. Sometimes it is clear from the nature of the message that is conveyed in a sequence of sentences and from the sequential order in which the sentences are arranged how the contributions of the successive sentences fit together. But this is by no means always so. In such cases it will be helpful, or even imperative, that the new sentence contains certain elements that make its connections with the preceding sentences clear. Given how important it is to get these connections right, it should not be surprising that natural languages include various types of such 'discourse linking' elements. In fact, there are many such elements and many of them are in constant use. (We know this to be the case at least for English and the range of other languages for which the question has been investigated, and we suspect that it is universal.)

Among the classical examples of sentence constituents which are capable of linking the sentences in which they occur to the preceding discourse are anaphoric pronouns. Anaphoric pronouns can have antecedents which occur at some earlier point in the same sentence, but as often as not their antecedents are not sentence internal. Often, but not necessarily, they occur in the immediately preceding sentence. In such cases the link between pronoun and antecedent also establishes a link between the content of the sentence containing the pronoun and that of the sentence that the antecedent belongs to. By way of illustration consider (2):

2. All these years Bill has kept in touch with one of the girls from his class in his final year in high school. He met her again last summer.

Here the pronoun 'her' can (in the absence of further context) only be construed as referring to the 'one of the girls from his class in his final year in high school' who is spoken of in the first sentence. This construal links the content of the second sentence to that of the first: the woman that, according to the second sentence, Bill met last summer is the same person as the girl that he has kept up with since his high school days. And in so linking the new content to the preceding one it also makes the former dependent on the other. It is a kind of 'add-on', i.e., an additional specification of the relation between Bill and the girl that the first sentence has already put on the interpretational map. This incrementality of discourse meaning, with the contributions by later parts building on those of earlier parts, is an aspect of linguistic meaning that substantially alters and complicates the picture of sentence meaning and utterance meaning sketched in section 3. Yet, as a feature of how natural languages work it is pervasive, and it comes in many different forms, the range of which is being uncovered only gradually.[61]

This is not the place to explore this range in depth, and we present just a few more examples that may give some flavour of what forms discourse linking can take. In example (3), the subject phrase of the second sentence, 'the other two', establishes a number of connected links with the subject of the first sentence. Because of the constraints that accompany these links (3) will be acceptable only if the number of students in the speaker's logic class was three. As a consequence,

[61]The systematic study of the effect of pronouns and other expressions with discourse linking effects is of comparatively recent date. It is one aspect of the approach to the study of meaning now widely known as 'dynamic semantics'. Cf., the references above, in footnote 36.

a recipient of (3), who assumes that the speaker has expressed herself in a way that is in keeping with what she is trying to convey, will conclude (in case he didn't already know that) that there were three students:

3. One of the students in my logic class flunked the final. The other two didn't turn up.

The constraints just spoken of are typical of expressions that establish this kind of 'anaphoric' discourse links.[62] Constraints of this kind, which utterances impose on the contexts in which they are made, are very common. In many (and presumably in all) languages there is a large variety of words and grammatical constructions which encode such constraints. The cover term that has come to be used for them is that of '(linguistic) presuppositions', or 'presuppositional constraints'.[63]

Presuppositional constraints do not always affect the content of the utterances which generate them in the way illustrated by (2) and (3). Two examples where this is not the case are the presuppositions triggered by the words 'too' and 'again' in (4a) and (4b), respectively.

4. (a) Yesterday John came too.
   (b) Yesterday John came again.
   (c) Yesterday John came.

An utterance of (4a) carries the presupposition that there was somebody else who came yesterday, and one of (4b) the presupposition that there was an occasion before yesterday when John came. Here the content that is asserted is in both cases the same as would have been conveyed by an utterance of (4c). Utterances of (4a) and (4b) differ from utterances of (4c) only with regard to the contexts in which they 'sound right', but not in the content they contribute when they do. But even so they, too, tend to produce discourse-linking effects. For instance,

[62] This is true also for the pronoun 'her', which requires that its referent be a female person (if we ignore special uses such as making reference to a ship or to a female animal to which the speaker feels or wants to imply a person-like relationship). This constraint is confirmed by the anaphoric link between the occurrence of 'her' in the second sentence of example (2) and the argument of 'with' in the first sentence. Had the 'with'-argument been 'classmate from his final year in high school' instead of 'one of the girls from his class in Bill's final year in high school', then the interpreter, seeing the 'with'-argument as the only possible antecedent for 'her', would have concluded that the classmate in question was a girl.

[63] The current tendency to subsume a large variety of context constraints under the term 'presupposition' is justified insofar as there is much that such constraints have in common, both in the ways in which they limit the contexts in which the expressions that generate them can be felicitously used and in the way they establish links to utterance contexts and thereby help to shape the content of connected discourse. But the term has the drawback that it tends to conceal some real differences which nevertheless exist between the various constraints that are subsumed under it. For the first clear recognition and articulation of the insight that anaphora and presupposition are closely related phenomena, and in fact that the terms 'anaphora' and 'presupposition' can be seen as each addressing one side of what is the same coin, cf., [van der Sandt, 1992]. A somewhat different, though also essentially dynamic perspective on presupposition and anaphora can be found in [Beaver, 2001].

the 'again' of (4b) draws attention to the fact that the described event — that of John coming (presumably his coming to an occasion of a certain, repeatable kind) was a repetition of something that had happened before. Such connections are often crucial to proper understanding. They belong to a dimension of discourse interpretation that lies beyond the reach of the notions of meaning and content outlined here, and which is only slowly becoming accessible to systematic, formally precise investigation.[64]

Presuppositions are usually described as 'constraints on the context'. That is a good way of describing their role and status, but it requires a certain understanding of the notion of context. As should be clear from our two anaphoric examples (2) and (3), the contexts that are the targets of anaphoric presuppositions are due to the preceding discourse. In fact, it is the content of the preceding discourse, as established by the interpretation that the reader or listener has made of it, that plays this role, and it is because of this double role — as content of what has been interpreted already and as context for what is being interpreted currently — that anaphoric constraints can engage with it in the way they do, relying on it for the satisfaction of the constraints they express and at the same time augmenting it (in its role as discourse content) with the content contribution derived from the current utterance. Thus discourse contexts — as contexts deriving from a discourse or discourse segment are usually called — function the way they do because they are content and context all in one. This unity of content and context is a direct reflection of the fact that the process of discourse interpretation is incremental, in that it modifies the discourse content/context step by step, adding each time the content contributed by the utterance or sentence that it has reached, after checking that the discourse context meets the current context constraints.

The incremental picture of interpretation throws an important new light on the nature of utterance content. The notion of content that we arrived at towards the end of section 3 was that of a set of possible worlds — those in which the current utterance is true. But for sentences whose interpretation requires linking one or more constituents to the discourse context this notion is no longer viable. Rather than determining a set of possible worlds in its own right all that an utterance of such a sentence can be said to identify by way of content is what it contributes to the discourse context established by the antecedent part of the discourse to which it belongs. In abstract terms this contribution can be characterised as a pair $\langle C, C' \rangle$ of discourse contexts, where C is provided by the antecedent discourse and C' is the discourse context that results from updating C with the contribution that is made by u, assuming that updating C with u is possible, i.e., that C is a discourse context which provides all that is needed for a proper interpretation of u.

In section 3.2 we have defined the meaning of a sentence S relative to an intensional model $\mathcal{M}$ as the function which maps each utterance u of S at a time t of

[64]The most ambitious current approach to this dimension of discourse interpretation that is familiar to us is the 'segmented discourse representation theory' developed by Asher and others. Cf., [Asher and Lascarides, 2003].

$\mathcal{M}$[65] to the propositional content of u in $\mathcal{M}$. In analogy to the relational characterisation of utterance content just given, we revise the notion of sentence meaning as follows. The meaning of a sentence S relative to an intensional model $\mathcal{M}$ is the partial function f_S which (i) maps each utterance u of S made in a discourse context C onto the pair $\langle C, C' \rangle$ in case C can be updated with the propositional content of u in $\mathcal{M}$ and the result of that update is C'; and (ii) is undefined otherwise. Such functions f_S are known as 'context change potentials' ('CCPs' for short), or 'update potentials'.

What CCPs are like depends first and foremost on how we identify discourse contexts. The first proposal that might come to mind is that discourse contexts can take over the role of our earlier utterance contents, and thus can be identified with sets of possible worlds. That is to say, one might think that, although the content of a sentence utterance can no longer be identified in those terms, it should still be possible to identify the content of a discourse in this way (and by the same token the content of any initial segment of it). On this assumption sentence meanings become functions from sets of possible worlds to such sets.

But this proposal won't work. It fails as soon as the question what are the available interpretation options for anaphoric pronouns is taken seriously. What issued may be involved in settling this issue is illustrated by the following example:

5. (a) One of the ten balls is missing from the bag. It has probably rolled behind the sofa.

   (b) Only nine of the ten balls are in the bag. *It has probably rolled behind the sofa.

The point of this example is this: The first sentence of (5a) and the first sentence of (5b) are true in precisely the same circumstances; an utterance of the first is true in precisely the same worlds as an utterance of the second.[66] So if the contents

[65]In the preceding section we emphasised the dependence of the propositional content of an utterance u on the time t at which u is uttered. The central topic of the present section is the dependence of utterance content on the discourse context. But of course, the second dependence does not abrogate the dependence on utterance time and other features of the utterance context, such as the identity of the speaker and that of her addressee(s). In other words, in general we find dependence both on the discourse context and on features of the utterance context. However, in the remainder of this chapter we will suppress explicit reference to features of the utterance context, including the utterance time, assuming that these are given with each individual utterance and could be recovered from these when necessary. So, when from now on we speak of the content of an utterance u in an intensional model $\mathcal{M}$ we mean the content of u at the time t of $\mathcal{M}$ at which u is made.

A few remarks about utterance context and discourse context as parts of a more inclusive notion of context can be found later in this section.

[66]Or, to put the point more pedantically, suppose that u is an utterance of the first sentence of (5a) that is made at some time t in some world w of a given intensional model $\mathcal{M}$, that u' is an utterance of the first sentence of (5b) that is made at the same time t in some other world w' and that in all other respects w and w' are exactly the same (so that in particular they contain exactly the same situation pertaining to the balls, bag and sofa that the utterances of (5a) in w and of (5b) in w' are targeted on. (It is a reasonable assumption that for all or most worlds w in which there is an utterance of the first sentence of (5a) at t there is such a world w' with a

of initial discourse segments are identified with utterance contents old style, then the discourse context determined by an utterance of the first sentence of (5a) will be indistinguishable from the discourse context established by an utterance of the first sentence of (5b). But then it becomes inexplicable why the anaphoric interpretation of the occurrence of 'it' in the second sentence of (5a) is possible but a similar interpretation of its occurrence in the second sentence of (5b) is not.

Examples like (5) indicate that pronoun interpretation is sensitive not just to what the described world must be like given what has been said about it — not just, to put the matter somewhat more abstractly, to what possible worlds are left as candidates for the described world — but also to certain aspects of how the discourse has described it. The first sentence of (5a) provides a description that permits interpreting it as referring to the missing ball, while the description provided by the first sentence of (5b) does not. A notion of discourse content that is to provide the basis for an account of these facts of pronoun interpretation must differentiate between the discourse contexts established by utterances of these two sentences; and utterance contents old style just don't do that.[67]

How can we modify the notion of discourse content so that this difference is captured? Not much reflection on examples like those in (5) is needed to see that the decisive difference between (5a) and (5b) is that the first sentence of (5a) 'introduces the missing ball into the discourse', as a kind of discourse entity, whereas the first sentence of (5b) does not do this. In (5b) the existence of the missing ball can only be inferred from the (old style) content, and apparently that is not good enough for the purpose of providing an antecedent for a singular pronoun. It may not be immediately clear how this idea can be turned into a formal definition of discourse content. To some extent this is brought out by the existing dynamic semantics literature, where a non-trivial number of different definitions can be found. We mention just one of these proposals, which seems to us a comparatively simple and natural realisation of the basic idea. According to this proposal each discourse content (relative to an intensional model $\mathcal{M}$) is based on a certain set X of so-called 'discourse referents' and consists of a set I of pairs $\langle \mathrm{w}, f\rangle$ where w is a world from $\mathcal{M}$ and f is a function from X to parts of w. (The common domain X of all the functions which occur as second components of pairs in I is called the referential base of I.) The discourse referents that make up the referential base X of a discourse content I should be thought of as the entities that are explicitly introduced by the discourse whose content is I. (Thus the referential base of the discourse content I_1 determined by an utterance of the first sentence of (5a) will consist of three discourse referents, one for the missing ball, one for the set of 10 balls of which the missing ball is a member and one for the bag. The referential base for the discourse content I_2 determined by an utterance of the first

corresponding utterance u' of the first sentence of (5b), and conversely.) Then the set of possible worlds of $\mathcal{M}$ in which u is true will be the same as the set of worlds of in which u' is true.

[67]Alternatively, one might attempt to analyse these examples in terms of descriptions, as proposed, e.g., by Neale [1990]; but cf, [Peregrin and von Heusinger, 2004] for arguments why this will not do.

sentence of (5b) will also contain three discourse referents, one for the set of nine balls in the bag, and further, as in the case of I_1, one for the set of ten balls and one for the bag. The base of I_1 contains an element that can serve as antecedent for it, the base for I_2 does not.)

Discourse contents of this type are usually referred to as 'information states'. Information states of this kind are about the simplest and most conservative refinement of our earlier notion of content as a set of possible worlds. They reflect only one aspect of the form of the discourse, viz., the set of discourse entities it introduces, which the information state captures through its referential base. Notwithstanding this conservativeness, information states thus defined can account for a remarkable variety of discourse linking mechanisms, some of which look at first glance quite different from the constraint on pronominal anaphora illustrated in (5). This is not to say that all such linking phenomena can be accounted for on the basis of discourse contents of this particular form. In fact, other, richer notions of discourse content have been proposed in order to deal with certain phenomena for which information states of the sort just defined do not seem adequate. At present the question what notions of discourse content are optimal for dealing with which aspects of discourse linking is far from settled.

We note for good measure that, obviously, the notion of a CCP co-varies with that of an information state, CCPs are always partial functions from information states to information states. Refinement of the notion of information state is automatically reflected in a similar refinement of the corresponding notion of a CCP.

There are two other fairly obvious points, which we record for further reference. The first is that an information state always determines a content in the sense of section 3. For an information state I of the kind defined above this content is the set $\text{prop}_{\mathcal{M}}(u)$ consisting of all worlds w such that for some f: $\langle \text{w}, f\rangle \in I$. In case the notion of information state is refined (in order to adapt it to the explanation of more complicated cases of discourse linking), the reduction to propositional contents may take a different form and be somewhat more involved. However, it is an essential ingredient to the notion of information state that each information state determines as propositional content, and thus that such reductions are always possible.

The second point is that among the sentences of a language such as English there are many whose content does not depend on the discourse context. Or, more accurately, there are many sentences such that any utterance of them has a content that is independent from the discourse context in which it is made. This is so whenever the sentence is free from anaphoric requirements and other presuppositional constraints. The content of an utterance of such a sentence s can still be identified in the manner of section 3 with the set of possible worlds in which the uttered sentence is true. Or, putting things more formally, given an intensional model $\mathcal{M}$ an utterance u of a sentence s (relative to a time t and a world w from $\mathcal{M}$) will have propositional content $\text{prop}_{\mathcal{M}}(u)$ that is independent of the discourse content C in which u is made. In such cases updating C with u will always be possible and will lead to a new discourse context C' of the propositional content is the intersection of the propositional content of C with $\text{prop}_{\mathcal{M}}(u)$. In particular, when C is identified with information state I, then the result of the update with u update is an information state I' such that $\text{prop}_{\mathcal{M}}(I') = \text{prop}_{\mathcal{M}}(I) \cap \text{prop}_{\mathcal{M}}(u)$. Of course this does not determine I' completely. In fact it is part of the point of example (5) that the first sentence of (5a) and the first sentence of (5b) can update the same information states I, that the resulting updated information states I_a and I_b have the same propositional content, but that they nevertheless differ in such a way that the same second sentence in (5a) and (5b) can serve as an update of I_a but not of I_b.

The question how much structure must be given to discourse contents so that they are suitably equipped for accounts of the various forms that discourse linking can take must be sharply distinguished from another question: should discourse contents be defined, as we have done so far, as model-theoretic constructs, i.e., as set-theoretic constructs built out of models and their components)? Or should they be characterised as semantic representations ('logical forms'), i.e., as structures that have their own syntax as well as a semantics determined by their syntactic structure (in the same sense in which, say, formulae of the predicate calculus have a semantics fully determined by their syntax)? Questions of the first type arise irrespective of how the second question is resolved, no less for those who opt for a representational approach than for those who prefer the 'non-representational' mode of analysis we have been following.[68]

Reasons for choosing between a representational and a non-representational approach should be looked for elsewhere. First of all, for the computational linguist, whose task is the design and implementation of algorithms for processing language on a computer, the representational approach is the only option. Only finite objects like representations can be computed and manipulated to further computational ends. The typically infinite structures which non-representational approaches use to model content and information are, as such, fundamentally non-computable, and even a computational linguist whose theoretical inclinations lean towards the non-representational perspective will be obliged to work with syntactic expressions which provide finitary descriptions of the infinitary objects he favours.

A second observation that might be seen as pointing towards the representational approach has to do more specifically with the cognitive dimension of language. Languages are used by people, whose processing capacities are, just like those of the computers they manufacture, finite; and when language is used, it is the finite minds of people which do the processing that is involved in both production and interpretation. Cognitive science is still at a stage where the most fundamental questions about how the mind works are a matter of debate, and this is true in particular with regard to the question how the mind processes language and represents the results of this processing. We still aren't in a position to assert

[68]This is not to say that the two approaches offer the same ranges of options for dealing with such questions. The question exactly how the representational and the non-representational approaches compare on this issue is still largely unanswered.

with full confidence that there is any sense in which the mind can be said to process language 'symbolically', in the sense of forming and manipulating representations similar to those assumed in theoretical linguistics. This goes for all levels of linguistic representation, including those of syntax and of semantics. Nevertheless, many of those concerned with the cognitive dimension of language work on the assumption that language processing is largely symbolic in a non-question begging sense, and that this is so in particular for processing at the syntactic and semantic level. When this general assumption is combined with the observations about discourse interpretation to which this section has been devoted, the inevitable conclusion seems to be that interpreting discourse involves representations of discourse content that can be incremented along the lines our discussion has indicated.[69]

Alternatively, one might regard the choice between a non-representational account that assumes infinite entities as semantic values and a representational one that offers finite 'reductions' of those entities as a false dilemma that in fact is a mere artifact of the specific model-theoretic assumptions that derive from the classical model, and ultimately of the classical logical approach to the treatment of formal languages. After all, there seems to be no principled reason why one would assume the various ingredients in a non-representational account (such as domains, sets of worlds, models, and even the language itself) to be infinite. In effect, it can be argued that the very notion of a language as an infinite object, which, when combined with semantic compositionality, brings along the conception of an infinite set of meanings, itself is a theoretical construct, the result of a set of assumptions we may make in order to facilitate the study of certain linguistic phenomena (such as studying syntactic productivity without worrying about actual performance limitations, or defining information update as elimination of possibilities), but that we may also discard again when they get in the way.[70]

In section 3 we drew attention to the role of the utterance context in the determination of the utterance content. In the present section we have focused on the role of the discourse context, and in doing so we have kept the dependence of content on the utterance context out of sight. This compartmentalisation is consistent with a practice that up to the present time has been quasi-universal: dependence on utterance context and dependence on discourse context are hardly ever discussed in the same breath. But it is a practice which has little to speak for itself. What we want is an integrated account of information content, which deals with dependence on utterance context and dependence on discourse context in tandem. As far as we can see there are no fundamental obstacles that stand in the way of such an integrated account. But to our knowledge none has yet been fully worked out.

---

[69]One of the long-term goals of the most representational version of dynamic semantics, discourse representation theory, is to uncover aspects of the semantic representation of content derived from linguistic input (i.e., the content that the recipients of linguistic input in spoken or written form extract from what they hear or read. Cf., [Kamp, 1984–1985; Kamp, 1990].

[70]For an early, philosophical argument along these lines, cf., [Davidson, 1986]; [Groenendijk and Stokhof, 2005] contains some thoughts in this direction regarding compositionality.

In an account of this kind utterance context and discourse context could be treated as completely separate and distinct, much as we have been presenting them in their respective sections 3 and 4. But from the perspective which provides the motivation for such an account it is tempting to see utterance context and discourse context as two components or aspects of a single comprehensive context, much as the terminology, which describes both as 'contexts', suggests. This second perspective seems especially compelling when we see utterance interpretation as something that takes place in the mind of the recipient. The recipient has to work with whatever information is available to him, and that information consists, apart from the expression uttered, of contextual information of various kinds, including information about the circumstances of the utterance event and the discourse context as the recipient has thus far constructed it.

As a matter of fact, the contextual information that the interpreter of an utterance relies on typically includes more than just his discourse context and information about the utterance context. For instance, world knowledge (including knowledge about both necessary as well as defeasible regularities that govern the events of the world in which we live) and the encyclopedic knowledge that competent speakers of a language associate with most of its words are notorious for being indispensable to interpretation, and thus it is natural to take them to be part of the context too. Along these lines we are led to a notion of an interpretation context as a complex structure, of which utterance context and discourse context are just two of the components. One way in which these components differ from each other is that some of them change in the course of a multisentence discourse while others, such as for instance the world knowledge component, normally remain fixed. But the dynamics of such integrated contexts will also involve interactions between different context components, leading for instance to information being transferred from the utterance context to the discourse context. Describing this dynamics correctly will be one of the major challenges for such an integrated context theory.

Let us take stock once more and see what the present section has taught us about content and information. We found that since discourse interpretation is incremental in nature, utterance content cannot in general be identified with sets of possible worlds, but rather has to be accounted for in terms of the updating effects that utterances produce and the context change potentials of sentences of which those updates are the manifestations. This led to a pair of two related conceptions of information, that of an information state as embodying the information content of a discourse, and that of a context change potential — a partial function from information states to information states — embodying the contributions that the utterances of sentences make to the discourses of which they are part.

It should be emphasised, however, that while these new notions cast a significantly different light on the nature of linguistic information, they are, just like the notions presented in section 3 they are meant to supplant, notions of information of a user-neutral sort, which abstracts away from the needs, interests and antecedent beliefs and convictions of those to whom information is imparted. They address the question what the information that is conveyed by an utterance or discourse actually means for the one that it is conveyed to only insofar as they identify what information he could get from the linguistic input if he interprets it in accordance with the rules of the language, including those rules that govern the incremental interpretation of multisentence discourse. But they have nothing to say about, for instance, the different effects that the same linguistic communication will have on two different recipients, either because of differences in prior information with which the newly acquired information can be inferentially combined, or because of differences in their respective needs and interests that the same new information may address to different extents or in different ways. In the next section we will have a brief look at some aspects of the cognitive perspective according to which information should be assessed in terms of what difference it makes to the one who receives it.

5 MODELLING THE RECIPIENT

Information content as we have explicated it so far is, as we noted at the end of section 4, independent of who is receiving it. When a speaker A makes a statement, using some discourse-context-independent sentence s, and B and C are among her audience, then, given how we have defined the notion, the information content of her utterance will be the same for B as it is for C. In an important sense this is right: on a natural understanding of the term 'information' B and C have obtained the same information. But how informative an utterance is, is not determined solely by the information it conveys, it also depends on what is in and on the recipient's mind at the moment he processes what the utterance has to say.

For one thing — this is the simplest but also the most telling distinction that can be made here — what is communicated to the recipient may be either new to him or it may be something that he knew already. For example, suppose that Anna tells Bernhard and Carl over dinner that Dorothea has gone to Paris. And suppose that this was news to Carl, but not to Bernhard, who had been informed about Dorothea's trip the day before. Then there is a sense in which the information which Anna's statement imparts to Bernhard is nil, while for Carl it may be significant news.

But this is not the only way in which one and the same communication may carry information of different significance to different recipients. It may be that the information is new to both Carl and Bernhard, but that Carl knew that Dorothea had applied for a job in Paris and accordingly infers from what Anna says that she must have been given the job, whereas to Bernhard, who knows nothing about Anna's job application, the question what she will be doing in Paris may not even occur. Or, yet another scenario, suppose that neither Bernhard nor Carl knew that Dorothea had left for Paris, but that their attitudes towards her are different. To Bernhard she is just someone he vaguely knows and that he has never paid much attention to, but to Carl she has been the object of deep and unrequited love. Again Anna's words will provide new information to both Bernhard and

Carl. But in Carl they are likely to trigger heated speculation about what reasons Dorothea may have had for going where she did, and about what she will be doing now that she is there. Bernhard, on the other hand, may be expected to file the new bit of information without giving it a second thought, and perhaps he will have forgotten it as soon as the dinner party is over.

The obvious fact these examples illustrate is that the significance that a piece of information has for the one who gets it is a function of his antecedent information,[71] and often also of his other mental attitudes — his concerns, convictions, affections, desires, goals and plans. This means that an account of the impact of information presupposes a systematic way of representing mental states, as composed of attitudes of these different kinds. At present no such theory of mental structure exists that has found general approval. But even if there were such a theory, this would not be the place to expound it. So we shall limit ourselves to a look at the first type of dependence mentioned above, viz., the dependence of the impact of newly conveyed information on prior information.[72]

The information that a person has at a given time must be represented in his mind in some way. For our present purposes it won't really matter in what way it is represented. It may be that the representation of at least some of the information takes the form of representations with a specific 'syntactic' structure — of formulae or terms from some 'language of thought' — but this assumption won't be essential for most of what we will have to say. What does matter is that the representations determine content. In view of what we have observed in section 4 this means that each such representation must determine an information state, where information states have at least the complexity of sets of world-assignment pairs. (For most of what we will say, however, it will suffice to assume that such representations determine propositional content.)

Let us consider, then, a person B who is the recipient of a statement u of some context independent sentence s. To make sense of u, B will have to process it in the light of what he knows. As was pointed out in section 3, this requires in the first place that B recognises the subject matter of u, i.e., the part of the world that u is about. And recognition means activating those elements of his information that pertain to this subject matter. It is this part that will be directly relevant to B's interpretation of u and that will be augmented with the information which his interpretation of u will produce.[73]

Evidently, where there is selection of elements there must be elements to be selected. So we must assume that B's information can be subdivided into elements that pertain to different subject matters. We leave open whether such divisions of a person's information into subject-related elements is always possible, and we doubt that the dividing lines can ever be entirely non-arbitrary and sharp. However, for

[71]Here and henceforth we use 'information' as an epistemically neutral term, covering both what a speech participant actually knows and what he only thinks he knows.

[72]Cf., section 7 for some references to work that takes dependencies on other factors, such as the action goals of speech participants, into account.

[73]The importance of this selection as part of the process of interpreting utterances has been underlined in particular in 'relevance theory'; cf., [Sperber and Wilson, 1995].

our present purposes a rough idealisation will suffice. Let us assume that these elements, bits of information, can be identified in the simplest and most abstract way possible, viz., as sets of possible worlds. Second, let a subject matter SM be the following equivalence relation between worlds: two worlds w and w' stand in the relation if they contain exactly the same facts pertaining to the subject matter, but otherwise they may be as different as you like.[74] Example: Suppose that the subject matter is what happens in Paris on the first of January 2006. Then w and w' stand in the corresponding relation if and only if what is the case in Paris in w coincides exactly with what happens in Paris on that day in w'. Furthermore, if SM is a subject matter and C a bit of information (i.e., a set of possible worlds), then we say that C has nothing to say about SM if and only if C has a non-empty intersection with each equivalence class of SM.[75] And finally, two distinct subject matters SM and SM', are said to be mutually independent if and only if each equivalence class of SM has a non-empty intersection with each equivalence class of SM'.[76]

About the selection of that part C_{SM} of his information that the recipient B of an utterance u activates as pertaining to the subject matter of u we make two assumptions. First, that C_{SM} is about a subject matter SM in the sense that it is the union of some set of equivalence classes of SM. And second, that B's total information can be decomposed into C_{SM} and some other part C_{RM} (the 'remainder' of B' information) in the sense that $C = C_{SM} \cap C_{RM}$, while at the same time C_{RM} has nothing to say about SM.

This puts us in a position to say something about the epistemic effect that u will have on B. Interpretation of u will lead to an augmentation of C_{SM} with the result of that interpretation. We denote this augmentation as $C_{SM} \otimes u$. It seems intuitively clear that the set-theoretic difference:

$$C_{SM} \setminus (C_{SM} \otimes u)$$

between C_{SM} and $C_{SM} \otimes u$ is a measure of the epistemic impact that u has on B.

On the assumption that B can represent thoughts about SM in some representation language L, there is also another, inference-related way of assessing the epistemic impact of u, viz., as the set:

$$\{\phi \in L : (C_{SM} \otimes u) \models \phi \;\&\; C_{SM} \models \phi\}$$

consisting of those representations belonging to L which B is in a position to deduce after he has obtained the information that is conveyed by u, but would not have been able to before. It is easily verified that the correlation between

[74]This means that the subject matter can be identified with a question in the so-called 'partition semantics of questions', with the equivalence classes representing the exhaustive answers; cf., [Groenendijk and Stokhof, 1984], [Groenendijk and Stokhof, 1997, section 4].

[75]In the terminology of the partition semantics of questions: if C is uninformative with regard to the question, i.e., if it does not exclude any of the possible exhaustive answers to the question.

[76]In terms of the analogy with questions once more: if no answer to either question implies or excludes an answer to the other.

this characterisation of epistemic impact and the set-theoretic one is monotone: suppose that u and u' are two utterances such that:

$$(C_{SM} \setminus (C_{SM} \otimes u)) \subseteq (C_{SM} \setminus (C_{SM} \otimes u'))$$

Then also:

$$\{\phi \in L : (C_{SM} \otimes u) \models \phi \ \& \ C_{SM} \not\models \phi\} \subseteq \{\phi \in L : (C_{SM} \otimes u') \models \phi \ \& \ C_{SM} \not\models \phi\}$$

(The converse implication doesn't hold without further assumptions about the expressive power of L.)

The information C_{SM} pertaining to *SM* can be seen as one component of the recipient's total information at the time of interpretation, but it is not the only one on which interpretation depends. In fact, the most important component, which plays a central part in every act of utterance interpretation, is his linguistic knowledge — his knowledge of the grammar of the language and of its lexicon — and, presumably, of a host of so-called 'encyclopedic knowledge'.[77] Let us represent this conglomerate of the recipient's linguistic and extra-linguistic knowledge as C_{LK}.

The way in which interpretation depends on C_{LK} is of course very different from the way in which it can depend on C_{SM}, and it may seem odd to mention these two almost in the same breath. But it is important to emphasise this second dependency as well, for it is the knowledge that goes into C_{LK} which is responsible for the very special character of information in natural language: linguistic expressions have the capacity to carry the information they do because of this very large package of knowledge that is shared (with close to total overlap) between the members of a speech community and thus in particular between any two members who use their language in an act of communication: the speaker uses her knowledge to encode a thought in words and her interlocutor makes use of the same knowledge to decode the verbal message to reconstruct the encoded thought. This is what makes linguistic information into the special thing it is and language into the uniquely powerful communication tool that it is.

Note that as a rule C_{SM} and C_{LK} will be quite different, and in fact can be assumed to be mutually independent in the sense defined above: any way that the given subject matter could have been is compatible with any way that the language could have been. For instance, suppose once more that the subject matter of an utterance u is the current whereabouts of Dorothea. Presumably that subject

[77]Encyclopedic knowledge is knowledge that isn't purely linguistic, but that nevertheless is important to interpretation, partly because it includes many of the preconditions of individual words — you cannot properly understand the meaning of 'levitate' without having some knowledge of the 'common sense physics' of gravity and its practical effects, or of the financial term 'futures' without knowing something about the stock market, or of the noun 'quark' without a substantial portion of knowledge about quantum physics (which is even harder), and so forth — and partly because it guides us in distinguishing plausible from implausible interpretations, and thereby helps us to deal with ambiguity and vagueness. To be sure, the boundary between what is linguistic knowledge sensu strictu and what counts as extra-linguistic, encyclopedic or world knowledge is itself rather vague.

matter has nothing to do with the conventions of the language; any possible fact about where Dorothea is, has moved from or is going to is compatible with any way the language of u could have been. Only in the special case when it is language itself that is the topic of discussion will C_{SM} and C_{LK} overlap, or even coincide.

Even in the special case where language itself is the subject matter, but certainly in all those where C_{SM} and C_{LK} are independent, the way in which interpretation depends on C_{LK} is clearly very different from the way it depends on C_{SM} — so different in fact that casting linguistic knowledge as part of the over-all context in which utterances are interpreted might seem rather artificial. But linguistic knowledge clearly is knowledge without which normal interpretation would be impossible, and, to repeat, for a proper appreciation of what makes linguistic information special the dependence of interpretation on this part of the interpreter's knowledge is crucial: that linguistic expressions have the capacity to carry the kind of information they do — and thus to carry as much information as they do — is due to the very large package of linguistic and encyclopedic knowledge on which the interpreter can and must rely, and which is common (largely, if not totally) to those who share knowledge of a given language.

This fact about human languages — that linguistic and paralinguistic knowledge is very extensive and that it is wholly or largely shared by those who can be said to speak them — is of particular importance for understanding the special nature of information as linguistically expressed and communicated. Part of the point here is not specific to natural language: in many contexts the information that is carried by a code belonging to a coding system of any kind is understood in terms of the coding and decoding algorithm that makes the system a coding system; and transmission of information using the system will function only if this algorithm is known to both sender and receiver. But of course, this is a notion of information that is derivative insofar as it presupposes some other language or medium in which information can be represented and with which the coding system is connected via its coding algorithm. It is a notion which passes, one might say, the question 'What is information?' on to that language or medium.

Human languages are special on the one hand because of the sheer quantity of knowledge that must be shared by those who use them to communicate. It is this which explains the possibility of packing as much information into an expression of modest size as we often manage to do and yet getting it across to our audience. But it isn't just the quantity of linguistic knowledge which makes the case of natural language special; it is also its quality. Our knowledge of our language isn't just a coding system that enables us to translate into and from it messages that are given in some other representation system (some 'language of thought'). It is in part genuinely semantic knowledge which links the expressions of our language directly to the world. It is these two properties of linguistic knowledge — that it is truly semantic and that it is shared between all speakers — which explain why linguistic expressions can be said to carry information in a non-derivative sense and at the same time be such remarkably effective information transmitters.

6 INFORMATION IN NATURAL LANGUAGE

In his chapter in this handbook Dretske reminds us of two distinct pretheoretical uses of the term 'information'. When we say that the puff of smoke we see in the distance means that there is a fire there, or that puffs of smoke normally carry the information that there is a fire, we use the term in a sense in which information by definition is true information. But this is not the sense we intend when we speak of 'linguistic information'. An utterance of the sentence 'There is a fire over there' can be described as meaning, or as carrying the information, that at the time of the utterance there is a fire somewhere in the direction that the speaker indicates. By itself this assessment does not entail that there is indeed a fire, in the relevant direction and at the relevant time, whenever someone makes a statement by uttering this sentence. Thus in cases such as this the term 'information' is used in such a way that its factual correctness is not assumed, i.e., in a way that allows for information that is true, but also for information that is false. It is this second sense in which we have been using the word 'information' throughout the present chapter. For it is this sense of 'information' that reflects the most fundamental characteristic of natural language meaning, viz., its ability to be about non-existent objects and non-obtaining situations.

Nevertheless , statements do carry a commitment to truth. It is constitutive of the practice of making statements that they are intended to convey true information, even if on occasion speakers fail to do so by mistake, or abuse the trust of their audience by lying. One situation in which this commitment to truth makes itself felt in the context of verbal communication is when the speaker makes statements which contradict the recipient's beliefs — in other words, statements u such that the propositional content of $C_{SM} \otimes u$ is the 'contradictory proposition' (the empty set of worlds). In such a situation the recipient may react in one of several different ways. He may conclude that if the speaker felt confident enough to make her statement, she must be right, and revise his beliefs to fit the opinion she has expressed. But he may also feel certain enough about his own beliefs to conclude that the speaker must be wrong; and in that case he may either keep his disbelief to himself or try to convince the speaker that she is wrong. And of course there are many shades between these two extremes. The recipient may feel strong enough about his own views not to accept the speaker's opinion without further ado, but not strong enough to dismiss her opinion out of hand. In such cases a discussion may ensue, perhaps ending in a joint view of whose opinion should be considered most likely.[78]

Even if we ignore the range of possibilities between the extremes of unquestioned acceptance and unconditional rejection, just the two extremes themselves show that our earlier binary distinction between old and new information is too simple. There are three basic relations that the speaker's statement u can stand in to the recipient's assumptions C_{SM} about the subject matter of u (or what he takes to be

[78] Assuming the discourse is a cooperative one. For a different perspective, cf., [Merin, 1999; van Rooij, 2004a].

its subject matter): (i) the interpretation he assigns to u may be entailed by C_{SM} — in this case the information carried by u is old for him; (ii) the interpretation may contradict C_{SM}; and (iii) it may be that the interpretation neither contradicts C_{SM} nor is entailed by it. So far we just counted cases (ii) and (iii) both as cases of new information, but as the remarks above should have made clear, from the recipient's point of view they are really very different.

This tripartite distinction is important not only in connection with the content of the statements speakers make, but also with the anaphoric and other presuppositions their statements carry. Suppose that A has just made the statement that 'Dorothea has just gone to Paris again.' This choice of words introduces the presupposition that Dorothea has been to Paris before. What is the interpreter to do with this presupposition? Again this will depend on how the presupposition is related to C_{SM}, and once again we have to distinguish between three possibilities — (i) C_{SM} entails the presupposition, (ii) the presupposition contradicts it, and (iii) neither. The first case, (i), is usually treated as the normal one: the antecedent information about the subject matter entails the presupposition; that is as it should be, and the interpreter can, detaching the presupposition, move straightaway to the non-presuppositional component of his interpretation. Cases of type (iii) are the ones that we believe presupposition theorists usually think of when they discuss accommodation. If a presupposition cannot be verified as following from the 'context', i.e., from what the interpreter currently knows or assumes, then he will adjust the context — accommodate it, as the technical vocabulary has it — so that it does entail the presupposition.[79]

And then there still is the second case, in which the presupposition contradicts C_{SM}. Once again there are several ways in which the interpreter could react: he could conclude that since the speaker was making a statement with this presupposition, she must have known the presupposition to be true, and adjust his own beliefs to fit; or he may conclude that the speaker's apparent belief in the truth of the presupposition is wrong; and in that case there are once again several options; he may try to point her mistake out to her or he may let the matter rest and take the non-presuppositional content of her statement as if no presupposition had been attached to it.

[79]By and large accommodation comes easily. In fact, it is one of the classical observations of presupposition theory that speakers will often exploit the readiness with which interpreters accommodate presuppositions by choosing wordings for their statements which trigger presuppositions of which they do not think that their interlocutors believe them already, but that they want them to adopt. And usually the ploy works: the interpreter will take the content of the presupposition on board much as he would have done if it had been asserted. In such cases, where the interlocutor reacts in accordance with the speaker's expectations, the effect on his beliefs will be the same as it would have been if the speaker had made the presupposition into a separate statement followed by the statement he actually made. The term 'accommodation', in the present technical use of it, can be traced back to [Lewis, 1979]. Some of Lewis' remarks can be read as suggesting that accommodation is always possible, but in the meantime we have learnt that there are presuppositions for which accommodation is subject to certain constraints (although it appears that the question which accommodation constraints apply to which presuppositions is still largely unanswered). Cf., [Beaver and Zeevat, 2007].

We note that these same observations also apply to presuppositions that are generated in speech acts other than assertions, such as questions or directives. Such non-assertive speech acts were mentioned in passing in the introduction, but since then nothing was said about them. This seems a suitable point to return to them. Our lack of attention to non-assertive speech acts throughout the chapter should not be construed as showing our lack of awareness of the crucial part they play in the normal use of language. (In any theory of the semantics and pragmatics of conversation their analysis is absolutely indispensable.) Rather it is a reflection of our conviction that in an account of linguistic information there is no need to make them the topic of a separate discussion. For by and large the information content of non-assertive utterances is determined according to the same principles as it is for assertions. The only difference is that non-assertive speech acts put their information content to different uses than assertions do. For the question what the information content of an expression is and how it gets transmitted that difference seems immaterial.

But the situation is different with regard to presuppositions. The presuppositions connected with non-assertive utterances have the same status as those connected with assertions. In either case their content must be verifiable in the context before the utterance can be accepted as a legitimate transmitter of its message. One consequence of this is that presuppositions are more markedly set aside from the non-presuppositional part of utterance content in the case of non-assertive speech acts than they are in the case of assertions. It is for this reason that non-assertive speech acts are particularly useful as presupposition tests: For instance, a presupposition triggered by a constituent of an interrogative sentence used in a yes-no question will often manifest itself more clearly as a presupposition there than it does in relation to an assertion involving the corresponding indicative sentence. Especially striking are those cases where you, the addressee of the question, think that the presupposition is false. It won't feel right to you to answer the question with either 'yes' or 'no'. For instance, suppose I ask you the question in (6a) and you know (i) that Fred did come to the session last night, but (ii) that he wasn't there either at any of the previous sessions. It wouldn't seem right for you in such a case to simply reply with 'yes', since that would imply that for all you know the contribution made by 'again' in (6a) — that Fred came to one or more earlier sessions — is true. Rather, you would feel it incumbent upon you to point out (in the words of (6b), say) that the presupposition was false, perhaps adding, after having got this matter out of the way, that as a matter of fact Fred was present at last night's session.

6. (a) Was Fred at the reading group again last night?
   (b) Well as a matter of fact he had never been there before. But yes, he was there yesterday.

In the above we have emphasised the importance of true information. The importance, we saw, shows up both in connection with presuppositional and with non-presuppositional information. Failure of a presupposition puts the whole communication process in jeopardy, something that can be observed with particular clarity in the case of non-assertive speech acts. But, as we also noted, truth is just as important for non-presuppositional content, in particular the content of assertions. (It is part of the conventions associated with speech acts of that kind, we observed, that the speaker commits herself to their content being true.) None of this is really surprising. For what people need and want first and foremost is true information about their world — information that makes it possible for them to plan their actions, by enabling them to make predictions about the consequences that the different lines of action open to them might have.

But none of this should blind us to the fact that it is nevertheless linguistic information as we have defined it — information about how the world might be, rather than information about how it actually is — that is the central notion in relation to human language; it is this kind of information that is language's principal commodity, not the kind of information that has truth built into it. One indication of this is that all we have said about the interpreter's handling of both non-presuppositional and presuppositional content that is motivated by the concern for truth is ultimately not about the truth as such but about what the interpreter thinks is true. It is because the interpreter can represent the world as being of a certain kind, and thus imagine it to be of that kind, that he is also capable of thinking that it is of that kind. But in the case of thought, as in that of language, the commitment to the world actually being of a certain kind is distinct and detachable from the conception of a world of such a kind as such. This distinction — between truth and mere possibility, or, if you prefer, between belief and imagination — is at the core of information both as a cognitive and a linguistic commodity.

The detachability of truth from linguistic information content comes into particularly clear view when we compare discourse about the real world with fiction. When we read a novel or listen to a story we assign information content to the words we hear or see in much the same way as we do when interpreting utterances that we take to be about the real world. By and large the same principles of interpretation apply, including those which regulate the resolution of anaphora or the contextual justification of presuppositions. But there is nevertheless one crucial difference: since the world that the fiction unfolds presents itself as one that is at the author's disposition, the interpreter has no basis for objecting to any bit of non-presuppositional or presuppositional content. For that would require detection of a conflict with what he knows (or thinks he knows) to be true on independent grounds, and in this case it is true by definition that there can be no independent grounds. (At best the interpreter could detect internal inconsistencies in the story or violations of the basic laws and regularities which any world, real or imagined, should obey.) In fictional discourse we see language at work as a means for providing pure information content, untrammelled by the concern that the world described might prove different from the world whose description is intended.

7 PROSPECTS AND CHALLENGES

In this chapter the focus has been on the information conveying function of natural languages. We have seen that the concepts of meaning and information have been related throughout much of the history of modern linguistics, though not always to the same extent and in the same sense. We have also described in some detail what kinds of concepts and ideas are needed to build a descriptively satisfactory and theoretically sound theory that models the information conveying capabilities of natural languages. To be sure, such a theory is still 'in the works', and no doubt other concepts and ideas will be needed for it to be developed further, but the basic contours of what natural language information is and how it is shaped by syntactic structure, the structure of the utterance context, the discourse context and the participants' doxastic states and strategic goals, seem reasonably clear. In what follows we mention a few current trends, and then end this chapter with some thoughts on the information exchange function of natural languages.

One aspect that currently gets a lot of attention relates to the strategic goals that language users have when they enter into a conversation, read a text, or communicate linguistically in some other form. Simple information exchange as such is hardly ever the ultimate goal: information is needed for certain purposes, e.g., in order to decide which action to undertake oneself, or to predict or influence actions of others, to explore possible courses of events, and so on. Language use then becomes part of a solution of a decision problem, and understanding the nature of the problem is essential since it determines what information (in terms of content and/or amount) is relevant in a given situation.[80]

Situations of information exchange become even more complex once one acknowledges that not only providing information, but also withholding information, or divulging information selectively (part of the information, to part of the other participants), may be a crucial element of an overall strategy. This type of information exchange is often analysed using tools from game theory.[81] So-called 'higher order effects' of information disclosure are a central topic here: if A tells B that p, then B potentially learns a lot of other things beside p: that A believes (knows) that p, that A wants B to believe (know) that p, that C, who happened to overhear the utterance, now also believes (knows) that p but not as a result of an intentional action of A, and so on.

A particular aspect of this problem set that has been studied quite extensively is how language users choose means of expressions, as speakers, and determine interpretations, as hearers. Given the fact that the relation between expression and content, form and meaning, is not one–one, but in general many–many, the problem how to express certain information, and how to decide what content a certain expression is used to convey, is a substantial one. In order to solve these problems, language users need general principles that they can use themselves and that they can assume the other users employ as well. Gricean pragmatics

[80]Cf., [Ginzburg, 1995; van Rooij, 2003].

[81]Cf., [van Benthem, 2006].

provides a first and partial indication. So-called 'bi-directional optimality theory' aims to generalise and systematise these ideas.[82] Although it greatly enhances the explanatory power of the classical Gricean framework, it stays within that framework in that it relies on an independent characterisation of the space of possible forms and the space of possible meanings. A framework in which meanings are not a precondition but a result of linguistic exchanges is provided by so-called 'signalling games',[83] which in that respect constitute a major step away from the classical model.

The integration of the various theories and paradigms that are currently being explored and that we can only mention, is still very much an open matter. No unified framework exists as yet, and developments are rapid. However, most approaches somehow build on the general principles that we have outlined in this chapter, which suggests that the current phase of diversification could be followed by one in which more comprehensive theories can be developed.

A question that we haven't addressed so far is to what extent information conveying is really natural language's 'core business', as many would claim. The reason we have not gone into that discussion is minimally this, that although not everyone agrees that information conveying is the function of natural language 'par excellence', no-one really wants to deny that it is one of the things natural languages are used for, and the how and why of that is really what this chapter is all about. Nevertheless, as a final reminder it may not be superfluous to indicate, albeit only very briefly, the possible limitations of this view on natural language.

First of all, in as much as the concepts and formal machinery that are put to use in theories of natural language semantics and pragmatics are taken from logic and theoretical computer science, where they were developed with the explicit purpose of providing tools for the description and analysis of processes of information exchange, it is hardly surprising that the resulting theories make natural language, too, appear as primarily concerned with that specific goal. Anything that doesn't fit simply disappears from sight, by being 'abstracted away' from. That by itself doesn't mean, of course, that there actually is something that doesn't fit, or that, if there is, it is of importance. But the fact remains that the tools used in the study of natural language are derived from the domain of formal systems and that whereas the latter are straightforwardly designed with a specific purpose in mind, the former can hardly be said to answer to such a description. So at least we need to allow the possibility of a certain one-sidedness, and concomitant distortion.

A second consideration pertains to the status of the use of natural language for information exchange. Of course nobody would deny that natural languages serve other purposes as well: we flatter and comfort (each other and ourselves), we sing songs and write poetry, we congratulate and curse. There is no denying that such utterances, too, convey information, if only of the 'higher order' type mentioned earlier on, but according to many it would be an unjustified generalisation to state

[82]Cf., [Dekker and van Rooij, 2000; Blutner *et al.*, 2006]. For a game-theoretical approach to the issue of ambiguity resolution, cf. [Parikh, 2002].

[83]Cf., [van Rooij, 2004b].

that their purpose is that of exchanging information. The real issue, they feel, is how these other uses and the information exchange use are weighed relative to each other. Is information exchange the core function of language, with other functions being somehow dependent on it? Or are the various uses we make of language relatively independent from each other? Or is there some other function than information exchange that is the primary one? These are difficult questions, and it would take us too far afield to discuss the various options that have been proposed and defended in the literature. Some seek the answer in an (often speculative) account of language evolution: did language evolve from the use of signals, to indicate foods, predators, and such?[84] Or are its origins rather to be found in a need for social cohesion, and did it start out as a way of maintaining social bonds within a group?[85] Others address these questions from a more systematic point of view. As a matter of fact, the history of western thinking about language displays quite a variety of opinions about what constitutes its inner nature, and the logical one, with its emphasis on reference, description and information exchange is but one of them. In modern times, the 'information oriented' views of, e.g., Locke and Leibniz, were balanced by, e.g., those of Rousseau, who saw the essence of language in the expressions of the passions, and of Herder and Humboldt, who emphasised its expressive role with regard to the spirit of a culture. In the twentieth century Wittgenstein explored the variety of the uses we make of language and emphasised their 'co-originality', and Austin and Searle further systematised certain elements of this view. And in other philosophical traditions, too, people have expressed yet other views on the nature of language, such as the hermeneutic perspective of Heidegger and Gadamer, the moral-political view of Habermas,[86] or the phenomenological one of Merleau-Ponty.[87]

What these alternative views have in common is that they reject, to some extent at least, the instrumentalism that seems inherent in the information exchange perspective. There language essentially is an instrument, a tool that is put to a use, viz., that of asking for and providing information. Again, the origins of the concepts and tools of modern semantics and pragmatics are conducive to such a view: in formal systems the languages and their semantics are defined according to predetermined specifications, which makes them instrumental through and through. In many of the alternative views, the distinction between the instrument and the use to which it is put is less clear, more difficult to make. One might say that the opposition really is a matter of whether 'language has use' or 'language is use'.

These observations, of course, merely scratch the surface of a very complicated, and still ongoing debate. We draw attention to them merely in order to balance the perspective, not to throw serious doubts on the view that information exchange is

[84] Cf., [Pinker and Bloom, 1990].

[85] Cf., [Dunbar, 1998]. And there are many other views, cf., various contributions in the already referred to [Christiansen and Kirby, 2003].

[86] Cf., [Lafont, 1999].

[87] Cf., [Edie, 1987].

a vital function of natural languages. The latter point is uncontroversial, as is the contention that theories exploring this perspective provide a real insight into the nature of natural languages, and have furthered our understanding of its structure, its meaning and its use immensely. The point to bear in mind is just that other perspectives, in their own way, contribute to such an understanding as well.

ACKNOWLEDGEMENTS

We would like to thank the editors for their patient support and Menno Lievers for his detailed critical comments, which have led to many improvements. Obviously, only the authors are to be held responsible for the end result.

BIBLIOGRAPHY

[Almog *et al.*, 1989] Joe Almog, John Perry, and Howard K. Wettstein, editors. *Themes from Kaplan.* Oxford University Press, 1989.

[Asher and Lascarides, 2003] Nicholas Asher and Alex Lascarides. *Logics of Conversation.* Cambridge University Press, Cambridge, 2003.

[Austin, 1962] John Longshaw Austin. *How To Do Things With Words.* Oxford University Press, Oxford, 1962.

[Baggio *et al.*, to appear] Giosuè Baggio, Michiel van Lambalgen, and Peter Hagoort. Language, linguistics and cognition. In Ruth Kempson and Tim Fernando, editors, *Handbook of Philosophy of Linguistics.* Elsevier, Amsterdam, to appear.

[Bartsch, 1996] Renate Bartsch. The myth of literal meaning. In Edda Weigand and Franz Hundschnurscher, editors, *Lexical Structures and Language Use*, pages 3–16. Niemeyer, Tübingen, 1996.

[Barwise and Perry, 1983] Jon Barwise and John Perry. *Situations and Attitudes.* MIT Press, Cambridge, Mass., 1983.

[Barwise and Seligman, 1997] Jon Barwise and Jerry Seligman. *Information Flow. The Logic of Distributed Systems*, volume 44 of *Cambridge Tracts in Theoretical Computer Science.* Cambridge University Press, Cambridge, 1997.

[Beaver and Zeevat, 2007] David I. Beaver and Henk Zeevat. Accommodation. In Gillian Ramchand and Charles Reiss, editors, *The Oxford Handbook of Linguistic Interfaces.* Oxford University Press, Oxford, 2007.

[Beaver, 2001] David I. Beaver. *Presupposition and Assertion in Dynamic Semantics.* CSLI, Stanford, 2001.

[Biletzki and Matar, 1998] Anat Biletzki and Anat Matar, editors. *The Story of Analytic Philosophy: Plots and Heroes.* Routledge, London, 1998.

[Blutner *et al.*, 2006] Reinhard Blutner, Helen de Hoop, and Petra Hendriks. *Optimal Communication.* CSLI, Stanford, 2006.

[Bruner, 1983] Jerôme Bruner. *Child's Talk. Learning to Use Language.* Norton, London, 1983.

[Burge, 1990] Tyler Burge. Wherein is language social? In C. Anthony Anderson and Joseph Owens, editors, *Propositional Attitudes*, pages 113–30. CSLI, Stanford, 1990.

[Cardona, 1988. 2nd ed 1997] George Cardona. *Pāṇini. His Work and Its Tradition.* Motilal Banarsidass, Delhi, 1988. 2nd ed 1997.

[Chomsky, 1955] Noam Chomsky. Logical syntax and semantics: Their linguistic relevance. *Language*, 31(1):36–45, 1955.

[Chomsky, 2005] Noam Chomsky. Three factors in language design. *Linguistic Inquiry*, 36(1):1–22, 2005.

[Christiansen and Kirby, 2003] Morton H. Christiansen and Simon Kirby, editors. *Language Evolution.* Oxford University Press, Oxford, 2003.

[Cresswell, 1973] Max J. Cresswell. *Logics and Languages.* Methuen, London, 1973.

[Davidson, 1967] Donald Davidson. Truth and meaning. *Synthese*, 17:304–23, 1967.

[Davidson, 1974] Donald Davidson. On the very idea of a conceptual scheme. *Proceedings and Adresses of The American Philosophical Association*, 47, 1974.

[Davidson, 1984] Donald Davidson. *Inquiries into Truth and Interpretation.* Oxford University Press, Oxford, 1984.

[Davidson, 1986] Donald Davidson. A nice derangement of epitaphs. In Ernest LePore, editor, *Truth and Interpretation. Perspectives on the Philosophy of Donald Davidson*, pages 433–46. Blackwell, Oxford, 1986. Reprinted in [Davidson, 2005].

[Davidson, 2005] Donald Davidson. *Truth, Language and History.* Clarendon Press, Oxford, 2005.

[Dekker and van Rooij, 2000] Paul Dekker and Robert van Rooij. Bi-directional optimality theory: An application of game theory. *Journal of Semantics*, 17:217–42, 2000.

[Dummett, 1996] Michael Dummett. *The Origins of Analytical Philosophy.* Duckworth, London, 1996.

[Dummett, 2004] Michael Dummett. *Truth and the Past.* Columbia University Press, 2004.

[Dunbar, 1998] Robin I.M. Dunbar. *Grooming, Gossip and the Evolution of Language.* Harvard University Press, Cambridge, Mass., 1998.

[Edie, 1987] James M. Edie. *Merleau-Ponty's Philosophy of Language.* University Press of America, Washington, 1987.

[Farkas, 2006] Katalin Farkas. Semantic internalism and externalism. In Ernest LePore and Barry Smith, editors, *The Oxford Handbook of Philosophy of Language.* Oxford University Press, Oxford, 2006.

[Fodor, 1987] Jerry A. Fodor. *Psychosemantics: The Problem of Meaning in the Philosophy of Mind.* MIT Press, Cambridge, Mass., 1987.

[Frege, 1879] Gottlob Frege. *Begriffsschrift. Eine der arithmetischen nachgebildete Formelsprache des reinen Denkens.* Louis Nebert, Halle a.S., 1879. English translation in van Heijenoort 1970.

[Frege, 1882] Gottlob Frege. Über die wissenschaftliche Berechtigung einer Begriffsschrift. *Zeitschrift für Philosophie und philosophische Kritik*, pages 48–56, 1882. English translation in [Frege, 1964].

[Frege, 1918–19] Gottlob Frege. Der Gedanke: eine logische Untersuchung. *Beiträge zur Philosophie des deutschen Idealismus*, 2:58–77, 1918–19. English translation in [Frege, 1977].

[Frege, 1964] Gottlob Frege. On the scientific justification of a concept-script. *Mind*, 73:155–60, 1964. Translated by J.M. Bartlett.

[Frege, 1977] Gottlob Frege. *Logical Investigations.* Blackwell, Oxford, 1977. Translated by P.T. Geach.

[Garton, 1992] A.F. Garton. *Social Interaction and the Development of Language and Cognition.* Erlbaum, Hillsdale, 1992.

[Ginzburg, 1995] Jonathan Ginzburg. Resolving questions, I & II. *Linguistics and Philosophy*, 18(5 & 6):459–527 & 567–609, 1995.

[Grice, 1957] H.P. Grice. Meaning. *The Philosophical Review*, 66:377–88, 1957.

[Grice, 1975] H.P. Grice. Logic and conversation. In Peter Cole and Jerry Morgan, editors, *Syntax and Semantics. Volume 3: Speech Acts*, pages 41–58. Academic Press, New York, 1975.

[Groenendijk and Stokhof, 1984] Jeroen Groenendijk and Martin Stokhof. *On the Semantics of Questions and the Pragmatics of Answers.* PhD thesis, Department of Philosophy, Universiteit van Amsterdam, Amsterdam, November 1984. (Available from: http://dare.uva.nl/en/record/123669).

[Groenendijk and Stokhof, 1990] Jeroen Groenendijk and Martin Stokhof. Dynamic Montague grammar. In László Kálmán and László Pólos, editors, *Papers from The Second Symposium on Logic and Language*, pages 3–48. Akadémiai Kiadó, Budapest, 1990.

[Groenendijk and Stokhof, 1991] Jeroen Groenendijk and Martin Stokhof. Dynamic predicate logic. *Linguistics and Philosophy*, 14(1):39–100, 1991.

[Groenendijk and Stokhof, 1997] Jeroen Groenendijk and Martin Stokhof. Questions. In Johan van Benthem and Alice ter Meulen, editors, *Handbook of Logic and Linguistics*, pages 1055–1124. Elsevier/MIT Press, Amsterdam/Cambridge Mass., 1997.

[Groenendijk and Stokhof, 2005] Jeroen Groenendijk and Martin Stokhof. Why compositionality? In Greg Carlson and Jeff Pelletier, editors, *Reference and Quantification: The Partee Effect*, pages 83–106. CSLI, Stanford, 2005.

[Groenendijk *et al.*, 1996] Jeroen Groenendijk, Martin Stokhof, and Frank Veltman. Coreference and modality. In Shalom Lappin, editor, *Handbook of Contemporary Semantic Theory*, pages 179–213. Blackwell, Oxford, 1996.

[Hacking, 1975] Ian Hacking. *Why Does Language Matter to Philosophy?* Cambridge. Cambridge University Press, 1975.

[Heim, 1983] Irene Heim. File change semantics and the familiarity theory of definiteness. In Rainer Bäuerle, Christoph Schwarze, and Arnim von Stechow, editors, *Meaning, Use, and Interpretation of Language*. De Gruyter, Berlin, 1983.

[Hendriks and de Hoop, 2001] Petra Hendriks and Helen de Hoop. Optimality theoretic semantics. *Linguistics and Philosophy*, 24(1):1–32, 2001.

[Hintikka, 1983] Jaakko Hintikka. *The Game of Language*. Reidel, Dordrecht, 1983.

[Jackendoff, 1990] Ray Jackendoff. *Semantic Structures*. MIT Press, Cambridge, Mass., 1990.

[Kamp and Reyle, 1993] Hans Kamp and Uwe Reyle. *From Discourse to Logic*. Kluwer, Dordrecht, 1993.

[Kamp, 1981] Hans Kamp. A theory of truth and semantic representation. In Jeroen Groenendijk, Theo Janssen, and Martin Stokhof, editors, *Formal Methods in the Study of Language*. Mathematical Centre, Amsterdam, 1981.

[Kamp, 1984–1985] Hans Kamp. Context, thought and communication. *Proceedings of the Aristotelian Society*, pages 239–61, 1984–1985.

[Kamp, 1990] Hans Kamp. Prolegomena to a structural account of belief and other attitudes. In C.A. Anderson and J. Owens, editors, *Propositional Attitudes: The Role of Content in Logic, Language, and Mind*. CSLI, Stanford, 1990.

[Kaplan, 1989] David Kaplan. Demonstratives. In Joe Almog, John Perry, and Howard K. Wettstein, editors, *Themes from Kaplan*, pages 481–563. Oxford University Press, Oxford, 1989.

[Kirby, 2000] Simon Kirby. Syntax without natural selection: How compositionality emerges from vocabulary in a population of learners. In C. Knight, editor, *The Evolutionary Emergence of Language: Social Function and the Origins of Linguistic Form*, pages 303–23. Cambridge University Press, Cambridge, 2000.

[Lafont, 1999] Christina Lafont. *The Linguistic Turn in Hermeneutic Philosophy*. MIT Press, Cambridge, Mass., 1999.

[Lakoff, 1987] George Lakoff. *Women, Fire and Dangerous Things*. The University of Chicago Press, Cambridge, 1987.

[Levinson, 1983] Stephen C. Levinson. *Pragmatics*. Cambridge University Press, Cambridge, 1983.

[Lewis, 1970] David K. Lewis. General semantics. *Synthese*, 22:18–67, 1970.

[Lewis, 1979] David K. Lewis. Scorekeeping in a language game. *Journal of Philosophical Logic*, 8:339–59, 1979.

[McGinn, 1989] Colin McGinn. *Mental Content*. Blackwell, Oxford, 1989.

[Merin, 1999] Arthur Merin. Information, relevance and social decisionmaking. In Larry Moss, Jonathan Ginzburg, and Maarten de Rijke, editors, *Logic, Language and Computation. Vol. 2*. CSLI, Stanford, 1999.

[Montague, 1970a] Richard Montague. English as a formal language. In Bruno Visentini, editor, *Linguaggi nella Società e nella Tecnica*, pages 189–224. Edizioni di Communità, Milano, 1970.

[Montague, 1970b] Richard Montague. Universal grammar. *Theoria*, 36:373–98, 1970.

[Montague, 1973] Richard Montague. The proper treatment of quantification in ordinary English. In Jaakko Hintikka, Julius Moravcsik, and Patrick Suppes, editors, *Approches to Natural Language*, pages 221–42. Reidel, Dordrecht, 1973.

[Montague, 1974] Richard Montague. *Formal Philosophy. Selected papers of Richard Montague. Edited and with an Introduction by Richmond H. Thomason*. Yale University Press, New Haven and London, 1974.

[Neale, 1990] Stephen Neale. *Descriptions*. MIT Press, Cambridge, Mass., 1990.

[Parikh, 2002] Prashant Parikh. *The Use of Language*. CSLI, Stanford, 2002.

[Partee, 1973] Barbara H. Partee. Some transformational extensions of Montague grammar. *Journal of Philosophical Logic*, 2:509–34, 1973.

[Peregrin and von Heusinger, 2004] Jaroslav Peregrin and von Heusinger. Dynamic semantics with choice functions. In Hans Kamp and Barbara H. Partee, editors, *Context Dependence in the Analysis of Linguistic Meaning*, pages 255–74. Elsevier, Amsterdam, 2004.

[Pinker and Bloom, 1990] Steven Pinker and Paul Bloom. Natural language and natural selection. *Behavioral and Brain Sciences*, 13(4):707–84, 1990.

[Plato, 1921] Plato. *Sophist.* Loeb Classical Library. Harvard University Press, Cambridge, Mass., 1921. Translated by H.N. Fowler.

[Pustejovsky, 1995] James Pustejovsky. *The Generative Lexicon.* MIT Press, Cambridge, Mass., 1995.

[Putnam, 1975] Hilary Putnam. The meaning of 'meaning'. In *Mind, Language, and Reality*, pages 215–71. Cambridge University Press, Cambridge, 1975.

[Quine, 1948] Willard Van Orman Quine. On what there is. *Review of Metaphysics*, 2:21–38, 1948. Reprinted in [Quine, 1953a].

[Quine, 1953a] Willard Van Orman Quine. *From a Logical Point of View.* Harper & Row, New York, 1953.

[Quine, 1953b] Willard Van Orman Quine. Two dogmas of empiricism. In *From a Logical Point of View*, pages 20–46. Harvard University Press, Cambridge, Mass., 1953.

[Recanati, 2004] Francois Recanati. *Literal Meaning.* Cambridge University Press, Cambridge, 2004.

[Robins, 1990] R. Robins. *A Short History of Linguistics.* Longman, London, 3rd edition, 1990.

[Rothstein, 2004] Susan Rothstein. *Structuring Events: A Study in the Semantics of Lexical Aspects*, volume 2 of *Explorations in Semantics.* Blackwell, Oxford, 2004.

[Searle, 1969] John R. Searle. *Speech Acts.* Cambridge University Press, Cambridge, 1969.

[Seuren, 1998] Pieter A.M. Seuren. *Western Linguistics: An Historical Introduction.* Blackwell, 1998.

[Soames, 2003] Scott Soames. *Philosophical Analysis in the Twentieth Century, Volumes 1 and 2.* Princeton University Press, Princeton, 2003.

[Sperber and Wilson, 1995] Dan Sperber and Deirdre Wilson. *Relevance: Communication and Cognition.* Blackwell, Oxford, 2nd edition, 1995.

[Stalnaker, 1974] Robert Stalnaker. Pragmatic presuppositions. In M. Munitz and P. Unger, editors, *Semantics and Philosophy.* New York University Press, New York, 1974.

[Stalnaker, 1979] Robert Stalnaker. Assertion. In Peter Cole, editor, *Syntax and Semantics 9 – Pragmatics.* Academic Press, New York, 1979.

[Stanley, 2005] Jason Stanley. Semantics in context. In Gerhard Preyer and Georg Peter, editors, *Contextualism in Philosophy.* Oxford University Press, Oxford, 2005.

[Steiner, 1975] George Steiner. *After Babel. Aspects of Language and Translation.* Oxford University Press, Oxford, 1975.

[Stenius, 1967] Eric Stenius. Mood and language-game. *Synthese*, 17:254–74, 1967.

[Stokhof, 2002] Martin Stokhof. Meaning, interpretation and semantics. In Dave Barker-Plummer, David Beaver, Johan F.A.K. van Benthem, and Patrick Scotto di Luzio, editors, *Words, Proofs and Diagrams*, pages 217–40. CSLI, Stanford, 2002.

[Sundholm, to appear] Göran Sundholm. A century of judgement and inference: 1837–1936. In L. Haaparanta, editor, *The Development of Logic.* Oxford University Press, Oxford, to appear.

[Szabó, 2005] Zoltán Gendler Szabó, editor. *Semantics versus Pragmatics.* Oxford University Press, 2005.

[Talmy, 2000] Leonard Talmy. *Towards a Cognitive Semantics.* MIT Press, Cambridge, Mass., 2000.

[Tomasello, 1999] Michael Tomasello. *The Cultural Origins of Human Cognition.* Harvard University Press, Cambridge, Mass., 1999.

[Tomasello, 2003] Michael Tomasello. *Constructing a Language: A Usage-Based Theory of Language Acquisition.* Harvard University Press, Cambridge, Mass., 2003.

[van Benthem, 2006] Johan F.A.K. van Benthem. One is a lonely number. In Z. Chatzidakis, P. Koepke, and W. Pohlers, editors, *Logic Colloquium '02*, volume 27 of *Lecture Notes in Logic.* Association for Symbolic Logic, 2006.

[van der Sandt, 1992] Rob van der Sandt. Presupposition projection as anaphora resolution. *Journal of Semantics*, 9(4):333–77, 1992.

[van Eijck and Stokhof, 2006] Jan van Eijck and Martin Stokhof. The gamut of dynamic logic. In Dov Gabbay and John Woods, editors, *Handbook of the History of Logic, Volume 6 – Logic and the Modalities in the Twentieth Century*, pages 499–600. Elsevier, Amsterdam, 2006.

[van Heijenoort, 1970] Jean van Heijenoort, editor. *Frege and Gödel. Two Fundamental Texts in Mathematical Logic.* Harvard University Press, Cambridge, Mass., 1970.

[van Rooij, 2003] Robert van Rooij. Asserting to resolve decision problems. *Journal of Pragmatics*, 35, 1161–79 2003.

[van Rooij, 2004a] Robert van Rooij. Cooperative versus argumentative communication. In Manuel Rebuschi and Tero Tulenheimo, editors, *Logique & Théorie des Jeux*, volume 8 of *Philosophia Scientiae*. Kimé, Paris, 2004.

[van Rooij, 2004b] Robert van Rooij. Signalling games select Horn strategies. *Linguistics and Philosophy*, 27:492–527, 2004.

[Veltman, 1996] Frank Veltman. Defaults in update semantics. *Journal of Philosophical Logic*, 25:221–61, 1996.

[Wittgenstein, 1958] Ludwig Wittgenstein. *Philosophical Investigations*. Blackwell, Oxford, 2nd edition, 1958.

[Wittgenstein, 1960] Ludwig Wittgenstein. *Tractatus Logico-Philosophicus*. Suhrkamp, Frankfurt a/M, 1960.

TRENDS IN THE PHILOSOPHY OF INFORMATION

Luciano Floridi

1 INTRODUCTION

"I love information upon all subjects that come in my way, and especially upon those that are most important." Thus boldly declares Euphranor, one of the defenders of Christian faith in Berkeley's *Alciphron.*[1] Evidently, information has been an object of philosophical desire and puzzlement for some time, well before the computer revolution, Internet or the dot.com pandemonium. Yet what does Euphranor love, exactly? *What is information*?

As with many other field-questions (consider for example "what is being?", "what is morally good?" or "what is knowledge?"), "what is information?" is to be taken not as a request for a dictionary definition, but as a means to demarcate a wide area of research. The latter has recently been defined as *the philosophy of information* (Floridi [2002; 2003b]). The task of this chapter is to review some interesting research trends in the philosophy of information (henceforth also PI). This will be achieved in three steps. We shall first look at a definition of PI. On this basis, we shall then consider a series of open problems in PI on which philosophers are currently working.[2] The conclusion will then highlight the innovative character of this new area of research.

2 DEFINING THE PHILOSOPHY OF INFORMATION

The philosophy of information may be defined as the philosophical field concerned with

a) the critical investigation of the conceptual nature and basic principles of information, including its dynamics, utilisation and sciences, and

b) the elaboration and application of information-theoretic and computational methodologies to philosophical problems.[3]

[1] Berkeley [1732], Dialogue 1, Section 5, Paragraph 6/10.

[2] For a longer and more detailed discussion see Floridi [2004b].

[3] The definition is first introduced in Floridi [2002]. The nature and scope of PI are further discussed in Floridi [2003b] and Floridi et al. [2005]. Floridi [2003c] provides an undergraduate level introduction to PI.

Handbook of the Philosophy of Science. Volume 8: Philosophy of Information
Volume editors: Pieter Adriaans and Johan van Benthem. General editors: Dov M. Gabbay, Paul Thagard and John Woods.

The first half of the definition concerns the philosophy of information as a new field. PI appropriates an explicit, clear and precise interpretation of the classic, Socratic question "ti esti...?" ("what is...?"), namely "What is the nature of information?". This is the clearest hallmark of a new field. PI provides critical investigations that are not to be confused with a quantitative theory of data communication (information theory). On the whole, we shall see that its task is to develop an integrated family of theories that analyse, evaluate and explain the various principles and concepts of information, their dynamics and utilisation, with special attention to systemic issues arising from different contexts of application and the interconnections with other key concepts in philosophy, such as knowledge, truth, meaning and reality.

By "dynamics of information" the definition refers to:

i) *the constitution and modelling of information environments*, including their systemic properties, forms of interaction, internal developments, applications etc.;

ii) *information life cycles*, i.e. the series of various stages in form and functional activity through which information can pass, from its initial occurrence to its final utilisation and possible disappearance;[4] and

iii) *computation*, both in the Turing-machine sense of *algorithmic processing*, and in the wider sense of *information processing*. This is a crucial specification. Although a very old concept, information has finally acquired the nature of a primary phenomenon only thanks to the sciences and technologies of computation and ICT (Information and Communication Technologies). Computation has therefore attracted much philosophical attention in recent years. Nevertheless, PI privileges "information" over "computation" as the pivotal topic of the new field because it analyses the latter as presupposing the former. PI treats "computation" as only one (although very important) of the processes in which information can be involved.

From an environmental perspective, PI is critical and normative about what may count as information, and how information should be adequately created, processed, managed and used. Methodological and theoretical choices in ICS (Information and Computer Sciences) are also profoundly influenced by the kind of PI a researcher adopts more or less consciously. It is therefore essential to stress that PI critically evaluates, shapes and sharpens the conceptual, methodological and theoretical basis of ICS, in short that it also provides a *philosophy of ICS*, as this has been plain since early work in the area of philosophy of AI [Colburn, 2000].

[4]A typical life cycle includes the following phases: occurring (discovering, designing, authoring, etc.), processing and managing (collecting, validating, modifying, organising, indexing, classifying, filtering, updating, sorting, storing, networking, distributing, accessing, retrieving, transmitting etc.) and using (monitoring, modelling, analysing, explaining, planning, forecasting, decision-making, instructing, educating, learning, etc.).

It is worth stressing here that an excessive concern with contemporary issues may lead one to miss the important fact that it is perfectly legitimate to speak of a philosophy of information even in authors who lived before the information revolution, and hence that it will be extremely fruitful to develop a historical approach and trace PI's diachronic evolution, as long as the technical and conceptual frameworks of ICS are not anachronistically applied, but are used to provide the conceptual method and privileged perspective to evaluate in full reflections that were developed on the nature, dynamics and utilisation of information before the digital revolution. This is significantly comparable with the development undergone by other philosophical fields like the philosophy of language, the philosophy of biology, or the philosophy of mathematics.[5]

The second half of the definition indicates that PI is not only a new field, but provides an innovative methodology as well. Research into the conceptual nature of information, its dynamics and utilisation is carried on from the vantage point represented by the methodologies and theories offered by ICS and ICT [Grim *et al.*, 1998] and [Greco *et al.*, 2005]. This perspective affects other philosophical topics as well. Information-theoretic and computational methods, concepts, tools and techniques have already been developed and applied in many philosophical areas,

- to extend our understanding of the cognitive and linguistic abilities of humans and animals and the possibility of artificial forms of intelligence (e.g. in the philosophy of AI; in information-theoretic semantics; in information-theoretic epistemology and in dynamic semantics);

- to analyse inferential and computational processes (e.g. in the philosophy of computing; in the philosophy of computer science; in information-flow logic; in situation logic; in dynamic logic and in various modal logics);

- to explain the organizational principles of life and agency (e.g. in the philosophy of artificial life; in cybernetics and in the philosophy of automata; in decision and game theory);

- to devise new approaches to modelling physical and conceptual systems (e.g. in formal ontology; in the theory of information systems; in the philosophy of virtual reality);

- to formulate the methodology of scientific knowledge (e.g. in model-based philosophy of science; in computational methodologies in philosophy of science);

- to investigate ethical problems (in computer and information ethics and in artificial ethics), aesthetic issues (in digital multimedia/hypermedia theory,

[5]See [Adams, 2003] for a reconstruction of the informational turn in philosophy and [Young, 2004] for an analysis of Wittgenstein's philosophy of information.

in hypertext theory and in literary criticism) and psychological, anthropological and social phenomena characterising the information society and human behaviour in digital environments(cyberphilosophy).

Indeed, the previous examples and the various chapters in this volume show that PI, as a new field, provides a unified and cohesive, theoretical framework that allows further specialisation.

3 OPEN PROBLEMS IN THE PHILOSOPHY OF INFORMATION

PI possesses one of the most powerful conceptual vocabularies ever devised in philosophy. This is because we can rely on informational concepts whenever a complete understanding of some series of events is unavailable or unnecessary for providing an explanation. In philosophy, this means that virtually any issue can be rephrased in informational terms. This semantic power is a great advantage of PI understood as a methodology (see the second half of the definition). It shows that we are dealing with an influential paradigm, describable in terms of an informational philosophy. But it may also be a problem, because a metaphorically pan-informational approach can lead to a dangerous equivocation, namely thinking that since any x can be described in (more or less metaphorically) informational terms, then the nature of any x is genuinely informational. And the equivocation obscures PI's specificity as a philosophical field with its own subject. PI runs the risk of becoming synonymous with philosophy. The best way of avoiding this loss of identity is to concentrate on the first half of the definition. PI as a philosophical discipline is defined by what a problem is (or can be reduced to be) *about*, not by *how* the latter is formulated. Although many philosophical issues seem to benefit greatly from an informational analysis, in PI one presupposes that a problem or an explanation can be legitimately and genuinely reduced to an informational problem or explanation. So the criterion to test the soundness of the informational analysis of x is not to check whether x *can* be formulated in informational terms but to ask what would be like for x not to have an informational nature at all. With this criterion in mind, we shall now review some of the most interesting problems in PI.

For reasons of space, only some research trends and issues could be included and even those selected are only briefly outlined and not represented with adequate depth, sophistication and significance. This is not only because of space, but also because the interested reader will find a wealth of further material in the other chapters of this Handbook. The issues included have been privileged because they represent macroproblems, that is, they are the hardest to tackle but also the ones that have the greatest influence on clusters of microproblems to which they can be related as theorems to lemmas. Some microproblems are mentioned whenever they seem interesting enough, but especially in this case the list is far from exhaustive. Some problems are new, others are developments of old problems, and in some cases philosophers have already begun to address them, but the review does not concern old trends and problems that have already received their due philosophical attention. There is also no attempt at keeping a uniform level of scope. Some problems are very general, others more specific. All of them have been chosen because they well indicate how vital and useful the new paradigm is in a variety of philosophical areas. Finally, whenever possible I have indicated which chapters in the Handbook are relevant to the problem under discussion.

4 THE NATURE OF INFORMATION

This is the hardest and most central question in PI. It has received many answers in different fields but, unsurprisingly, several surveys do not even converge on a single, unified definition of information (see for example [Braman, 1989; Losee, 1997; Machlup and Mansfield, 1983; Debons and Cameron, 1975; Larson and Debons, 1983]). Information is notoriously a polymorphic phenomenon and a polysemantic concept so, as an explicandum, it can be associated with several explanations, depending on the level of abstraction adopted and the cluster of requirements and desiderata orientating a theory. Claude E. Shannon, for one, was very cautious: "The word 'information' has been given different meanings by various writers in the general field of information theory. It is likely that at least a number of these will prove sufficiently useful in certain applications to deserve further study and permanent recognition. *It is hardly to be expected that a single concept of information would satisfactorily account for the numerous possible applications of this general field.* (italics added)" [Shannon, 1993, p. 180]. Thus, following Shannon, Weaver [1949] supported a tripartite analysis of information in terms of (1) technical problems concerning the quantification of information and dealt with by Shannon's theory; (2) semantic problems relating to meaning and truth; and (3) what he called "influential" problems concerning the impact and effectiveness of information on human behaviour, which he thought had to play an equally important role. And these are only two early examples of the problems raised by any analysis of information.

Indeed, the plethora of different analyses can be confusing. Complaints about misunderstandings and misuses of the very idea of information are frequently expressed, even if to no apparent avail. Sayre [1976], for example, already criticised the "laxity in use of the term 'information"' in [Armstrong, 1968] (see now [Armstrong, 1993]) and in Dennett [1969] (see now [Dennett, 1986]), despite appreciating several other aspects of their work. More recently, Harms [1998] pointed out similar confusions in Chalmers [1996], who "seems to think that the information theoretic notion of information [see section 3, my addition] is a matter of what possible states there are, and how they are related or structured [...] rather than of how probabilities are distributed among them" (p. 480).

Information remains an elusive concept. This is a scandal not by itself, but because so much basic theoretical work, both in science and in philosophy, relies on a clear grasping of the nature of information and of its cognate concepts. We know that information ought to be quantifiable (at least in terms of partial ordering), additive, storable and transmittable. But apart from this, we still do not seem to have a much clearer idea about its specific nature.

Information is often approached from three perspectives: information as reality (e.g. as patterns of physical signals, which are neither true nor false), also known as *ecological* information; information about reality (semantic information, which is alethically qualifiable and an ingredient in the constitution of knowledge); and information for reality (instruction, like genetic information, algorithms and recipes). Many extensionalist approaches to the definition of information as/about reality provide different starting points. The following list contains only some of the most philosophically interesting or influential, and I shall say a bit more about each of them presently. They are not to be taken as necessarily alternative, let alone incompatible:

1. the communication theory approach (mathematical theory of codification and communication of data/signals (Shannon and Weaver [1949 rep. 1998]; see also the chapter by Topsøe and Harremoës) defines information in terms of probability space distribution;

2. the algorithmic approach (also known as Kolmogorov complexity, [Li and Vitányi, 1997]; see also the chapters by Grunwald and Vitányi and by Adriaans) defines the information content of X as the size in bits of the smallest computer program for calculating X [Chaitin, 2003];

3. the probabilistic approach [Bar-Hillel and Carnap, 1953; Bar-Hillel, 1964; Dretske, 1981]; see also the chapter by Dretske), is directly based on (1) above and defines semantic information in terms of probability space and the inverse relation between information in p and probability of p;

4. the modal approach defines information in terms of modal space and in/consistency (the information conveyed by p is the set of possible worlds excluded by p);

5. the systemic approach (situation logic, [Barwise and Perry, 1983; Israel and Perry, 1990; Devlin, 1991]) defines information in terms of states space and consistency (information tracks possible transitions in the states space of a system);

6. the inferential approach defines information in terms of inferences space (information depends on valid inference relative to a person's theory or epistemic state);

7. the semantic approach [Floridi, 2004c; 2005b] defines information in terms of data space (semantic information is well-formed, meaningful and truthful data).

Each extentionalist approach can be given an intentionalist reading by interpreting the relevant space as a doxastic (i.e. belief-related) space, in which information is seen as a reduction in the degree of uncertainty or level of surprise given a state of knowledge of the informee (see the chapters by Baltag, Moss and van Ditmarsch and by Rott).

Communication theory in (1) approaches information as a physical phenomenon, syntactically. It is not interested in the usefulness, relevance, meaning, interpretation or reference of data, but in the level of detail and frequency in the uninterpreted data (signals or messages). It provides a successful mathematical theory because its central question is whether and how much data, not what information is conveyed.

The algorithmic approach in (2) is equally quantitative and solidly based on theory of computation. It interprets information and its quantities in terms of the computational resources needed to specify it.

The remaining approaches all address the question "what is *semantic* information?". They seek to give an account of information as semantic content, usually adopting a propositional orientation (they analyse examples like "The earth has only one moon"). Do (1) or (2) provide the necessary conditions for any theory of semantic information in (3)–(7)? Are all the remaining semantic approaches mutually compatible? Is there a logical hierarchy? Do any of the previous approaches provide a clarification of the notion of data as well? Most of the problems in PI acquire a different meaning depending on how we answer this cluster of questions. Indeed, positions might be more compatible than they initially appear owing to different interpretations of the concept(s) of information involved.

Once the concept of information is clarified, each of the previous approaches needs to address the following question.

5 THE DYNAMICS OF INFORMATION

The question does not concern the nature of management processes (information seeking, data acquisition and mining, information harvesting and gathering, storage, retrieval, editing, formatting, aggregation, extrapolation, distribution, verification, quality control, evaluation, etc.) but, rather, information processes themselves, whatever goes on between the input and the output phase. Communication theory, as the mathematical theory of data transmission, provides the necessary conditions for any physical communication of information, but is otherwise of only marginal help. The information flow — understood as the carriage and transmission of information by some data about a referent, made possible by regularities in a distributed system — has been at the centre of logical studies for some time [Barwise and Seligman, 1997; van Benthem, 2003], but still needs to be fully explored. How is it possible for something to carry information about something else? The problem here is not yet represented by the "aboutness" relation, which needs to be discussed in terms of meaning, reference and truth. The problem here concerns the nature of data as vehicles of information. In this version, the problem plays a central role in semiotics, hermeneutics and situation logic. It is closely related to the problem of the naturalisation of information. Various other logics, from classic first order logic to epistemic, erotetic and dynamic logic, provide useful approaches with which to analyse the logic of information, but there is still much work to be done [van Benthem and van Rooy, 2003; Allo, forthcoming; Allo and Floridi, forthcoming; Floridi, forthcoming].

Information processing, in the general sense of information states transitions, includes at the moment effective computation (*computationalism*, [Newell, 1980; Pylyshyn, 1984; Fodor, 1975; 1987; Dietrich, 1990]), distributed processing (*connectionism*, [Smolensky, 1988; Churchland and Sejnowski, 1992]), and dynamical-system processing (*dynamism*, [van Gelder, 1995; van Gelder and Port, 1995; Eliasmith, 1996]). The relations between the current paradigms remain to be clarified (Minsky [1990], for example, argues in favour of a combination of computationalism and connectionism in AI, as does Harnad [1990] in cognitive science), as do the specific advantages and disadvantages of each, and the question as to whether they provide complete coverage of all possible *internalist* information processing methods.

The two previous questions in §§ 4 and 5 and are closely related to a third, more general problem.

6 THE CHALLENGE OF A UNIFIED THEORY OF INFORMATION

The reductionist approach holds that we can extract what is essential to understanding the concept of information and its dynamics from the wide variety of models, theories and explanations proposed. The non-reductionist argues that we are probably facing a network of logically interdependent but mutually irreducible concepts. The plausibility of each approach needs to be investigated in detail. Both approaches, as well as any other solution in between, are confronted by the difficulty of clarifying how the various meanings and phenomena of information are related, and whether some concepts of information are more central or fundamental than others and should be privileged. Waving a Wittgensteinian suggestion of family resemblance means only acknowledging the problem, not solving it. The reader interested in a positive answer the question may wish to read the essays collected in Hofkirchner [1998]. A defence of a more skeptical view, following Shannon, can be found in [Floridi, 2003a].

7 THE DATA GROUNDING PROBLEM: HOW DATA ACQUIRE THEIR MEANING

We have seen that most analyses of the nature of information tend to concentrate on its semantic features, quite naturally. So it is useful to carry on our review of problem areas in PI by addressing next the cluster of issues arising in informational semantics. Their discussion is bound to be deeply influential in several areas of philosophical research. But first, a warning. It is hard to formulate problems clearly and in some detail in a completely theory-neutral way. So in what follows, the semantic frame will be adopted (see above § 4, (7)), namely the view that semantic information can be satisfactorily analysed in terms of well-formed, meaningful and veridical data. This semantic approach is simple and powerful enough for the task at hand. If the problems selected are sufficiently robust, it is reasonable to expect that their general nature and significance are not relative to the theoretical vocabulary in which they are cast but will be exportable across conceptual platforms.

We have already encountered the issue of the nature of data. Suppose data are intuitively described as uninterpreted differences (symbols or signals). How do they become meaningful? This is the data grounding problem.

Searle [1990] refers to a specific version of the data grounding problem as the problem of intrinsic meaning or "intentionality". Harnad [1990] defines it as the symbols grounding problem and unpacks it thus: "How can the semantic interpretation of a formal symbol system be made intrinsic to the system, rather than just parasitic on the meanings in our heads? How can the meanings of the meaningless symbol tokens, manipulated solely on the basis of their (arbitrary) shapes, be grounded in anything but other meaningless symbols?" (p. 335).

Arguably, the frame problem (how a situated agent can represent, and interact with, a changing world satisfactorily) and its sub-problems are a consequence of the data grounding problem (Harnad [1993], Taddeo and Floridi [2005]). In more metaphysical terms, this is the problem of the semanticisation of being and it is further connected with the problem of whether information can be naturalised.

8 THE SEMANTIC PROBLEM: HOW MEANINGFUL DATA ACQUIRE THEIR TRUTH VALUE

Once grounded, meaningful data can acquire different truth values, the question is how. The question then gains new dimensions when asked within epistemology and the philosophy of science. It also interacts with the way in which we approach both a theory of truth and a theory of meaning, especially a truth-functional one (see the chapter by Kamp and Stokhof). Are truth and meaning understandable on the basis of an informational approach, or is it information that needs to be analysed in terms of non-informational theories of meaning and truth? To call attention to this important set of issues it is worth asking two more place-holder questions:

1. can information explain truth?

   In this, as in the following question, we are not asking whether a specific theory could be couched, more or less metaphorically, in some informational vocabulary. This would be a pointless exercise. What is in question is not even the mere possibility of an informational approach. Rather, we are asking

(a) could an informational theory explain truth more satisfactorily than other current approaches? And

(b) should (1a) be answered in the negative, could an informational approach at least help to clarify the theoretical constraints to be satisfied by other approaches?

The second major question mentioned above is:

2. can information explain meaning?

Several informational approaches to semantics have been investigated in epistemology ([Dretske, 1981; 1988]), situation semantics ([Seligman and Moss, 1997]), discourse representation theory ([Kamp, 1984]) and dynamic semantics ([Muskens *et al.*, 1997]). Is it possible to analyse meaning not truth-functionally but as the potential to change the informational context? Can semantic phenomena be explained as aspects of the empirical world? Since the problem is whether meaning can at least partly be grounded in an objective, mind- and language-independent notion of information (naturalisation of intentionality), it is strictly connected with the problem of the naturalisation of information.

9 INFORMATION PROCESSING AND THE STUDY OF COGNITION

Information and its dynamics are central to the foundations of AI and of cognitive science (see the chapters by McCarthy and Boden). Both discipline study cognitive agents as informational systems that receive, store, retrieve, transform, generate and transmit information. This is the *information processing* view. Before the development of connectionist and dynamic-system models of information processing and the IT revolution, it was also known as the computational view. The latter expression was acceptable when a Turing machine [Turing, 1936] and the machine involved in the Turing test [Turing, 1950] were inevitably the same. It has recently become misleading, however, because computation, when used as a technical term (effective computation), refers now to the specific class of algorithmic and symbolic processes that can be performed by a Turing machine, that is recursive functions [Turing, 1936; Minsky, 1967; Floridi, 1999; Boolos *et al.*, 2002]. Not all information processing is computational in this precise sense, and in the literature one can now find approaches that use the expression more loosely.

The information/computational processing view of cognition, intelligence and mind provides the oldest and best-known cluster of significant problems in PI.[6] Some of their formulations, however, have long been regarded as uninteresting.

[6]In 1964, introducing his influential anthology, Anderson wrote that the field of philosophy of AI had already produced more than a thousand articles [Anderson, 1964, p. 1]. No wonder that (sometimes overlapping) editorial projects have flourished. Among the available titles, the reader may wish to keep in mind [Ringle, 1979] and [Boden, 1990], which provide two further good collections of essays, and [Haugeland, 1981], which was expressly meant to be a sequel to [Anderson, 1964] and was further revised in [Haugeland, 1997].

Turing [1950] considered "can machines think?" a meaningless way of posing the otherwise interesting problem of the functional differences between AI and NI (natural intelligence). Searle [1990] has equally dismissed "is the brain a digital computer?" as ill-defined. The same holds true of the unqualified question "are naturally intelligent systems information processing systems?". Such questions are vacuous. Informational concepts are so powerful that, given the right level of abstraction (LoA; [Floridi and Sanders, 2004; Floridi and Sanders, forthcoming]), anything can be presented as an information system, from a building to a volcano, from a forest to a dinner, from a brain to a company, and any process can be simulated informationally — heating, flying and knitting. So pancomputationalist views have the hard task of providing a credible answer to the question: what would it mean for a physical system *not* to be an informational system (that is, a computational system, if computation is used to mean information processing, see [Chalmers, 1996] and [Chalmers, online]. The task is hard because pancomputationalism does not seem vulnerable to a refutation, in the form of a realistic token counterexample in a world nomically identical to the one to which pancomputationalism is applied.[7] A good way of posing the problem is not: "is 'x is y' adequate?", but rather "if 'x is y' at some specified Level of Abstraction z, is z adequate?".

10 SCIENCE AND INFORMATION MODELLING

In many contexts (probability or modal states and inferential spaces), we often adopt a conditional, laboratory view. We analyse what happens in "as being (of type, or in state) F is correlated to b being (of type, or in state) G, thus carrying for the observer the information that b is G"(Barwise and Seligman [1997] provide a similar analysis based on Dretske [1981]) by assuming that $F(a)$ and $G(b)$. In other words, we assume a given model. The question asked here is: how do we build the original model? Many approaches seem to be ontologically over-committed. Instead of assuming a world of empirical affordances and constraints to be designed, they assume a world already well-modelled, ready to be discovered. The semantic approach to scientific theories [Suppes, 1960; Suppes, 1962; van Fraassen, 1980; Giere, 1988; Suppe, 1989], on the other hand, argues

[7]Chalmers [online] seems to believe that pancomputationalism is empirically falsifiable, but what he offers is not (a) a specification of what would count as an instance of x that would show how x is not to be qualified computationally (or information-theoretically, in the language of this paper) given the nomic characterisation N of the universe, but rather (b) just a re-wording of the idea that pancomputationalism might be false, i.e. a negation of the nomic characterisation N of the universe in question: "To be sure, there are some ways that empirical science might prove it to be false: if it turns out that the fundamental laws of physics are noncomputable and if this noncomputability reflects itself in cognitive functioning, for instance, or if it turns out that our cognitive capacities depend essentially on infinite precision in certain analog quantities, or indeed if it turns out that cognition is mediated by some non-physical substance whose workings are not computable." To put it simply, we would like to be told something along the lines that a white raven would falsify the statement that all ravens are black, but instead we are told that the absence of blackness or of ravens altogether would, which it does not.

that "scientific reasoning is to a large extent model-based reasoning. It is models almost all the way up and models almost all the way down." [Giere, 1999, p. 56].

Theories do not make contact with phenomena directly, but rather higher models are brought into contact with other, lower models. These are themselves theoretical conceptualisations of empirical systems, which constitute an object being modelled as an object of scientific research. Giere [1988] takes most scientific models of interest to be non-linguistic abstract objects. Models, however, are the medium, not the message. Is information the (possibly non-linguistic) content of these models? How are informational models (semantically, cognitively and instrumentally) related to the conceptualisations that constitute their empirical references? What is their semiotic status, e.g. structurally homomorphic or isomorphic representations or data-driven and data-constrained informational constructs? What levels of abstraction are involved? Is science a social (multi-agents), information-designing activity? Is it possible to import, in (the philosophy of) science, modelling methodologies devised in information system theory? Can an informational view help to bridge the gap between science and cognition? Answers to these questions are closely connected with the discussion of the problem of an informational theory of truth see above. The reader interested in some specific applications will find them in the chapters by Devlin and Rosenberg, and by Collier.

The possibility of a more or less informationally constructionist philosophy of science leads to our next cluster of problems, concerning the relation between information and the natural world.

11 THE ONTOLOGICAL STATUS OF INFORMATION

Barwise and Seligman [1997] have remarked that "If the world were a completely chaotic, unpredictable affair, there would be no information to process. Still, the place of information in the natural world of biological and physical systems is far from clear." (p. xi). This lack of clarity prompts a whole family of problems.

It is often argued that there is no information without (data) representation. Following Landauer and Bennett [1985]; Landauer [1987; 1991; 1996], this principle is usually interpreted materialistically, as advocating the impossibility of physically disembodied information, through the equation "representation = physical implementation". The view that there is no information without physical implementation is an inevitable assumption, when working on the physics of computation, since computer science must necessarily take into account the physical properties and limits of the carriers of information. It is also the ontological assumption behind the Physical Symbol System Hypothesis in AI and cognitive science [Newell and Simon, 1976]. However, the fact that information requires a representation does not entail that the latter ought to be physically implemented. Arguably, environments in which there are only noetic entities, properties and processes (e.g. Berkeley, Spinoza), or in which the material or extended universe has a noetic or non-extended matrix as its ontological foundation (e.g. Pythagoras,

Plato, Leibniz, Hegel), seem perfectly capable of upholding the representationalist principle without also embracing a materialist interpretation (see [Floridi, 2004a] for a defence of this view). The relata giving rise to information could be monads, for example. So the problem here becomes: is the informational an independent ontological category, different from the physical/material and (assuming one could draw this Cartesian distinction) the mental? Wiener, for example, thought that "Information is information, not matter or energy. No materialism which does not admit this can survive at the present day" [Wiener, 1948, p. 132].

If the informational is not an independent ontological category, to which category is it reducible? If it is an independent ontological category, how is it related to the physical/material and the mental? Answers to these questions determine the orientation a theory takes with respect to the following problem.

12 NATURALISED INFORMATION

The problem is connected with the semanticisation of data. It seems hard to deny that information is a natural phenomenon, so this is not what one should be asking here. Even elementary forms of life such as sunflowers survive because they are capable of some chemical data processing. The problem here is whether there is information in the world independently of forms of life capable to extract it and, if so, what kind of information is in question (an informational version of the teleological argument for the existence of God argues both that information is a natural phenomenon and that the occurrence of environmental information requires an intelligent source). If the world is sufficiently information-rich, perhaps an agent may interact successfully with it by using "environmental information" directly, without being forced to go through a representation stage in which the world is first analysed informationally. "Environmental information" still presupposes (or perhaps is identical with) some physical support but it does not require any higher-level cognitive representation or computational processing to be immediately usable. This is argued, for example, by researchers in AI working on animats (artificial animals, either computer simulated or robotic). Animats are simple reactive agents, stimulus-driven. They are capable of elementary, "intelligent" behaviour, despite the fact that their design excludes the possibility of internal representations of the environment and any effective computation (Mandik [2002] for an overview, the case for non-representational intelligence is famously made by Brooks [1991]). So, are cognitive processes continuous with processes in the environment? Is semantic content (at least partly) external (Putnam)? Does "natural" or "environmental" information pivot on natural signs (Peirce) or nomic regularities? Consider the typical example provided by the concentric rings visible in the wood of a cut tree trunk, which may be used to estimate the age of the plant. The externalist/extensionalist, who favours a positive answer (e.g. Dretske and Barwise), is faced by the difficulty of explaining what kind of information and how much of it saturates the world, what kind of access to, or interaction with "information in the world" an informational agent can enjoy, and how information dynamics is possible. The internalist/intentionalist (e.g. Fodor and Searle), who privileges a negative answer, needs to explain in what specific sense information depends on intelligence and whether this leads to an anti-realist view.

The location of information is related to the question whether there can be information without an informee, or whether information, in at least some crucial sense of the word, is essentially parasitic on the meanings in the mind of the informee, and the most it can achieve, in terms of ontological independence, is systematic interpretability. Before the discovery of the Rosetta Stone, was it legitimate to regard Egyptian hieroglyphics as information, even if their semantics was beyond the comprehension of any interpreter? Admitting that computers perform some minimal level of proto-semantic activity works in favour of a "realist" position about "information in the world".

Before moving to the next problem, it remains to be clarified whether the previous two ways of locating information might not be restrictive. Could information be neither here (intelligence) nor there (natural world) but on the threshold, as it were, as a special relation or interface between the world and its intelligent inhabitants (constructionism)? Or could it even be elsewhere, in a third world, intellectually accessible by intelligent beings but not ontologically dependent on them (Platonism)? The reader interested in the physics of information is adviced to read the chapter by Bais and Farmer.

13 THE IT FROM BIT HYPOTHESIS

Can nature be informationalised? The neologism "informationalised" is ugly but useful to point out that this is the converse of the previous problem. Here too, it is important to clarify what the problem is not. We are not asking whether the metaphorical interpretation of the universe as a computer is more useful than misleading. We are not even asking whether an informational description of the universe, as we know it, is possible, at least partly and piecemeal. This is a challenging task, but formal ontologies already provide a promising answer [Smith, 2004]. We are asking whether the universe in itself could essentially be made of information, with natural processes, including causation, as special cases of information dynamics (e.g. information flow and algorithmic, distributed computation and forms of emergent computation). Depending on how one approaches the concept of information, it might be necessary to refine the problem in terms of digital data or other informational notions.

Answers to this problem deeply affect our understanding of the distinction between virtual and material reality, of the meaning of artificial life in the ALife sense [Bedau, 2004], and of the relation between the philosophy of information and the foundations of physics: if the universe is made of information, is quantum physics a theory of physical information? Moreover, does this explain some of its paradoxes? If nature can be informationalised, does this help to explain how life emerges from matter, and hence how intelligence emerges from life? "Can we build a gradualist bridge from simple amoeba-like automata to highly purposive intentional systems, with identifiable goals, beliefs, etc.?" [Dennett, 1998, p. 262].

14 CONCLUSION

Our brief survey ends here. We have had a quick look to many questions of a wide variety of nature and scope. This should not be disheartening. On the contrary, we saw at the beginning of this chapter that Berkeley-Euphranor loved "information upon all subjects". It has required several scientific, technological and social transformations, but philosophers have finally begun to address the new intellectual challenges arising from the world of information and the information society. Michael Dummett recently acknowledged that "Evans had the idea that there is a much cruder and more fundamental concept than that of knowledge on which philosophers have concentrated so much, namely the concept of information. Information is conveyed by perception, and retained by memory, though also transmitted by means of language. One needs to concentrate on that concept before one approaches that of knowledge in the proper sense. Information is acquired, for example, without one's necessarily having a grasp of the proposition which embodies it; the flow of information operates at a much more basic level than the acquisition and transmission of knowledge. I think that this conception deserves to be explored. It's not one that ever occurred to me before I read Evans, but it is probably fruitful. That also distinguishes this work very sharply from traditional epistemology" [Dummett, 1993, p. 186]. Dummett is arguably correct. PI evolves out of the analytic movement, but does not seem to belong to it. It attempts to expand the frontier of philosophical research, not by putting together pre-existing topics, and thus reordering the philosophical scenario, but by enclosing new areas of philosophical inquiry?which have been struggling to be recognised and may not yet found room in the traditional philosophical syllabus?and by providing innovative methodologies to address traditional problems from new perspectives. Clearly, PI promises to be one of the most exciting and fruitful areas of philosophical research of our time. As this volume proves, it is already affecting the overall way in which new and old philosophical problems are being addressed, bringing about a substantial innovation of the philosophical system. This represents the information turn in philosophy.

ACKNOWLEDGEMENTS

This chapter is based on Floridi [2002; 2004b; 2005a]. I wish to acknowledge the kind permission by Blackwell and by the Stanford Encyclopedia of Philosophy to reproduce parts of the texts from those publications.

BIBLIOGRAPHY

[Adams, 2003] F. Adams. The Informational Turn in Philosophy, *Minds and Machines*, 13(4), 471-501, 2003.

[Allo, forthcoming] P. Allo. Being Informative, *Lecture Notes in Artificial Intelligence*, forthcoming.

[Allo and Floridi, forthcoming] P. Allo and L. Floridi. Logic and the Philosophy of Information, *Special Issue of Logique et Analyse*, forthcoming.

[Anderson, 1964] A. R. Anderson. *Minds and Machines* (Englewood Cliffs: Prentice-Hall), 1964.

[Armstrong, 1968] D. M. Armstrong. *A Materialist Theory of the Mind* (London: Routledge & Kegan Paul), 1968.

[Armstrong, 1993] D. M. Armstrong. *A Materialist Theory of the Mind* 2nd edition (London: Routledge), 1993.

[Bar-Hillel, 1964] Y. Bar-Hillel. *Language and Information : Selected Essays on Their Theory and Application* (Reading, Mass ; London: Addison-Wesley), 1964.

[Bar-Hillel and Carnap, 1953] Y. Bar-Hillel and R. Carnap. An Outline of a Theory of Semantic Information, 1953; repr. in Bar-Hillel [1964], pp. 221-74.

[Barwise and Perry, 1983] J. Barwise and J. Perry. *Situations and Attitudes* (Cambridge, Mass.: MIT Press), 1983.

[Barwise and Seligman, 1997] J. Barwise and J. Seligman. *Information Flow: The Logic of Distributed Systems* (Cambridge: Cambridge University Press), 1997.

[Bedau, 2004] M. Bedau. Artificial Life in *The Blackwell Guide to the Philosophy of Computing and Information*, edition, edited by L. Floridi (New York - Oxford: Blackwell), chap. 16, 2004.

[Berkeley, 1732] G. Berkeley. *Alciphron: Or the Minute Philosopher* (Edinburgh: 1948-57: Thomas Nelson), 1732.

[Boden, 1990] M. A. Boden. *The Philosophy of Artificial Intelligence* (Oxford: Oxford University Press), 1990.

[Boolos *et al.*, 2002] G. Boolos, J. P. Burgess, and R. C. Jeffrey. *Computability and Logic* 4th ed. (Cambridge: Cambridge University Press), 2002.

[Braman, 1989] S. Braman. Defining Information, *Telecommunications Policy*, 13, 233-242, 1989.

[Brooks, 1991] R. Brooks. Intelligence without Representation, *Artificial Intelligence*, 47, 139-159, 1991.

[Chaitin, 2003] G. Chaitin. Two Philosophical Applications of Algorithmic Information Theory, *Proceedings DMTCS'03 - Springer Lecture Notes in Computer Science*, 2731, 1-10, 2003.

[Chalmers, online] D. J. Chalmers. *A Computational Foundation for the Study of Cognition*, online.

[Chalmers, 1996] D. J. Chalmers. *The Conscious Mind : In Search of a Fundamental Theory* (New York: Oxford Univ. Press), 1996.

[Churchland adn Sejnowski, 1992] P. S. Churchland and T. J. Sejnowski. *The Computational Brain* (Cambridge MA: MIT/Bradford Press), 1992.

[Colburn, 2000] T. R. Colburn. *Philosophy and Computer Science* (Armonk, N.Y.: M.E. Sharpe), 2000.

[Debons and Cameron, 1975] A. Debons and W. J. Cameron, ed. *Perspectives in Information Science : Proceedings of the Nato Advanced Study Institute on Perspectives in Information Science, Held in Aberystwyth, Wales, Uk, August 13-24, 1973* (Leiden: Noordhoff), 1975.

[Dennett, 1998] D. C. Dennett. *Brainchildren* (Cambridge Ma: MIT Press), 1998.

[Dennett, 1969] D. C. Dennett. *Content and Consciousness* (London: Routledge & Kegan Paul), 1969.

[Dennett, 1986] D. C. Dennett. *Content and Consciousness* 2nd edition (London: Routledge & Kegan Paul), 1986.

[Devlin, 1991] K. J. Devlin. *Logic and Information* (Cambridge: Cambridge University Press), 1991.

[Dietrich, 1990] E. Dietrich. Computationalism, *Social Epistemology*, 4, 135-154, 1990.

[Dretske, 1988] F. I. Dretske. *Explaining Behavior* (Cambridge, Ma: MIT Press), 1988.

[Dretske, 1981] F. I. Dretske. *Knowledge and the Flow of Information* (Oxford: Blackwell), 1981. Reprinted in 1999 (Stanford, CA: CSLI Publications).

[Dummett, 1993] M. Dummett. *Origins of Analytical Philosophy* (London: Duckworth), 1993.
[Eliasmith, 1996] C. Eliasmith. The Third Contender: A Critical Examination of the Dynamicist Theory of Cognition, *Journal of Philosophical Psychology*, 9(4), 441-463, 1996.
[Floridi, 1999] L. Floridi. *Philosophy and Computing: An Introduction* (London; New York: Routledge). 1999.
[Floridi, 2002] L. Floridi. What Is the Philosophy of Information? *Metaphilosophy*, 33(1-2), 123-145, 2002.
[Floridi, 2003a] L. Floridi. Information in *The Blackwell Guide to the Philosophy of Computing and Information*, edition, edited by L. Floridi (Oxford - New York: Blackwell), 40-61. 2003.
[Floridi, 2003b] L. Floridi. Two Approaches to the Philosophy of Information, *Minds and Machines*, 13(4), 459-469, 2003.
[Floridi, 2004a] L. Floridi. Informational Realism, *ACS - Conferences in Research and Practice in Information Technology (Computers and Philosophy 2003 - Selected Papers from the Computer and Philosophy conference CAP 2003)*, 37, 7-12, 2004.
[Floridi, 2004b] L. Floridi. Open Problems in the Philosophy of Information, *Metaphilosophy*, 35(4), 554-582, 2004.
[Floridi, 2004c] L. Floridi. Outline of a Theory of Strongly Semantic Information, *Minds and Machines*, 14(2), 197-222, 2004.
[Floridi, 2005a] L. Floridi. Information, Semantic Conceptions Of, *Stanford Encyclopedia of Philosophy*. Edward N. Zalta (ed.), `http://plato.stanford.edu/entries/information-semantic/`
[Floridi, 2005b] L. Floridi. Is Information Meaningful Data? *Philosophy and Phenomenological Research*, 70(2), 2005.
[Floridi, forthcoming] L. Floridi. The Logic of Being Informed, *Logique et Analyse*, forthcoming.
[Floridi and Sanders, 2004] L. Floridi and J. W. Sanders. The Method of Abstraction in *Yearbook of the Artificial. Nature, Culture and Technology. Models in Contemporary Sciences*, edition, edited by M. Negrotti (Bern: Peter Lang), 2004.
[Floridi and Sanders, forthcoming] L. Floridi and J. W. Sanders. Levellism and the Method of Abstraction, forthcoming.
[Fodor, 1975] J. A. Fodor. *The Language of Thought* (New York: Thomas Y. Crowell), 1975.
[Fodor, 1987] J. A. Fodor. *Psychosemantics* (Cambridge, Ma: MIT/Bradford), 1987.
[Giere, 1988] R. N. Giere. *Explaining Science: A Cognitive Approach* (Chicago: University of Chicago Press), 1988.
[Giere, 1999] R. N. Giere. Using Models to Represent Reality in *Model-Based Reasoning in Scientific Discovery*, edition, edited by L. Magnani, N. J. Nersessian, and P. Thagard (Dordrecht: Kluwer), 1999.
[Greco *et al.*, 2005] G. M. Greco, G. Paronitti, M. Turilli, and L. Floridi. How to Do Philosophy Informationally, *Lecture Notes in Computer Science*, 3782, 623-634, 2005.
[Grim *et al.*, 1998] P. Grim, G. Mar, and P. St. Denis. *The Philosophical Computer* (Cambridge Mass.: MIT Press), 1998.
[Harms, 1998] W. F. Harms. The Use of Information Theory in Epistemology, *Philosophy of Science*, 65(3), 472-501, 1998.
[Harnad, 1990] S. Harnad. The Symbol Grounding Problem, *Physica Scripta*, D(42), 335-346, 1990.
[Harnad, 1993] S. Harnad. Problems, Problems: The Frame Problem as a Symptom of the Symbol Grounding Problem, *Psycoloquy*, 4(34), 1993.
[Haugeland, 1981] J. Haugeland. *Mind Design : Philosophy, Psychology, Artificial Intelligence* (Cambridge, Mass ; London: MIT Press), 1981.
[Haugeland, 1997] J. Haugeland. *Mind Design Ii : Philosophy, Psychology, Artificial Intelligence* Rev. and enl. (Cambridge, Mass. ; London: MIT Press), 1997.
[Hofkirchner, 1998] W. Hofkirchner, ed. *The Quest for a Unified Theory of Information : Proceedings of the Second International Conference on the Foundations of Information Science* (Amsterdam: Gordon & Breach), 1998.
[Israel and Perry, 1990] D. Israel and J. Perry. What Is Information? in Hanson [1990], pp. 1-28, 1990.
[Kamp, 1984] H. Kamp. A Theory of Truth and Semantic Interpretation in *Truth, Interpretation and Information*, edition, edited by J. Groenendijk, T. M. V. Janssen, and M. Stokhof (Dordrecht: Foris), 1984.

[Landauer, 1987] R. Landauer. Computation: A Fundamental Physical View, *Physica Scripta*, 35, 88-95, 1987.

[Landauer, 1991] R. Landauer. Information Is Physical, *Physics Today*, 44, 23-29, 1991.

[Landauer, 1996] R. Landauer. The Physical Nature of Information, *Physics Letter*, A 217, 188, 1996.

[Landauer and Bennett, 1985] R. Landauer and C. H. Bennett. The Fundamental Physical Limits of Computation, *Scientific American*, July, 48-56, 1985.

[Larson and Debons, 1983] A. G. Larson and A. Debons, ed. *Information Science in Action : System Design. Proceedings of the Nato Advanced Study Institute on Information Science, Crete, Greece, August 1-11, 1978* (The Hague: M. Nijhoff), 1983.

[Li and Vitaányi, 1997] M. Li and P. M. B. Vitányi. *An Introduction to Kolmogorov Complexity and Its Applications* 2nd ed. (New York: Springer), 1997.

[Losee, 1997] R. M. Losee. A Discipline Independent Definition of Information, *Journal of the American Society for Information Science*, 48(3), 254-269, 1997.

[Machlup and Mansfield, 1983] F. Machlup and U. Mansfield, ed. *The Study of Information : Interdisciplinary Messages* (New York: Wiley), 1983.

[Mandik, 2002] P. Mandik. Synthetic Neuroethiology in *Cyberphilosophy: The Intersection of Philosophy and Computing*, edition, edited by T. W. Bynum and J. H. Moor (New York - Oxford: Blackwell), 11-29, 2002.

[Minsky, 1967] M. L. Minsky. *Computation: Finite and Infinite Machines* (Englewood Cliffs, NJ: Prentice Hall), 1967.

[Minsky, 1990] M. L. Minsky. Logical Vs. Analogical or Symbolic Vs. Connectionist or Neat Vs. Scruffy in *Artificial Intelligence at Mit, Expanding Frontiers*, edition, edited by P. H. Winston (Cambridge, Ma: MIT Press), 1990.

[Muskens *et al.*, 1997] R. Muskens *et al.* Dynamics in *Handbook of Logic and Language*, edition, edited by J. van Benthem and A. Ter Meulen (Amsterdam: Elsevier), chap. 10, 1997.

[Newell, 1980] A. Newell. Physical Symbol Systems, *Cognitive Science*, 4, 135 - 183, 1980.

[Newell and Simon, 1976] A. Newell and H. A. Simon. Computer Science as Empirical Inquiry: Symbols and Search, *Communications of the ACM*, 19 113-126, 1976.

[Pylyshyn, 1984] Z. W. Pylyshyn. *Computation and Cognition* (Cambridge, Ma: MIT/Bradford), 1984.

[Ringle, 1979] M. Ringle. *Philosophical Perspectives in Artificial Intelligence* (Atlantic Highlands N.J.: Humanities Press), 1979.

[Sayre, 1976] K. M. Sayre. *Cybernetics and the Philosophy of Mind* (London: Routledge & Kegan Paul), 1976.

[Searle, 1990] J. R. Searle. Is the Brain a Digital Computer? *Proceedings and Addresses of the American Philosophical Association*, 64, 21-37, 1990.

[Seligman and Moss, 1997] J. Seligman and L. S. Moss. Situation Theory in *Handbook of Logic and Language*, edition, edited by Van Benthem J. and Ter Meulen A. (Amsterdam: Elsevier), chap. 4, 1997.

[Shannon, 1993] C. E. Shannon. *Collected Papers.*, edited by N. J. A. Sloane and A. D. Wyner (New York: IEEE Press), 1993.

[Shannon and Weaver, 1949] C. E. Shannon and W. Weaver, *The Mathematical Theory of Communication* (Urbana: University of Illinois Press). Foreword by Richard E. Blahut and Bruce Hajek, 1949, repringed 1988.

[Smith, 2004] B. Smith. Ontology in *The Blackwell Guide to the Philosopjhy of Computing and Information*, edition, edited by L. Floridi (New York - Oxford: Blackwell), chap. 12, 2004.

[Smolensky, 1988] P. Smolensky. On the Proper Treatment of Connectionism, *Behavioral and Brain Sciences*, 11(1), 1-23, 1988.

[Suppe, 1989] F. Suppe. *The Semantic Conception of Theories and Scientific Realism* (Urbana, Ill.: University of Illinois Press), 1989.

[Suppes, 1960] P. Suppes. A Comparison of the Meaning and Uses of Models in Mathematics and the Empirical Sciences, *Synthese*, 12, 287-301, 1960.

[Suppes, 1962] P. Suppes. Models of Data in *Logic, Methodology and Philosophy of Science: Proceedings of the 1960 International Congress*, edition, edited by E. Nagel, P. Suppes, and A. Tarski (Stanford: Stanford University Press), 252-261, 1962.

[Taddeo and Floridi, 2005] M. Taddeo and L. Floridi. Solving the Symbol Grounding Problem: A Critical Review of Fifteen Years of Research, *Journal of Experimental and Theoretical Artificial Intelligence*, 17(4), 419–445, 2005.

[Turing, 1936] A. M. Turing. On Computable Numbers, with an Application to the Entscheidungsproblem, *Proceedings of the London Mathematics Society, 2nd series*, 42, 230-265. Correction published in Vol. 43 (1936), pp. 544-546.

[Turing, 1950] A. M. Turing. Computing Machinery and Intelligence, *Minds and Machines*, 59, 433-460, 1950.

[van Benthem, 2003] J. van Benthem. Logic and the Dynamics of Information, *Minds & Machines*, 13(4), 503-519, 2003.

[van Benthem and van Rooy, 2003] J. van Benthem and R. van Rooy. Connecting the Different Faces of Information, *Journal of Logic, Language and Information*, 12(4), 375-379, 2003.

[van Fraassen, 1980] B. van Fraassen. *The Scientific Image* (Oxford: Clarendon Press), 1980.

[van Gelder, 1995] T. van Gelder. What Might Cognition Be, If Not Computation? *Journal of Philosophy*, 92, 345-381, 1995.

[van Gelder and Port, 1995] T. van Gelder and R. Port, ed. *Mind as Motion: Explorations in the Dynamics of Cognition* (Cambridge, Ma: MIT Press), 1995.

[Weaver, 1949] W. Weaver. The Mathematics of Communication, *Scientific American*, 181(1), 11-15, 1949.

[Wiener, 1948] N. Wiener. *Cybernetics or Control and Communication in the Animal and the Machine* 2nd ed. (Cambridge, Ma: MIT Press), 1948.

[Young, 2004] R. A. Young. Wittgenstein's Tractatus Project as Philosophy of Information, *Minds and Machines*, 14(1), 119-132, 2004.

LEARNING AND THE COOPERATIVE COMPUTATIONAL UNIVERSE

Pieter Adriaans

1 INTRODUCTION

In the summer of 1956, a number of scientists gathered at the Dartmouth College in Hanover, New Hampshire. Their goal was to study human intelligence with the help of computers. Their central hypothesis was: "that every aspect of learning or any other feature of intelligence can in principle be so precisely described that a machine can be made to simulate it." During that conference, where amongst others John McCarthy, Claude Shannon and Marvin Minsky were present, the new discipline of Artificial Intelligence was born. It is striking that 'learning' was considered to be an important aspect of human intelligence from the start. A better understanding of the phenomenon of learning was high on the agenda of the emerging young science.

Now, fifty years later, the study of learning is one of the success stories of AI. There is a multitude of learning techniques for the computer. Data mining techniques are being used for marketing, stock management, production optimization and fraud detection in the commercial domain. Biologically inspired learning models such as neural networks and genetic algorithms are being used to simulate human cognition and evolution. In disciplines like computer vision and computational linguistics, machine learning is in the center of interest [Kearns and Vazirani, 1994; Mitchell, 1997; Adriaans and Zantinge, 1997; Cornuéjols and Miclet, 2003].

But, researchers do not have much reason to sit back and rest, because there is still a whole list of questions that are begging for answers. One of the biggest embarrassments is that we still do not know what learning is exactly. The toolbox of a machine learner looks like a haphazardly collected bunch of screwdrivers, hammers en chisels of dubious origin. For some jobs they work, but we do not understand why, for others they do not work and we also do not understand why. One thing is certain. If we understand learning as data compression then there will never be a general theory that explains what learning is exactly.

1.1 *Philosophy of information*

It is clear that with the adventure of artificial intelligence we have hit upon a problem domain that has much wider repercussions than the creation of intelligent

Handbook of the Philosophy of Science. Volume 8: Philosophy of Information
Volume editors: Pieter Adriaans and Johan van Benthem. General editors: Dov M. Gabbay, Paul Thagard and John Woods.

computers. Recently a new discipline has emerged: the philosophy of information [Floridi, 2004].[1] This discipline reformulates central questions of philosophy from the perspective of modern insights from computer science. Developments like these, urge us to formulate the question of the relation between philosophy on one side and logic, mathematics, theory of information and computation on the other.

First of all philosophy, in my view, is not science. It takes a meta-position and is always at most *a reflection on science* and scientific results. It is not the primary task of the philosopher to formulate and prove theorems. It is his task to reflect on the consequences of theorems and theories. On the other hand *philosophy can not claim to have any form of privileged access to reality.* There is no fixed Archimedean position from which the philosopher can judge the results of scientific endeavors.[2] Philosophy and science therefore are doomed to live permanently in each other's shadow without any possibility of a final reconciliation. Any scientific result can be made the object of philosophical analysis, but ... only, or predominantly, in terms of the concepts that the sciences have constructed themselves. Philosophy therefore is at its best when it is in dialogue with foundational programs of science and the humanities. The more it removes itself from these central issues, the more substance it loses and the more it deteriorates into a (possibly brilliant) literary exercise at best. In this sense, philosophical reflection may be seen as an inherent and necessary aspect of scientific heuristics. It provides us with a rich historical context of 2500 years of reflection on foundational programs and invites us to investigate the more extreme consequences of our theories and models.

The study of theory of knowledge, theory of information and computation, methodology of science, theory of induction and meta-mathematics share a common history in which related questions have been analyzed in different guises. The work of Solomonoff and Kolmogorov provides direct answers to questions about the nature of knowledge and induction proposed by Carnap and the Wiener Kreis and much earlier Kant and Hume. In this light, one has to interpret the reflections on theory of information and learning that I present below.

1.2 Philosophy of learning

First, I show that the question of the essence of learning is embedded in fundamental epistemological questions. The old philosophical problem of the essence of knowledge is fundamentally associated with learning. The notion of efficiency of learning plays an essential role in this context. Our models of learning show us that tasks, like learning a language, that human beings perform without too much difficulty, are from a formal point of view extremely complex and next to impossible. This leaves us with the riddle of human efficiency. I show how the contours of

---

[1] See the chapter by Floridi in this book

[2] Specifically: no privileged direct access to ones own consciousness, no Husserlian epoche, no historical laws of materialism, no recourse to immediately given sense data, no special rapport with Being itself, etc.

an analysis of this mysterious efficiency of human learning takes shape in the light of recent insights from complexity theory and thermodynamics. Central questions in this respect are:

QUESTION 1. What is learning?

QUESTION 2. What are data sets from which we can learn?

QUESTION 3. What kind of systems produce those data sets?

The answer to the first question is: learning is algorithmic compression of data sets. Not all forms of learning are caught by this definition, but a broad class of philosophically relevant learning phenomena fall under this description.[3] The answer to the second question is: data sets that can be compressed by a computer algorithm without too much effort.[4] An answer to the third question is - quite naturally - systems with relatively low entropy: i.e. self-organizing systems, systems that are not in a state of thermal equilibrium and systems that redirect energy from their environment in order to keep their internal entropy lower than that of the environment. This kind of self-organization is typical for life and for computational processes. The picture that emerges is that those systems in nature that produce data sets from which something can be learned are by necessity systems with a relatively low entropy. The data sets themselves consequently have low entropy and are easy to decipher. This seems to be the solution to the problem of the efficiency of our learning algorithms. A deep analysis of the idea that the universe can be interpreted as a computational process shows that nature necessarily acts as a cooperative teacher. This is a philosophical insight that transcends the local context of Artificial Intelligence. At the same time these insights help us to develop new algorithms that solve problems from every day life. Learning in the form of data compression helps us to classify viruses, analyze music [Cilibrasi and Vitanyi, 2005] and to learn languages [Adriaans, 2001].

1.3 A short historical digression

The notion that knowing something implied knowing its 'form' goes back to Plato's theory of ideas as forms. Aristotle's more empirical doctrine of the four causes (causalis, finalis, formalis and efficiens) also distinguishes the notion of form as a crucial element of knowledge. The original technical notion of the Latin word 'in-formare' (giving form to something, impressing ideas/forms in the mind in the

[3]Neural networks, genetic algorithms, decision tree induction, clustering, nearest neighbor, support vector machines, association rules, to name a few. As a counter example: simple rote learning of a finite set of facts does not necessarily involve compression of data.

[4]Technically: data sets that can be compressed by means of constructive resource bounded compression. The 'without too much effort' restriction is added because it actually is possible to construct highly compressible data sets that from the outside look random, e.g. encrypted data or expansions of very special real numbers like π and e. There are no general algorithms to compress these sets. It is highly unlikely that these data sets occur frequently in nature. Anyhow, we would not notice them.

Platonic sense) that is found in the writings of Cicero[5] and Augustine seems to have played no role in the emergence of the modern concept of information. The word 'idea' seems the true modern heir of the classical term 'information' [Capurro, 1978; Capurro and Hjørland, 2003].

In the 15th century, the French term 'information' finds its way into the colloquial vocabulary of European languages with various subtle differences in meaning, clustering around meanings like 'investigation', 'education', 'the act of informing or communicating knowledge', 'intelligence' etc. After Descartes the technical term seems to vanish from the philosophical debate. It does not play any specific role in the work of a broad philosopher like Kant. There is no lemma on information in Windelbands famous 'Lehrbuch der Geschichte der Philosophie' from 1889 [Windelband, 1921]. Even Edward's Encyclopedia of Philosophy from 1967 does not have a separate lemma on information [Edwards, 1967]. The same holds for the well-known History of Logic written by Kneale and Kneale that first appeared in 1962 [Kneale and Kneale, 1988]. In short the term 'information' seems to have been absent from the philosophical dialogue for hundreds of years.

In the history of philosophy the phenomenon of learning has long been studied implicitly, because it is related to knowledge, but since circa 1700 AD the problem of learning is placed explicitly on the philosophical agenda. A key insight in the study of the history of the concept of information is formulated in this book by Devlin and Rosenberg in their chapter on information in the social sciences, where information is described as an abstract notion that is the natural byproduct resulting from the advent of modern media. When human communication was transformed from a direct dialogue between individuals to an interaction that was mediated by technology (telescopes, microscopes, books, newspapers, the telephone, television, internet etc.) the need to create an abstract umbrella term to denote the 'stuff' that was transmitted from a sender to a receiver of a message emerged. In this respect, the emergence of the empirical sciences in the 17th century is a central period in history of the conceptualization of information.

Descartes (1596–1650) formulated a firm mathematical framework for the description of the material world, but his dualism prevented him from understanding the interplay between language and the growth of knowledge. For Descartes, man's rationality was equivalent to mastering language and was an innate quality. The communication between the res extensa and the res cogitans remained a central problem. Descartes is important because he is the first philosopher who formulated a theoretical framework in which the mediation between mind and body, between the knower and the known becomes problematic. With hindsight one could say that in the work of Descartes the need for *an abstract concept of mediation between knower and the known*, i.e. a concept of information, is identified for the first

[5]Cicero used the word information as a translation of the Epicurean notion of 'prolepsis', i.e. a representation in the mind. A notion that can be compared to the later use of the word 'idea' by Descartes and Locke. See 'On the nature of the Gods", I, 43. Also Greek terms like 'hypothesis' and 'eidos' were translated with the term 'information' by Latin authors [Capurro, 1978].

time. Descartes' metaphysics can not describe such a mediation. Because of this lack, he was incapable of developing an adequate philosophical theory of language and thus of an adequate conceptualization of the interplay between language and knowledge.

The next philosopher to take up this challenge was Locke (1632–1704) who developed a psychological version of Cartesian dualism in the "Essay concerning human understanding" (1690) [Locke, 1961]. The Cartesian cogito becomes a epistemological subject that starts as a tabula rasa and is gradually filled up with 'ideas' that find their origin in experience. Descartes had formulated the notion of ideas as innate forms of thought but Locke is quite liberal in his concept of an 'idea': "*whatsoever is the object of understanding when a man thinks ... whatever is meant by phantasm, notion, species, or whatever it is which the mind can be employed about when thinking*". (Essay, I,i,8) This abstract notion of an idea, as a qualitative building block of knowledge, can be interpreted as a philosophical precursor of the modern concept of information. Ideas emerge in the mind as a result of sensory experience, they can be isolated and combined into new knowledge. When we receive ideas our knowledge grows.

This conceptualization of the growth of knowledge in terms of the combination of 'chunks' of knowledge implied a reformulation of a number of central problems in philosophy that would dominate the discussion for the next centuries. Central questions are:

- Can we validate general statements about the properties of a class on the basis of a finite number of observations of members of that class? Can we derive the statement "All swans are white" on the basis of "All swans we have seen so far are white"?

- Can we generalize from the past to the future?

- What part of knowledge is a priori, what part a posteriori?

In *An Enquiry Concerning Human Understanding*, (par. 4.1.20-27, par. 4.2.28-33) the philosopher Hume (1711–1776) argued that there is no logical necessity that the future will resemble the past. The insight that it is impossible to select the best theory to explain a set of observations with absolute certainty, is known as the induction problem since Hume [1909, 1914]. It denies science the possibility to formulate universal laws with absolute certainty. Several philosophers have tried to deal with this problem. It was the main motivation for the development of Kant's transcendental philosophy in the Kritik der reinen Vernunft. Kant's attempt is the last major effort to bridge the gap between empirical science and traditional philosophy striving at the formulation of absolute truths.

The empiricist program was revived by the so-called Vienna circle in the beginning of the 20th century. The ambition was to seek the foundation of science in the analysis of elementary phenomena that could be observed empirically. Needless to say that, with this methodology, the induction problem is a major obstacle for science. Popper, who occasionally attended meetings of the Vienna circle, formulated a solution in terms of the asymmetry between verification and falsification [Popper, 1952]. Although this solved part of the problem, the issue of heuristics remained open (Context of discovery versus context of justification).

One solution to the induction problem is to view scientific knowledge as being essentially statistical. The concept of probability is far from harmless from a philosophical point of view [Hájek, 2002]. Carnap [1950] has argued that there exist two very distinct forms of probability: a priori probability or "Rational credibility" and empirical probability in the sense of "limiting relative frequency of occurrence". Indeed there seems to be a distinct difference between the use of the notion of probability in observations like: "It is highly probable that an English sentence contains more *e*s than *q*s" and "It is highly probable that life on earth originated from outer space". The first is a statement about the frequency of letters in English. It can be corroborated by a sequence of experiments. The second statement seems different. It has prima facie nothing to do with limiting frequency. It can not be corroborated by experiments. Even if our planet was the only planet in the universe with life, the statement still could be true. It seems to express a rational belief that somebody could have after carefully examining the evidence.

Black [1967] has criticized Carnap: different modes of verification for probability statements do not imply that there necessarily exist different notions of probability. The fact remains that we sometimes make judgements about the probability of individual isolated structures. This seems to involve a notion of a priori probability. If we can assign a priori probabilities to theories and data sets and conditional probabilities to a data set given a theory, then we can calculate the probability of a theory given a data set. The formulation of an exact answer to these theoretical questions is one of the great achievements of computer science in the 20th century. Solomonoff defined the idea of algorithmic complexity of a binary object as the shortest program that computes this object on a universal reference Turing machine [Solomonoff, 1997].[6] He showed that the algorithmic or Kolmogorov complexity of an object is associated with an a priori probability of this object. It allows us in theory to assign an a priori probability as well as a complexity to an individual binary object (universal distribution). These measures exist, but can not be computed. This is the basis for modern theories about learnability and studies of methodology of science.

A central concept that ties information theory and learning together is the so-called Minimum Description Length Principle (MDL) [Rissanen, 1999]. Below I will give a formal treatment of the principle, but the main idea is that formal representations of scientific theories can be used to compress data sets with empirical observations. The shortest adequate MDL code explaining a data set will be the one that minimizes the sum of a description, in bits, of the theory, plus a description, in bits, of the set of observations given the theory. One could think of the observations of Tycho Brahe and Kepler's laws as theory. The laws of Kepler

[6]The same concept was somewhat later discovered independently by Kolmogorov and Chaitin.

explain the observations of Tycho Brahe, because these observations can be represented concisely using these laws. Kepler's laws are much simpler than the rules of the cosmology of Ptolemy based on celestial spheres and they also do a good job of predicting the motions of the planets. One of the main ambitions of this paper is to study the philosophical implications of this concept. The theory of Kolmogorov complexity provides us with an excellent framework for a philosophical analysis of the concepts behind MDL. This is, in my view, the form in which the problem of induction should be studied in the current context of philosophy of information.

The MDL principle is often described as being equivalent to Ockham's razor (entia non sunt multiplicanda preater necessitate, William of Ockham, ca. 1290–1349). An association that is debatable, since Ockham's razor is related to a specific nominalistic critique of Plato's theory of ideas (as defended by Duns Scotus, 1266–1308) that is quite far removed from the general problem of induction. In fact, the idea of explaining a certain set of observations in terms of an optimized two-part code (Theory + Data encoded with the theory) could as well be interpreted as a Platonic ambition, where the Theory is the *ideal* description of the data and the Data encoded with the theory is a description of the noise, or *faults*, in the data. The underlying problem seems to have a different nature: the question of the regularity of nature, or in other words the notion of a cooperative universe.

2 AN UNEASY MARRIAGE BETWEEN LEARNING AND KNOWING: PARTICIPATION VERSUS CONSTRUCTION

A theory of learning has consequences in at least three areas:

- Theory of knowledge: how do we gather knowledge?
- Cognition: how does our brain work?
- Methodology of science: how do we construct scientific knowledge?

Knowledge and learning have always had a rather uneasy relationship in philosophy. The subject easily could fill a book in itself. A clear picture emerges if we try to develop a simple logic of learning and knowing. We can adopt two axioms:

1. Priority of knowing: I know everything that I have learned.
2. Priority of learning: I have learned everything that I know.

The first axiom seems obvious. Learning would not really be learning if it did not lead to knowledge. Yet, this is not unproblematic. Learning has a temporal aspect. It involves a transformation from not knowing to knowing. If we simply learn a finite number of facts, this is straight forward. If somebody tells me that Amsterdam is the capital of the Netherlands and I did not know that, then I have learned something. Of course, I trust my source of information to speak the truth. He must be a trustworthy teacher. Even if that is the case, things get more complicated if I try to learn an infinite number of facts in a finite time. Since Hume, philosophers know that this is logically impossible. One can never learn a general law on the basis of a finite number of observations. Even if I have seen millions of white swans, this does not allow me to draw the conclusion that the statement "All swans are white" is true. I only need to observe one black swan and my general law can be scrapped [Popper, 1952]. The conclusion seems clear. Logically, it is impossible to learn an infinite set on the basis of a finite number of observations. To put it in other words: we can learn facts, but we can not learn general laws. This would mean the end of science. Philosophers that endorse the first axiom implicitly sweep the problem of learning under the carpet: learning actually is remembering what you already know (Plato), you can only learn if knowledge is innate (Descartes, Chomsky), mathematical research is the discovery of what is already there (Hilbert, Gödel). Under axiom 1) scientific knowledge is only possible if one has what I call a participation theory of truth. The amount of knowledge of the human subject grows in time, but not by means of learning. The human mind seems to participate in the realm of truth and this participation allows us to separate true from untrue insights. It is clear that this theory of learning is less satisfactory.

So let's have a look at axiom 2) the priority of learning. From this perspective we seem to loose our grip on the concept of knowledge. Results that we have learned are preliminary: they can change, they have a statistical nature. In most cases, learning leads to a hypothesis that only has a certain degree of plausibility. It does not seem to be a good idea to accept the derivation "The hypothesis P is very probable, therefore I know P" as valid. Knowing seems to be an absolute concept. The situation in which I testify in court that I know that John has killed Mary is very different from the situation in which I testify that it is very probably that John is the killer. Nevertheless we are willing to sentence somebody, even if we are not completely sure that he is guilty. Beyond reasonable doubt is a phrase that finds its philosophical roots in the work of Hume, who has chosen the second axiom as his starting point. This position leads to what I call a construction theory of truth. A supporter of this theory has two options. Either he admits that knowledge is a statistical phenomenon or he limits himself to knowledge that can be constructed out of elementary observations. This last option leaves very little room for science. Yet this position has been defended vigorously in the philosophy of mathematics by Brouwer and the early Wittgenstein. Traces of the first solution can be found in the works of Aristotle, Euclid, Locke, Hume and the members of the Wiener Kreis.

This short analysis shows that one could rewrite the history of philosophy with learning as a central theme. For a long time such a history would not contain much more than what I summarized above. Both axioms lead to unfortunate conclusions. A good choice is not really possible: a real philosophical problem. In the second half of the 20th century theoretical ideas developed rapidly mainly as a result of the application of insights from mathematical model theory and thermodynamics to an analysis of the phenomenon of learning.

3 THE RIDDLE OF HUMAN EFFICIENCY

The mathematics of learning starts with the conception of learning as a game that is played between a student and a teacher. The game theoretical model of learning was first introduced by Gold in Information and Control in 1967. The problem that Gold studies is learning a language. The form of the game is as follows:

1. There is background knowledge. The teacher and the student agree beforehand on a(n) (infinite) class of possible languages, one of which is to be learned.
2. The teacher chooses one language from this class that he is going to teach.
3. A move of the teacher consists of the presentation of an example sentence from the language he has chosen. The teacher must be faithful. He is obliged to produce all possible sentences of the language in the limit at least once.
4. A move of the pupil consists of a guess of the language (a hypothesis) that the teacher has selected.
5. The game continues indefinitely. The pupil learns the language (wins the game) when he does not need to update his hypothesis anymore.

We can suggest the following practical interpretations of this abstract model:

- Theory of knowledge: the student is any human being, experience is the teacher, the class of languages is the set of possible theories about the world.
- Cognition: the student is the brain, the teacher is perception, the class of languages is the number of concepts that the human brain can learn.
- Methodology of science: the student is the scientist, the teacher is nature, the class of languages is the set of possible laws of nature.

For our purpose, the abstract model is rich enough. The surprise of Gold's paper was that he could prove that under these conditions, even if the game could go on for ever, the student could not learn classes of languages of any interest with absolute certainty. This holds a fortiori for all natural languages that we all learn as children without much difficulty. Here we find an interesting problem that has not been solved adequately until this day and really only has become more urgent. One could baptize this problem the riddle of human efficiency. All our formal models of learning tasks indicate that learning, from a formal point of view, is next to impossible or at least extremely hard. The central issue here is that learning in Gold's model is distribution free, i.e. the only constraint is that every sentence of the language has a positive probability of being produced by the teacher. This allows for highly non-standard distributions on which one cannot expect general learning algorithms to converge.

In the last 40 years, we have seen an overwhelming number of amendments and adaptations of Gold's model and theory construction certainly is not finished (See e.g. [Angluin, 1988]). The research concentrates on a number of issues: a restriction on the class of languages, using statistical techniques to select the hypothesis, richer interaction between the student and the teacher and the attitude of the teacher. In the original model of Gold, the teacher only has to be reliable. He gives all the examples in a random sequence. It is easy to imagine that the teacher helps the student a bit, for instance by selecting simple examples first or by adapting the information content of the examples to the progress of the student. In this case, we have a cooperative teacher. In its simplest form the cooperative teacher is nothing but a probability distribution over the set of examples that gives a higher probability to simpler examples. A student that studies under the guidance of a cooperative teacher has a much higher chance of selecting the right hypothesis with the help of statistical reasoning. Here, we distinguish the contours of an interesting solution to the riddle of human efficiency in learning. Our efficiency might not be an achievement of human intelligence but more a reflection of the structure of the world in which we live. Nature is not completely random, it is organized and works as a cooperative teacher. Before we explore this concept further, we need to develop a formal framework to study these concepts.

3.1 Learning as data compression

Suppose you switch on your television set and there are three different channels from which you can choose: random noise, a picture of a forest and a test image. From a computational point of view, we can analyze these three data sets in the following way:

1. **Random noise**: this data set has a high complexity and therefore contains from a theoretical point of view a lot of information. Because the data set is the result of a random process it cannot be compressed into a shorter description. This means that it does not contain any meaningful information. No part of the data set contains any information about any other part. There is no self-information. Nothing can be learned from it. These data sets are typical for systems that are in thermal equilibrium and thus have maximal entropy.

2. **The picture of a forest**: this data set has high complexity, but it also contains structure (the forms of the branches, leaves and trees repeat themselves: there is self-information). Therefore the image can be compressed into a shorter description. We can extract meaningful information from the picture (e.g. the fact that we can distinguish 10 trees in the picture). We can learn a lot from this data set. These data set are typical for self-organizing

systems that extract energy from the environment to create some form of order, e.g. living things, computational processes.

3. **The test image**: this data set looks very simple with regular geometrical shapes. It can easily be compressed and thus contains little information at all. Nothing much can be learned from it.

From these examples it is clear that we can learn the most interesting things from data sets that show a mix of structure and random elements. This is exactly the sort of data that one would expect in a computationally cooperative universe. Modern learning theory focuses on the analysis of this kind of data sets. The ambition is to find an optimal short description of the data set in terms of two new data sets:

- A structural part that described the regularities in the data set.
- An ad hoc part that describes the random elements of the data set.

Such a description is technically adequate if the length of the new description in terms of two data sets is (much) shorter than that of the original data set. In the literature this principle is known as the Minimum Description Length principle [Rissanen, 1999], sometimes interpreted as two part code optimization [Vereshchagin and Vitányi, 2004]. Suppose that the picture of the forest has a size of 1280 $\times$ 800 pixels of 256 colors, than the uncompressed file will have a size of about 31 Mb. This is the number of bytes we need to send via a communication channel if we want to communicate the contents of the file. As soon as we have an analysis of the meaningful content of the picture at our disposal we can summarize the content. In this way we get a sequence of interpretations of the picture in which more and more of the content is revealed:

Ad Hoc	**Structural**
A forest	A general description of forests
A set of 10 trees	A general description of the structure of a tree
A set of 3 birches, 4 willows and 3 oaks	A description of the specific structure of birches, willows and oaks
Etc.	Etc.

An important part of the research in learning theory concentrates itself on the development of algorithms that can separate a data set in an ad hoc and a structural part. Many scientific problems can be reformulated in terms of a two part code optimization problem. I give a number of examples:

Data Set	Ad Hoc	Structural
Description of our solar system	Trajectories and size of the planets	Kepler's laws
Reuters Database	Structure and sequence of the individual sentences	English grammar
A composition by Bach	Structure and sequence of themes	Specifics of Bach's style
Human DNA	Structure and sequence of regions that code genes	A description of genes

Finding such a two part code optimization is usually not an easy task. One can formally prove that there is no universal learning algorithm for such a task. For some data sets we have good algorithms, for others not (yet). It is possible with a learning technique called genetic programming to derive the laws of Kepler from the observations of Tycho Brahe, but a good algorithm for learning a grammar on the basis of a corpus is not yet available [Adriaans and van Zaanen, 2004]. In the following paragraphs we will develop a deeper understanding of learning as compression.

4 LEARNING, COMPUTATION, INFORMATION AND ENTROPY

In this section we will develop a formal framework that helps us to understand learning better. The crucial step is the definition of the concept of information as something that could be objectively quantified. It is immediately clear that the concepts of information and learning are related. It seems impossible to learn without gaining information and impossible to gain information without learning. A discussion of the technical issues concerning the concept of information is not possible without an understanding of the concept of a Turing machine. In the next paragraphs we will first describe this basic notion and then turn our attention to the definition of information.

The Turing machine

In its simplest form, a Turing machine is a device with a read-write head, an infinite working tape on which symbols can be read and written and a finite deterministic program for the manipulation of symbols. The only symbols needed are '1', '0' and 'b' (blank). The machine starts its calculation by reading input from the tape, and stops when a certain predefined final state is reached. Not all programs will stop. In fact, Turing proved that there does not exist a program that decides in all cases whether a certain machine will stop given a certain input (undecidability). The combination of machines and programs that stop in finite time is known as the *Halting Set.* This set could be seen as a transcendent object in computer science: we know it exists, but it can not be constructed. There are a number of reasons why Turing's device can claim to be associated with a universal scientific language.

First of all, the set of all possible programs for a Turing machine is the set of all possible binary strings $\{0,1\}^*$, which is equivalent to the set of natural numbers. Secondly, one can define a 'universal' Turing machine, that emulates all possible computations of all possible Turing machines by first reading a definition of a machine from the tape followed by the definition of the program and the execution of the program on the emulated machine. This allows us to interpret the Turing machine as a universal computing device. Thirdly, all the current definitions of the concept of computation (Lambda calculus, combinatorial logic, recursive functions, etc.) are known to be Turing equivalent, i.e. can be emulated on a Turing machine. This fact has lead to the formulation of the so-called Church-Turing thesis, which states everything computable is computable on a Turing machine. It is hard to imagine how this claim could ever be verified. In the worst case it is destined to be an unproven metaphysical claim for ever. The thesis could easily be falsified by a conception of calculation that can not be emulated on a Turing machine, but so far, these conceptions of computation escape our imagination.

From a transcendental point of view, the Turing machine encapsulates fundamental notions: *The local physical storage and processing of a finite set of discrete symbols as a sequential finite discrete process in time according to a finite set of (deterministic) rules.* The apparent universality of these notions lead to what one might call the central working hypothesis of modern computer science:

CONJECTURE 4. Any finite discrete system or process can be described in terms of a program for a Turing machine.

Personally I expect this claim to be disproven (or at least amended) somewhere in the future, but for the moment it gives the foundation for a methodological research program that is rich in perspectives and far from exhausted. It defines a universal scientific methodology. For any system X, we have to ask ourselves the fundamental question: is X a finite discrete system? If so, we can apply our methodology and try to construct an adequate program to model it. The decision to consider a certain phenomenon X (say a financial administration, turbulence around a sail, human consciousness, the human cell, a black hole or the universe as a whole) to be a finite discrete system can be controversial from a philosophical point of view and require a separate philosophical motivation. These questions are not part of our current analysis. For the moment, my aim is the clarification of the central concepts and not an analysis of their applicability.

The association with the old philosophical ambition of a mathesis universalis is immediately clear from the Turing equivalence of recursive functions, which lead to the following corollary:

COROLLARY 5. *Any finite discrete system or process can be described in terms of operations on natural numbers.*[7]

This analysis of Turing machines does not lead to a theory of information. It is a

[7]Wolfram states a related notion that he calls the Principle of Computational Equivalence: "... whenever one sees behavior that is not obviously simple ... it can be thought of as computation of equivalent sophistication" [Wolfram, 2001, p. 5].

theory-neutral conception of manipulation of binary strings. In order to determine what kind of information, and how much of it, is contained in these strings we need separate definitions. Even within this context, there are a number of competing conceptualizations of the notions of information that need to be treated here.

Shannon Information and optimal codes

The idea that the frequency of a letter is associated with the information it contains (or its value) is well known to any person who solves a crossword puzzle or plays Scrabble. If one knows that a word contains a 'z' this is more informative than an 'e' because there are less words with a 'z'. This 'information' about the 'z' implies a bigger reduction of the search space. The crucial insight that has lead to a mathematical theory of information is formulated by Shannon [Weaver and Shannon, 1949]. Here the information content of a message is defined in terms of its probability:

DEFINITION 6. The Shannon information contained in a message x is $I(x) = \log 1/P(x) = -\log P(x)$,

where $I(x)$ is the number of bits of information contained in x and $P(x)$ is a probability distribution ($0 \leq P(x) \leq 1$). Note that[8]: If $P(x) = 1$ then $I(x) = 0$. $I(x \text{ and } y) = I(x) + I(y)$.

From a philosophical point of view, it is important to note that Shannon information says nothing about the meaning of the messages, nor about their epistemological status. One bit is the maximal amount of information that can be stored in a binary symbol. A bit can simply be used as a physical unit. Alternative notions are nat, based on the natural logarithm, and hartley, based on log base 10. One nat corresponds to about 1.44 bits ($1/(ln2)$), or 0.434 hartleys ($1/(ln10)$). If x is a message and $P(x) = 2^{-3}$, then the amount of information contained in x is three bits and an optimal code for x would use three bits, say 001. Apart from this, x could have any meaning, varying from "John has passed his exam" to "Goldbach's conjecture is true". In itself, this is strange. We are inclined to say that if we get the information that John passed his exam from a reliable source we consequently *know* that John passed his exam. A simple bit code like 001 does not convey this information. Apparently there are meanings of the term 'information' that are not fully covered by Shannon's definitions. Shannon himself, by the way, would be the first to acknowledge this. Also there is no straightforward translation of Shannon's definitions into a theory of knowledge. A valuable attempt to fill this gap is made by Dretske [Dretske, 1981]. The least one can say is that, on top of the formal definitions that are offered by Shannon, the information that is received by an agent is dependent on the context of the dialogue and on the background knowledge shared by parties involved in the exchange of messages.

A second observation that is philosophically relevant is that Shannon information, as such, is independent of the notion of a Turing machine. Shannon defines

[8] log is used for log_2

information in terms of bits and Turing machines operate on strings of zeros and ones that could be interpreted as bit strings. In these terms Turing machines could be seen as information processing devices, but this is only a very weak connection. Shannon's notion of information and Turing's definition of computation seem to be orthogonal. Shannon uses the notion of a bit to measure amounts of information, but his theory does not say anything about the amount of information that is stored in a string of bits itself.

The concept of Shannon information only makes sense in the context of a set of potential messages that are sent between a sender and a receiver and a probability distribution over this set. If we have such a setting, we can design an optimal code system. Suppose X is a set of messages $x_i(I = 1, \dots n)$ the **communication entropy** of X is:[9]

$$H(X) = -\sum_{i=1,n} P(x_i) \log P(x_i)$$

The **Maximal entropy** of a set of n messages, if $P(x_i) = 1/n$ for each I:

$$H_{max}(X) = -n(1/n)\ log\ (1/n) = \log n$$

The **Optimal code** (that minimizes the expected message length) assigns $-logP(x_i)$ bits to encode message x_i. One finds an extensive discussion of these definitions in the chapter by Harremoës and Topsøe. The notion of optimality of a code system is associated with the idea of compression of a set of messages. Suppose, for the sake of argument, that we want to develop an optimal code for a certain book, say Dickens' "A Tale of Two Cities", and that we simplify the task to finding an optimal code for an alphabet of 26 letters.[10] We can code each of the 26 letters with a standard length of 5 bits. A set of messages in which the frequency of each letter would be equal (e.g. 1/26) has maximal entropy. Of course, such a set would contain only nonsense. It could not be normal English since the frequency of letters in English varies greatly. Therefore a standard 5 bit code is redundant and can be optimized. We can assign shorter codes to more frequent letters. Giving up the fixed code length implies that our code has to be *prefix free*: no code can be a prefix of any other code. Standard Huffman code provides an optimal solution for this problem. Using Huffman code one can compress "A Tale of Two Cities" 0.81 bit per character comparison with the 5 bit code. We can ask ourselves if Huffman code is the best solution for compressing a book. In a sense it is, if one sticks to compression of characters, but there is no reason to do this. One could try to compress words instead or maybe one could use an analysis of idiosyncrasies of Dickens' style. This poses an interesting theoretical problem: what would be the theoretical shortest code for "A Tale of Two Cities"? In order to find an answer for this question we have to turn our attention to a different

[9]This definition is exactly equal to the definition of Gibbs entropy in thermodynamics. See the chapter by Bais and Farmer in this book.

[10]This example is discussed extensively by Harremoës and Topsøe.

definition of the concept of information that is intricately related to the notion of a Turing machine: Algorithmic information.

Algorithmic information

We have seen that with the theory developed by Turing we can define a universal Turing machine. In fact, there is an infinite number of such universal Turing machines, so let us select a standard (small) one and call it U. The input of U consists of two parts: a definition of a special Turing machine T_i in prefix code, followed by the input code, or data D for T_i. Observe that, using Huffman code, we can create a program that reproduces "A Tale of Two Cities" as output on U. The crucial insight is that it is easy to construct a Turing machine that decodes Huffman code. Let $D_{ToTC,Huf}$ be the Huffman code for "A Tale of Two Cities" and let T_{Huf} be a Turing machine that decodes Huffman code in the standard prefix free input format of U. The text of "A Tale of Two Cities" can be coded as

$$U(T_{Huf} + D_{ToTC,Huf})$$

When confronted with the input $T_{Huf} + D_{ToTC,Huf}$ our universal machine U will first read the definition of T_{Huf}, reconfigure itself as an interpreter for Huffman code and then start to interpret $D_{ToTC,Huf}$ resulting in the text of "A Tale of Two Cities" as output. The bit string $T_{Huf} + D_{ToTC,Huf}$ can be seen as a program for the text of "A Tale of Two Cities". Let $|D|$ be the length in bits of the data set D and let $D_{ToTC,5bit}$ be the 5 bit code for "A Tale of Two Cities. We will have:

$$|T_{Huf} + D_{ToTC,Huf}| < |D_{ToTC,5bit}|$$

Given the fact that a Turing machine for interpreting Huffman code is not complicated, the set $T_{Huf} + D_{ToTC,Huf}$ will be shorter than the original 5 bit code for "A Tale of Two Cities". In this way, we have created a computer program that generates the text of "A Tale of Two Cities" on a universal Turing machine. The bit code of this program is shorter than the original text. We could go on and try to find more clever code systems that compress the text even more. Such a code system, say $T_{CodeSystem_i}$ could make use of the frequency of words in the text, knowledge about the grammar of English and idiosyncrasies in the style of the author. Such a code system would be 'better' than the Huffman code if:

$$|T_{CodeSystem_i} + D_{ToTC:i}| < |T_{Huf} + D_{ToTC,Huf}|$$

where $D_{ToTC:i}$ is the text encoded in the new code.

We can now answer the theoretical challenge from the previous paragraph: the theoretical shortest code for "A Tale of Two Cities" would be the shortest program that generates this text on U. In order to find this program ideally, what we have to do is enumerate all possible programs for U, test them, and select the shortest that generates "A Tale of Two Cities". Alas this is impossible because of the uncomputability of the halting set. We know that such a program exists, but it remains an intensional object.

This fact gives rise to a different definition of the concept of information [Li and Vitányi, 1997]. The descriptive complexity of a string x relative to a Turing machine T and a binary string y is defined as the shortest program that gives output x on input y:

$$K_T(x|y) = \min\{|p| : p \in \{0,1\}^*, T(p,y) = x\}$$

One can prove that there is a universal Turing machine U, such that for each Turing machine T there is a constant c_T, such that for all x and y, we have $K_U(x|y) \leq K_T(x|y) + c_T$.[11] This definition is invariant up to a constant with respect to different universal Turing machines. Hence we fix a reference universal Turing machine U, and drop the subscript U by setting $K(x|y) = K_U(x|y)$. We define:

DEFINITION 7. The Prefix Kolmogorov complexity of a binary string x is $K(x) = K(x|\epsilon)$. That is the shortest prefix free program that produces x on an empty input string.

Kolmogorov complexity is a competing notion of information. It allows us to assign a complexity to individual strings and data sets.

A unified view on Shannon information and Kolmogorov complexity

We are now in a position to evaluate the difference between Shannon information and Algorithmic information, i.e. Kolmogorov complexity. Suppose we have a data set encoded in bits, say a five bit code of the text of "A Tale of Two Cities". We can analyze this set from two perspectives:

- From a Shannon perspective as a *collection of messages*. In this we can construct an optimal code using variation in frequency of the messages. This leads to a relative compression of the set of messages that can be computed. More frequent messages get shorter codes and contain less information.[12] We could call this concept of information *relative to the probability of a message.*

- From a Kolmogorov perspective as a *single message*. In this case, relative frequency has no meaning, but there exists an optimal compression of the message in terms of the shortest program on a Turing machine. The length of this program is an absolute measure for the amount of information contained in the message. This program is an intensional object and can not be computed as such. Messages that are highly compressible contain little information. This could be seen as a concept of information *relative to a Turing machine.*

[11]For an extensive discussion of these definitions, see the chapter by Grünwald and Vitányi in this book.

[12]This would work equally well in a case where frequency is an actual count, a probability in a Platonic world or a Bayesian belief.

As an example, suppose we have a bit string 01010101010101010101010101. We can *recode* this string in Shannon's sense as '01'=1;1111111111111, or we can *reprogram* it in Kolmogorov's sense as `for x = 1 to 13 write '01'`. Both structures are shorter than the original code reflecting the fact that the string shows a regular pattern. In this case, both the Shannon and the Kolmogorov compression do their work. In my view, both algorithmic information and Shannon information are different mathematical guises of one and the same concept of information that is associated with entropy of data sets.

CLAIM 8. Information is associated with the entropy of data sets. Data sets with low entropy can be compressed and contain less information than data sets with maximal entropy, which cannot be compressed and contain exactly themselves as information. There are various ways to explain these relations mathematically.

Shannon information starts with a segmentation of the set. In the limiting case where we have very few segments, or only one, Shannon's theory collapses into Kolmogorov's conception of information. Kolmogorov's conception of information is more powerful, but the price we have to pay is threefold: it is non-constructive, therefore it can only be approximated and it is asymptotic.

LEMMA 9. *The concepts of Kolmogorov complexity and Shannon information are equivalent in terms of predicting incompressibility of data sets with maximal entropy.*

Proof. In Shannon's conception a set of messages can not be compressed if they all have equal probability. Suppose we have a sequence of k messages with maximal entropy based on a code system of 2^n code words of n bits, then this is equivalent to a random string of $l = kn$ bits and thus it can not be compressed in Kolmogorov's sense. Suppose, conversely, that we have a random bit string $l = kn$ bits with l fixed, then for each segmentation of l in k messages the entropy is maximal thus it can not be compressed in Shannon's sense. ■

Note that the difference between Shannon information and Kolmogorov information can be seen as a difference in granularity. Kolmogorov complexity is coarse grained giving the whole set of messages a complexity in one shot. Shannon information is fine grained, it calculates the information for individual messages first and then establishes an entropy for the whole set. Given the equivalence of Shannon information and Kolmogorov complexity, one would expect that also in the limiting case of considering a bit string as one unsegmented message it is possible to assign a probability to it. This is indeed the case. In Shannon's case we reason from probabilities to entropies, in the Kolmogorov world we derive probabilities from entropies. Using results of Solomonoff [1997; 2003] and Levin we can define an a priori probability of a finite binary string.

DEFINITION 10 (Solomonoff, Levin). The universal a priori probability $P_U(x)$ of a binary string x is

$$P_U(x) = \sum_{U(p)=x} 2^{-|p|}$$

This is the sum of the probabilities of all the programs that generate x on a universal Turing machine on an empty input string. Thus strings with a low Kolmogorov complexity, i.e. the ones that are compressible, get a higher a priori probability. Associated with a universal a priori probability, we expect to get a universal distribution. We can define a semi-measure along these lines. A recursively enumerable semi-measure μ on N is called universal if it multiplicatively dominates every other enumerable semi-measure μ' i.e. $\mu(x) \geq c\mu'(x)$ for a fixed positive constant c independent of x. Levin proved that such a universal enumerable semi-measure exists. Since there might be more, we fix a universal semi-measure $\mathbf{m}(\mathrm{x})$. The semi-measure $\mathbf{m}(\mathrm{x})$ converges to 0 slower than any positive recursive function which converges to 0. Of course, $\mathbf{m}(\mathrm{x})$ itself is not recursive. We now give without proof a theorem that relates all these concepts with each other:

THEOREM 11 (Levin).

$$-\log \mathbf{m}(x) = -\log P_U(x) + O(1) = K(x) + O(1)$$

The universal distribution has quite wonderful qualities and its philosophical relevance has hardly been explored up till now.

4.1 Thermodynamics, Information and Computation

It is clear that the study of information and computation is related to concepts of thermodynamics on a fundamental level. The first law of thermodynamics states that energy in a closed system is conserved. The second law states that the entropy of a closed system can never decrease. After a certain time a closed system will reach an equilibrium in which the entropy is maximal. Another way of phrasing the second law is that self-organization is not possible without external energy.

As the entropy of a set of messages grows, so does the set of accessible states and so does the number of bits that we need to identify those states (according to Boltzmann the formula entropy was simply $S = \ln w$, where w is the number of accessible states, this is equal to the maximum entropy in Shannon's definition). Consequently in a closed system, when the entropy grows, the amount of information stored in the system grows. A closed system can increase its internal information without exchange of heat with the environment.

A thought experiment can help here. Think of a bit string as a gas in a one dimensional container (say 0s are spaces and 1s molecules). If the bits are allowed to move freely through the space, starting from any configuration they will eventually reach an equilibrium state in which the Kolmogorov complexity of the accessible states is maximal. These states are exactly the ones in which the bits contain maximal information (in terms of Kolmogorov complexity). Random bit strings contain the most information, have the highest entropy and correspond to a thermal equilibrium.[13]

[13]It is possible to develop a thermodynamics of bit strings along these lines.

All this is quite counter intuitive. If we dissolve milk in coffee, or we spill sugar in sand, we feel we lose possibilities. It seems strange to assume that noise on a channel is actually the richest source of information possible. The reason for our unease seems to be the fact that high entropy is the normal situation in the universe. Order (i.e. low entropy) is more interesting since is needs to have a specific cause. High entropy does not point at specific causal processes of any interest. Low entropy is a sign that somebody or something redirected energy to a system. That is the reason why, when we want to detect life in outer space, we scan the sky for signals with less then maximal entropy. In order to be meaningful to us, a set of messages has to have some structure and consequently have less than maximal entropy. This concept of meaningful information in a system is from a thermodynamical point of view related to the free energy in the system and from a learning view to two part code optimization.

Thermodynamics therefore has interesting consequences for the physics of computing. A universe in which we can calculate has to obey the following conditions:

- It must be stable enough to **store information**. Structures should have a certain stability; identity over a certain period of time should be guaranteed. This points to relatively low entropy. In a system that is in a perfect thermodynamic equilibrium, structures would not be robust enough to store information at all.

- There must be enough free energy to **process information**. There must be reversible processes that facilitate the transition between stable states: i.e. there must be mechanisms to flip bits. This condition implies more than minimal entropy. Computation can not exist in systems with extremely low entropy, e.g. computation at zero degrees Kelvin is not possible.

Computation seems to presuppose some kind of state of intermediate non equilibrium entropy.[14] Luckily, we live in a universe that satisfies these conditions exactly. This is no surprise, because in a universe that does not offer these possibilities; intelligent life would not be possible. This is a variant of the anthropic principle [Hawking, 1988]. The hypothesis of the cooperative universe however goes deeper because it states that such a universe would be easy to learn. It is a number of random processes, but these processes are necessarily of limited complexity.

Out of these observations the following picture emerges: A deterministic computer is simply a Laplacian system that, in itself, cannot add information to the universe. Its future is completely determined by its initial conditions. Still a deterministic computer can easily use energy to erase information and thereby reduce the amount of information in the subsystem (say its tape). The total entropy in the universe will still grow as a result of this action. For a subjective observer,

[14]This goes against the interpretation of Lloyd and Ng [Lloyd and Ng, 2004] who consider almost any physical process as a computer, e.g. black holes and pure plasma. In these cases it is better to speak of computational processes than of computers.

however, the situation is different. He might not know whether a certain computation will finish. If he observes that the computational process comes to a halt this certainly adds to his information, even if he lives in a Laplacian universe.

Suppose, on the other hand, that a statistical observer can only make measurements of a certain granularity. He can, for instance, measure the local density of bits on the tape with a certain accuracy, but not observe individual bits. In such a case, the subjective entropy generated by a deterministic computing process can be much bigger than the entropy of the initial conditions. Suppose that the computer writes the binary expansion of the number e on the tape. This is a data set with very low entropy, but, for such a statistical observer, it cannot be distinguished from random noise (since he cannot identify the individual bits). Here, we seem to cross the border from theory of computation to thermodynamics. Very much the same thing happens if we see the generation of a fractal. This is a data set of very low entropy, but to our subjective eye full of interesting details. A non-deterministic computer adds information to the universe with each randomized computing step it takes.

As a last note, observe that thermodynamics only works for systems in a state of equilibrium. Computing systems tend to specifically stay out of equilibrium so the applicability of classical thermodynamics for the understanding of computing processes is limited. At the moment, we are missing a theory that helps us to understand these matters adequately. The following theoretical observations give an initial outline of such a theory.

4.2 A universal a priori near optimal Shannon code based on Kolmogorov complexity

Levin's theorem allows us to explore the relation between Shannon information and Kolmogorov complexity at a more fundamental level. We define the standard bijection b between the set of binary strings $\{0,1\}^*$ and the set of natural numbers N as

$$b(0,\epsilon), b(1,0), b(2,1), b(3,00), b(4,01), \ldots$$

Where ϵ denotes the empty word. We can define the function $S : \{0,1\}^* \rightarrow \{0,1\}^*$ as:

DEFINITION 12. $S(x) = \min_{i \in N}\{p : b(i,p), U(p,\epsilon) = x\}$

Here U is a universal Turing machine. S associates each binary object x with the first program that produces x on U with empty input.

COROLLARY 13. *S is a universal a priori near optimal code associated with* $\mathbf{m}$ *for binary strings in Shannon's sense.*

Proof. According to Shannon an optimal code for x given $\mathbf{m}$ would be $-\log \mathbf{m}(x)$ bits long. According to Levin we have $-\log \mathbf{m}(x) = K(x) + O(1)$. But then $S(x)$ is such an optimal Shannon code, because by definition $|S(x)| = K(x)$ since $S(x)$ is the first, and thus the shortest, program that produces x on U. The code is near optimal, because of the factor $O(1)$ in Levin's theorem. $S(x)$ will always be maximally $O(1)$ removed from the factual optimal code. ■

The function S is interesting because it brings the concepts of Shannon information and Kolmogorov complexity together. On one hand $|S(x)|$ is the Kolmogorov complexity of x, on the other $S(x)$ is an optimal a priori code for x. Of course, S can never be computed, but suppose that some Platonic oracle would give us S. In that case we would have a universal a priori solution to the problem of induction. $S(x)$ reflects *any regularity (e.g. deviation from maximal entropy, i.e. compressibility) that can be expressed solely in terms of the internal structure* x. Observe that $S(x)$ will itself always be random (and thus incompressible) because it is the *first* program that computes x. If $S(x)$ would be compressible, it would itself have been identified much earlier by S. It is important to note that, although S can not be constructed, it nevertheless exists. S is the closest we can get to a universal language of science, given the current state of research in computer science.

To give some examples. S would make it easy to find binary expansions of transcendent numbers like π and e. There are simple programs for these extensions. In fact, S would identify almost *any* discrete object of *any* mathematical interest for us. On top of that S would give us an optimal code for the text of "A Tale of Two Cities" and indeed of any other conceivable poem, novel, piece of music, movie or any work of art in digital code. The same would hold for any digital data set that scientific inquiry could produce. S would 'explain' the regularities and idiosyncrasies of these data sets in so far as they can be expressed in terms of deviation of maximal entropy.

4.3 Intensive and extensive data sets

A very interesting consequence of having S would be that we are capable of measuring the scale invariance of complexities and entropies. A little thought experiment will help. Suppose that we study some segment L of length l, starting at the p-th bit, of the binary expansion of a transcendental number, say π. Since we are studying an expansion of π the Kolmogorov complexity of the sequence is low. In the sense of lemma 9 we could analyze this as a sequence of $l = kn$ bits, i.e. k messages based on a code system of 2^n code words of n bits. The total measured complexity of L using S with granularity n could be defined as:

$$K(L)_{S,n} = \sum_{i=0}^{k-1} S(x_{(i\times n)+1}, x_{(i\times n)+2}, \ldots x_{(i\times n)+(n-1)})$$

If we plot the size of $K(L)_{S,n}$ in terms of the size of n we will see the following effect: for small n the function $K(L)_{S,n}$ will show a slow decrease that will be linear in n. This is because of the diminished overhead of S per segment. For small n all segments will be random for S, because of the transcendentality of π. At a certain point, 'close' to $\log p + \log l/n + O(1)$, the value of $K(L)_{S,n}$ will

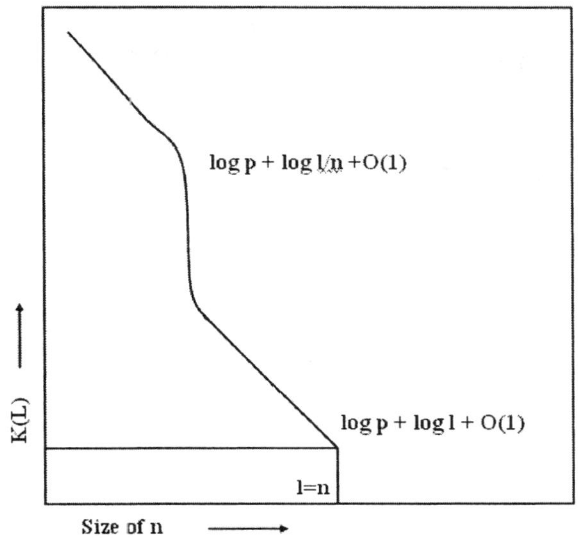

Figure 1. The size of $K(L)_{S,n}$ in relation to the granularity n while sampling a segment of π

drop suddenly.[15] This is exactly the point where n is big enough so that S starts to 'sense' the compressibility of L. For $n = l$ the function $K(L)_{S,n}$ will land at the value $\log p + \log l + O(1)$. What this amounts to is that for certain data sets, e.g. bit representations of transcendental numbers (but there are many others), complexity (and consequently entropy) is *non-extensive*. Another way of putting this is that the Shannon entropy of the collection of messages diverges from the Kolmogorov complexity as a measure of entropy for the set as a whole. Local estimates of the complexity do not tell us anything about global complexity and consequently complexities of various regions of the data set can not be added to get a global complexity estimate. The complexity of these data sets is not robust under statistical operations and under re-scaling of the code system.[16] Clearly for the application of efficient learning algorithms the non-extensive complexity of such data sets is an insurmountable barrier. No algorithm can compress data sets that look random from the outside but are in fact highly compressible, e.g. encrypted data or expansions of very special real numbers like π and e.

Uncompressibility and extensiveness are in fact the same notions, as is clear

[15]The $\log p$ gives us an index in L, $\log l/n$ code the length of the individual segment and the $O(1)$ term contains the program for π. This information is sufficient to describe any substring in L.

[16]The custom in thermodynamics to take the averages of values in the sample regions is just one specific form of recoding.

from the following analysis. A data set D is extensive if the sum of the complexity of two arbitrary disjoint subsets A and B equals that complexity of the union of that set: $K(A) + K(B) = K(A \cup B) + O(1)$. This is only the case if D does not contain any redundancy i.e. if D is random. On the other hand, suppose that D is very compressible. If we know A already, then B would add no information, i.e. $K(A) + K(B) = K(A) + \log|B| + O(1)$. In other words B would only add its own size to our knowledge. This is for instance the case when D contains extremely simple regular patterns. This suggests the following definitions:

DEFINITION 14. A bit string D is **extensive** for a sample granularity g if for each substring $A \in D$ such that $|A| \geq g$ we have $K(A) > |A| - O(1)$. A bit string D is **intensive** if for each substring $A \in D$ such that $|A| \geq g$ we have $K(A) < \log|A| + \log|D| + O(1)$. **Sub-extensive** data strings have $|A| \gg K(A) + O(1)$ and **super-intensive** strings have $K(A) \gg \log|A| + \log|D| + O(1)$.

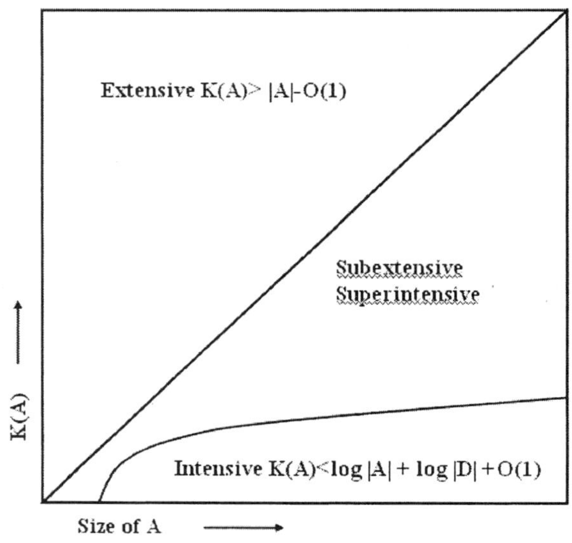

Figure 2. The relation between extensive, sub-extensive, super-intensive and intensive strings

Sub-extensive data sets are the ones from which we can learn something. The borderline between extensive, sub-extensive, super-intensive and intensive data sets is blurry, but the general idea stands. If we sample an extensive data set we really get value for money, every bit counts. But there is a price to pay. The information is completely random. Nothing can be learned from this set. This corresponds with the picture of random noise at the television set that was discussed earlier in this chapter. On the other end of the spectrum we find the picture of the test image: this data set is almost totally intensive. It is a simple repeatable pattern for which we need only the information about the number of repetitions to encode it. Extensiveness corresponds to maximal randomness, intensiveness to maximal redundancy. Figure 1 shows that we can make each string extensive by taking a small granularity. This corresponds to the fact that, even if a data set is very regular, there is a learning phase in which we have to analyze the pattern itself. At this time the data set cannot be distinguished from a random one. A finite program producing an infinite data set has to go through loops. If we cannot compress the data set on the basis of samples that are in the order of the complexity of a loop of the program that generates the data we are in trouble. Because the increase in information after this phase will be only logarithmic. So if we have not spotted the regularity after, say 10, loops then we will probably never spot it because the only new information we get from x repetitions is of size $\log x$. This gives rise to the following claim:

CLAIM 15. From the point of view of intelligent systems of a certain complexity, nature is by necessity shallow. Intensive data sets can either be learned by an intelligent system (a resource bounded learning algorithm) that is of the order of the complexity of the algorithm generating the data set, or not at all.

From completely intensive strings we can learn only their generating program and their size. One could call this the *self information* of a data set. The program generating an intensive string can be seen as its **intension.**[17] Intensive data sets asymptotically have their size as their most defining characteristic. Extensive data sets do not have an intension, or to say this in other words: they only describe themselves. Their extension is their intension. Super-intensive data sets contain more information, but this might be just noise. They are non random, but not completely regular either. From a physical point of view they are associated with systems that are in a non equilibrium state. It is the kind of information that we find in the picture of the forest on our television screen. The trees are generated by a program and thus have regular specific features. But the program is not completely deterministic. Individual trees show random variation. It is interesting to characterize sciences in terms of the nature of their data sets. Data sets of mathematicians and physicists are close to intensive. Data sets of the humanities are super-intensive. The eternal question whether history repeats itself, can be answered by stating that history is sub-extensive and super-intensive. There are patterns but they will never repeat themselves exactly. In physics we have explanation and prediction exactly because the data sets are intensive.

A consequence of this analysis is that the amount of randomness we observe is dependent on the granularity of our measurements. In one sweeping statement one might say: randomness has a scale. Suppose we are looking at a movie of a hand flipping a coin.[18] At normal speed we are looking at a random (or at least a very

[17]Here we have a computational equivalent of Platos notion of an idea. The intension of an object is the program generating it.

[18]Suppose also that this hand does not belong to Persi Diaconis, the well known mathemati-

complex) process. This data set certainly has extensive elements. Note that the data set itself in this case is not random. It is a movie of coin flipping that contains a lot of information. We could for instance learn a lot about Newtonian mechanics if we analyze it at an appropriate scale. Now suppose that we slow the movie down extremely, say we stretch out one second to a million years. In this case, the movie will be rather dull on a human scale. It will be close to a intensive process that contains very little information. On the other hand if we speed the movie up so that a million years is compressed into one second. Then again the movie would on a human scale be reduced to a meaningless grey blur that contains no information. On this scale the data set would again be intensive. The important thing to notice is that the data set contains the most information if we sample it at a granularity where the extensiveness is maximal. Both at a larger and at a smaller granularity we will lose information. In short: even randomness has a scale. Every form of randomness necessarily can only be observed at a granularity in which it is in equilibrium. When we see smoke dissolve in the air, then on a human scale we observe increase of entropy, on a molecular scale the increase does not exist and on the scale of, say the solar system, the effect is too small to notice. An optimal analysis of a data set involves finding a granularity that optimizes the randomness of the data.[19]

Researchers in machine learning are familiar with the idea that certain phenomena can only be explained at certain scales. Some structures can only be learned when the data set is sampled with a certain granularity.[20] This can also be observed in the text of "A Tale of Two Cities". When we only sample individual bits of this data set no useful information emerges. When we sample letters, we can make good statistical estimates based on frequency. This is already somewhat harder for words and next to impossible for sentences, leave alone paragraphs or

cian/magician that has proved that coin flipping is actually a deterministic process. Some of the material in this paragraph is influenced by the lecture that Professor Diaconis gave on the occasion of receiving the Van Wijngaarden award at CWI in 2006.

[19]This insight is related to Jaynes' maximal entropy principle and the minimal randomness deficiency principle to be discussed later. There is a further analogy with thermodynamics, where we find exactly the same scaling issues. Suppose that we have a number of gas particles in a isolated container at low entropy. After some time, an equilibrium will be reached. On a micro scale the entropy can not have increased because the evolution of particles in the container is determined by simple deterministic Newtonian physics. Macroscopic measurements however will show an increase in entropy. Just like our example of the binary expansion of π, the data set will have low complexity at micro level and appear to be random at lager scales. In a strictly deterministic universe randomness takes the form of coarse grained undecidability.

[20]This was one of the more interesting results of the Robosail project, an attempt to use machine learning techniques to learn to sail automatically that I started in 1998 [van Aartrijk *et al.*, 2002]. Measurements of almost all relevant human concepts like 'wave', 'gust of wind', 'change of wind direction' and 'wind strength' were dependent on selecting an adequate granularity for the measurements. What you subjectively experience as a wave is dependent on the size of your boat. Some of the conceptual distinctions used by sailors depend on sophisticated phase transitions in chaotic media that were only observable at certain scales. This holds for instance for the distinction between light air (laminar flow) and breeze (turbulent flow). In the final system we implemented learning agents that were living in a variety of time scales: 10 Hz, 1 Hz, 10^{-3} Hz, etc.

chapters. There is a certain granularity that reveals the structure of the text optimally.

A deeper analysis of these kind of phase transitions and their meaning for learning algorithms is necessary, but it is clear from this short analysis that the analogy between information and thermodynamics can be carried further than is commonly accepted.

4.4 Induction and Minimum Description Length

Let us have a closer look at the relation between S and the problem of induction. In one special guise induction amounts to selecting the most probable hypothesis to explain a given data set. In terms of Bayesian learning this task can be formulated as follows [Mitchell, 1997]. The **prior probability** of a hypothesis h is $P(h)$. Probability of the data D is $P(D)$. The **Posterior probability** of the hypothesis given the data is:

$$P(h|D) = \frac{P(h)P(D|h)}{P(D)}$$

THEOREM 16. *Suppose that* $h, D \in \{0,1\}^*$, *i.e. both the data set and the hypothesis range over the full class of finite binary strings. Selecting the* **Maximum A Posteriori hypothesis (MAP)** *to explain* D, *amounts to selecting the hypothesis that minimizes the length in bits of*

$$S(h) + S(D|h)$$

Here $S(h)$ is the universal optimal Shannon code for the hypothesis and $S(D|h)$ is the universal optimal Shannon code for the data set given the hypothesis.

Proof.

$$h_{MAP} \equiv argmax_{h \in H}\ P(h|D)$$

$$= argmax_{h \in H}\ (P(h)P(D|h))/P(D)$$

(since D is constant)

$$= argmax_{h \in H}\ (P(h)P(D|h))$$

$$= argmax_{h \in H} \log P(h) + \log P(D|h)$$

$$= argmin_{h \in H} - \log P(h) - \log P(D|h)$$

(Since $h, D \in \{0,1\}^*$ and according to Shannon $-\log P(h)$ is the optimal code for the hypothesis and $-\log P(D|h)$ is the optimal code for the data given the hypothesis.)

$$= argmin_{h \in H} S(h) + S(D|h)$$

■

This result is closely related to the so-called:

DEFINITION 17. **The Minimum Description Length principle (MDL)**: The best theory to explain a set of data is the one which minimizes the sum of

- the length, in bits, of the description of the theory and
- the length, in bits, of the data when encoded with the help of the theory

This principle was first formulated by Rissanen [1999]. Research in this domain is far from finished and these concepts are still the object of fierce debate [Domingos, 1998; Domingos, 1999]. A common misconception is the idea that the minimum description length principle can be transformed into a methodology for the construction of a sequence of improving theories by means of an incremental compression of the data set. Suppose that S_i, h_j, Sp and h_q are arbitrary coding schemes and hypotheses such that:

$$|S(h) + S(D|h)| < |S_i(h_j) + S_i(D|h_j)| < |S_p(h_q) + S_p(D|h_q)| < |D|$$

Although h is the best theory it is not necessarily the case that h_i is better than h_q. This could for instance be guaranteed if $S = S_i = S_p$, i.e. when the code is optimal [Adriaans and Vitányi, 2005]. Translating these observations to the domain of methodology of science gives us a number of interesting insights: Given the fact that entropy in nature tends to increase the regularity of the world we observe around us is extremely improbable, when we suppose that the world started from a state of thermal equilibrium. The process of reducing a set of observations to a general theory explaining these observations can be described as a process of data-compression. A universal methodology of science would have the following form:

- Represent your data set D in binary format.
- Select a hypothesis h in binary format such that $|S(h) + S(D|h)|$ is minimal.

This program fails because of the uncomputability of S but it can serve as a regulative ideal for the study of methodology of science. In certain cases the theoretical results allow us to solve real life problems and to develop more efficient algorithms [Li and Vitányi, 1997]. Note that we have characterized learnable data sets as non- and sub-extensive, they contain a mix of random and deterministic elements. MDL aims at finding a compression for such a set that exactly separates the random (extensive) elements ($S(D|h)$) from the non-random (intensive) ones ($S(h)$). For intensive data sets the two part code will simply consist of a description of the program generating the data set ($S(h)$) and the length of the data set ($S(D|h)$).

Another way to look at this is from the perspective of the so-called *randomness deficiency* [Vereshchagin and Vitányi, 2004]:

$$(1)\quad \delta(D|M,d) = \log\binom{m}{d} - K(D|M,d),$$

Here M is a model of size m and $D \subseteq M$ is a data set of size d. The expression $\log\binom{m}{d}$ is the measure of the maximum information in a subset of M of size d. The expression $K(D|M,d)$ is the actual entropy of the data set D in the model, i.e. conditional Kolmogorov complexity of D given M and d. If the actual entropy is much smaller than the maximal entropy of an average set of size d in M then D still contains a lot of regularity that is not explained by M. In other words M is not an optimal model. A model would be optimal if the randomness deficiency is minimal. In such a case D would be a typical element (extensive) of M and M would explain all that is worth knowing about D, i.e. its intension. The principle of minimal randomness deficiency is very close to Jaynes' maximal entropy principle: in order to explain a set D try to find the set M for which the entropy is maximal under a set of constraints observed in D.[21]

5 THE COOPERATIVE COMPUTATIONAL UNIVERSE

From this discussion it is clear that the philosophy of learning touches on a number of philosophical issues: To name a few: entropy, information, computation, objective and subjective probability. In order to study these issues let's define a thought experiment. For the sake of argument we will restrict ourselves to the case in which we observe a string of bits from an unknown source. Even in this simple setting there are some fundamental philosophical issues to be dealt with.

Suppose that we reserve a room at the University of Amsterdam for the purpose of this experiment. The room has no windows and the door is closed. In the room there is a black box. The black box produces a bit every minute. If the bit is '1' the light is switched on, if it is '0' the light is switched off. This bit is published on a web site. Of course, nobody knows the contents of the black box, but, for the sake of argument, we choose three possible configurations. The box could contain:

1. A *random process* that generates bits (e.g. a person flipping a coin, a quantum process or some other ergodic process.).

2. A *deterministic computer program* generating bits.

3. An *infinite database* with a list of bits.

These three definitions represent radically different views on the phenomenon of a source of information. The first is an objective random process associated with an objective form of probability. It generates an extensive data set. All the information that is contained in the sequence can be measured in terms of its fundamental statistical characteristics: mean, variance, autocorrelation function etc. The second is a deterministic process with a definition of finite length. The

[21]See the paper of Bais and Farmer in this book.

maximal amount of information in a string produced by the program is limited to the length of the definition of the program. It is an intensive data set. It could lead to a sequence of bits with a certain statistical bias (e.g. repeating patterns), but this is not necessary. Some transcendental numbers have short definitions (e.g. e and π) but lead after a bit of twisting to bit patterns that cannot be recognized as non-random. The third is a deterministic process with a definition of infinite length. The generating data set itself could be in- or extensive. It potentially contains an infinite amount of information that can never be learned in a finite amount of time.

THEOREM 18. *The three sources of information, (a random process, a deterministic computer program and an infinite database) cannot be distinguished from each other by a receiver of the information.*

Proof. Each of the three sources can produce a sequence of bits that cannot be distinguished from a random sequence. 1) The case of the random process is trivial 2) A deterministic program can generate strings that cannot be recognized as non-random. The non-computability of Kolmogorov complexity tells us that there will always be compressible strings for which no compression can be computed. 3) An infinite database can continue a random set of bits or a set of non-random bits that cannot be recognized as such. ■

The philosophical importance of this result is obvious. We cannot make a distinction between a source of information that is random and a source of information that has high complexity. This makes the traditional controversy between determinism and indeterminism from the point of view of informatics senseless. It reveals the famous dictum by Einstein "God does not play dice" as a real metaphysical position. It is not a question that can be settled by any argument. It also shows that it is impossible to assign any form of objective probability to a source of information. In this context one might ask to which extent randomness is in any sense a scientific concept. We can define randomness of strings in terms of incompressibility, but we do not need the concept of randomness to study incompressibility. The notion of flipping a coin or throwing a dice are real scientific paradigms in the original Kuhnian sense, but au fond they are deterministic processes that in most cases are simply too complex to predict and therefore can act as place holders for supposedly real random processes. They serve as anecdotic topoi in the scientific discourse, nothing more. The notions of extensiveness and incompressibility still have an exact meaning in a deterministic Laplacian universe, so they seem to be more fundamental than the concept of randomness. Macroscopic measurements of microscopic deterministic processes might subjectively be interpreted as random. Even in a Laplacian universe there are data sets that are both strictly deterministic and extensive (e.g. the Halting set).

In such a world however there is a form of subjective probability that is relevant. Suppose that we want to form a hypothesis about the internal structure of the black box and the black box produces a string that shows some regularity. In that case it is extremely unlikely that the source of bits is random. Suppose that our black box produces a string of n ones $1_1 1_2 \dots 1_n$. The probability of creating this string with n flips of a perfect coin is 2^{-n}. So, intuitively, with each one that is produced by our black box the hypothesis that it contains a random process becomes more unlikely in favor of the hypothesis that the bits are produced by some deterministic process. Yet this argument is flawed because *any* bit string of length n produced by flipping a perfect coin has probability 2^{-n} and therefore is extremely unlikely. We have no clear ground to favor any regular string over a random one as a ground for selecting between hypotheses about the content of the black box. As we have seen, the theory of Kolmogorov complexity allows us to define the concept of *randomness deficiency* of a string. The idea is the following. A string like, say, 11100101000100 is *typical* for a random source. Such a string is produced by a source that is perfectly compatible with the hypothesis that the source is random. A string like 11111111111111 is *atypical* for a random source. When produced by a source it makes the hypothesis that the source is random unlikely. A high randomness deficiency corroborates the theory that the process in the black box is non-random.

This analysis suggests that the best thing we can do in science is: observe a set of phenomena, estimate the randomness deficiency and formulate a theory. Unfortunately in the case of the Amsterdam room the situation is more complicated. This becomes clear if we analyze the following claims.

CLAIM 19. We get exactly one bit of objective information each minute.

It is clear that each bit that is published on the web by the black box contains real information about the actual binary situation in the room: the light is on or off.

CLAIM 20. The meaning of the message contained in the bit and the knowledge generated as a consequence of receiving the message is not dependent on the content of the black box.

Yet there is a subtle interplay between the growth of our subjective information and our theories about the nature of the black box.

CLAIM 21. The objective amount of information we get is dependent upon our interpretation of the nature of the source of information.

The three possible interpretations of the content of the box could be seen as three different types of senders of messages. I will define three possible receivers along the same line:

1. A forgetful receiver that determines the statistical characteristics of the sequence: mean, variance, autocorrelation function etc. Here our subjective information grows incrementally at a very slow rate with each objective bit that is received. This observer corresponds with an interpretation of the source as a system in equilibrium. The statistical (macroscopic) qualities of the system are all that we can know about the system.

2. A machine learning program with bounded computing time and memory, that tries to reconstruct the finite structure of the black box. Here our subjective information grows in an irregular but monotone way with each bit of objective information that is received. This observer corresponds to an interpretation of the data set as intensive. After some finite point in time our information will only grow with the factor $logx$ where x is the number of bits we have seen so far.

3. An infinite database with a list of bits recording every bit that is received. Here our subjective information grows with exactly 1 bit per bit that is received, if the data set itself is considered to be extensive.

This example shows that we can not restrict ourselves to a purely subjective interpretation of information when we analyze a source of messages. We need to make an a priori decision about the nature of our source.

Our analysis shows that nature and science play an asymmetrical game. Non-random strings are very rare. To make this more specific: in the limit the density of compressible strings x in the set $\{0,1\}^{\leq k}$ for which we have $K(x) < |x|$ is zero. Data sets that appear to be random may be actually compressible, but the occurrence of such objects in nature is extremely unlikely. If a data set looks random, we may with high probability assume that it is random. On the other hand if a data set from the point of view of an intelligent agent appears to be regular then it is with extremely high probability not random and can be learned because of the shallowness claim 15. Therefore a learning system that simply scans the environment for areas of low entropy and tries to compress the data sets it finds there will be successful with high probability, if the complexity of data sets is of the same order of magnitude as the agent. Local low entropy data sets correspond with energy consuming non-equilibrium systems that with high probability can be described in terms of computational models. Learning is not as hopeless as our formal models seem to imply. We are computational processes of limited complexity analyzing computational processes of limited complexity in a universe that generates computational processes of limited complexity. In this sense, we live in a cooperative computational universe. This is as close as we can get to the solution of certain philosophical problems in terms of information and computer science.

So why is this the case? Why do we live in a world that is intelligible at all? This question pervades philosophy from its early conception on (Herakleitos vs Parmenides). In form of a sweeping statement: prima facie, the God of Leibniz might very well have created a universe in which the Minimum Description Length principle would not hold. There seems to be no theoretical necessity to favor simplicity. The extreme regularity of the universe could be a 'local' condition accidentally observed by us. In terms of modern information theory: every infinite random string has an infinite number of regions of extreme regularity. If we transpose this idea to the analysis of our world we might just accidentally live in such a regular region in a purely random universe [Li and Vitányi, 1992]. A rather horrifying thought.

On the other hand imagine the following thought experiment: an infinite set of universal Turing machines working in parallel with input tapes that are created by means of some random process (e.g. flipping a coin). The set of input tapes is infinite so every finite prefix free program will occur an infinite number of times. Yet the density of 'shorter' programs will be exponentially higher than that of 'longer' ones. Some programs will run for ever, others will stop in finite time. After n time steps a number of 'simple' programs will have stopped and produced a fixed output. This means that the set of outputs we observe in this thought experiment will have a strong bias for simplicity. In other words even a universe that consists of purely random computational processes has a strong bias for simplicity. The distribution of phenomena it produces is cooperative in the sense that we get examples of the simple structures first. This is the hypothesis of the cooperative universe in another guise: nature produces the information that we need to interpret her in such a way that hypotheses we form are right with high probability. In such a universe MDL therefore will be a viable methodological principle. It coincides with another well known dictum of Einstein: Subtle is the Lord, but malicious He is not. The exact relation between various computational models of the universe, cooperative distributions, the universal distribution $\mathbf{m}$ and the problem of induction is, in my view, one of the most important open problems in the philosophy of information.

These issues (subjective versus objective probability, regularity versus randomness, information versus meaning) are far from resolved and should be at the center of a philosophical research program of a philosophy of information.

6 CONCLUSION

The research on learning and induction that has emerged because of the growing interest in artificial intelligence is still developing. The results do not only lead to useful industrial applications, but also influence the way we think about fundamental philosophical questions about the origin of human knowledge, the structure of our brain and methodology of science. A formal analysis of the mathematics of learning helps us to understand the efficiency of human learning. Human beings can only learn complex structure like language and the laws of nature if the underlying probabilities are 'benign'. The hypothesis of the cooperative universe is an attempt to explain why we live in a world that can be learned efficiently.

Finally, a tongue in cheek observation: Our human brain can contain about 10^{14} bits of information. The total storage capacity of the known universe is estimated to be about 10^{92} bits [Lloyd and Ng, 2004]. The old philosophical ambition of understanding the universe as a whole amounts to the wish to find a compression of the universe of the following nature: a structural description of less than 10^{14} bits (the laws of nature) and an ad hoc description of more than 10^{78} bits (the actual structure given the laws of nature) . There is only one conclusion possible. The universe can only be understood by human beings if it is extremely compressible -in other words- if almost nothing of any significance happens.

BIBLIOGRAPHY

[Adriaans and van Zaanen, 2004] Pieter W. Adriaans and Menno M. van Zaanen. Computational grammar induction for linguists. *Grammars*, 7:57–68, 2004.

[Adriaans and Vitányi, 2005] P. Adriaans and P.M.B. Vitányi. The power and perils of MDL. Technical report, Human Computer Studies Lab, Universiteit van Amsterdam, 2005.

[Adriaans and Zantinge, 1997] Pieter Adriaans and Dolf Zantinge. *Data mining*. Addison-Wesley Longman Publishing Co., Inc., Boston, MA, USA, 1997.

[Adriaans, 2001] Pieter Adriaans. Learning shallow context-free languages under simple distributions. In Ann Copestake and Kees Vermeulen (eds.), editors, *Algebras, Diagrams and Decisions in Language, Logic and Computation*. CSLI/CUP, 2001.

[Angluin, 1988] D. Angluin. Queries and concept learning. *Machine Learning*, 1988.

[Black, 1967] M. Black. Probability. *The Encyclopedia of Philosophy, Paul Edwards (ed.)*, 6:464–479, 1967.

[Capurro and Hjørland, 2003] R. Capurro and B. Hjørland. The concept of information. *Annual Review of Information Science and Technology*, 37(8):343–411, 2003.

[Capurro, 1978] R. Capurro. *Information. Ein Beitrag zur etymologischen und ideengeschichtlichen Begründung des Informationsbegriffs*. München, New York, London, Paris: Saur, 1978.

[Carnap, 1950] Rudolf Carnap. *Logical foundations of probability*. The University of Chicago Press, 1950.

[Cilibrasi and Vitanyi, 2005] R. Cilibrasi and P. Vitanyi. Clustering by compression. *IEEE Transactions on Information Theory*, 2005.

[Cornuéjols and Miclet, 2003] A. Cornuéjols and L. Miclet. *Apprentissage artificial, concepts et algorithmes*. Eyrolles, 2003.

[Domingos, 1998] Pedro Domingos. Occam's two razors: The sharp and the blunt. In *Proceedings of the Fourth International Conference on Knowledge Discovery and Data Mining*. AAAI Press, 1998.

[Domingos, 1999] Pedro Domingos. The role of occam's razor in knowledge discovery. *Data Mining and Knowledge Discovery*, 3(4):409–425, 1999.

[Dretske, 1981] F. Dretske. *Knowledge and the Flow of Information*. Cambridge MA: MIT Press, 1981.

[Edwards, 1967] Paul Edwards. *The Encyclopedia of Philosophy*. Macmillan Publishing Company, 1967.

[Floridi, 2004] L. Floridi. Open problems in the philosophy of information. *Metaphilosophy*, 2004.

[Hájek, 2002] A. Hájek. Interpretations of probability: http://plato.stanford.edu/entries/probability-interpret/. Stanford Encyclopedia of Philosophy, ed. E. Zalta, 2002.

[Hawking, 1988] S.W. Hawking. *A brief history of time: from the big bang to black holes*. Toronto; New York: Bantam Books, 1988.

[Hume, 1909, 1914] David Hume. *An Enquiry Concerning Human Understanding*, Vol. XXXVII, Part 3 of *The Harvard Classics*. P.F. Collier & Son, 1909, 1914.

[Kearns and Vazirani, 1994] M.J. Kearns and U.V. Vazirani. *An introduction to computational learning theory*. 1994.

[Kneale and Kneale, 1988] William Kneale and Marthe Kneale. *The Development of Logic*. Oxford, Clarendon Press, 1988.

[Li and Vitányi, 1992] M. Li and P. M. B. Vitányi. Philosophical issues in Kolmogorov complexity. *Automata, Languages and Programming: Proc. of the 19th International Colloquium*, pages 1–15, 1992.

[Li and Vitányi, 1997] M. Li and P.M.B. Vitányi. *An introduction to Kolmogorov complexity and its applications*. Springer-Verlag, 2 edition, 1997.

[Lloyd and Ng, 2004] S. Lloyd and Y.J. Ng. Black hole computers. *Scientific American*, 2004.

[Locke, 1961] John Locke. *An Essay Concerning Human Understanding*. London : Dent ; New York : Dutton, 1961.

[Mitchell, 1997] Tom M. Mitchell. *Machine Learning*. McGraw-Hill, New York, 1997.

[Popper, 1952] K.R. Popper. *The Logic of Scientific Discovery*. London: Hutchinson & Co. Postman, L., & Brown, D.R., 1952.

[Rissanen, 1999] J. Rissanen. Hypothesis selection and testing by the MDL principle. *The Computer Journal*, 42:60–269, 1999.

[Solomonoff, 1997] Ray J. Solomonoff. The discovery of algorithmic probability. *Journal of Computer and System Sciences*, 55(1):73–88, 1997.

[Solomonoff, 2003] Ray J. Solomonoff. The Kolmogorov lecture. The universal distribution and machine learning. *Computer Journal*, 46(6):598–601, 2003.

[van Aartrijk *et al.*, 2002] Martijn L. van Aartrijk, Claudio P. Tagliola, and Pieter W. Adriaans. AI on the ocean: the robosail project. In Frank van Harmelen, editor, *ECAI*, pages 653–657. IOS Press, 2002.

[Vereshchagin and Vitányi, 2004] N.K. Vereshchagin and P.M.B. Vitányi. Kolmogorov's structure functions and model selection. *IEEE Trans. Information Theory*, 50(12):3265–3290, 2004.

[Weaver and Shannon, 1949] W. Weaver and C.E. Shannon. *The Mathematical Theory of Communication.* University of Illinois Press, Urbana, Illinois, 1949. republished in paperback 1963.

[Windelband, 1921] Wilhelm Windelband. *Lehrbuch der Geschichte der Philosophie.* Tübingen, Verlag von J.C.B. Mohr (Paul Siebeck), 9,10 edition, 1921.

[Wolfram, 2001] Stephen Wolfram. *A new Kind of Science.* Wolfram Media Inc., 2001.

Part C

Three Major Foundational Approaches

THE QUANTITATIVE THEORY OF INFORMATION

Peter Harremoës and Flemming Topsøe

1 BASIC CONCEPTS OF INFORMATION THEORY

Information theory as developed by Shannon and followers is becoming more and more important for a number of sciences. The concepts appear to be just the right ones with intuitively appealing operational interpretations. Furthermore, the information theoretical quantities are connected by powerful identities and inequalities. In this section we introduce *codes, entropy, divergence, redundancy* and *mutual information* which are considered to be the most important concepts.

1.1 Shannon's break-through

Shannon's 1948 paper [Shannon, 1948]: "A mathematical theory of communication" marks the birth of modern information theory. It immediately caught the interest of engineers, mathematicians and other scientists. Naturally, one had speculated before Shannon about the nature of information but mainly at the qualitative level. Precise and widely applicable notions and tools did not exist before Shannon.

Shannon focused on engineering-type problems of communication. Because of the great impact for the economy, this is where the main interest from society lies. But information theory captures fundamental aspects of many other phenomena and has implications at the philosophical level regarding our understanding of the world of which we are part. More applied areas include the interrelated fields *communication theory, coding theory, signal analysis* and *cryptography.*

1.2 Coding

Information is always *information about something.* The *description* of information must be distinguished from this "something", just as the words used to describe a dog are different from the dog itself. Description of information in precise technical terms is important since, in Shannon's words it will allow *"reproducing at one point either exactly or approximately a message selected at another point"*. The descriptions in information theory are called *codes.*

Handbook of the Philosophy of Science. Volume 8: Philosophy of Information
Volume editors: Pieter Adriaans and Johan van Benthem. General editors: Dov M. Gabbay, Paul Thagard and John Woods.

vowel	code-word	code-word length
a	11	2
e	00	2
i	01	2
o	100	3
u	1010	4
y	1011	4

Table 1. Codebook for vowels in English.

An *information source* is some device or mechanism which generates elements from a certain set, the *source alphabet* $\mathbb{A}$. Table 1 shows a *code-book* related to a source which generates a vowel of the English alphabet. The various *code-words* may be taken as a way to *represent*, indeed to *code*, the vowels. Or we may conceive the code-book as a strategy for obtaining information about the actual vowel from a knowledgeable "guru" via a series of yes/no questions. In our example, the first question will be "is the letter one of *a, o, u* or *y*?" . This corresponds to a "1" as the first *binary digit* — or *bit* as we shall say — in the actual code-word. Continuing asking questions related to the further bits, we end up by knowing the actual vowel. The number of bits required in order to identify a vowel is the *code-word length*, i.e. the number of bits in the corresponding code-word.

The term "bit" is used in two ways, as a rather loose reference to 0 or 1 (as above) and then, as a more precisely defined *unit of information*: *A bit is the maximal amount of information you can obtain from a yes/no question* . To clarify, consider questions posed as above but with respect to a modified code-book where 11, the code-word for *a*, is replaced by 111. If the two first questions are both answered by "yes", then, according to the new code-book, you should ask a new question which you can of course do, but it gives no further information as you already know that the actual letter must be *a*. The definition points to classical logic with its reference to "yes/no" (or "1/0" or "true/false"). In Section 1.3 we shall follow up with a more precise mathematical treatment of the concepts "*amount of information*".

To ensure unambiguous identification, we require that a code is *prefix-free*, i.e. no code-word in the code-book is allowed to be the beginning of another. Denoting code-word lengths by l_x , $x \in \mathbb{A}$, *Kraft's Inequality*

$$(1) \quad \sum_{x\in\mathbb{A}} 2^{-l_x} \leq 1$$

must hold — indeed, the binary subintervals of the unit interval that correspond, via successive bisections, to the various code-words must be pairwise disjoint, hence have total length at most 1. And, in the other direction, if numbers l_x are given satisfying (1) then there exists a prefix-free code with the prescribed l_x's as code-word lengths.

We may express Kraft's Inequality differently, as the property that any length function $x \curvearrowright l_x$ must satisfy the lower bound restriction

(2) $l_x \geq -\log_2 p_x$ for all $x \in \mathbb{A}$

for some probability distribution $P = (p_x)_{x \in \mathbb{A}}$. Here and below, "$\log_2$" denotes logarithm to the base 2.

The case of equality in (1) corresponds to *compact codes*, i.e. codes where no code-word can be added to the code-book without breaking the prefix-free property.

A guiding principle is to design codes that achieve efficient *compression*, i.e. which have as short code-word lengths as possible, understood in some appropriate way. Design criteria depend on the type of knowledge one has about the source. If, in the example, we actually know nothing about the source, then "minimax" is a suitable design criterion (and the code in Table 1 is not optimal as it is easy to design a code with maximal code-word lengths equal to 3 rather than 4).

Consider another extreme where very detailed knowledge about the source is available. We have chosen to look at Charles Dickens' "A Tale of Two Cities". It generates individual letters, spaces, punctuation marks etc. To simplify, we ignore the finer details and only pay attention to the standard letters. We may then summarize our knowledge about the source by listing the frequencies of letters, cf. Table 2. It can be proved that the code listed in the table as a *Huffman code* is optimal in the sense that it requires the smallest number of bits to *encode* the entire novel. This smallest number is 2.444.253 bits or in average 4.19 bits for each of the 583.426 letters.

We stress that above we have only aimed at efficient coding of single letters. Our success in compression can then be expressed by the one number 4.19 (bits/letter). We can also consider the optimal code as a *reference code* and measure the performance of other codes in relation to it. For instance, for the *fixed length code* which is also shown in Table 2, there is a *redundancy* of 0.81 bits/letter, expressing that these bits are superfluous when we compare with the optimally achievable compression.

The situation could also be that originally, before we had detailed knowledge about the statistics of the letters in the novel, we used the fixed length code and then the redundancy tells us how much we can save by switching to an optimal code once we have obtained more detailed knowledge.

If we code the entire novel using the optimal code in Table 2, the coded string starts off with

$$1010010011101110100101010000010111100$$
$$0100101011001111000101010001110001001$$

which is *decoded* as "itwasthebestoftimes" corresponding to the opening words in Dickens' novel.

What we have considered above is *noiseless coding*. If, however, errors can occur, many new problems turn up. For instance, if the 19th bit (0) and the 52nd

Letter	frequency		fixed length		Huffman code		ideal
	count	in %	word	length	word	length	length
a	47064	8.07	00000	5	1110	4	3.63
b	8140	1.40	00001	5	101111	6	6.16
c	13224	2.27	00010	5	01111	5	5.46
d	27485	4.71	00011	5	0110	4	4.41
e	72883	12.49	00100	5	000	3	3.00
f	13155	2.25	00101	5	111100	6	5.47
g	12120	2.08	00110	5	111101	6	5.59
h	38360	6.57	00111	5	1000	4	3.93
i	39786	6.82	01000	5	1010	4	3.87
j	622	0.11	01001	5	1111111110	10	9.87
k	4635	0.79	01010	5	11111110	8	6.98
l	21523	3.69	01011	5	10110	5	4.76
m	14923	2.56	01100	5	00111	5	5.29
n	41310	7.08	01101	5	1101	4	3.82
o	45118	7.73	01110	5	1100	4	3.69
p	9453	1.62	01111	5	101110	6	5.95
q	655	0.11	10000	5	1111111100	10	9.80
r	35956	6.16	10001	5	0010	4	4.02
s	36772	6.30	10010	5	1001	4	3.99
t	52396	8.98	10011	5	010	3	3.48
u	16218	2.78	10100	5	00110	5	5.17
v	5065	0.87	10101	5	1111110	7	6.85
w	13835	2.37	10110	5	01110	5	5.40
x	666	0.11	10111	5	1111111101	10	9.77
y	11849	2.03	11000	5	111110	6	5.62
z	213	0.04	11001	5	1111111111	10	11.42
total = 583.426		**100**	**mean = 5.00**		**mean = 4.19**		**H = 4.16**

Table 2. Statistics of letters in "A Tale of Two Cities" and two codebooks.

bit (1) in the above string are transmitted incorrectly, decoding leads to the string "itwalierfttltotimes" with an irritating period out of synchronization. We realize the need to develop tools for *detection* and *correction* of errors. There is a huge literature on these aspects. Here we only note that some redundancy is needed to prevent corruption of the whole message caused by a few accidental errors. Indeed, if we use the fixed length code of Table 2 instead of the optimal code, we are much better protected against occasional bit flip errors.

Coding is partly of a combinatorial nature due to the requirement of integers as code-word lengths. For theoretical discussions it is desirable to take the combinatorial dimension out of coding. This can be done by allowing arbitrary real numbers as code-word lengths. We therefore define an *idealized code over the alphabet* $\mathbb{A}$ as a map $x \curvearrowright l_x$ of $\mathbb{A}$ into the positive real numbers such that *Kraft's Inequality* holds, i.e. such that

$$(3)\quad \sum_{x\in\mathbb{A}} 2^{-l_x} \leq 1.$$

The l_x's are thought of as code-word lengths and the idealization lies in accepting arbitrary real values for the l_x's. If equality holds in Inequality 3 then the code is said to be *compact*. Apparently, there is a one-to-one relationship between compact codes and probability distributions. It is given by the formulas

$$(4)\quad l_x = -\log_2 p_x \ ; \ p_x = 2^{-l_x}.$$

When these formulas hold, we say that the code κ is *adapted to* P or that P *matches* κ.

We can then consider *optimal idealized codes*, in analogy with the notion of ordinary (combinatorial) optimal codes. It turns out that an optimal idealized code is unique. For the example chosen, the idealized code shown in Table 2 in two-decimal precision is in fact the optimal one. If we use this code, and accept the interpretation as lengths of idealized code-words, we should use 2.426.739,10 bits to encode the entire novel. If we allow idealized coding, the performance of other codes should be measured relative to the optimal idealized code. Hence the redundancy of the fixed length code in Table 2 should be 0.84 rather than 0.81 bits/letter and the redundancy of the Huffman code is 0.03 bits/letter.

1.3 Entropy

The relative frequencies in Table 2 are formally defining a probability distribution over the 26-letter alphabet. For many considerations it is not important whether a distribution describes observed relative frequencies or unobserved random events. Therefore assume that an alphabet $\mathbb{A}$ is given with a known probability distribution $P = (p_x)_{x\in\mathbb{A}}$.

The compression problem of the previous section gives rise to the definition of the *entropy* $H(P)$ of P as:

$$(5)\quad H(P) = \min_{\kappa} \sum_{x \in \mathbb{A}} p_x l_x ,$$

it being understood that the minimum is over all idealized codes κ (with the l_x's denoting the idealized code-word lengths). Thus, *entropy is minimal average code-word length* understood in an idealized sense. A key result is the analytical identification of entropy :

THEOREM 1 (First main theorem of information theory). *The entropy of P defined by (5) can be expressed analytically as follows:*

$$(6)\quad H(P) = -\sum_{x \in \mathbb{A}} p_x \log_2 p_x \,.$$

The relation of entropy to coding was emphasized by introducing the concept of idealized codes. By Theorem 1, the idealized code adapted to P is the optimal idealized code of a source governed by P. We will return to the *duality* expressed by (4) in Section 3.

The idealization in Theorem 1 is a great convenience and no serious restriction. To emphasize this, let us insist, for a moment, to use codes with integer lengths. Then we can choose code-lengths l_x close to $-\log_2 p_x$ and ensure in this way that $H(P) \leq \sum p_x l_x < H(P) + 1$. Moreover, if we consider a source generating sequences of letters independently according to the distribution P, then the minimum average code-word length per letter when we consider longer and longer sequences of letters converges to $H(P)$.

Often, entropy is measured in *natural units* ("nats") rather than in bits. In (6) then, $\log_2$ should be replaced by ln and exponentiation should be with respect to e rather than 2. Clearly, H in nats equals H in bits multiplied by $\ln 2 \approx 0.6931$.

1.4 Divergence and redundancy

Assume that you use an idealized code κ with code-word lengths $l_x\,;\, x \in \mathbb{A}$ to represent data but realize — due to new information obtained or otherwise — that it is better to change to another idealized code, κ' with code-word lengths $l'_x\,;\, x \in \mathbb{A}$. *Redundancy* or *divergence*, which we denote $D(\kappa' \| \kappa)$, measures the gain in bits that can be obtained by changing to the new idealized code. The idea behind the definition is that the preference for κ' reflects the belief that this idealized code could be optimal, i.e. the distribution matching it, $P = (p_x)_{x \in \mathbb{A}}$, could be the "true" distribution. This suggests the definition

$$(7)\quad D(\kappa' \| \kappa) = \sum_{x \in \mathbb{A}} p_x l_x - \sum_{x \in \mathbb{A}} p_x l'_x \,.$$

If $Q = (q_x)_{x \in \mathbb{A}}$ denotes the distribution matching κ (thus Q is the distribution which you originally found best represented the data) we can express $D(\kappa' \| \kappa)$ in terms of P and Q and write $D(P \| Q)$ instead. This is the notation mainly found in the literature. It is the *Kullback-Leibler divergence*, or just the *divergence*, from P to Q. We find that

$$(8)\quad D(P\|Q) = D(\kappa'\|\kappa) = \sum_{x\in\mathbb{A}} p_x \log_2 \frac{p_x}{q_x}\,.$$

The quantity is of great significance for many theoretical studies and for applications. The interpretation focuses on a situation where you start with partial knowledge and then, somehow, obtain information which makes you change behaviour. The properties of the logarithmic function implies that $0 \le D(P\|Q)$ with equality if and only if $P = Q$. This is the most basic inequality of information theory.

We find that

$$(9)\quad \sum p_x l_x = H(P) + D(P\|Q)\,,$$

i.e. *actual average code length is the sum of minimal average code length and divergence.* We refer to (9) as the *linking identity.*

For several applications it is important that divergence makes sense also for continuous distributions. Formally this can be achieved via a limiting process based on the discrete case or one may define divergence directly as an integral. For the present text we will base the exposition on the discrete case and rely on an intuitive understanding when we comment on the continuous case.

1.5 Mutual information

It is important that key notions such as entropy can be extended from dealing only with distributions to incorporate also random elements. The *entropy* of a random element is defined as the entropy of the corresponding distribution. If the random element X is defined on a sample space governed by the probability measure $\mathbb{P}$ and X takes values in $\mathbb{A}$, then, denoting the distribution of X by P_X, we define the *entropy* of X by $H(X) = H(P_X)$, i.e.

$$(10)\quad H(X) = -\sum_{x\in\mathbb{A}} P_X(x)\log_2 P_X(x) = -\sum_{x\in\mathbb{A}} \mathbb{P}(X=x)\log_2 \mathbb{P}(X=x)\,.$$

As $H(X)$ only depends on X through its distribution and as it is the actual values of X which carry semantic information, one must admit that the extension only contributes moderately to incorporate semantic aspects.

If several random elements are defined on the same probability space, *joint entropy* such as $H(X,Y)$ makes good sense. So does *conditional entropy*, $H(X|Y)$, defined in the natural way as the average of the entropies of the conditional distributions (here indicated by $X|Y=y$ or by $P_{X|y}$):

$$(11)\quad H(X|Y) = \sum_y \mathbb{P}(Y=y)H(X|Y=y) = \sum_y P_Y(y)H(P_{X|y})\,.$$

The conditional entropy $H(X|Y)$ is also called the *equivocation of X given Y*. It represents the uncertainty that remains about X after having obtained information about Y.

Information theory operates with a number of intuitive identities and inequalities. Here we mention what is often referred to as *Shannon's Identity*, (12), and *Shannon's Inequality*, either (13) or (14) below:

$$H(X,Y) = H(X) + H(Y|X)\,, \tag{12}$$

$$H(X,Y) \leq H(X) + H(Y)\,, \tag{13}$$

$$H(Y|X) \leq H(Y)\,. \tag{14}$$

Equality holds in (13) and (14) if and only if X and Y are independent (assuming that the involved entropies are finite). Regarding (13) and (14), a simple proof depends on the basic inequality $D \geq 0$ in connection with (17) and (18) below.

The availability of notions of entropy for random elements is a great help in many situations. For instance, one may express development in time through a series $X_1, X_2, \cdots$ of random elements which could represent bits, letters, words or other entities.

Consider two random elements, X and Y with our interest attached to X. To begin with we have no information about X. Assume now that we can obtain information, not about X, but about Y. *Mutual information*, $I(X;Y)$, measures the amount of information in bits we can obtain about X by knowing Y. At least three different ideas for a sensible definition are possible: Firstly, as *uncertainty removed*, secondly, as *average redundancy* and thirdly, admittedly less intuitive, as *divergence related to a change of joint distributions*. It is a surprising fact that all suggested definitions give the same quantity. In more detail:

$$I(X;Y) = H(X) - H(X|Y) \tag{15}$$

$$= \sum_y \mathbb{P}(Y=y) D(X|Y=y \| X) = \sum_y P_Y(y) D(P_{X|y} \| P_X) \tag{16}$$

$$= D(P_{X,Y} \| P_X \otimes P_Y)\,. \tag{17}$$

In (17), $P_X \otimes P_Y$ denotes the distribution $(x,y) \curvearrowright P_X(x) \cdot P_Y(y)$ corresponding to independence of X and Y.

Rewriting (15) as

$$H(X) = H(X|Y) + I(X;Y) \tag{18}$$

and combining with (15) and (12) we realize that

$$I(X;Y) = I(Y;X)\,. \tag{19}$$

This *symmetry of mutual information* has puzzled many authors as it is not intuitively obvious that information about X, knowing Y quantitatively amounts to the same as information about Y, knowing X.

Another significant observation is that we may characterize entropy as *self-information* since, for $Y = X$, (15) shows that

$$H(X) = I(X;X)\,. \tag{20}$$

Previously we emphasized that information is always information about something. So entropy of a random variable is a measure of information in the seemingly weak sense that this "something" is nothing but the variable itself. Although this interpretation is self-referential it has turned out to be very useful.

1.6 Data reduction and side information

If, when studying a certain phenomenon, you obtain extra information, referred to as *side information*, this results in a *data reduction* and you will expect quantities like entropy and divergence to decrease. Sometimes the extra information can be interpreted as information about the *context* or about the *situation.*

Shannon's Inequality (14) can be viewed as a data reduction inequality. There, the side information was given by a random element. Another way to model side information is via a *partition* of the relevant sample space. Recall that a partition of a set A is a collection of non-empty, non-overlapping subsets of A with union A; the subsets are referred to as the *classes* of the partition.

As an example, consider prediction of the two first letters x_1, x_2 in an English text and assume that, at some stage, you obtain information about the first letter, x_1. As a model you may use the random element X_1, X_2 with X_1 expressing the side information. Or you may consider modeling based on the partition of the original set of all $26 \times 26 = 676$ two-letter words into the 26 classes defined by fixing the first letter.

Consider distributions over a general alphabet $\mathbb{A}$ and let θ denote a partition of $\mathbb{A}$. Denote the classes of θ by A_i (with i ranging over some appropriate index set) and denote the set of classes by $\partial\mathbb{A}$. In mathematics this is the *quotient space* $\mathbb{A}/\theta$. If P is a source over $\mathbb{A}$, ∂P denotes the *derived source* over $\partial\mathbb{A}$ given by $\partial P(A_i) = P(A_i)$. By the *conditional entropy of P given the side information θ* we understand the quantity

$$(21)\quad H^\theta(P) = \sum_i P(A_i) H(P|A_i)$$

with summation over all indices (which could be taken to be summation over $\partial\mathbb{A}$). Similarly, if two sources over $\mathbb{A}$ are considered, *conditional divergence under the side information θ* is defined by

$$(22)\quad D^\theta(P\|Q) = \sum_i P(A_i) D(P|A_i \| Q|A_i)\,.$$

Simple algebraic manipulations show that the following *data reduction identities* hold:

$$H(P) = H(\partial P) + H^\theta(P)\,, \tag{23}$$

$$D(P\|Q) = D(\partial P\|\partial Q) + D^\theta(P|Q)\,. \tag{24}$$

Immediate corollaries are the *data reduction inequalities*

$$H(\partial P) \leq H(P)\,, \tag{25}$$

$$D(\partial P \| \partial Q) \leq D(P\|Q)\,, \tag{26}$$

as well as the *inequalities under conditioning*

$$H^{\theta}(P) \leq H(P)\,, \tag{27}$$

$$D^{\theta}(P\|Q) \leq D(P\|Q)\,. \tag{28}$$

As a more special corollary of (26) we mention *Pinsker's Inequality*

$$D(P\|Q) \geq \frac{1}{2}V^2(P,Q) \tag{29}$$

where $V(P,Q) = \sum |p_x - q_x|$ denotes *total variation* between P and Q. This inequality is important as the basic notion of convergence of distributions in an information theoretical sense, called *convergence in information* and defined by the requirement $D(P_n\|P) \to 0$, is then seen to imply convergence in total variation, $V(P_n, P) \to 0$ which is an important and well-known concept.

1.7 Mixing

Another important process, which applies to distributions is that of *mixing*. Intuitively one should think that mixing results in more "smeared out" distributions, hence should result in an increase in entropy. Regarding divergence, the "smearing out" should have a tendency to bring distributions closer together, hence in diminishing divergence.

To be precise, consider a mixture, say a finite mixture

$$P_0 = \sum_{n=1}^{N} \alpha_n P_n \tag{30}$$

of N distributions over $\mathbb{A}$ (thus, the α's are non-negative and add to 1).

Just as in the case of data reduction, certain natural inequalities suggest themselves and these can be derived from simple identities. In fact, from the linking identity (9), you easily derive the following identities:

$$H\left(\sum_{n=1}^{N} \alpha_n P_n\right) = \sum_{n=1}^{N} \alpha_n H(P_n) + \sum_{n=1}^{N} \alpha_n D(P_n\|P_0)\,, \tag{31}$$

$$\sum_{n=1}^{N} \alpha_n D(P_n\|Q) = D\left(\sum_{n=1}^{N} \alpha_n P_n \| Q\right) + \sum_{n=1}^{N} \alpha_n D(P_n\|P_0)\,. \tag{32}$$

As corollaries we see that entropy $P \curvearrowright H(P)$ is *concave* and divergence $P \curvearrowright D(P\|Q)$ *convex* for fixed Q:

$$H\left(\sum_{n=1}^{N} \alpha_n P_n\right) \geq \sum_{n=1}^{N} \alpha_n H(P_n)\,, \tag{33}$$

$$D\left(\sum_{n=1}^{N} \alpha_n P_n \| Q\right) \leq \sum_{n=1}^{N} \alpha_n D(P_n \| Q)\,. \tag{34}$$

The common term which appears in (31) and in (32) is of importance in its own right, and has particular significance for an even mixture $P_0 = \frac{1}{2}P_1 + \frac{1}{2}P_2$ when it is called *Jensen-Shannon divergence.* Notation and definition is as follows:

$$JSD(P_1, P_2) = \frac{1}{2}D(P_1 \| P_0) + \frac{1}{2}D(P_2 \| P_0)\,. \tag{35}$$

Jensen-Shannon divergence is a smoothed and symmetrized version of divergence. In fact, it is the square of a metric, which metrizes convergence in total variation.

1.8 Compression of correlated data

A basic theme has been *compression of data.* This guided us via coding to key quantities of information theory. The simplest situation concerns a single source, but the concepts can be applied also in more complicated cases when several sources interact and produce correlated data. This already emerged from the definitions involving conditioning.

As a more concrete type of application we point to compression of data in a *multiple access channel.* To simplify, assume that there are only two senders and one receiver. Sender 1 knows the value of the random variable X and Sender 2 the value of Y. The random variables may be correlated. The same channel, assumed noiseless, is available to both senders. There is only one receiver. If there were no collaboration between the senders, Sender 1 could, optimally compress the data to the rate $R_1 = H(X)$ bits and Sender 2 to the rate $R_2 = H(Y)$ bits, resulting in a joint rate of $R_1 + R_2 = H(X) + H(Y)$ bits needed for the receiver to know both X and Y. This should be compared to the theoretically optimal joint compression of the joint variable (X, Y), which is

$$\begin{aligned} H(X,Y) &= H(X) + H(Y) - I(X;Y) \\ &= H(X) + H\left(Y \mid X\right) = H\left(X \mid Y\right) + H\left(Y\right). \end{aligned} \tag{36}$$

In fact, in a remarkable paper [Slepian and Wolf, 1973], Slepian and Wolf showed that it is possible for Sender 1 to compress to $H(X)$ bits and independently for Sender 2 to compress to $H(Y \mid X)$ bits, in such a way that the receiver is able to recover X and Y. Similarly, Sender 1 can compress to $H(X \mid Y)$ bits and Sender 2 to $H(Y)$ bits, and the receiver is still able to recover X and Y. As it is possible

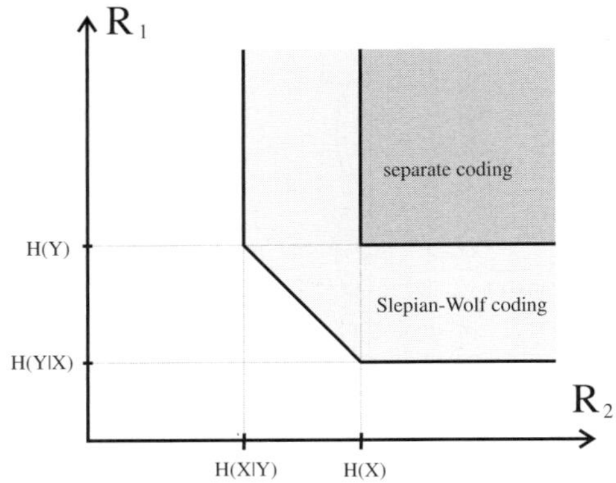

Figure 1. Compression region obtained by Slepian-Wolf coding.

to introduce timesharing between the two protocols described this leads to the following result: The rates of compression R_1 and R_2 are achievable if and only if

$$\begin{aligned} R_1 &\geq H(X \mid Y) \\ R_2 &\geq H(Y \mid X) \\ R_1 + R_2 &\geq H(X, Y). \end{aligned} \tag{37}$$

For a technically correct result, one has to consider multiple outcomes of X and Y and also to allow a small probability of error when X and Y are recovered.

Note that the result does not tell which of the two protocols is the best one or whether it is one of the timesharing protocols.

1.9 Other definitions of basic information theoretical quantities

The key definitions of information theory are those rooted in Shannon's work. There are, however, many other ways of defining entropy and related quantities. Here we shall introduce certain entropy and divergence measures going back to Rényi [Rényi, 1961]. These measures appear in many studies, cf. [Cambell, 1965], [Csiszár, 1995] and [Arndt, 2001]. Moreover, they have operational definitions which relate directly to coding and as such may be considered to be members of the "Shannon family" of information measures.

Previously, much attention was given to the axiomatic approach. In our opinion this often hides essential aspects. When possible, an approach based on operational definitions is preferable.

Consider two probability distributions P and Q over the discrete alphabet $\mathbb{A}$ and a parameter $\alpha \in]0, 1[$. Let λ and γ be the compact codes adapted to P and Q, respectively. If we want to express belief in P as well as in Q, a possibility is to consider the convex mixture $\kappa = \alpha\lambda + (1 - \alpha)\gamma$. Then κ is also an idealized code but it is not compact except when $\lambda = \gamma$. However, $\kappa - d$ is a compact code with $d \geq 0$ defined by

$$(38)\quad d = -\log_2 \left(\sum_{x \in \mathbb{A}} 2^{-\kappa(x)} \right) .$$

The constant d is a measure of *discrepancy* between P and Q. We define the *Rényi divergence of order α between P and Q*, denoted $D_\alpha(P\|Q)$, to be $\frac{1}{1-\alpha}d$ or, in terms of P and Q,

$$(39)\quad D_\alpha(P\|Q) = \frac{1}{\alpha - 1} \log_2 \left(\sum_{x \in \mathbb{A}} p_x^\alpha q_x^{1-\alpha} \right) .$$

The chosen normalization ensures that we regain the usual Kullback-Leibler divergence as the limit of D_α for $\alpha \to 1$. Formally, (39) makes sense for all real α.

One may consider divergence as the most fundamental concept of information theory. Then mutual information and entropy appear as derived concepts. For a finite alphabet $\mathbb{A}$, *entropy differences* may be defined directly from divergence using the guiding equation

$$(40)\quad D_\alpha(P\|U) = H_\alpha(U) - H_\alpha(P) ,$$

with U the uniform distribution over $\mathbb{A}$. Then *Rényi's entropy* of P of *order* α is obtained if one adds the assumption that the entropy of a uniform distribution for any sensible notion of entropy must be the *Hartley entropy*, the logarithm of the size of the alphabet. Doing that, one finds that (40) leads to the quantity

$$(41)\quad H_\alpha(P) = \frac{1}{1 - \alpha} \log_2 \sum_{x \in \mathbb{A}} p_x^\alpha .$$

It is arguably more satisfactory first to define mutual information and then to define entropy as self-information, cf. (20). If one bases mutual information on (16) one will end up with the Rényi entropy of order α, whereas, if one uses (17) as the basis for mutual information, one ends up with Rényi entropy, not of order α though, but of order $2 - \alpha$. Thus, leaving the classical Shannon case, it appears that entropy "splits up" in H_α and $H_{2-\alpha}$.

In certain parts of non-classical statistical physics the quantity obtained from (41) by using the approximation $\ln u \approx u - 1$ has attached much interest, but a direct operational definition is not yet clear. For more on this form of entropy, , the *Tsallis entropy* see the contribution on physics in this handbook.

The considerations in this section point to some difficulties when leaving purely classical grounds. A complete clarification must depend on operational definitions and has to await further progress.

2 BEYOND YES AND NO

Coding is used for storing, transmission and reconstruction of information. If the information is carried by a continuous variable, such as a 2-dimensional image or the result of a measurement of a physical quantity, perfect storage is not possible in a digital medium. This poses serious technical problems for which there is no universal solution. These problems are handled in *rate-distortion theory*. The interest for this Handbook lies in the fundamental problem of the nature of the world. Discrete or continuous? Does modeling with continuous quantities make sense? Though rate-distortion theory does not contribute to answer the philosophical questions it does give a clue to what *is* possible if you use modeling by the continuous.

2.1 Rate distortion theory

Consider a continuous random variable X with values in the source alphabet $\mathbb{A}$ and with distribution P_X. In simple examples, $\mathbb{A}$ is one of the Euclidean spaces $\mathbb{R}^n$ or a subspace thereof but more complicated settings may arise, for instance in image analysis. The continuous character means that $\sum_{x\in\mathbb{A}} P_X(x) < 1$ (typically, this sum is 0).

The treatment of problems of coding and reconstruction of continuous data builds on a natural idea of *quantization*. Abstractly, this operates with a finite *reconstruction alphabet* $\mathbb{B}$, and a *quantizer* $\phi : \mathbb{A} \to \mathbb{B}$ which maps $a \in \mathbb{A}$ into its *reconstruction point* $b = \phi(a)$. Considering, for each $b \in \mathbb{B}$, the set of $a \in \mathbb{A}$ with $\phi(a) = b$ we realize that this defines a partition of $\mathbb{A}$. For simplicity we shall only consider the case when $\mathbb{B}$ is a subset of $\mathbb{A}$ and $\phi(b) = b$ for each $b \in \mathbb{B}$. The idea is illustrated by Figure 2.

A *rate-distortion code* is an idealized code over $\mathbb{B}$. Associated with a rate-distortion code we consider the *length function,* which maps $x \in \mathbb{A}$ to the length of the "code-word" associated with $\phi(x)$. The reconstruction points are used to define the decoding of the code in an obvious manner. If we ignore the requirement to choose reconstruction points, this construction amounts to the same as a data reduction, cf. Section 1.6.

In order to study the *quality* of reconstruction we introduce a *distortion function* d defined on $\mathbb{A}$ (formally on $\mathbb{A} \times \mathbb{B}$). This we may also think of as an expression of the *relevance* — with a high degree of relevance corresponding to a small distortion. The quantity of interest is the *distortion* $d(x, \hat{x})$ with $\hat{x} = \phi(x)$. Maximizing over $\mathbb{A}$ or taking mean values over $\mathbb{A}$ with respect to P_X we obtain the *maximal distortion* and the *mean distortion*. In practice, e.g. in image analysis, it is often difficult to specify sensible distortion functions. Anyhow, the set-up in rate distortion theory, especially the choice of distortion function, may be seen as one way to build semantic elements into information theory.

As examples of distortion measures on $\mathbb{R}$ we mention *squared error distortion* $d(x, \hat{x}) = (x - \hat{x})^2$ and *Hamming distortion*, which is 0 if $\hat{x} = x$ and 1 other-

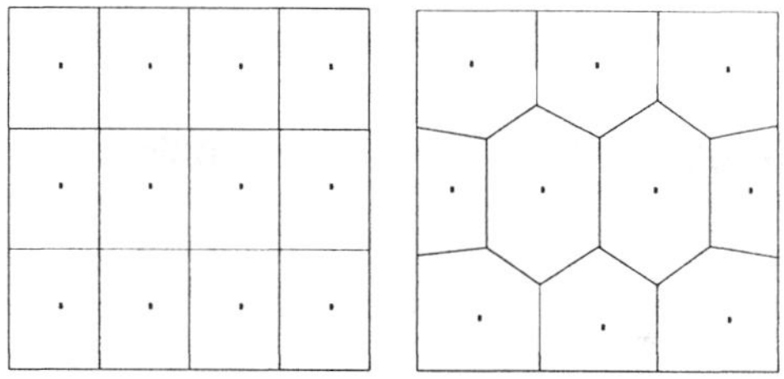

Figure 2. Two quantizers with partitions and reconstruction points shown. It is far from obvious which of the quantizers is the best one.

wise. Thus Hamming distortion tells whether a reproduction is perfect or not whereas squared error distortion weighs small geometric errors as being of small significance. Hamming distortion is the distortion function used in ordinary information theory and corresponds to the situation where one only distinguishes between "yes" and "no" or "black" and "white".

By $B(x,\varepsilon)$ we denote the *distortion ball* around x with radius ε, i.e. the set of y such that $d(x,y) \leq \varepsilon$. The following result is analogous to Kraft's inequality as expressed by (2):

THEOREM 2. *Let $l : X \to \mathbb{R}_+$ be the length function of a rate distortion code with maximal distortion ε. Then there exists a probability distribution P such that, for all $x \in \mathbb{A}$,*

$$(42)\quad l(x) \geq -\log_2 (P(B(x,\varepsilon))).$$

The converse is only partially true, but holds asymptotically if one considers average length of length functions corresponding to long sequences of inputs. We see that a small ε corresponds to large code lengths. The inequality should be considered as a distortion version of Kraft's inequality, and it extends the duality (4) to cover also rate-distortion.

If a probability distribution on the source alphabet $\mathbb{A}$ is given, then the quantizer induces a probability distribution on the reconstruction alphabet $\mathbb{B}$. The *rate* of the quantizer is defined as the entropy of the induced probability distribution, i.e. as $R = H(\phi(P_X))$ (here, $\phi(P_X)$ denotes the distribution of ϕ). A high rate reflects a fine resolution. Consider, as above, a fixed continuous random variable with distribution P_X. In order to characterize the performance of any quantization method as described above it is reasonable to use two quantities, the rate R and the mean distortion $D = E\left(d(X,\hat{X})\right)$. The set of feasible values of (D,R) forms the *rate-distortion region* for the distribution P_X. If distortion is small, the rate must be large. Therefore, not all points in $\mathbb{R}^2$ are feasible. The borderline between feasible and infeasible points is called the *rate-distortion curve* and is most often expressed as the *rate-distortion function*, cf. Figure 3. It describes the optimal trade-off between distortion and rate.

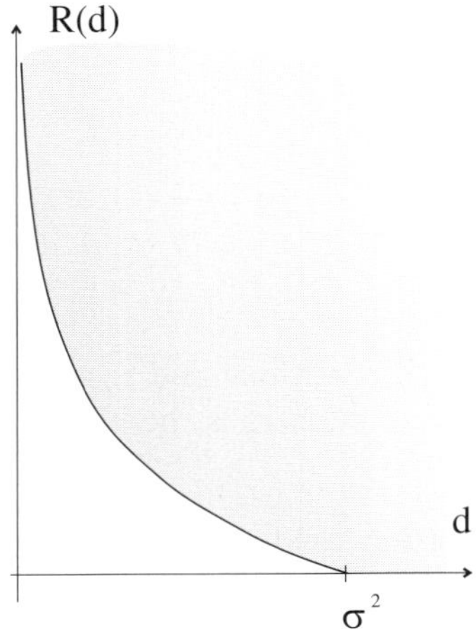

Figure 3. Rate distortion function for a Gaussain distribution.

In special cases it is possible to calculate the rate distortion function exactly using Shannon's celebrated *Rate Distortion Theorem.* For instance, let X be Gaussian with variance σ^2. Then the rate-distortion function is given by

$$(43)\quad R(d) = \begin{cases} \frac{1}{2}\log_2\left(\frac{\sigma^2}{d}\right) & d \le \sigma^2 \\ 0 & d > \sigma^2 \end{cases} .$$

In other cases the rate-distortion function can be approximated using numerical methods. In cases where the rate-distortion function can be determined the results from the previous sections can be extended to a continuous setting. In practice it has turned out to be quite difficult to implement these theoretical ideas. The reason is that practical problems typically involve a high number of variables, and it is very difficult to specify distortion measures and probability distributions on these high-dimensional spaces.

Let X be a random variable with probability density f. The *differential entropy* of X is given by the formula

$$(44)\quad h(X) = -\int f(x)\log_2 f(x)\ dx.$$

If we use squared error distortion, the rate-distortion function is given, approximately, by

$$(45)\quad R(d) \approx h(X) - \frac{1}{2}\log_2(2\pi e \cdot d)$$

for small values of d. This also gives an interpretation of the differential entropy as

(46) $h(X) \approx R(d) + \frac{1}{2}\log_2(2\pi e \cdot d)$.

In fact, the right hand side converges to $h(X)$ for d tending to zero.

2.2 *Aspects of quantum information theory*

Classical information theory is based on natural concepts and tools from analysis and probability theory. The first many years one did not take the physical dimension into consideration. It was believed that the nature of the physical devices used as carriers of information would not have any impact on the theory itself. In particular, it was expected that the classical theory would carry over and apply to quantum systems without essential changes as soon as the appropriate concepts had been identified. In the 70'ties and 80'ties studies looking into these questions were initiated and a number of preliminary results established. However, it was not until the 90'ties that the new *quantum information theory* really took off and gained momentum. This was partly due to progress by experimental physicists.

Today, quantum information theory is a thriving field, but still containing controversies and basic open questions. The theory is fundamentally different from the classical theory. The new aspects are interesting from a mathematical, a physical as well as a purely philosophical point of view. The theory brings us beyond the "yes" and "no" tied to the classical theory and bound to the fundamental unit of a bit.

A *quantum experiment* provides a connection between the *preparation* of the system and the possible *measurements* on the system. The focus on measurements forms an extra layer between the system and the observer which is necessary in order to enable meaningful statements about the system. The set-up may be conceived as a "black box", a "coupling" or an "information channel" between the preparation and the measuring device. Two preparations represent the same *state* of the system if the preparations cannot be distinguished by any available measurement. Defined in this way, the set of all states, the *state space*, depends on the set of possible measurements. If, therefore, an experiment involves a preparation and a measurement on an electron and the state found is S, it will be misleading to say that "the electron is in state S". Instead, you may say that "our knowledge about the electron is completely described by the state S".

Usually, in quantum physics, the state space can be identified with a set of *density matrices* (or operators). For the simplest quantum systems, the state space consists of 2×2 *density matrices*, matrices of the form

(47) $\begin{pmatrix} \frac{1}{2}+\alpha & \beta+i\gamma \\ \beta-i\gamma & \frac{1}{2}-\alpha \end{pmatrix}$,

where the real numbers α, β and γ satisfy the relation

(48) $\alpha^2 + \beta^2 + \gamma^2 \leq \frac{1}{4}$

(with i the complex imaginary unit)[1]. Geometrically, this state space is a ball. States on the boundary of the state space are *pure states* whereas states in the interior are *mixed states*. The principle behind *mixing* is the following: Consider two possible preparations. Construct a new preparation by flipping a coin and choose the first preparation if the coin shows "head" and the second preparation if the coin shows "tail". In this way, the resulting preparation is constructed by mixing. A mixed state can always be represented as a mixture of pure states. In classical physics, the normal situation is that any state is a unique mixture of pure states. A special feature of quantum physics is that a mixed state can always be obtained in several ways as a mixture of pure states. This implies that, if one observes a mixed state, it is theoretically impossible to infer which preparations were involved in the mixing. This is a fundamental new feature of quantum information theory.

The fact that the state space has a high degree of symmetry — as was the case with the ball above — is no coincidence. In general, symmetries in the state space reflect that physical operations like rotations have to leave the state space invariant.

A simple system as described by matrices of the form (47) is called a *qubit*. Physically, a qubit may be implemented by a particle of spin $\frac{1}{2}$ with α, β and γ indicating direction of the spin.

The qubit is the unit of quantum information theory. This is a natural choice of unit as one can device a protocol which, with high fidelity, transforms any quantum information system into a system involving only qubits. Quite parallel to the classical theory, main tasks of quantum information theory are then to represent complicated quantum systems by qubits and to consider representation, transmission and reconstruction of states.

It is easy to encode a bit into a qubit. By orthogonality of spin up and spin down, one can perform a measurement which recovers the bit perfectly. In this way a preparation described by the probability distribution $(\frac{1}{2}+\alpha, \frac{1}{2}-\alpha)$ is mapped into the density matrix

$$\begin{pmatrix} \frac{1}{2}+\alpha & 0 \\ 0 & \frac{1}{2}-\alpha \end{pmatrix}. \tag{49}$$

This shows how bits and, more generally, any classical information system can be embedded in quantum systems. Thus quantum information theory contains classical information theory. The two theories are not equivalent as there is no way in which a qubit can be represented by classical bits.

In order to manipulate quantum information, we need a quantum computer. Recall that a classical computer is based on *gates* which operates on one or two bits. Similar gates can be constructed also for the manipulation of qubits but there is an important restriction of *reversibility* on the gates in a quantum computer.

[1] A description in terms of vectors in Hilbert space is also possible, but the density matrices express in a better way essential aspects related to mixing and measurements.

According to this restriction, to each quantum gate, there should correspond a reverse gate which transforms the output into the input. For instance it is not possible to transform two qubits into one qubit. Similarly it is not possible to transform one qubit into two qubits. This is called the *No-cloning Principle.* Thus quantum information cannot be created, copied or destroyed. In this sense quantum information is physical and behaves somewhat like a liquid.

2.3 Entanglement

In order to explain, if only briefly, the important notion of *entanglement*, consider a system composed of initially independent subsystems, with an associated observer who can prepare a quantum state. If the observers are allowed to manipulate the states by local quantum operations and classical communication, the states of the total system which are achievable in this way are said to be *separable*. If the observers are allowed also to exchange quantum information (via qubits or other non-local quantum operations) then the joint system may be described by states which are not separable. These states are said to be *entangled*.

The electrons in a Helium atom have total spin 0. This means that if one of the electrons is measured to have spin up, the other must have spin down (if measured in the same direction). The two electrons behave like one and such a pair is called an *Einstein-Podolsky-Rosen pair*, an EPR-pair for short. This is the simplest example of an entangled system.

Above, we saw that bits can be encoded into qubits, but qubits cannot be encoded into bits with only classical resources available. If entanglement is available to Alice and Bob in a quantum communication system, this leads to special possibilities. In this case two bits may be encoded into one qubit. This is called *super-dense coding*. The two bits are encoded into two qubits in the sense that the decoder (Bob) receives two qubits. The new thing is that the first qubit (which is one of the particles in an EPR-pair) may be received by both Alice and Bob before Alice knows which bit to send. Although the sharing of an EPR-pair does not represent classical communication, it is a kind of communication that makes the measurement apparatus more sensitive and enables measurements which would not otherwise be possible.

If Alice and Bob share an EPR-pair it is also possible to encode a qubit into two bits. This process is called *quantum teleportation*. The reason for this name is that our entire knowledge about the quantum particle is contained in the density matrix and at the output we receive a particle with exactly the same density matrix. One may say that the particle was destroyed at the input and reconstructed at the output, but nothing is lost by the destruction and reconstruction, so many physicists use the terminology that the particle was teleported from the input to the output. This leads to the physically and philosophically interesting question: Can a particle be identified with the knowledge we have about the particle? Mathematically this is not of significance because all calculations concern the knowledge we have about the system as represented by its density matrix.

3 DUALITY BETWEEN TRUTH AND DESCRIPTION

It is important to distinguish between ontology, how the world is, and epistemology, observations of the world. Niels Bohr said that physics deals with what can be said about nature, not how nature is. The positivists take another position: Physics should uncover objective knowledge about nature. Ontology and epistemology are usually considered as opposed, but information theory offers a position in between. Truth and description are different, but there is a duality between the concepts. To any "true" model there exists an optimal description and, to any description, there exists a model of the world such that the description is optimal if the model is "true". Here the word *true* is in quotation marks because it makes associations to ontology though objective truth is disputable. Instead of speaking about "truth" we shall focus on observations — those already made and observations planned for the future.

3.1 Elements of game theory

As a prelude to the subsections to follow we provide a short introduction to certain parts of game theory.

In game theory situations are modeled where "players" interact in such a way that the satisfaction of each player (or group of players) depends on actions, *strategies*, chosen by all players. Typically, the players are individuals, but animals, machines or other entities could also be considered. We shall only deal with static games, games with no succession of strategic choices. The many variants of the theory operates with different rules regarding the possible actions of the players and the flow of information among them.

A central theme is the investigation of possibilities for rational behaviour of the players. Here, the notion of *equilibrium* comes in. The idea is that if, somehow, the players can decide under the rules of the game to choose specific strategies this is a sign of stability and features associated with such a collective choice can be expected to be observed. For our treatment of game theory it is immaterial how the decisions of the players are arrived at.

Assume that there are n players and that the *cost* or *loss* for player i is given by a real-valued *loss function* $(x_1, \cdots, x_n) \curvearrowright c_i(x_1, \cdots, x_n)$ where $x_1, \cdots, x_n$ represents the strategic choices by the players. The set of strategies $x_1, \cdots, x_n$ defines a *Nash equilibrium* if no player can benefit from a change of strategy provided the other players stick to their strategies. For example, for Player 1, no strategy x_1^* different from x_1 will yield a lower loss, so $c_1(x_1^*, x_2, \cdots, x_n) \geq c_1(x_1, x_2, \cdots, x_n)$ must hold in a Nash equilibrium. This notion of equilibrium is related to *non-cooperation* among the players. It may well be that, for strategies which obey the criteria of a Nash equilibrium, two or more of the players may jointly benefit from a change of their strategies whereas no single player cannot benefit from such a change.

	scissors	paper	stone
scissors	0	−1	1
paper	1	0	−1
stone	−1	1	0

Table 3. Loss function in the scissors-paper-stone game

A Nash equilibrium may not exist. However, a general result guarantees that a, often unique, Nash equilibrium exists if certain convexity assumptions regarding the loss functions are fulfilled. These conditions normally reflect acceptance of *mixed strategies* or *randomization.*

EXAMPLE 3. Consider the two-person *scissors-paper-stone* game. The loss function for, say, Player 1 is shown in Table 3. We assume that $c_2 = -c_1$. This is an instance of a *two-person zero-sum game*, reflecting that what is good for the one player is bad — and equally much so — for the other.

Clearly, there is no Nash equilibrium for this game, no set of strategies you can expect the players to agree on. The game is psychological in nature and does not encourage rational considerations. However, if the game is repeated many times and we allow randomization and use averaging to define the new loss functions, we find that there is a unique choice of strategies which yields a Nash equilibrium, viz. for both players to choose among the three "pure strategies" with equal probabilities.

Games such as the psychologically thrilling scissors-paper-stone game are often best treated by invoking methods of artificial intelligence, learning theory, non-classical logic and psychology. We note that by allowing randomization, an initial game of hazard is turned into a conflict situation which encourages rational behaviour, hence opens up for quantitative statements.

3.2 Games of information

Many problems of information theory involve optimization in a situations that can be modelled as conflicts. Among the relevant problems we mention *prediction, universal coding, source coding, cryptography* and, as the key case we shall consider, the *maximum entropy principle.* The relevant games for these problems are among the simplest of game theory, the *two-person zero-sum games*, cf. Example 3 above.

For these *games of information* one of the players represents "you" as a person seeking information and the other represents the area you are seeking information about. We choose to refer to the players as *Observer* and *Nature*, respectively. In any given context you may prefer to switch to other names, say statistician/model, physicist/system, mother/child, investor/market or what the case may be. Strategies available to Observer are referred to as *descriptors* and strategies available to

Nature are called *worlds*. The set of strategies available to the two players are denoted $\mathcal{D}$, respectively $\mathcal{W}$. We refer to $\mathcal{W}$ as the *set of possible worlds*. Our preferred generic notation for descriptors and worlds are, respectively κ and P which, later, will correspond to, respectively, idealized codes and probability distributions.

Seen from the point of view of Observer, the loss function $(P, \kappa) \curvearrowright c(P, \kappa)$ represents the cost in some suitable sense when the world chosen by Nature is P and the descriptor chosen by Observer is κ. The zero-sum character of the game dictates that we take $-c$ as the loss function for Nature. Then, the Nash equilibrium condition for a pair of strategies (P^*, κ^*) amounts to the validity of the *saddle-value inequalities*

(50) $c(P, \kappa^*) \leq c(P^*, \kappa^*) \leq c(P^*, \kappa)$ for all $P \in \mathcal{W}, \kappa \in \mathcal{D}$.

The *risk* associated with Observers choice $\kappa \in \mathcal{D}$ is defined as the maximal possible cost:

(51) $r(Q) = \max_{P \in \mathcal{W}} c(P, \kappa)$,

and the *minimal risk* is defined by

(52) $r_{min} = \min_{\kappa \in \mathcal{D}} r(Q)$.

A descriptor $\kappa \in \mathcal{D}$ is *optimal* if $r(Q) = r_{min}$.

Similar quantities for Nature are the *gain* (more accurately, the *guaranteed gain*)

(53) $h(P) = \min_{\kappa \in \mathcal{D}} c(P, \kappa)$,

and the *maximal gain*

(54) $h_{max} = \max_{P \in \mathcal{W}} h(P)$.

The requirement of optimality for Nature therefore amounts to the equality $h(P) = h_{max}$.

Quite generally, the *mini-max inequality*

(55) $h_{max} \leq r_{min}$

holds. If there is equality in (55), the common value (assumed finite) is simply called the *value* of the game. Existence of the value is a kind of equilibrium:

THEOREM 4. *If a game of information has a Nash equilibrium, the value of the game exists and Observer and Nature both have optimal strategies.*

In fact, the existence of a Nash equilibrium is also necessary for the conclusion of the theorem. The search for a Nash equilibrium is, therefore, quite important. In some special cases, Nash equilibria are related to *robust descriptors* by which we mean descriptors $\kappa \in \mathcal{D}$ such that, for some finite constant h, $c(P, \kappa) = h$ for all possible worlds P [2].

[2]These strategies correspond closely to the *exponential families* known from statistics.

We now introduce an additional assumption of *duality* by requiring that every world has a best descriptor. In more detail we require that to any possible world P_0, there exists a descriptor κ_0, the *descriptor adapted to* P_0, such that

$$(56)\quad \min_{\kappa \in \mathcal{D}} c(P_0, \kappa) = c(P_0, \kappa_0)\,,$$

and further, we assume that the minimum is only attained for $\kappa = \kappa_0$ (unless $c(P_0, \kappa_0) = \infty$). The condition implies that the gain associated with P_0 is given by $h(P_0) = c(P_0, \kappa_0)$. Also note that the right hand inequality of the saddle value inequalities (50) is automatic under this condition (with κ^* the descriptor adapted to P^*). It is easy to establish the following simple, yet powerful result:

THEOREM 5. *Assume that P^* is a possible world and that the descriptor κ^* adapted to P^* is robust. Then the pair (P^*, κ^*) is the unique Nash equilibrium pair.*

Thus, in the search for Nash equilibrium strategies, one may first investigate if robust descriptors can be found.

3.3 The maximum entropy principle

Consider the set $\mathcal{D}$ of all idealized codes $\kappa = (l_x)_{x \in \mathbb{A}}$ over the discrete alphabet $\mathbb{A}$ and let there be given a set $\mathcal{W}$ of distributions over $\mathbb{A}$. Take average code length as cost function, i.e.

$$(57)\quad c(P, \kappa) = \sum_{x \in \mathbb{A}} p_x l_x\,.$$

By the linking identity (9), the duality requirements related to (56) are satisfied and also, we realize that the gain associated with $P \in \mathcal{W}$ is nothing but the entropy of P. Therefore, h_{max} is the *maximum entropy value* given by

$$(58)\quad H_{max} = H_{max}(\mathcal{W}) = \sup_{P \in \mathcal{W}} H(P)$$

and an optimal strategy for Nature is the same as a *maximum entropy distribution*, a distribution $P^* \in \mathcal{W}$ with $H(P^*) = H_{max}$. In this way, game theoretical considerations have led to a derivation of the *maximum entropy principle* — which encourages the choice of a maximum entropy distribution as the preferred distribution to work with.

EXAMPLE 6. Assume that the alphabet $\mathbb{A}$ is finite with n elements and let $\mathcal{W}$ be the set of *all* distributions over $\mathbb{A}$. Clearly, the constant descriptor $\kappa = (log_2 n)_{x \in \mathbb{A}}$ is robust and hence, by Theorem 5 this descriptor is optimal for Observer and the associated distribution, i.e. the uniform distribution, is the maximum entropy distribution.

EXAMPLE 7. Let $\mathbb{A} = \{0, 1, 2, \cdots\}$, let $\lambda > 0$ and consider the set $\mathcal{W}$ of all distributions with mean value λ. Let $\kappa = (l_n)_{n \geq 0}$ be an idealized code. Clearly, if κ is of the form

(59) $\kappa_n = \alpha + \beta n$

then $\langle P, \kappa \rangle = \alpha + \beta\lambda$ for all $P \in \mathcal{W}$, hence κ is robust. The constant α can be determined from (3) and by a proper choice of β one finds that the associated distribution is one of the possible worlds. This then, again by Theorem 5, must be the maximum entropy distribution. Going through the calculations one finds that for this example, the maximum entropy distribution is the *geometric distribution* with mean value λ, i.e. the distribution $P^* = (p_n^*)_{n\geq 0}$ given by

(60) $p_n^* = pq^n \text{ with } p = 1 - q = \dfrac{1}{\lambda + 1}\,.$

The length function for the optimal descriptor is given by

(61) $l_n = \log_2(\lambda + 1) + n \log_2 \dfrac{\lambda + 1}{\lambda}$

and the maximum entropy value is

(62) $H_{max} = \log_2(\lambda + 1) + \lambda \log_2 \dfrac{\lambda + 1}{\lambda}\,.$

The overall philosophy of information theoretical inference can be illuminated by the above example. To do so, consider a dialogue between the statistician (S) and the information theorist (IT):

S: Can you help me to identify the distribution behind some interesting data I am studying?
IT: OK, let me try. What do you know?
S: All observed values are non-negative integers.
IT: What else?
S: Well, I have reasons to believe that the mean value is 2.3.
IT: What more?
S: Nothing more.
IT: Are you sure?
S: I am!
IT: This then indicates the geometric distribution.
S: What! You are pulling my leg! This is a very special distribution and there are many, many other distributions which are consistent with my observations.
IT: Of course. But I am serious. In fact, any other distribution would mean that *you would have known something more.*
S: Hmmm. So the geometric distribution is the true distribution.
IT: I did not say that. The true distribution we cannot know about.
S: But what then did you say — or mean to say?
IT: Well, in more detail, certainty comes from observation. Based on your information, the best descriptor for you, until further observations are made, is the one adapted to the geometric distribution. In case you use any other descriptor there is a risk of a higher cost.

S: This takes the focus away from the phenomenon I am studying. Instead, you make statements about my behaviour.
IT: Quite right. "Truth" and "reality" are human imaginations. All you can do is to make careful observations and reflect on what you see as best you can.
S: Hmmmm. You are moving the focus. Instead of all your philosophical talk I would like to think more pragmatically that the geometric distribution is indeed the true one. Then the variance should be about 7.6. I will go and check that.
IT: Good idea.
S: But what now if my data indicate a different variance?
IT: Well, then you will know something more, will you not? And I will change my opinion and point you to a better descriptor and tell you about the associated distribution in case you care to know.
S: But this could go on and on with revisions of opinion ever so often.
IT: Yes, but perhaps you should also consider what you are willing to know. Possibly I should direct you to a friend of mine, expert in complexity theory.
S: Good heavens no. Another expert! You have confused me sufficiently. But thanks for your time, anyhow. Goodbye!

There are interesting models which cannot be handled by Theorem 5. For some of these, a Nash equilibrium is unattainable though the value of the game exists. For these games Observer, typically, has a unique optimal strategy, say the idealized code κ^*. Further, the world associated with κ^*, P^*, is an *attractor* for Nature in the sense that any attempt to define a maximum entropy distribution must converge to P^*. One will expect that $H(P^*) = H_{\max}$ but an interesting phenomenon of *collapse of entropy* with $H(P^*) < H_{\max}$ may occur.

Models with collapse of entropy appear at a first glance to be undesirable. But this is not the case.

Firstly, for such models Nature may well have chosen the strategy P^* (even though a better match to the choice κ^* by Observer is possible). Since why should Nature be influenced by actions available for the Observer, a mere human? Thus, the circumstances do not encourage a change of strategies and may therefore be conceived as stable. A second reason why such models are interesting is that they allow approximations to the attractor at a much higher entropy level than the level of the attractor itself. This is a sign of *flexibility*. Thus, we do not only have *stability* as in more classical models but also a desirable flexibility. An instance of this has been suggested in the modeling of natural languages at the lowest semantic level, that of words, cf. [Harremoës and Topsøe, 2001; Harremoës and Topsøe, 2006].

We may summarize by saying that Nature and Observer have different roles and the game is not so much a conflict between the two players understood in the usual common sense but rather *a conflict governed by duality considerations between Observer and Observers own thoughts about Nature.*

3.4 Universal coding

Consider again the problem of coding the letters of the English alphabet. If the source is Dickens "A Tale of Two Cities" and if we consider idealized coding, we know how to proceed, viz. to adapt the idealized code to the known data as shown in Table 2. But if we want to design an idealized code so as to deal with other sources, perhaps corresponding to other types of texts, it is not so clear what to do. We shall now show how the game theoretical approach can also be used to attack this problem.

Let $P_1, \cdots, P_N$ be the distributions related to the possible sources. If we take $\{P_1, \cdots, P_N\}$ as the set of possible worlds for Nature, we have a situation of hazard similar to the scissors-paper-stone game, Example 3. We therefore randomize and take instead the set of all distributions $\alpha = (\alpha_n)_{n \le N}$ over $\{P_1, \cdots, P_N\}$ as the set $\mathcal{W}$ of possible worlds. As the set $\mathcal{D}$ of descriptors, we here find it convenient, instead of idealized codes, to consider the corresponding set of distributions. Thus, $\mathcal{D}$ is the set of all distributions Q over the alphabet. Finally, as cost function we take c defined by

$$(63)\quad c(\alpha, \kappa) = \sum_{n \le N} \alpha_n D(P_n \| Q)\,.$$

This time, the duality requirements related to (56) are satisfied due to the identity (32) which also identifies $h(\alpha)$ with a certain mutual information. Of special interest for this game is the identification of r_{min} as the *mini-max redundancy*

$$(64)\quad r_{min} = \min_{Q \in \mathcal{D}} \max_{n \le N} D(P_n \| Q)\,.$$

The identification of Nash equilibrium strategies can sometimes be based on Theorem 5 but more often one has to use a more refined approach based on (50).

3.5 Other games of information

The game theoretical approach applies in a number of other situations. Of particular interest perhaps are games where, apart from a descriptor as considered up to now, a *prior world* is also known to Observer. The goal then is to find a suitable *posterior world* and in so doing one defines an appropriate measure of the gain associated with *updating* of the prior. For these games it is thus more appropriate to work with an objective function given as a gain rather than a cost. The games indicated adopt a *Bayesian view*, well known from statistics.

3.6 Maximum entropy in physics

The word *entropy* in information theory comes from physics. It was introduced by Clausius in thermodynamics. In thermodynamics the definition is purely operational:

(65) $dS = \frac{dQ}{T}$.

It is a macroscopic quantity you can measure, which is conserved during reversible processes, but increases during irreversible processes in isolated systems. If the entropy has reached its maximum, no more irreversible processes can take place. Often one says that "entropy increases to its maximum" but the process may be extremely slow so that the validity of this statement is of limited interest. Classical equilibrium thermodynamics is only able to describe reversible processes in detail, and irreversible processes are considered as a kind of black boxes. This presents a paradox because reversible processes have speed zero and hence the entropy is constant. In practice equilibrium thermodynamics is a good approximation to many real world processes. Equilibrium thermodynamics can be extended to processes near equilibrium, which solves some of the subtleties but not all.

EXAMPLE 8. An ideal gas is enclosed in a cylinder at an absolute temperature T. The volume of the the cylinder is increased to j times the original volume using a piston, and the temperature is kept fixed. In order to measure the change in entropy, the piston should be moved very slowly. If the system had been isolated this would result in a decrease in temperature. Therefore you have to slowly add heat. This will result in a entropy increase proportional to $\ln j$.

Bolzmann and Gibbs invented statistical mechanics. In statistical mechanics one works with two levels of description. The macroscopic level corresponding to thermodynamics and the microscopic level corresponding to Newtonian (or quantum) mechanics. For instance absolute temperature (a macroscopic quantity) is identified with average kinetic energy. The main task then is to deduce macroscopic properties from microscopic ones or the other way round. This works quite well but also introduces new complications. Typically, the macroscopic quantities are identified as average values of microscopic ones. Thus thermodynamic variables that were previously considered as deterministic quantities have to be replaced by random variables. The huge number of molecules (typically of the order 10^{23}) implies that the average is close to the mean value with high probability. Boltzmann observed that

(66) $S \sim \ln(N)$

where S denotes the entropy of a macro state and N denotes the number of micro states that give exactly that macro state. Thus the maximum entropy distribution corresponds to the macrostate with the highest number of microstates. Normally one assigns equal probability to all micro states. Then the maximum entropy distribution corresponds to the most probable macro state.

EXAMPLE 9. Consider Example 8 again. In the j-fold expansion, each of the n molecules is now allowed in k times as many states as before. Therefore the difference in entropy is proportional to

(67) $\ln j^n = n \ln j$.

EXAMPLE 10. Assume that we know the temperature of a gas, hence the mean kinetic energy. The energy of a molecule is $1/2\ m\left\|v\right\|^2$ where $\|v\|$ is the length of the 3-dimensional velocity vector v. The maximum entropy distribution on velocity vectors with given mean length is a 3-dimensional Gaussian distribution. Then the probability distribution of the length $\|v\|$ is given by the Maxwell distribution with density

$$(68)\quad \frac{4\beta^{3/2}}{\pi^{1/2}}x^2e^{-\beta x^2}.$$

Often it is convenient to work with (Helmholtz) *free energy* A instead of entropy. One can prove that

$$(69)\quad A - A_{eq} = kT \cdot D\left(P\|P_{eq}\right),$$

where P is the actual state and P_{eq} is the corresponding equilibrium state. Hence the amount of information we know about the actual state being different from the equilibrium state can be extracted as energy. The absolute temperature tells how much energy can be extracted if we have one bit of information.

Jaynes introduced the maximum entropy principle as a general principle [Jaynes, 1957]. Previously, the physicists tried to explain why entropy is increasing. Jaynes turned the arguments upside down. Maximum entropy is a fundamental principle, so if we know nothing else, we better describe a system as being in the maximum entropy state. If we do not describe the system as being in its maximum entropy state this would correspond to knowing something more, cf. Section 3.3. Then, the system will be governed by the maximum entropy distribution among all distributions that also satisfy these extra conditions. In a closed thermodynamical system we only know the initial distribution. If the system undergoes a time evolution then our knowledge about the present state will decrease. Thus, the number of restrictions on the distribution will decrease and the set of feasible distributions will increase, resulting in an increase of the entropy.

3.7 Gibbs conditioning principle

Apart from the considerations of Section 3.3, there are some theorems, which support Jaynes' maximum entropy principle. Assume that we have a system which can be in one of k states. As a prior distribution on the k states we use the uniform distribution. Let X be a random variable with values in the set. Somehow we get the information that the mean value of X is λ which is different from the mean value when the uniform distribution is used. We are interested in a new distribution that takes the new information into account. Let C denote the set of feasible distributions, i.e. distributions for which the mean value of X is λ. Jaynes suggests to use the maximum entropy distribution as the new distribution.

One can also argue as follows. How can we actually know the mean value of X? Somehow we must have measured the average value of X. Consider a number

Number of eyes	Prior probability	Simulations				Max. ent. distribution
		1	10	100	1000	
1	0.167	0	0	12	102	0.103
2	0.167	0	2	14	125	0.123
3	0.167	0	2	11	147	0.146
4	0.167	1	3	15	172	0.174
5	0.167	0	0	21	205	0.207
6	0.167	0	3	27	249	0.247

Table 4. Simulation of 1, 10, 100 and 1000 outcomes of a die under the condition that the average number of eyes is exactly 4.

of independent identically distributed variables $X_1, X_2, ..., X_n$. Consider the set of events such that

$$(70)\quad \frac{X_1 + X_2 + ... + X_n}{n} = \lambda.$$

Now consider the distribution of X_1 given that (70) holds. If n is large, then the distribution is close to the maximum entropy distribution. This result is called the *conditional limit theorem*, *Gibbs conditioning principle* or *the conditional law of large numbers*.

EXAMPLE 11. The mean number of eyes on a regular die is 3.5. Take a large number of dice and throw them. Assume that the average number of eyes in the sample is 4 and not 3.5 as expected. If one counts the number of ones, twos, etc. then with high probability the relative frequency of the different outcomes will be close to the maximum entropy distribution among all distributions on the set $\{1, 2, 3, 4, 5, 6\}$ for which the mean value is 4 (see Table 4).

EXAMPLE 12. Assume that all velocity vectors of n molecules are equally probable. Let v_i denote the velocity of molecule i. Then the mean kinetic energy is proportional to

$$(71)\quad \frac{1}{n}\sum \|v_i\|^2 .$$

We can measure the mean kinetic energy as the absolute temperature. Assume that we have measured the temperature. If n is huge as in macroscopic thermodynamic systems then the probability distribution of $\|v_1\|$ is approximately the Maxwell distribution.

Example 11 can be used to analyze to which extent our assumptions are valid. The first condition is that the uniform distribution is used as prior distribution. Hence we cannot use the maximum entropy principle to argue in favor of the uniform distribution. Some symmetry considerations are needed in order to single out the uniform distribution at first hand. Next, according to our prior distribution it is highly unlikely to observe that the empirical average is 4. From a classical statistical point of view one should use the high value of the average to reject the uniform distribution, but if the uniform distribution is rejected as being false then we will not be able to calculate the a posteriori distribution. Hence if the conditional limit theorem is used as an argument in favor of the maximum entropy principle then we are forced to use a Bayesian interpretation of the prior probability distribution. Many physicists find this problematic. Thermodynamic entropy increases, they argue, independently of how we assign prior distributions of the system.

In order to single out the physical problems from the statistical ones, the concept of *sufficiency* is useful. Consider an ideal gas in an isolated container of a specific volume. At equilibrium the gas can be described by the number of molecules and the temperature. Using the maximum entropy formalism we can calculate for instance the velocity distribution and all other quantities and distributions of interest. We say that the number of molecules and the temperature are *sufficient*. Then one may ask: "why are number and temperature sufficient?" If the container has an isolating division we have to know the number of molecules and the temperature on each side of the division, and four numbers will be sufficient in this case. Only the experienced physicists should be able to tell which statistics are sufficient for the specific setup. Thus, we can formulate the following result:

The maximum entropy principle may be used as a general formalism, but it tells little or nothing about which statistics are sufficient.

The conditional limit theorem can also be formulated for a prior distribution different from the uniform distribution. Consider a distribution P and a (mathematically well behaved) set C of probability distributions. Then the probability of observing the empirical distribution in C satisfies

$$(72)\quad P^n(C) \leq 2^{-nD(Q\|P)}$$

where Q is the information projection of P into C, i.e. the distribution Q in C that minimizes the divergence $D(Q\|P)$. Furthermore there is a high probability that the empirical distribution is close to Q given that it belongs to C. If P is the uniform distribution then the information projection equals the maximum entropy distribution.

3.8 Applications in statistics

Statistical analysis is based on *data* generated by random phenomena. Actual data are used to make *inference* about the statistical nature of the phenomena studied. In this section we assume that $X_1, X_2, \cdots, X_n$ are independent random variables, distributed according to a common, unknown *law* (probability distribution) Q.

Assume that Q is discrete with point probabilities $q_1, q_2, \cdots, q_m$. If the *observed frequencies* in a sample ω of size n are $n_1, n_2, \cdots, n_m$, then the *empirical distribution of size* n, $Emp_n(\omega)$, is the distribution with point probabilities $\frac{n_1}{n}, \frac{n_2}{n}, \cdots, \frac{n_m}{n}$. The *likelihood ratio*, a quantity of central importance in statistics, is the ratio between the probability of the actually observed data, measured with respect to $Emp_n(\omega)$, respectively the theoretical distribution Q. For the *log-likelihood ratio* we find the expression

$$(73)\ \ln \frac{(\frac{n_1}{n})^{n_1} \cdot (\frac{n_2}{n})^{n_2} \cdots (\frac{n_m}{n})^{n_m}}{q_1^{n_1} \cdot q_2^{n_2} \cdots q_m^{n_m}}$$

which we easily recognize as n times the information divergence $D(Emp_n(\omega)\|Q)$. This simple observation is indicative of the relevance for statistics of information theory, especially regarding the concept of information divergence.

Let us have a closer look at . Typically, the statistician considers two hypothesis, denoted H_0 and H_1, and called, respectively, the *null hypothesis* and the *alternative hypothesis.* In classical statistics these hypothesis are treated quite differently. According to Karl Popper, one can never verify a hypothesis. Only falsification is possible. Therefore, if we want to give statistical evidence for an alternative hypothesis — typically that something "special" is going on, the coin is irregular, the drug has an effect or what the case may be — one should try to falsify a suitably chosen null hypothesis, typically expressing that everything is "normal" .

Consider a test of the alternative hypothesis H_1 against the null hypothesis H_0. In order to decide between H_0 and H_1, the statistician chooses a partition of the simplex of all probability distributions over the possible outcomes into two classes, A_0 and A_1, called *acceptance regions*. If the observed empirical distribution $Emp_n(\omega)$ belongs to A_0, one accepts H_0 (or rather, one does not reject it) whereas, if $Emp_n(\omega) \in A_1$, one rejects H_0 (and, for the time being, accepts H_1).

The acceptance regions generate in a natural way a decomposition of the n-fold sample space of possible sequences $\omega = (x_1, x_2, \cdots, x_n)$ of observed values of $X_1, X_2, \cdots, X_n$. The sets in this decomposition we denote by A_0^n and A_1^n. For example, A_0^n consists of all $\omega = (x_1, x_2, \cdots, x_n)$ for which $Emp_n(\omega) \in A_0$.

A *type-I error* occurs when you accept H_1 though H_0 is true (everything is "normal") and a *type-II error* occurs when you accept H_0 though H_1 is true (something "special" is happening).

In case H_0 and H_1 are both *simple*, i.e. of the form $H_0 : Q = P_0$ and $H_1 : Q = P_1$ with P_0 and P_1 fixed, known distributions, we can use the product distributions P_0^n and P_1^n to calculate the *error probabilities*, i.e. the probabilities of a type-I, respectively a type-II error. With natural notation for these error probabilities, we find the expressions

$$(74)\ Pr(A_1|H_0) = P_0^n(A_1^n)\,,\ Pr(A_0|H_1) = P_1^n(A_0^n)\,.$$

The quantity $Pr(A_1|H_0)$ is called the *significance level* of the test and $1 - Pr(A_0|H_1)$ the *power* of the test.

Under the simplifying assumptions we have made, the *Neyman-Pearson Lemma* often leads to useful tests. To formulate this result, consider, for any $t \geq 0$, the test defined by the region

$$(75)\ A_1 = \{P|D(P\|P_1) \leq D(P\|P_0) + t\}$$

as acceptance region of H_1. Then this test is a *best test* in the sense that any other test at the same (or lower) significance level has power at most that of this special test.

Hypothesis testing is used to gradually increase ones knowledge about some stochastic phenomenon of interest. One starts with a null hypothesis everyone can accept. Then, as one gains experience through observation, one reconsiders the hypothesis and formulates an alternative hypothesis. If, some day, the null hypothesis is falsified, you take the alternative hypothesis as your new null hypothesis. The process is repeated until you find that you have extracted as much information about the nature of the phenomenon as possible, given the available time and resources.

Note the significance of quantitative information theory as a guide in the subtle process of selection and falsification of hypothesis until you end up with a hypothesis you are either satisfied with as final expression of your knowledge about the phenomenon or else you do not see how to falsify this hypothesis, given the available resources.

We now turn to more subtle applications of information divergence. We consider fixed hypothesis $H_0 : Q = P_0$ and $H_1 : Q = P_1$ (with $D(P_0\|P_1) < \infty$) and a series A_n of acceptance regions for H_0. The index n indicates that testing is based on a sample of size n. Then, for mathematically well behaved regions,

$$(76)\quad Pr(A_n|H_1) \leq \exp(-nD(Q_n\|P_1))$$

where Q_n is the information projection of P_1 on A_n. This upper bound on the type-II error probability is asymptotically optimal for a fixed significance level. Indeed, if all tests are at the same significance level, then

$$(77)\quad \lim_{n\to\infty} -\frac{1}{n} Pr(A_n|H_1) = D(P_0\|P_1)$$

as illustrated in Figure 4.

Note that this limit relation gives an interesting interpretation of information divergence in statistical terms. The result was found by Chernoff [1952], but is normally called *Stein's Lemma.* In 1947 Wald [1947] proved a similar but somewhat weaker result. This was the first time information divergence appeared, which was one year before Shannon published his basic paper and five years before Kullback and Leibler defined information divergence as an independent quantity.

Among other applications of information theoretical thinking to statistics, we point to the *Minimum Description Length principle* (MDL), due to J. J. Rissanen [1978], which is a variant of the principle that among different possible descriptions one shall choose the shortest one. Thus the parameters in a statistical model shall be chosen such that coding according to the resulting distribution gives the shortest total length of the coded message. So far all agree. The new idea is to incorporate not only the data but also the description of the statistical model. In general, a model with three parameters will give a better description than a model with only two parameters. On the other hand the three-parameter model

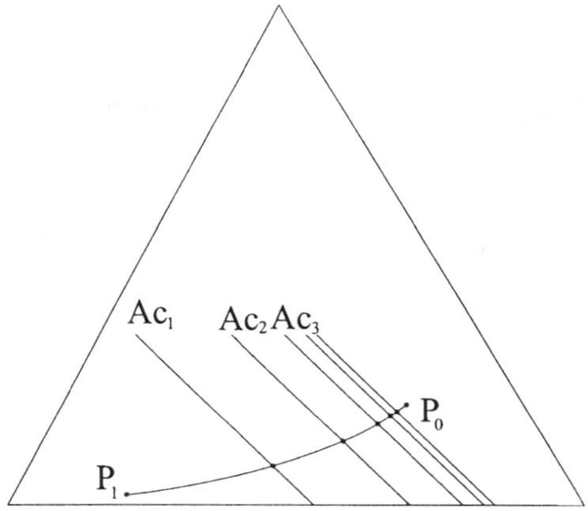

Figure 4. Decreasing sequence of acceptance regions in the probability simplex.

is more complicated, so there is a trade-off between complexity of the model and the coding of the data according to the model.

A simple and well-known example is the description of a single real parameter. How many digits shall be given? A rule of thumb states that the uncertainty shall be at the last digit. The Minimum Description Length principle tries to justify or to modify such rules.

We refer to [Csiszár and Shields, 2004] for a review of the relations to statistics and further references. The most thorough treatment of the Minimum Description Length principle in statistics can be found in [Grünwald, 2007].

3.9 Law of Large Numbers and Central Limit Theorems

Inequality (76) states that the probability of observing an empirical distribution far from the theoretical distribution is small. As a consequence we immediately get a Law of Large Numbers:

THEOREM 13. *Let P be a probability distribution. Let A be a convex set of probability distributions not containing P. Then the probability that the empirical distribution belongs to A converges to zero when the number of observations tends to infinity.*

We can also formulate this result for random variables.

THEOREM 14. *Let $X_1, X_2, \ldots$ be a sequence of independent and identically distributed random variables. Assume that X_i has mean value μ. Then if n is chosen sufficiently large,*

$$(78) \quad \frac{X_1 + X_2 + \ldots + X_n}{n}$$

is close to μ with high probability.

Inequality (76) gives more. The probability of getting a deviation from the mean decreases exponentially. Therefore the sum of the probabilities of deviations is finite. This has important applications. Let A be a set of probability measures such that $D(Q\|P) \geq 1/2$ for all $Q \notin A$. Then the probability that the empirical distribution belongs to A is upper bounded by $1/2^n$. The probability that at least one of the empirical distributions belong to A for $n \geq N$ is upper bounded by

$$\frac{1}{2^N} + \frac{1}{2^{N+1}} + \frac{1}{2^{N+1}} + \dots = \tfrac{1}{2^N}\left(1 + \tfrac{1}{2} + \tfrac{1}{4} + \dots\right) = 2 \cdot \tfrac{1}{2^N}. \tag{79}$$

If N is large then this is small. The law of large numbers states that there is a high probability that $Emp_N(\omega) \in A$, but we even have that there is a high probability that $Emp_n(\omega) \in A$ for all $n \geq N$. Thus most sequences will never leave A again. This is formulated as the *strong law of large numbers*:

THEOREM 15. *Let P be a probability distribution. Then the empirical distribution converges to P with probability one.*

For random variables the theorem states that:

THEOREM 16. *Let $X_1, X_2, \dots$ be a sequence of independent and identically distributed random variables. Assume that X_i has mean value μ. Then*

$$\frac{X_1 + X_2 + \dots + X_n}{n} \tag{80}$$

converges to μ with probability one.

We have seen that $\frac{X_1+X_2+\dots+X_n}{n}$ is close to μ with high probability. Equivalently,

$$\frac{(X_1-\mu) + (X_2-\mu) + \dots + (X_n-\mu)}{n} \tag{81}$$

is close to zero. If we divide with a number smaller than n we get a quantity not as close to zero. In order to keep the variance fixed we divide by $n^{1/2}$ instead. Put

$$S_n = \frac{(X_1-\mu) + (X_2-\mu) + \dots + (X_n-\mu)}{n^{1/2}}. \tag{82}$$

Thus $E(S_n) = 0$ and $Var(S_n) = Var(X_1)$. Let P_n be the distribution of S_n. Let Φ denote the distribution of a centered Gaussian random variable. The differential entropy of P_n satisfies

$$h(P_n) = h(\Phi) - D(P_n\|\Phi). \tag{83}$$

Thus we see that the differential entropy of P_n is less than or equal to the differential entropy of the Gaussian distribution. The Central Limit Theorem in its standard formulation states that P_n converges to a Gaussian distribution.

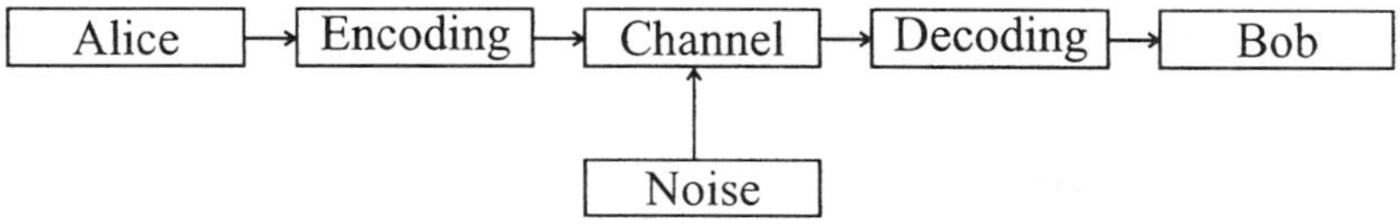

Figure 5. Shannon's model of a noisy channel.

THEOREM 17. *If there exists n such that $h(P_n) < \infty$ then $h(P_n)$ increases and converges to its maximum, which equals $h(\Phi)$. Equivalently, $D(P_n \| \Phi)$ decreases to zero.*

In this formulation the Central Limit Theorem corresponds to the second law of thermodynamics, which states that the entropy of a physical system increases and converges to its maximum. Here the variance turns out to be sufficient. We see that addition of random variables gives a "dynamics" which supports the maximum entropy principle in that it explains a mechanism behind entropy increase. It turns out that all the major theorems of probability theory can be formulated as maximum entropy results or minimum information divergence results.

4 IS CAPACITY ONLY USEFUL FOR ENGINEERS?

4.1 Channel coding

We consider a situation where Alice sends information to Bob over a noisy information channel. Alice attempts to encode the information in such a way that it is tolerant to noise, yet at the same time enabling Bob to recover the original message.

A simple error-correcting protocol is to send the same message several times. If the message is sent three times and a single error has occurred, then two of the received messages are still identical and Bob concludes that these must be identical to the original message. Another simple protocol is possible when feedback is allowed. Alice sends the message. Bob sends the received message back again. If Alice receives what she sent, she can be quite certain that Bob received the original message without error, and she can send a new message. If she receives a different message from the one sent, she sends the original message again. These protocols are simple but they are not always efficient. More complicated codes are possible.

EXAMPLE 18. In this example a message consisting of three bits is encoded into seven bits. Let X_1, X_2 and X_3 be the three bits. We shall use the convention that $1 + 1 = 0$. Put

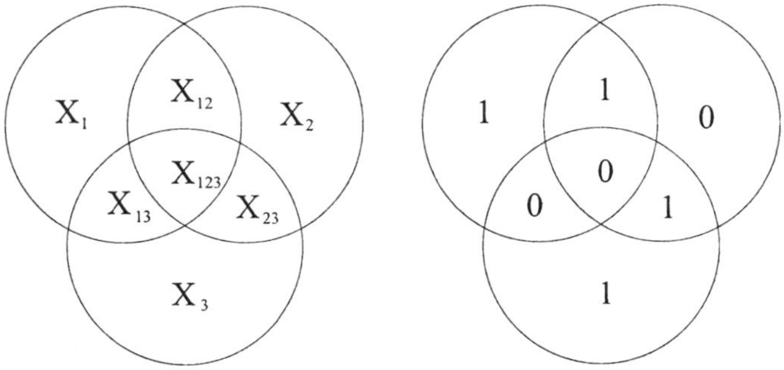

Figure 6. The code in Example 18 is constructed such that the sum of bits inside any circle is zero. The right diagram corresponds to the codeword 101.

$$
\begin{aligned}
X_{12} &= X_1 + X_2 \\
X_{23} &= X_2 + X_3 \\
X_{13} &= X_3 + X_1 \\
X_{123} &= X_1 + X_2 + X_3.
\end{aligned} \tag{84}
$$

See Figure 6. Now transmit the code-word $X_1X_2X_3X_{12}X_{23}X_{13}X_{123}$. If the received code-word $Y_1Y_2Y_3Y_{12}Y_{23}Y_{13}Y_{123}$ is identical with $X_1X_2X_3X_{12}X_{23}X_{13}X_{123}$, then the received code-word satisfies the following parity check equations

$$
\begin{aligned}
Y_1 + Y_{12} + Y_{13} + Y_{123} &= 0 \\
Y_2 + Y_{12} + Y_{23} + Y_{123} &= 0 \\
Y_3 + Y_{13} + Y_{23} + Y_{123} &= 0.
\end{aligned} \tag{85}
$$

If a single bit has been corrupted then one or more of the parity check equations will not hold. It is then easy to identify the corrupted bit and recover the original message. Indeed, as the reader will realize, this can be done by considering the faulty equations and the domains they represent.

4.2 Capacity

Let $X \in \mathbb{A}$ denote the input to an information channel and $Y \in \mathbb{B}$ the output. Then, if X can be (almost perfectly) reconstructed from the output Y, $H(X \mid Y)$ is small and then by (15),

$$H(X) \approx I(X;Y). \tag{86}$$

Hence, if Alice wants to send a lot of information through the information channel she wants $I(X;Y)$ to be big. Alice can choose which input symbols to send frequently and which to send less frequently. As Alice controls the distribution of the input letters, we define the *capacity* C of the information channel to be the maximum of the mutual information $I(X;Y)$ over all distributions on the input letters.

Consider the *binary symmetric channel* where $\mathbb{A} = \mathbb{B} = \{0, 1\}$ and where $Q(Y = 1 \mid X = 0) = Q(Y = 0 \mid X = 1) = \varepsilon \in [0; 1/2]$ is called the transmission error. Here, the *uniform input distribution* $P^*(0) = P^*(1) = \frac{1}{2}$ is optimal and the capacity is

$$\begin{aligned} C &= \log_2 2 - H(Q(\cdot \mid 0)) = \log_2 2 - H(Q(\cdot \mid 1)) \\ &= D((\varepsilon, 1-\varepsilon) \,\|\, (1/2, 1/2)) \,. \end{aligned} \tag{87}$$

As is natural, capacity is the largest, 1 bit, if $\varepsilon = 0$ and the smallest, 0 bits, if $\varepsilon = \frac{1}{2}$.

We note that determination of the capacity of a channel can be viewed as a game. In fact, the game is identical — but with different interpretations and emphasis — to the game related to universal coding, cf. Section 3.4.

Before Shannon most people believed that a lot of redundancy or feedback is needed in order to ensure a high probability of correct transmission. Shannon showed that this is not the case.

THEOREM 19 (Second main theorem of information theory). *If X is an information source and $H(X) < C$ then the source can be transmitted almost perfectly if the channel is used many times and complicated coding schemes are allowed.*

Shannon also showed that feedback does not increase capacity. In order to prove the theorem Shannon introduced the concept of *random coding* where code-words are assigned to messages at random. A code-book containing all these code-words is enormous, and Alice has to provide Bob with the code-book before the transmission starts. A lot of bits are thus used just to transmit the code-book, but Alice only needs to transmit the code-book once. Therefore, even if a large code-book is used and this code-book saves just one bit compared to a simpler code-book then, if sufficiently many transmissions are performed, the saved bits will exceed the number of extra bits in the big code-book. Since Shannon published the second main theorem of information theory it has been a challenge to construct codes which are both simple and efficient.

It turns out that the repetition code is inefficient except if the capacity is very small. It also turns out that feedback does not increase capacity. One may ask why these codes are so widely used when they, according to the first main theorem of information theory, are inefficient. Actually, Shannon-type coding does not seem to be used among humans or animals. Instead much more primitive codes are used. There are, apparently, several reasons for this.

The first is that efficient coding is complicated. Thus efficient coding schemes will only evolve if there is a high selective pressure on efficient communication. Often there is a high selective pressure on getting the message across, but if the transmission cost is low there is no reason to develop sophisticated coding schemes. It is known that the simple coding schemes are efficient in a very noisy environment, so if there is uncertainty about the actual noise level it may be better to be on the safe side and transmit “too much”.

The human language is highly structured. In logic, semantics and linguistics one studies the relation between the more formal structures inside the language and the world outside the language. Many grammatical structures work to some extent as a kind of error correction in the language (but may have other functions as well). But we know that it is very hard to learn a language with a complicated grammar. If the language used some of the coding techniques used by engineers, a lot of new "grammatical rules" had to be introduced. In a sentence like "The man has a box" the word "man" can be replaced with "woman", "boy", "girl", "friend" etc. and the word "box" can, independently, be replaced by "ball", "pen", "stick" etc. Each of the sentences would make sense and correspond to a scenario which is true or false, possible or impossible, probable or improbable. In our simple example the sentence may be compressed to "man box" and we can still replace the words and recover the original structure. If the sentence was coded using Shannon coding there would not be the same possibility of restructuring the sentence, because error correcting codes introduce dependencies which were not there before. In this sense:

Data compression emphasize structure, and channel coding smudges structure.

4.3 Transmission of quantum information

The key to the success of Shannon's theory lies to a great extent in the quantitative results regarding possibilities for faithful transmission of classical information. When we turn to the similar problems regarding transmission of quantum information, new phenomena occur. Technically, it is even difficult to get started on the investigations as it is not clear what the proper notion of a channel should be in the quantum setting. This concerns questions about the type of input and output allowed (classical and/or quantum), the necessary attention to the handling of sequential input (where entanglement has to be taken into consideration) and finally, it concerns questions about feedback.

Considering the various options, one is lead to more than twenty different types of quantum channels, and even for the simplest of these, basic properties are not yet fully developed[3]. The many possibilities indicate that quantum information theory is not just a simple extension of the classical theory. For instance, when sender and receiver in a quantum communication share an EPR-pair, then, though this in itself cannot be used for transfer of information, it can facilitate such transfer and raise the capacity of the communication system. Thinking about it, such new possibilities raise qualitative philosophical questions about the nature of information. New emerging ideas, which are only partly developed today, may well change our understanding of the very concept of information in radical ways.

5 MULTI-USER COMMUNICATION

In the kind of problems we have discussed information is something Alice sends to Bob. Thus there have only been *one sender* and *one receiver*. In many situations

[3]This concerns, in particular, the so-called *additivity conjecture* related to properties of one of the notions of *quantum capacity*.

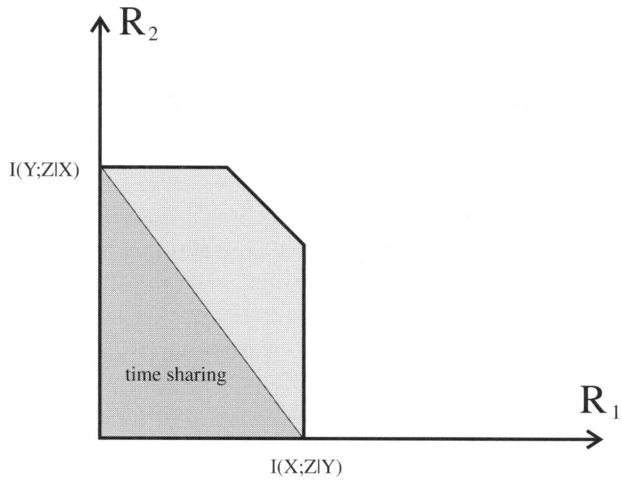

Figure 7. Capacity region of multiple access channel.

there are more senders and receivers at the same time. A television signal is sent from an antenna to a large number of receivers. This is a so-called *broadcast system*. In a *multiple access system* there are many senders and only one receiver. An example of a multiple access system is a class room where the teacher wants some information from the pupils. If all pupils speak at the same time the teacher will just receive a lot of noise. Timesharing, a common solution to this problem, dictates that one pupil speaks at a time. An important example of a multi-user system is the internet where the servers send signals to each other. Timesharing for the whole internet is possible but very inefficient. The main problem of multi-user information theory is to find more efficient protocols than timesharing, and to determine theoretical bounds on the efficiency of the protocols. A special example of a multiuser system is a cryptographic system where Alice sends a message to Bob, but a second potential receiver is Eve who wiretaps the system or tries to disturb the message.

The engineers have developed many sensible protocols, but there are only few theoretical results, so, in general, it is not known if the protocols are optimal. Here we shall describe some well understood problems and indicate the more general ones. We shall see the kind of results one may dream of for more complicated systems.

5.1 The multiple access channel

Consider a noisy multiple access channel with two senders. The senders send variables X and Y and the receiver receives a variable Z. The channel is given in the sense that we know the distribution of Z given the input (X, Y). Consider a specific input distribution on (X, Y). We are interested in which pairs (R_1, R_2) have the property that Sender 1 can send at rate R_1 and Sender 2 can send at rate R_2. Assume that Sender 1 and the receiver knows Y. Then Sender 1 can send information at a rate

(88) $R_1 \leq I(X; Z \mid Y)$,

which gives the rate pair $(I(X; Z \mid Y), 0)$. If X is known to Sender 2 and to the receiver then Sender 2 can send information at a rate

(89) $R_2 \leq I(Y; Z \mid X)$,

which gives the rate pair $(0, I(Y; Z \mid X))$. By timesharing, the senders can send at rates which are combinations of $(I(X; Z \mid Y), 0)$ and $(0, I(Y; Z \mid X))$. But one can achieve a better performance. If the two senders both know X and Y they can send at rate

(90) $R_1 + R_2 \leq I((X, Y); Z)$.

It turns our that the three conditions (88), (89) and (90) are necessary and sufficient for the rate pair to be achievable.

Therefore, correlated variables can be sent over a multiple access channel if and only if the compression region and the capacity region intersect. In order to achieve a rate pair in this intersection, the source coding should be adapted to the channel and the channel coding should be adapted to the correlations in the source. Thus source and channel coding cannot be separated in multi user information theory.

5.2 *Network coding*

We shall start with an example. Consider a network with two senders A_1 and A_2 and two receivers B_1 and B_2 and intermediate nodes C and D as illustrated in Figure 9. Assume that A_1 want to send one bit of information to B_2 and A_2 wants to send one bit of information to B_1. Assume that each edge has capacity one bit. If A_1 sends her bit along the path A_1CDB_2 then it is not possible at the same time for A_2 to send her bit along the path A_2CDB_1. The solution is that A_1 sends her bit to B_1 and C, and A_2 sends her bit to B_2 and C. Then C should send the sum of the received bits to D, which should send the received bit to B_1 and B_2. Now, B_1 will be able to reconstruct the bit sent from A_2 from the two received bits and, similarly, B_2 will be able to reconstruct the message sent from A_1. This is the simplest example of *network coding*, and was given in [Ahlswede *et al.*, 2000].

Since 2000 many remarkable results in network coding have been obtained. The theory works well as long as the noise is by deletions, i.e. a symbol can disappear during transmission, but it cannot be altered. A simple protocol is obtained when each node transmits a random mixture of the received signals. The original message is reconstructed by comparing the received mixed noisy signals. If transmission of a message from one node to another is possible by any protocol,

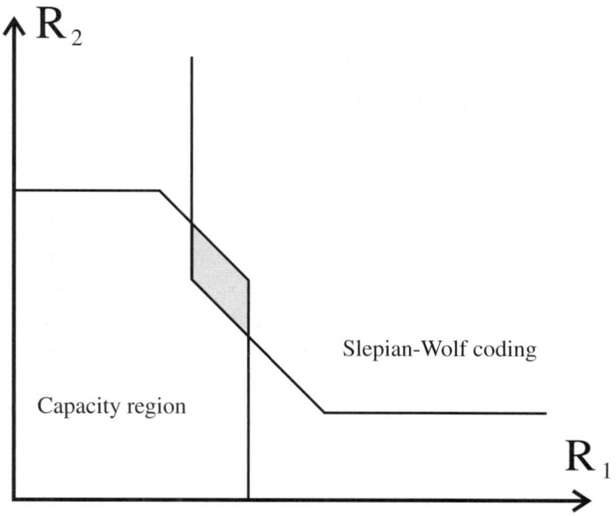

Figure 8. Intersection of capacity and compression region.

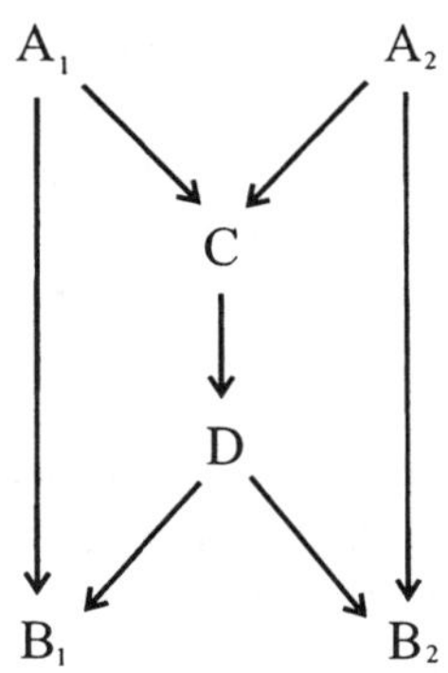

Figure 9. Network where network coding makes it possible for A_1 to send one bit to B_2 and for A_2 to send one bit to B_1 though each of the edges only has capacity one bit.

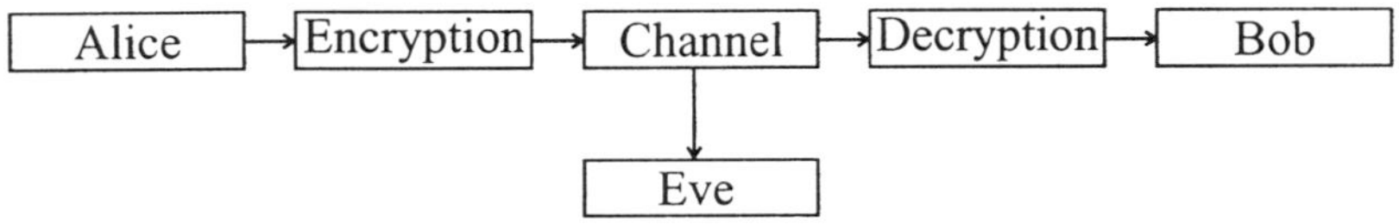

Figure 10. Channel with an eavesdropper.

then it is also possible with this simple random protocol, if the transmission is repeated sufficiently many times. These new results should both have practical and philosophical implications.

A review of the subject and further references can be found in [Yeung *et al.*, 2005].

5.3 Broadcast problems

In a broadcast system there is one sender and a number of receivers. The broadcast problem is to determine the capacity region, assuming the distributions of the received signals given the sent signal are known. There would be a tremendous number of applications of such a result, and therefore it is considered as one of the major open problems in information theory.

A special kind of broadcast system is an *identification system.* An example is call-outs in an airport. There is a special message for a single passenger. The speaker can address the message to all passengers, but this is clearly inefficient because most passengers are not interested. Therefore the speaker starts saying "A message for Mr. Bob Johnson..." After hearing this introduction all passengers except Bob Johnson can choose not to listen to the last part. The speaker may even choose to say "Mr. Bob Johnson, please, go to the information desk". If there is a lot of noise the speaker may choose to repeat the sentence or introduce error-correction by some other method. This is called an *identification problem*, because the main problem is to identify who should receive the message. One may argue that this is not transmission of information. First of all there is no message in the ordinary sense. Secondly it is hard to call the passenger Mrs. Alice Brown a receiver. After hearing the word "Mr." she knows that there is no reason to the listen to the rest. The situation is sometimes termed *information transfer* rather than information transmission.

5.4 Cryptography

Consider a crypto-system where Alice wants to send a message to Bob but at the same time she wants to prevent an eavesdropper Eve from picking up the message. This can sometimes be done if Alice and Bob shares a secret code-word, Z, called the *key*. Using the key Z, Alice encrypts the *plain text* X into a *cipher text* Y.

For this to work consider the following three conditions:

1. X is independent of Z,
2. Y is determined by X and Z,
3. X is determined by Y and Z.

The first condition is that the key is chosen independently of the message Alice wants to communicate to Bob. The second condition is the possibility of encryption and the third condition is the possibility of decryption.

A crypto-system is said to be *unconditionally secure* if X is independent of Y, i.e. knowledge of the cipher-text gives no information about the plain-text.

EXAMPLE 20 (The one-time pad). Consider a plain-text $X_1X_2...X_n$ of bits. Alice and Bob share a secret key $Z_1Z_2...Z_n$ consisting of bits generated in such a way that the bits are independent and each of them with a uniform distribution. Alice constructs a cipher-text $Y_1Y_2...Y_n$ by adding the key, i.e. by putting $Y_j = X_j + Z_j$. Here she uses the convention that $1+1=0$. Bob decrypts the received cipher-text by subtracting the key. Here he uses the convention that $0-1=1$. Thus Bob recovers the plain-text. Remark that with the conventions used adding a key or subtracting the key gives the same result. The method is called the one-time pad because each bit in the key is used only once during the encryption procedure.

The one-time pad requires very long keys. If a file of size 1 Gb has to be encrypted the key has to be 1 Gb as well. One may ask if a key can be used in a more efficient way such that shorter keys can be used.

Various inequalities can be derived from these conditions. The most important is the following result:

THEOREM 21. *For an unconditionally secure crypto system,* $H(X) \leq H(Z)$ *where* X *denotes the plain text and* Z *the key.*

If $H(X)$ is identified with the length of the (compressed) plain text and $H(Z)$ is identified with the length of the (compressed) key, we see that the key must be at least as long as the plain-text if we want unconditional security. In everyday life much shorter keys and passwords are used. The theorem shows that they cannot be unconditionally secure. If Eve had a sufficiently strong and fast computer she would in principle be able to recover most of the plain-text from the cipher-text. This was exactly what happened to the ciphers used during the second world war. When modern ciphers using short keys are said to be (conditionally) secure there is always a condition/assumption that the eavesdropper has limited computational power.

One of the most important problems in implementing cryptographic systems is key distribution as it involves both technical and social problems.

Both Alice and Bob have to know the key, but it shall be secret to Eve. Hence we have to introduce a secret channel used to send the key to Bob. This may for instance be a courier. Then Theorem 21 states that the amount of secret information that Alice can send to Bob is bounded by the capacity of the secret channel.

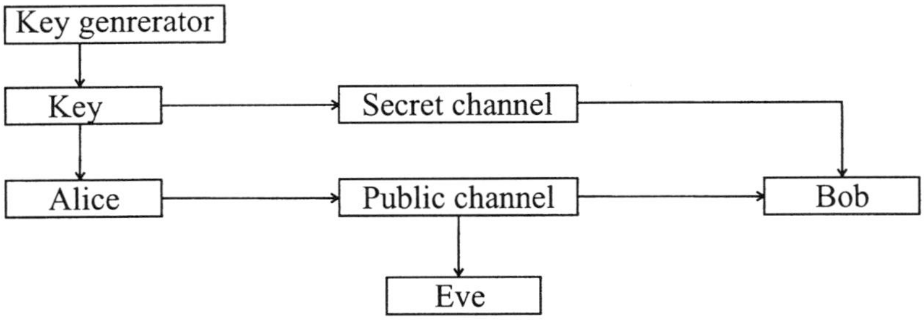

Figure 11. A crypto system with a public and a secret channel.

This kind of thinking may be extended to scenarios where the information channels are noisy and Eve is only able to wiretap part of the communication between Alice and Bob. We are interested in how many secret bits Alice is able to transmit to Bob and we can define the least upper bound as the secrecy capacity of the system. Even in systems involving only three users there are open mathematical problems.

6 CONCLUSIONS

The quantitative theory of information as developed by Shannon and his successors, provides powerful tools that allow modeling of a wide range of phenomena where information in one sense or another plays the central role. Modeling is rooted in interpretations, which captures basic philosophical aspects of information. This is especially apparent in the duality between truth and description, which we have put much emphasis on.

Duality allows you to switch back and forth between modeling based on distributions and modeling based on codes. Though formally a one-to-one correspondence, the importance lies in the asymmetries, and the different points of view attached to the two possibilities. This interplay is important technically as well as for a proper understanding.

A technical development of information theory is under way, which will put concepts related to uncertainty, information and knowledge on a more firm theoretical footing and, apart from the philosophical impact, this is believed to result in a change of paradigm and a better understanding of certain parts of science, especially probability theory and statistics.

BIBLIOGRAPHY

[Ahlswede *et al.*, 2000] R. Ahlswede, N. Cai, S.-Y. R. Li, and R. W. Yeung. Network information flow. *IEEE Trans. Inform. Theory*, IT-46:1204–1216, 2000.

[Amari and Nagaoka, 2000] S. Amari and H Nagaoka. *Methods of Information Geometry*, volume 191 of *Translations of Mathematical monographs.* Oxford University Press, 2000.

[Amari, 2002] S. I. Amari. Information geometry of statistical inference - an overview. In *Proceedings of 2002 IEEE Information Theory Workshop, Bangalore*, pages 86–89, 2002.

[Arimoto, 1972] S. Arimoto. An algorithm for computing the capacity of arbitrary discrete memoryless channels. *IEEE Trans. Inform. Theory*, 18:14–20, 1972.

[Arndt, 2001] C. Arndt. *Information Measures.* Springer, Berlin, 2001.

[Aubin, 1993] J. P. Aubin. *Optima and equilibria. An introduction to nonlinear analysis.* Springer, Berlin, 1993.

[Barron *et al.*, 1998] A. Barron, J. Rissanen, and B. Yu. The minimum description length principle in coding and modeling. *IEEE Trans. Inform. Theory*, 44:2743–2760, 1998.

[Barron, 1986] A. R. Barron. Entropy and the Central Limit Theorem. *Annals Probab. Theory*, 14(1):336–342, 1986.

[Blahut, 1972] R. E. Blahut. Computation of channel capacity and rate-distortion functions. *IEEE Trans. Inform. Theory*, 18:460–473, 1972.

[Boltzmann, 1872] L. E. Boltzmann. Weitere Studien ber das Wrmegleichgewicht unter Gasmoleklen; *Wiener Berichte*; vol.66; p.275-370; 1872.

[Cambell, 1965] L. L. Cambell. A coding theorem and Rényi's entropy. *Informat. Contr.*, 8:423–429, 1965.

[Chernoff, 1952] H. Chernoff. A measure of asymptotic efficiency for tests of a hypothesis based on a sum of observations. *Ann. Math. Statist.*, 23:493–507, 1952.

[Clausius, 1865] R. Clausius. *The Mechanical Theory of Heat - with its Applications to the Steam Engine and to Physical Properties of Bodies*; London: John van Voorst; 1865.

[Cover and Thomas, 1991] T. Cover and J. A. Thomas. *Elements of Information Theory.* Wiley, 1991.

[Csiszár and Körner, 1981] I. Csiszár and J. Körner. *Information Theory: Coding Theorems for Discrete Memoryless Systems.* Academic, New York, 1981.

[Csiszár and Shields, 2004] I. Csiszár and P. Shields. *Information Theory and Statistics: A Tutorial.* Foundations and Trends in Communications and Information Theory. Now Publishers Inc., 2004.

[Csiszár, 1975] I. Csiszár. I-divergence geometry of probability distributions and minimization problems. *Ann. Probab.*, 3:146–158, 1975.

[Csiszár, 1995] I. Csiszár. Generalized cutoff rates and Rényi information measures. *IEEE Trans. Inform. Theory*, 41(1):26–34, Jan. 1995.

[Dembo and Zeitouni, 1993] A. Dembo and O. Zeitouni. *Large Deviations Techniques and Applications.* Jones and Bartlett Publishers International, Boston, 1993.

[Einstein *et al.*, 1935] A. Einstein, B. Podolsky and N. Rosen. Can Quantum-Mechanical Description of Physical Reality Be Considered Complete? *Phys. Rev.*; vol. 47; p.777-780; 1935.

[Gibbs, 1961] J.W. Gibbs. *On the Equilibrium of Heterogeneous Substances*; 1876 and 1878; reprinted in Scientific papers, vol. 1, p.184ff; New York, Dover; 1961

[Goldie and Pinch, 1991] C. M. Goldie and R. G. E. Pinch. *Communication Theory.* Cambridge University Press, Cambridge, 1991.

[Grünwald and Dawid, 2004] P. D. Grünwald and A. P. Dawid. Game theory, maximum entropy, minimum discrepancy, and robust Bayesian decision theory. *Annals of Mathematical Statistics*, 32(4):1367–1433, 2004.

[Grünwald, 2007] P. Grünwald. *the Minimum Description Length principle.* MIT Press, 2007.

[Harremoës and Topsøe, 2001] Peter Harremoës and Flemming Topsøe. Maximum entropy fundamentals. *Entropy*, 3(3):191–226, Sept. 2001.

[Harremoës and Topsøe, 2006] P. Harremoës and F. Topsøe. Zipf's law, hyperbolic distributions and entropy loss. In R. Ahlswede et al., editor, *Information Transfer and Combinatorics*, volume 4123 of *Lecture Notes in Computer Science*, pages 788–792. Springer-Verlag, Berlin Heidelberg, 2006.

[Holevo, 2002] A. S. Holevo. *An introduction to quantum information theory.* MCCME (Publishing House of Moscow Independent University), Moscow, 2002.

[Huffman, 1952] D. A. Huffman. A method for the construction of minimum redundancy codes. In *Proc. IRE 40*, pages 1098–1101, 1952.

[Jaynes, 1957] E. T. Jaynes. Information theory and statistical mechanics, I and II. *Physical Reviews*, 106 and 108:620–630 and 171–190, 1957.

[Jaynes, 2003] E. T. Jaynes. *Probability Theory — The Logic of Science.* Cambridge University Press, Cambridge, 2003.

[Johnson and Barron, 2004] O. Johnson and A. R. Barron. Fisher information inequalities and the central limit theorem. *Probability Theory and Related Fields*, 129(3):391–409, April 2004.

[Kelly, 1956] J. L. Kelly. A new interpretation of information rate. *Bell System Technical Journal*, 35:917–926, 1956.

[Kontoyiannis *et al.*, 2005] I. Kontoyiannis, P. Harremoës, and O. Johnson. Entropy and the law of small numbers. *IEEE Trans. Inform. Theory*, 51(2):466–472, Feb. 2005.

[Kraft, 1949] L. G. Kraft. A device for quantizing, grouping, and coding amplitude modulated pulses. Master Thesis, Electrical Engineering Department, Massachusetts Institute of Technology, 1949.

[Kullback and Leibler, 1951] S. Kullback and R. Leibler. On information and sufficiency. *Ann. Math. Statist.*, 22:79–86, 1951.

[Lin, 1991] J. Lin. Divergence Measures Based on the Shannon Entropy, *IEEE Trans. Information Theory*, vol.37, p. 145-151, 1991.

[Linnik, 1959] Yu. V. Linnik. An information-theoretic proof of the central limit theorem with Lindebergcondition. *Theory Probab. Appl.*, 4:288–299, 1959.

[Maxwell, 1871] J. C. Maxwell. *Theory of Heat*; Longmans and Co.; 1871

[Nash, 1951] J.F. Nash, Non-cooperative games. *Annals of Mathematics*; vol.54; p.286-295: 1951.

[Neyman and Pearson, 1933] J. Neyman and E. S. Pearson. On the Problem of the Most Efficient Tests of Statistical Hypothesis; *Philosophical Transactions of the Royal Society of London*, Serias A; vol. 231; p. 289-337; 1933.

[Ohya and Petz, 1993] M. Ohya and D. Petz. *Quantum Entropy and Its Use.* Springer, Berlin Heidelberg New York, 1993.

[Pinsker, 1960] M. S. Pinsker. *Information and Information Stability of Random Variables and Processes.* Izv. Akad. Nauk, Moskva, 1960. in Russian.

[Rényi, 1961] A. Rényi. On measures of entropy and information. In *Proc. 4th Berkeley Symp. Math. Statist. and Prob.*, volume 1, pages 547–561, Berkely, 1961. Univ. Calif. Press.

[Rissanen, 1978] J. J. Rissanen. Modelling by shortest data description. *Automatica*, 14:465–471, 1978.

[Shannon, 1948] C. E. Shannon. A mathematical theory of communication. *Bell Syst. Tech. J.*, 27:379–423 and 623–656, 1948.

[Slepian and Wolf, 1973] D. Slepian and J. K. Wolf. Noiseless coding of correlated information sources. *IEEE Trans. Inform. Theory*, IT-19:471–480, 1973.

[Tsallis, 1988] C. Tsallis. Possible generalization of Boltzmann-Gibbs statistics, *J. Statist. Physics*; vol. 52, p.479; 1988.

[Topsøe, 1979] F. Topsøe. Information theoretical optimization techniques. *Kybernetika*, 15(1):8 – 27, 1979.

[Topsøe, 2001] F. Topsøe. Basic concepts, identities and inequalities – the toolkit of information theory. *Entropy*, 3:162–190, 2001.

[Wald, 1947] A. Wald. *Sequensial Analysis.* Wiley, 1947.

[Yeung *et al.*, 2005] R. W. Yeung, S.-Y. R. Li, N. Cai, and Z. Zhang. Theory of network coding. *Foundations and Trends in Communications and Information Theory.*, 2(4 and 5):241–381, 2005.

THE STORIES OF LOGIC AND INFORMATION

Johan van Benthem and Maricarmen Martinez

1 INTRODUCTION AND SUMMARY

Information is a notion of wide use and great intuitive appeal, and hence, not surprisingly, different formal paradigms claim part of it, from Shannon channel theory to Kolmogorov complexity. Information is also a widely used term in logic, but a similar diversity repeats itself: there are several competing logical accounts of this notion, ranging from semantic to syntactic. In this chapter, we will discuss three major logical accounts of information.

Information as range

The first notion is semantic, associated with sets of possible worlds, taken in a relaxed light sense, and we call it *information as range.* The greater one's range of options for what the real world is like, the less information one has. This setting reflects common sense ideas about information and uncertainty, but it also brings a more technical agenda of what information is good for — and we will develop some epistemic logic showing this. In particular, ranges of options change as agents make observations, or engage in communication. This process of 'update' high-lights a key feature of information in logic, and also in general: information is always to be understood in connection with some *dynamic process* using and transforming it. We will look at epistemic dynamic logics for this purpose, which form a natural calculus of changes in information ranges triggered by events of observation or communication.

Information as correlation

The second major logical strand in information high-lights another semantic feature, viz. that information tends to be *about something* that is relevant to us, and hence it crucially involves connections between different situations: my own, and others. This notion of *information as correlation* has been developed in situation theory, starting from a naturalist theory of meaning for agents living in information-rich physical environments, and moving to a more abstract view of components of a distributed system whose parts show dependencies. The correlation paradigm brings with it a further agenda of the structure of situations

— non-well-founded or 'circular' — ways of classifying them, constraints on their joint behaviour, and channels which allow for information flow. As for dynamic processes accessing this correlational structure, one can again think of observation or measurement or discourse, even though these have not been modeled in traditional situation theory per se.

Information as code

Finally, there is a third major logical sense of information, oriented toward syntax, inference, and computation. This may be the primary sense for students learning logic and being told that valid conclusions 'add no information' to the premises. Thinking of information as encoded in sentences at some abstraction level, we come to the idea of *information as code.* In this concrete combinatorial setting, the major dynamic processing paradigm is 'inference' in some general sense, and the relevant logical sub-discipline is no longer model theory, but *proof theory* and theories of computation. Again dynamic processes are of the essence here, as both deduction and computation are stepwise activities of 'elucidation' which manipulate syntactic representations.

In all, then, we will see several static representations of information in logic, and several dynamic processes transforming them. In addition, further basic features come up in the relevant logical systems – in particular, the notions of 'aboutness': information is *about something*, and 'agency': information is *for someone.* As to the latter, logic has a rich account of attitudes and activities of agents, including their knowledge, beliefs, and activities of learning and revision over time. In an unpretentious diagram, a total view might be pictured as follows:

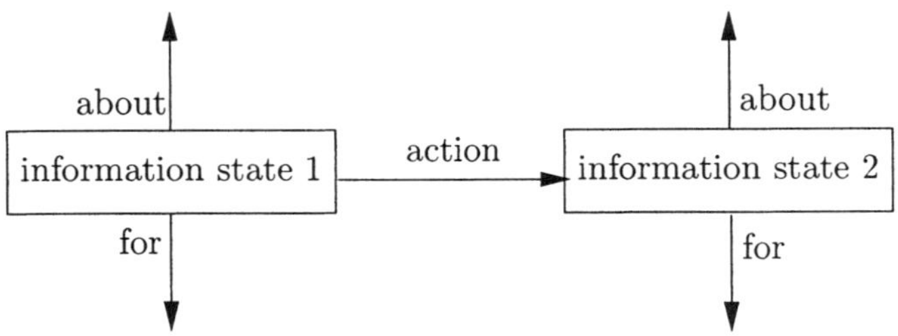

Co-existence versus unification

This picture suggests that information as range and information as correlation are compatible semantic notions, and we show later how they can be merged, in the spirit of modern logics of dependence and interaction. Likewise, the co-existence of semantic and syntactic perspectives invites comparison. Semantic and syntactic perspectives have always co-existed in logic, and their interplay in capturing the same set of valid consequences is at the heart of the celebrated *completeness theorems.* We will show how this harmony also suggests links between more proof-theoretic information as code and semantic views of information and the processes transforming it, especially when we bring in the agents involved in these activities. Even so, this chapter provides no full-fledged grand unification of all bona fide logical notions of information. We even doubt whether one such unified notion is desirable, if it exists at all.

Beyond our horizon

Much more could be said about logic and information, and we briefly list a few other topics at the end, including the increasing role of non-linguistic visual (graphic) information carriers in logic, as well as the border-line between logic and quantitative approaches such as probability theory, which is spawning new hybrids today. But the above is our main agenda, and we refer to the literature for further historical and systematic background.

After this top-level sketch, let's plunge into concrete logical matters.

2 INFORMATION IN LOGIC

Just a shaky metaphor?

Information has a somewhat precarious status in logic. One uses it colloquially to explain to beginning students what logic is all about, and a favourite metaphor is then that deduction is useful for the purpose of 'extracting information' from the data at our disposal. Say, you know that $A \vee B$ and you learn that $\neg A$. Then logic tells you that you now know B, since the following inference schema is valid:

$$A \vee B, \neg A / \neg B.$$

By the way, this classic is also the schema behind *Sudokus* and other logic puzzles sweeping the planet — so the reality of these phenomena of learning through logic is well-attested. But what one metaphor gives, another one takes. Further on in the course, we cheerfully tell students that the hall-mark of logical validity is its 'analyticity' in the sense of the classical philosophical tradition. And that says that the conclusions 'do not add information' to what was already there in the premises.[1] Indeed, there are many different views of information in logic, but strikingly, all of them are *implicit*. The field has official definitions for its concepts of proof, computation, truth, or definability, but not of information! Indeed, many logicians will feel that this is significant. We do not need this notion in the mechanics or even the foundations of the formal theory — or, as Laplace once said to Napoléon, who inquired into the absence of God in his *Mécanique Céleste*: "Sire, je n'avais pas besoin de cette hypothèse".

[1]This is close to validity in traditional logic, revived in the 'information-theoretic' account of Corcoran [1998]. Saguillo [1996] discusses three views of consequence with their historical and philosophical background, casting the information-based one as a Third Way in addition to the received proof-theoretic and semantic views of validity.

Information, inference, and computation

Still, more can be said. There are several areas in modern logic where notions of information emerge naturally. We have already noticed the connection between information and inference, i.e., the *proof-theoretic* stance toward validity. Information states are then stages in a dynamic process of deduction, and the corresponding informational moves are proof steps, or more general computation steps. For a concrete illustration, think of successive stages in the solution of a 3×3 'Sudokoid':

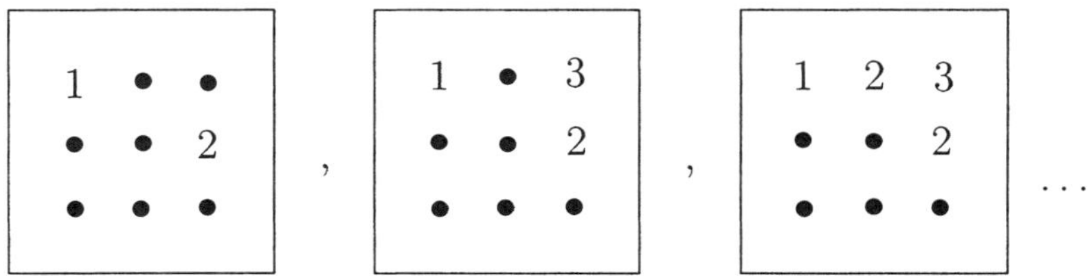

Intuitively, each successive diagram displays a bit more information about the eventual solution. Thus, on this view, information is brought to light in logical deduction or computation, and inferences are actions transforming the current state, i.e., the current representation structure appropriate to the topic of investigation. In the sequel, to coin a neutral phrase, we will sometimes refer to this dynamic process as *elucidation.*

Information range and update

To find further logical notions of information beyond inference, however, consider just the simplest realistic scenario. Premises do not drop out of the sky! In a café, your friend has ordered Applejack, while you asked for a Brandy. A new waiter comes back from the kitchen with two glasses. What we see around us every day is that the new waiter asks who has the Applejack (say), puts it down, and then puts the other glass without asking. This scenario shows two informational processes intertwined. The final inference by the waiter produces information in the earlier sense, but so did your answer to his question, which was crucial to the whole information flow, be it in another sense. Your response did not involve deductive steps on your or his part, but rather changes in the waiter's information *through observation*, giving him new facts.

How should we model this second kind of information flow? Before learning the premises of the earlier inference, your information state contains 4 options. Receiving the first premise $A \vee B$ cuts this down to 3, and receiving the second premise $\neg A$ cuts down to just one option: $\neg A$ & B is the actual situation. The rest seems indeed a matter of deductive inference, since a further update, say with the remaining fact B, would not change the final state any more:

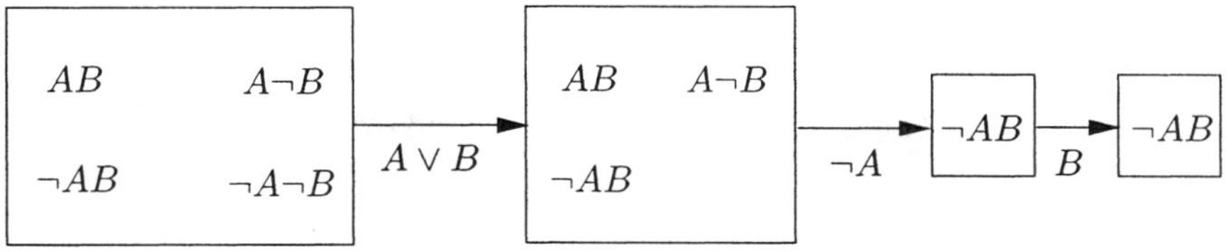

Again there is a process of information flow, this time through update. The resulting process is related to the *semantic view of validity*, which says that the conclusion must be true in all models for the premises. We do not know what the real situation is, but we have some information about it, encoded in a current *range of models*. And the relevant informational actions this time are updates: each subsequent assertion that we learn decreases that range, to a first approximation. This notion of information is found in epistemic logic and related semantic paradigms which we will discuss below in Sections 3 and 4.

Incidentally, note that in this second informational perspective, we are also extending the descriptive scope of logic. As the Restaurant example shows, inference is now just one of several information-producing processes entangled in even the simplest scenarios of information flow, and it seems tedious, and untrue to cognitive reality, to insist that 'logic' would just be about the inferential part, and not the rest. Indeed, there is even more to the scenario, as questions and answers involve processes of communication, another natural domain for logic.

Information and correlation

But there are still further perspectives in our panorama! The research program which put information on the map perhaps most forcefully as a major theme for logical study per se was *Situation Theory*, the topic of Section 5 below. Here, flow of information is not seen primarily as driven by events of observation or communication between agents. It rather rests on the fact that we often learn about some situation that is not directly accessible to us, via information at our disposal which is about that situation. Making sense of this *aboutness* calls attention to the 'constraints' in our environment, which correlate behaviour of spatio-temporally different situations. These constraints provide *information channels* between situations, and we can avail ourselves of these to draw conclusions about what we may not be able to observe directly.

Again, this correlational view may be tied in with notions of inference — once we are willing to step away from the usual varieties. E.g., one recurrent 'syllogism' in the ancient Indian logic tradition runs roughly as follows [Staal, 1988]. I am standing at the foot of a mountain, and cannot inspect directly what is going on there. But I can make observations in my current situation. Then, a useful inference might work as follows:

> "I see smoke right here. Seeing smoke here indicates fire on the mountain. So, there is a fire on the mountain top."[2]

[2]This version is simplified: Indian syllogisms also had other interesting features.

Compare this with the Aristotelean syllogism, which is about one fixed situation. The main new idea will be clear: logical inference can also cross between situations. Given suitable information channels between situations, observations about one may give reliable information concerning another. Indeed, the Indian example is almost the running example in Barwise & Seligman [1997] on seeing a flash-light on the mountain by some observer safely in the valley.[3]

Incidentally, on the Indian view, reflected in parts of Western logic, inference is a sort of last resort, when other informational processes have failed. If I can see for myself what is happening in the room, that suffices. If I can ask some reliable person who knows, then that suffices as well. But if no direct or indirect observation is possible, we must resort to reasoning. Again, we see the entanglement of different informational processes mentioned earlier.

Information and interaction

In this context of informational activities, it is worth noting that logic also has *pragmatic* perspectives beyond syntactic and semantic ones. Lorenzen [1955] famously explained a valid consequence as a dialogical claim in an argumentation game, whose proponent has a *winning strategy* against any opponent granting the premises. This is the world of Plato's "Dialogues" and public debate in the Greek polis, rather than the abstract proofs of Euclid's "Elements". On this pragmatic view we are after strategies which agents can use to win debating game or perform other interactive tasks. Going back in history, this stance also fits early views in Western and Chinese traditions on the role of inference in explanation to others, and the law of non-contradiction as the engine of debate [Liu & Zhang, 2007]. This view is also in line with the earlier-mentioned scenario with questions and answers, which typically involved more than one agent. Thus, logic can also cast information as what flows in communication, and more generally, as the lubricant of intelligent interaction. We refer to the chapters by Devlin & Rosenberg and by Walliser in this Handbook for more on this take, which fits well with current views of logic as a theory of interaction, with game theory as a natural ally. As with the earlier proof-theoretic and semantic views, we find information arising in a process, not as an absolute commodity by itself.

The goal of this chapter

The major purpose of this chapter is to bring out information as a theme running all through logic, even when it is usually left implicit. The resulting take on logic today and some of its major features is shown in the following sections. This might seem mainly a matter of presentation and re-telling existing stories. But there are genuine conceptual problems to be addressed as well. The many notions of information in logic pose a problem because they are so different — and yet

[3]Other *très* Indian examples include observing a coiled object in a dark room: using logic, rather than touch, to find out if it is a piece of rope or a cobra.

they each clearly make some valid point. Thus the second purpose of this chapter is to provide a more unified perspective, pointing out some new connections, and raising some further questions. These issues have not been much addressed in the philosophy of logic, where the agenda still seems largely concerned with somewhat fossilized questions from the past.

To see that there is a potential for, if not unification, at least successful integration, we note that the above logical information theories have clear similarities. In particular, all involve an interplay between *statics* and *dynamics*. There are structures representing the information, but these only make sense as vehicles for various processes of information flow. Yet epistemic logic, situation theory, proof theory, and dialogue systems all approach this information dynamics from different stances. We think this diversity may be a blessing rather than a curse. We will elaborate on the stances, and discuss links between communication and correlation, or between observation and deduction. But we do not (yet) have one unified notion of information coming out of all this — and thus logic shows the same diversity one finds with the notion of information in general in this Handbook. Indeed, we will also high-light some connections to adjoining fields.

Finally, related issues to those discussed here occur elsewhere in this Handbook. The reader will find analogies in the chapters by Baltag, van Ditmarsch & Moss on epistemic dynamics, Rott on belief revision, Kamp & Stokhof on linguistic communication, and Abramsky on the information flow in computation.

3 INFORMATION AS RANGE: STATE SPACES, EPISTEMIC, AND DOXASTIC LOGIC

In this section, we develop the notion of information as range. In our view, current epistemic logic — broadly understood — is the 'information theory' that goes with this. We show how this brings along an agenda of further themes showing how this information functions, how it can be computed with, and what basic methodological issues arise. The headings in our text identify what these are.

3.1 Information, state spaces, and sentences

Successive assertions inform us about a situation which the current discourse is about. Here is the natural semantic picture. It takes an 'inverse view' where extra information does not add things: it rather shrinks something, viz. the current range of options for what the real situation might be. We saw this with the earlier update pictures for the inference $A \vee B, \neg A / \neg B$, where the initial state of ignorance had 4 possible options, of which 3 remained after the input $A \vee B$, and only 1 after the further input of the premise $\neg A$. The inverse relationship is as follows, with T for sets of formulas, and $MOD(T)$ for the class of models making all of T true:

$$T \subseteq T' \text{ iff } MOD(T) \supseteq MOD(T')$$

Using sets of models as information ranges is like the 'state spaces' used in Bar-Hillel & Carnap [1953] for describing the information associated with an assertion. But it would be silly to award a patent for this view to specific individuals. As we have said before, it seems close to the common sense through history.

Semantic sets of models are rather rough information states. A more finely-grained syntactic alternative would use languages describing properties of situations or worlds. Assembling assertions over time creates an ever-growing 'book' of sentences, the total current information — perhaps including inferential connections between sentences, and rankings as to relevance or plausibility. This syntactic view of the information at our disposal is natural, too, and it may be the most attractive from a computational point of view. Against this background, information as range provides only rough counterparts to information as sets of sentences, since $MOD(T) = MOD(T')$ for logically equivalent sets of assertions T, T', even when these are vastly different syntactically. To most logicians, this is a virtue, as they find 'details of syntax' irrelevant to content (as long as they are, one hopes, not reading love letters). Nevertheless, one can seek finer intermediate representations, and Carnap himself used syntactic 'state descriptions' for computations in his inductive logic. Lewis [1970] and Moore [1989] have lively discussions, going back to Russell, of how these issues play in logical semantics — but they have really exercised about every major philosophical logician, including Hintikka, Kripke, and Stalnaker. Indeed, much of the discussion of 'propositions' and 'meanings' in the philosophical literature (cf. the chapter by Kamp & Stokhof in this Handbook) might be seen as the search for a level of information in between mere sets of models and every last detail of syntax.

3.2 Knowledge and epistemic logic

We now discuss a setting which combines the semantic and syntactic perspectives. The best-known paradigm incorporating information as semantic range is *epistemic logic,* a subject proposed in Hintikka 1962, and developed by many authors since, across different disciplines such as philosophy, computer science, and economics. Its main ideas are easy to describe: models of the system describe information ranges for agents, while the matching language describes a notion of knowledge that can be paraphrased as "to the best of the agent's information". Here is how this works in more formal detail.

Language and models

The syntax has proposition letters $p, q, \ldots$, Boolean operators $\neg, \vee$, and modal operators $K_i\phi$. The latter say that agent i *knows that* ϕ, while the dual $\langle i\rangle\phi = \neg K_i \neg\phi$ says that i *considers* ϕ *possible.* The following semantics provides a precise underpinning for this intuitive reading. Models $\boldsymbol{M}$ for the language are triples

$$(W, \{\sim_i | i \in G\}, V),$$

where W is a set of worlds, the $\sim_i$ are binary accessibility relations between worlds, and V is a propositional valuation recording in which world which atomic propositions are true. The worlds ('states', 'situations', ...) in the set W represent the options for how the actual situation might be, while the relations $\sim_i$ encode the uncertainty, or alternatively, the current information of the agents:

$x \sim_i y$ says that, at world x,
i considers y an option for being the actual world.

These accessibility relations may be different for different agents, who evidently need not all have the same information. One often takes the $\sim_i$ to be equivalence relations, but this is not crucial to epistemic logic.[4] Working with equivalence relations does validate some much-debated epistemic features of agents like 'positive' and 'negative introspection' concerning one's own knowledge. Now, the semantic truth condition for the knowledge operator makes the most evident stipulation in this setting, using a universal quantifier over the current information:

Agents know what is true throughout their current range of uncertainty:
$\boldsymbol{M}, s \models K_i\phi$ iff for all t with $s \sim_i t : \boldsymbol{M}, t \models \phi$.

The dual $\langle i \rangle \phi = \neg K_i \neg \phi$ is then the existential quantifier 'in some currently accessible world', stating that agent i holds it possible that ϕ is the case.

We follow the usual 'knowledge' terminology for the operator $K_i\phi$ of epistemic logic in much of what follows. Even so, as we have already said above, the more neutral and less ambitious term

'to the best of i's information' for $K_i\phi$

states much better what the universal quantification over the current information range really achieves, and how epistemic logic can then serve as an information theory. We will briefly address this point of intended meaning again when discussing connections with formal epistemology.

Digression: alternative semantics for knowledge

Possible worlds models have been under constant attack from critics.[5] While some criticism seems merely a misguided response to the unfortunate 'worlds' metaphor for the situations represented in our models, there are respectable alternative traditions. Indeed, the earliest semantics for modal logic in the 1930s used *topological models*, reading $K_i\phi$ as 'ϕ is true throughout some open neighbourhood of the current point'. This stipulation represents another broad geometrical intuition about the structure of knowledge as range, generalizing the above graph-based models with accessibility arrows to include topological structures like the real numbers, or Euclidean space. Topological semantics for knowledge is undergoing a modest

[4]Hintikka himself favoured mere reflexive transitive orderings.

[5]This may explain why the paradigm is so vibrant and vigorous today.

renaissance these days [van Benthem & Sarenac, 2005; van Benthem & Bezhanishvili, 2007], and generalized versions blossom in 'neighbourhood semantics' (cf. [Chellas, 1980; Arlo-Costa & Pacuit, 2005]).[6]

Factual and higher-order information

The epistemic language can formulate non-trivial scenarios. Consider a model for two agents $\boldsymbol{Q}, \boldsymbol{A}$ (say, 'questioner' and 'answerer') with one world where the atomic fact P holds, and another where it fails. We assume that the real world (there always *is* one!) is the one indicated by the black dot — though this is of course an outsider's annotation, rather than the agents' own information. Labeled lines linking worlds indicate uncertainties. In particular, $\boldsymbol{Q}$ does not know which world is the actual one, while $\boldsymbol{A}$ is better informed: if the actual world has P then she knows that is the case, and if it does not, then she knows that, too:

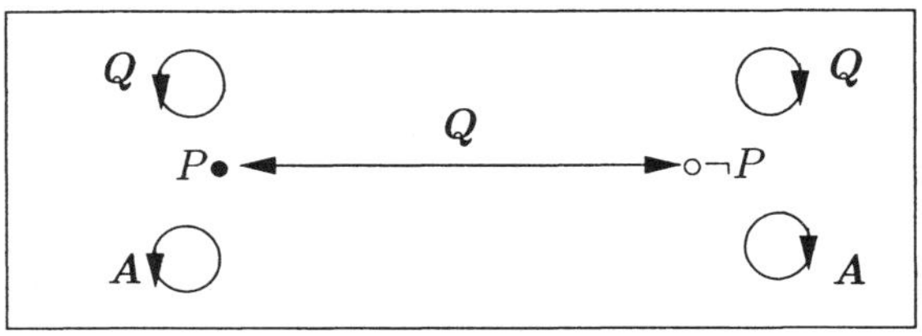

$\boldsymbol{M}$

This diagram also encodes 'higher' information that agents have about their own and each others' information. For instance, in a precise sense to be defined below, $\boldsymbol{Q}$ knows that she does not know if P while $\boldsymbol{A}$ does. Hence, it would be a good idea for $\boldsymbol{Q}$ to *ask a question* to $\boldsymbol{A}$, and find out. But before getting to that in Section 4 below, let us take stock of the model $\boldsymbol{M}$, representing the current state of the informational process. In formulas of the epistemic language, here are a few things which are true in the world to the left:

> $P, K_A P, \neg K_Q P, K_Q \neg K_Q P, K_Q(\neg K_Q P \wedge \neg K_Q \neg P)$ ($\boldsymbol{Q}$ knows that she does not know if P), $K_Q(K_A P \vee K_A \neg P)$ ($\boldsymbol{Q}$ knows that $\boldsymbol{A}$ knows whether P)

Thus, this language of information can express complicated epistemic patterns.

In particular, the preceding *iterations* of knowledge about oneself and others reveal something essential about the notion of information as used by humans. 'Higher-order information' about (lack of) information of ourselves and others is just as important as ground-level factual information! By providing such iterations, our static language encodes crucial patterns of information in interaction. E.g., to see that mere knowledge about others' ignorance can be helpful, consider the following model, with the actual world at the black dot:

[6] Moss, Parikh & Steinsvold [2007] develop another type of epistemic logic based on topology.

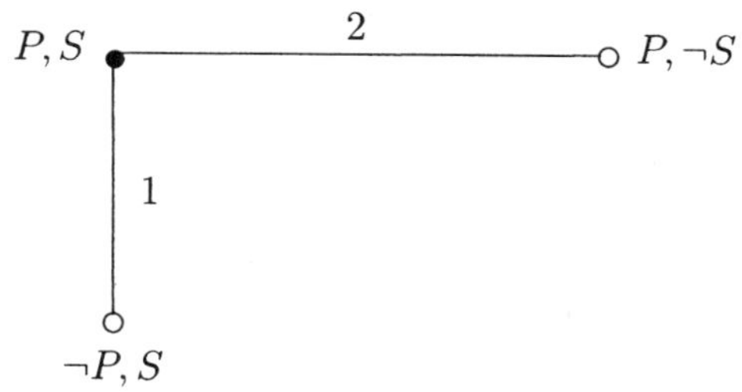

Given the uncertainty pattern, neither agent 1 nor 2 knows the real situation. But if 2 were to say she does not know, this is informative to 1, as it rules out the bottom-most world. Hence, 1 learns the real situation, and can inform 2.[7]

There is more to be said in logic about the information dynamics in these examples; and we shall do so in Section 3 below. For the moment, we just note that the above formal definitions alone do not really show the merits of a framework. Only a practical *art of modeling* can do that. And indeed, in the hands of skilled practitioners, epistemic logic is an information theory which handles much more complex scenarios than the baby examples here. Cf. [Fagin *et al.*, 1995; van der Hoek and Meijer, 1995; van Ditmarsch *et al.*, 2007].

Model-theoretic digression: information and invariance

In addition to its uses as a practical modeling device, epistemic logic raises general model-theoretic issues of expressive power. As in other formalisms, the crucial issue here is the harmony between expressive power and matching *semantic invariances* between models. The latter style of thinking raises the conceptual issue of

When are two given information models the same?

We cannot just take things at face value, and assume that geometrically different diagrams automatically model different information. To see this, consider the following variant $\boldsymbol{M}_1$ of the epistemic model in the above scenario, with the actual world again indicated in bold-face:

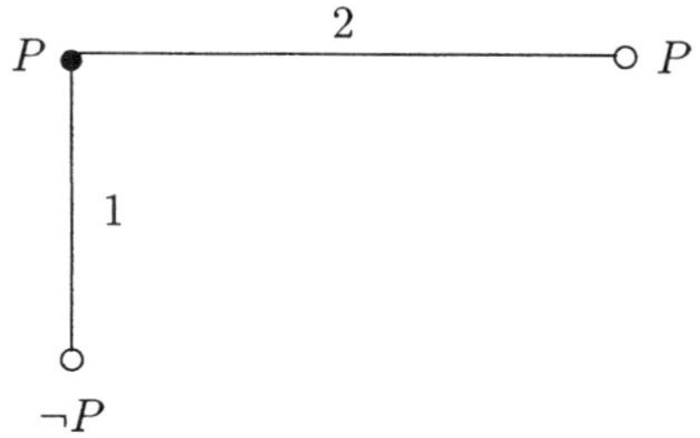

[7]In fact, in this setting, if both agents say *simultaneously* that they don't know, an epistemic model containing just the actual situation is reached at once.

Now, let agent 2 announce the true fact that she does not know $\neg P$. The updated model will lose the lower-left world to become $\boldsymbol{M}_2$:

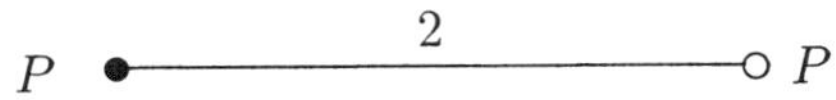

But clearly, $\boldsymbol{M}_2$ is essentially *the same* as the following one-world model $\boldsymbol{M}_3$:

as can be seen by identifying both worlds in $\boldsymbol{M}_3$ with the single one in $\boldsymbol{M}_3$. Indeed, understanding what makes different diagrams carry 'the same' information is a good test of our understanding of the notion of information itself.

The standard semantic answer of epistemic logic is clear [van Benthem, 1996; van Benthem and Blackburn, 2006]. Two information models $(\boldsymbol{M}, s)$ and $(\boldsymbol{N}, t)$ satisfy the same epistemic formulas[8] iff there exists a *bisimulation* between $\boldsymbol{M}$ and $\boldsymbol{N}$ connecting world s to world t. We will not pursue the issue whether this is the only possible account of 'same information structure', but on our view, a true understanding of information should definitely come with a 'criterion of identity' allowing us to recognize the same information under different representations.

Having said this, we return to more standard and better-known issues.

Epistemic logic as information theory: two functions

Any logical system has at least two major functions. First, its language can state properties of situations, and hence it has *descriptive and communicative* uses, just like a natural language. And once we have stated such properties, we can check computationally whether they hold in specific models, or we can compute concrete updates of the current model.

But there is also a second use for a logic, as a description of valid inference, supplying in our current case a 'calculus of information'. This second function comes ideally in the form of complete syntactic calculi describing all valid principles of inference with knowledge, such as multi-agent 'K' or '$S5$' or other well-known axiomatic systems. E.g., here is the complete set of principles for $S5$, on top of any complete classical propositional logic:

$K_j(\phi \to \psi) \to (K_j\phi \to K_j\psi)$	Knowledge Distribution
$K_j\phi \to \phi$	Veridicality
$K_j\phi \to K_jK_j\phi$	Positive Introspection
$\neg K_j\phi \to K_j\neg K_j\phi$	Negative Introspection

The complete logic with all the above principles is decidable, be it computationally more complex than its propositional base. It is '*Pspace*-complete', rather than '*NP*-complete', giving it the same complexity as many *games* [Papadimitriou, 1994] — again an indication of its interactive character. These laws of inference can be applied in every concrete scenario, just like the principles of probability

[8]In an infinitary version of the epistemic language whose technical details are irrelevant here.

theory. This axiomatic system describes both the agents' own reasoning, and our external reasoning as theorists about them – though the distinction has been played up in the recent literature. We will ignore that issue here, but we do briefly discuss a few of these axioms below, since they are not just rock-bottom truths: they reflect a particular view of powers of epistemic agents.

Altogether then, epistemic logic can be used to describe both information update and inference, making it serve the broader purposes of logic as a theory of different kinds of information advocated in Section 2.

3.3 Other attitudes: doxastic and conditional logic

In the above, we have noted a separation between information per se coming from some source, and agents' attitudes and responses to it. Indeed, there are many attitudes that agents can have toward propositions beyond knowledge. Our natural language has a fine-grained repertoire, from knowing propositions to believing, or even just 'entertaining' them. Moreover, we can also doubt propositions, maybe on the basis of new information — and more generally, change our allegiance from one epistemic attitude to another. Some of these ubiquitous attitudes toward information have received formal treatment in logic.

Basic doxastic logic

In particular, 'doxastic logics' analyze assertions $B_i\varphi$ standing for *'agent i believes that φ'*. The semantics for the belief operator used in this chapter seems folklore. It adds a new idea to the mere ranges for agents in our epistemic models so far. We now assume further semantic structure, in the form of a *plausibility ordering* of worlds as seen from some vantage point:

$\leq_{i,s} xy$ in world s, agent i considers y at least as plausible as x.

Thus, while the earlier ranges of alternatives corresponded to the strict information that agents have, these same ranges ordered by plausibility give much finer gradations. In particular, we can now define belief semantically as less demanding than knowledge, in the sense of *'truth in the most plausible options'*:

$$\boldsymbol{M}, s \models B_i\phi \text{ iff } \boldsymbol{M}, t \models \phi \text{ for all } t \text{ maximal in the ordering } x \leq_{i,s} y.$$

Here is an elementary example. Consider a model $\boldsymbol{M}$ with two worlds that are mutually epistemically accessible, but the one with $\neg P$ is considered more plausible than the other with P (which happens to be the actual world):

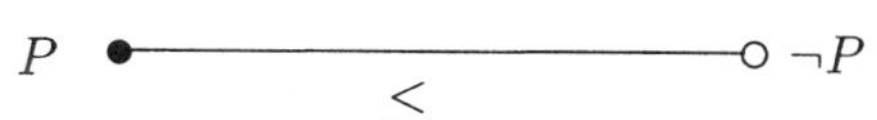

In this model, at the actual world where P holds, the agent does not know if P, but she does (mistakenly!) believe that $\neg P$. It is crucial to any realistic account of informational agents that our beliefs can be false.[9]

[9]There are some technical complications in making this semantics work in infinite models.

As with epistemic logic, there are complete doxastic logics, and a further theory around them [Fagin *et al.*, 1995]. In general the resulting logics also analyze the *interplay* between knowledge and belief, with knowledge implying belief, and more controversially, taking a stance on such issues as whether one knows one's beliefs, and so on. While inter-relations between attitudes toward information are an important topic in logic, we omit it here as tangential to our main concerns.

Digression: information or 'attitude'?

We can think of these epistemic-doxastic information models with two relations $\sim_i, \leq_j$ in two ways. The orderings $\leq_{i,s}$ may just encode an agent's private response to the strict information received, making the belief *subjective*. But we can also view them more *objectively* as encoding intrinsic gradations of plausibility in the incoming information — with beliefs the mere registering of this by sensitive logical observers. Both interpretations occur in the literature, like with probability — and both have their uses.

Conditional doxastic logic

In doxastic logic as used in dynamic informational processes, one soon finds that mere beliefs are not sufficient for explaining agents' behaviour over time. We want to know what they would do in certain scenarios where they receive new information. This requires *conditional belief*:

> $\boldsymbol{M}, s \models B_i^{\psi}\phi$ iff $\boldsymbol{M}, t \models \phi$ for all worlds t which are *maximal* for $x \leq_{i,s} y$ in the set $\{u \mid \boldsymbol{M}, u \models \psi\}$.

Conditional beliefs $B_i^{\psi}\phi$ are like logical conditionals in general, expressing what might happen under different circumstances from where we are now. Thus they *pre-encode* beliefs in ϕ we would have if we were to learn new things ψ. The analogy is so close that conditional belief on reflexive transitive plausibility models satisfies exactly the principles of the minimal conditional logic [Burgess, 1981; Veltman, 1985]. We will sharpen things up when discussing the more detailed mechanics of belief revision in Section 4.

3.4 Agents: powers, attitudes, and collective information

Epistemic logic highlights three aspects of our schematic diagram of information flow in Section 1. It is about *information states* (the above semantic models) and *information dynamics* — but crucially, always as performed by information-handling *agents*. We have already seen several aspects of this agent-orientation. Information is relative to agents, and they can have a spectrum of different attitudes toward it: knowledge, belief, and others. In addition, agents can also have very different abilities — with some choice-points high-lighted in the axioms of epistemic logic. For instance, the Distribution Axiom $K_j(\phi \to \psi) \to (K_j\phi \to K_j\psi)$

says that knowledge is closed under known implications, giving agents unlimited powers of stepwise inference from what they already know. Likewise, the Introspection Axioms give them unlimited powers of self-reflection on what they know or do not know. These abilities of 'omniscience' and 'introspection' have been questioned in much of the philosophical and computational literature — though no alternative consensus model of 'bounded rationality' has emerged so far. Still, epistemic systems today provide for agent variation. Now this raises an issue, and perhaps also a fact of life.

Agent relativity and agent diversity

Agent-relativity seems to dilute the absoluteness of a bona fide information theory. But it may be a genuine insight. Our account of 'knowledge' as 'to the best of an agent's information' really shows that there is an unavoidable interplay between two notions which needs to be made explicit: (a) the information in a situation per se, and (b) the *powers of the agents* having access to it. Put as a general principle, to *see what information is available, one must take the informational nature of the agents into account.*

Existing epistemic logics differ on what they ascribe to agents [van Benthem and Liu, 2004; Liu, 2006] in terms of deductive powers, introspective abilities, observational powers, and memory capacity. Such assignments are reflected in correspondences between such powers of agents and axioms of the logic. Examples are the much-discussed link between the 'KK principle' $K\phi \to KK\phi$ and *transitivity* of accessibility, or between *commutation laws* for communication (cf. Section 5) and memory assumptions of *perfect recall.* These correspondences provide a refined logical view of what information can flow in a given setting, given the nature of a source and that of the recipient.

Group knowledge and social structure

Our final observation about agent orientation in epistemic logic is its social character. This showed in the interactive nature of knowledge about others, but it goes much further than this. Agents also form groups, as in the above question-answer scenario, where communication involves a group consisting of the participants, which can have knowledge of its own. In particular, epistemic logic also has notions of group knowledge which are sui generis, with pride of place going to

$C_G\phi$ ('ϕ is *common knowledge in group* G'),

which is read in the above information models as follows:

$\boldsymbol{M}, s \models C_G\phi$ iff for all worlds t reachable from s by some finite sequence of $\sim_i$ steps $(i \in G) : \boldsymbol{M}, t \models \phi$.

In the logic, the additional axioms for C_G bring out the 'reflexive' fixed-point character which common knowledge has intuitively:

$C_G\phi \leftrightarrow \phi \ \& \ E_G C_G \phi$ Equilibrium Axiom
$(\phi \ \& \ C_G(\phi \rightarrow E_G\phi)) \rightarrow C_G\phi$ Induction Axiom[10]

Another natural form of group knowledge is 'distributed knowledge' referring to what a group might come to know if it pooled the information available to individual members (cf. [Fagin *et al.*, 1995; van Benthem, 2006b]). The same social group notions occur in the literature for belief.

3.5 Connections with other fields

Philosophy

Epistemic-doxastic logic is a somewhat austere account of qualitative information. But ever since its birth, it has triggered discussions in philosophical epistemology. It was soon found that the universal quantifier in information as range provides a rather poor analysis of knowledge in the philosopher's demanding sense, where the quest for a satisfactory definition of *knowing that* P involves finding the right sort of 'robustness' in addition to the obvious features of *truth of* P and *belief in* P. Plato famously proposed 'justified true belief', but sophisticated new definitions have kept appearing until today, such as Dretske's 'true belief grounded in correct information', or Nozick's 'true belief with counterfactual tracking: if P had been false, we would have believed that $\neg P$'. Even though epistemic logic has never offered a definition of knowledge of comparable sophistication, the very mismatch with the richer philosophical tradition has been much more exciting than many happy marriages, leading to clarification and fruitful debate. For instance, the distribution axiom $K(\phi \rightarrow \psi) \rightarrow (K\phi \rightarrow K\psi)$ has sparked debates about Logical Omniscience, and the axiom $K\phi \rightarrow KK\phi$ about Positive Introspection. We refer the reader to [Hendricks, 2006; Williamson, 2000; van Benthem, 2006a; Baltag *et al.*, 2007] for a wide array of interface topics, such as definitions of knowledge, skeptical arguments (see also Dretske's chapter in this Handbook), different sources of information (language, senses, etc.), omniscience, bounded rationality, and reflection.[11]

Computer science, and economics

But maybe the philosophers set their standards of knowledge too high. Who searches for the Perfect Knight might never get married. For many practical purposes, the picture of knowledge as range seems quite sufficient. Think of scenarios like this. The cards have been dealt. I know there are 52 of them, and I know their colours. The possible worlds are just the possible deals. Of course, I could be wrong about this,[12] but this worry seems morbid, and not useful in understanding normal information flow. What is true is that less well-founded attitudes will

[10] Here $E_G\phi$ says that everyone in the group G knows that ϕ.

[11] Many of these issues are high-lighted by persistent puzzles in the epistemological literature, such as infelicitous Moore sentences, the Fitch Paradox, or the Ramsey Test.

[12] Perhaps someone replaced the King of Hearts by Bill Clinton's visiting card.

come in as well. I may only have ephemeral beliefs about who holds which card, or about how the other agents will play. And indeed, we are sensitive to such distinctions. 'Knowledge' may then be a context-dependent term for 'the strictest attitude in the current setting'.

Notions of knowledge based on epistemic logic have penetrated other areas, notably computer science, witness the interdisciplinary TARK conferences since the early 1980s [Fagin *et al.*, 1995], with their current companion event LOFT. Originally, knowledge was ascribed here metaphorically to processors in distributed systems, with accessibility arising as these can only distinguish global system states through their own local state. But in modern intelligent agent-based computing systems, the difference with humans seems slight. Also, and even somewhat earlier, epistemic logic entered economic game theory in the 1970s through the work of Aumann (cf. [Osborne and Rubinstein, 1994; de Bruin, 2004]), as a way of stating what players know about each other, in an account of the reasoning about rationality underpinning notions like Backward Induction and Nash Equilibrium. Cf. [van der Hoek and Pauly, 2006] for further details, as well as the chapter by Walliser in this Handbook.

In applications such as these, knowledge often occurs intermingled with moves or actions. Players reason using their knowledge of what certain moves will bring about, and also, after observing a move by other players, they readjust their current information. This natural combination will be the topic of the next section.

4 INFORMATION FLOW AND DYNAMIC LOGIC

Communication and information flow

We start by summarizing a point which pervaded much of the preceding section. As in other information theories, such as Shannon's quantitative account of channel transmission (cf. the chapters by Harremoes and Topsoe and by Grunwald and Vitanyi in this Handbook), information really comes into its own only in a dynamic setting of *communication* and *information flow.* As a simplest case of this phenomenon in epistemic logic, consider the following conversational scenario:

(a) $\boldsymbol{Q}$ asks $\boldsymbol{A}$ the question "P?",

(b) $\boldsymbol{A}$ gives the true answer "Yes".

Then the situation depicted in the model $\boldsymbol{M}$ *changes*, since information will now come to flow. If the episode is one of simple cooperative Gricean communication, the question (a) itself conveys the information that $\boldsymbol{Q}$ does not know the answer, but also, that she thinks $\boldsymbol{A}$ might know. In general, this iteration can be highly informative and useful to the answerer. Next, the answering event (b) conveys that $\boldsymbol{A}$ knows that P, and its public announcement in the group $\{\boldsymbol{Q}, \boldsymbol{A}\}$ makes sure that $\boldsymbol{Q}$ now also knows that P, that both agents know this about each other, and so on to every depth of iteration. In terms of the philosophical classic Lewis [1969], after the episode, the agents have achieved *common knowledge* of P.

Structure and process

Summing up, our earlier point returns. It is hard to think of information in isolation from the processes which create, modify, and convey it. This combination of structure and process is quite natural in many disciplines. In computer science, one designs data structures in tandem with the processes that manipulate them, and the tasks which the latter should perform. And the same point is familiar from philosophy, as in David Lewis' famous dictum that 'Meaning Is what Meaning Does'. We can only give good representations of meanings for linguistic expressions when we state how they are going to be *used*: in communication, disambiguation, inference, and so on. In a slogan:

Structure should always come in tandem with a process!

So, which dynamic processes drive the notion of information? Many candidates vie for this honour in the chapters of this Handbook, including computation, inference, update, revision, correction, question answering, communication, games, and learning. Some items in this list are activities of single agents, while others are intrinsically interactive 'social' phenomena — with interesting connections between the two. We will not investigate all these links in detail, or even the issue whether one notion of information serves them all.[13] Instead, we will highlight one instance of this Dynamic Turn, in terms of informational processes for epistemic-doxastic logic. The chapters by Baltag, van Ditmarsch and Moss and by Rott in this Handbook provide much further elaboration.

Information update and model change

One of the simplest events providing new information is a *public announcement* $!P$ of some fact P. One can think of this as a statement coming from some authoritative source — or more generally, as a totally reliable observation based on one's senses. If I see that the Ace of Spades is played on the table, I now come to know that no one of us holds it any more.[14] These events of new *hard information* in the earlier sense change what I know.

Formally, this happens by their triggering changes in the current epistemic model. In particular, public announcements $!P$ work as described in Section 1 above. They eliminate all worlds incompatible with P, thereby zooming in on the actual situation. Thus the current model $(\boldsymbol{M}, s)$ with actual world s changes into its sub-model $(\boldsymbol{M}|P, s)$ whose domain is the new restricted set $\{t \in \boldsymbol{M} | \boldsymbol{M}, t \models P\}$. In a picture, one goes

[13] Cf. the general Editorial to this Handbook for further thoughts on this issue.

[14] Of course, we can be mistaken about what we hear or see, but much worse than that would be to let overly morbid worries block the progress of logic.

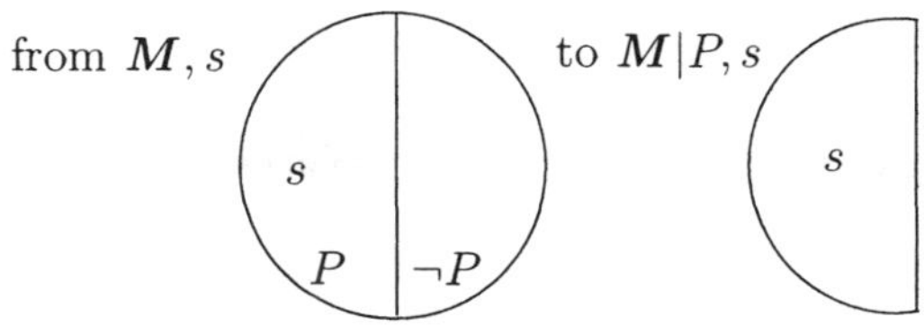

Dynamic logic of public announcement

Crucially, truth values of epistemic formulas may change in such an update step: typically, agents who did not know that P now do after the announcement. One can keep track of this epistemic dynamics in *public announcement logic PAL*, extending the epistemic language with action expressions in a two-level syntax:

Formulas	P:	$p \mid \neg\phi \mid \phi \vee \psi \mid K_i\phi \mid C_G\phi \mid [A]\phi$
Action expressions	A:	$!P$

The semantic clause for the dynamic action modality is as follows:

$$\boldsymbol{M}, s \models [!P]\phi \quad \text{iff} \quad \textit{if } \boldsymbol{M}, s \models P, \textit{ then } \boldsymbol{M}|P, s \models \phi$$

As all this is less known than standard epistemic logic, we state the principles of the complete calculus of information flow under public announcement. It has the usual laws of epistemic logic over one's chosen base models (as discussed earlier) plus, crucially, the following *reduction axioms*:

(a)	$[!P]q$	$\leftrightarrow$	$P \to q$	*for atomic facts q*
(b)	$[!P]\neg\phi$	$\leftrightarrow$	$P \to \neg[!P]\phi$	
(c)	$[!P](\phi \wedge \psi)$	$\leftrightarrow$	$[!P]\phi \wedge [!P]\psi$	
(d)	$[!P]K_i\phi$	$\leftrightarrow$	$P \to K_i(P \to [!P]\phi)$	

Taken together, these principles allow for compositional analysis of what agents know after new hard information has come in. In particular, the equivalence (d) is a dynamic 'recursion equation' relating knowledge *before* and *after* the event $!P$. This is the earlier idea of *pre-encoding*. There is already sufficient information about the effects of events $!P$ in the current state through the *relativized knowledge* of the form $K_i(P \to$. In order to find similar recursion equations for other informational events, and other attitudes of agents such as belief, similar expressive demands will have to be met by the static language.[15]

Agent powers again

Logical axioms high-light basic issues in concentrated form. The dynamic knowledge axiom $[!P]K_i\phi \leftrightarrow (P \to K_i(P \to [!P]\phi))$ also involves an *interchange* between the dynamic operator $[!P]$ and the epistemic K_i. This commutativity is not obvious, witness its possible failures in general logics of knowledge and action [Moore,

[15]As an extended modal logic, *PAL* still has a bisimulation-based model theory, and it raises many new issues of expressive power and computational complexity (cf. [van Benthem, 2006b]).

1985]. I know now that after drinking I am boring. But the tragedy of my life is that, after drinking, I do not know that I am boring... The reason is that drinking, unlike mere observation, impairs my epistemic abilities. Knowledge action interchanges presuppose abilities of agents such as Perfect Recall [Halpern and Vardi, 1989; van Benthem, 2001]. Again we see that studying information flow immediately raises issues of agent ability.

General observation and event update

Public announcement or observation is just one process which produces information. More complex changes occur in the current epistemic model for a group of agents when parlour games are played, or emails are sent with 'hidden buttons' like *bcc*. The relevant changes high-light another power of agents which admits of variation. This time, the issue is not inferential zeal, introspection, or memory, but agents' potentially limited and diverse *powers of observation*. Say, I see you draw a card from the stack, but unlike you, without seeing exactly *which one*. So, how to model this varying observational access? The answer is a non-trivial generalization of public update:

In general *dynamic-epistemic logics* [Baltag *et al.*, 1998; Gerbrandy, 1999; van Benthem *et al.*, 2006; van Ditmarsch *et al.*, 2007], so-called *event models* $\boldsymbol{A}$ describe complex scenarios where not all agents have the same observational access to what is happening. This leads to a mechanism of *product update*, turning the current epistemic model $\boldsymbol{M}$ into a model $\boldsymbol{M} \times \boldsymbol{A}$ which can even be larger than $\boldsymbol{M}$ itself, recording information of different agents about base facts and what others know. Product update redefines the universe of relevant possible worlds, and also the epistemic accessibility relations between them. Conversations, games, internet transactions, and other real activities are like this. For this natural continuation of our present analysis, we refer to the chapter by Baltag, van Ditmarsch and Moss in this Handbook.[16]

Hard and soft information

But there is more to information flow than *hard information*, whether observed totally or partially, and the resulting knowledge. As we saw in Section 3, further fine-structure may be present on epistemic models through plausibility orderings. These can be viewed as representing agents' attitudes, or as the result of receiving what might be called *soft information*. This plausibility ordering is reflected in the informational attitude of belief, absolute or conditional, telling us — roughly — what agents would believe when confronted with new hard information. And in the end, it is the total interplay of all these attitudes which would describe

[16]Systems with similar aims, but with richer languages allowing for process description in the long run over time, are *epistemic-temporal* logics, developed by Gupta and Thomason, Belnap, Perlof and Ming Xu, Fagin, Halpern, Moses and Vardi, Parikh and Ramanujam, and many other authors. Van Benthem and Pacuit [2006] and van Benthem, Gerbrandy and Pacuit [2007] give some up-to-date surveys and comparisons.

our stances toward information, and the way they are affected by new incoming information. Thus, we get a mixture of information per se, and the ways in which agents take it — and this logical entanglement is such it is hard to say where one notion ends and the other begins.[17]

Changing beliefs

Now we are ready for the next obvious step. The dynamic process perspective on information change explained here also applies to the doxastic attitude of our *beliefs*, and how to *revise* these on the basis of incoming information [Gärdenfors, 1987]. This involves changes, not in the range of available worlds or in their epistemic accessibility patterns, but rather in the earlier *plausibility orderings* $x \leq_{i,s} y$ among worlds.

How this works precisely, depends on the incoming signal. When we receive hard information $!P$, update will proceed by world elimination as before. We then get new beliefs related to our earlier conditional beliefs, and the counterpart to the above reduction axiom (d) is the following recursion equation saying which new beliefs — and indeed, conditional beliefs — are acquired [van Benthem, 2007a]:

$$\begin{array}{lll} \text{(e)} & [!P]B_i\phi & \leftrightarrow \quad P \rightarrow B_i^P[!P]\phi \\ \text{(f)} & [!P]B_i^{\psi}\phi & \leftrightarrow \quad P \rightarrow B_i^{P\wedge[!P]\psi}[!P]\phi \end{array}$$

But often, we just get *soft information* signals of the sort mentioned before. These increase our 'preference' for P-worlds, but without telling us to abandon the others. A typical 'belief revision policy' in this spirit is *lexicographic upgrade* $\Uparrow P$ [Rott, 2006] which replaces the current ordering relation $\leq$ between worlds by this:

all P-worlds become better than all ¬P-worlds,
and within those two zones, the old ordering remains.

Belief changes under such policies can again be axiomatized completely (cf. again [van Benthem, 2007a]). Just for illustration, here is the recursion equation for conditional beliefs following the $\Uparrow P$ revision policy. It looks somewhat forbidding — but perhaps necessarily so: after all, we are now describing a more complex informational process than mere epistemic update:

$$\begin{array}{lll} \text{(g)} & [\Uparrow P]B^{\psi}\phi & \leftrightarrow \quad (\Diamond(P \wedge [\Uparrow P]\psi) \wedge B^{P\wedge[\Uparrow P]\psi}[\Uparrow P]\varphi) \;\vee \\ & & \quad (\neg(\Diamond(P \wedge [\Uparrow P]\psi) \wedge B^{[\Uparrow P]\psi}[\Uparrow P]\varphi)^{18} \end{array}$$

Still richer dynamic doxastic logics use the above event models as triggers for belief revision, with 'signal events' ordered by plausibility [Baltag and Smets, 2006]. This relocates revision policies in the structure of the incoming signal, while sticking to one product update rule for the new plausibility ordering. We refer to the

[17]Other views distinguish between a base level of pure information processing and a higher level of beliefs on the basis of this information, where belief changes occur in a sort of 'reflective dynamics'. Cf. the chapter by Rott in this Handbook.

[18]Here the existential epistemic modality $\Diamond$ in the consequent says that the assertion is true 'in some world of the current range'.

chapters by Baltag, van Ditmarsch and Moss and by Rott in this Handbook for further policies, and deeper discussion of the issues raised by all this.[19]

Information-processing agents: the general setting

The over-all view of information dynamics which emerges from our discussion of information update and belief revision in Sections 3, 4 is quite elaborate. Agents have powers of inference, observation, introspection, and self-correction, maybe not all to the same degree — and moreover, they exercise these powers interactively, in groups of agents, leading to behavioral equilibria that naturally invite study per se. This picture is in line with broader current trends. These include 'logical dynamics' [van Benthem, 1996], 'information dynamics' in computer science (cf. Abramsky's chapter in this Handbook) and 'interactive epistemology' in philosophy and game theory (cf. the chapter by Walliser). There are also connections with mathematical *learning theory* ([Kelly, 1996], and also his chapter in this Handbook).

Here we note once more that our logical systems have the following features as 'information theories'. They are *dynamic*, treating informational actions and events on a par with static information structures. They are also *social*, in that basic scenarios have more than one agent together. [20] Thus *interaction* between agents, and even irreducible group action and group knowledge, become crucial to a logical understanding of information flow.

5 INFORMATION AS CORRELATION: THE WORLD OF SITUATION THEORY

We now turn to the only program in modern logic that treats information as its central notion, shaking up the usual agenda of research. It originated in the 1980s [Barwise and Perry, 1983] as a foundational framework for 'situation semantics', but in its underlying *situation theory*, the program has come to study the notion of information from a much more general point of view (cf. also [Devlin, 1991; Allo, 2007], and the chapter by Devlin and Rosenberg in this Handbook).

According to situation semantics, meaning arises from the interaction of organisms and their information-rich environment. As a formal tool originally devised for modeling this view of meaning and associated phenomena, situation theory goes beyond possible-worlds models by providing richer structure. Among other things, this yields a natural treatment of 'the problem of grain': different *false* claims about the world are represented by different mathematical objects. In addition, the theory provides an account of important context effects that are part of everyday inference and natural language.

The central role of the notion of information in situation theory, and the way it was formalized, reflects several influences. For example, like the Gibsonian view

[19] There are also strong analogies between processes of plausibility re-ordering and recent dynamic logics of *preference change* [van Benthem and Liu, 2007].

[20] Even observation is really a matter between 'me' and 'my source'.

of reality and its role in visual perception [Gibson, 1979], in situation theory everything is part of as a structured reality which is full of 'uniformities'. Organisms are 'attuned' to those regularities, and that allows them to survive. Information is a pervasive aspect of reality, prior to cognitive action. Other important influences came from philosophy. For example, Putnam's Twin Earth thought experiment [Putnam, 1981] has convinced many that meaning cannot be just in the mind, and hence external reality must play a crucial role. But it is Dretske's theory of information flow [Dretske, 1981] that conspicuously provided key ideas about the basic informational notion to study. Dretske builds on Shannon's theory of information transmission, but his account is qualitative and his focus is squarely on semantic *content*, not on the *amount* of information transmitted. It uses the following variant of the usual notion of confirmation:

> *Information Content* To an agent with prior knowledge k, signal r carries the information that s is F iff Pr (s is $F \mid r$ and k) $= 1$, but Pr(s is $F \mid k$) < 1.

Dretske then defines knowledge as belief caused by information flow. One reason why the definition includes the strict equality Pr(s is $F \mid r$ and k) $= 1$,[21] is the following intuitively appealing consequence:

> *Xerox Principle* If A carries the information that B and B carries the information that C, then A carries the information that C.

Situation theory adopts the same guiding idea — though dropping the probabilities — and it encompasses not only natural but also conventional signals and others.

Furthermore, in situation theory we distinguish between *having* and *carrying* information. This distinction is reflected in the study of two kinds of reports in natural language. *Epistemic attitude reports* state the information an agent *has:*

1. Elwood knows that Santa Cruz is east of Berkeley
2. Elwood sees that Gretchen is limping.

In contrast to this, *information reports* tell what is shown or indicated by an event or the state of an object ('the signal'):

3. The x-ray's being of pattern ϕ shows that Gretchen has a broken leg.
4. The array of pixels on the screen being of its specific pattern carried the information that the Forty-niners had won the championship.

In (3) the x-ray having its specific pattern is the *indicating fact;* the x-ray itself is the *carrier* of the information.

Treatments of information as *correlation* as mentioned in Section 1 focus on information reports, in which one state of affairs carries information about another;

[21]This stipulation has been the target of some criticisms, which we disregard here.

the way one thing indicates how another thing is; the way the x-ray turned out carries information about Gretchen's leg. More generally, one situation carries information about another situation *connected* to it in some way, by some *channel* of information, in virtue of a regularity or *constraint:* perhaps a natural law, a convention, or something else. By contrast, epistemic attitude reports about having information are closer to the epistemic logic *range* view discussed before.

We now come to the basic apparatus, which we describe in enough detail to contrast it with the preceding sections. Still, the thrust will differ somewhat. Epistemic logic in its static and dynamic versions has a well-developed model theory and proof calculus, but it provides little information at a practical level about what might be called the 'art of modeling'. Situation theory does a kind of converse. It offers a rich apparatus for modeling challenging phenomena — but so far, it has (with a few exceptions to be noted below) not yet generated canonical calculi of reasoning with the same sweep and wide application as epistemic logic.

5.1 Basic concepts of situation theory

Situation theory starts from a concrete reality, with concrete parts but no concrete alternatives. This reality can be thought about, perceived, studied and analyzed in a variety of different ways, from different perspectives, for different purposes. But ultimately everything that exists, everything that happens, everything that is true, has its status because of the nature of this reality. The parts of reality are what we call *situations*. Situation theory is committed to there being situations, but not to there being a largest total situation of which the rest are parts.

States of affairs

When we think or talk about reality, we need some way of analyzing it. This we call a *system of classification and individuation*. It consists of situations, relations, locations and individuals. The commonplace that different schemes can be used to study the same reality is one to which situation theory subscribes. But this does not make situations structure-less, with properties projected onto them by language or thought. Rather, situations are rich in structure, and support a variety of schemes, suited (or unsuited) to various needs.

Each relation R comes with a set of argument roles. For example, the relation of *eating* comes with the roles of eater, eaten, and the location of eating. Objects of appropriate sorts play these roles. A relation, together with objects assigned to its roles, gives rise to an *issue*, namely, whether or not the objects stand in the relation. There are two possibilities, and each of these we call a *state of affairs*. For example, if *eating* is the relation, Bush is the eater, a certain quantity of succotash is the eaten, and the White House at a certain time is the location (call it *loc*), then there are the following two states of affairs:

$$\langle\langle \text{ Eats}; \textit{loc}, \text{Bush, the succotash}; 1\rangle\rangle\ ,$$
$$\langle\langle \text{ Eats}; \textit{loc}, \text{Bush, the succotash}; 0\rangle\rangle$$

The issue is whether Bush eats the succotash at the location; the possible answers are 'yes' (1) and 'no' (0). The first state of affairs resolves the issue positively, the second, negatively. We say the first has a positive and the second a negative *polarity*. Each of these two is the *dual* of the other.

Although we don't assume that the argument roles of a relation have a natural order, we often use the order suggested by English to identify argument roles, without explicitly mentioning the roles themselves, as we did above.

Facts and the partiality of situations

Situations determine whether a given state of affairs or its dual is a fact. In a familiar notation,

$s \models \sigma$ means that s *makes* σ factual, or s *supports* σ

and

$\models \sigma$ means that σ *is* factual (i.e., some real situation supports it)

Given a state of affairs σ, these are uncontroversial theses about the $\models$ relation:

1. Some situation will make σ or its dual factual,
2. No situation will make both σ and its dual factual,
3. Some situations will leave the relevant issue unresolved, making neither σ nor its dual factual.

In contrast, the following is a controversial thesis about this important relation:

4. Some situation resolves all issues (*i.e.*, there is a largest total situation).

The third thesis says that situations are partial. Hence, there are two ways a situation s can fail to make a given state of affairs σ factual. Namely, s may make the dual of σ factual, or s may fail to resolve the σ-issue one way or the other.

Parameters, anchors, and infons

For theoretical purposes, it is useful, though not strictly necessary, to extend our ontology with *parameters*. Parameters are constructs resembling the well-known operator of λ-abstraction in the λ-calculus [Barendregt, 1984]. We start by adding some infinite set of parameters $\boldsymbol{a}_1, \boldsymbol{a}_2, \ldots$ for individuals, $\boldsymbol{r}_1, \boldsymbol{r}_2$, for relations, etc. With these, we can now work with an abstract form of states of affairs, that we call *infons*. Infons are just like states of affairs, but some roles may be filled with parameters instead of concrete objects. The term 'infon' was coined by Keith Devlin (cf. [Devlin, 1991]) to suggest that parametric states of affairs are theoretical entities that serve as the basic units of information. One can also think of infons as corresponding to properties. For example, the infon

$$\sigma = \langle\langle kisses,\ \boldsymbol{a}_1, \boldsymbol{a}_2; 1\rangle\rangle$$

captures the property shared by all situations where someone is kissing someone.

The step from states of affairs to infons is a sort of abstraction. To get from the infons back to states of affairs, we need *anchors*. An anchor is a partial function from the domain of parameters to appropriate objects. We will always assume here that f gives a value to all parameters involved. Where f is an anchor and $\langle\langle \ldots \boldsymbol{a}, \ldots\rangle\rangle$ is an infon, $\langle\langle \ldots \boldsymbol{a}, \ldots\rangle\rangle[f] = \langle\langle \ldots, f[\boldsymbol{a}], \ldots\rangle\rangle$. For instance, with σ as above, we have $\sigma[\boldsymbol{a}_1 \to\ Anne, \boldsymbol{a}_2 \to\ Bill] = \langle\langle kisses,\ Anne,\ Bill; 1\rangle\rangle$.

Notice that states of affairs are special infons: those whose set of parameters is empty. We can extend the earlier support relation $\models$ to infons, by saying that an anchor f *satisfies* an infon σ relative to a situation s if $s \models \sigma[f]$. An anchor f satisfies an infon i *simpliciter* if $\models \sigma[f]$, i.e., if there is a situation s such that $s \models \sigma[f]$. Finally, we say that $s \models \sigma$ (s *supports* σ) if there is some anchor f such that $s \models \sigma[f]$. For example, if s is a situation where Anne is kissing Bill, then

$$s \models \langle\langle kisses,\ Anne,\ \boldsymbol{a}_2; 1\rangle\rangle$$

because $s \models \langle\langle kisses,\ Anne, \boldsymbol{a}_2; 1\rangle\rangle[\boldsymbol{a}_1 \to\ Bill]$, so

$$s \models \langle\langle kisses,\ Anne,\ Bill;\ 1\ \rangle\rangle.$$

Compound infons

Infons so far are abstractions of single states of affairs. But there are natural ways to construct others. A guiding principle here is the Principle of Persistence: all information supported by a situation s_1 must be supported by any other situation s_2 that extends s_1.[22] Two important constructs that comply with persistence are the meet of a set of infons and the existentialization of an infon with respect to a parameter. Where Φ is a set of infons, we can form the infon $\Lambda\Phi$ such that an anchor f satisfies $\Lambda\Phi$ relative to situation s if and only if $s \models \sigma[f]$ for all σ in Φ. As for existentials, given an infon σ and a parameter $\boldsymbol{x}$, we can construct the infon $\exists\boldsymbol{x}\sigma$ such that an anchor f satisfies $\exists\boldsymbol{x}\sigma$ relative to situation s if and only if for some object a, we have that $f_{x/a}$ satisfies σ relative to s.

Of course, once we have constructs like these, we would like to have an algebra and logic of infons. We will not go in details here (see [Devlin, 1991; Moss and Seligman, 1997]), but just point out that the usual logical intuitions cannot be taken for granted here, due to the fact that *information is fine-grained*: logically equivalent compound infons may still be different pieces of information.

[22]This is like in intuitionistic logic, and partial logics: cf. the corresponding chapters in the *Handbook of Philosophical Logic*: [van Dalen, 2002; Blamey, 2002].

Situation types

Once we have infons, we can abstract over situation parameters,[23] and thus obtain a rich collection of *abstract situation types*, of the form

$$T = [\boldsymbol{s} | \boldsymbol{s} \models \sigma]$$

where $\boldsymbol{s}$ is a situation parameter and σ the conditioning infon of the type, *cond(T)*. This infon may be a state of affairs, in which case we call T a nonparametric type. If T is a parametric type, we refer to it just as *a type*, and sometimes write $T(\boldsymbol{p})$ to stress that its conditioning infon σ uses parameters $\boldsymbol{p}$. If f is an anchor with domain $\boldsymbol{p}, T[f]$ denotes the type determined by the state of affairs $\sigma[f]$. We will overload the notation $\models$ and use $s \models T$ to mean that situation s is of type T. Notice that this happens precisely when there is an anchor such that $s \models \sigma[f]$.

Constraints

Now we are ready to introduce the notions which are at the heart of the situation theoretic account of information. The crucial idea is that reality is full of regularities, lawlike *constraints* between types of situations, and it is the existence of those regularities that makes it possible for one situation to carry information about another. Constraints correspond to natural phenomena such as natural laws of physics, social conventions like those in language and traffic signs, and many other kinds of dependence. Formally, a *constraint* is an infon

$$\langle\langle \textit{Involves}, T[\boldsymbol{p}], T'; 1\rangle\rangle.$$

This infon is factual if the following existence condition is satisfied:

> for every anchor f with domain $\boldsymbol{p}$, whenever s is a situation of type $T[f]$, then there is a situation s' of type $T'[f]$.

If *smoke* is the type of situations where there is smoke and *fire* the type of those where something is burning, then $\langle\langle$*Involves*, *smoke*, *fire*;1$\rangle\rangle$ is factual.

Simple constraints do not make the kind of *connection* explicit that exists between the constituents of the two types involved. For that, we need *relative constraints*, infons of the form

$$\langle\langle \textit{Involves}_R, T[\boldsymbol{p}], T', T''[\boldsymbol{q}]\rangle\rangle$$

Such a relative constraint is factual if for every anchor f with domain $\boldsymbol{p} \cup \boldsymbol{q}$, if s, s'' are two situations of types $T[f]$ and $T''[f]$, respectively, then there is also a situation s' of type $T'[f]$.[24]

[23]There is a collection of that kind of parameters, too.

[24]Intuitively, T involves T' relative to T'' if for any pair of situations of the first and third types, there is a situation of the second type.

5.2 *Information flow*

Next we come to the fundamental issue of *how one situation can carry information about another*, which may be far away in time or space? We present two ways of formalizing this claim. The first is given here, in terms of the above ontology. The second requires the theory of classifications and channels [Barwise and Seligman, 1997], and we leave it for the next subsection.

Propositions

In situation theory, the bearers of truth are *propositions,* nonlinguistic abstract objects that have absolute truth values. There are two kinds of these. *Russellian propositions* are characterized just by a type T, and they are true if there exists a situation of type T. *Austinian propositions* are claims of the form $s \models T$, thus involving a type and a situation.[25] This distinction was high-lighted by Barwise and Etchemendy (cf. [Barwise and Etchemendy, 1987]), where it plays a key role in the treatment of semantic paradoxes.

Propositions are not infons. Infons characterize situations; propositions are truth bearers. We assume that for each situation type and each situation there is an Austinian proposition true just in case the situation is of that type. For Russellian propositions, we shall assume that for each type, there is a proposition true just in case some situation is of that type. This last strong assumption can lead to paradox [Barwise and Etchemendy, 1987], but it will not affect us here.

Information

Now, while propositions are the bearers of truth, it is particular situations that act as carriers of information. More precisely, it is the fact that some proposition $s \models T$ is true, plus the existence of factual constraints relating T with other types, that allows for s to carry information about other parts of reality. The basic kind of informational statement for this reads:

> the fact that $s \models T$ carries the information that *Prop.*

Depending on whether the proposition *Prop* is Russellian or Austinian, this tells us whether the information carried by s is 'signal-bound' or 'incremental' [Israel and Perry, 1990]. In the Russellian case,

> $s \models T$ carries the *signal-bound* information that there is some situation of type T' if there is an anchor f and a factual constraint $\langle\langle$Involves, $T_1, T_2; 1\rangle\rangle$ such that $T_1[f] = T$ and $T_2[f] = T'$.

If s is a situation where there is an x-ray on the desk of a vet, we may say that

> The x-ray's being φ-ish indicates that there is a dog, of whom this is an x-ray, and that dog has a broken leg.

[25] If we adopted the controversial fourth thesis above that there is a total situation, then Russellian propositions could be taken as Austinian propositions determined by that total situation.

This is a case of signal-bound information, since the information carried is *about* the signal itself (the x-ray). We can represent this regularity scenario as follows. First, the relevant factual constraint is $C = \langle\langle \text{Involves}, T_1, T_2; 1\rangle\rangle$, where

$$T_1 = [s|s \models \langle\langle X\text{-ray}, \boldsymbol{x}, \boldsymbol{t} : 1\rangle\rangle \wedge \langle\langle \text{Has-pattern-}\phi, \boldsymbol{x}, \boldsymbol{t}; 1\rangle\rangle]$$
$$T_2 = [s|s \models \langle\langle \text{Is-}X\text{-ray-of}, \boldsymbol{x}, \boldsymbol{y}, \boldsymbol{t}; 1\rangle\rangle \wedge \langle\langle \text{Has-broken-leg}, \boldsymbol{y}, \boldsymbol{t}; 1\rangle\rangle]$$

The indicating situation s will support the infon

$$\langle\langle X\text{-ray}, a, t'; 1\rangle\rangle \wedge \langle\langle \text{Has-pattern-}\phi, a, t'; 1\rangle\rangle$$

where a is the particular x-ray on the desk and t' is the time. This infon is an instantiation of $cond(T_1)$ via the anchor $f(\boldsymbol{x}) = a, f(\boldsymbol{t}) = t'$, so s is of type $T_1[f]$. The signal-bound information carried by s is the proposition that there is some situation of type $T_2[f]$, that is, one that supports the infon

$$\langle\langle \text{Is-}X\text{-ray-of, } a, \boldsymbol{y}, t'; 1\rangle\rangle \wedge \langle\langle \text{Has-broken-leg, } \boldsymbol{y}, t'; 1\rangle\rangle$$

This says there is a dog with a leg depicted by a at t' and that leg is broken at t'.

But to guide a vet's action appropriately, it is not enough for her to be acquainted with the fact that the x-ray's is φ-ish and to be aware of the relevant constraint. She must also know that the information the x-ray carries is about Gretchen — i.e., the incremental information the x-ray carries, given the additional fact that it was taken of the specific dog Gretchen. We say that

> the fact that $s \models T$ carries the *incremental* information that $s' \models T'$ (relative to T'') if there is an anchor f and a factual relative constraint $\langle\langle \textit{Involves}_R, T_1, T_2, T_3; 1\rangle\rangle$ with $T_1[f] = T, T_2[f] = T'$, and $T_3[f] = T''$.

Notice that the informational content in this case is the Austinian proposition that $s' \models T'$. Incremental information therefore, is information about a *concrete* situation s' via the indicating fact $s \models T$, in virtue of how s is connected to s'.

Our example turns on the relative constraint that, if an x-ray is of this type, and it is of a dog, then that dog had a broken leg at the time the x-ray was taken. That the x-ray was of Gretchen is the *connecting fact*, and the incremental information content is the proposition that Gretchen has a broken leg. The latter is about Gretchen, but not about the x-ray. The relevant factual relative constraint is:

$$C' = \langle\langle \text{Involves}_R, T_1, T_2, T_3; 1\rangle\rangle$$

with indicating type T_1 as before, and indicated and connecting types T_2 and T_3:

$$T_2 = [s|s \models \langle\langle \text{Has-broken-leg}, \boldsymbol{y}, \boldsymbol{t}; 1\rangle\rangle]$$
$$T_3 = [s|s \models \langle\langle \text{Is-}X\text{-ray-of}, \boldsymbol{x}, \boldsymbol{y}, \boldsymbol{t}; 1\rangle\rangle]$$

As before, we assume that the indicating situation s supports the state of affairs

$$\sigma = \langle\langle X\text{-ray}, a, t'; 1\rangle\rangle \wedge \langle\langle \text{Has-pattern-}\phi; a, t'; 1\rangle\rangle$$

Further, we assume that the connecting state of affairs σ' is factual — where b stands for Gretchen, σ' is $\langle\langle \text{Is } X\text{-ray-of}, a, b, t'; 1\rangle\rangle$.

Any anchor f of the right kind here, with $\sigma = cond(T_1)[f]$ and $\sigma' = cond(T_2)$, must be defined on the parameter $\boldsymbol{y}$ of the connecting type: in particular, it must anchor $\boldsymbol{y}$ to Gretchen. For any such anchor f, the proposition carried incrementally by the fact $s \models T_1$ (relative to σ') states that there is a situation s'' with $s'' \models \langle\langle \text{Has-broken-leg}, b, t'; 1\rangle\rangle$. This is a proposition about Gretchen, not at all about the x-ray. And it is, after all, our real dog Gretchen that we are concerned about.

Agents using information

Typically, information is used by *agents* to guide their actions, and thus, it is relative to the manner in which an agent is adapted to the world. This means the information an agent can get out of a situation depends on the constraints to which it is *attuned*. With the x-ray, what information is available to the vet depends on the regularities about x-rays to which she is attuned. We will now proceed to make this next set of notions more precise.

5.3 Information flow and distributed systems

So far, we have shown how the machinery of situation theory models the fact that one situation can carry information about another. Examples of this flow abound beyond those already discussed. The manner in which the diaphragm of a microphone is vibrating carries information about what the announcer is saying. The modulation of the electromagnetic signal arriving at some antenna carries information about the way that diaphragm is vibrating. And finally, the modulation of the electromagnetic signal arriving at the antenna carries information about what the announcer is saying, for instance, "Hillary Clinton is irritated".

Situation theory provides tools for answering the question: how does the modulation of the electromagnetic signal at the antenna carry information about the words that the announcer spoke? The theory relies on a number of fundamental principles about the nature of information as it flows across distributed systems, that is, systems that can be analyzed in sub-parts:

1. The availability of information depends on the existence of regularities between connected parts of a distributed system.

2. These regularities are relative to how the system is analyzed into parts.

3. Flow of information crucially involves types and their concrete instances. It is in virtue of constraints (relations on types) that concrete situations, being of certain types, can act as carriers of (concrete or general) information.

We will now present a second, more mathematical way of formalizing the idea of one situation carrying information about another. This is done via the theory of *classifications and channels* [Barwise and Seligman, 1997]. This paradigm, based on category theory, is an elegant formal distillation of the ideas presented so far.

5.4 Classifications and channel theory

The notions of information flow presented in the previous section involve facts of the form $s \models T$, where s is a situation and T a type. These facts are about how situations are *classified*. The notion of classification, independently discovered around 1990 as 'Chu Spaces' [Gupta, 1994] and 'Formal Contexts' [Ganter and Wille, 1997], is the basic notion on which the theory of classifications and channels is built. *Classifications* are triples, often depicted as follows:

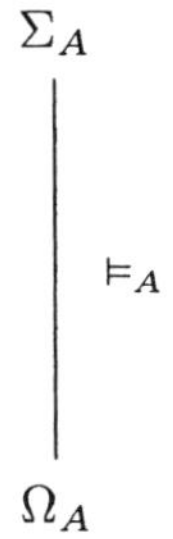

Here Ω_A is a set of *tokens* (for example, situations), Σ_A is a set of *types* (conceived of as anything that can classify tokens), and $\models$ is a relation between tokens and types. If s is a token and T a type, then $s \models_A T$ reads as 's is of type T'.

The natural 'part-of' relationships that exist between parts of a system are called *infomorphisms*. An infomorphism $f : \boldsymbol{A} \to \boldsymbol{C}$ between two classifications is a pair $\langle f^\wedge, f^\vee \rangle$ of functions

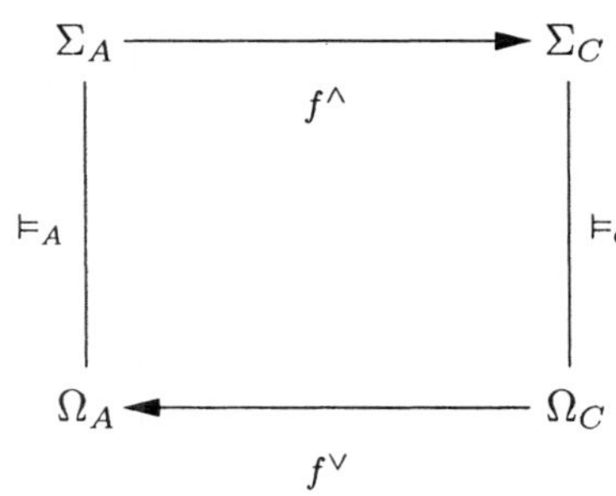

such that for all tokens $b \in \Omega_C$ and all types $T \in \Sigma_A$

$$f^\vee(c) \models_A T \text{ if and only if } c \models_C f^\wedge(T) \qquad (*)$$

Infomorphisms are of independent interest as an abstract invariance behind translation between theories, and that of general category-theoretic adjunctions. But here we look at their concrete uses as an intuitive model for information flow.

A concrete scenario

The Smoke-on-the-Mountain scenario of Indian logic, mentioned in Section 2, involves at least two classifications $\boldsymbol{A}$ and $\boldsymbol{C}$. Tokens of $\boldsymbol{A}, \Omega_A$ might be situations where somebody is facing a mountain, while the relevant types Σ_A might include SeesSmoke, LookingUp, LookingDown, Blind etc. On the other hand, the classification $\boldsymbol{C}$ might correspond to the overall setting including the observer and the mountain. Its tokens are situations extending those of Ω_A, and types in Σ_C might include ObserverSeesSmoke, ThereIsAFireOnTop, etc. The map $f^\vee$ maps each large situation to the sub-situation capturing just the point of view of the observer. The map $f^\wedge$ sends SmokeObserved to ObserverSeesSmoke, LookingDown to ObserverLookingDown, etc. Thus type T of $\boldsymbol{A}$ is mapped to a type of $\boldsymbol{C}$ intended to mean that 'the observing situation' is of type T. Condition (*) ensures that things work out just this way.

As before, it is the existence of constraints, in the form of regularities between types that makes information flow within a distributed system. In this more general abstract setting, a *constraint* $\Sigma_1 \vdash \Sigma_2$ of classification $\boldsymbol{A}$ consists of two sets of types such that for all a in Ω_A, if $a \models_A \wedge\Sigma_1$, then $a \models_A \vee\Sigma_2$. If $\boldsymbol{A}$ were the classification of observers facing a mountain, SeesSmoke, Blind $\vdash \varnothing$ would be a constraint of $\boldsymbol{A}$, saying that no blind observer sees smoke on top of the mountain.

Channels and information flow

Let us now add to our observer and his mountain a third classification $\boldsymbol{B}$ for what is happening at the mountain top. Its tokens are situations located on mountain tops, and its types include ThereIsFire, ThereIsFog, etc. $\boldsymbol{B}$ is also a 'part' of the big component $\boldsymbol{C}$ — say, via the infomorphism g depicted here:

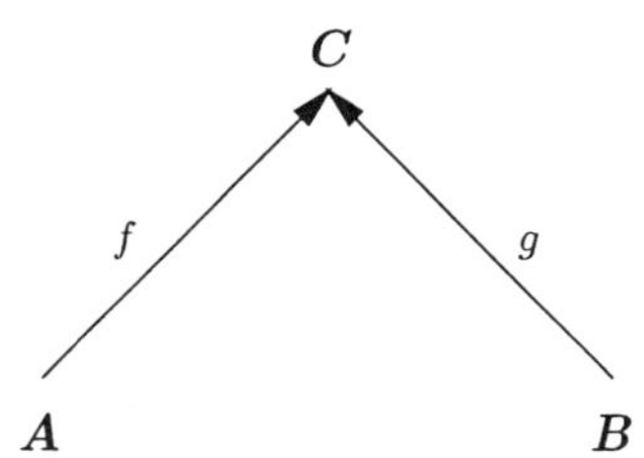

A collection of infomorphisms sharing codomain C is called a *channel* with *core* C. Tokens of the core are called *connections*, because they connect subparts into a whole. Tokens a from Ω_A and b from Ω_B are *connected* in channel $\boldsymbol{C}$ if there is a token $c \in \Omega_C$ such that $f^\vee(c) = a$ and $g^\vee(c) = b$. In the example, an observing situation and a mountain top are connected if they belong to the same overall situation. We can now formulate a notion of (incremental!) information flow:

> $a \models_A T$ *carries the information that* $b \models_B T'$ (relative to $\boldsymbol{C}$) if a, b are connected in $\boldsymbol{C}$ and $f^\wedge(T) \vdash g^\wedge(T')$ is a constraint of C.

This notion of information flow is relative to a channel — and hence, to an analysis of a whole distributed system into its parts. Again we see that 'carrying information' is not an absolute property: the mere fact that a token or situation is of a certain type does not completely determine what information it carries.[26]

Here is our observer-mountain example in these terms. Let $s \in A$ be a situation of type SEESSMOKE: the observer in it sees smoke on top of the mountain. Let $s' \in B$ be the top-of-the mountain observed from the base of the mountain. Our choice of $\boldsymbol{C}$ makes s and s' be linked by some connection in $\mathbf{C}$. In addition,

$$f^\wedge(\text{SEESSMOKE}) = \text{OBSERVERSEESSMOKE}$$
$$g^\wedge(\text{THEREISFIRE}) = \text{THEREISAFIREONTOP}.$$

Since OBSERVERSEESSMOKE $\vdash$ THEREISAFIREONTOP is a constraint of $\boldsymbol{C}$, s being of type SEESSMOKE carries the information that s' is of type THEREISFIRE.

Xerox Principle revisited

The current information flow complies with Dretske's Xerox Principle, in the sense that we can compose channels. If we have two channels with cores $\boldsymbol{C}_1$ and $\boldsymbol{C}_2$, as those shown in the lower part of this diagram:

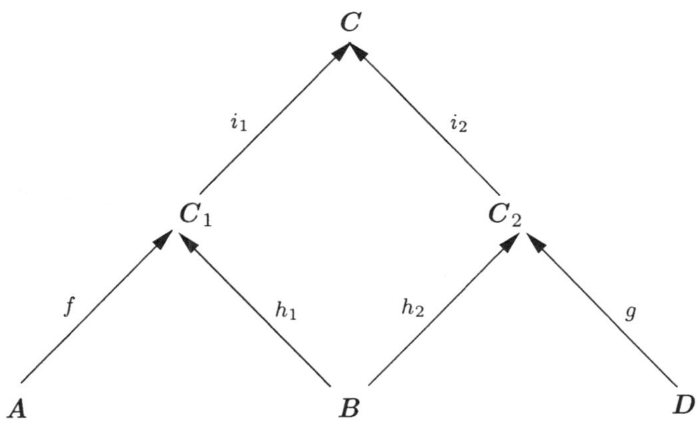

then we can always complete the diagram by taking the classification $\boldsymbol{C}$ to be the one whose tokens are ordered pairs of tokens of $\boldsymbol{C}_1$ and $\boldsymbol{C}_2$ and whose types belong to the disjoint union of those of $\boldsymbol{C}_1$ and $\boldsymbol{C}_2$. Here the type component of infomorphisms i_1 and i_2 is the identity, and tokens go via left and right projections. Within this setting, if $a \models_A T$ carries the information that $b \models_B T'$ (relative to $\boldsymbol{C}_1$) and $b \models_B T'$ carries the information that $d \models_D T''$ (relative to $\boldsymbol{C}_2$), then $a \models_A T$ carries the information that $d \models_D T''$ relative to the channel with core $\boldsymbol{C}$ and infomorphisms $i_1 f : \boldsymbol{A} \to \boldsymbol{C}$ and $i_2 g : \boldsymbol{D} \to \boldsymbol{C}$.

[26]This feature links up with general context dependency in logic (cf. Section 6).

Digression: model-theoretic relations between situations

So far, we just *assumed* that a given distributed system has dependencies. But *why* does one situation correlate with another? Sometimes, this is a mere accident, just as funds in the stock market may have fluke correlations. But often, there are more intimate links. A watch keeps time because its stepwise operation mirrors the unfolding of time. Infomorphism is one elegant mathematical notion supporting such harmony. Following Barwise and Seligman [1997], van Benthem [2000] has studied them model-theoretically, determining the precise first-order properties of channel-theoretic structures which they preserve. But other model-theoretic relations between situations make sense, too, such as *model extension* to larger situations. For a study of model-crossing notions of inference and *logics of model change*, cf. [van Benthem, 2007b]. Following earlier work by Lindström, Barwise and van Benthem [1999] study general 'information links', and entailment along model-crossing relations.

Local constraints and reasoning

As we have seen, an agent's ability to get information from one part of a system based on what it can observe at another, depends on what constraints she is attuned to. Now, most regularities to which agents are attuned hold only within some regions of reality. If a massive object is released in the air, it will fall, *given that* we are on Earth, not in a satellite. The notion of *local constraint* captures this idea. A local constraint of classification $\boldsymbol{A}$ has the form $\Gamma \vdash_A \Sigma$ (on S), where S is a collection of tokens from $\boldsymbol{A}$, while Γ, Σ are two sets of types from $\boldsymbol{A}$, and for all $a \in$ S, if $a \models_A \wedge\Gamma$, then $a \models_A \vee\Sigma$.

A this point, it becomes natural to consider explicit constraint-based *reasoning.* Now in general, modeling agents that can reason about their world would require a good account of 'inferential information', which is not an easy notion to capture in a satisfactory manner (cf. Sections 7, 8 below). For instance, Barwise and Seligman [1997] propose the following, following Dretske:

> *Inferential information*: To an agent with prior knowledge $k, r \models F$ carries the information that $s \models G$ if and only if the agent can legitimately infer that $s \models G$ from $r \models F$ and k, but not from k alone.

But what does 'legitimately infer' mean in this setting? So far this question has not been satisfactorily answered. However, the notion of a local constraint has suggested some first approaches, which we survey here, following Martinez [2004]. For a start, our setting validates local versions of basic inference rules, such as

Weakening

$$\frac{\Sigma_1 \vdash_A \Sigma_2 \text{ (on } S)}{\Sigma_1, \Gamma_1 \vdash_A \Sigma_2, \Gamma_2 \text{ (on } S)}$$

Adding to the rule of Weakening all obvious identity axioms plus a strong form of the classical Cut rule, one can prove an abstract completeness theorem: If we fix

S and a collection of constraints on S, then the collection of theorems is precisely the theory (i. e., the set of all the constraints) of some classification.

But simple 'localization' of the usual rules does not allow for inferences where knowledge about one part of the system drives inferences about another. To achieve that, we need rules for shifting classifications (changing from $\vdash_A$ to $\vdash_B$) and conditions (from S to another S'). Martinez [2004] has rules for this, including:

S-Weakening

$$\frac{\Sigma_1 \vdash_A \Sigma_2 \text{ (on } S)}{\Sigma_1, \Gamma_1 \vdash_A \Sigma_2, \Gamma_2 \text{ (on } S')} \qquad \text{whenever } S' \subseteq S.$$

Enlargement

$$\frac{\Sigma_1, T \vdash_A \Sigma_2 \text{ (on } Nec(T, S))}{\Sigma_1, T \vdash_A \Sigma_2 \text{ (on } S)}$$

where *Nec* is a function mapping each pair (*type* T, *set of tokens* S) to a subset of S that includes all tokens of type T in S. Natural modifications of other rules from Barwise and Seligman [1997] would also yield principles such as

f-Intro:

$$\frac{\Gamma \vdash_A \Sigma \text{ (on } S)}{f[\Gamma] \vdash_C f[\Sigma] \text{ (on } (f^\vee)^{-1}[S])}$$

where $f[\Gamma]$ is the set of types obtained by applying $f^\wedge$ to all types in Γ.

It is still an open question if calculi based on these rules are enough to formalize the notion of 'legitimate inference' in this setting in a fully satisfactory manner.

5.5 *Harnessing information*

How agents *use* the information available to them is crucial in cognitive modeling. An agent's ability to make *inferences* is just one aspect here. A more fundamental (and general) reason why agents are able to use information is structural: the architecture of the agent is geared towards doing so [Israel and Perry, 1991]. This turn fits well with the general Tandem View in this chapter, found also in other approaches, that notions of information must eventually be understood in terms of the dynamic processes which that information is used for. Now to specifics!

Clearly, for information to be of use to some agent, or to enable a device to do what it is supposed to, it is not enough that the states of the agent or device *carry* the information. The agent or device must in some sense *have* the information. We conceive of an agent having information as meaning that the architecture of the agent 'harnesses' the information to trigger action that advances the agent's goals. This architecture can be due to Nature or to a designer, and the goal can be a natural goal of the agent, or a goal for which the designer creates the agent.

For simplicity, assume an agent with one available action A and one goal G. Relative to these, circumstances can be divided into a range of possibilities, P and

$\neg P$, those in which the action achieves the goal, and those in which it does not.[27] The circumstances in P are the success conditions of A relative to G. The idea of harnessing information is now simple: some state of the agent that carries the information that P should trigger the action A.[28] An old-fashioned *mousetrap* provides a simple example of these ideas:

Cheese is placed on a small lever. The trap is placed somewhere, behind the refrigerator say, where the only likely cause of the lever moving is the presence of a mouse directly in front of it. The trap has one action: a wire-blade is released. The goal is killing a mouse; this will occur when there is a mouse in the path of the blade. The trap is constructed so that a mouse causing the lever to move will be in the path of the blade, and the movement of the lever will release the blade.

What we want then is an account of having information as being in a state that plays two roles (Israel and Perry 1991). First, the agent's being in the state carries certain information relative to a constraint. Second, an agent's being in that state has an effect, i.e., it triggers an action relative to some other constraint, that is appropriate given the information. In that case, the agent not only carries but *has* the information. More elaborate formalizations of the relevant events occurring in such scenarios might involve a mixture of ideas from situation theory and dynamic logic — witness the brief discussion in Section 6.5 below.

5.6 *Some further developments*

Circularity and set theory

Situation theory has been applied to information flow involving self-reference. Natural language as a medium for communication sometimes involves utterances about the very same situation in which they are made. Now, in set-theoretic approaches to situation theory, situations are usually modeled as sets of tuples (standing for the infons supported by the situation), but self-referring situations cannot be naturally modeled as sets in the standard Zermelo universe. This is because the Axiom of Foundation bans the existence of chains of sets of the form $u \in u_1 \in \ldots \in u$. Thus, a theory of sets without Foundation would fit situation theory better. Indeed, the study of circularity and its applications has been an important contribution of situation theory.

Technically, most of this work has taken place in *AFA* set theory, which consists of *ZFC* with the Foundation Axiom replaced by this Antifoundation Axiom (cf. [Forti and Honsell, 1983; Aczel, 1988, Barwise and Moss, 1996]):

[27] This is, by the way, the standard story from Decision Theory, except that we are ignoring risk or uncertainty with respect to the state of the reality.

[28] Here is a more general way of thinking about this. Any event will carry all sorts of information, relative to different constraints and connecting facts. It is the range of possibilities determined by the goals and actions that determines the information that is relevant. This again links information as range — now basically as having information about success conditions — and information as correlation — carrying information about the world external to the agents.

Given any graph (W, R) whose childless nodes are labeled with urelements, there is a unique way to label the other nodes with sets such that, if $d(v)$ is the label of v, then $d(v) = \{d(u) \mid vRu\}$ (i.e., $d(v)$ is the set of labels of v's children).

A graph with its childless nodes labeled by urelements represents a set, and each set can be represented as a labeled graph. In fact, the same set may be represented by different, but *bisimilar* labeled graphs.[29] E.g., there is a unique set with the self-referential or circular specification $\Omega = \{\Omega\}$, and it can be represented by the following two graphs (among many other bisimilar ones):

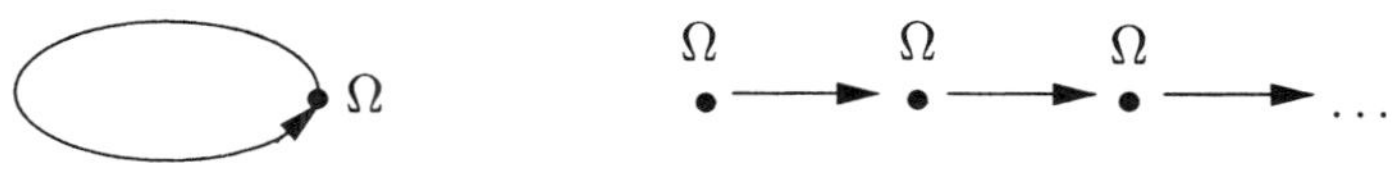

Bisimilarity enters the picture here, because in the *AFA* universe, equality of elements is not enough as a criterion for deciding whether two sets are equal.

Circularity, modal logic, and co-algebra

The preceding facts lead to interesting connections with modal logic and the use of recent *co-algebraic* methods.[30] Given a monotone operator F on sets, an *F-coalgebra* is just a mapping $f : X \to F(X)$. As a simple but important example, when we take $F(X) = P(X)$, an F-coalgebra turns out to be just a modal relational Kripke frame. Main themes in this new direction are the duality between algebras, inductive definitions and smallest fixed points on the one hand, and coalgebras, coinductive definitions and largest fixed points on the other hand [Jacobs and Rutten, 1997]. For the modal logic connection, see [Venema, 2006]. Barwise and Moss [1996] is a pioneering study of the general theory, starting from work by Aczel, with a mixture of *AFA* set theory and early co-algebra, and Baltag [1999] is a further study of *AFA*-classes.[31]

Summary and further applications

Situation theory is a logical account of information in distributed systems (including reality itself as parsed by agents) which model information as correlation. It brings along its own agenda of technical issues, which include formal theories of circularity in modeling and the associated reasoning apparatus, and a general account of relations between situations and of the channels along which information

[29] We already encountered bisimulation in a different guise in Section 3, as a measure of identity for information structure in epistemic logic.

[30] Cf. [SanGiorgi, 2007] for an excellent history of the history of the notion of bisimulation since the 1970s, through modal logic, theories of computation, and set theory.

[31] This last issue has been important beyond purely technical reasons. Although the initial intuition sees situations as 'small', situation theory also deals with large set-theoretic objects. For instance, the treatment of paradoxes via Austinian propositions allows for models of reality resolving all possible issues, which are proper classes [Barwise and Etchemendy, 1987].

can flow. For further applications, we refer to the more linguistic/social-science chapter by Devlin and Rozenberg in this Handbook, as well as more computational studies such as Akman and Tin [1997] and Martinez [2004]. Finally, [Devlin, 1991] is a good introduction to the general ideas and state of the art in the early 1990s.

6 MERGING RANGE AND CONSTRAINT VIEWS

Sections 3 and 5 have told two separate logical semantic stories of information, once as range and once as correlation. In this section, we compare the two, and propose some merges, in line with parts of the computational literature. While we cannot do justice to all aspects of this comparison here, we do show that the two styles of thinking are very congenial.

Many readers might rather have expected a polemical shoot-out here. It is often thought that epistemic logic and situation theory are hostile paradigms, witness discussions between, e.g., Perry and Stalnaker in the 1980s. For instance, much has been made of the *partiality* of situations versus the 'totality' of possible worlds. But in practice, many of these differences are minor. In many models of epistemic logic, possible worlds are small and partial. In the card examples in Section 2, 'worlds' are just possible hands, and hence 'no big deal'. Conversely, some entities cited as 'situations' can be pretty complex, such as countries, or world-wars. Also, situation theory has been cast as 'realist' and non-modal, in that only chunks of the real world are needed in its semantics. But this difference, too, evaporates, once one tries to give a situation-theoretic account of belief and being wrong (cf. [Kratzer, 2006]). Significantly, Barwise 1997 eventually introduces some sort of possible worlds and modal languages into situation theory. Indeed, the more striking historical development has been the steady discovery of significant analogies between situation theory and modal logic. This has happened to such an extent that, in the definitive mathematical framework of Barwise and Moss [1996], *modal logic* has become the vehicle for developing a bisimulation-based situation theory! Indeed, the two paradigms seem compatible and congenial both technically and conceptually — and they share a common cause.

This section emphasizes three themes, following van Benthem 2005B:

(a) modal and first-order logics of dependence as an account of constraints,

(b) merges between constraint-based models and dynamic epistemic logic,

(c) the addition of events to create dynamic versions of situation theory.

This topic is more a digression in this chapter, but it serves two purposes. First, since both aspects of information are relevant in practice, we need to understand how epistemic logic and situation theory can live together in modeling actual phenomena. Moreover, juxtaposing these systems raises some interesting new questions, and allows us to see where both lie in the larger picture of logic today.

6.1 *Correlation through gaps in state spaces*

In a world of one-shot events, no significant information can flow. Genuine constraints only arise in situations with different 'states' that can be correlated. To make this more precise, consider two situations s_1, s_2, where s_1 can have some proposition letter p either true or false, and s_2 a proposition letter q. There are four possible configurations:

$s_1 : p, s_2 : q$	$s_1 : p, s_2 : \neg q$
$s_1 : \neg p, s_2 : q$	$s_1 : \neg p, s_2 : \neg q$

With all these present, one situation does not carry information about another, as p and q do not correlate in any way. A significant constraint on the total system arises only when we *leave out* some possible configurations. For instance, let the system have just the following two states:

$s_1 : p, s_2 : q, \quad s_1 : \neg p, s_2 : \neg q$

Now, the truth value of p in s_1 will determine that of q in s_2, and vice versa. In a formula with some obvious notation, we have the truth of the following constraint:

$$s_1 : p \leftrightarrow s_2 : q$$

But even a less constrained system with three instead of just two global configurations still allows for significant information flow:

$s_1 : p, s_2 : q \quad s_1 : \neg p, s_2 : q, \quad s_1 : \neg p, s_2 : \neg q$

Presence of p in s_1 still conveys the information that q in s_2, but absence of p does not convey information about s_2.Again in a formula, we have the implication:

$$s_1 : p \rightarrow s_2 : q$$

Contraposing the implication, absence of q in s_2 tells us about absence of p in s_1, but presence of q has no immediate informative value about s_1.

Thus, correlation between different situations amounts to restrictions on the total state space of possible simultaneous behaviors. The more 'gaps' in that state space, the more information there is in the system, ready to be used in principle by potential observers.[32]

[32]Recall that in epistemic logic, gaps in the set of worlds encode common knowledge.

The bare bones of this setting are brought out by *constraint models*

$$\boldsymbol{M} = (\mathit{Sit},\ \mathit{State},\ \boldsymbol{C}, \mathit{Pred})$$

with a set *Sit* of situations, a set *State* of possible valuations, a predicate *Pred* recording which atomic predicates hold where, and crucially, a 'constraint relation' C stating which assignments of states to situations are possible in the system.[33]

6.2 Modal logics of constraints and correlation

Constraint models suggest a logic — in the simplest case, a modal one. To see this, take a language with names x for situations (a tuple $\boldsymbol{x}$ names a tuple of situations), and atomic assertions $P\boldsymbol{x}$ for properties of or relations between situations. We take Boolean operations, plus a universal modality $U\phi$ ('ϕ is true everywhere'):

$$P\boldsymbol{x} \mid \neg \mid \vee \mid U$$

The semantic interpretation has obvious clauses for the following notion:

$$\boldsymbol{M}, \boldsymbol{s} \models \phi \qquad \phi \text{ is true in global state } \boldsymbol{s} \text{ of model } \boldsymbol{M}$$

In particular, $P\boldsymbol{x}$ holds at $\boldsymbol{s}$ if the tuple of local states assigned by $\boldsymbol{s}$ to the tuple $\boldsymbol{x}$ satisfies the predicate denoted by P. This language defines basic constraints across situations such as $U(Px \to Qy)$. The resulting logic is classical propositional logic plus the modal logic $S5$ for the universal modality U.

But we can go one step further. Intuitively, a situation x which satisfies p 'settles' the truth of p, plus all repercussions this has for other situations in the system. Accordingly, define the following new relation between global states:

$$\boldsymbol{s} \sim_x \boldsymbol{t} \text{ iff } \boldsymbol{s}(x) = \boldsymbol{t}(x).$$

This generalizes to a relation $\sim_{\boldsymbol{x}}$ for sets or tuples of situations $\boldsymbol{x}$ by requiring equality of $\boldsymbol{s}$ and $\boldsymbol{t}$ for all coordinates in $\boldsymbol{x}$. Thus, there are modalities $\Box_{\boldsymbol{x}}\phi$ for each such tuple, which say intuitively that the situations in $\boldsymbol{x}$ settle the truth of ϕ in the current system:

$$\boldsymbol{M}, \boldsymbol{s} \models \Box_{\boldsymbol{x}}\phi \text{ iff } \boldsymbol{M}, \boldsymbol{t} \models \phi \text{ for each global state } \boldsymbol{t} \sim_{\boldsymbol{x}} \boldsymbol{s}.$$

This language can express more subtle properties of information.

---

[33]We are not claiming any originality for this set-up. Constraint models are also much like 'context models' or 'local semantics' in the style of Ghidini and Giunchiglia [2001], and they also resemble the local state models of 'interpreted systems' in the style of Fagin *et al.* [1995] — be it with a non-epistemic interpretation, for now.

Digression

One vivid metaphor here views situations themselves as *agents*. Operators $\Box_x, \Box_{\boldsymbol{x}}$ then express what single situations or groups *know* by inspecting their own local properties. This epistemic interpretation of constraint models is strengthened by one more analogy. The tuple modalities $\Box_{\boldsymbol{x}}$ involve an *intersection* of accessibility relations $\Box_x$ for single situations x. This is like 'distributed knowledge' for groups in epistemic logic, describing what whole sets of agents may be said to 'know implicitly'.[34]

As to valid laws, constraint models satisfy persistence axioms for atomic facts:

$$P\boldsymbol{x} \to \Box_{\boldsymbol{x}} P\boldsymbol{x}, \quad \neg P\boldsymbol{x} \to \Box_{\boldsymbol{x}} \neg P\boldsymbol{x}^{35}$$

More generally, the extended modal constraint language has a decidable complete logic with modal $S5$ for each tuple modality, plus all axioms of the forms $U\phi \to \Box_{\boldsymbol{x}}\phi$, and $\Box_{\boldsymbol{x}}\phi \to \Box_{\boldsymbol{y}}\phi$ whenever $\boldsymbol{y} \subseteq \boldsymbol{x}$.

6.3 *Digression: constraint logic and logic of dependence*[36]

A natural alternative to our story of situations and constraints involves another major notion in current logical theory, viz. *dependence*. Dependence fits well with information as correlation, and as we shall see, also with information as range. To bring this out, we can stay close to standard first-order logic.

For a start, think of the earlier situations as *variables* x, y, ... which store values. A global state $\boldsymbol{s}$ is then a *variable assignment* in the usual sense: a function assigning an object to each variable. Now first-order logic has no genuine dependencies between variables. In any assignment $\boldsymbol{s}$, we can shift the value of x to some object d to obtain a new assignment $\boldsymbol{s}[x := d]$, where all other variables have retained their $\boldsymbol{s}$-value. This is the reason why first-order logic typically has validities like the following commutation of quantifiers:

$$\exists x \exists y \phi \leftrightarrow \exists y \exists x \phi$$

The order of assigning values to the variables x and y is completely independent. But in many natural forms of reasoning, e.g., in probability theory, variables x, y *can* be dependent, in the sense that changes of value for one must co-occur with changes of value for the other. Such cases of genuine dependence can be modeled in a first-order setting [van Benthem, 1996, Chapters 9, 10]. A *general assignment model* is a pair $(\boldsymbol{M}, \mathbb{V})$ of a first-order model $\boldsymbol{M}$ with domain D and interpretation

[34] Similar points have been made by Baltag and Smets [2007] in their dynamic logic analysis of information structure and measurement-driven information flow in quantum mechanics.

[35] These implications do not hold for all formulas. E.g., $\Box_x Px \to \Box_y \Box_x Px$ is invalid, since accessible global states for x may change after a shift in the y coordinate.

[36] This section may be skipped without loss of continuity, but the main ideas are really simple, and they connect the preceding analysis with the heartland of logic.

function I, and $\mathbb{V}$ a non-empty set of assignments on $\boldsymbol{M}$, i.e., a subset of the total space D^{VAR}. The first-order language is now interpreted as usual, but using triples $\boldsymbol{M}, \mathbb{V}, s$ with $s \in \mathbb{V}$ — with the following clause for quantifiers:

$\boldsymbol{M}, \mathbb{V}, s \models \exists x\phi$ iff *for some* $t \in \mathbb{V} : s =_x t$ *and* $\boldsymbol{M}, \mathbb{V}, t \models \phi$.
Here $=_x$ relates assignments identical up to x-values.

The analogy with the earlier constraint language will be clear. Moreover, in this broader semantic setting, we get new dependence operators, such as *polyadic quantifiers* $\exists \boldsymbol{x}$ binding tuples of variables $\boldsymbol{x}$:

$$\boldsymbol{M}, \mathbb{V}, s \models \exists \boldsymbol{x}\phi \text{ iff } \textit{for some } t \in \mathbb{V} : s =_{\boldsymbol{x}} t \text{ and } \boldsymbol{M}, \mathbb{V}, t \models \phi^{37}$$

In first-order logic, $\exists xy \bullet \phi$ is just short-hand for $\exists x \exists y \phi$ or $\exists y \exists x \phi$ in any order. But in general assignment models, these expressions are no longer equivalent, as not all 'intermediate assignments' for x- or y-shifts need be present — and both fail to capture $\exists xy \bullet$ as defined here.[38]

The complete logic of general assignment models is a decidable subsystem of standard predicate logic. It contains those valid first-order laws which hold even when variables may be correlated [Németi, 1995; van Benthem, 1996; 2005a]. Beyond this decidable core logic, further axioms express special features of constraint models. E.g., the commutativity law $\exists x \exists y \phi \rightarrow \exists y \exists x \phi$ says that the following *Diamond Property* should hold:

If $s \sim_x t \sim_y u$, then there is another available assignment v with $s \sim_y v \sim_x u$.

Imposing such special conditions makes models much like full function spaces, and the complete first-order logic (of independence) becomes *undecidable.*

Van Benthem [2005] shows how the above modal constraint logic can be faithfully embedded into the first-order logic of dependent variables, and also vice versa. Thus, constraints and dependence are the same topic in two different guises! Dependence of various kinds is a major theme in foundations of first-order logic these days [Abramsky, 2006; Väänänen, 2007]. We have shown this is also a move toward a logic of information and constraints in the situation-theoretic sense.

6.4 *Combining epistemic logic and constraint logic*

Intuitively, information as correlation and information as range seem different notions. The difference is one of *agency.* A blinking dot on my radar screen carries information about some airplane approaching. But it does so whether or not I observe it. I may 'have' the information about the airplane, when I am in a situation at the screen, but unless I *know* that there is a blinking dot, it will not

[37] Here, $=_{\boldsymbol{x}}$ is identity between assignments up to values for all variables in $\boldsymbol{x}$.

[38] In these models, one can also interpret single or polyadic *substitution operators* in their own right: $\boldsymbol{M}, \mathbb{V}, s \models [\boldsymbol{y}/\boldsymbol{x}]\phi$ iff $s[\boldsymbol{x} := s(\boldsymbol{y})] \in \mathbb{V}$ & $\boldsymbol{M}, \mathbb{V}, s[\boldsymbol{x} := s(\boldsymbol{y})] \models \phi$.

do me much good. That knowledge arises from an additional event, my observing the screen. If I make that observation S, and I know the right constraint $S \Rightarrow A$, I will indeed also know that there is an airplane A.

Though distinct, the two phenomena obviously form a natural and compatible pair. To bring them together technically in a simple setting, we can use a *combined modal logic* of constraints and knowledge, first static and eventually also dynamic. We give one system, just to show how easy this is, viz. a combined *epistemic constraint language* with syntax

$$P\boldsymbol{x} \mid \neg \mid \vee \mid U \mid \Box_{\boldsymbol{x}} \mid K_i$$

Epistemic constraint models are then bi-modal structures of the form

$$\boldsymbol{M} = \ (Sit,\ State,\ \boldsymbol{C},\ Pred,\ \sim_i)$$

where global states have the earlier component-wise relations modeling constraints, while there are additional abstract epistemic accessibility relations $\sim_i$ for each agent i. Specifics of the latter depend on how the relevant scenario specifies agents' access to the situational structure.[39] We now have a simple language combining correlation and range talk. E.g., suppose that our model $\boldsymbol{M}$ satisfies the constraint $s_1 : p \rightarrow s_2 : q$. Then the agent knows this, as the implication is true in all worlds in $\boldsymbol{M}$. Now suppose the agent knows that $s_1 : p$. In that case, the agent also knows that $s_2 : q$, by the Distribution law of epistemic logic:

$$(Ks_1 : p \wedge K(s_1 : p \rightarrow s_2 : q)) \rightarrow Ks_2 : q$$

The converse requires more thought. The point is not that the agent already knows that $s_1 : p$, but that, if she were to *learn* this fact, she would also know that $s_2 : q$. In our earlier dynamic-epistemic terms, this would read as follows:

$$[!s_1 : p]Ks_2 : q.$$

This formula is equivalent to the truth of the constraint — by the axioms for dynamic-epistemic logic (cf. Section 4). Next, what do agents know about the informational content of specific situations x? If $\Box_{\boldsymbol{x}}\phi$ holds at some world $\boldsymbol{s}$, must the agent know this fact: $\Box_x\phi \rightarrow K\Box_x\phi$? Not so: $\Box_x\phi$ can be true at some worlds, and false at epistemically accessible ones. What a situation x 'knows' in the earlier impersonal sense of correlation need not be known to an external agent, unless this agent makes an observation about x. Thus, a combined modal-epistemic logic brings out the interaction between our two senses of information.

Digression: interpreted systems and a uniform vector view

In the paradigm of 'interpreted systems' [Fagin, *et al.*, 1995], epistemic worlds are vectors of 'local states' for individual agents, and epistemic accessibility between

[39] The general logic of epistemic constraint models is a mere fusion of that for constraint models and a poly-$S5$ epistemic logic of knowledge. In the case of one single agent, the earlier universal modality U then serves as an epistemic K.

worlds $\boldsymbol{s}, \boldsymbol{t}$ for agent i is just component-wise equality $(\boldsymbol{s})_i = (\boldsymbol{t})_i$. This structured view of possible worlds, extends to $\sim_{\boldsymbol{i}}$ for groups of agents $\boldsymbol{i}$, just as we did in constraint models with the move from $\sim_x$ to $\sim_{\boldsymbol{x}}$. On such a view, epistemic accessibility for agents and constraint accessibility for situations as part of the relevant physical world become formally analogous. Accordingly, we can then also use one uniform vector format for our combined models. Consider once again essentially the earlier example

$s_1: p, s_2: q$ $\quad$ $s_1: \neg p, s_2: \neg q$ $\quad$ $s_1: \neg p, s_2: \neg q$

Let some agent i have an accessibility structure indicated by the black dotted line:

$s_1: p, s_2: q$ $\;\ldots\;$ $s_1: \neg p, s_2: \neg q$ $\quad$ $s_1: \neg p, s_2: \neg q$

We can bring this into vector format by casting the agent itself as a further component, which can be in one of two states, as in the following picture:

$s_1: p, s_2: q, i:\ state_1$ $\quad$ $s_1: \neg p, s_2: \neg q, i:\ state_1$ $s_1: \neg p, s_2: \neg q, i:\ state_2$

The component-wise accessibility is the same as in the preceding picture.[40]

To summarize, information as correlation and information as range co-exist happily inside one formal modeling, and that even in more than one way.

6.5 *Explicit dynamics: events, information change, and correlation*

But the compatibility extends to the *dynamics* of both. Dynamic epistemic logic made informational events a key feature. Likewise, situation theory involved event scenarios for making use of ('harnessing') information, such as The Mousetrap in Section 5. Indeed, the constraint models of this section have a dynamic aspect in the first place. Constraints relate different situations, and the most useful correlations (think of the ground station and the mountain top) involve *events over time*, leading to different global states of some evolving system.

It seems of interest to bring out this temporal dynamics explicitly in some logic. Many situation-theoretic scenarios may be viewed as specifying automata consisting of many parts which can undergo correlated changes, through actions or events e. The Mousetrap and other running examples are of this kind: a complex system with a finite number of components cycles through a finite number of possible states, according to some fixed scenario. This style of analysis may be represented in combined *dynamic constraint models*

[40]Interpreted systems also involve stepwise action by agents, but more on this below.

$$\boldsymbol{M} = (\textit{Sit, State, } \boldsymbol{C}, \textit{ Pred, Event})$$

where events e are binary transition relations between global states. For instance, we may have had absence of a fire and a smoke signal, and then a combustion event takes place, changing the global state in our Mountain Top setting to *(smoke, fire)*. The matching language simply combines the earlier modal constraint operators with dynamic event modalities:

$$P\boldsymbol{x} \mid \neg \mid \vee \mid U \mid \Box_{\boldsymbol{x}} \mid [e]$$

Here, we interpret the dynamic modality in the usual modal style:

$$\boldsymbol{M}, \boldsymbol{s} \models [e]\phi \text{ iff } \textit{for all } \boldsymbol{t} \textit{ with } \boldsymbol{s}R_e\boldsymbol{t} : \boldsymbol{M}, t \models \phi$$

This language describes constraints on the current state, but its event modalities also record what happens as the system moves in the space of all its possible developments.[41] These merged logics seem easy to use and perspicuous.[42]

More sophisticated scenarios combining epistemic information, correlations, and informational events are discussed in Baltag and Smets [2007] on the structure of quantum information states and measurement actions. Its state spaces seem quite close to the view presented in this section, but now in Hilbert spaces.

If we want to describe longer-term system properties describing fixed and perhaps even shifting correlations over time, an earlier richer option is available. *Epistemic temporal languages* for multi-component systems in the sense of [Fagin *et al.*, 1995], or [Parikh and Ramanujam, 2003] describe the long-term behaviour of processes and agents over time, and they include intertwined accounts of external events and internal message passing.[43] Van Benthem *et al.* [2007] discuss how to explicitly model the implicit dynamics in the core situation-theoretic scenarios for 'harnessing information' in dynamic and temporal logics.

6.6 Conclusion: semantic co-existence

Information as range gave rise to both static and dynamic epistemic logics. Likewise, information as correlation leads to static and dynamic constraint logics. The two can be merged in a modal setting, implementing the schema in Section 1, where systems change over time, adapting their connections to agents who are informed about them, and to the reality they are about. Thus information as range

[41] A more realistic language would have local events at specific locations, making global events tuples of simultaneous local ones. This would be a form of 'concurrent' dynamic logic.

[42] *A connection with DEL.* Suppose a system has 2 states, and we do not know if it is in P or $\neg P$. Now we perform an observation, and learn that the state is P. This is not an internal 'system event', as it affects an outside agent's view of the system. In Section 4, public observations $!P$ changed epistemic models $\boldsymbol{M}$ to sub-models $\boldsymbol{M}|P$ with domain P^M. Both internal events and external observations can be implemented in dynamic constraint models by making agents components of the global state as before. The agent is trying to find out the current state $\boldsymbol{s}$, but the language can also express that $\boldsymbol{s}$ may change to $\boldsymbol{t}$ after system-internal events e.

[43] This also seems the proper setting for a logical account of the crucial issue of how our information and the beliefs based on it 'track' the world over time (cf. [Roush, 2006]).

and as correlation are intimately related. The two agendas merge naturally into one logical semantic view of information structure and dynamics.

7 INFORMATION AS CODE: SYNTAX, PROOF AND COMPUTATION

7.1 Information as code and elucidation

Our analysis so far has been largely semantical in orientation, and it seems successful in fitting logical notions of information into one perspective. But as noted at the start of this chapter, logic has another broad take on information structure and its processing, viz. inference on formulas. The relevant focus is then *syntax* and *proof theory* rather than model theory. This is a vast area, and we only discuss a few basic issues here, fleshing out a few relevant features of what might be called *information as code* – with a matching dynamics of *elucidation*.

For a start, the precise notion of 'information' behind the many calculi of inference in modern logic – and an optimal level of abstraction for it –, seems even less established than in logical semantics. Inference by its nature is more tied to details of syntactic representation, either on formulas per se, or some abstraction thereof. This allows for more variation in what counts as information states than our earlier semantic models. In line with this, there is a great variety of logical systems for deduction and proof, with very different formats. Standard mathematical Proof Theory [Troelstra and Schwichtenberg, 1996; Feferman, 2000; Tennant, 1978] uses natural deduction and associated calculi of type theory. Other well-established formats that have been linked with modeling information flow are logic programming and resolution theorem proving (cf. [Kowalski, 1979; Doets, 1994]), and more general unification-based inferential mechanisms (cf. [Rounds, 1997]). Less syntax-laden mathematical formats employ Category Theory [Lambek and Scott, 1994].

Proof theory is a rich field, and a whole chapter parallel to the present one might be written in its terms. Moreover, it has its own natural levels of abstraction for dealing with information in type theory or category theory, making room for a variety of logics: classical, intuitionistic, or sub-structural [Prawitz, 1965; Belnap, 1982; Restall, 2000]. We will not go into this here. Instead, we discuss a few general issues linking up with our earlier sections. We refer to the chapter by Abramsky in this Handbook for a state-of-the-art account of proof structure as it relates to the information flow in computation, another key process of elucidation.

7.2 How can inference be informative?

Let us start off with a somewhat surprising question, to which no definitive answer appears to be known. Simply put, our earlier semantic theories of information have the following problem. *Semantic approaches do not account for the informative nature of inference!* Depending on where one puts the blame, this problem may be formulated in various terms. Jaakko Hintikka has called it 'the scandal of deduction'. This is a real problem when we look at logical practice. For, deduction is clearly informative. Natural tasks show a clear interplay of update through observation, or recollection, or whatever source, with more combinatorial inferential steps. When solving a puzzle, we do not just update information spaces, we also make appropriate *deductions* which highlight some important aspect of the solution, or at least, the road toward it.

Example: update and/or inference

In practice, reasoning often trumps update. Here is a simple toy example. You are throwing a party subject to the following constraints: (a) John comes if Mary or Ann does, (b) Ann comes if Mary does not, (c) If Ann comes, John does not. Is this a feasible event? In principle, one can update as in Section 2 here, starting with 8 possible sets of invitees, and then ruling out 3 by (a), 2 more by (b), and finally, using (c) to cut down to the only solution remaining. But in reality, you would probably do something more like this (with logicians' annotation):

> By (c), if Ann comes, John does not. But by (a), if Ann comes, John does: a contradiction, so *Ann does not come.* Therefore, by (b), *Mary comes.* Then by (a) once more, *John comes.* Indeed, {John, Mary} satisfies all requirements.

There is a clear sense in which these successive inferential steps add relevant information. Indeed, cognitive science tells us (Knauff 2007) how in cognitive tasks, the brain continually mixes model inspection and inferential steps. But there is a difficulty in saying just how the latter process can be informative. An inference step does not shrink the current information range, and it does not add to correlational constraints. It rather adds information in a more combinatorial sense, which may depend on the agent's coded representations and even purposes. Other versions of this problem are known in the philosophy of science. E.g., much has been written about the way deductive consequences of a physical theory, such as Einstein's deduction for the perihelion perturbation of Mercury from the General Theory of Relativity [Fitelson, 2006], can uncover startling new information which a scientific community did not have before.

Information in computation

This problem is not unique to logic: witness several chapters of this Handbook. Inference is much like computation, and one issue running through the chapter by Abramsky in this Handbook is how computation can provide information, and how abstract, often category-theoretic semantics can provide a handle on this. In a more standard mathematical framework, the chapter by Adriaans analyzes information growth through computation in terms of numerical measures such as Kolmogorov complexity. There, too, inference or computation can increase information, but only by a little, since a shortest description of the theorem generator plus the allotted running length suffices. Thus, the divide between semantical and code-based approaches to information occurs outside of logic, too. In what follows we discuss a few strands at the syntax/semantics interface.

7.3 Logical syntax as information structure

In response to the challenge of information flow through deduction, various answers have been given. For instance, Hintikka [1973] proposed a way inside predicate logic for distinguishing levels of information in the syntactic analysis of a formula. First-order formulas ϕ describe the existence of objects with certain properties while excluding others, and these enumerations of object types can be unpacked level by level until the full quantifier nesting depth of ϕ. Inferences that stay close to the original syntax contain only 'surface information', those requiring full processing into normal form carry 'depth information' — notions which have been elaborated in some numerical detail.[44] [45] All this is clever, and it suggests that logical syntax might have natural information content where syntactic manipulation can increase information. But it is also very system-dependent. So far, no coherent and generally useful notion of syntax-based information has evolved – partly because we have no general theory of syntax, abstracting away from details, which would provide the vehicle for this. Frankly, we do not know how to remedy this. What we will do instead, in Section 8 below, is look at some ways in which semantic and syntactic accounts of information can be *merged*.

7.4 Proofs and information dynamics

By far the most sophisticated paradigm bringing out information flow in inference and computation, as we have noted already, is logical *Proof Theory.* Proof systems are a form of 'discourse syntax', generating ordered tree structures of sentences with their inferential links indicated, and hence they are a natural candidate for representing information structure at a syntactic level.

Natural deduction and type theory

In particular, the widely used method of *natural deduction* is an elegant calculus putting together, linking, and re-packaging pieces of evidence with their dependencies. A paradigmatic illustration are its proof rules for the basic logical operation of conditional assertion:

[44]Similar distinctions have even been used for defining the classical philosophical 'analytic/synthetic' distinction as a matter of first-order logic.

[45]Hintikka's specific numerical proposals have been criticized ever since they appeared.

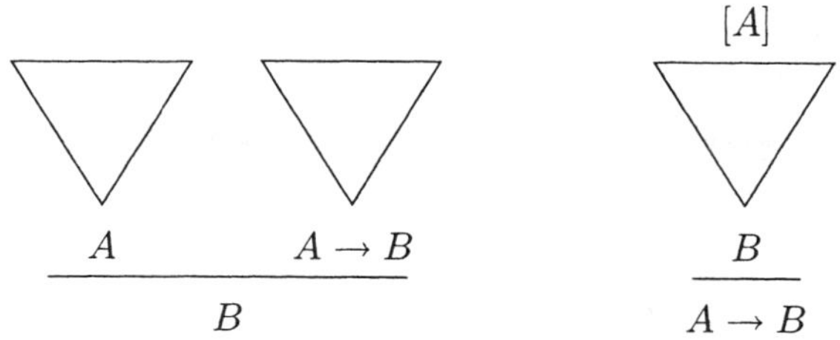

The view of mathematical proof calculi as general systems of *combining evidence for claims* has been put forward forcefully since the 1970s by authors such as Dummett, Martin-Löf, and Prawitz (cf. [Sundholm, 1986]). Moreover, there are natural notions of proof equivalence backing this up, for instance, through the valid identities of the lambda calculus and related abstract category-theoretic notions of equivalence (cf. [Lambek and Scott, 1994]). The resulting normal forms provide an attractive abstract level of information structure beyond brute details of syntax, and in the setting of linear logic, Girard [1987; 1989] has emphasized how proof normalization steps may be seen as communication moves, and hence as a form of information flow. We will not elaborate natural deduction, type theory, or category-theoretic proof paradigms in this chapter, but we refer to the chapter by Abramsky in this Handbook for an approach very much in the same spirit.

Proof construction as evidence dynamics

Inference steps construct proofs, and hence they transform information at a syntactic level in a process of elucidation. This dynamics seems in line with the constructivist, or *intuitionist* slant of many proof theories [Dummett, 1977; Troelstra and van Dalen, 1988], where one thinks of proof as performed by human agents creating objects and moving through successive information stages in the course of a mathematical enquiry. And even more dynamically, calculi of constructive proof can also be interpreted as models for multi-agent interaction. For instance, in the earlier-mentioned *dialogue games* of Lorenzen [1955], proof rules are *game moves*, and proofs are winning strategies for players. Viewed that way, proofs suddenly seem close to the dynamics we have placed at centre stage in this chapter. In this spirit, current 'logic of proofs' (starting from Artemov [1994]) is developing a general take on evidence combination, and one can view the 'labeled deductive systems' of Gabbay [1996] as a program with similar ambitions. We will briefly return to both in Section 8.

Even so, claims about the philosophical relevance of proof theory have a problem of 'transfer'. Mathematical proof is a very special way of establishing knowledge – and proofs, however elegant, hardly seem a paradigm for all the different sorts of evidence that humans manipulate, or the different ways in which they do so. Clearly, we are not just playing what Hintikka 1973 called 'indoor games' of proof or argumentation. Impeccable evidence also comes from our senses, our memory, from being told, and so on. And these other sources were described rather well in our earlier semantic accounts of information, without any obvious recasting as proof or computation. Therefore, a more promising line seems to be how to *merge* proof-theoretic and model-theoretic views of information into one story about rational agents capable of obtaining and integrating both.

Digression: logic programs

As we said before, there are other proof formats which can represent information flow in inference: cf. the extensive discussion in [Allo, 2007]. In particular, Jago [2006] shows how logic programming, too, provides attractive models for inferential information flow. Here information states are sets of propositional literals $(\neg)p$, and information generating events are applications of Horn-clause rules $p_1 \& \ldots \& p_n \rightarrow q$, which let these states grow step by step. This dynamics can be made explicit by taking proof rules themselves as action expressions, and describing inference steps with modal operators $\langle rule \rangle \phi$.[46]

In Section 8 we will discuss how proof-theoretic or other syntactic formats for information link up with our earlier semantic views. That there can be a significant connection seems clear from Gödel's *completeness theorem* which started off modern logic by showing how, in major systems like first-order logic, a formula ϕ is true in all semantic models if and only if ϕ has a syntactic proof.

7.5 Agent orientation once more!

Proof theory is often considered the most upper-class part of logic, dealing with crystalline mathematical structures where all traces of real human reasoning have been removed. But a return to the real world occurs as soon as we ask *what good* the information does which is created and transformed by inference. Syntax is clearly a representation of information as used *by some agent*, including possible idiosyncracies of formulation, and useful only given particular powers for performing elucidation on her data. What is informative for me in this manner need not be informative for you. Thus, deduction seems informative only for people engaged in certain tasks, and we are back with the central role of agency which also emerged in the earlier sections on agents construed semantically. This agent-orientation seems the right setting for understanding how an inference can inform us, and sometimes even surprise us, contradicting previous beliefs.

8 SEMANTICS, SYNTAX AND PROOF: MERGING PERSPECTIVES

Even though syntactic and semantic views of information seem far apart in spirit, the recent literature has seen a number of attempts toward combining them. While there is no consensus on how to do this, or what can be achieved, we group a few approaches in this section, to introduce the reader to at least some challenges.

[46] Like Égre [2006], Jago also views Fregean Morning Star / Evening Star-type puzzles involving substituting identicals in intensional contexts as further sources of non-omniscience and true inferential information.

8.1 *Two-level syntax-semantics formats*

One kind of proposal, in various guises, combines semantic and syntactic views. Suppose that information states of agents really have *two different levels*: (a) the current range of models or possible worlds, and (b) *some current set of sentences* true in these models which the agent has available right now. Like the worlds in epistemic logic, and the situations of situation theory, the syntax domain, too, may be structured — say by an ordering of formulas as to relative importance in our current set of beliefs, or our current 'theory' [Ryan, 1991; Gärdenfors, 1987]. Then we can think of informational processes as operating on two levels:

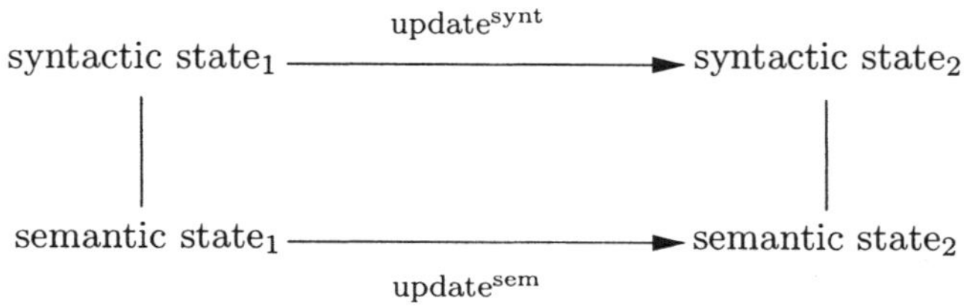

Range-changing updates modify the lower level, while inference modifies the upper syntax component of the current state, and hence the information 'immediately present' to the agent. This process can be described in the same sort of dynamic logic format used for epistemic logic in Section 3. Examples of this sort of approach include [Dung, 1995] on dynamic logic of argumentation, [Wassermann, 2001] on belief revision over syntactic knowledge bases, [Gabbay *et al.*, 2001] on dynamic logic of grammars, and the earlier-mentioned [Jago, 2006].[47] Two- and many-level views also occur at the interface of logic and *cognitive science* [Castelfranchi and Lorini, 2007] in realistic modeling of the many cognitive processes involved when humans process information.

8.2 *Merging calculi of proofs and evidence*

There are several ways of merging epistemic and proof-theoretic paradigms. Section 7 mentioned current attempts at broadening standard proof-theoretic accounts of internal elucidation with external observation of new information. We first discuss one strand in the recent literature which stays close to proof theory.

Range and evidence: a case of $\forall$ *versus* $\exists$

For a start, as we have said earlier, proof-theoretic and semantic perspectives on information are not in conflict: they complement each other. But how? Let us first high-light a prima facie logical difference. As we saw in Section 2, epistemic logic of information as range revolves around a *universal quantifier*: $K_i\phi$ says that

[47]Also relevant are current syntax-flavoured 'neighbourhood models' for epistemic notions, such as the treatment of beliefs over time in [Arlo-Costa and Parikh, 2005].

> ϕ *is true* in all situations *agent i considers as candidates for the real s.*

But there is also an existential quantifier, occurring in the famous traditional account of knowledge that ϕ as 'justified true belief. The first phrase "justified" in this definition says intuitively that

> *there exists a justification* for the formula ϕ.

In this sense, knowledge consists in having evidence for a proposition: sometimes, on an exceptionally clear Cartesian day, even a mathematical proof. But as we saw, the two views co-exist in logic! Completeness theorems establish an equivalence between validity (an $\forall$-type notion) and provability (an $\exists$-type notion).

Combined calculi

To have *both views together* then, van Benthem [1993] proposed merging epistemic logic with a 'calculus of evidence'. Candidates range from *intuitionistic logic* with binary type-theoretic assertions of the form

> *x is a proof for* ϕ.

to the much more general 'labeled deductive systems' of Gabbay 1996, which were designed as a general calculus of statements $x : \phi$, where the label x can stand for a proof, an observation, or indeed any kind of evidence from any source, possibly even with further information attached about reliability, probability, and so on, of the support that x offers to ϕ.

Even closer to the epistemic logics of Sections 2, 3, the 'logic of proofs' of Artemov [1994; 2005] is a modal language enriched with proof- or evidence-labeled assertions $[x]\phi$ then, e.g., the ubiquitous (though philosophically controversial; [Egré, 2004]) epistemic Distribution Axiom $K_i(\phi \to \psi) \to (K_i\phi \to K_i\psi)$ becomes the more informative statement that

$$[x](\phi \to \psi) \ \& \ [y]\phi \to [x\#y]\psi,$$

where # is some natural *sum operation* on proofs, or pieces of evidence generally. Incidentally, van Benthem [2006a] shows how this also works for contextualist views of information — as discussed in the chapter by Dretske in this Handbook. In that case the crucial law is

$$[c_1]K_i(\phi \to \psi) \ \& \ [c_2]K_i\phi \to [c_1\#c_2]K_i\psi,$$

where # is now an operation of 'context merge' reminiscent of situation theory. This is all fine, but it has to be admitted that one coherent philosophical interpretation of these combined systems still remains to be found.[48]

[48]For a merge of logic of proofs with dynamic epistemic logic, cf. [Renne, 2007]. Also relevant is [Artemov, 2007] on connections with general epistemology.

8.3 Merging proof and observation: two-level dynamic logics

But integrating inferential information flow with observational update can also be done in formats that stay close to the dynamic logic of our Section 4. Here is a toy illustration from van Benthem [2007c], joining two simple components.

Consider the earlier logic programs for a propositional language with sets of literals that can be modified by inferential steps. Let observations be incoming hard information, as in the logic *PAL* of public announcements. We need to hook up the syntactic information states with semantic ones. Here is one way of doing this. A semantic model $\boldsymbol{M}$ has valuations V accepting or rejecting each proposition letter, i.e., complete worlds for our simple language. Next, $\boldsymbol{M}$ specifies a family of inference rules $\boldsymbol{R}$ whose corresponding conditionals are valid in the given set of valuations. Finally, it has partial sets of literals Xwhich do not yet decide every proposition letter. Each such set X stands in an obvious 'compatibility relation' with all total valuations V containing it. More precisely then, we can say that

information states are *pairs* (V, X) with X a set of literals true in V.

One can think of these as an agent's current semantic range about the empirical facts in the actual world *plus* its current inferential approximation to the worlds in that range. A matching logical language will access both components. Here a knowledge operator $K\varphi$ will operate on V components only, referring to the whole semantic range as before. Next, one could access the X component by an operator $I\varphi$ saying that the agent 'realizes that φ' if the literal φ is actually present in X. In this setting, there are then several natural kinds of informative action:

Internal *inference steps* do not change the V component of the current state, but take some rule and add literals to the current X component (if the premises are there). For external observation we take *PAL*-style *public announcements* $!P$, removing all states (V, X) where V fails to satisfy P. But the interesting point of the combination is that there may be other 'intermediate' actions with something of both. In particular, think of the act of *seeing* in some conscious sense. We can model this with a third kind of update action on our hybrid information states. *Explicit observations* $+q$ of literals q operate like announcements $!q$, but then also place q directly into all X components of pairs (V, X) in the remaining model. Putting all this into a logical system with both static and dynamic modalities, we can make combined assertions which describe the fine-structure of all these informational processes, both inferential and observational, within one setting.[49]

8.4 Abstract viewpoints: toward a deeper unification?

The preceding two approaches shows that evidence based $\exists$-type and range-based $\forall$-type accounts of information can live together. But this compatibility does

[49] Jago [2007] presents an alternative take on this, however, enriching the set of possible worlds with worlds carrying 'non-standard valuations' which are eliminated successively from the set of candidates through inferential steps. This seems a full-fledged alternative to the above pair-approach, based on ideas from relevant and paraconsistent logic (cf. [Priest, 1997]).

not answer the more ambitious question if there is some *deeper identity* between evidence combination in an epistemic sense and that of syntax with proof theory. Stated as an observation about information and computation in general, *proof theory* and *dynamic logics* often address very similar issues and provide attractive solutions, whether in process theory or information dynamics, but the precise analogy between these two broad styles of thinking remains a bit of a mystery.

Abstract information theories: relevant and categorial logics

Indeed, there are several abstract perspectives merging the two perspectives. One is that of Scott Information Systems, discussed by Michael Dunn in this Handbook. Another line are logics in the tradition of *categorial* and *relevant logic*, which have often been given an informational interpretation. One example is the 'Gaggle Theory' of Dunn 1991, inspired by the algebraic semantics for relevant logic, which provides an abstract framework that can be specialized to combinatory logic, lambda calculus and proof theory, but on the other hand to relational algebra and dynamic logic, i.e., the modal approach to informational events. Van Benthem [1991] arrives at a similar duality starting from categorial grammars for natural language, which sit at the interface of parsing-as-deduction and dynamic semantics. Indeed, he points out how the basic laws of the categorial 'Lambek Calculus' for product and its associated directed implications have both dynamic and informational interpretations:

$$A \cdot B \Rightarrow C \quad \text{iff} \quad B \Rightarrow A \rightarrow C$$
$$A \cdot B \Rightarrow C \quad \text{iff} \quad A \Rightarrow C \leftarrow B$$

Here, the product can be read dynamically as *composition* of binary relations modeling transitions of some process, and the implications as the corresponding right- and left-inverses. But these laws can be read equally well as describing a universe of information pieces which can be *merged* by the product operation. E.g., we can read $A \rightarrow B$ as the directed implication denoting $\{X \mid \forall y \in A : y \cdot x \in B\}$, with $B \leftarrow A$ read in the obvious corresponding left-adjoining manner. On both interpretations, the principles of the Lambek Calculus hold (cf. [van Benthem, 1991] for further theory). Beyond that, however, the usual structural rules of classical inference turn out to fail,[50] and thus, there is a strong connection between *substructural logics* and what might be called abstract information theory [Mares, 1996; 2003; Restall, 2000]. This dynamic/informational interpretation also makes sense for Gabbay's earlier-mentioned paradigm of 'labeled deductive systems'.[51]

[50] In particular, the rules of Contraction and Permutation would express highly questionable assumptions about procedural or informational resources, which have no appeal in general.

[51] Van Benthem [1998] points out how semantic and proof-theoretic strands come together in labeled deductive systems. He unifies both in terms of a ternary calculus of *abstract combination*, with the following two rules: (a) $x : A, y : A \rightarrow B, Rz, xy \vdash z : B$, where Rz, xy is some *ternary* condition relating z, x, y. Here the atom Rz, xy can be read as 'z is the sum of the information pieces x, y', 'z results from composing transitions x, y', 'z results from applying function x to argument y', etc. Principle (a) is then the abstract analogue of the earlier Modus Ponens in

Sequoiah-Grayson [2007] is a spirited modern defense of the Lambek calculus as a minimal core system of information structure and information flow. While this is appealing, it has to be said that the above axioms merely encode the minimal properties of mathematical adjunctions, and these are so ubiquitous that they can hardly be seen as a substantial theory of information.[52]

Remark: Situation Theory after all

Interestingly, authors in the relevant logic tradition also tend to see their systems as a direct continuation of the abstract analysis of information flow in situation theory — a link which we do not pursue here [Mares, 2003; Restall, 2000]. The doctoral thesis [Allo, 2007] is an illuminating discussion of logical models underpinning the 'philosophy of information' set forth in the chapter by Floridi in this Handbook, joining up epistemic logic with sub-structural logics [Restall, 2000], and Batens' 'adaptive logic' program (cf. [Primero, 2007]).

But we are far from having exhausted all existing abstract information theories.

Other abstract models for unification are found in the work on logical omniscience in Parikh [1987; 1995], and most recently Parikh [2007]. These papers propose an abstraction level that tries again to explain how deduction can be informative. [53] Other relevant work is found in the Czech Tradition, of which we mention [Tichy, 2004; Materna, 2004; Duzi *et al.*, 2005]. But we conclude with a few strands from the grand tradition of foundational research.

8.5 Coda: intuitionistic logic and modal information models

Let us step back for a moment from latter-day abstractions. Maybe the oldest and still quite attractive joint proof-theoretic and semantic view of information occurs in *intuitionistic logic* [Dummett, 1977 ; Troelstra and van Dalen, 1988; cf. Section 7]. Unlike epistemic logic, it is a calculus of 'implicit knowledge' [van Benthem, 1993], where the very meanings of the standard logical constants are tied up with being 'known' or 'provable'. Intuitionistic logic found its original proof-theoretic explanation in natural deduction style through the so-called Brouwer-Heyting-Kolmogorov interpretation. But it also has modal-style semantic models due to Tarski, Beth, and Kripke, where 'worlds' are information stages in some ongoing enquiry. This long-standing harmony suggests that in logical systems like this, inferential and semantic information are very close. This strong connection comes out even very forcefully in the 'full completeness' theorems of Abramsky

natural logic. Here is the principle for the dual rule of Conditionalization: (b) $\tau: X, x: A, Rz, xy \vdash z: B$ implies $\tau: X \vdash y: A \to B$. So far, this is still an abstract format, however, without a corresponding intuitive unification.

[52] Just to mention some quite different sources, the same laws arise in matrix algebra, theories of vector spaces in mathematical morphology, and 'spatial implications' in logics of security.

[53] Parikh [2007] also has an extremely interesting account of knowledge as that which is revealed in observable Nash equilibria of a new kind of evaluation games for testing assertions.

and Jagadeesan [1994], which do not just link provability to validity, but specific proofs to concrete objects in (category-theoretic) semantics.

From intuitionistic to modal logics of information

Intuitionistic logic has a very specific view on valid inference, far beyond what one would want to endorse in a general information theory. Thus, attempts have been made at abstracting a more general view of information, though retaining its attractive features. Van Benthem 1996B proposes a 'modal theory of information' over models of abstract information stages ordered by inclusion (cf. also the 'data semantics' of Veltman 1984, Landman 1986), and there are related recent approaches to abstract information modeling in Sequoiah-Grayson 2006. These modal views seem related to the earlier-mentioned categorial and relevant approaches, but much remains to be clarified.

Digression: information that and information how

Merging semantic and proof-theoretic perspectives is of interest in other ways, too. Proof theory has a clear *algorithmic* aspect, which we have ignored here (but cf. the chapters by Abramsky, Adriaans, and Grunwald and Vitanyi in this Handbook on various algorithmic aspects of information). Indeed, ever since Euclid's "Elements", constructive proofs are both ways of seeing that a conclusion is true and methods for constructing the relevant objects. In type-theoretic calculi, proofs are definable function terms in some mathematical universe through the Curry-Howard isomorphism. This duality ties in with the famous distinction between *knowledge that* and *knowledge how* [Gochet, 2006] in philosophy and computer science. The latter is about procedures and general cognitive skills. This also makes sense for information. As pointed out in Israel and Perry 1991, this notion is not just about what things are like, but also how to get thing things done.

8.6 Conclusion

Despite the abundance of ideas recorded in Sections 7 and 8, there is no consensus on the integration of inferential and semantic views of information structure and information dynamics. Indeed, the chapter by Samson Abramsky in this Handbook highlights the contrast even more, distinguishing logic as *describing* informational dynamics (the main thrust of our semantic approaches) from logic as *embodying* dynamics, the more proof-theoretical view that logical denotations are themselves informational processes. In such a light, our earlier Tandem View of representation and process would reflect two fundamental faces of logic itself. However this may be, given the history of logic, semantic and inferential views of information may be irreducibly complementary stances, and it may be precisely their interplay which is needed to grasp the notion of information in its entirety.

9 CONCLUSION: LOGIC AS INFORMATION THEORY

This chapter has painted logic as a theory of information, with an actual or potential impact on many disciplines. This is not at all the standard outside view or self-image of the field, but it does seem a legitimate perspective. We have woven a story which puts many logical notions and results into one new line, but one rather different from a standard textbook presentation. Here are our major conclusions, or rather: our main claims, in doing so.

First, to get the real picture of logical information, one needs to address the *statics and dynamics in parallel*, with intertwined accounts of information structure and the dynamic processes which manipulate the latter. Next, there are many such dynamic processes, which need not all reduce to one primitive: be it inference or update, and hence logical views of information come in complementary kinds. More specifically, we have re-interpreted epistemic logic as an information theory of *range, knowledge*, and *observation-based update*. Like classical information theories, this is not just a model of information, but also a calculus for computing information flow, witness the concrete procedures for model update, and the axiomatic dynamic-epistemic laws governing this. We also emphasized the essential role of agents, and the ways in which *agents take information*: alone, but much more importantly, in *interaction* with others. Information is studied best in a setting of many agents, communication, and perhaps even interactive games.

Next, we have high-lighted another major perspective on information, viz. its 'aboutness', and the matching situation-theoretic account of *correlations and dependence* between different parts of distributed systems. This provides a much richer view of how agents can actually access information, and why it is there for them to pick up and communicate in the first place. We have shown that, despite some common misconceptions, this situation-theoretic perspective is not in conflict with the epistemic one, but rather forms its natural complement — and we have provided a number of merged systems doing justice to all aspects so far.

Finally, logic offers another major view of *information as code* through its proof systems and other syntactic calculi which elucidate implicitly available semantic information by computing on suitable data structures. Our view is that this is another informational stance in its own right, close to the worlds of algorithmics and computation. We have surveyed some ways in which the syntactic stance and its natural processes can live in harmony with the semantic ones.

Re-interpreting logical systems in this way raises further questions. What does the usual logical agenda mean in this light? Does it provide a calculus for information comparable to that of the powerful Shannon-style or Kolmogorov-style information theories in this Handbook? Most standard results in logic are concerned with (a) expressive power and *definability*, (b) *axiomatic completeness* for proof systems, or (c) computational *complexity* of tasks like model checking or proof search. We are not claiming that all of these are equally important as general concerns about 'information', but we do think that our re-interpretation might shed some fresh light on what we are doing in our standard grooves.

Finally, there is the issue of *Grand Unification*. Can all stances toward information be unified in one framework? We do not think so. Some abstract unification of all logical information theories may always exist, ascending far enough into mathematical Heaven. But what is the distance to Earth, i.e., our cognitive practices? Indeed, we doubt if a Grand Unification is necessary. Having several *complementary stances* in a field is fruitful in itself. Just compare the interplay of semantic and algorithmic views of information running throughout this Handbook. What is good enough, instead of unification, both in practice and in theory, is a rigorous way of analyzing differences and possible merges between stances. And that the latter is feasible should be clear from our text.

10 FURTHER TOPICS AND BROADER OUTREACH

Our presentation has focused on a few major senses of information in logic, and some systems showing how they work. This is by no means an exhaustive survey of all topics and issues, and several alternative chapters could have been written mapping out the territory differently. Just think of analogies and differences with *Probability Theory*, logic's neighbour, but also its closest rival in the analysis of information. Of the omissions that weigh most on our conscience, besides logic and probability (cf. [Loewe *et al.*, 2007] for some recent interfaces), we mention a few, with some highly unsystematic references:

- Information content and the methodology of science (cf. [Kuipers, 2000])
- Visual and other information carriers beyond language [Allwein and Barwise, 1996; Kerdiles, 2001, Aiello *et al.*, 2007]
- Information and context [Buvac and McCarthy, 2004]
- Information, interaction, and games [Sevenster, 2006]

We console ourselves with the thought that many of these themes are found in other guises in other chapters of this Handbook, in particular, the ones on language, philosophy, economic game theory, computation, and cognition.

ACKNOWLEDGMENTS

This chapter was written by two authors. Maricarmen Martinez was responsible for the parts on situation theory, with some valuable textual contributions by John Perry — and by Johan van Benthem for the epistemic logic and the other parts. We thank our commentators David Israel and John Perry for their critical comments throughout. In addition, we thank the following people for their stimulating responses to various versions of this chapter: Pieter Adriaans, Patrick Allo, Marie Duzi, Branden Fitelson, Paolo Mancosu, Edwin Mares, Ron van der Meyden, Larry Moss, Rohit Parikh, José Saguillo, and Sebastian Sequoiah-Grayson.

BIBLIOGRAPHY

[Abramsky, 2006] S. Abramsky. Socially Responsive, Environmentally Friendly Logic, in T. Aho and A-V Pietarinen, eds., *Truth and Games: Essays in Honour of Gabriel Sandu*, Acta Philosophica Fennica, 17–45, 2006.

[Abramsky and Jagadeesan, 1994] S. Abramsky and R. Jagadeesan. Games and Full Completeness for Multiplicative Linear Logic, *Journal of Symbolic Logic* 59, 543–57, 1994.

[Aczel, 1988] P. Aczel. *Non-well-founded Sets*, CSLI Publications, Stanford, 1988.

[Aiello *et al.*, 2007] M. Aiello, I. Pratt-Hartmann and J. van Benthem, eds., *Handbook of Spatial Logics*, Springer, Dordrecht, 2007.

[Akman and Tin, 1997] V. Akman and E. Tin. Situated Non-Monotonic Temporal Reasoning with BABY-SIT, *AI Communications*, 10; 93-109, 1997.

[Allo, 2007] P. Allo. *On Logics and Being Informative*, Ph.D. Thesis, Faculty of Arts, Vrije Universiteit Brussel, 2007.

[Allwein and Barwise, 1996] G. Allwein and J. Barwise, eds., *Logical Reasoning with Diagrams,* Oxford University Press, New York, 1996.

[Arlo-Costa and Pacuit, 2005] H. Arlo-Costa and E. Pacuit. First-Order Classical Modal Logic: Applications in Logics of Knowledge and Probability. In R. van der Meyden, ed., *Theoretical Aspects of Rationality and Knowledge:* Proceeding of TARK X, 262–278, 2005.

[Arlo-Costa and Parikh, 2005] H. Arlo-Costa and R. Parikh. Some Non-Normal Logics of Knowledge without Logical Omniscience, Department of Philosophy, Carnegie-Mellon University, Pittsburgh, 2005.

[Artemov, 1994] S. Artemov. Logic of Proofs, *Annals of Pure and Applied Logic* 67, 29–59, 1994.

[Artemov, 2005] S. Artemov. Evidence-Based Common Knowledge, CUNY Graduate Center, New York, 2005.

[Artemov, 2007] S. Artemov. Justification Logic, CUNY Graduate Center, New York, 2007.

[Baltag, 1999] A. Baltag. STS: A Structural Theory of Sets, *Logic Journal of the ILPG* 7 (4), 481-515, 1999.

[Baltag *et al.*, 2007] A. Baltag, J. van Benthem and S. Smets. A Dynamic-Logical Approach to Interactive Epistemology, Working Paper, Institute for Logic, Language and Computation, Amsterdam, 2007.

[Baltag *et al.*, 1998] A. Baltag, L. Moss and S. Solecki. The Logic of Public Announcements, Common Knowledge and Private Suspicions, *Proceedings TARK 1998*, 43–56, Morgan Kaufmann Publishers, Los Altos, 1998.

[Baltag and Smets, 2006] A. Baltag and S. Smets. Dynamic Belief Revision over Multi-Agent Plausibility Models, *Proceedings LOFT 2006*, Department of Computing, University of Liverpool, 2006.

[Baltag and Smets, 2007] A. Baltag and S. Smets. Logics for Quantum Information Flow, Workshop Philosophy of Information, Oxford University. To appear in *Knowledge, Rationality and Action.*

[Barendregt, 1984] H. Barendregt. *The Lambda Calculus, its Syntax and Semantics*, North-Holland, Amsterdam, 1984.

[Bar-Hillel and Carnap, 1953] Y. Bar-Hillel and R. Carnap. Semantic Information, *The British Journal for the Philosophy of Science* 4:14, 147–157, 1953.

[Barwise, 1997] J. Barwise. Information and Impossibilities, *Notre Dame Journal of Formal Logic* 38, 488–515, 1997.

[Barwise and van Benthem, 1999] J. Barwise and J. van Benthem. Interpolation, Preservation, and Pebble Games, *Journal of Symbolic Logic* 64, 881–903, 1999.

[Barwise and Etchemendy, 1987] J. Barwise and J. Etchemendy. *The Liar: An Essay on Truth and Circularity*, Oxford University Press, New York, 1987.

[Barwise and Moss, 1996] J. Barwise and L. Moss. *Vicious Circles: On the Mathematics of Non-Well-founded Phenomena*, CSLI Publications, Stanford, 1996.

[Barwise and Seligman, 1997] J. Barwise and J. Seligman. *Information Flow: The Logic of Distributed Systems*, Cambridge University Press, Cambridge, 1997.

[Barwise and Perry, 1983] J. Barwise and J. Perry. *Situations and Attitudes*, The MIT Press, Cambridge (Mass.), 1983.

[Belnap, 1982] N. Belnap. Display Logic, *Journal of Philosophical Logic* 11, 375–417, 1982.

[van Benthem, 1991] J. van Benthem. *Language in Action: Categories, Lambdas, and Dynamic Logic*, North-Holland Amsterdam and MIT Press, Cambridge (Mass.), 1991.

[van Benthem, 1993] J. van Benthem. Reflections on Epistemic Logic, *Logique et Analyse* 34, 5–14, 1993.

[van Benthem, 1996a] J. van Benthem. *Exploring Logical Dynamics*, CSLI Publications, Stanford, 1996.

[van Benthem, 1996b] J. van Benthem. Modal Logic as a Theory of Information, in J. Copeland, ed., *Logic and Reality. Essays on the Legacy of Arthur Prior*, Clarendon Press, Oxford, 135–168, 1996.

[van Benthem, 1998] J . van Benthem. Proofs. Labels, and Dynamics in Natural Language, in U. Reyle and H-J Ohlbach, eds., *Festschrift for Dov Gabbay*, 31–41, Kluwer Academic Publishers, Dordrecht, 1998.

[van Benthem, 2000] J. van Benthem. Information Transfer Across Chu Spaces, *Logic Journal of the IGPL 8*, 719–731, 2000.

[van Benthem, 2001] J. van Benthem. Games in Dynamic Epistemic Logic, *Bulletin of Economic Research* 53:4, 219–248, 2001.

[van Benthem, 2005a] J. van Benthem. Guards, Bounds, and Generalized Semantics, *Journal of Logic, Language and Information* 14, 263–279, 2005.

[van Benthem, 2005b] J. van Benthem. Information as Correlation versus Information as Range, Research Report, Institute for Logic, Language and Computation, University of Amsterdam. To appear in L. Moss, ed., *Logic and Cognition*, Memorial Volume for Jon Barwise, 2005.

[van Benthem, 2006a] J. van Benthem. Epistemic Logic and Epistemology: the state of their affairs, *Philosophical Studies* 128, 49–76, 2006.

[van Benthem, 2006b] J. van Benthem. One is a Lonely Number: on the logic of communication, in Z. Chatzidakis, P. Koepke and W. Pohlers, eds., *Logic Colloquium 02,* ASL and A.K. Peters, Wellesley MA, 96–129, 2006.

[van Benthem, 2007a] J. van Benthem. Dynamic Logic of Belief Revision, *Journal of Applied Non-Classical Logics* 17, 129–155, 2007.

[van Benthem, 2007b] J. van Benthem. Inference in Action, Research Report, Institute for Logic, Language and Computation, University of Amsterdam. To appear in *Logic in Serbia*, Serbian Academy of Sciences, Beograd, 2007.

[van Benthem, 2007c] J. van Benthem. Tell It Like It Is: Information Flow in Logic, *Philosophical Review*, Beijing, 2007.

[van Benthem and Bezhanishvili, 2007] J. van Benthem and G. Bezhanishvili. Modal Logics of Space, in M. Aiello, I. Pratt and J. van Benthem, eds., *Handbook of Spatial Logics*, Springer, Dordrecht, 217–298, 2007.

[van Benthem and Blackburn, 2006] J. van Benthem and P. Blackburn, Modal Logic, a Semantic Perspective, in J. van Benthem, P. Blackburn and F. Wolter, eds., *Handbook of Modal Logic*, Elsevier, Amsterdam, 1–84, 2006.

[van Benthem *et al.*, 2006] J. van Benthem, J. van Eijck and B. Kooi. Logics of Communication and Change, *Information and Computation* 204(11), 1620–1662, 2006.

[van Benthem *et al.*, 2007a] J. van Benthem, J. Gerbrandy and E. Pacuit. Merging Frameworks for Interaction: *DEL* and *ETL*, ILLC Amsterdam and Informatics Torino. *Proceedings TARK* 2007, University of Namur, 2007.

[van Benthem *et al.*, 2007b] J. van Benthem, D. Israel and J. Perry. Situation Theory and Dynamic Logic, Department of Philosophy, Stanford University, 2007.

[van Benthem and Liu, 2004] J. van Benthem and F. Liu. Diversity of Logical Agents in Games, *Philosophia Scientiae 8:2,* 163–178, 2004.

[van Benthem and Liu, 2007] J. van Benthem and F. Liu. Dynamic Logics of Preference Upgrade, *Journal of Applied Non-Classical Logics* 17, 157–182, 2007.

[van Benthem and Pacuit, 2006] J. van Benthem and E. Pacuit. The Tree of Knowledge in Action, *Proceedings Advances in Modal Logic,* ANU Melbourne, 2006.

[van Benthem and Sarenac, 2005] J. van Benthem and D. Sarenac. The Geometry of Logic, in J-Y Béziau, A. Costa Leite and A. Facchini, eds., *Aspects of Universal Logic*, Centre de Recherches Sémiologiques, Université de Neuchatel, 1– 31, 2005.

[Blamey, 2002] S. Blamey. Partial Logic, in D. Gabbay and F. Guenthner, eds., *Handbook of Philosophical Logic*, Vol. 5 (2nd ed.), Kluwer, Dordrecht, 261-353, 2002.

[de Bruin, 2004] B. de Bruin. *Explaining Games*, Dissertation. ILLC, University of Amsterdam, 2004.

[Burgess, 1981] J. Burgess. Quick Completeness Proofs for some Logics of Conditionals, *Notre Dame Journal of Formal Logic* 22, 76–84, 1981.

[Buvac and McCarthy, 2004] S. Buvac and J. McCarthy. Formalizing Context (Expanded Notes), Technical Report CS-TN-94-13, Department of Computer Science, Stanford University, 2004.

[Carnap, 1952] R. Carnap. *The Continuum of Inductive Methods*, The University of Chicago Press, Chicago, 1952.

[Chellas, 1980] B. Chellas. *Modal Logic: an Introduction*, Cambridge University Press, Cambridge, 1980.

[Corcoran, 1998] J. Corcoran. Information-Theoretic Logic, in C. Martínez, U. Rivas, L. Villegas–Forero, eds., *Truth in Perspective*, Ashgate Publishing, Aldershot, 113–135, 1998.

[van Dalen, 2002] D. van Dalen. Intuitionistic Logic, in D. Gabbay and F. Guenthner, eds., *Handbook of Philosophical Logic*, Vol. 5 (2nd ed.), Kluwer, Dordrecht, 1-114, 2002.

[Devlin, 1991] K. Devlin. *Logic and Information*, Cambridge University Press, 1991.

[van Ditmarsch *et al.*, 2007] H. van Ditmarsch, W. van der Hoek and B. Kooi. *Dynamic Epistemic Logic*, Springer. Dordrecht, 2007.

[Doets, 1994] K. Doets. *From Logic to Logic Programming*, The MIT Press, Cambridge (Mass.), 1994.

[Dretske, 1981] F. Dretske. *Knowledge and the Flow of Information*, The MIT Press, Cambridge (Mass.), 1981.

[Dummett, 1977] M. Dummett. *Elements of Intuitionism*, Oxford University Press, Oxford, 1977.

[Dung, 1995] P. Dung. An Argumentation-Theoretic Foundation for Logic Programming, *Journal of Logic Programming* 22, 151–177, 1995.

[Dunn, 1991] M. Dunn. Gaggle Theory: An abstraction of Galois connections and Residuation, with applications to negation, implication, and various logical operators, in J. van Eijck, ed., *Logics in AI (Amsterdam,* 1990), Springer, Berlin, 31–51, 1991.

[Duzi *et al.*, 2005] M. Duzi, B. Jespersen and B. Müller. Epistemic Closure and Inferable Knowledge, in L. Behounek and M. Bilkova, eds., *The Logica Yearbook 2004*, Filosofia, Prague, 125–140, 2005.

[Egré, 2004] P. Egré. *Attitudes Propositionnelles et Paradoxes Épistémiques*, Thése de Doctorat, Université Paris 1 Panthéon-Sorbonne, IHPST, 2004.

[Fagin *et al.*, 1995] R. Fagin, J. Halpern, Y. Moses and M. Vardi. *Reasoning about Knowledge,* The MIT Press, Cambridge (Mass.), 1995.

[Feferman, 2000] S. Feferman. Highlights in Proof Theory, in V. Hendricks et al., eds., *Proof Theory*, pages 11–31. Kluwer Academic Publishers, Dordrecht, 2000.

[Fitelson, 2006] B. Fitelson. Old Evidence, Logical Omniscience and Bayesianism, Lecture ILLC Workshop Probability and Logic, Department of Philosophy, University of California at Berkeley, 2006.

[Forti and Honsell, 1983] M. Forti and F. Honsell. Set Theory with Free Construction Principles, *Annali della Scuola Normale Superiore di Pisa*, Classe di Scienze, Serie IV:10, 493–522, 1983.

[Gabbay, 1996] D. Gabbay. *Labeled Deductive Systems*, Oxford University Press, Oxford, 1996.

[Gabbay *et al.*, 2001] D. Gabbay, R. Kempson, and W. Meyer-Viol. *Dynamic Syntax: The Flow of Language Understanding*, Blackwell, Oxford, 2001.

[Ganter and Wille, 1997] B. Ganter and R. Wille. *Formal Context Analysis: Mathematical Foundations*, Springer Verlag, New York, 1997.

[Gärdenfors, 1987] P. Gärdenfors. *Knowledge in Flux, on the Dynamics of Epistemic States*, The MIT Press, Cambridge (Mass.), 1987.

[Gärdenfors and Rott, 1995] P. Gärdenfors and H. Rott. Belief Revision, in D. M. Gabbay, C. J. Hogger and J. A. Robinson, eds., *Handbook of Logic in Artificial Intelligence and Logic Programming* 4, Oxford University Press, Oxford 1995.

[Gerbarndy, 1999] J. Gerbrandy. *Bisimulations on Planet Kripke*, Dissertation, Institute for Logic, Language and Computation, University of Amsterdam.

[Ghidini and Giunchiglia, 2001] C. Ghidini and F. Giunchiglia. Local Model Semantics, or Contextual Reasoning = Locality + Compatibility, *Artificial Intelligence* 127, 221–259, 2001.

[Gibson, 1986] J. Gibson. *The Ecological Approach to Visual Perception*, Lawrence Erlbaum Associates, Hillsdale, NJ, 1986.

[Girard, 1987] J-Y Girard. Linear Logic, *Theoretical Computer Science* 50, 1–102, 1987.

[Girard, 1989] J-Y Girard. Towards a Geometry of Interaction, in J. W. Gray and A. Scedrov, eds., *Categories in Computer Science and Logic*, American Mathematical Society, 69–108, 1989.

[Gochet, 2006] P. Gochet. La Formalisation du Savoir-Faire, Lecture at Pierre Duhem Colloquium IPHRST Paris, Philosophical Institute, Université de Liege, 2006.

[Gupta, 1994] V. Gupta. *Chu Spaces: A Model of Concurrency*, PhD Thesis, Stanford University, September 1994.

[Halpern and Vardi, 1989] J. Halpern and M. Vardi. The Complexity of Reasoning about Knowledge and Time, *Journal of Computer and System Sciences* 38, 195–237, 1989.

[Hendricks, 2006] V. Hendricks. *Mainstream and Formal Epistemology*, Cambridge University Press, New York, 2006.

[Hintikka, 1962] J. Hintikka. *Knowledge and Belief,* Cornel University Press, Ithaca, 1962.

[Hintikka, 1963] J. Hintikka. *Logic, Language-Games and Information*, Clarendon Press, Oxford, 1963.

[van der Hoek and Meijer, 1995] W. van der Hoek and J-J Meijer. *Epistemic Logic for AI and Computer Science*, Cambridge University Press, Cambridge, 1995.

[van der Hoek and Pauly, 2006] W. van der Hoek and M. Pauly. Modal Logic for Games and Information, in P. Blackburn, J. van Benthem and F. Wolter, eds., *Handbook of Modal Logic*, Elsevier, Amsterdam, 1077–1148, 2006.

[Israel and Perry, 1990] D. Israel and J. Perry. What is Information? , In: P. Hanson (ed.). *Information, Language and Cognition.* University of British Columbia Press, Vancouver, 1990.

[Israel and Perry, 1991] D. Israel and J. Perry. Information and Architecture. In: J. Barwise, J. M. Gawron, G. Plotkin, and S. Tutiya, eds., *Situation Theory and Its Applications*, vol. 2. CSLI Publications, Stanford, 1991.

[Jacobs and Rutten, 1997] B. Jacobs and J. Rutten. A Tutorial on (Co)Algebras and (Co)Induction, *EACTS Bulletin* 62, 222-259, 1997.

[Jago, 2006] M. Jago. *Logics for Resource-Bounded Agents*, Dissertation, Department of Philosophy, University of Nottingham, 2006.

[Jago, 2007] M. Jago. Logical Information is Vague, Department of Philosophy, University of Nottingham, to appear in *Knowledge, Rationality and Action*, 2007.

[Kelly, 1996] K. Kelly. *The Logic of Reliable Inquiry*, Oxford University Press, New York, 1996.

[Kerdiles, 2001] G. Kerdiles. *Saying It with Pictures: a Logical Landscape of Conceptual Graphs*, Dissertation, Institute for Logic, Language and Computation, University of Amsterdam, 2001.

[Knauff, 2007] M. Knauff. How our Brains Reason Logically, *Topoi: an International Review of Philosophy* 26, 19–36, 2007.

[Kowalski, 1979] R. Kowalski. *Logic for Problem Solving*, North-Holland, New York, 1979.

[Kratzer, 2006] A. Kratzer. Situations in Natural Language Semantics, *Stanford Encyclopedia of Philosophy*, 2006. `http://www.plato.stanford.edu`

[Kuipers, 2000] Th. Kuipers. *From Instrumentalism to Constructive Realism*, Kluwer, Dordrecht, 2000.

[Lambek and Scott, 1994] J. Lambek and Ph. Scott. *Introduction to Higher-Order Categorical Logic*, Cambridge University Press, Cambridge, 1994.

[Landman, 1986] F. Landman. *Towards a Theory of Information, the Status of Partial Objects in Semantics*, Foris, Dordrecht, 1986.

[Lewis, 1969] D. Lewis. *Convention, A Philosophical Study*, Harvard University Press, Harvard, 1969.

[Lewis, 1970] D. Lewis. General Semantics, *Synthese* 22, 18–67, 1970.

[Liu, 2006] F. Liu. Diversity of Agents, Research Report, Institute for Logic, Language and Computation, Amsterdam. To appear in *Journal of Logic, Language and Information*, 1006.

[Liu and Zhang, 2007] F. Liu and J. Zhang. Some Thoughts on Mohist Logic, in J. van Benthem, S. Ju and F. Veltman, eds., *A Meeting of the Minds*, Proceedings LORI Beijing 2007, College Publications, London, 79–96, 2007.

[Loewe *et al.*, 2007] B. Loewe, E. Pacuit and J-W Romeijn. Foundations of the Formal Sciences VI: *Reasoning about Probabilities and Probabilistic Reasoning*, Institute for Logic, Language and Computation, University of Amsterdam, 2007.

[Lorenzen, 1955] P. Lorenzen. *Einführung in die Operative Logik und Mathematik,* Springer, Berlin, 1955.

[Lorini and Castelfranchi, 2007] E. Lorini and C. Castelfranchi. The Cognitive Structure of Surprise: looking for basic principles, *Topoi* 2007;, 133–149, 2007.
[Mares, 1996] E. Mares. Relevant Logic and the Theory of Information, *Synthese* 109, 345–360, 1996.
[Mares, 2003] E. Mares. *Relevant Logic, a Philosophical Interpretation*, Cambridge University Press, Cambridge, 2003.
[Martinez, 2004] M. Martinez. *Commonsense Reasoning via Product State Spaces*, Ph.D. Thesis, Indiana University, Bloomington, USA, 2004.
[Materna, 2004] P. Materna. *Conceptual Systems*, Logos Verlag, Berlin, 2004.
[Moore, 1985] R. Moore. A Formal Theory of Knowledge and Action. In J. Hobbs and R. Moore, eds., *Formal Theories of the Commonsense World*, Ablex Publishing Corp, pp. 319-358, 1985.
[Moore, 1989] R. Moore. Propositional Attitudes and Russellian Propositions, in R. Bartsch, J. van Benthem and P. van Emde Boas, eds., *Semantics and Contextual Expression*, Foris, Dordrecht, 147–174, 1989.
[Moss *et al.*, 2007] L. Moss, R. Parikh and Ch. Steinsvold. Topology and Epistemic Logic, in M. Aiello, I. Pratt and J. van Benthem, eds., *Handbook of Spatial Logics*, Springer, Dordrecht, 299–341, 2007.
[Moss and Seligman, 1997] L. Moss and J. Seligman. Situation Theory, in J. van Benthem and A. ter Meulen, eds., *Handbook of Logic and Language*, North Holland, Amsterdam, 1997.
[Németi, 1995] I. Németi. Decidable Versions of First-Order Logic and Cylindric- Relativized Set Algebras, in L. Csirmaz, D. Gabbay and M. de Rijke, eds., 1995, *Logic Colloquium 92. Veszprem, Hungary*, Studies in Logic, Language and Information, CSLI Publications, Stanford, 177-241, 1995.
[Osborne and Rubinstein, 1994] M. Osborne and A. Rubinstein. *A Course in Game Theory*, The MIT Press, Cambridge (Mass.), 1994.
[Papadimitriou, 1994] Ch. Papadimitriou. *Computational Complexity*, Addison-Wesley, Reading (Mass.), 1994.
[Parikh, 1987] R. Parikh. Knowledge and the Problem of Logical Omniscience, ISMIS-87, *International Symposium on Methodology for Intelligent Systems*, North-Holland, 432–439, 1987.
[Parikh, 1995] R. Parikh. Logical Omniscience, in D. Leivant, ed., *Logic and Computational Complexity*, Springer, Heidelberg, 22–29, 1995.
[Parikh, 2007] R. Parikh. Sentences, Propositions, and Group Knowledge, Lecture at *First Synthese Annual Conference Copenhagen*, CUNY Graduate Center, New York, 2007.
[Parikh and Ramanujam, 2003] R. Parikh and R. Ramanujam. A Knowledge Based Semantics of Messages, *Journal of Logic, Language and Information* 12, 453–467, 2003.
[Prawitz, 1965] D. Prawitz. *Natural Deduction. A Proof-Theoretical Study*, Almqvist and Wiksell, Stockholm, 1965.
[Priest, 1997] G. Priest. Impossible Worlds - Editors Introduction, *Notre Dame Journal of Formal Logic* 38, 481–487, 1997.
[Primero, 2007] G. Primero. Belief Merging Based on Adaptive Interaction, in J. van Benthem, S. Ju and F. Veltman, eds., *A Meeting of the Minds*, Proceedings LORI Beijing 2007, College Publications London, 315–320, 2007.
[Putnam, 1981] H. Putnam. *Reason, Truth and History*, Cambridge University Press, 1981.
[Renne, 2007] B. Renne. Public Communication in Justification Logic, CUNY Graduate Center, New York, 2007.
[Restall, 2000] G. Restall. *An Introduction to Substructural Logics*, Routledge, London.
[Rott, 2006] H. Rott. Shifting Priorities: Simple Representations for Twenty-Seven Iterated Theory Change Operators, Philosophical Institute, University of Regensburg, 2006.
[Rounds, 1997] W. C. Rounds. Feature Logics, in J. van Benthem and A. ter Meulen, eds., *Handbook of Logic and Language*, Elsevier, Amsterdam, 475–533, 1997.
[Roush, 2006] Sh. Roush. *Tracking Truth: Knowledge, Evidence and Science*, Oxford University Press, Oxford, 2006.
[Ryan, 1991] M. Ryan. *Defaults and Revision in Structured Theories*, Dissertation, Department of Computing, Imperial College, London.
[Saguillo, 1997] J. Saguillo. Logical Consequence Revisited, *The Bulletin of Symbolic Logic* 3:2, 216-241, 1997.
[SanGiorgi, 2004] D. SanGiorgi. Bisimulation: From The Origins to Today, *Proceedings LICS*, 298-302, Extended new version, 2007, Department of Informatics, University of Bologna, 2004.

[Schwichtenberg and Troelstra, 1996] H. Schwichtenberg and A. Troelstra. *Basic Proof Theory,* Cambridge University Press, Cambridge, 1996.

[Sequoiah-Grayson, 2006] S. Sequoiah-Grayson. Information Flow and Impossible Situations, *Logique et Analyse* 196, 371–398, 2006.

[Sequoiah-Grayson, 2007] S. Sequoiah-Grayson. Information Gain from Inference, Philosophical Institute, Oxford. To appear in *Knowledge, Rationality and Action*, 2007.

[Sevenster, 2006] M. Sevenster. *Branches of Imperfect Information*, Dissertation, Institute for Logic, Language and Computation, University of Amsterdam, 2006.

[Staal, 1988] F. Staal. *Universals: Studies in Indian Logic and Linguistics,* University of Chicago Press, Chicago and London, 1988.

[Sundholm, 1986] G. Sundholm. Proof Theory and Meaning, in D. Gabbay and F. Guenthner, eds., *Handbook of Philosophical Logic* III, Reidel, Dordrecht, 471–506, 1986.

[Tennant, 1978] N. Tennant. *Natural Logic*, Edinburgh University Press, Edinburgh, 1978.

[Tichy, 2004] P. Tichy. *Collected Papers in Logic and Philosophy*, V. Svoboda, B. Jespersen, and C. Cheyne, eds., Filosofia, Prague and University of Otago Press, Dunedin, 2004.

[Troelstra and van Dalen, 1988] A. Troelstra and D. van Dalen. *Constructivism in Mathematics: An Introduction*, North-Holland Publishing, Amsterdam, 1988.

[Väänänen, 2007] J. Väänänen. *Dependence Logic: A New Approach to Independence Friendly Logic*, Cambridge University Press, Cambridge, 2007.

[Veltman, 1984] F. Veltman. Data Semantics, in J. Groenendijk, Th. Janssen and M. Stokhof, eds., *Truth, Interpretation and Information*, Foris, Dordrecht, 43–63, 1984.

[Veltman, 1985] F. Veltman. *Logics for Conditionals*, Dissertation, Philosophical Institute, University of Amsterdam, 1985.

[Venema, 2006] Y. Venema. Algebras and Co-algebras, in J. van Benthem, P. Blackburn and F. Wolter, eds., *Handbook of Modal Logic*, Elsevier, Amsterdam, 331–426, 2006.

[Wassermann, 2001] R. Wassermann. *Resource Bounded Belief Revision*, Dissertation, Institute for Logic, Language and Computation, University of Amsterdam, 2001.

[Williamson, 2000] T. Williamson. *Knowledge and its Limits*, Oxford University Press, Oxford, 2000.

ALGORITHMIC INFORMATION THEORY

Peter D. Grünwald and Paul M. B. Vitányi

1 INTRODUCTION

How should we measure the amount of information about a phenomenon that is given to us by an observation concerning the phenomenon? Both 'classical' (Shannon) information theory (see the chapter by [Harremoës and Topsøe, 2008]) and algorithmic information theory start with the idea that this amount can be measured by *the minimum number of bits needed to describe the observation.* But whereas Shannon's theory considers description methods that are optimal relative to some given probability distribution, Kolmogorov's algorithmic theory takes a different, nonprobabilistic approach: any computer program that first computes (prints) the string representing the observation, and then terminates, is viewed as a valid description. The amount of information in the string is then defined as the size (measured in bits) of the *shortest* computer program that outputs the string and then terminates. A similar definition can be given for infinite strings, but in this case the program produces element after element forever. Thus, a long sequence of 1's such as

(1)
$$\overbrace{11\ldots1}^{10000 \text{ times}}$$

contains little information because a program of size about log 10000 bits outputs it:

$$\texttt{for } i := 1 \texttt{ to } 10000 \text{ ; } \texttt{print } 1.$$

Likewise, the transcendental number $\pi = 3.1415...$, an infinite sequence of seemingly 'random' decimal digits, contains but a few bits of information (There is a short program that produces the consecutive digits of π forever).

Such a definition would appear to make the amount of information in a string (or other object) depend on the particular programming language used. Fortunately, it can be shown that all reasonable choices of programming languages lead to quantification of the amount of 'absolute' information in individual objects that is invariant up to an additive constant. We call this quantity the 'Kolmogorov complexity' of the object. While regular strings have small Kolmogorov complexity, random strings have Kolmogorov complexity about equal to their own length. Measuring complexity and information in terms of program size has turned out to be a very powerful idea with applications in areas such as theoretical computer science, logic, probability theory, statistics and physics.

Handbook of the Philosophy of Science. Volume 8: Philosophy of Information
Volume editors: Pieter Adriaans and Johan van Benthem. General editors: Dov M. Gabbay, Paul Thagard and John Woods.

This Chapter Kolmogorov complexity was introduced independently and with different motivations by R.J. Solomonoff (born 1926), A.N. Kolmogorov (1903–1987) and G. Chaitin (born 1943) in 1960/1964, 1965 and 1966 respectively [Solomonoff, 1964; Kolmogorov, 1965; Chaitin, 1966]. During the last forty years, the subject has developed into a major and mature area of research. Here, we give a brief overview of the subject geared towards an audience specifically interested in the philosophy of information. With the exception of the recent work on the Kolmogorov structure function and parts of the discussion on philosophical implications, all material we discuss here can also be found in the standard textbook [Li and Vitányi, 1997]. The chapter is structured as follows: we start with an introductory section in which we define Kolmogorov complexity and list its most important properties. We do this in a much simplified (yet formally correct) manner, avoiding both technicalities and all questions of motivation (why this definition and not another one?). This is followed by Section 3 which provides an informal overview of the more technical topics discussed later in this chapter, in Sections 4– 6. The final Section 7, which discusses the theory's philosophical implications, as well as Section 6.3, which discusses the connection to inductive inference, are less technical again, and should perhaps be glossed over before delving into the technicalities of Sections 4– 6.

2 KOLMOGOROV COMPLEXITY: ESSENTIALS

The aim of this section is to introduce our main notion in the fastest and simplest possible manner, avoiding, to the extent that this is possible, all technical and motivational issues. Section 2.1 provides a simple definition of Kolmogorov complexity. We list some of its key properties in Section 2.2. Knowledge of these key properties is an essential prerequisite for understanding the advanced topics treated in later sections.

2.1 Definition

The Kolmogorov complexity K will be defined as a function from finite binary strings of arbitrary length to the natural numbers $\mathbb{N}$. Thus, $K : \{0,1\}^* \to \mathbb{N}$ is a function defined on 'objects' represented by binary strings. Later the definition will be extended to other types of objects such as numbers (Example 3), sets, functions and probability distributions (Example 7).

As a first approximation, $K(x)$ may be thought of as the length of the shortest computer program that prints x and then halts. This computer program may be written in Fortran, Java, LISP or any other *universal programming language*. By this we mean a general-purpose programming language in which a universal Turing Machine can be implemented. Most languages encountered in practice have this property. For concreteness, let us fix some universal language (say, LISP) and define Kolmogorov complexity with respect to it. The *invariance theorem* discussed below implies that it does not really matter which one we pick. Computer programs often make use of data. Such data are sometimes listed inside the program. An example is the bitstring `"010110..."` in the program

(2) `print"01011010101000110...010"`

In other cases, such data are given as additional input to the program. To prepare for later extensions such as conditional Kolmogorov complexity, we should allow for this possibility as well. We thus extend our initial definition of Kolmogorov complexity by considering computer programs with a very simple input-output interface: programs are provided a stream of bits, which, while running, they can read one bit at a time. There are no end-markers in the bit stream, so that, if a program p halts on input y and outputs x, then it will also halt on any input yz, where z is a continuation of y, and still output x. We write $p(y) = x$ if, on input y, p prints x and then halts. We define the Kolmogorov complexity relative to a given language as the length of the shortest program p *plus input* y, such that, when given input y, p computes (outputs) x and then halts. Thus:

(3) $$K(x) := \min_{y,p:p(y)=x} l(p) + l(y),$$

where $l(p)$ denotes the length of input p, and $l(y)$ denotes the length of program y, both expressed in bits. To make this definition formally entirely correct, we need to assume that the program p runs on a computer with unlimited memory, and that the language in use has access to all this memory. Thus, while the definition (3) can be made formally correct, it does obscure some technical details which need not concern us now. We return to these in Section 4.

2.2 Key Properties of Kolmogorov Complexity

To gain further intuition about $K(x)$, we now list five of its key properties. Three of these concern the size of $K(x)$ for commonly encountered types of strings. The fourth is the invariance theorem, and the fifth is the fact that $K(x)$ is uncomputable in general. Henceforth, we use x to denote finite bitstrings. We abbreviate $l(x)$, the length of a given bitstring x, to n. We use boldface $\mathbf{x}$ to denote an infinite binary string. In that case, $x_{[1:n]}$ is used to denote the initial n-bit segment of $\mathbf{x}$.

1(a). Very Simple Objects: $K(x) = O(\log n)$. $K(x)$ must be small for 'simple' or 'regular' objects x. For example, there exists a fixed-size program that, when input n, outputs the first n bits of π and then halts. As is easy to see (Section 4.2), specification of n takes $O(\log n)$ bits. Thus, when x consists of the first n bits of π, its complexity is $O(\log n)$. Similarly, we have $K(x) = O(\log n)$ if x represents the first n bits of a sequence like (1) consisting of only 1s. We also have $K(x) = O(\log n)$ for the first n bits of e, written in binary; or even for the first n bits of a sequence whose i-th bit is the i-th bit of $e^{2.3}$ if the $i-1$-st bit was a one, and the i-th bit of $1/\pi$ if the $i-1$-st bit was a zero. For certain 'special' lengths n, we may have $K(x)$ even substantially smaller than $O(\log n)$. For example, suppose $n = 2^m$ for some $m \in \mathbb{N}$. Then we can describe n by first describing m and then describing a program implementing the function $f(z) = 2^z$. The description of m takes $O(\log m)$ bits, the description of the program takes a constant number of bits not depending on n. Therefore, for such values of n, we get $K(x) = O(\log m) = O(\log \log n)$.

1(b). Completely Random Objects: $K(x) = n + O(\log n)$. A *code* or *description method* is a binary relation between source words — strings to be encoded — and code words — encoded versions of these strings. Without loss of generality, we can take the set of code words to be finite binary strings [Cover and Thomas, 1991]. In this chapter we only consider *uniquely decodable* codes where the relation is one-to-one or one-to-many, indicating that given an encoding $E(x)$ of string x, we can always reconstruct the original x. The Kolmogorov complexity of x can be viewed as the code length of x that results from using the *Kolmogorov code* $E^*(x)$: this is the code that encodes x by the shortest program that prints x and halts.

The following crucial insight will be applied to the Kolmogorov code, but it is important to realize that in fact it holds for *every* uniquely decodable code. For any uniquely decodable code, there are no more than 2^m strings x which can be described by m bits. The reason is quite simply that there are no more than 2^m binary strings of length m. Thus, the number of strings that can be described by less than m bits can be at most $2^{m-1} + 2^{m-2} + \ldots + 1 < 2^m$. In particular, this holds for the code E^* whose length function is $K(x)$. Thus, the fraction of strings x of length n with $K(x) < n - k$ is less than 2^{-k}: the overwhelming majority of sequences cannot be compressed by more than a constant. Specifically, if x is determined by n independent tosses of a fair coin, then all sequences of length n have the same probability 2^{-n}, so that with probability at least $1 - 2^{-k}$,

$$K(x) \geq n - k.$$

On the other hand, for arbitrary x, there exists a program '`print x; halt`'. This program seems to have length $n + O(1)$ where $O(1)$ is a small constant, accounting for the 'print' and 'halt' symbols. We have to be careful though: computer programs are usually represented as a sequence of bytes. Then in the program above x cannot be an arbitrary sequence of bytes, because we somehow have to mark the end of x. Although we represent both the program and the string x as bits rather than bytes, the same problem remains. To avoid it, we have to encode x in a prefix-free manner (Section 4.2) which takes $n + O(\log n)$ bits, rather than $n + O(1)$. Therefore, for all x of length n, $K(x) \leq n + O(\log n)$. Except for a fraction of 2^{-c} of these, $K(x) \geq n - c$ so that for the overwhelming majority of x,

$$K(x) = n + O(\log n). \tag{4}$$

Similarly, if x is determined by independent tosses of a fair coin, then (4) holds with overwhelming probability. Thus, while for very regular strings, the Kolmogorov complexity is small (sublinear in the length of the string), *most* strings have Kolmogorov complexity about equal to their own length. Such strings are called *(Kolmogorov) random*: they do not exhibit any discernible pattern. A more precise definition follows in Example 4.

1(c). Stochastic Objects: $K(x) = \alpha n + o(n)$. Suppose $\mathbf{x} = x_1 x_2 \ldots$ where the individual x_i are realizations of some random variable X_i, distributed according to some distribution P. For example, we may have that all outcomes $X_1, X_2, \ldots$ are independently identically distributed (i.i.d.) with for all i, $P(X_i = 1) = p$ for some $p \in [0, 1]$. In that case, as will be seen in Section 5.3, Theorem 10,

(5) $K(x_{[1:n]}) = n \cdot H(p) + o(n)$,

where log is logarithm to the base 2, and $H(p) = -p \log p - (1 - p) \log(1 - p)$ is the binary entropy, defined in Section 5.1. For now the important thing to note is that $0 \leq H(p) \leq 1$, with $H(p)$ achieving its maximum 1 for $p = 1/2$. Thus, if data are generated by independent tosses of a fair coin, (5) is consistent with (4). If data are generated by a biased coin, then the Kolmogorov complexity will still increase linearly in n, but with a factor less than 1 in front: the data can be compressed by a linear amount. This still holds if the data are distributed according to some P under which the different outcomes are dependent, as long as this P is 'nondegenerate'.[1] An example is a *k-th order Markov chain*, where the probability of the i-th bit being a 1 depends on the value of the previous k bits, but nothing else. If none of the 2^k probabilities needed to specify such a chain are either 0 or 1, then the chain will be 'nondegenerate' in our sense, implying that, with P-probability 1, $K(x_1, \ldots, x_n)$ grows linearly in n.

2. Invariance It would seem that $K(x)$ depends strongly on what programming language we used in our definition of K. However, it turns out that, for any two universal languages L_1 and L_2, letting K_1 and K_2 denote the respective complexities, for all x of each length,

(6) $|K_1(x) - K_2(x)| \leq C$,

where C is a constant that depends on L_1 and L_2 *but not on x or its length.* Since we allow *any* universal language in the definition of K, $K(x)$ is only defined up to an additive constant. This means that the theory is inherently *asymptotic*: it can make meaningful statements pertaining to strings of increasing length, such as $K(x_{[1:n]}) = f(n) + O(1)$ in the three examples 1(a), 1(b) and 1(c) above. A statement such as $K(a) = b$ is not very meaningful.

It is actually very easy to show (6). It is known from the theory of computation that for any two universal languages L_1 and L_2, there exists a compiler, written in L_1, translating programs written in L_2 into equivalent programs written in L_1. Thus, let L_1 and L_2 be two universal languages, and let Λ be a program in L_1

[1] This means that there exists an $\epsilon > 0$ such that, for all $n \geq 0$, all $x^n \in \{0,1\}^n$, for $a \in \{0,1\}$, $P(x_{n+1} = a \mid x_1, \ldots, x_n) > \epsilon$.

implementing a compiler translating from L_2 to L_1. For concreteness, assume L_1 is LISP and L_2 is Java. Let (p, y) be the shortest combination of Java program plus input that prints a given string x. Then the LISP program Λ, when given input p followed by y, will also print x and halt.[2] It follows that $K_{\text{LISP}}(x) \leq l(\Lambda) + l(p) + l(y) \leq K_{\text{Java}}(x) + O(1)$, where $O(1)$ is the size of Λ. By symmetry, we also obtain the opposite inequality. Repeating the argument for general universal L_1 and L_2, (6) follows.

3. Uncomputability Unfortunately $K(x)$ is not a recursive function: the Kolmogorov complexity is not computable in general. This means that there exists no computer program that, when input an arbitrary string, outputs the Kolmogorov complexity of that string and then halts. We prove this fact in Section 4, Example 3. Kolmogorov complexity can be computably approximated (technically speaking, it is *upper semicomputable* [Li and Vitányi, 1997]), but not in a practically useful way: while the approximating algorithm with input x successively outputs better and better approximations $t_1 \geq t_2 \geq t_3 \geq \ldots$ to $K(x)$, it is (a) excessively slow, and (b), it is in general impossible to determine whether the current approximation t_i is already a good one or not. In the words of [Barron and Cover, 1991], (eventually) "You know, but you do not know you know".

Do these properties make the theory irrelevant for practical applications? Certainly not. The reason is that it is possible to approximate Kolmogorov complexity after all, in the following, weaker sense: we take some existing data compression program C (for example, gzip) that allows every string x to be encoded and decoded computably and even efficiently. We then approximate $K(x)$ as the number of bits it takes to encode x using compressor C. For many compressors, one can show that for "most" strings x in the set of all strings of interest, $C(x) \approx K(x)$. Both *universal coding* [Cover and Thomas, 1991] and the *Minimum Description Length (MDL) Principle* (Section 6.3) are, to some extent, based on such ideas. Universal coding forms the basis of most practical lossless data compression algorithms, and MDL is a practically successful method for statistical inference. There is an even closer connection to the *normalized compression distance* method, a practical tool for data similarity analysis that can explicitly be understood as an approximation of an "ideal" but uncomputable method based on Kolmogorov complexity [Cilibrasi and Vitányi, 2005].

3 OVERVIEW AND SUMMARY

Now that we introduced our main concept, we are ready to give a summary of the remainder of the chapter.

Section 4: Kolmogorov Complexity — Details We motivate our definition of Kolmogorov complexity in terms of the theory of computation: the Church–

[2]To formalize this argument we need to setup the compiler in a way such that p and y can be fed to the compiler without any symbols in between, but this can be done; see Example 2.

Turing thesis implies that our choice of description method, based on universal computers, is essentially the only reasonable one. We then introduce some basic coding theoretic concepts, most notably the so-called *prefix-free codes* that form the basis for our version of Kolmogorov complexity. Based on these notions, we give a precise definition of Kolmogorov complexity and we fill in some details that were left open in the introduction.

Section 5: Shannon vs. Kolmogorov Here we outline the similarities and differences in aim and scope of Shannon's and Kolmogorov's information theories. Section 5.1 reviews the *entropy*, the central concept in Shannon's theory. Although their primary aim is quite different, and they are functions defined on different spaces, there is a close relation between entropy and Kolmogorov complexity (Section 5.3): if data are distributed according to some computable distribution then, roughly, *entropy is expected Kolmogorov complexity.*

Entropy and Kolmogorov complexity are concerned with information in a single object: a random variable (Shannon) or an individual sequence (Kolmogorov). Both theories provide a (distinct) notion of *mutual information* that measures the information that *one object gives about another object.* We introduce and compare the two notions in Section 5.4.

Entropy, Kolmogorov complexity and mutual information are concerned with *lossless* description or compression: messages must be described in such a way that from the description, the original message can be completely reconstructed. Extending the theories to *lossy* description or compression enables the formalization of more sophisticated concepts, such as 'meaningful information' and 'useful information'.

Section 6: Meaningful Information, Structure Function and Learning The idea of the Kolmogorov Structure Function is to encode objects (strings) in two parts: a *structural* and a *random* part. Intuitively, the 'meaning' of the string resides in the structural part and the size of the structural part quantifies the 'meaningful' information in the message. The structural part defines a 'model' for the string. Kolmogorov's structure function approach shows that the meaningful information is summarized by the *simplest* model such that the corresponding two-part description is not larger than the Kolmogorov complexity of the original string. Kolmogorov's structure function is closely related to J. Rissanen's *minimum description length principle*, which we briefly discuss. This is a practical theory of learning from data that can be viewed as a mathematical formalization of Occam's Razor.

Section 7: Philosophical Implications Kolmogorov complexity has implications for the foundations of several fields, including the foundations of mathematics. The consequences are particularly profound for the foundations of *probability* and *statistics*. For example, it allows us to discern between *different forms* of randomness, which is impossible using standard probability theory. It provides a precise prescription for and justification of the use of Occam's Razor in statistics, and leads to the distinction between *epistemological* and *metaphysical* forms of Occam's Razor. We discuss these and other implications for the philosophy of information in Section 7, which may be read without deep knowledge of the technicalities described in Sections 4–6.

4 KOLMOGOROV COMPLEXITY: DETAILS

In Section 2 we introduced Kolmogorov complexity and its main features without paying much attention to either (a) underlying motivation (why is Kolmogorov complexity a useful measure of information?) or (b) technical details. In this section, we first provide a detailed such motivation (Section 4.1). We then (Section 4.2) provide the technical background knowledge needed for a proper understanding of the concept. Based on this background knowledge, in Section 4.3 we provide a definition of Kolmogorov complexity directly in terms of Turing machines, equivalent to, but at the same time more complicated and insightful than the definition we gave in Section 2.1. With the help of this new definition, we then fill in the gaps left open in Section 2.

4.1 Motivation

Suppose we want to describe a given object by a finite binary string. We do not care whether the object has many descriptions; however, each description should describe but one object. From among all descriptions of an object we can take the length of the shortest description as a measure of the object's complexity. It is natural to call an object "simple" if it has at least one short description, and to call it "complex" if all of its descriptions are long. But now we are in danger of falling into the trap so eloquently described in the Richard-Berry paradox, where we define a natural number as "the least natural number that cannot be described in less than twenty words." If this number does exist, we have just described it in thirteen words, contradicting its definitional statement. If such a number does not exist, then all natural numbers can be described in fewer than twenty words. We need to look very carefully at what kind of descriptions (codes) D we may allow. If D is known to both a sender and receiver, then a message x can be transmitted from sender to receiver by transmitting the description y with $D(y) = x$. We may define the descriptional complexity of x under specification method D as the length of the shortest y such that $D(y) = x$. Obviously, this descriptional complexity of x depends crucially on D: the syntactic framework of the description language determines the succinctness of description. Yet in order to objectively compare descriptional complexities of objects, to be able to say "x is more complex than z," the descriptional complexity of x should depend on x alone. This complexity can be viewed as related to a universal description method that is a priori assumed by all senders and receivers. This complexity is optimal if no other description method assigns a lower complexity to any object.

We are not really interested in optimality with respect to all description methods. For specifications to be useful at all it is necessary that the mapping from y to $D(y)$ can be executed in an effective manner. That is, it can at least in principle be performed by humans or machines. This notion has been formalized as that of "partial recursive functions", also known simply as *computable* functions. According to generally accepted mathematical viewpoints — the so-called 'Church-Turing thesis' — it coincides with the intuitive notion of effective computation [Li and Vitányi, 1997].

The set of partial recursive functions contains an optimal function that minimizes description length of every other such function. We denote this function by D_0. Namely, for any other recursive function D, for all objects x, there is a description y of x under D_0 that is shorter than any description z of x under D. (That is, shorter up to an additive constant that is independent of x.) Complexity with respect to D_0 minorizes the complexities with respect to all partial recursive functions (this is just the invariance result (6) again).

We identify the length of the description of x with respect to a fixed specification function D_0 with the "algorithmic (descriptional) complexity" of x. The optimality of D_0 in the sense above means that the complexity of an object x is invariant (up to an additive constant independent of x) under transition from one optimal specification function to another. Its complexity is an objective attribute of the described object alone: it is an intrinsic property of that object, and it does not depend on the description formalism. This complexity can be viewed as "absolute information content": the amount of information that needs to be transmitted between all senders and receivers when they communicate the message in absence of any other a priori knowledge that restricts the domain of the message. This motivates the program for a general theory of algorithmic complexity and information. The four major innovations are as follows:

1. In restricting ourselves to formally effective descriptions, our definition covers every form of description that is intuitively acceptable as being effective according to general viewpoints in mathematics and logic.

2. The restriction to effective descriptions entails that there is a universal description method that minorizes the description length or complexity with respect to any other effective description method. Significantly, this implies Item 3.

3. The description length or complexity of an object is an intrinsic attribute of the object independent of the particular description method or formalizations thereof.

4. The disturbing Richard-Berry paradox above does not disappear, but resurfaces in the form of an alternative approach to proving Gödel's famous result that not every true mathematical statement is provable in mathematics (Example 4 below).

4.2 Coding Preliminaries

Strings and Natural Numbers Let $\mathcal{X}$ be some finite or countable set. We use the notation $\mathcal{X}^*$ to denote the set of finite *strings* or *sequences* over $\mathcal{X}$. For example,

$$\{0,1\}^* = \{\epsilon, 0, 1, 00, 01, 10, 11, 000, \ldots\},$$

with ϵ denoting the *empty word* '' with no letters. We identify the natural numbers $\mathbb{N}$ and $\{0,1\}^*$ according to the correspondence

(7) $(0, \epsilon), (1, 0), (2, 1), (3, 00), (4, 01), \ldots$

The *length* $l(x)$ of x is the number of bits in the binary string x. For example, $l(010) = 3$ and $l(\epsilon) = 0$. If x is interpreted as an integer, we get $l(x) = \lfloor \log(x+1) \rfloor$ and, for $x \geq 2$,

(8) $\lfloor \log x \rfloor \leq l(x) \leq \lceil \log x \rceil$.

Here, as in the sequel, $\lceil x \rceil$ is the smallest integer larger than or equal to x, $\lfloor x \rfloor$ is the largest integer smaller than or equal to x and log denotes logarithm to base two. We shall typically be concerned with encoding finite-length binary strings by other finite-length binary strings. The emphasis is on binary strings only for convenience; observations in any alphabet can be so encoded in a way that is 'theory neutral'.

Codes We repeatedly consider the following scenario: a *sender* (say, A) wants to communicate or transmit some information to a *receiver* (say, B). The information to be transmitted is an element from some set $\mathcal{X}$. It will be communicated by sending a binary string, called the *message*. When B receives the message, he can decode it again and (hopefully) reconstruct the element of $\mathcal{X}$ that was sent. To achieve this, A and B need to agree on a *code* or *description method* before communicating. Intuitively, this is a binary relation between *source words* and associated *code words*. The relation is fully characterized by the *decoding function*. Such a decoding function D can be any function $D : \{0,1\}^* \to \mathcal{X}$. The domain of D is the set of *code words* and the range of D is the set of *source words*. $D(y) = x$ is interpreted as "y is a code word for the source word x". The set of all code words for source word x is the set $D^{-1}(x) = \{y : D(y) = x\}$. Hence, $E = D^{-1}$ can be called the *encoding* substitution (E is not necessarily a function). With each code D we can associate a *length function* $L_D : \mathcal{X} \to \mathbb{N}$ such that, for each source word x, $L_D(x)$ is the length of the shortest encoding of x:

$$L_D(x) = \min\{l(y) : D(y) = x\}.$$

We denote by x^* the shortest y such that $D(y) = x$; if there is more than one such y, then x^* is defined to be the first such y in lexicographical order.

In coding theory attention is often restricted to the case where the source word set is finite, say $\mathcal{X} = \{1, 2, \ldots, N\}$. If there is a constant l_0 such that $l(y) = l_0$

for all code words y (equivalently, $L(x) = l_0$ for all source words x), then we call D a *fixed-length* code. It is easy to see that $l_0 \geq \log N$. For instance, in teletype transmissions the source has an alphabet of $N = 32$ letters, consisting of the 26 letters in the Latin alphabet plus 6 special characters. Hence, we need $l_0 = 5$ binary digits per source letter. In electronic computers we often use the fixed-length ASCII code with $l_0 = 8$.

Prefix-free code In general we cannot uniquely recover x and y from $E(xy)$. Let E be the identity mapping. Then we have $E(00)E(00) = 0000 = E(0)E(000)$. We now introduce *prefix-free codes*, which do not suffer from this defect. A binary string x is a *proper prefix* of a binary string y if we can write $y = xz$ for $z \neq \epsilon$. A set $\{x, y, \ldots\} \subseteq \{0,1\}^*$ is *prefix-free* if for any pair of distinct elements in the set neither is a proper prefix of the other. A function $D : \{0,1\}^* \to \mathbb{N}$ defines a *prefix-free code*[3] if its domain is prefix-free. In order to decode a code sequence of a prefix-free code, we simply start at the beginning and decode one code word at a time. When we come to the end of a code word, we know it is the end, since no code word is the prefix of any other code word in a prefix-free code. Clearly, prefix-free codes are uniquely decodable: we can always unambiguously reconstruct an outcome from its encoding. Prefix codes are not the only codes with this property; there are uniquely decodable codes which are not prefix-free. In the next section, we will define Kolmogorov complexity in terms of prefix-free codes. One may wonder why we did not opt for general uniquely decodable codes. There is a good reason for this: It turns out that every uniquely decodable code can be replaced by a prefix-free code without changing the set of code-word lengths. This follows from a sophisticated version of the Kraft inequality [Cover and Thomas, 1991, Kraft-McMillan inequality, Theorem 5.5.1]; the basic Kraft inequality is found in [Harremoës and Topsøe, 2008], Equation 1.1. In Shannon's and Kolmogorov's theories, we are only interested in code word *lengths* of uniquely decodable codes rather than actual encodings. The Kraft-McMillan inequality shows that without loss of generality, we may restrict the set of codes we work with to prefix-free codes, which are much easier to handle.

Codes for the integers; Pairing Functions Suppose we encode each binary string $x = x_1 x_2 \ldots x_n$ as

$$\bar{x} = \underbrace{11\ldots1}_{n \text{ times}} 0 x_1 x_2 \ldots x_n.$$

The resulting code is prefix-free because we can determine where the code word $\bar{x}$ ends by reading it from left to right without backing up. Note $l(\bar{x}) = 2n + 1$; thus, we have encoded strings in $\{0,1\}^*$ in a prefix-free manner at the price of doubling their length. We can get a much more efficient code by applying the

[3]The standard terminology [Cover and Thomas, 1991] for such codes is 'prefix codes'. Following [Harremoës and Topsøe, 2008], we use the more informative 'prefix-free codes'.

construction above to the length $l(x)$ of x rather than x itself: define $x' = \overline{l(x)}x$, where $l(x)$ is interpreted as a binary string according to the correspondence (7). Then the code that maps x to x' is a prefix-free code satisfying, for all $x \in \{0,1\}^*$, $l(x') = n + 2\log n + 1$ (here we ignore the 'rounding error' in (8)). We call this code the *standard prefix-free code for the natural numbers* and use $L_{\mathbb{N}}(x)$ as notation for the codelength of x under this code: $L_{\mathbb{N}}(x) = l(x')$. When x is interpreted as a number (using the correspondence (7) and (8)), we see that $L_{\mathbb{N}}(x) = \log x + 2\log\log x + 1$.

We are often interested in representing a pair of natural numbers (or binary strings) as a single natural number (binary string). To this end, we define the *standard 1-1 pairing function* $\langle\cdot,\cdot\rangle : \mathbb{N} \times \mathbb{N} \to \mathbb{N}$ as $\langle x, y\rangle = x'y$ (in this definition x and y are interpreted as strings).

4.3 Formal Definition of Kolmogorov Complexity

In this subsection we provide a formal definition of Kolmogorov complexity in terms of Turing machines. This will allow us to fill in some details left open in Section 2. Let $T_1, T_2, \ldots$ be a standard enumeration of all Turing machines [Li and Vitányi, 1997]. The functions implemented by T_i are called the *partial recursive* or *computable* functions. For technical reasons, mainly because it simplifies the connection to Shannon's information theory, we are interested in the so-called prefix complexity, which is associated with Turing machines for which the set of programs (inputs) resulting in a halting computation is prefix-free[4]. We can realize this by equipping the Turing machine with a one-way input tape, a separate work tape, and a one-way output tape. Such Turing machines are called prefix machines since the halting programs for any one of them form a prefix-free set.

We first define $K_{T_i}(x)$, the prefix Kolmogorov complexity of x relative to a given prefix machine T_i, where T_i is the i-th prefix machine in a standard enumeration of them. $K_{T_i}(x)$ is defined as the length of the shortest input sequence y such that $T_i(y) = x$; that is, the i-th Turing machine, when run with input y, produces x on its output tape and then halts. If no such input sequence exists, $K_{T_i}(x)$ remains undefined. Of course, this preliminary definition is still highly sensitive to the particular prefix machine T_i that we use. But now the 'universal prefix machine' comes to our rescue. Just as there exists universal ordinary Turing machines, there also exist universal prefix machines. These have the remarkable property that they can simulate every other prefix machine. More specifically, there exists a prefix machine U such that, with as input the concatenation $i'y$ (where i' is the standard encoding of integer i, Section 4.2), U outputs $T_i(y)$ and then halts. If U gets any other input then it does not halt.

DEFINITION 1. Let U be our reference prefix machine, i.e. for all $i \in \mathbb{N}, y \in \{0,1\}^*$, $U(\langle i, y\rangle) = U(i'y) = T_i(y)$. The *prefix Kolmogorov complexity* of x is

---

[4]There exists a version of Kolmogorov complexity corresponding to programs that are not necessarily prefix-free, but we will not go into it here.

defined as $K(x) := K_U(x)$, or equivalently:

$$\begin{aligned} K(x) &= \min_z \{l(z) : U(z) = x, z \in \{0,1\}^*\} = \\ &= \min_{i,y} \{l(i') + l(y) : T_i(y) = x, y \in \{0,1\}^*, i \in \mathbb{N}\}. \end{aligned} \tag{9}$$

We can alternatively think of z as a program that prints x and then halts, or as $z = i'y$ where y is a program such that, when T_i is input program y, it prints x and then halts.

Thus, by definition $K(x) = l(x^*)$, where x^* is the lexicographically first shortest self-delimiting (prefix-free) program for x with respect to the reference prefix machine. Consider the mapping E^* defined by $E^*(x) = x^*$. This may be viewed as the encoding function of a prefix-free code (decoding function) D^* with $D^*(x^*) = x$. By its definition, D^* is a very parsimonious code.

EXAMPLE 2. In Section 2, we defined $K(x)$ as the shortest program for x in some standard programming language such as LISP or Java. We now show that this definition is equivalent to the prefix Turing machine Definition 1. Let L_1 be a universal language; for concreteness, say it is LISP. Denote the corresponding Kolmogorov complexity defined as in (3) by K_{LISP}. For the universal prefix machine U of Definition 1, there exists a program p in LISP that simulates it [Li and Vitányi, 1997]. By this we mean that, for all $z \in \{0,1\}^*$, either $p(z) = U(z)$ or neither p nor U ever halt on input z. Run with this program, our LISP computer computes the same function as U on its input, so that

$$K_{\text{LISP}}(x) \leq l(p) + K_U(x) = K_U(x) + O(1).$$

On the other hand, LISP, when equipped with the simple input/output interface described in Section 2, is a language such that for all programs p, the set of inputs y for which $p(y)$ is well-defined forms a prefix-free set. Also, as is easy to check, the set of syntactically correct LISP programs is prefix-free. Therefore, the set of strings py where p is a syntactically correct LISP program and y is an input on which p halts, is prefix-free. Thus we can construct a prefix Turing machine with some index i_0 such that $T_{i_0}(py) = p(y)$ for all $y \in \{0,1\}^*$. Therefore, the universal machine U satisfies for all $y \in \{0,1\}^*$, $U(i_0'py) = T_{i_0}(py) = p(y)$, so that

$$K_U(x) \leq K_{\text{LISP}}(x) + l(i_0') = K_{\text{LISP}}(x) + O(1).$$

We are therefore justified in calling $K_{\text{LISP}}(x)$ a version of (prefix) Kolmogorov complexity. The same holds for any other universal language, as long as its set of syntactically correct programs is prefix-free. This is the case for every programming language we know of.

EXAMPLE 3. [**$K(x)$ as an integer function; uncomputability**] The correspondence between binary strings and integers established in (7) shows that Kolmogorov complexity may equivalently be thought of as a function $K : \mathbb{N} \rightarrow \mathbb{N}$

where $\mathbb{N}$ are the nonnegative integers. This interpretation is useful to prove that Kolmogorov complexity is uncomputable.

Indeed, let us assume by means of contradiction that K is computable. Then the function $\psi(m) := \min_{x \in \mathbb{N}}\{x : K(x) \geq m\}$ must be computable as well (note that x is interpreted as an integer in the definition of ψ). The definition of ψ immediately implies $K(\psi(m)) \geq m$. On the other hand, since ψ is computable, there exists a computer program of some fixed size c such that, on input m, the program outputs $\psi(m)$ and halts. Therefore, since $K(\psi(m))$ is the length of the shortest program plus input that prints $\psi(m)$, we must have that $K(\psi(m)) \leq L_{\mathbb{N}}(m) + c \leq 2 \log m + c$. Thus, we have $m \leq 2 \log m + c$ which must be false from some m onwards: contradiction.

EXAMPLE 4. [**Gödel's incompleteness theorem and randomness**] We say that a formal system (definitions, axioms, rules of inference) is *consistent* if no statement which can be expressed in the system can be proved to be both true and false in the system. A formal system is *sound* if only true statements can be proved to be true in the system. (Hence, a sound formal system is consistent.)

Let x be a finite binary string of length n. We write 'x is c-random' if $K(x) > n - c$. That is, the shortest binary description of x has length not much smaller than x. We recall from Section 2.2 that the fraction of sequences that can be compressed by more than c bits is bounded by 2^{-c}. This shows that there are sequences which are c-random for every $c \geq 1$ and justifies the terminology: the smaller c, the more random x.

Now fix any sound formal system F that is powerful enough to express the statement 'x is c-random'. Suppose F can be described in f bits. By this we mean that there is a fixed-size program of length f such that, when input the number i, outputs a list of all valid proofs in F of length (number of symbols) i. We claim that, for all but finitely many random strings x and $c \geq 1$, the sentence 'x is c-random' is not provable in F. Suppose the contrary. Then given F, we can start to exhaustively search for a proof that some string of length $n \gg f$ is random, and print it when we find such a string x. This procedure to print x of length n uses only $\log n + f + O(1)$ bits of data, which is much less than n. But x is random by the proof and the fact that F is sound. Hence F is not consistent, which is a contradiction.

Pushing the idea of Example 4 much further, [Chaitin, 1987] proved a particularly strong variation of Gödel's theorem, using Kolmogorov complexity but in a more sophisticated way, based on the number Ω defined below. Roughly, it says the following: there exists an *exponential Diophantine equation*,

$$(10) \quad A(n, x_1, \ldots, x_m) = 0$$

for some finite m, such that the following holds: let F be a formal theory of arithmetic. Then for all F that are sound and consistent, there is only a finite number of values of n for which the theory determines whether (10) has finitely or infinitely many solutions $(x_1, \ldots, x_m)$ (n is to be considered a parameter rather than a variable). For all other, infinite number of values for n, the statement '(10) has a finite number of solutions' is logically independent of F.

Chaitin's Number of Wisdom Ω An axiom system that can be effectively described by a finite string has limited information content — this was the basis for our proof of Gödel's theorem above. On the other hand, there exist quite short strings which are mathematically well-defined but uncomputable, which have an astounding amount of information in them about the truth of mathematical statements. Following [Chaitin, 1975], we define the *halting probability* Ω as the real number defined by

$$\Omega = \sum_{U(p)<\infty} 2^{-l(p)},$$

the sum taken over all inputs p for which the reference machine U halts. We call Ω the halting probability because it is the probability that U halts if its program is provided by a sequence of fair coin flips. It turns out that Ω represents the *halting problem* very compactly. The following theorem is proved in [Li and Vitányi, 1997]:

THEOREM 5. *Let y be a binary string of length at most n. There exists an algorithm A which, given the first n bits of Ω, decides whether the universal machine U halts on input y; i.e. A outputs 1 if U halts on y; A outputs 0 if U does not halt on y; and A is guaranteed to run in finite time.*

The halting problem is a prime example of a problem that is *undecidable* [Li and Vitányi, 1997], from which it follows that Ω must be uncomputable.

Knowing the first 10000 bits of Ω enables us to solve the halting of all programs of less than 10000 bits. This includes programs looking for counterexamples to Goldbach's Conjecture, Riemann's Hypothesis, and most other conjectures in mathematics which can be refuted by a single finite counterexample. Moreover, for all axiomatic mathematical theories which can be expressed compactly enough to be conceivably interesting to human beings, say in less than 10000 bits, $\Omega_{[1:10000]}$ can be used to decide for every statement in the theory whether it is true, false, or independent. Thus, Ω is truly the number of Wisdom, and 'can be known of, but not known, through human reason' [C.H. Bennett and M. Gardner, *Scientific American*, 241:11(1979), 20–34].

4.4 Conditional Kolmogorov complexity

In order to fully develop the theory, we also need a notion of *conditional* Kolmogorov complexity. Intuitively, the conditional Kolmogorov complexity $K(x|y)$ of x given y can be interpreted as the shortest program p such that, when y is given to the program p as input 'for free', the program prints x and then halts. Based on conditional Kolmogorov complexity, we can then further define Kolmogorov complexities of more complicated objects such as functions and so on (Example 7).

The idea of providing p with an input y is realized by putting $\langle y, p\rangle$ rather than just p on the input tape of a universal *conditional* prefix machine U. This is a prefix machine U such that for all y, i, q, $U(\langle y, \langle i, q\rangle\rangle) = T_i(\langle y, q\rangle)$, whereas for any input not of this form, U does not halt. Here $T_1, T_2, \ldots$ is some effective enumeration of prefix machines. It is easy to show that such a universal conditional prefix machine U exists [Li and Vitányi, 1997]. We now fix a reference conditional universal prefix machine U and define $K(x|y)$ as follows:

DEFINITION 6. **[Conditional and Joint Kolmogorov Complexity]** The *conditional prefix Kolmogorov complexity* of x given y (for free) is

$$K(x|y) = \min_p \{l(p) : U(\langle y, p\rangle) = x, p \in \{0,1\}^*\}. \tag{11}$$

$$= \min_{q,i} \{l(\langle i, q\rangle) : U(\langle y, \langle i, q\rangle\rangle) = x, q \in \{0,1\}^*, i \in \mathbb{N}\} \tag{12}$$

$$= \min_{q,i} \{l(i') + l(q) : T_i(y'q) = x, q \in \{0,1\}^*, i \in \mathbb{N}\}. \tag{13}$$

We define the *unconditional complexity* $K(x)$ as $K(x) = K(x|\epsilon)$. We define the *joint* complexity $K(x, y)$ as $K(x, y) = K(\langle x, y\rangle)$.

Note that we just redefined $K(x)$ so that the unconditional Kolmogorov complexity is *exactly* equal to the conditional Kolmogorov complexity with empty input. This does not contradict our earlier definition: having chosen some reference conditional prefix machine U, we can always find an effective enumeration T_1', T_2' and a corresponding unconditional universal prefix machine U' such that for all p, $U(\langle \epsilon, p\rangle) = U'(p)$. Then we automatically have, for all x, $K_{U'}(x) = K_U(x|\epsilon)$.

EXAMPLE 7. **[K for general objects: functions, distributions, sets, ...]** We have defined the Kolmogorov complexity K of binary strings and natural numbers, which we identified with each other. It is straightforward to extend the definition to objects such as real-valued functions, probability distributions and sets. We briefly indicate how to do this. Intuitively, the Kolmogorov complexity of a function $f : \mathbb{N} \rightarrow \mathbb{R}$ is the length of the shortest prefix-free program that computes (outputs) $f(x)$ to precision $1/q$ on input $x'q'$ for $q \in \{1, 2, \ldots\}$. In terms of conditional universal prefix machines:

$$K(f) = \min_{p \in \{0,1\}^*} \big\{l(p) : \text{for all } q \in \{1, 2, \ldots\}, x \in \mathbb{N}\text{: } |U(\langle x, \langle q, p\rangle\rangle) - f(x)| \leq 1/q\big\}. \tag{14}$$

The Kolmogorov complexity of a function $f : \mathbb{N} \times \mathbb{N} \rightarrow \mathbb{R}$ is defined analogously, with $\langle x, \langle q, p\rangle\rangle$ replaced by $\langle x, \langle y, \langle q, p\rangle\rangle\rangle$, and $f(x)$ replaced by $f(x, y)$; similarly for functions $f : \mathbb{N}^k \times \mathbb{N} \rightarrow \mathbb{R}$ for general $k \in \mathbb{N}$. As a special case of (14), the Kolmogorov complexity of a probability distribution P is the shortest program that outputs $P(x)$ to precision q on input $\langle x, q\rangle$. We will encounter $K(P)$ in Section 5.

The Kolmogorov complexity of *sets* can be defined in various manners [Gács, Tromp, and Vitányi, 2001]. In this chapter we only consider finite sets S consisting of finite strings. One reasonable method of defining their complexity $K(S)$ is as the length of the shortest program that sequentially outputs the elements of S (in an arbitrary order) and then halts. Let $S = \{x_1, \ldots, x_n\}$, and assume that $x_1, x_2, \ldots, x_n$ reflects the lexicographical order of the elements of S. In terms of conditional prefix machines, $K(S)$ is the length of the shortest binary program p such that $U(\langle \epsilon, p \rangle) = z$, where

(15) $z = \langle x_1, \langle x_2, \ldots, \langle x_{n-1}, x_n \rangle \ldots \rangle \rangle.$

This definition of $K(S)$ will be used in Section 6. There we also need the notion of the Kolmogorov complexity of a string x given that $x \in S$, denoted as $K(x|S)$. This is defined as the length of the shortest binary program p from which the (conditional universal) U computes x from input S given literally, in the form of (15).

This concludes our treatment of the basic concepts of Kolmogorov complexity theory. In the next section we compare these to the basic concepts of Shannon's information theory.

5 SHANNON AND KOLMOGOROV

In this section we compare Kolmogorov complexity to Shannon's [1948] information theory, more commonly simply known as 'information theory'. Shannon's theory predates Kolmogorov's by about 25 years. Both theories measure the amount of information in an object as the length of a description of the object. In the Shannon approach, however, the method of encoding objects is based on the presupposition that the objects to be encoded are outcomes of a known random source—it is only the characteristics of that random source that determine the encoding, not the characteristics of the objects that are its outcomes. In the Kolmogorov complexity approach we consider the individual objects themselves, in isolation so-to-speak, and the encoding of an object is a computer program that generates it. In the Shannon approach we are interested in the minimum expected number of bits to transmit a message from a random source of known characteristics through an error-free channel. In Kolmogorov complexity we are interested in the minimum number of bits from which a particular message can effectively be reconstructed. A little reflection reveals that this is a great difference: for *every* source emitting but two messages the Shannon information is at most 1 bit, but we can choose both messages concerned of arbitrarily high Kolmogorov complexity. Shannon stresses in his founding article that his notion is only concerned with *communication*, while Kolmogorov stresses in his founding article that his notion aims at supplementing the gap left by Shannon theory concerning the information in individual objects. To be sure, both notions are natural: Shannon ignores the object itself but considers only the characteristics of the random source of which the object is one of the possible outcomes, while Kolmogorov considers only the object itself to determine the number of bits in the ultimate compressed version irrespective of the manner in which the object arose.

These differences notwithstanding, there exist very strong connections between both theories. In this section we given an overview of these. In Section 5.1 we recall the relation between probability distributions and codes, and we review Shannon's fundamental notion, the *entropy*. We then (Section 5.2) indicate how Kolmogorov complexity resolves a lacuna in the Shannon theory, namely its inability to deal with information in individual objects. In Section 5.3 we make precise and explain the important relation

$$\text{Entropy} \approx \text{expected Kolmogorov complexity.}$$

Section 5.4 deals with Shannon and algorithmic *mutual information*, the second fundamental concept in both theories.

5.1 Probabilities, Codelengths, Entropy

We now briefly recall the two fundamental relations between probability distributions and codelength functions, and indicate their connection to the entropy, the fundamental concept in Shannon's theory. These relations are essential for understanding the connection between Kolmogorov's and Shannon's theory. For (much) more details, we refer to [Harremoës and Topsøe, 2008]'s chapter in this handbook, and, in a Kolmogorov complexity context, to [Grünwald and Vitányi, 2003]. We use the following notation: let P be a probability distribution defined on a finite or countable set $\mathcal{X}$. In the remainder of the chapter, we denote by X the random variable that takes values in $\mathcal{X}$; thus $P(X = x) = P(\{x\})$ is the probability that the event $\{x\}$ obtains. We write $P(x)$ as an abbreviation of $P(X = x)$, and we write $E_P[f(X)]$ to denote the expectation of a function $f : \mathcal{X} \rightarrow \mathbb{R}$, so that $E_P[f(X)] = \sum_{x\in\mathcal{X}} P(x)f(x)$.

The Two Relations between probabilities and code lengths

1. For every distribution P defined on a finite or countable set $\mathcal{X}$, there exists a code with lengths $L_P(x)$, satisfying, for all $x \in \mathcal{X}$, $L_P(x) = \lceil -\log P(x)\rceil$. This is the so-called *Shannon-Fano* code corresponding to P. The result follows directly from the Kraft inequality [Harremoës and Topsøe, 2008, Section 1.2].

2. If X is distributed according to P, then the Shannon-Fano code corresponding to P is (essentially) the optimal code to use in an expected sense.

   Of course, we may choose to encode outcomes of X using a code corresponding to a distribution Q, with lengths $\lceil -\log Q(x)\rceil$, whereas the outcomes are actually distributed according to $P \neq Q$. But, as expressed in the *noiseless coding theorem* or, more abstractly, in [Harremoës and Topsøe, 2008, Section 1.3] as the *First main theorem of information theory*, such a code cannot be significantly better, and may in fact be much worse than the code with lengths $\lceil -\log P(X)\rceil$: the noiseless coding theorem says that

(16) $E_P[-\log P(X)] \leq$

$$\min_{C:\ C \text{ is a prefix-free code}} E_P[L_C(X)] \leq E_P[-\log P(X)] + 1,$$

so that it follows in particular that the expected length of the Shannon-Fano code satisfies

$$E_P\lceil -\log P(X) \rceil \leq E_P[-\log P(X)]+1 \leq \min_{C:\ C \text{ is a prefix-free code}} E_P[L_C(X)]+1.$$

and is thus always within just bit of the code that is optimal in expectation.

In his 1948 paper, Shannon proposed a measure of information in a distribution, which he called the 'entropy', a concept discussed at length in the chapter by [Harremoës and Topsøe, 2008] in this handbook. It is equal to the quantity appearing on the left and on the right in (16):

DEFINITION 8. [**Entropy**] Let $\mathcal{X}$ be a finite or countable set, let X be a random variable taking values in $\mathcal{X}$ with distribution P. Then the (Shannon-) *entropy* of random variable X is given by

(17) $H(P) = -\sum_{x \in \mathcal{X}} P(x) \log P(x),$

Entropy is defined here as a functional mapping a distribution on $\mathcal{X}$ to real numbers. In practice, we often deal with a pair of random variables (X, Y) defined on a joint space $\mathcal{X} \times \mathcal{Y}$. Then P is the joint distribution of (X, Y), and P_X is its corresponding marginal distribution on X, $P_X(x) = \sum_y P(x, y)$. In that case, rather than writing $H(P_X)$ it is customary to write $H(X)$; we shall follow this convention below.

Entropy can be interpreted in a number of ways. The noiseless coding theorem (16) gives a precise coding-theoretic interpretation: it shows that the entropy of P is essentially equal to the average code length when encoding an outcome of P, if outcomes are encoded using the optimal code (the code that minimizes this average code length).

5.2 A Lacuna in Shannon's Theory

EXAMPLE 9. Assuming that x is emitted by a random source X with probability $P(x)$, we can transmit x using the Shannon-Fano code. This uses (up to rounding) $-\log P(x)$ bits. By Shannon's noiseless coding theorem this is optimal *on average*, the average taken over the probability distribution of outcomes from the source. Thus, if $x = 00\ldots0$ (n zeros), and the random source emits n-bit messages with equal probability $1/2^n$ each, then we require n bits to transmit x (the same as transmitting x literally). However, we can transmit x in about $\log n$ bits if we ignore probabilities and just describe x individually. Thus, the optimality with respect to the average may be very sub-optimal in individual cases.

In Shannon's theory 'information' is fully determined by the probability distribution on the set of possible messages, and unrelated to the meaning, structure or content of individual messages. In many cases this is problematic, since the distribution generating outcomes may be unknown to the observer or (worse), may not exist at all[5]. For example, can we answer a question like "what is the information in this book" by viewing it as an element of a set of possible books with a probability distribution on it? This seems unlikely. Kolmogorov complexity provides a measure of information that, unlike Shannon's, does not rely on (often untenable) probabilistic assumptions, and that takes into account the phenomenon that 'regular' strings are compressible. Thus, it measures the information content of an *individual finite object.* The fact that such a measure exists is surprising, and indeed, it comes at a price: unlike Shannon's, Kolmogorov's measure is asymptotic in nature, and not computable in general. Still, the resulting theory is closely related to Shannon's, as we now discuss.

5.3 Entropy and Expected Kolmogorov Complexity

We call a distribution P computable if it can be computed by a finite-size program, i.e. if it has finite Kolmogorov complexity $K(P)$ (Example 7). The set of computable distributions is very large: it contains, for example, all Markov chains of each order with rational-valued parameters. In the following discussion we shall restrict ourselves to computable distributions; extensions to the uncomputable case are discussed by [Grünwald and Vitányi, 2003].

If X is distributed according to some distribution P, then the optimal (in the average sense) code to use is the Shannon-Fano code. But now suppose it is only known that $P \in \mathcal{P}$, where $\mathcal{P}$ is a large set of computable distributions, perhaps even the set of all computable distributions. Now it is not clear what code is optimal. We may try the Shannon-Fano code for a particular $P \in \mathcal{P}$, but such a code will typically lead to very large expected code lengths if X turns out to be distributed according to some $Q \in \mathcal{P}, Q \neq P$. We may ask whether there exists another code that is 'almost' as good as the Shannon-Fano code for P, no matter what $P \in \mathcal{P}$ actually generates the sequence? We now show that (perhaps surprisingly) the answer is yes.

Let X be a random variable taking on values in the set $\{0,1\}^*$ of binary strings of arbitrary length, and let P be the distribution of X. $K(x)$ is fixed for each x and gives the shortest code word length (but only up to a fixed constant). It is *independent* of the probability distribution P. Nevertheless, if we weigh each individual code word length for x with its probability $P(x)$, then the resulting P-expected code word length $\sum_x P(x)K(x)$ almost achieves the minimal average code word length $H(P) = -\sum_x P(x) \log P(x)$. This is expressed in the following theorem (taken from [Li and Vitányi, 1997]):

[5] Even if we adopt a Bayesian (subjective) interpretation of probability, this problem remains [Grünwald, 2007].

THEOREM 10. *Let P be a computable probability distribution on $\{0,1\}^*$. Then*

$$0 \leq \left(\sum_x P(x)K(x) - H(P) \right) \leq K(P) + O(1).$$

The theorem becomes interesting if we consider sequences of P that assign mass to binary strings of increasing length. For example, let P_n be the distribution on $\{0,1\}^n$ that corresponds to n independent tosses of a coin with bias q, where q is computable (e.g., a rational number). We have $K(P_n) = O(\log n)$, since we can compute P_n with a program of constant size and input n, q with length $l(n') + l(q') = O(\log n)$. On the other hand, $H(P_n) = nH(P_1)$ increases linearly in n (see, e.g., the chapter by [Harremoës and Topsøe, 2008] in this handbook; see also paragraph 1(c) in Section 2.2 of this chapter). So for large n, the optimal code for P_n requires on average $nH(P_1)$ bits, and the Kolmogorov code E^* requires only $O(\log n)$ bits extra. Dividing by n, we see that the additional number of bits needed per outcome using the Kolmogorov code goes to 0. Thus, remarkably, whereas the entropy is the expected codelength according to P under the optimal code for P (a code that will be wildly different for different P), there exists a single code (the Kolmogorov code), which is asymptotically almost optimal for *all* computable P.

5.4 Mutual Information

Apart from entropy, the *mutual information* is perhaps the most important concept in Shannon's theory. Similarly, apart from Kolmogorov complexity itself, the *algorithmic mutual information* is one of the most important concepts in Kolmogorov's theory. In this section we review Shannon's notion, we introduce Kolmogorov's notion, and then we provide an analogue of Theorem 10 which says that essentially, Shannon mutual information is averaged algorithmic mutual information.

Shannon Mutual Information How much information can a random variable X convey about a random variable Y? This is determined by the (Shannon) *mutual information* between X and Y. Formally, it is defined as

$$\begin{aligned} I(X;Y) &:= H(X) - H(X|Y) \\ &= H(X) + H(Y) - H(X,Y) \end{aligned} \tag{18}$$

where $H(X|Y)$ is the conditional entropy of X given Y, and $H(X,Y)$ is the joint entropy of X and Y; the definition of $H(X,Y), H(X|Y)$ as well as an alternative but equivalent definition if $I(X;Y)$, can be found in [Harremoës and Topsøe, 2008]. The equality between the first and second line follows by straightforward rewriting. The mutual information can be thought of as the expected (average) reduction in the number of bits needed to encode X, when an outcome of Y is given for free. In accord with intuition, it is easy to show that $I(X;Y) \geq 0$, with equality if and only if X and Y are independent, i.e. X provides no information about Y. Moreover, and less intuitively, a straightforward calculation shows that this information is *symmetric*: $I(X;Y) = I(Y;X)$.

Algorithmic Mutual Information In order to define algorithmic mutual information, it will be convenient to introduce some new notation: We will denote by $\stackrel{+}{<}$ an inequality to within an additive constant. More precisely, let f, g be functions from $\{0,1\}^*$ to $\mathbb{R}$. Then by '$f(x) \stackrel{+}{<} g(x)$' we mean that there exists a c such that for all $x \in \{0,1\}^*$, $f(x) < g(x) + c$. We write '$f(x) \stackrel{+}{>} g(x)$' if $g(x) \stackrel{+}{<} f(x)$. We denote by $\stackrel{+}{=}$ the situation when both $\stackrel{+}{<}$ and $\stackrel{+}{>}$ hold.

Since $K(x,y) = K(x'y)$ (Section 4.4), trivially, the symmetry property holds: $K(x,y) \stackrel{+}{=} K(y,x)$. An interesting property is the "Additivity of Complexity" property

(19) $K(x,y) \stackrel{+}{=} K(x) + K(y \mid x^*) \stackrel{+}{=} K(y) + K(x \mid y^*)$.

where x^* is the first (in standard enumeration order) shortest prefix program that generates x and then halts. (19) is the Kolmogorov complexity equivalent of the entropy equality $H(X,Y) = H(X) + H(Y|X)$ (see Section I.5 in the chapter by [Harremoës and Topsøe, 2008]). That this latter equality holds is true by simply rewriting both sides of the equation according to the definitions of averages of joint and marginal probabilities. In fact, potential individual differences are averaged out. But in the Kolmogorov complexity case we do nothing like that: it is quite remarkable that additivity of complexity also holds for individual objects. The result (19) is due to [Gács, 1974], can be found as Theorem 3.9.1 in [Li and Vitányi, 1997] and has a difficult proof. It is perhaps instructive to point out that the version with just x and y in the conditionals doesn't hold with $\stackrel{+}{=}$, but holds up to additive logarithmic terms that cannot be eliminated.

To define the algorithmic mutual information between two individual objects x and y with no probabilities involved, it is instructive to first recall the probabilistic notion (18). The algorithmic definition is, in fact, entirely analogous, with H replaced by K and random variables replaced by individual sequences or their generating programs: The *information in y about x* is defined as

(20) $I(y : x) = K(x) - K(x \mid y^*) \stackrel{+}{=} K(x) + K(y) - K(x,y)$,

where the second equality is a consequence of (19) and states that this information is symmetric, $I(x : y) \stackrel{+}{=} I(y : x)$, and therefore we can talk about *mutual information.*[6]

Theorem 10 showed that the entropy of distribution P is approximately equal to the expected (under P) Kolmogorov complexity. Theorem 11 gives the analogous result for the mutual information.

[6]The notation of the algorithmic (individual) notion $I(x : y)$ distinguishes it from the probabilistic (average) notion $I(X;Y)$. We deviate slightly from [Li and Vitányi, 1997] where $I(y : x)$ is defined as $K(x) - K(x \mid y)$.

THEOREM 11. *Let P be a computable probability distribution on $\{0,1\}^* \times \{0,1\}^*$. Then*

$$(21)\quad I(X;Y) - K(P) \stackrel{+}{<} \sum_x \sum_y p(x,y) I(x:y) \stackrel{+}{<} I(X;Y) + 2K(P).$$

Thus, analogously to Theorem 10, we see that the expectation of the algorithmic mutual information $I(x:y)$ is close to the probabilistic mutual information $I(X;Y)$.

Theorems 10 and 11 do not stand on their own: it turns out that just about *every* concept in Shannon's theory has an analogue in Kolmogorov's theory, and in all such cases, these concepts can be related by theorems saying that *if* data are generated probabilistically, then the Shannon concept is close to the expectation of the corresponding Kolmogorov concept. Examples are the probabilistic vs. the algorithmic sufficient statistics, and the probabilistic rate-distortion function [Cover and Thomas, 1991] vs. the algorithmic Kolmogorov structure function. The algorithmic sufficient statistic and structure function are discussed in the next section. For a comparison to their counterparts in Shannon's theory, we refer to [Grünwald and Vitányi, 2004].

6 MEANINGFUL INFORMATION

The information contained in an individual finite object (like a finite binary string) is measured by its Kolmogorov complexity—the length of the shortest binary program that computes the object. Such a shortest program contains no redundancy: every bit is information; but is it meaningful information? If we flip a fair coin to obtain a finite binary string, then with overwhelming probability that string constitutes its own shortest program. However, also with overwhelming probability all the bits in the string are meaningless information, random noise. On the other hand, let an object x be a sequence of observations of heavenly bodies. Then x can be described by the binary string *pd*, where p is the description of the laws of gravity and the observational parameter setting, while d accounts for the measurement errors: we can divide the information in x into meaningful information p and accidental information d. The main task for statistical inference and learning theory is to distill the meaningful information present in the data. The question arises whether it is possible to separate meaningful information from accidental information, and if so, how. The essence of the solution to this problem is revealed as follows. As shown by [Vereshchagin and Vitányi, 2004], for all $x \in \{0,1\}^*$, we have

$$(22)\quad K(x) = \min_{i,p}\{K(i) + l(p) : T_i(p) = x\} + O(1),$$

where the minimum is taken over $p \in \{0,1\}^*$ and $i \in \{1,2,\ldots\}$.

To get some intuition why (22) holds, note that the original definition (1) expresses that $K(x)$ is the sum of the description length $L_{\mathbb{N}}(i)$ of some Turing machine i when encoded using the standard code for the integers, plus the length of a program such that $T_i(p) = x$. (22) expresses that the first term in this sum may be replaced by $K(i)$, i.e. the *shortest* effective description of i. It is clear that (22) is never larger than (9) plus some constant (the size of a computer program implementing the standard encoding/decoding of integers). The reason why (22) is also never smaller than (9) minus some constant is that there exists a Turing machine T_k such that, for all i, p, $T_k(i^*p) = T_i(p)$, where i^* is the shortest program that prints i and then halts, i.e. for all i, p, $U(\langle k, i^*p \rangle) = T_i(p)$ where U is the reference machine used in Definition 1. Thus, $K(x)$ is bounded by the constant length $l(k')$ describing k, plus $l(i^*) = K(i)$, plus $l(p)$.

The expression (22) shows that we can think of Kolmogorov complexity as the length of a *two-part code*. This way, $K(x)$ is viewed as the shortest length of a two-part code for x, one part describing a Turing machine T, or *model*, for the *regular* aspects of x, and the second part describing the *irregular* aspects of x in the form of a program p to be interpreted by T. The regular, or "valuable," information in x is constituted by the bits in the "model" while the random or "useless" information of x constitutes the remainder. This leaves open the crucial question: How to choose T and p that together describe x? In general, many combinations of T and p are possible, but we want to find a T that describes the meaningful aspects of x. Below we show that this can be achieved using the *Algorithmic Minimum Sufficient Statistic.* This theory, built on top of Kolmogorov complexity so to speak, has its roots in two talks by Kolmogorov [1974a; 1974b]. Based on Kolmogorov's remarks, the theory has been further developed by several authors, culminating in [Vereshchagin and Vitányi, 2004], some of the key ideas of which we outline below.

Data and Model We restrict attention to the following setting: we observe data x in the form of a finite binary string of some length n. As models for the data, we consider finite sets S that contain x. In statistics and machine learning, the use of finite sets is nonstandard: one usually models the data using probability distributions or functions. However, the restriction of sets is just a matter of convenience: the theory we are about to present generalizes straightforwardly to the case where the models are arbitrary computable probability density functions and, in fact, other model classes such as computable functions [Vereshchagin and Vitányi, 2004]; see also Section 6.3.

The intuition behind the idea of a set as a model is the following: informally, 'S is a good model for x' or equivalently, S captures all structure in x, if, in a sense to be made precise further below, it summarizes all simple properties of x. In Section 6.1 below, we work towards the definition of the algorithmic minimal sufficient statistic (AMSS) via the fundamental notions of 'typicality' of data and 'optimality' of a set. Section 6.2 investigates the AMSS further in terms of the important *Kolmogorov Structure Function.* In Section 6.3, we relate the AMSS to the more well-known *Minimum Description Length Principle.*

6.1 *Algorithmic Sufficient Statistic*

We are now about to formulate the central notions 'x is typical for S' and 'S is optimal for x'. Both are necessary, but not sufficient requirements for S to precisely capture the 'meaningful information' in x. After having introduced optimal sets, we investigate what further requirements we need. The development will make heavy use of the Kolmogorov complexity of sets, and conditioned on sets. These notions, written as $K(S)$ and $K(x|S)$, were defined in Example 7.

Typical Elements

Consider a string x of length n and prefix complexity $K(x) = k$. We look for the *structure* or *regularity* in x that is to be summarized with a set S of which x is a *random* or *typical* member: given S containing x, the element x cannot be described significantly shorter than by its maximal length index in S, that is, $K(x \mid S) \geq \log|S| + O(1)$. Formally,

DEFINITION 12. Let $\beta \geq 0$ be an agreed-upon, fixed, constant. A finite binary string x is a *typical* or *random* element of a set S of finite binary strings, if $x \in S$ and

(23) $K(x \mid S) \geq \log|S| - \beta.$

We will not indicate the dependence on β explicitly, but the constants in all our inequalities ($O(1)$) will be allowed to be functions of this β.

This definition requires a finite S. Note that the notion of typicality is not absolute but depends on fixing the constant implicit in the O-notation.

EXAMPLE 13. Consider the set S of binary strings of length n whose every odd position is 0. Let x be an element of this set in which the subsequence of bits in even positions is an incompressible string. Then x is a typical element of S. But x is also a typical element of the set $\{x\}$.

Note that, if x is not a typical element of S, then S is certainly not a 'good model' for x in the intuitive sense described above: S does not capture all regularity in x. However, the example above ($S = \{x\}$) shows that even if x is typical for S, S may still not capture 'all meaningful information in x'.

EXAMPLE 14. If y is not a typical element of S, this means that it has some simple special property that singles it out from the vast majority of elements in S. This can actually be proven formally [Vitányi, 2005]. Here we merely give an example. Let S be as in Example 13. Let y be an element of S in which the subsequence of bits in even positions contains two times as many 1s than 0s. Then y is not a typical element of S: the overwhelming majority of elements of S have about equally many 0s as 1s in even positions (this follows by simple combinatorics). As shown in [Vitányi, 2005], this implies that $K(y|S) \ll |\log S|$, so that y is not typical.

Optimal Sets

Let x be a binary data string of length n. For every finite set $S \ni x$, we have $K(x) \leq K(S) + \log|S| + O(1)$, since we can describe x by giving S and the index of x in a standard enumeration of S. Clearly this can be implemented by a Turing machine computing the finite set S and a program p giving the index of x in S. The size of a set containing x measures intuitively the number of properties of x that are represented: The largest set is $\{0,1\}^n$ and represents only one property of x, namely, being of length n. It clearly "underfits" as explanation or model for x. The smallest set containing x is the singleton set $\{x\}$ and represents all conceivable properties of x. It clearly "overfits" as explanation or model for x.

There are two natural measures of suitability of such a set as a model for x. We might prefer either (a) the simplest set, or (b) the smallest set, as corresponding to the most likely structure 'explaining' x. Both the largest set $\{0,1\}^n$ [having low complexity of about $K(n)$] and the singleton set $\{x\}$ [having high complexity of about $K(x)$], while certainly statistics for x, would indeed be considered poor explanations. We would like to balance simplicity of model vs. size of model. Both measures relate to the optimality of a two-stage description of x using a finite set S that contains it. Elaborating on the two-part code described above,

$$\begin{aligned} K(x) &\leq K(S) + K(x \mid S) + O(1) \\ &\leq K(S) + \log|S| + O(1), \end{aligned} \tag{24}$$

where the first inequality follows because there exists a program p producing x that first computes S and then computes x based on S; if p is not the shortest program generating x, then the inequality is strict. The second substitution of $K(x \mid S)$ by $\log|S| + O(1)$ uses the fact that x is an element of S. The closer the right-hand side of (24) gets to the left-hand side, the better the two-stage description of x is. This implies a trade-off between meaningful model information, $K(S)$, and meaningless "noise" $\log|S|$. A set S (containing x) for which (24) holds with equality,

(25) $K(x) = K(S) + \log|S| + O(1)$,

is called *optimal.* The first line of (24) implies that if a set S is optimal for x, then x must be a typical element of S. However, the converse does not hold: a data string x can be typical for a set S without that set S being optimal for x.

EXAMPLE 15. It can be shown that the set S of Example 13 is also optimal, and so is $\{x\}$. Sets for which x is typical form a much wider class than optimal sets for x: the set $\{x, y\}$ is still typical for x but with most y it will be too complex to be optimal for x. A less artificial example can be found in [Vereshchagin and Vitányi, 2004].

While 'optimality' is a refinement of 'typicality', the fact that $\{x\}$ is still an optimal set for x shows that it is still not sufficient by itself to capture the notion of 'meaningful information'. In order to discuss the necessary refinement, we first need to connect optimal sets to the notion of a 'sufficient statistic', which, as its name suggests, has its roots in the statistical literature.

Algorithmic Sufficient Statistic

A *statistic* of the data $x = x_1 \dots x_n$ is a function $f(x)$. Essentially, every function will do. For example, $f_1(x) = n$, $f_2(x) = \sum_{i=1}^{n} x_i$, $f_3(x) = n - f_2(x)$, and $f_4(x) = f_2(x)/n$, are statistics. A "sufficient" statistic of the data contains all information in the data about the model. In introducing the notion of sufficiency in classical statistics, Fisher [1922] stated: "The statistic chosen should summarize the whole of the relevant information supplied by the sample. This may be called the Criterion of Sufficiency ... In the case of the normal distributions it is evident that the second moment is a sufficient statistic for estimating the standard deviation." For example, in the Bernoulli model (repeated coin flips with outcomes 0 and 1 according to fixed bias), the statistic f_4 is sufficient. It gives the mean of the outcomes and estimates the bias of the Bernoulli process, which is the only relevant model information. For the classic (probabilistic) theory see, for example, [Cover and Thomas, 1991]. [Gács, Tromp, and Vitányi, 2001] develop an algorithmic theory of sufficient statistics (relating individual data to individual model) and establish its relation to the probabilistic version; this work is extended by [Grünwald and Vitányi, 2004]. The algorithmic basics are as follows: Intuitively, a model expresses the essence of the data if the two-part code describing the data consisting of the model and the data-to-model code is as concise as the best one-part description. In other words, we call a shortest program for an optimal set with respect to x an *algorithmic sufficient statistic* for x.

EXAMPLE 16. **(Sufficient Statistic)** Let us look at a coin toss example. Let k be a number in the range $0, 1, \dots, n$ of complexity $\log n + O(1)$ given n and let x be a string of length n having k 1s of complexity $K(x \mid n, k) \geq \log \binom{n}{k}$ given n, k. This x can be viewed as a typical result of tossing a coin with a bias about $p = k/n$. A two-part description of x is given by first specifying the number k of 1s in x, followed by the index $j \leq \log |S|$ of x in the set S of strings of length n with k 1s. This set is optimal, since, to within $O(1)$, $K(x) = K(x, \langle n, k \rangle) = K(n, k) + K(x \mid n, k) = K(S) + \log |S|$. The shortest program for S, which amounts to an encoding of n and then k given n, is an algorithmic sufficient statistic for x.

The optimal set that admits the shortest possible program (or rather that shortest program) is called *algorithmic minimal sufficient statistic* of x. In general there can be more than one such set and corresponding program:

DEFINITION 17 (Algorithmic minimal sufficient statistic). An *algorithmic sufficient statistic* of x is a shortest program for a set S containing x that is optimal, i.e. it satisfies (25). An algorithmic sufficient statistic with optimal set S is *minimal* if there exists no optimal set S' with $K(S') < K(S)$.

The algorithmic minimal sufficient statistic (AMSS) divides the information in x in a relevant structure expressed by the set S, and the remaining randomness with respect to that structure, expressed by x's index in S of $\log |S|$ bits. The shortest program for S is itself alone an algorithmic definition of structure, without a probabilistic interpretation.

EXAMPLE 18. **(Example 13, Cont.)** The shortest program for the set S of Example 13 is a minimum sufficient statistic for the string x mentioned in that example. The program generating the set $\{x\}$, while still an algorithmic sufficient statistic, is not a minimal sufficient statistic.

EXAMPLE 19. **(Example 16, Cont.)** The S of Example 16 encodes the number of 1s in x. The shortest program for S is an algorithmic minimal sufficient statistic for *most* x of length n with k 1's, since only a fraction of at most 2^{-m} x's of length n with k 1s can have $K(x) < \log |S| - m$ (Section 4). But of course there exist x's with k ones which have much more regularity. An example is the string starting with k 1's followed by $n - k$ 0's. For such strings, S is not optimal anymore, nor is S an algorithmic sufficient statistic.

To analyze the minimal sufficient statistic further, it is useful to place a constraint on the maximum complexity of set $K(S)$, say $K(S) \leq \alpha$, and to investigate what happens if we vary α. The result is the *Kolmogorov Structure Function*, which we now discuss.

6.2 The Kolmogorov Structure Function

The *Kolmogorov structure* function [Kolmogorov, 1974a; 1974b; Vereshchagin and Vitáyni, 2004] h_x of given data x is defined by

$$(26)\quad h_x(\alpha) = \min_S\{\log |S| : S \ni x,\ K(S) \leq \alpha\},$$

where $S \ni x$ is a contemplated model for x, and α is a non-negative integer value bounding the complexity of the contemplated S's. Clearly, the Kolmogorov structure function is nonincreasing and reaches $\log |\{x\}| = 0$ for $\alpha = K(x) + c_1$ where c_1 is the number of bits required to change x into $\{x\}$. For every $S \ni x$ we have (24), and hence $K(x) \leq \alpha + h_x(\alpha) + O(1)$; that is, the function $h_x(\alpha)$ never decreases more than a fixed independent constant below the diagonal *sufficiency line* L defined by $L(\alpha) + \alpha = K(x)$, which is a lower bound on $h_x(\alpha)$ and is approached to within a constant distance by the graph of h_x for certain α's (e.g., for $\alpha = K(x) + c_1$). For these α's we thus have $\alpha + h_x(\alpha) = K(x) + O(1)$; a model corresponding to such an α (witness for $h_x(\alpha)$) is a sufficient statistic, and it is *minimal* for the least such α [Cover and Thomas, 1991; Gács, Tromp, and Vitányi, 2001]. This is depicted in Figure 1. Note once again that the structure function is defined relative to given data (a single sequence x); different sequences result in different structure functions. Yet, all these different functions share some properties: for all x, the function $h_x(\alpha)$ will lie above the diagonal sufficiency line for all $\alpha \leq \alpha_x$. Here α_x is the complexity $K(S)$ of the AMSS for x. For $\alpha \geq \alpha_x$, the function $h_x(\alpha)$ remains within a constant of the diagonal. For stochastic strings generated by a simple computable distribution (finite $K(P)$), the sufficiency line will typically be first hit for α close to 0, since the AMSS will grow as $O(\log n)$. For example, if x is generated by independent fair coin flips, then, with probability 1, one AMSS will be $S = \{0, 1\}^n$ with complexity $K(S) = K(n) = O(\log n)$. One

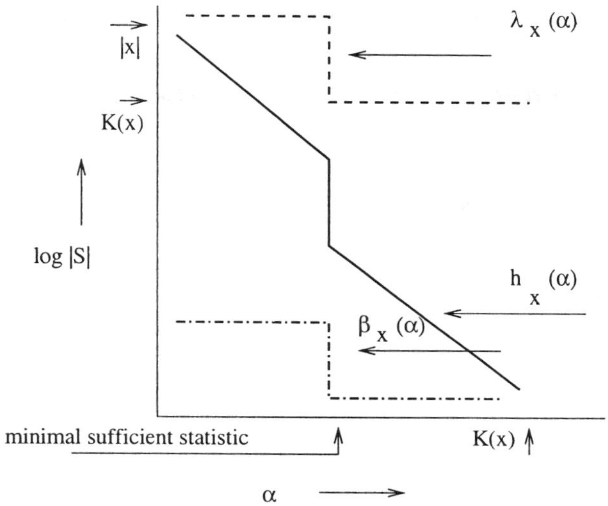

Figure 1. Structure functions $h_x(i), \beta_x(\alpha), \lambda_x(\alpha)$, and minimal sufficient statistic.

may suspect that all intuitively 'random' sequences have a small sufficient statistic of order $O(\log n)$ or smaller. Surprisingly, this turns out not to be the case, as we show in Example 21.

EXAMPLE 20. **(Lossy Compression)** The Kolmogorov structure function $h_x(\alpha)$ is relevant to lossy compression (used, e.g., to compress images). Assume we need to compress x to α bits where $\alpha \ll K(x)$. Of course this implies some loss of information present in x. One way to select redundant information to discard is as follows: let S_0 be the set generated by the Algorithmic Minimum Sufficient Statistic S_0^* (S_0^* is a shortest program that prints S_0 and halts). Assume that $l(S_0^*) = K(S_0) \leq \alpha$. Since S_0 is an optimal set, it is also a typical set, so that $K(x|S_0) \approx \log |S_0|$. We compress x by S_0^*, taking α bits. To reconstruct an x' close to x, a decompressor can first reconstruct the set S_0, and then select an element x' of S_0 uniformly at random. This ensures that with very high probability x' is itself also a typical element of S_0, so it has the same properties that x has. Therefore, x' should serve the purpose of the message x as well as does x itself. However, if $l(S_0^*) > \alpha$, then it is not possible to compress all meaningful information of x into α bits. We may instead encode, among all sets S with $K(S) \leq \alpha$, the one with the smallest $\log |S|$, achieving $h_x(\alpha)$. But inevitably, this set will not capture all the structural properties of x.

Let us look at an example. To transmit a picture of "rain" through a channel with limited capacity α, one can transmit the indication that this is a picture of the rain and the particular drops may be chosen by the receiver at random. In this interpretation, the complexity constraint α determines how "random" or "typical"

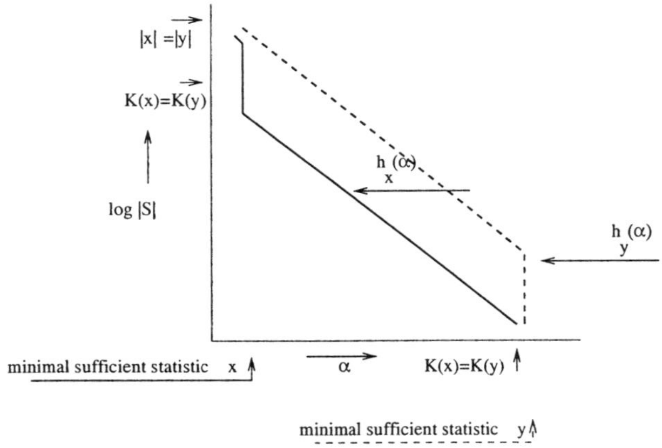

Figure 2. Data string x is "positive random" or "stochastic" and data string y is just "negative random" or "non-stochastic".

x will be with respect to the chosen set S —and hence how "indistinguishable" from the original x the randomly reconstructed x' can be expected to be.

We end this section with an example of a strange consequence of Kolmogorov's theory:

EXAMPLE 21. **"Positive" and "Negative" Individual Randomness:** [Gács, Tromp, and Vitányi, 2001] showed the existence of strings for which essentially the singleton set consisting of the string itself is a minimal sufficient statistic (Section 6.1). While a sufficient statistic of an object yields a two-part code that is as short as the shortest one part code, restricting the complexity of the allowed statistic may yield two-part codes that are considerably longer than the best one-part code (so that the statistic is insufficient). In fact, for every object there is a complexity bound below which this happens; this is just the point where the Kolmogorov structure function hits the diagonal. If that bound is small (logarithmic) we call the object "stochastic" since it has a simple satisfactory explanation (sufficient statistic). Thus, Kolmogorov [1974a] makes the important distinction of an object being random in the "negative" sense by having this bound high (it has high complexity and is not a typical element of a low-complexity model), and an object being random in the "positive, probabilistic" sense by both having this bound small and itself having complexity considerably exceeding this bound (like a string x of length n with $K(x) \geq n$, being typical for the set $\{0,1\}^n$, or the uniform probability distribution over that set, while this set or probability distribution has complexity $K(n) + O(1) = O(\log n)$). We depict the distinction in Figure 2.

6.3 The Minimum Description Length Principle

Learning The main goal of statistics and machine learning is to learn from data. One common way of interpreting 'learning' is as a search for the structural, regular properties of the data — all the patterns that occur in it. On a very abstract level, this is just what is achieved by the AMSS, which can thus be related to learning, or, more generally, inductive inference. There is however another, much more well-known method for learning based on data compression. This is the Minimum Description Length (MDL) Principle, mostly developed by J. Rissanen [1978; 1989] — see [Grünwald, 2007] for a recent introduction; see also [Wallace, 2005] for the related MML Principle. Rissanen took Kolmogorov complexity as an informal starting point, but was not aware of the AMSS when he developed the first, and, with hindsight, somewhat crude version of MDL [Rissanen, 1978], which roughly says that the best theory to explain given data x is the one that minimizes the sum of

1. The length, in bits, of the description of the theory, plus

2. The length, in bits, of the description of the data x when the data is described with the help of the theory.

Thus, data is encoded by first encoding a theory (constituting the 'structural' part of the data) and then encoding the data using the properties of the data that are prescribed by the theory. Picking the theory minimizing the total description length leads to an automatic trade-off between complexity of the chosen theory and its goodness of fit on the data. This provides a principle of inductive inference that may be viewed as a mathematical formalization of 'Occam's Razor'. It automatically protects against overfitting, a central concern of statistics: when allowing models of arbitrary complexity, we are always in danger that we model random fluctuations rather than the trend in the data [Grünwald, 2007].

The MDL Principle has been designed so as to be practically useful. This means that the codes used to describe a 'theory' are not based on Kolmogorov complexity. However, there exists an 'ideal' version of MDL [Li and Vitányi, 1997; Barron and Cover, 1991] which does rely on Kolmogorov complexity. Within our framework (binary data, models as sets), it becomes [Vereshchagin and Vitányi, 2004; Vitányi, 2005]: pick a set $S \ni x$ minimizing the two-part codelength

(27) $K(S) - \log |S|$.

In other words: any "optimal set" (as defined in Section 6.1) is regarded as a good explanation of the theory. It follows that every set S that is an AMSS also minimizes the two-part codelength to within $O(1)$. However, as we already indicated, there exist optimal sets S (that, because of their optimality, may be selected by MDL), that are not minimal sufficient statistics. As explained by [Vitányi, 2005], these do not capture the idea of 'summarizing all structure in x'. Thus, the AMSS may be considered a refinement of the idealized MDL approach.

Practical MDL The practical MDL approach uses probability distributions rather than sets as models. Typically, one restricts to distributions in some model class such as the set of all Markov chain distributions of each order, or the set of all polynomials f of each degree, where f expresses that $Y = f(X) + Z$, and Z is some normally distributed noise variable (this makes f a 'probabilistic' hypothesis). These model classes are still 'large' in that they cannot be described by a finite number of parameters; but they are simple enough so that admit efficiently computable versions of MDL — unlike the ideal version above which, because it involves Kolmogorov complexity, is uncomputable. The Kolmogorov complexity, set-based theory has to be adjusted at various places to deal with such practical models, one reason being that they have uncountably many elements. MDL has been successful in practical statistical and machine learning problems where overfitting is a real concern [Grünwald, 2007]. Technically, MDL algorithms are very similar to the popular Bayesian methods, but the underlying philosophy is very different: MDL is based on finding structure in *individual data* sequences; distributions (models) are viewed as *representation languages for expressing useful properties of the data*; they are neither viewed as objectively existing but unobservable objects according to which data are 'generated'; nor are they viewed as representing subjective degrees of belief, as in a mainstream Bayesian interpretation.

In recent years, ever more sophisticated refinements of the original MDL have developed [Rissanen, 1996; Rissanen and Tabus, 2005; Grünwald, 2007]. For example, in modern MDL approaches, one uses *universal codes* which may be two-part, but in practice are often *one-part* codes.

7 PHILOSOPHICAL IMPLICATIONS AND CONCLUSION

We have given an overview of algorithmic information theory, focusing on some of its most important aspects: Kolmogorov complexity, algorithmic mutual information, their relations to entropy and Shannon mutual information, the Algorithmic Minimal Sufficient Statistic and the Kolmogorov Structure Function, and their relation to 'meaningful information'. Throughout the chapter we emphasized insights that, in our view, are 'philosophical' in nature. It is now time to harvest and make the philosophical connections explicit. Below we first discuss some implications of algorithmic information theory on the philosophy of (general) mathematics, probability theory and statistics. We then end the chapter by discussing the philosophical implications for 'information' itself. As we shall see, it turns out that nearly all of these philosophical implications are somehow related to *randomness*.

Philosophy of Mathematics: Randomness in Mathematics In and after Example 4 we indicated that the ideas behind Kolmogorov complexity are intimately related to Gödel's incompleteness theorem. The finite Kolmogorov complexity of any effective axiom system implied the existence of bizarre equations like (10), whose full solution is, in a sense, random: *no* effective axiom system can fully determine the solutions of this single equation. In this context, Chaitin writes: "This is a region in which mathematical truth has no discernible structure or pattern and appears to be completely random [...] Quantum physics has shown that there is randomness in nature. I believe that we have demonstrated [...] that randomness is already present in pure Mathematics. This does not mean that the universe and Mathematics are completely lawless, it means that laws of a different kind apply: statistical laws. [...] Perhaps number theory should be pursued more openly in the spirit of an experimental science!".

Philosophy of Probability: Individual Randomness The statement 'x is a random sequence' is essentially meaningless in classical probability theory, which can only make statements that hold for ensembles, such as 'relative frequencies converge to probabilities *with high probability*, or *with probability 1*'. But in reality we only observe one sequence. What then does the statement 'this sequence is a typical outcome of distribution P' or, equivalently, is 'random with respect to P' tell us about the sequence? We might think that it means that the sequence satisfies all properties that hold with P-probability 1. But this will not work: if we identify a 'property' with the set of sequences satisfying it, then it is easy to see that the intersection of all sets corresponding to properties that hold 'with probability 1' is empty. The Martin-Löf theory of randomness [Li and Vitányi, 1997] essentially resolves this issue. Martin-Löf's notion of randomness turns out to be, roughly, equivalent with Kolmogorov randomness: a sequence x is random if $K(x) \approx l(x)$, i.e. it cannot be effectively compressed. This theory allows us to speak of the randomness of single, individual sequences, which is inherently impossible for probabilistic theories. Yet, as shown by Martin-Löf, his notion of randomness is entirely consistent with probabilistic ideas. It opens up a whole new area, which is illustrated by Example 21, in which we made distinctions between different *types* of random sequences ('positive' and 'negative') that cannot be expressed in, let alone understood from, a traditional probabilistic perspective.

Philosophy of Statistics/Inductive Inference: Epistemological Occam's Razor There exist two close connections between algorithmic information theory and inductive inference: one via the algorithmic sufficient statistic and the MDL Principle; the other via Solomonoff's induction theory, which there was no space to discuss here [Li and Vitányi, 1997]. The former deals with finding structure in data; the latter is concerned with sequential prediction. Both of these theories implicitly employ a form of Occam's Razor: when two hypotheses fit the data equally well, they prefer the simplest one (with the shortest description). Both the MDL and the Solomonoff approach are theoretically quite well-behaved: there exist several convergence theorems for both approaches. Let us give an example of such a theorem for the MDL framework: [Barron and Cover, 1991] and [Barron, 1985] show that, if data are distributed according to some distribution in a contemplated model class (set of candidate distributions) $\mathcal{M}$, then two-part MDL will eventually find this distribution; it will even do so based on a reasonably small sample. This holds both for practical versions of MDL (with restricted model classes) as well as for versions based on Kolmogorov complexity, where $\mathcal{M}$ consists of the huge class of all distributions which can be arbitrarily well approximated by finite computer programs. Such theorems provide a justification for MDL. Looking at the proofs, one finds that the preference for simple models is crucial: the convergence occurs precisely because complexity of each probabilistic hypotheses P is measured by its codelength $L(P)$, under some prefix-code that allows one to encode all P under consideration. If a complexity measure $L'(P)$ is used that does *not* correspond to any prefix code, then, as is easy to show, in some situations one will not converge at all, and, no matter how many data are observed, will keep selecting overly complex, suboptimal hypotheses for the data. In fact, even if the world is such that data are generated by a very complex (high $K(P)$) distribution, it is wise to prefer simple models at small sample sizes [Grünwald, 2007]! This provides a justification for the use of MDL's version of Occam's razor in inductive inference. It should be stressed that this is an *epistemological* rather than a *(meta-) physical* form of Occam's Razor: it is used as an effective *strategy*, which is something very different from a belief that 'the true state of the world is likely to have a short description'. This issue, as well as the related question to what extent Occam's Razor can be made representation-independent, is discussed in great detail in [Grünwald, 2007].

A further difference between statistical inference based on algorithmic information theory and almost all other approaches to statistics and learning is that the algorithmic approach focuses on individual data sequences: there is no need for the (often untenable) assumption of classical statistics that there is some distribution P according to which the data are distributed. In the Bayesian approach to statistics, probability is often interpreted subjectively, as a degree of belief. Still, in many Bayesian approaches there is an underlying assumption that there exists 'states of the world' which are viewed as probability distributions. Again, such assumptions need not be made in the present theories; neither in the form which explicitly uses Kolmogorov complexity, nor in the restricted practical form. In both cases, the goal is to find regular patterns in the data, no more. All this is discussed in detail in [Grünwald, 2007].

Philosophy of Information On the first page of the chapter on Shannon information theory in this handbook [Harremoës and Topsøe, 2008], we read "information is always *information about something.*" This is certainly the case for Shannon information theory, where a string x is always used to communicate some state of the world, or of those aspects of the world that we care about. But if we identify 'amount of information in x' with $K(x)$, then it is not so clear anymore what this 'information' is about. $K(x)$, the algorithmic information in x looks at the information in x itself, independently of anything outside. For example, if x consists of the first billion bits of the binary expansion of π, then its information content is the size of the smallest program which prints these bits. This sequence does not describe any state of the world that is to be communicated. Therefore, one may argue that it is meaningless to say that 'x carries information', let alone to measure its amount. At a workshop where many of the contributors to this handbook were present, there was a long discussion about this question, with several participants insisting that "algorithmic information misses "aboutness" (sic), and is therefore not really information". In the end the question whether algorithmic information should really count as "information" is, of course, a matter of definition. Nevertheless, we would like to argue that there exist situations where intuitively, the word "information" seems exactly the right word to describe what is being measured, while nevertheless, "aboutness" is missing. For example, $K(y|x)$ is supposed to describe the amount of "information" in y that is not already present in x. Now suppose y is equal to $3x$, expressed in binary, and x is a random string of length n, so that $K(x) \approx K(y) \approx n$. Then $K(y|x) = O(1)$ is much smaller than $K(x)$ or $K(y)$. The way an algorithmic information theorist would phrase this is "x provides nearly all the *information* needed to generate y." To us, this seems an eminently reasonable use of the word information. Still, this "information" does not refer to any outside state of the world.[7]

Let us assume then that the terminology "algorithmic *information* theory" is justified. What lessons can we draw from the theory for the philosophy of information?

First, we should emphasize that the amount of 'absolute, inherent' information in a sequence is only well-defined asymptotically and is in general uncomputable. Thus, an objective measure of information without 'aboutness' is possible, but at an (unavoidable) price. If we want a nonasymptotic and efficiently computable measure, we are forced to use a restricted class of description methods. Such restrictions naturally lead one to universal coding and practical MDL. The resulting notion of information is always defined *relative* to a class of description methods and can make no claims to objectivity or absoluteness. Interestingly though, unlike Shannon's notion, it is still meaningful for individual sequences, independently of any outside *probabilistic* assumptions: this is an aspect of the general theory that can be retained in the restricted forms [Grünwald, 2007].

Second, the algorithmic theory allows us to formalize the notion of 'meaningful information' in a distinctly novel manner. It leads to a separation of the meaningful information from the noise in a sequence, once again without making any probabilistic assumptions. Since learning can be seen as an attempt to find the meaningful information in data, this connects the theory to inductive inference.

Third, the theory re-emphasizes the connection between measuring amounts of information and data compression, which was also the basis of Shannon's theory. In fact, algorithmic information has close connections to Shannon information after all, and *if* the data x are generated by some probabilistic process P, so that the information in x is actually really 'about' something, then the algorithmic

[7]We may of course say that x carries information "about" y. The point, however, is that y is not a state of any imagined external world, so here "about" does not refer to anything external. Thus, one cannot say that x contains information about some external state of the world.

information in x behaves very similarly to the Shannon entropy of P, as explained in Section 5.3.

Further Reading Kolmogorov complexity has many applications which we could not discuss here. It has implications for aspects of physics such as the second law of thermodynamics; it provides a novel mathematical proof technique called the *incompressibility method*, and so on. These and many other topics in Kolmogorov complexity are thoroughly discussed and explained in the standard reference [Li and Vitányi, 1997]. Additional (and more recent) material on the relation to Shannon's theory can be found in Grünwald and Vitányi [2003; 2004]. Additional material on the structure function is in [Vereshchagin and Vitányi, 2004; Vitányi, 2005]; and additional material on MDL can be found in [Grünwald, 2007].

ACKNOWLEDGMENTS

Paul Vitányi was supported in part by the EU project RESQ, IST-2001-37559, the NoE QIPROCONE +IST-1999-29064 and the ESF QiT Programme. Both Vitányi and Grünwald were supported in part by the IST Programme of the European Community, under the PASCAL Network of Excellence, IST-2002-506778. This publication only reflects the authors' views.

BIBLIOGRAPHY

[Barron and Cover, 1991] A. R. Barron and T. Cover. Minimum complexity density estimation. *IEEE Transactions on Information Theory 37*(4), 1034–1054, 1991.

[Barron, 1985] A. R. Barron. *Logically Smooth Density Estimation.* Ph. D. thesis, Department of EE, Stanford University, Stanford, Ca, 1985.

[Chaitin, 1966] G. Chaitin. On the length of programs for computing finite binary sequences. *Journal of the ACM 13*, 547–569, 1966.

[Chaitin, 1975] G. Chaitin. A theory of program size formally identical to information theory. *Journal of the ACM 22*, 329–340, 1975.

[Chaitin, 1987] G. Chaitin. *Algorithmic Information Theory.* Cambridge University Press, 1987.

[Cilibrasi and Vitányi, 2005] R. Cilibrasi and P. M. Vitányi. Clustering by compression. *IEEE Transactions on Information Theory 51*(4), 1523–1545, 2005.

[Cover and Thomas, 1991] T. Cover and J. Thomas. *Elements of Information Theory.* New York: Wiley Interscience, 1991.

[Gács, 1974] P. Gács. On the symmetry of algorithmic information. *Soviet Math. Dokl. 15*, 1477–1480, 1974. Correction, Ibid., 15:1480, 1974.

[Gács, Tromp, and Vitányi, 2001] P. Gács, J. Tromp, and P. M. Vitányi. Algorithmic statistics. *IEEE Transactions on Information Theory 47*(6), 2464–2479, 2001.

[Grünwald and Vitányi, 2003] P. D. Grünwald and P. M. Vitányi. Kolmogorov complexity and information theory; with an interpretation in terms of questions and answers. *Journal of Logic, Language and Information 12*, 497–529, 2003.

[Grünwald, 2007] P. D. Grünwald. *The Minimum Description Length Principle.* Cambridge, MA: MIT Press, 2007.

[Grünwald and Vitányi, 2004] P. D. Grünwald and P. M. Vitányi. Shannon information and Kolmogorov complexity, 2004. Submitted for publication. Available at the Computer Science CoRR arXiv as `http://de.arxiv.org/abs/cs.IT/0410002`.

[Harremoës and Topsøe, 2008] P. Harremoës and F. Topsøe. The quantitative theory of information. In J. van Benthem and P. Adriaans (Eds.), *Handbook of the Philosophy of Information*, Chapter 6. Elsevier, 2008.

[Kolmogorov, 1965] A. Kolmogorov. Three approaches to the quantitative definition of information. *Problems Inform. Transmission 1*(1), 1–7, 1965.

[Kolmogorov, 1974a] A. Kolmogorov. Talk at the Information Theory Symposium in Tallinn, Estonia, 1974.

[Kolmogorov, 1974b] A. Kolmogorov. Complexity of algorithms and objective definition of randomness. A talk at Moscow Math. Soc. meeting 4/16/1974. A 4-line abstract available in *Uspekhi Mat. Nauk* 29:4, 155, 1974 (in Russian).

[Li and Vitányi, 1997] M. Li and P. M. Vitányi. *An Introduction to Kolmogorov Complexity and Its Applications* (revised and expanded second ed.). New York: Springer-Verlag, 1997.

[Rissanen, 1978] J. Rissanen. Modeling by the shortest data description. *Automatica 14*, 465–471, 1978.

[Rissanen, 1989] J. Rissanen. *Stochastic Complexity in Statistical Inquiry.* World Scientific Publishing Company, 1989.

[Rissanen, 1996] J. Rissanen. Fisher information and stochastic complexity. *IEEE Transactions on Information Theory 42*(1), 40–47, 1996.

[Rissanen and Tabus, 2005] J. Rissanen and I. Tabus. Kolmogorov's structure function in MDL theory and lossy data compression. In P. D. Grünwald, I. J. Myung, and M. A. Pitt (Eds.), *Advances in Minimum Description Length: Theory and Applications.* MIT Press, 2005.

[Solomonoff, 1964] R. Solomonoff. A formal theory of inductive inference, part 1 and part 2. *Information and Control 7*, 1–22, 224–254, 1964.

[Vereshchagin and Vitányi, 2004] N. Vereshchagin and P. M. Vitányi. Kolmogorov's structure functions and model selection. *IEEE Transactions on Information Theory 50*(12), 3265–3290, 2004.

[Vitányi, 2005] P. M. Vitányi. Algorithmic statistics and Kolmogorov's structure function. In P. D. Grünwald, I. J. Myung, and M. A. Pitt (Eds.), *Advances in Minimum Description Length: Theory and Applications.* MIT Press, 2005.

[Wallace, 2005] C. Wallace. *Statistical and Inductive Inference by Minimum Message Length.* New York: Springer, 2005.

Part D

Major Themes in Transforming and Using Information

OCKHAM'S RAZOR, TRUTH, AND INFORMATION

Kevin T. Kelly

1 INTRODUCTION

Suppose that several or even infinitely many theories are compatible with the information available. How ought one to choose among them, if at all? The traditional and intuitive answer is to choose the "simplest" and to cite *Ockham's razor* by way of justification. Simplicity, in turn, has something to do with minimization of entities, description length, causes, free parameters, independent principlies, or ad hoc hypotheses, or maximization of unity, uniformity, symmetry, testability, or explanatory power.

Insofar as Ockham's razor is widely regarded as a rule of scientific inference, it should help one to select the true theory from among the alternatives. The trouble is that it is far from clear how a fixed bias toward simplicity could do so [Morrison, 2000]. One wishes that simplicity could somehow indicate or inform one of the true theory, the way a compass needle indicates or informs one about direction. But since Ockham's razor always points toward simplicity, it is more like a compass needle that is frozen into a fixed position, which cannot be said to to indicate anything. Nor does it suffice to respond that a prior bias toward simplicity can be corrected, eventually, to allow for convergence to the truth, for alternative biases are also correctable in the limit.

This paper reviews some standard accounts of Ockham's razor and concludes that not one of them explains successfully how Ockham's razor helps one find the true theory any better than alternative empirical methods. Thereafter, a new explanation is presented, according to which Ockham's razor does not indicate or *inform* one of the truth like a compass but, nonetheless, keeps one on the straightest possible route to the true theory, which is the best that any inductive strategy could possibly guarantee. Indeed, no non-Ockham strategy can be said to guarantee so straight a path. Hence, a truth-seeker always has a good reason to stick with Ockham's razor even though simplicity does not indicate or inform one of the truth in the short run.

2 STANDARD ACCOUNTS

The point of the following review of standard explanations of Ockham's razor is just to underscore the fact that they do not connect simplicity with selecting the true

Handbook of the Philosophy of Science. Volume 8: Philosophy of Information
Volume editors: Pieter Adriaans and Johan van Benthem. General editors: Dov M. Gabbay, Paul Thagard and John Woods.

theory. For the most part, the authors of the accounts fairly and explicitly specify motives other than finding the true theory — e.g., coherence, data-compression, or accurate estimation. But the official admonitions are all too easily forgotten in favor of a vague and hopeful impression that simplicity is a magical oracle that somehow extends or amplifies the information provided by the data. None of the following accounts warrants such a conclusion, even though several of them invoke the term "information" in one way or another.

2.1 Simple Virtues

Simple theories have attractive aesthetic and methodological virtues. Aesthetically, they are more unified, uniform and symmetrical and are less ad hoc or messy. Methodologically, they are more severely testable [Popper, 1968; Glymour, 1981; Friedman, 1983; Mayo, 1996], explain better [Kitcher, 1981], predict better [Forster and Sober, 1994], and provide a compact summary of the data [Li and Vitanyi, 1997; Rissanen, 1983].[1] However, if the truth happens not to be simple, then the truth does not possess the consequent virtues, either. To infer that the truth is simple because simple worlds and the theories that describe them have desirable properties is just wishful thinking, unless some further argument is given that connects these other properties with finding the true theory [van Fraassen, 1981].

2.2 Bayesian Prior Probabilities

According to Bayesian methodology, one should update one's degree of belief $P(T)$ in theory T in light of evidence e according to the rule:

$$p(T|e) = \frac{p(T) \cdot p(e|T)}{p(e)}.$$

Subjective Bayesians countenance any value whatever for the prior probability $p(T)$, so it is permissible to start with a prior probability distribution biased toward simple theories [Jeffreys, 1985]. But the mere adoption of such a bias hardly explains how finding the truth is facilitated better by that bias than by any other.

A more subtle Bayesian argument seems to avoid the preceding circle. Suppose that S is a simple theory that explains observation e, so that $p(e|S) \approx 1$ and that $C = \exists\theta C(\theta)$ is a competing theory that is deemed more complex due to its free parameter θ, which can be tuned to a small range of "miraculous" values over which $p(e|C(\theta)) \approx 1$. Strive, this time, to avoid any prior bias for or against simplicity. Ignorance between S and C implies that $p(S) \approx p(C)$. Hence, by the

[1] Rissanen is admirably explicit that finding short explanations is an end-in-itself, rather than a means for finding the true theory.

standard, Bayesian calculation:

$$\frac{p(S|e)}{p(C|e)} = \frac{p(S) \cdot p(e|S)}{p(C) \cdot p(e|C)} \approx \frac{p(e|S)}{p(e|C)} \approx \frac{1}{\int p(e|C(\theta)) \cdot p(C(\theta)|C)\, d\theta}.$$

Further ignorance about the true value of θ given that C is true implies that $p(C(\theta)|C)$ is flattish. Since $p(e|C(\theta))$ is high only over a very small range of possible values of θ and $p(C(\theta)|C)$ is flattish, the integral assumes a value near zero. So the posterior probability of the simple theory S is sharply greater than that of C [Rosenkrantz, 1983]. It seems, therefore, that simplicity is "truth conducive", starting from complete ignorance.

The magic evaporates when the focus shifts from theories to ways in which the alternative theories can be true. The S world carries prior probability 1/2, whereas the prior probability of the range of worlds $C(\theta)$ in which θ is tuned to explain e is vanishingly small. That sharp, prior bias in favor of the S world is merely passed along through the Bayesian computation, accounting entirely for the sharp "confirmation" of S over C. More generally, Bayesian "ignorance" with respect to one partition of possibilities implies a strong prejudice with respect to another — e.g., "ignorance" between blue and non-blue together with ignorance between non-blue hues implies a strong bias against yellow — and that is all that is going on here. The point is not that science should be entirely free from biases. It is, rather, that direct appeal to one's bias hardly explains how that bias is better for finding the truth than alternative biases might be — every bias flatters itself.

2.3 Objective Prior Probabilities

One way to avoid the subjectivity of the preceding arguments is to select some particular prior probability distribution as special and to show that Ockham's razor follows. For example, R. Carnap [1950] viewed confirmation as a generalized notion of logical consequence in which $p(T|e)$ supposedly represents the degree to which premise e *partially entails* conclusion T. This putative degree of entailment is understood in terms of the total weight of possibilities satisfying $T\&e$ divided by the total weight of possibilities satisfying e. "Weight" is explicated in terms of probability, so there is the usual, Bayesian question of which prior probability measure to impose. Carnap imposed prior probabilities favoring uniform sequences of observable outcomes, with higher degrees of confirmation for predictions that resemble the past as a not-so-surprising result.

The trouble with Carnap's logical defense of Ockham's razor is that its prior bias toward uniformity is not preserved under linguistic translation and, hence, cannot be logical. On Carnap's proposal, a long run of green observations strongly confirms at stage n that the next observation will be green, rather than blue, because an invariantly green world is more uniform. N. Goodman [1955] responded that one can translate green/blue into grue/bleen, where grue means "green through n and blue thereafter" and bleen means "blue through n and green thereafter". A sequence of observations is uniform with respect to green/blue if and only if it is non-uniform with respect to grue/bleen, so uniformity and, hence, confirmation, is not preserved under translation. Against the objection that green/blue are "natural" predicates whereas grue/bleen involve a "magic time n", the predicates green/blue equally involve a magic time n in the grue/bleen language, so the situation is *logically* symmetrical. Therefore, Ockham's razor must be sought outside of logic.

Goodman, himself, proposed to rule out "grue-like" predicates by appealing to success in past inductions, which is a matter of history, rather than of logic. However, is hard to see how that can help if the "magic" time n still lies in the future, since then grue and green would have yielded identical success rates. A currently popular approach, called *algorithmic information theory* [Li and Vitanyi, 1997], seeks uniformity not in pure logic, but in the presumably objective nature of computation. The algorithmic complexity of a string corresponds (roughly) to the length of the shortest computer program (in some fixed computer language) that generates the string. The intuitive idea is that a simple string has structure that a short program can exploit to reproduce it, whereas a complex or "random" string does not. This gives rise to the notion that good explanations are short theories that compress the data and that Ockham's razor is a matter of minimizing the sum of the lengths of the theory and of the compressed data. The proposal that one should infer the best explanation in this sense is called the *minimum description length* principle or MDL for short [Rissanen, 1983]. Algorithmic information theorists have developed the notion of a *universal* prior probability over bit strings with the property that more compressible strings tend to have higher prior probability. It can be shown that under certain conditions the MDL approach approximates Bayesian updating with the universal prior probability [Vitanyi and Li, 2000].

Algorithmic complexity may help to explicate some slippery but important methodological concepts, such as interest, beauty, or emergence [Adriaans, 2007]. The focus here, however, is on the putative connection, if any, between data-compression and finding the true theory. Some proponents of the approach (e.g., Rissanen, himself) deny that there is one and urge data-compression as an alternative aim. One reason for doubt is that program length depends heavily upon the particular programming language assumed in the definition of program length. In algorithmic complexity theory, a computer language is identified with a *universal machine*, which simulates an arbitrary program p, step by step, to produce the output of p. Suppose that, in a "natural" programming language L, the shortest program p that generates a random-looking string σ is almost as long as σ itself. But now one can specify a new programming language L' whose universal machine I' is just like the universal machine I for L except that, when presented with a very short program p', I' simulates I on the long program p, generating σ. In other words, the complexity of p can be "buried" inside of I' so that it does not show up in the L' program p' that generates σ. This arbitrariness makes it hard to take program length seriously as an indicator of how simple the world really is unless a theory of "natural" programming languages is provided — but the theory of algorithmic complexity is stated in terms of an arbitrary, Turing-equivalent programming language.[2]

Quite aside from the relativity of program length to one's choice of computer language, there is a further question about the process by which observations are encoded or transduced into the bit-strings presupposed by algorithmic complexity theory. One transducer could encode green wavelengths as 0 and blue wavelengths as 1, whereas another, grue-like transducer could reverse these assignments at some random-looking times. Algorithmic complexity judges the same world to be either extremely complex or extremely simple depending upon which transducer is employed, but no bias that depends upon mere conventions about how the data are passed along to the scientist could plausibly be an indicator of truths lying behind the data-reporting process.

Finally, and most importantly, insofar as there is any theoretical connection between simplicity and truth in the MDL story, it amounts to the selection of a universal (i.e., simplicity-biased) prior probability measure, which adds nothing to the standard, circular, Bayesian account already discussed (cf. [Mitchell, 1997]). Therefore, it is important not to be confused by talk of bits and nats into believing that simplicity somehow provides *information* about the true theory.

2.4 Over-fitting and Empirical Estimation

Classical statisticians have an alternative account of the connection between simplicity and truth based on the concept of "over-fitting" (cf. [Wasserman, 2003]). Since this explanation does not invoke prior probabilities at all, it is free from the shadow of circularity characteristic of Bayesian explanations. However, the underlying aim is not to choose the true theory, but to find a false theory that yields accurate empirical estimates at small sample sizes. One might expect that no theory predicts more accurately than the true theory, but that is emphatically not how "accuracy" is understood in the over-fitting literature. Hence, the over-fitting explanation of Ockham's razor avoids circular appeal to a prior simplicity bias only by adopting a skeptical or instrumentalistic stance toward theories [Forster and Sober, 1994].

To see how false theories can predict more "accurately" than true ones, imagine a marksman firing a rifle at a target from a tripod that can be locked in both the vertical and the horizontal dimensions. When both locks are off, the best marksman produces a cloud of shots centered on the bull's eye. Suppose that the

[2]Algorithmic complexity theorists respond to the preceding concern as follows. The first universal machine I has a program $p_{I'}$ that simulates universal machine I'. Let p' be the shortest program producing some string σ according to I'. Then the result p of chaining together the programs $p_{I'}$ and p' generates σ in L. Chaining $p_{I'}$ onto p' adds only constant length to p', so there exists a constant k that bounds the difference in length of the shortest program in L from the length the shortest program in L' that generates an arbitrary string σ. But that is scant comfort when one applies Ockham's razor in a particular instance, for it is still the case that an arbitrarily complex theory in the first universal machine could be the simplest possible theory for the second. The constants connecting systems can be arbitrarily large, so no matter how many reversals of simplicity ranking one wishes to effect, one could fish for an alternative universal machine that effects them.

inaccuracy of a marksman is measured in terms of the expected distance from the bull's eye of a single shot. Call this the marksman's "risk" (of missing the bull's eye). A good marksman's risk is due entirely to the spread or *variance* of his shots around the bull's eye. Now consider a lazy marksman, who locks the tripod in both dimensions, so every shot hits at the same point at a distance b from the bull's eye. The lazy marksman has no variance, but has *bias* b, because his average shot hits at distance b from the bull's eye. There is a critical bias $b > 0$ below which the lazy marksman is more "accurate" than the good marksman as measured by risk. Think of the bull's eye as the true value of an empirical parameter and of a shot as an empirical estimate of the parameter based on a random sample. Free aim corresponds to an empirical estimate using a complex theory. The locked tripod corresponds to a fixed empirical estimate based on a simple theory with no free parameters. The bias of the simple theory implies its falsehood (it rules out the true sampling distribution). So even if the true theory is very complex and is known in advance, risk minimization argues for using a false, over-simplified theory for estimation purposes. Hence, over-fitting hardly explains how Ockham's razor helps one find the true theory. That conclusion may sound odd in light of popular glosses of over-fitting such as the following:

> It is overwhelmingly probable that any curve that fits the data perfectly is false. Of course, this negative remark does not provide a recipe for disentangling signal from noise. We know that any curve with perfect fit is probably false, but this does not tell us which curve we should regard as true. What we would like is a method for separating the *trends* in the data from the random deviations from those trends generated by error. A solution to the curve fitting problem will provide a method of this sort [Forster and Sober, 1994].

One might naturally conclude that the *trend* in the data is the *true signal* and that the aim is to strike the *true* balance between signal and noise, which only the true theory can do. However, as the authors of the passage later explain with care, over-fitting and under-fitting are defined in terms of estimation risk at a given sample size, rather than in terms of the true curve: "under-fitting" occurs when sub-optimal risk is due to bias and "over-fitting" occurs when sub-optimal risk is due to variance. Thus, as discussed above, if the sample size is small and the truth is not as simple as possible, risk minimization recommends selection of an over-simplified theory that falsely explains true signal as noise.

In the scientific case, one does not know the true sampling distribution a priori, so one does not know the bias and, hence, the risk, of using a given theory for estimation purposes. One can estimate the risk from the sample by calculating the average squared distance of data points from predictions by the theory. But the estimated risk of a complex theory is biased toward optimism because risk is estimated as fit to the data and a sufficiently complex theory can fit the data exactly, even if the true risk of estimation is considerable due to noise. To assuage this systematic estimation bias, the risk estimate must incorporate a tax on free parameters. Then one can choose, for estimation purposes, a theory whose corrected estimated risk is minimal. This is the basic logic between such standard, classical estimation procedures as *Akaike's information criterion (AIC))* (1973), cross-validation, and *Mallow's statistic* (cf. [Wasserman, 2003]).

Structural risk minimization (SRM) is an interesting generalization and extension of the over-fitting perspective [Vapnik, 1998]. In the SRM approach, one does not merely construct an (approximately) unbiased estimate of risk; one solves for objective, worst-case bounds on the chance that estimated risk differs by a given amount from actual risk. A crucial term in these bounds is called the *Vapnik Chervonenkis* dimension or VC dimension for short. The VC dimension is a measure of the range of possible samples the theory in question has the "capacity" to accommodate, which suggests a connection to simplicity and Ockham's razor. As in the over-fitting account, one can seek the "sweet spot" between simplicity (low VC-dimension) and fit (estimated risk) that minimizes the worst-case bound on the error of the risk estimate. Then one can choose the parameter setting that minimizes estimated risk within that theory.

Again, the aim is not to find the true theory. And yet, the SRM approach can explain other approaches (e.g., MDL and Bayesianism) as respectable ways to control worst-case estimation risk, eliminating the circular appeals to prior simplicity biases [Vapnik, 1998]. The moral is skeptical. If risk minimization is the last word on Ockham's razor, then the apparent rhetorical force of simplicity is founded upon a fundamental confusion between theories as true propositions and theories as useful instruments for controlling variability in empirical estimates.

It is tempting, at this point, to ask whether theoretical truth really matters — accurate predictions should suffice for all practical purposes. That is true so far as passive prediction is concerned. But beliefs are for guiding action and actions can alter the world so that the sampling distribution we drew our conclusions from is altered as well — perhaps dramatically. Negligible relativistic effects are amplified explosively when a sufficient quantity of uranium ore is processed. A crusade to eliminate ash trays breaks the previously observed, strong correlation between ash trays and cancer, undermining the original motivation for the policy. Theories that guide action are supposed to provide accurate *counterfactual* estimates about what would happen if the world (and, hence, the sampling distribution) were altered in various ways [Spirtes *et al.*, 2000]. An accurate estimate of the true sampling distribution is not enough in such cases, because distributions corresponding to complex theories can be arbitrarily similar to distributions corresponding to simple theories, that have very different counterfactual import. This point will be sharpened below, when the details of the contemporary literature on causal discovery are discussed.

Finally, it is clear that the over-fitting story depends, essentially, upon noise in the data and, hence, in the shots at the truth taken by the estimator, since non-noisy estimates involve no variance and, hence, no bias-variance balance. However, Ockham's razor seems no less compelling in deterministic settings. One would prefer that the connection between simplicity and theoretical truth not depend essentially upon randomness.

2.5 *Convergence*

The preceding explanations promise something in the short run, but theoretical truth cannot be guaranteed in the short run, even with high chance, because complex effects in nature may be two small or subtle to notice right away. Bayesians address this difficulty by circular appeal to the very bias to be explained. Risk minimization responds by shifting the focus from theoretical truth to predictive risk. A third option is to relax the demand for immediate success. Arbitrarily small, complex effects requiring free parameters to explain them can be detected eventually, as more data are collected, as more regions of the universe are explored, and as observational technology improves, so if it is assumed in advance (as in polynomial curve fitting) that there are at most finitely many such effects to be found, then at some point all the effects are noticed and Ockham's razor converges to the true theory. For example, it can be shown that, in a wide range of cases, Bayesian updating armed with a simplicity-biased prior probability does converge to the true theory in the limit. However, if indication or pointing to the true theory is too stringent to be feasible, mere convergence to the true theory is too weak to single out Ockham's razor as the best truth-finding policy in the short run. Convergence requires merely that a prior simplicity bias "wash out", eventually, in complex worlds. But the question is not how to overcome a prior simplicity bias; it is, rather, how such a bias helps one find the truth better than alternative biases. Convergence, alone, cannot answer that question, since if a method converges to the truth, so does every finite variant of that method [Salmon, 1967]. Hence, mere convergence says nothing about how the interests of truth-finding are *particularly* furthered by choosing the simplest theory *now.* But that is what the puzzle of simplicity is about.

3 DIAGNOSIS

To recapitulate, the two standard notions of finding truth are (1) *indication* or *informing* of the truth in the short run and (2) *convergence* in the long run. The former aim is too strong to support an a priori explanation of Ockham's razor, since an arbitrarily complex world can appear arbitrarily simple in the short run, before the various dimensions of complexity have been detected. The latter aim is too weak to support an a priori explanation of Ockham's razor, since a prior bias toward complexity can also be washed out by further information. Therefore, if the apparent connection between simplicity and theoretical truth has an explanation, it should be sought somewhere between these two extremes: Ockham's razor should somehow help one converge to the true theory better or more efficiently than alternative strategies. Just such an account will now be presented. The basic idea is that a bias toward simplicity neither points at the truth nor merely converges to it, but converges to it in the most efficient or direct manner possible, where efficiency is measured in terms of errors, reversals of opinion, and the time delay to such reversals.[3]

4 TRAVELER'S AID

To state the simplicity puzzle in its most basic terms, how could fixed, one-size-fits-all advice be guaranteed to help one find something that might be anywhere — in a sense stronger than merely guaranteeing that one will find it by exhaustive search? It happens every day. Suppose that a city dweller is lost in a small town on a long automobile journey. He asks a local resident for directions. The resident directs him to the freeway entrance ramp. The traveler follows the advice and travels as directly as possible to the freeway, which is by far the most direct route home — in spite of a few, unavoidable curves around major geographical features.

Now suppose that the traveler stubbornly ignores the resident's advice. Indeed, suppose that, in so doing, the traveler follows a road on the true compass heading to his destination, whereas getting on the freeway requires a short jog in the opposite direction. The chosen route narrows and begins to meander through the mountains. The traveler finally concedes that it wasn't a good idea and retraces his route back to the resident. He then follows the resident's directions to the freeway and proceeds home via the best possible route. The traveler's reward for ignoring the resident's advice is a humiliating U-turn right back to where he started, followed by all the unavoidable twists and turns encountered on the freeway over the mountains. Had he heeded the advice, he would have encountered only the unavoidable curves along the freeway. So he should have heeded it.

In connection with the simplicity puzzle, this unremarkable tale has some remarkable features.

1. The resident's advice is the *best possible* advice in the sense that it puts one on the most direct route to the goal, for violating it incurs at least one extra, initial U-turn, regardless of one's destination.

2. The advice is the best possible even if it aims the traveler in the wrong direction initially.

3. The resident can give precisely the same, *fixed* advice to every stranger who asks, even though she does not know where they are headed — no Ouija board or other occult channel of information is required.

[3]The basic idea of counting mind-changes is originally due to H. Putnam (1965). It has been studied extensively in the computational learning literature — for a review cf. [Jain *et al.*, 1999]. But in that literature, the focus is on categorizing the complexities of problems rather than on singling out Ockham's razor as an optimal strategy. I viewed the matter the same way in [Kelly, 1996]. Schulte [1999a; 1999b] derives short-run constraints on strategies from retraction minimization. Kelly [2002] extends the idea, based on a variant of the ordinal mind-change account due to [Freivalds and Smith, 1993], but that approach does not apply to cases like curve fitting, in which theory complexity is unbounded. Subsequent steps toward the present approach may be found in [Kelly, 2004; 2006] and in [Kelly and Glymour, 2004].

So directions to the nearest freeway entrance ramp satisfy all the apparently arcane and paradoxical demands that a successful explanation of Ockham's razor must satisfy. It remains to explain what the freeway to the truth is and how Ockham's razor keeps one on it.

5 SOME EXAMPLES

For some guidance in the general developments that follow, consider some familiar examples.

Polynomial structures. Let S be a finite set of natural numbers and suppose that the truth is some unknown polynomial law:

$$y = f(x) = \sum_{i \in S} a_i x^i,$$

where for each $i \in S$, $a_i \neq 0$. Say that S is the *structure* of the law, as it determines the form of the law as it would be written in a textbook. Suppose that the problem is to infer the true structure S of the law. It is implausible to suppose that for a given value of the independent variable x one could observe the exact value of the dependent variable y, so suppose that for each queried value of x at stage k of inquiry, the scientist receives an arbitrarily small, open interval around the corresponding value of y and that repeated queries of x result in an infinite sequence of open intervals converging to $\{y\}$.

It is impossible to be sure that one has selected S correctly by any finite time, since there may be some $i \in S$ such that a_1 is set to a very small value in f, making it appear that the monomial $a_i x^i$ is missing from f. Ockham's razor urges the conclusion that $i \notin S$ until the corresponding monomial is noticed in the data.

There is a connection between the complexity of the true polynomial structure and what scientists and engineers call *effects*. Suppose that $S_0 = \{0\}$, so for some $a_i > 0$, $f_0(x) = a_i$. Let experience e_0 present a finite sequence of interval observations of the sort just described for f_0. Then there is a bit of wiggle room in each such interval, so that for some suitably small $a_1 > 0$, the curve $f_1(x) = a_1 x + a_0$ of form $S_1 = \{0, 1\}$ is compatible with e_0. Eventually, some open interval around $y = a_0$ is presented that excludes f_0. Call such information a first-order *effect*. If e_1 extends that information and presents an arbitrary, finite number of shrinking, open intervals around f_1 then, again, there exists suitably small $a_2 > 0$ such that $f_2(x) = a_2 x^2 + a_1 x + a_0$ of form $S_2 = \{0, 1, 2\}$ passes through each of the intervals presented in e_1. Eventually, the intervals tighten so that no linear curve passes between them. Call such information a second-order effect, and so forth. The number of effects presented by a world corresponds to the cardinality of S, so there is a correspondence between empirical effects and empirical complexity. A general account of empirical effects is provided in Section 16 below.

Linear dependence. Suppose that the truth is a multivariate linear law

$$y = f(x) = \sum_{i \in S} a_i x_i,$$

where for each $i \in S$, $a_i \neq 0$. Again, the problem is to infer the structure S of f. Let the data be presented as in the preceding example. As before, it seems that complexity corresponds with the cardinality of S which is connected, in turn, to the number of effects presented by nature if f is true.

Conservation laws. Consider an idealized version of explaining reactions with conservation laws, as in the theory of elementary particles [Schulte, 2001; Valdez-Perez, 1996]. Suppose that there are n observable types of particles, and it is assumed that they interact so as to conserve n distinct quantities. In other words, each particle of type p_i carries a specific amount of each of the conserved quantities and for each of the conserved quantities, the total amount of that quantity going into an arbitrary reaction must be the total amount that emerges. Usually, one thinks of a reaction in terms of inputs and outputs; e.g.,

$$r = (p_1, p_1, p_1, p_2, p_2, p_3 \to p_1, p_1, p_2, p_3, p_3).$$

One can represent the inputs by a vector in which entry i is the number of input particles of type p_i in r, and similarly for the output:

$$\begin{aligned} \mathbf{a} &= (3,2,1); \\ \mathbf{b} &= (2,1,2); \\ r &= (\mathbf{a} \to \mathbf{b}). \end{aligned}$$

A *quantity* $\mathbf{q}$ (e.g., mass or spin) is an assignment of real numbers to particle types, as in $\mathbf{q} = (1,0,1)$, which says that particles a_1, a_3 both carry a unit of $\mathbf{q}$ and a_2 carries none. Quantity $\mathbf{q}$ is *conserved* in r just in case the total $\mathbf{q}$ in is the total $\mathbf{q}$ out. That condition is just:

$$\sum_{i=1}^{3} q_i a_i = \sum_{i=1}^{3} q_i b_i,$$

or, in vector notation,

$$\mathbf{q} \cdot \mathbf{a} = \mathbf{q} \cdot \mathbf{b},$$

which is equivalent to:

$$\mathbf{q} \cdot (\mathbf{a} - \mathbf{b}) = 0.$$

Since reaction r enters the condition for conservation solely as the vector difference $\mathbf{a} - \mathbf{b}$, there is no harm, so far as conservation is concerned, in identifying reaction r with the difference vector:

$$\mathbf{r} = \mathbf{a} - \mathbf{b} = (1, 1, -1).$$

Then the condition for $\mathbf{r}$ conserving $\mathbf{q}$ can be rewritten succinctly as:

$$\mathbf{q} \cdot \mathbf{r} = 0,$$

which is the familiar condition for geometrical orthogonality of $\mathbf{q}$ with $\mathbf{r}$. Thus, the reactions that preserve quantity $\mathbf{q}$ are precisely the integer-valued vectors orthogonal to $\mathbf{q}$. In this example, $\mathbf{r}$ does conserve $\mathbf{q}$, for:

$$(1, 0, 1) \cdot (1, 1, -1) = 1 + 0 - 1 = 0.$$

But so do reactions $\mathbf{u} = (1, 0, -1)$ and $\mathbf{v} = (0, 1, 0)$, which are linearly independent. Since the subspace of vectors orthogonal to $\mathbf{q}$ is two-dimensional, every reaction that conserves $\mathbf{q}$ is a linear combination of $\mathbf{u}$ and $\mathbf{v}$ (e.g., $\mathbf{r} = \mathbf{u} + \mathbf{v}$). If the only conserved quantity were $\mathbf{q}$, then it would be strange to observe only scalar multiples of $\mathbf{r}$. In that case, one would expect that the possible reactions are constrained by some other conserved quantity linearly independent of $\mathbf{q}$, say $\mathbf{q}' = (0, 1, 1)$. Now the possible reactions lie along the intersection of the planes respectively orthogonal to $\mathbf{q}$ and $\mathbf{q}'$, which are precisely the scalar multiples of $\mathbf{r}$. Notice that any two linearly independent quantities orthogonal to $\mathbf{r}$ would suffice — the quantities, themselves, are not uniquely determined.

Now suppose that the problem is to determine how many quantities are conserved, assuming that some conservation theory is true and that every possible reaction is observed, eventually. Let an "effect" be the observation of a reaction linearly independent of the reactions seen so far. As in the preceding applications, effects may appear at any time but cannot be taken back after they occur and the correct answer is uniquely determined by the (finite) number of effects that occur.

In this example, favoring the answer that corresponds to the fewest effects corresponds to positing the greatest possible number of conserved quantities, which corresponds to physical practice (cf. [Ford, 1963]). In this case, simplicity intuitions are consonant with testability and explanation, but run counter to minimization of free parameters (posited conserved quantities).

Discovering causal structure. If one does not have access to experimental data, due to cost, feasibility, or ethical considerations, one must base one's policy recommendations on purely observational data. In spite of the usual advice that correlation does not imply causation, sometimes it does. The following setup is based upon [Spirtes *et al.*, 2000]. Let V be a finite set of empirical variables. A *causal structure* associates with each unordered pair of variables $\{X, Y\}$ one of the following statements:

$$X \rightarrow Y; \quad X \leftarrow Y; \quad X \| Y;$$

interpreted, respectively, as X is a direct cause of Y, Y is a direct cause of X, and X, Y have no direct causal connection. The first two cases are *direct causal connections* and the fourth case denies such a connection. A causal structure can, therefore, be presented as a directed, acyclic *graph* (DAG) in which variables are *vertices* and arrows are direct causal connections. The notation $X - Y$ means that there is a direct connection in either direction between X and Y without specifying which. A partially oriented graph with such ambiguous edges is understood, for present purposes, to represent the disjunction of the structures that result from specifying them in each possible way.

At the core of the approach is a rule for associating causal structures with probability distributions. Let p be a joint probability distribution on variables V. If S is a subset of V, let $(X \amalg Y)|S$ abbreviate that X is statistically independent of Y conditional on S in p. A sequence of variables is a *path* if each successive pair is immediately causally connected. A *collision* on a path is a variable with arrows coming in from adjacent variables on the path (e.g., variable Y in path $X \to Y \leftarrow Z$). A path is *activated* by variable set S just in case the only variables in S that occur on the path are collisions and every collision on the path has a descendent in S. Then the key assumption relating probabilities to causal structures is simply:

$$(X \amalg Y)|S \text{ if and only if no path between } X \text{ and } Y \text{ is activated by } S.$$

Let T_p denote the set of all causal structures satisfying this relation to probability measure p.

To see why it is intuitive to associate T_p with p, suppose that $X \to Y \to Z$ and that none of these variables are in conditioning set S. Then knowing something about Z tells one something about X and knowing something about the value of X tells one something about Z. But the ultimate cause X yields no further information about Z when the intermediate cause Y is known (unless there is some other activated path between X and Y). On the other hand, suppose that the path is $X \to Y \leftarrow Z$ with collision Y. If there is no further path connecting X with Z, knowing about X says nothing about Z (X and Z are independent causes of Y), but since X and Z may cooperate or compete in a systematic way to produce Y, knowing the value of Y together with the value of X yields some information about the corresponding setting of Z. The dependency among causes given the state of the common effect turns out to be an important clue to causal orientation.

It follows from the preceding assumption that there is a direct connection $X - Y$ just in case X and Y are dependent conditional on each set of variables not including X, Y. There is a collision ($X \to Y \leftarrow Z$) if ($X - Y - Z$) holds (by the preceding rule) and ($X - Z$) does not hold (by the preceding rule) and, furthermore, X, Z are dependent given every set of variables including Y but not X, Z [Spirtes *et al.*, 2000, theorem 3.4]. Further causal orientations may be entailed in light of background assumptions. The preceding rules (actually, more computationally efficient heuristic versions thereof) have been implemented in "data-mining" software packages that search for causal structures governing large sets of observational variables. The key points to remember are that (1) a direct causal connection is implied by the appearance of some set of statistical dependencies and (2) edge orientations depend both on the appearance of some statistical dependencies and on the non-appearance in the future of further statistical dependencies.

The above considerations are taken to be general. However, much of the literature on causal discovery focuses on two special cases. In the *discrete multinomial* case, say that $G \in D_g$ if and only if $G \in T_p$ and p is a discrete, joint distribution over a finite range of possible values for each variable in G. In the *linear Gaussian* case, say that $G \in L_p$ if and only if $G \in T_p$ and p is generated from G as follows: each variable in G is assumed to be a linear function of its parents, together with an extra, normally distributed, unobserved variable called an *error term* and the error terms are assumed to be uncorrelated. For brevity, say that G is *standard* for p if and only if $G \in D_p$ or $G \in L_p$. The following discussion is restricted to the standard cases because that is where matters are best understood at present.

In practice, not all variables are measured, but assume, optimistically, that all causally relevant variables are measured. Even then, in the standard cases, the DAGs in T_p cannot possibly be distinguished from one another from samples drawn from p, so one may as well require only convergence to T_p in each p compatible with background assumptions.[4]

Statistical dependencies among variables must be inferred from finite samples, which can result in spurious causal conclusions because finite samples cannot reliably distinguish statistical independence from weak statistical dependence. Idealizing, as in the preceding examples, suppose that one receives the outputs of a data-processing laboratory that merely informs one of the dependencies[5] that have been verified so far (at the current, growing sample size) by a standard statistical dependency test, where the null hypothesis is independence.[6] Think of an *effect* as data verifying that a partial correlation is non-zero. Absence of an effect is compatible with noticing it later (the correlation could be arbitrarily small). If it is required only that one infer the true indistinguishability class $T(p)$ for arbitrary p representable by a DAG, then effects determine the right answer.

What does Ockham say? In the light of the preceding examples, something like: assume no more dependencies than one has seen so far, unless background knowledge and other dependencies entail them. It follows, straightforwardly, that direct causal connections add complexity, and that seems intuitively right. Causal orientation of causal connections is more interesting. It may seem that causal orientation does affect complexity, because, with binary variables, a common effect depends in some manner that must be specified upon four states of the joint causes whereas a common cause affects each effect with just two states. Usually, free

[4]It is known that in the linear, non-Gaussian case, causal structure can be recovered uniquely if there are no unobserved variables [Shimizu *et al.*, 2006]. The same may be true in the non-linear Gaussian case.

[5]In the standard cases, it is known that all of the over-identifying constraints follow from conditional independence constraints [Richardson and Spirtes, 2002]. That is known to be false in the linear, non-Gaussian case [Shimizu *et al.*, 2006], so in that case simplicity must be relativized to a wider range of potential effects. Indeed, in the linear, non-Gaussian case, the set of possible empirical effects is so rich that there are no proper inclusion relations among the sets of effects corresponding to alternative causal models, so the simplicity ranking is flat.

[6]Also, the significance level is tuned down at a sufficiently slow rate to ensure that the test converges in probability to the right answer. At the end of the paper, some of the issues that arise in a serious application to statistical model selection are raised.

parameters contribute to complexity, as in the curve-fitting example above. But given the overall assumptions of causal discovery, a result due to Chickering [2003] implies that these extra parameters do not correspond to potential empirical effects and, hence, do not really contribute to empirical complexity. In other words, given that no further edges are coming, one can afford to wait for data that *decide* all the discernable facts about orientation [Schulte, 2007]. Standard MDL procedures that tax free parameters can favor non-collisions over collisions before the data resolve the issue, risking extra surprises.[7]

For example, when there are three variables X, Y, Z and $(X - Y - Z)$ is known, then, excluding unobserved causes, there are two equivalence classes of graphs, the collision orientation $(X \rightarrow Y \leftarrow Z)$ in one class C and all the other orientations in the complementary class $\neg C$. Looking at the total set of implied dependencies for C, C', it turns out that the only differences are that C entails $\neg((X \amalg Z)|Y)$ but not $\neg(X \amalg Z)$, whereas $\neg C$ entails $\neg(X \amalg Z)$ but not $\neg((X \amalg Z)|Y)$, so there is no inclusion relationship between the dependencies characterizing C and the dependencies characterizing $\neg C$. Therefore, both hypotheses are among the simplest compatible with the data, so Ockham's razor does not choose among them. Moreover, given that the truth is $(X - Y - Z)$, nature must present either $\neg(X \amalg Z)$ or $\neg((X \amalg Z)|Y)$ eventually (given that the causal truth can be represented by some graph over the observable variables) so it seems that science can and should wait for nature to resolve the matter instead of racing ahead — and that is just how Ockham's razor is interpreted in the following discussion. Regardless of which effect nature elects to present, it remains possible, thereafter, to present the other effect as well, in which case each variable is connected immediately to every other and one can infer nothing about causal directionality. This situation involves more effects than either of the two preceding cases, but another direct causal connection is also added, reflecting the increase in complexity.

The preceding evolution can result in spectacular reversals of causal conclusions as experience increases, not just in terms of truth, but in terms of practical consequences as well. Suppose that it is known that $(X \rightarrow Y - Z)$ and none of these variables has yet exhibited any dependence with W. Then discovery of $\neg((X \amalg Z)|Y)$, background knowledge, and Ockham's razor unambiguously imply $(X \rightarrow Y \leftarrow Z)$, a golden invitation to exploit Z to control Y. Indeed, the connections may be obvious and strong, inviting one to invest serious resources to exploit Z. But the conclusion rests entirely on Ockham's razor, for the further discovery of $\neg(X \amalg Z)$ is incompatible with $(X \rightarrow Y \leftarrow Z)$ and the new Ockham answer is $(X \rightarrow Y - Z)$ with edge $(X - Z)$ added. Further discovery that $\neg((Z \amalg W)|X, Y)$ and that $\neg((Y \amalg W)|Z)$ results in the conclusion $Y \rightarrow Z \leftarrow W$, reversing the origi-

[7] A similar issue arises in the inference of regular sets from positive examples. The most liberal automaton is a one-state universal acceptor with a loop for each input character. But assuming that the language is learned from positive examples only, that is the most complex hypothesis in terms of empirical effects. In typical scientific applications, such as curve fitting, extra parameters imply extra effects. But not always, and then it is the effects, rather than the parameters, that determine retraction efficiency.

nal conclusion that Y can be controlled by Z.[8] The orientation of the direct causal connection $Y - Z$ can be flipped n times in sequence by assuming causes $X_0, \ldots X_n$ of Y in the role of X and potential collisions $W_0, \ldots, W_n$ in the role of W. There is no way that a convergent strategy can avoid such discrete flips of $Y - Z$; they are an ineluctable feature of the problem of determining the efficacy of Z on Y from non-experimental data, no matter how strong the estimate of the strength of the cause $Y \rightarrow Z$ is prior to the reversal. Indeed, standard causal discovery algorithms exhibit the diachronic retractions just discussed in computer simulations. The practical consequences of getting the edge orientation wrong are momentous, for if Z does not cause Y, the policy of manipulating Z to achieve results for Y will have no benefits at all to justify its cost. Indeed, in the case just described, sample size imposes no non-trivial bound on arbitrarily large mis-estimates of the effectiveness of Y in controlling Z (cf. [Robins *et al.*, 2003; Zhang and Spirtes, 2003]). Therefore a skeptical stance toward causal inference is tempting:

> We could try to learn the correct causal graph from data but this is dangerous. In fact it is impossible with two variables. With more than two variables there are methods that can find the causal graph under certain assumptions but they are large sample methods and, furthermore, there is no way to ever know if the sample size you have is large enough to make the methods reliable [Wasserman, 2003, p. 275].

This skepticism is one more symptom of the unrealizable demand that simplicity should reliably point toward or inform one of the true theoretical structure, a popular — if infeasible — view both in statistics and philosophy [Goldman, 1986; Mayo, 1996; Dretske, 1981]. The approach developed below is quite different: insofar as finding the truth makes reversals of opinion unavoidable, they are not only justified but laudable — whereas, insofar as they are avoidable, they should be avoided. So the best possible strategies are those that converge to the truth with as few course-reversals as possible. That is what standard causal inference algorithms tend to do, and it is the best they could possibly do in the standard cases.

To summarize, an adequate explanation of Ockham's razor should isolate what is common to the simplicity intuitions in examples like the preceding ones and should also explain how favoring the simplest theory compatible with experience helps one find the truth more directly or efficiently than competing strategies when infallibility or even probable infallibility is hopeless. Such an explanation, along the lines of the freeway metaphor, will now be presented. First, simplicity and efficient convergence to the truth must be defined with mathematical rigor and then a proper proof must be provided that Ockham's razor is the most efficient possible strategy for converging to the truth.

---

[8] I am indebted to Richard Scheines for suggesting this example.

6 INFERENCE OF THEORETICAL STRUCTURE

In light of the preceding examples, say that an empirical *effect* is experience that (1) may take arbitrarily long to appear due to its subtlety or difficulty to produce and that (2) never disappears once it has been seen. Furthermore, (3) at most finitely many effects appear for eternity and (4) the correct theoretical structure is uniquely determined by the (finite) set of effects one encounters for eternity. In light of (4), one may as well understand the problem of finding the true theory as a matter of inferring which finite set of effects (corresponding to some structure or other) one will encounter for eternity.

Accordingly, let E be a countable set of potential effects satisfying (1–4) which, for the time being, will not be analyzed further (a deeper analysis, explaining what effects are, is provided below). Let Ω denote the set of all finite subsets of E. It may happen that one knows *a priori* that some theoretical structures are impossible (e.g., not every finite set of statistical dependencies corresponds to a causal graph). Let $\Gamma \subseteq \Omega$ be the set of possible sets of effects compatible with background knowledge. An empirical *world* w is an infinite sequence of mutually disjoint, finite subsets of E that converges to $\emptyset$, where the finite set $w(i)$ corresponds to the set of as-yet unobserved effects encountered for the first time at stage i of inquiry. Let W denote the set of all such worlds. If no new effects are encountered at i, then $w(i)$ is empty. Let $w|k = (w_0, \ldots, w_{k-1})$, the finite initial segment of w of length k. The finite set of all effects presented by w (or by finite sequence $e = w|k$) is given by:

$$S_w = \bigcup_{i=0}^{\infty} w(i); \quad S_e = \bigcup_{i=0}^{k-1} w(i).$$

For each $w \in W$ define the *modulus* of w to be the first moment from which no more new effects appear:

$$\mu(w) = \text{ the least } k \text{ such that } S_w = S_{w|k}.$$

The background restriction $\Gamma \subseteq \Omega$ on sets of effects can be viewed as a material restriction on empirical worlds as follows:

$$K_\Gamma = \{w \in W : S_w \in \Gamma\}.$$

Recall that each theoretical structure T corresponds uniquely to some finite set S of effects. Let theoretical structure T_S corresponding to finite set $S \subseteq E$ be identified with the set of all worlds in which T_S is correct — namely, the set of all worlds that present exactly S:

$$T_S = \{w \in W : S_w = S\}.$$

The set:

$$\Pi_\Gamma = \{T_S : S \in \Gamma\}$$

partitions W into mutually exclusive and exhaustive alternative propositions called *potential answers* and will be referred to as the *question* posed by the problem of inferring theoretical structures. Then T_{S_w} is the unique answer in Π_Γ that contains (is true of) w. Finally, the *theoretical structure inference problem* with *possible structures* Γ is represented by the ordered pair:

$$\mathcal{P}_\Gamma = (K_\Gamma, \Pi_\Gamma),$$

where Π_Γ is the empirical *question* and K_Γ is the empirical *background presupposition.*

Every concept and proposition that follows is relative to Γ so, to eliminate some symbolic clutter, think of Γ as a "global variable" held fixed in the background, to be referred to as clarity demands.

7 EMPIRICAL STRATEGIES AND CONVERGENT SOLUTIONS

What makes science unavoidably fallible is that one does not get to see the entire empirical world w all at once; rather, one sees incrementally longer, finite, initial segments of w as time passes. The set of all possible finite sequences the scientist might see as time passes is given by:

$$F_\Gamma = \{w|i : w \in K_\Gamma \text{ and } i \in N\}.$$

When e is a finite, initial segment of e' (i.e., there exists i such that $e = e'|i$), say that $e \leq e'$. When e is a sub-sequence but not necessarily an initial segment of e', then abuse notation by writing $e \subseteq e'$. Let $e * e'$ denote sequence concatenation, where it is always understood that e is finite and that e' may be finite or infinite. Finally (in the proofs in the Appendix), if x is some generic set-theoretic object, let x^∞ denote the infinite sequence in which only x occurs.

An empirical *strategy* M for problem $\mathcal{P}_\Gamma$ is a mapping of type:

$$M : F_\Gamma \rightarrow \Pi \cup \{\text{`?'}\}.^9$$

In other words, M maps each finite sequence $e \in F_\Gamma$ either to an answer $T_S \in \Pi$ or to '?', indicating refusal to choose an answer. Then in world $w \in K_\Gamma$, M produces the unending sequence of outputs:

$$M[w] = (M(w|0), M(w|1), M(w|2), \ldots),$$

where the square brackets are a reminder that M does not get to see w "all at once".

After seeing finite input sequence e, background presupposition K_Γ entails that one must live in a world $w \in K_\Gamma$ that extends e, so let:

$$K_\Gamma|e = \{w \in K_\Gamma : w \geq e\}$$

[9] In a more realistic setup, M could output disjunctions of answers in Π_Γ or degrees of belief distributed over Π_Γ. The ideas that follow extend to both situations.

denote the set of all such extensions. Then one may resrict Π_Γ to the answers compatible with e as follows:

$$\Pi_\Gamma|e = \{T \in \Pi_\Gamma : T \cap K_\Gamma|e \neq \emptyset\}.$$

Say that M *solves* $\mathcal{P}_\Gamma$ *in the limit* given e if and only if for each $w \in K_\Gamma|e$,

$$\lim_{i \to \infty} M(w|i) = T_{S_w},$$

in which case, say that M is a *convergent solution* to $\mathcal{P}_\Gamma$ given e. A *convergent solution* to $\mathcal{P}_\Gamma$ is just a convergent solution given the empty sequence ().

One obvious, convergent solution to $\mathcal{P}_\Omega$ (i.e., no finite set of effects is ruled out *a priori*) is just:

$$M(e) = T_{S_e},$$

for if $w \in K_\Gamma$, new effects stop appearing, eventually — say by stage n — so for all $m \geq n$, $M(w|m) = T_{S_{w|m}} = T_w$. But there are infinitely many alternative, convergent solutions as well — each finite variant of the obvious, convergent solution is a convergent solution — and it is not trivial to say how and in what sense the obvious strategy helps one to find the truth better than these do. That is the question answered by the following argument.

8 EMPIRICAL COMPLEXITY DEFINED IN TERMS OF EFFECTS

If $\Gamma = \Omega$, as in the polynomial structure problem, then an obvious definition of the *empirical complexity* of world w given e is

$$c(w, e) = |S_w| - |S_e|,$$

the number of new effects presented by w after the end of e (cf. [Kelly, 2007]). When $\Gamma \subset \Omega$, as in the causal inference problem (some finite sets of partial correlations correspond to no causal graph), a slightly more general approach is required.[10] The basic idea is that effects, relative to a problem, correspond to successive opportunities for nature to force the scientist to switch from one answer to another. Restrict Γ to those sets of effects compatible with e:

$$\Gamma|e = \{S \in \Gamma : S_e \subseteq S\}.$$

This set includes all the possible theoretical structures that might serve as potential interpretations of what has been presented by e. Say that a *path* in $\Gamma|e$ is a finite, non-repetitive, ascending sequence of elements of $\Gamma|e$. If $S, S' \in \Gamma|e$, let $\pi_e(S, S')$ denote the set of all paths in π_e that start with S and terminate with S'. Then $\pi_e(*, S')$ denotes all paths in $\Gamma|e$ that terminate with S' and $\pi_e(S, *)$ denotes all

[10] E.g., suppose that $\Gamma = \{\emptyset, \{a, b\}\}$. Then seeing a implies that one will see b, so a and b are not independent effects. They are more like correlated aspects of one effect, so they should not be counted separately.

paths in $\Gamma|e$ that start with S. So $\pi_e(*, S)$ represents all the possible paths nature might have taken to S from some arbitrary starting point in $\Gamma|e$. Then for $e \in F_\Gamma$, $w \in K_\Gamma|e$, and $P \subseteq K_\Gamma$, define *empirical complexity* as follows:

$$\begin{aligned} c(w, e) &= \max\{\text{length}(p) : p \in \pi_e(*, S_w)\} - 1; \\ c(P, e) &= \min\{c(w, e) : w \in P \cap K_\Gamma|e\}. \end{aligned}$$

Then since $(S) \in \pi_e(*, S)$ if $S \in \Gamma|e$ and lengths are discrete, it is immediate that:

PROPOSITION 1 (Empirical complexity is non-negative). *If $w \in K_\Gamma|e, P \in \Pi|e$, then $c(w, e), c(P, e)$ assume values in the natural numbers.*

Hence, answers with complexity zero are simplest. Define:

$$(\Gamma|e)_{\min} = \{S \in \Gamma|e : \text{for all } S' \in \Gamma|e, S' \not\subset S\},$$

and say that S is *minimally compatible* with e if and only if $S \in (\Gamma|e)_{\min}$.

PROPOSITION 2 (Characterization of zero complexity). *Let $w \in K_\Gamma|e$ and $e \in F_\Gamma$ and $T_S \in \Pi_\Gamma|e$. Then:*

1. *$c(w, e) = 0$ if and only if $S_w \in (\Gamma|e)_{min}$;*
2. *$c(T_S, e) = 0$ if and only if $S \in (\Gamma|e)_{min}$.*

Maximum simplicity is minimum complexity. Borrowing a standard re-scaling trick from information theory, one can convert complexity degrees to *simplicity degrees* in the unit interval as follows:

$$s(P, e) = \exp(-c(P, e)).$$

Unconditional complexity and simplicity are definable as:

$$\begin{aligned} c(P) &= c(P, ()); \\ s(P) &= s(P, ()). \end{aligned}$$

9 OCKHAM'S RAZOR

The *Ockham* answer given e, if it exists, is the unique answer $T \in \Pi_\Gamma|e$ such that $c(T, e)$ is minimal over all alternative theories $T' \in \Pi_\Gamma|e$. In light of Proposition 1, the Ockham answer is the unique answer in $T \in \Pi_\Gamma|e$ such that $c(T, e) = 0$. Empirical strategy M *satifies Ockham's razor* (or is *Ockham*, for short) at e iff

$$M(e) \text{ is Ockham given } e \text{ or } M(e) = \text{‘?’}.^{11}$$

[11] If M is allowed to output disjunctions of answers in Π_Γ, then Ockham's razor requires that $\bigcup\{T_S : S \in (\Gamma|e)_{\min}\} \subseteq M(e)$.

Furthermore, M is *Ockham* from e onward iff M is Ockham at each e' extending e; and M is *Ockham* if M is Ockham at each $e \in F_\Gamma$.

When S is in $\Gamma|e$ and S is a subset of each $R \in \Gamma|e$, say that S is the *minimum* in $\Gamma|e$. The Ockham answer, if it exists, can be characterized both in terms of uniquely minimal compatibility and in terms of being minimum.

PROPOSITION 3 (Ockham answer characterization). *Let $e \in F_\Gamma$ and $T_S \in \Pi_\Gamma|e$. Then the following statements are equivalent:*

1. *T_S is Ockham given e;*
2. *$(\Gamma|e)_{min} = \{S\}$;*
3. *S is the minimum in $\Gamma|e$.*

10 STALWARTNESS AND EVENTUAL INORMATIVENESS

Ockham's razor does not constrain suspension of judgment in any way, but it would be odd to adopt the Ockham answer T at e and then to drop T later, even though T is still the Ockham answer — further effect-free experience would only seem to "confirm" the truth of T. Accordingly, let $e \in F_\Gamma$ and let $e * S$ denote the extended, finite input sequence along which finite $S \subseteq E$ is reported right after the end of e. Say that strategy M is *stalwart* at $e * S$ if and only if for each answer $T \in \Pi_\Gamma$, if $M(e) = T$ and $M(e * S) \neq T$ then T is not the Ockham answer at $e * S$ (i.e., an answer is dropped only if it is not the Ockham answer when it is dropped). As with the Ockham property, itself, one may speak of M being stalwart from e onward or as just being stalwart, which means that M is stalwart at each e.

Similarly, it would be too skeptical never to conclude that no more effects are forthcoming, no matter how much effect-free experience has been collected. Accordingly, say that a strategy is *eventually informative* from e onward if there is no world $w \in K_\Gamma|e$ on which M converges to '?'. Then M is *eventually informative* if M is eventually informative from the empty input sequence onward.

Finally, a *normal* Ockham strategy from e onward is an eventually informative, stalwart, Ockham strategy from e onward and a normal Ockham strategy is normally Ockham from the empty sequence onward. The normal Ockham strategies are intuitively quite plausible. Such a strategy M may wait for a while but eventually chooses the Ockham answer and retains it until it is no longer Ockham. Furthermore, after each new effect is encountered, there is some finite amount of effect-free experience that lulls M to plump for the simplest theory once again. That is pretty much what people and animals do, and also describes, approximately, the behavior of a simplicity-biased Bayesian agent who selects only the theory whose posterior probability is above some high threshold. But plausibility and rhetoric are not the points at issue — finding the true theory is — so it is more pertinent to observe that normally Ockham strategies are, at least, guaranteed to converge to the truth.

PROPOSITION 4 (Normal Ockham Convergence). *If M is normally Ockham for $\mathcal{P}_\Gamma$ from e onward, then M is a solution to P_Γ from e onward.*

Furthermore, eventual informativeness is a necessary condition for being a solution, for a strategy that is not eventually informative evidently fails to converge to any theory in some world w:

PROPOSITION 5 (Convergence implies eventual informativeness). *If M solves $\mathcal{P}_\Gamma$ from e onward, then M is eventually informative from e onward.*

So there is always a motive to be eventually informative, if one wishes to find the truth at all. The same is not clear, yet, for Ockham's razor and stalwartness, since there are infinitely many eventually informative, non-Ockham solutions. For example, an alternative solution favors some set S of size fifty until the anticipated fifty effects fail to appear for ten thousand stages, after which it concedes defeat and reverts back to Ockham's razor. So it remains to determine how, if at all, Ockham strategies are better at finding the true theory than these variants are.

11 EPISTEMIC COSTS OF CONVERGENCE

As in the parable of the traveler, the aim is to show that normal Ockham strategies are the properly most *efficient* strategies for finding the truth, where efficiency is a matter of minimizing epistemic costs en route to convergence to the truth.

(1) Since the aim is to find the truth, an evident cost of inquiry is the number of times one selects a false answer prior to convergence.

(2) Nobody likes it when science changes its tune, but the intrinsic fallibility of theory choice makes some reversals of course unavoidable. Therefore, the best one can demand of an optimally truth-conducive strategy for theory choice is that it not reverse course more than necessary. A method *retracts* its previous answer whenever its current answer fails to entail its previous answer.[12] In the narrow context of methods that produce answers in $\Pi \cup \{`?'\}$ (where '?' is interpreted as the most uninformative answer $W$), strategy $M$ *retracts* at $e * S$ if and only if $M(e) \neq `?'$ and $M(e * S) \neq M(e)$.

Retractions have been studied as an objective feature of the complexity of problems, both computational and empirical. H. Putnam [1965] noticed that the concept of computability can be extended by allowing Turing machines to "take back" their answers some fixed number of times and called properties having such generalized decision procedures *n-trial predicates*. In a similar spirit, computational learning theorists speak of *mind-changes* and have studied bounds on the number of mind-changes required to find the truth in various empirical questions [Jain *et*

[12] In belief revision theory, a belief change that adds content is an *expansion*, a belief change that removes content is a *contraction* and a belief change that does any of the above is a *revision* [Gärdenfors, 1988]. In that terminology, a retraction is any revision in which content is lost and, hence, may be expressed as a non-trivial contraction followed by an expansion. In spite of this connection, belief revision theorists have not begun to examine the normative consequences of minimizing contractions (or of finding the truth).

al., 1999]. The body of results obtained makes it clear that mind-changes are an invariant feature both of empirical and of purely formal inquiry. The idea here is to shift the focus from problems back to methods.

(3) A third cost of inquiry is elapsed time to each retraction. Theories are used to derive conclusions that tend to accumulate through time. When the theory is retracted, all of these subsidiary conclusions are called into question with it. The accompanying angst is not merely practical but cognitive and theoretical, and it should be minimized by getting retractions over with as soon as possible. Also, aside from such subsidiary conclusions, there is a tragic aspect of unwittingly "living a lie" when one is destined to retract in the future, even if the retracted theory happens to be true. The insouciance is all the worse if one is destined to retract many times. It would be better to relieve the hubris as soon as possible.[13]

Taken together, errors, retractions, and retraction times paint a fairly representative picture of what might be termed the quality or directness of a strategy's connection with or route to the truth. If e is an input stream, let the *cumulative cost* or *loss* of strategy M on $e \in K_\Gamma$ be given by the pair $\lambda(M, w) = (b, \tau)$, where b is the total number of false answers produced by M along e and τ is the sequence of times at which the successive retractions performed by M along e occur. The length of τ (which is finite for convergent strategies) is, then, the total number of retractions performed.

It would be a shame if Ockham's razor were to rest upon some idiosyncratic, subjective weighting of errors, retractions, and retraction times but, happily, the proposed argument for Ockham's razor rests only on comparisons that agree in all dimensions (i.e. on *Pareto* comparisons). First, consider retractions and retraction times. If σ, τ are finite, ascending sequences of natural numbers, define:[14]

$$\sigma \leq \tau \quad \text{iff} \quad \text{there exists a subsequence } \gamma \text{ of } \tau \text{ such that for each } i \leq \text{length}(\sigma), \sigma(i) \leq \gamma(i).$$

For example, $(1, 3, 7) \leq (2, 3, 4, 8)$ in virtue of sub-sequence $(2, 3, 8)$. Then if (b, σ) and (c, τ) are both cumulative costs, define:

$$(b, \sigma) \leq (c, \tau) \quad \text{iff} \quad b \leq c \text{ and } \sigma \leq \tau;$$
$$(b, \sigma) \equiv (c, \tau) \quad \text{iff} \quad (b, \sigma) \leq (c, \tau) \text{ and } (c, \tau) \leq (b, \sigma);$$
$$(b, \sigma) < (c, \tau) \quad \text{iff} \quad (b, \sigma) \leq (c, \tau) \text{ and } (c, \tau) \not\leq (b, \sigma).$$

12 WORST-CASE COST BOUNDS

Non-Ockham strategies do not necessarily incur greater costs prior to convergence: nature could be so kind as to present the extra effects posited by a non-Ockham strategy immediately, in which case it would beat all Ockham competitors in

[13] Elimination of hubris as soon as possible is a Platonic theme, arising, for example, in the *Meno*.

[14] Context will distinguish whether $\leq$ denotes this relation or the initial segment relation.

the race to the truth. The same point is familiar in the theory of computational complexity: if an inefficient algorithm is optimized for speed on a single input, even the best algorithms will fail to dominate it in terms of computational resources expended before the answer is found. For that reason, algorithmic efficiency is ordinarily understood to be a matter of optimizing worst-case cost [Garey and Johnson, 1979]. Adoption of a similar approach to empirical strategies and to Ockham's razor requires some careful attention to worst-case bounds on total costs of inquiry. Let ω denote the first infinite ordinal number. A *potential cost bound* is a pair (b, σ), where $b \leq \omega$ and σ is a finite or infinite, non-descending sequence of entries $\leq \omega$ in which no finite entry occurs more than once. If (b, σ) is a cost vector and (c, τ) is a cost bound, then $(b, \sigma) \leq (c, \tau)$ can be defined just as for cost vectors, themselves. Cost bounds $(c, \tau), (d, \gamma)$ may now be compared as follows:

$$(c, \tau) \leq (d, \gamma) \text{ iff for each cost vector } (b, \sigma), \text{ if } (b, \sigma) \leq (c, \tau) \text{ then } (b, \sigma) \leq (d, \gamma);$$
$$(c, \tau) \equiv (d, \gamma) \text{ iff } (c, \tau) \leq (d, \gamma) \text{ and } (d, \gamma) \leq (c, \tau);$$
$$(c, \tau) < (d, \gamma) \text{ iff } (c, \tau) \leq (d, \gamma) \text{ and } (d, \gamma) \not\leq (c, \tau).$$

Thus, for example, $(4, (2)) < (\omega, (2, \omega)) < (\omega, (0, 1, 2, \ldots)) \equiv (\omega, (\omega, \omega, \omega, \ldots))$. Now, each set C of cost vectors has a unique (up to equivalence) least upper bound $\sup(C)$ among the potential upper bounds [Kelly, 2007]. Suppose that finite input sequence e has already been seen. Then one knows that possibilities incompatible with e cannot happen, so define the *worst-case cost* of M at e as:

$$\lambda_e(M) = \sup_{w \in K_\Gamma | e} \lambda(M, w).$$

13 RELATIVE EFFICIENCY

The final hurdle in arguing for the efficiency of Ockham's razor is the triviality of worst-case cost bounds: for each e and for each convergent solution M to $\mathcal{P}_\Omega$, the worst-case cost bound achieved by M at e is just:

$$\lambda_e(M) = (\omega, (\omega, \omega, \omega, \ldots)).$$

For let m be an arbitrary, natural number and let $\{a_0, \ldots a_n, \ldots\}$ be an arbitrary enumeration of the set E of possible effects. Nature can present $\emptyset$ until, on pain of not converging to the truth, M produces answer $T_\emptyset$ at least m times consecutively. Then Nature can present $\{a_0\}$ followed by repetitions of $\emptyset$ until, on pain of not converging to the truth, M produces $T_{\{e_0\}}$ at least m times, consecutively, etc. Hence, normal Ockham strategies are not distinguished from alternative, convergent solutions in terms of worst-case efficiency.

Again, a similar difficulty is familiar in the assessment of computer algorithms: typically, the number of steps required by an algorithm is not finitely bounded across all possible inputs since larger inputs require more steps of computation.

The problem disappears if worst-case bounds are taken over problem instances (inputs) of a given size, rather than over all possible problem instances, for there are at most finitely many such inputs, so the worst-case performance of an algorithm over inputs of a given size is guaranteed to exist [Garey and Johnson, 1979]. Then one compares the worst-case cost bounds over each instance size as instance size increases. In $\mathcal{P}_\Omega$, every problem instance (input stream) is of infinite length, so length is no longer a useful notion of instance size. But *empirical complexity* $c(w, e)$, defined above, is such a notion. Furthermore, each normal Ockham strategy is a convergent solution that retracts at most n times over instances of empirical complexity n, so non-trivial cost bounds are achievable. Accordingly, define the nth *empirical complexity class* $C_e(n)$ of worlds in $K_\Gamma|e$ as:

$$C_e(n) = \{w \in K_\Gamma|e : c(w, e) = n\}.$$

Then one may define the *worst-case cost* of strategy M given e over $C_e(n)$ as follows:

$$\lambda_e(M, n) = \sup_{w \in C_e(n)} \lambda(M, w).$$

Now it is possible to compare strategies in terms of their worst-case costs over problem instances of various sizes.

$$\begin{array}{lll} M \leq_e M' & \text{iff} & (\forall n)\ \lambda_e(M,n) \leq \lambda_e(M',n); \\ M <_e M' & \text{iff} & M \leq_e M' \text{ and } M' \not\leq_e M; \\ M \prec_e M' & \text{iff} & (\forall n) \text{ if } C_e(n) \neq \emptyset \text{ then } \lambda_e(M,n) < \lambda_e(M',n). \end{array}$$

When $M \leq_e M'$, say that M is *as efficient as* M' given e. If $M <_e M'$ say that M is (weakly) *more efficient than* M' given e. Finally, when $M \prec_e M'$, say that M is *strongly* more efficient than M'.

The concept "more efficient than" is a hybrid, lying between dominance (doing as well in each world and better in some world) and worst-case (minimax) reasoning (doing better in the worst case overall). The hybrid character of "more efficient than" is just what is required to expose the superiority of normal Ockham strategies: dominance is too strict because non-Ockham strategies can get lucky and the worst-case overall is too loose because even normal Ockham strategies guarantee no nontrivial, worst-case bound.

14 OPTIMALITY

Suppose that a scientist facing problem $\mathcal{P}_\Gamma$ has been using strategy M for a while and that the final datum in finite input sequence $e = (x_1, \ldots, x_n)$ has just been presented. Let the data e_- observed just prior to the end of e be defined by:

$$\begin{array}{rcl} e_-(()) & = & (); \\ e_-((S_0, \ldots, S_n, S_{n+1})) & = & (S_0, \ldots, S_n). \end{array}$$

At e, past actions along e_- directed by the scientist's strategy M can no longer be "taken back" at e. Hence, an alternative strategy M' cannot possibly be adopted and implemented at e unless its outputs agree with those of M all along e_-, in which case write $M \asymp_{e_-} M'$. Define:

> M is *optimal* at e iff M is a solution at e and for each strategy $M' \asymp_{e_-} M$ that is a solution at e, $M \leq_e M'$.

It is not enough for strategy M to be optimal at e. If the user of M is not to have reason to dispense with M later, it had best be the case that M is always optimal:

> M is *always optimal* iff for each $e \in F_\Gamma$, M is optimal at e.

When e is empty, say simply that M is *optimal*.

15 UNIQUE OPTIMALITY OF NORMAL OCKHAM STRATEGIES

Here is the promised, non-circular argument, based entirely on truth-finding efficiency, for always following Ockham's razor. The results are relative to a fixed problem $\mathcal{P}_\Gamma$.

THEOREM 6 (Optimality). *If M is a normal Ockham strategy, then M is always an optimal solution.*

But that is not enough. One trouble with much of the standard literature on Ockham's razor is that it shows only that Ockham's razor is sufficient for, say, convergence to the truth, but what is required is an argument that Ockham's razor is *necessary* for optimal truth-conduciveness. Here is such an argument:

THEOREM 7 (Unique optimality). *Let $e \in F_\Gamma$. If M' is a convergent solution that violates Ockham's razor for the first time at e, then every strategy $M \asymp_{e_-} M'$ that is always normally Ockham is a more efficient solution than M' at e;*

The proof (cf. the Appendix) is closely analogous to the parable of the traveler discussed above, with extra cases added to allow for the possibility of branching freeway ramps. The two theorems jointly imply the following corollary, which summarizes the proposed argument for Ockham's razor.

COROLLARY 8 (Ockham efficiency characterization). *The following statements are equivalent:*

1. *M is always a normal, Ockham strategy;*
2. *M is always an optimal solution;*
3. *no solution M' is ever a more efficient solution than M.*

In other words, the normally Ockham methods are coextensive with the always efficient strategies and with the strategies such that no alternative strategy is ever more efficient.[15]

[15] It is a further question whether it is always better to follow Ockham's razor even after

16 A GENERAL DEFINITION OF EMPIRICAL COMPLEXITY

In the preceding development, empirical effects were stipulated, by appeal to intuition, for each of the examples considered and the effects appealed to were quite different from case to case. Empirical effects will now be defined in a general way that explains the apparently ad hoc choices in the examples.[16] The idea is to locate effects and, hence, empirical complexity, in the power of nature to force an arbitrary, convergent method to change its answer to the problem to be solved. Thus, empirical complexity is a structural, semantic feature of the problem to be solved, rather than a matter of syntactic or computational brevity. As such, it is invariant under grue-like translations.

An empirical *problem* is a pair $\mathcal{P} = (K, \Pi)$ where K is now an arbitrary set of infinite sequences of *inputs* drawn from some arbitrary set I and Π is an arbitrary partition of K. The elements of I are just inputs (e.g., boolean bits in a binary coding scheme for meter readings or what-not). Answers are just arbitrary, mutually exclusive and exhaustive propositions over K. This is a very general conception of empirical problems. An empirical strategy takes finite, initial segments of elements of K as inputs and outputs potential answers in $\Pi \cup \{\text{`?'}\}$ in response. Convergence and many other concepts like $M[w]$, $F|e$, $K|e$ extend to this more general setting in an obvious way. It remains to reconstruct Γ and the concepts that presuppose it.

Let some problem $\mathcal{P} = (K, \Pi)$ be understood to be fixed in the background. An *answer pattern* is a finite sequence of elements of Π without immediate repetitions (non-immediate repetitions are allowed). Pattern s is *forcible* by nature given e of length k if and only if for each convergent solution M, there exists $w \in K|e$ such that from stage k onward, M produces a sequence of answers of which s is a sub-sequence. In other words, no convergent solution can avoid producing the successive conclusions in s in the worst case, given e. Let Δ_e denote the set of all patterns forcible by nature given e. Restrict attention to problems $\mathcal{P}$ such that:

AXIOM 9 (Forcible path convergence). for each $w \in K$, $\lim_{i \to \infty} \Delta_{w|i}$ exists, in the sense that the sequence $\{\Delta_{w|i} : i \geq 0\}$ stabilizes to a fixed set eventually.

Define:

$$\begin{aligned} \Delta_w &= \lim_{i \to \infty} \Delta_{w|i}; \\ \Gamma|e &= \{\Delta_w : w \in K|e\}; \end{aligned}$$

violating it. The answer is negative: let $\Gamma = \{\{a\}, \{b\}, \{b, c\}\}$, let $M((\emptyset)) = M'((\emptyset)) = T_{\{b\}}$, and let $M((\emptyset, \emptyset)) = \text{`?'}$ whereas $M'((\emptyset, \emptyset)) = T_{\{b\}}$. Then M uses one extra retraction in reaching theory $T_{\{b,c\}}$ after seeing $e = (\emptyset, \emptyset)$, so $\lambda_e(M, 2) \not\leq \lambda_e(M', 2)$.

There is still something to say in favor of Ockham, however. Method M is *strongly Ockham* at e if M never favors an answer T_S such that some alternative S' compatible with e has a longer path through $\Gamma|e$. Then one can argue, along the same lines, that at each strong Ockham violation and at each violation of stalwartness by M, some alternative, convergent M' is more efficient.

[16] Preliminary versions of the following ideas can be found in [Kelly, 2007].

$$\Gamma = \Gamma|().$$

Elements of $\Gamma|e$ serve the same purpose as before — they are the possible, permanent stopping places for nature given e because each element Δ_w of Γ is converged to in world w and world w is compatible with e. Call the elements of Γ *empirical problem states.* Define *epistemic accessibility* among states in the following way:

> $X \leq Y$ if and only if for each $e \in F$ such that $\Delta_e = X$, there exists $e' \in F$ such that $e' \geq e$ and $\Delta_{e'} = Y$.

Let $\pi_e(X, Y)$ denote the set of all $\leq$-paths between two states in Γ with respect to order $\geq$. Finally, define $c(w, e)$ and $c(P, e)$ in terms of these paths, as before, with Δ_w in place of S_w. Let the new, more general concepts so defined be marked with a prime, as in $c'(w, e)$, to distinguish them from the notions defined in terms of stipulated effects.[17]

The general concept of empirical complexity just defined agrees with the effect-based definition.

PROPOSITION 10 (Recovery). *Let* $(\Gamma', \leq')$ *be constructed from problem* (K_Γ, Π_Γ) *in the manner just described. Then for each* $e \in F$, *the mapping* $\phi(S_w) = \Delta_w$ *is well-defined and witnesses:*

$$(\Gamma|e, \subseteq) \textit{ is order-isomorphic to } (\Gamma'|e, \leq').$$

Thus, for each $w \in K$:

$$\begin{aligned} c'(w, e) &= c(w, e); \\ c'(P, e) &= c(P, e). \end{aligned}$$

Something far more interesting is also true. Let (K, Π) be any one of the examples considered above (e.g., the conservation law problem) *prior* to being represented in the form (K_Γ, Π_Γ). The problem (K, Π) does not wear its empirical effects "on its sleeve" — the reactions may be presented in some obscure or even grue-like code that is highly misleading. But it is still the case that applying the preceding construction directly to (K, Π) results in $(\Gamma', \leq')$ order-isomorphic to $(\Gamma, \subseteq)$ and, hence, to the same empirical complexity concept $c(w, e)$. Depending on the structure of the problem $\mathcal{P}$, empirical complexity reflects extra parameters, extra conserved quantities, extra causes, etc., regardless of how gerrymandered the data-gathering process happens to be.[18] Therefore, the set Γ assumed in each case reflects more than mere notation, convention, or whim — it is an intrinsic, structural feature of the original problem that survives every sort of re-description that preserves the meanings of the background presupposition K and of the question Π.

[17] The structure $(\Gamma'|e, \leq')$ can be viewed as a model of an epistemic logic, in which elements of $\Gamma'|e$ are worlds and increasing information e "chops down" the set of worlds, in accordance with what is known as *dynamic epistemic logic* [van Benthem, 2006]. What is new is a motivated constraint on accessibility and the idea that empirical complexity is a matter of maximum accessibility path length into a world.

[18] Contrast this result with the preceding discussion of algorithmic complexity, which is relative both to the choice of a computer language and to the encoding of observations.

17 A WORD ON STOCHASTIC APPLICATIONS

In real curve-fitting and causal discovery problems, the data are not merely inexact, but random. The above treatment of these problems in terms of collapsing open intervals around the true observations is intended only as an indication of how a fully statistical story might go (think of the intervals as idealizations of high probability quantiles). This section sketches some promising pieces of a fully statistical version of the theory.

A *world* is an objective probability distribution of interest (e.g., the distribution induced by a polynomial curve with normally distributed measurement error). A *question* is a partition of worlds. A *method* maps samples of arbitrary size to answers to the question. A method is *consistent* just in case the probability that the method produces the true answer converges to unity as sample size increases. The *retraction in chance* of answer T by method M at sample size $n+1$ in distribution p is definable as the drop in chance that M outputs T from sample size n to sample size $n+1$:

$$p^n(M = T) - p^{n+1}(M = T).$$

The total retractions in chance in p are the sums of the retractions in chance for all $T \in \Pi$, and for all sample sizes n.

A sequence of answers is *forcible in chance* if and only if nature can force an arbitrary, consistent method to produce the first answer in the sequence with arbitrarily high chance followed by the second answer in the sequence with arbitrarily high chance, etc. For a simplistic illustration of how this works (a similar argument applies in causal discovery), let K consist of independent, bivariate normal means of fixed, known variance and let the possible answers correspond to the number of non-zero components of the true mean vector $\mu = (\mu_X, \mu_Y)$, so answer T_i is the set of all $p \in K$ such that exactly i components of μ are non-zero. Let M be a consistent method. Let $p_0 \in T_0$ and let $\epsilon > 0$ be as small as desired. Since M is consistent, there is a sample size n_0 such that

$$p_0^{n_0}(M = T_0) > 1 - \epsilon.$$

Since the chance of a fixed measurable event is continuous in μ, there exists $p_1 \in T_1$ such that

$$p_1^{n_0}(M = T_0) > 1 - \epsilon.$$

Since M is consistent, there exists sample size $n_1 > n_0$ such that

$$p_1^{n_1}(M = T_1) > 1 - \epsilon.$$

Again, by continuity, there exists $p_2 \in T_2$ such that:

$$\begin{aligned} p_2^{n_0}(M = T_0) &> 1 - \epsilon; \\ p_2^{n_1}(M = T_1) &> 1 - \epsilon. \end{aligned}$$

Again, by consistency, there exists $n_2 > n_1$ such that:

$$p_2^{n_1}(M = T_2) > 1 - \epsilon.$$

Hence, the sequence of answers (T_0, T_1, T_n) is forcible by nature in chance. The only premise required for this forcing argument is convergence to the truth, so a Bayesian's degree of belief in each successive answer can also be forced arbitrarily high. A Bayesian's retraction in chance of T at $n + 1$ in p can be measured in terms of the drop in his expected degree of belief in T at sample size $n + 1$ in p.[19] Simplicity can be defined in terms of statistically forcible sequences of answers, just as in the deterministic case. It remains to recover a suitable analogue of Corollary 8 in the setting just described.

18 OCKHAM, FALLIBILITY, AND "INFORMATION"

Like it or not, we do infer theoretical forms, they are subject to the problem of induction, and we may have to take them back. Indeed, there is no bound on the number of times science might have to change its tune as new layers of complexity are successively revealed in nature. Ockham's razor merely keeps science on the straightest path to the truth, crooked as it may be. For millennia, fallibility has been thought to *undermine* the justification of science, resulting in the usual, circular, metaphysical, or skeptically evasive justifications of Ockham's razor. The proposed account reverses the traditional reasoning — Ockham's razor is justified not because it points straight at the truth, but because its path to the truth, albeit crooked, is uniquely straightest. The Ockham path is straightest because its unavoidable kinks are due to the intrinsic fallibility of theory choice. Therefore, the ineluctable fallibility of theory choice justifies, rather than undermines, Ockham's razor. That is why the proposed account is not circular, metaphysical, or evasive of the connection between method and true theoretical structure.

Ockham's razor is, nonetheless, so firmly anchored in our animal spirits that it feels as if, somehow, simplicity *informs* us about the true theory in a way that the data alone do not, just as a compass needle augments the information provided by one's native sense of direction. Then there must be some benevolent cosmic cause behind the correlation of simplicity and truth — a mysterious, undetected agency that operates across evolutionary time and across domains from subatomic particles to cell metabolism to social policy — the irony of defending Ockham's razor with such hidden, metaphysical fancies notwithstanding [Koons, 2000].

Therein lies a concern about the association of information-theoretic terminology with Ockham's razor, as in the MDL approache. When information theory is applied to a telephone line, as originally intended, it really has something to do with informative signals from a source. If one wishes to minimize expected message length to maximize the line's capacity, it makes sense to adopt shorter codes

[19]Bayesians are sub-optimal: moving from ignorance (.5/.5) to knowledge (.99/.01) implies a retraction of nearly one half that could have been avoided by modeling ignorance as (0/0), as Schafer [1976] proposed.

for more frequently sent words. But applications of information theory to theory choice are not about sending information over a line. They are a formal recipe either for constructing short codes for plausible explanations or (contrariwise) for assigning high plausibility to short explanations. Either way, the ultimate connection between simplicity and truth is stipulated rather than explanatory. But since the stipulated connection is formulated in the language of "information", it is all too readily confused, in the popular mind, with a deep theoretical revelation that simplicity *does* provide a magical communication channel to the truth that amplifies the only real information available — the data. Better not to mention "information" at all than to kindle that perennial wish.

19 ACKNOWLEDGEMENTS

This work has benefitted from helpful comments from and patient discussions with (in alphabetical order), Pieter Adriaans, Seth Casana, Stephen Fancsali, Clark Glymour, Conor Mayo-Wilson, Joe Ramsey, Richard Scheines, Cosma Shalizi, Peter Spirtes, Wolfgang Spohn, my very helpful commentator, Larry Wasserman, and Jiji Zhang. The errors, of course, are my responsibility alone.

BIBLIOGRAPHY

[Adriaans, 2007] P. Adriaans. The philosophy of learning, the cooperative computational universe, in *Handbook of the Philosophy of Information*, P. Adriaans, and J. van Benthem, eds. Dordrecht: Elsevier, 2007.

[Akaika, 1973] H. Akaike. Information theory and an extension of the maximum likelihood principle, *Second International Symposium on Information Theory*. pp. 267-281, 1973.

[Carnap, 1950] R. Carnap. *Logical Foundations of Probability*, Chicago: University of Chicago Press, 1973.

[Chickering, 2003] D. Chickering. Optimal Structure Identification with Greedy Search, JMLR, 3: 507-554, 2003.

[Dretske, 1981] F. Dretske. *Knowledge and the Flow of Information*, Cambridge: MIT Press, 1981.

[Forster and Sober, 1994] M. R. Forster and E. Sober. How to Tell When Simpler, More Unified, or Less Ad Hoc Theories will Provide More Accurate Predictions, *The British Journal for the Philosophy of Science* 45: 1-35, 1994.

[Freivalds and Smith, 1993] R. Freivalds and C. Smith. On the Role of Procrastination in Machine Learning, *Information and Computation* 107: pp. 237-271, 1993.

[Ford, 1963] K. Ford. *The World of Elementary Particles*, New York: Blaisdell, 1963.

[Friedman, 1983] M. Friedman. *Foundations of Space-Time Theories*, Princeton: Princeton University Press.), 1983.

[Gärdenfors, 1988] P. Gärdenfors. *Knowledge in Flux*, Cambridge: MIT Press, 1988.

[Garey and Johnson, 1979] M. Garey and D. Johnson. *Computers and Intractability*, New York: Freeman, 1979.

[Glymour, 1980] C. Glymour. *Theory and Evidence*, Princeton: Princeton University Press, 1980.

[Goldman, 1986] A. Goldman. *Epistemology and Cognition*, Cambridge: Harvard University Press, 1986.

[Goodman, 1983] N. Goodman. *Fact, Fiction, and Forecast*, fourth edition, Cambridge: Harvard University Press, 1983.

[Jeffreys, 1985] H. Jeffreys. *Theory of Probability*, Third edition, Oxford: Clarendon Press, 1985.

[Jain *et al.*, 1999] S. Jain, D. Osherson, J. Royer, and A. Sharma. *Systems That Learn: An Introduction to Learning Theory*. Camabridga: MIT Press, 1999.

[Kelly, 1996] K. Kelly. *The Logic of Reliable Inquiry*, New York: Oxford, 1996.

[Kelly, 2002] K. Kelly. Efficient Convergence Implies Ockham's Razor, *Proceedings of the 2002 International Workshop on Computational Models of Scientific Reasoning and Applications*, Las Vegas, USA, June 24-27, 2002.

[Kelly, 2004] K. Kelly. Justification as Truth-finding Efficiency: How Ockham's Razor Works, *Minds and Machines* 14: 485-505, 2004.

[Kelly and Glymour, 2004] K. Kelly and C. Glymour. Why Probability Does Not Capture the Logic of Scientific Justification, forthcoming, C. Hitchcock, ed., *Contemporary Debates in the Philosophy of Science*, Oxford: Blackwell, 2004 pp. 94-114, 2004.

[Kelly, 2007] K. Kelly. Ockham's Razor, Empirical Complexity, and Truth-finding Efficiency, *Theoretical Computer Science* 317: 227-249, 2007.

[Kitcher, 1981] P. Kitcher. Explanatory Unification, *Philosophy of Science*, 48, 507-31, 1981.

[Koons, 2000] R. Koons. The Incompatibility of Naturalism and Scientific Realism, In *Naturalism: A Critical Appraisal*, edited by J. Moreland and W. Craig, London: Routledge, 2000.

[Kuhn, 1962] T. Kuhn. *The Structure of Scientific Revolutions*, Chicago: University of Chicago Press, 1962.

[Mayo, 1996] D. Mayo. *Error and the Growth of Experimental Knowledge*, Chicago: University of Chicago Press, 1996.

[Morrison, 2000] M. Morrison. *Unifying Scientific Theories: Physical Concepts and Mathematical Structures*, Cambridge: Cambridge University Press, 2000.

[Li and Vitanyi, 1997] M. Li and P. Vitanyi. *An Introduction to Kolmogorov Complexity and Its Applications*, New York: Springer, 1997.

[Mitchell, 1997] T. Mitchell. *Machine Learning*. New York: McGraw-Hill, 1997.

[Putnam, 1965] H. Putnam. Trial and Error Predicates and a Solution to a Problem of Mostowski, *Journal of Symbolic Logic* 30: 49-57, 1965.

[Popper, 1968] K. Popper. *The Logic of Scientific Discovery*, New York: Harper, 1968.

[Richardson and Spirtes, 2002] T. Richardson and R. Spirtes. Ancestral Graph Markov Models, *Annals of Statistics* 30: pp. 962-1030, 2002.

[Rissanen, 1983] J. Rissanen. A universal prior for integers and estimation by inimum description length, *The Annals of Statistics*, 11: 416-431, 1983.

[Robins *et al.*, 1999] J. Robins, R. Scheines, P. Spirtes, and L. Wasserman. Uniform Consistency in Causal Inference, *Biometrika* 90:491-515, 1999.

[Rosenkrantz, 1983] R. Rosenkrantz. Why Glymour is a Bayesian, in *Testing Scientific Theories*, J. Earman ed., Minneapolis: Universitiy of Minnesota Press, 1983.

[Salmon, 1967] W. Salmon. *The Logic of Scientific Inference*, Pittsburgh: University of Pittsburgh Press, 1967.

[Schulte, 1999a] O. Schulte. The Logic of Reliable and Efficient Inquiry, *The Journal of Philosophical Logic*, 28:399-438, 1999.

[Schulte, 1999b] O. Schulte. Means-Ends Epistemology, *The British Journal for the Philosophy of Science*, 50: 1-31, 1999.

[Schulte, 2001] O. Schulte. Inferring Conservation Laws in Particle Physics: A Case Study in the Problem of Induction, *The British Journal for the Philosophy of Science*, 51: 771-806, 2001.

[Schulte *et al.*, 2007] O. Schulte, W. Luo, and R. Griner. Mind Change Optimal Learning of Bayes Net Structure. Unpublished manuscript, 2007.

[Schwarz, 1978] G. Schwarz. Estimating the Dimension of a Model, *The Annals of Statistics*, 6:461-464, 1978.

[Shimizu *et al.*, 2006] S. Shimizu, P. Hoher, A. Hyvärinen, and A. Kerminen. A Linear Non-Gaussian Acyclic Model for Causal Discovery, *Journal of Machine Learning Research* 7: pp. 2003-2030, 2006.

[Spirtes *et al.*, 2000] P. Spirtes, C. Glymour, and R. Scheines. *Causation, Prediction, and Search*. Cambridge: MIT Press, 2000.

[Valdez-Perez and Zytkow, 1996] R. Valdez-Perez, and J. Zytkow. Systematic Generation of Constituent Models of Particle Families, *Physical Review*, 54:2102-2110, 1996.

[van Benthem, 2006] J. F. A. K. van Benthem. Epistemic Logic and Epistemology, the state of their affairs, *Philosophical Studies* 128: 49–76, 2006.

[van Fraassen, 1980] B. van Fraassen. *The Scientific Image*, Oxford: Clarendon Press, 1980.

[Vapnik, 1998] V. Vapnik. *Statistical Learning Theory*, New York: John Wiley and Sons, Ltd, 1998.

[Vitanyi and Li, 2000] P. Vitanyi and M. Li. Minimum Description Length Induction, Bayeisanism, and Kolmogorov Complexity, *IEEE Transactions on Information Theory* 46: 446-464, 2000.

[Wasserman, 2003] L. Wasserman. *All of Statistics: A Concise Course in Statistical Inference.* New York: Springer, 2003.

[Zhang and Spirtes, 2003] J. Zhang and P. Spirtes. Strong Faithfulness and Uniform Consistency in Causal Inference, in *Proceedings of the Nineteenth Conference on Uncertainty in Artificial Intelligence*, 632-639. Morgan Kaufmann, 2003.

APPENDIX

A PROOFS AND LEMMAS

Proof of Proposition 1. Immediate. ■

Proof of Proposition 2. For (1), suppose that $w \in K_\Gamma$ and that $c(w, e) = 0$. Then each $p \in \pi_e(*, S_w)$ has unit length and terminates with S_w, so since $S_w \in \Gamma|e$, $\pi_e(*, S_w) = \{(S_w)\}$. Hence, $S_w \in (\Gamma|e)_{\min}$. Conversely, suppose that $S_w \in (\Gamma|e)_{\min}$. Then for each $R \in \Gamma|e$, $R \not\subset S_w$. So $\pi_e(*, S_w) = \{(S_w)\}$, so $c(w, e) = 0$.

For (2), suppose that $T_S \in \Pi_\Gamma|e$ and that $c(T_S, e) = 0$. Then there exists $w \in T_S$ such that $c(w, e) = 0$. So by (1), $S = S_w \in (\Gamma|e)_{\min}$. Conversely, suppose that $S \in (\Gamma|e)_{\min}$. Let $w = e * (S \setminus S_e) * \emptyset^\infty$. Then $w \in T_S$ and $S_w = S \in (\Gamma|e)_{\min}$. So by part (1), $c(w, e) = 0$. So $c(T_S, e) = 0$. ■

Proof of Proposition 3. The equivalence (1) $\Leftrightarrow$ (2) is by part 2 of Proposition 2. Equivalence (2) $\Leftrightarrow$ (3) is an elementary property of $\subseteq$ over a collection of finite sets. ■

Proof of Proposition 4. Suppose that $w \in K_\Gamma|e$. Let $k \geq \mu(w)$ so that $S_w = S_{w|k}$. Then $(\Gamma|(w|k))_{\min} = \{S_w\}$, so by Proposition 3, T_{S_w} is Ockham given e. Since M is eventually informative from e onward, M produces some answer from Π_Γ after e in w. Since M is Ockham from e onward, the answer M chooses is T_{S_w}. Since M is stalwart from e onward, M never drops T_{S_w} thereafter. So $\lim_{i \to \infty} M(w|i) = T_{S_w}$. ■

Proof of Proposition 5. Immediate. ■

Proof of Theorem 6. Let M be a strategy that is always normally Ockham. Hence, M is a solution, by Proposition 4. Let $e \in F_\Gamma$ have length k. Let M' be an arbitrary solution given e such that $M' \asymp_{e_-} M$. Let d denote the maximum, over all $w \in C_e(0)$, of the number of errors committed by both M and M' along e_- and let the retraction times for M, M' along e_- be $(r_1, \ldots, r_m)$. Consider the case in which M retracts at e. Since M is always stalwart and Ockham, it follows that $T_S = M(e_-)$ is Ockham at e_- but not at e, so by Proposition 3, $(\Gamma|e_-)_{\min} = \{S\}$ and $(\Gamma|e)_{\min} \neq \{S\}$. So by Lemma 19, $S \notin (\Gamma|e)_{\min}$.

Suppose that $w \in C_e(0)$. By parts (1) and (2) of Lemma 12, M never retracts or commits an error from stage $k+1$ onward in w. Hence:

$$\lambda_e(M, 0) \leq (d, (r_1, \ldots, r_m, k)).$$

Since M retracts at e, there exists $S \subseteq E$ such that

$$M(e_-) = M'(e_-) = T_S \neq M(e).$$

Since $e \in F_\Gamma$, Lemma 16 implies that there exists $w' \in C_e(0)$ such that $S_{w'} \neq S$. Since $M'(e_-) = T_S$ and M' is a solution, it follows that M' retracts after e_- along w'. So since M' commits at least d errors in some world in $C_0(e)$ (they do not have to be committed in w' to affect the worst-case lower bound):

$$\lambda_e(M, 0) \leq (d, (r_1, \ldots, r_m, k)) \leq \lambda_e(M', 0).$$

Now suppose that $w \in C_e(n+1)$. By part 1 of Lemma 12, M retracts at most $n+1$ times in w from $k+1$ onward, so allowing for the extra retraction at k:

$$\lambda_e(M, n+1) \leq (\omega, (r_1, \ldots, r_m, k, \underbrace{\omega, \ldots, \omega}_{n+1}))).$$

Suppose that there exists $w \in C_e(n+1)$. By Lemma 11, there exists path $(S_0, \ldots, S_{n+1})$ and $w' \in C_e(n)$ such that (1) $S_0 \in (\Gamma|e)_{\min}$ and (2) $S_{w'} = S_w$ and for each $i \leq n+1$, M' produces T_{S_i} at least b times in immediate succession in w' after the end of e. It was shown above that $S \notin (\Gamma|e)_{\min}$. So, since $S_0 \in (\Gamma|e)_{\min}$, it follows that $S_0 \neq S$. So M' retracts from T_S to T_{S_0} no sooner than stage k. By incrementing b, w can be chosen so that the retractions of M' between answers $T_{S_0}, \ldots, T_{S_{n+1}}$ along w' occur arbitrarily late and M' produces arbitrarily many errors along w', so:

$$\lambda_e(M', n+1) \geq (\omega, (r_1, \ldots, r_m, k, \underbrace{\omega, \ldots, \omega}_{n+1}))) \geq \lambda_e(M, n).$$

For the case in which M does not retract at e, simply erase the k's from the bounds in the preceding argument. ■

Proof of Theorem 7. Let $e \in$ be of length k. Let M' be given and let $M \asymp_e M'$ be a strategy that is normally Ockham from e' onward. Hence, M is a solution given e', by Proposition 4. Let $b \geq 0$. Let d denote the maximum, over $w \in C_e(0)$ of the number of errors committed by both M and M' along e_- and let the retraction times for M, M' along e_- be $(r_1, \ldots, r_m)$.

Suppose that solution M' violates Ockham's razor at $e \in F_\Gamma$ of length k but not at any proper, initial segment of e. So $T_S = M'(e)$ is not Ockham at e. By Proposition 3, $(\Gamma|e)_{\min} \neq \{S\}$. Thus, there exists $S' \neq S$ such that $S' \in (\Gamma|e)_{\min}$. Let $w \in C_e(0)$. Since M is Ockham from e onward, if $M(e) = T_{S'}$ then $T_{S'}$ is Ockham at e so, by Proposition 3, $(\Gamma|e)_{\min} = \{S'\}$ and, hence, $w \in T_{S'}$ so M commits no error at e in w. So by Lemma 12:

$$\lambda_e(M, 0) \quad \leq \quad (d, (r_1, \ldots, r_m, k)).$$

Let $w \in T_{S'}$. Then $w \in C_0(e)$, since $S' \in (\Gamma|e)_{\min}$. Since M' is a solution, M' converges to $T_{S'}$ in w, so M' retracts T_S properly later than stage k in w. Since M' commits at least d errors in some world in $C_0(e)$ (they do not have to be committed in w' to affect the worst-case lower bound):

$$\lambda_e(M', 0) > (d, (r_1, \ldots, r_m, k)) \geq \lambda_e(M, 0).$$

Suppose that there exists $w \in C_e(n+1)$. As in the proof of Proposition 6,

$$\lambda_e(M, n+1) \leq \lambda_e(M', n+1).$$

Next, suppose that M' violates stalwartness at e of length k. Then $T_S = M'(e_-) = M(e_-)$ is Ockham at e, so by Proposition 3, $(\Gamma|e)_{\min} = \{S\}$. Since M is stalwart from e onward, $M(e) = T_S$, so M does not retract at e. Let $w \in C_e(0)$. Then, by Proposition 2, $S_w \in (\Gamma|e)_{\min}$, so $S_w = S$. So M commits no error at e. So by Lemma 12:

$$\begin{aligned} \lambda_e(M, 0) &\leq (d, (r_1, \ldots, r_m)); \\ \lambda_e(M, n+1) &\leq (\omega, (r_1, \ldots, r_m, \underbrace{\omega, \ldots, \omega}_{n+1}))). \end{aligned}$$

By Lemma 16, there exists $w \in C_e(0)$. Since M' retracts at e:

$$\lambda_e(M', 0) \geq (d, (r_1, \ldots, r_m, k)) > \lambda_e(M, 0).$$

Suppose that there exists $w \in C_e(n+1)$. By Lemma 11, there exists $w' \in C_e(n+1)$ such that $S_{w'} = S_w$ and in w', M' produces $n+2$ distinct blocks of answers in $K_\Gamma|e$ after the end of e, each block having length at least b. So in w', M' retracts at e and at the end of each block prior to the terminal block. By incrementing b, w can be chosen so that the retractions occur arbitrarily late and there are arbitrarily many errors, so including the assumed retraction at k,

$$\lambda_e(M', n+1) \quad \geq \quad (\omega, (r_1, \ldots, r_m, k, \underbrace{\omega, \ldots, \omega}_{n+1})) > \lambda_e(M, n+1).$$

■

Proof of Corollary 8. The equivalence (1) $\Rightarrow$ (2) is by Proposition 4 and Theorem 6. Equivalence (2) $\Rightarrow$ (3) is immediate from the definitions. (3) $\Rightarrow$ (1) is by Proposition 5 and Theorem 7. ■

Proof of Proposition 10. When the problem is (K_Γ, Π_Γ), the following relations hold:

$$\text{i.} \quad S_w = S_{w'} \quad \text{if and only if} \quad \Delta_w = \Delta_{w'};$$
$$\text{ii.} \quad S_w \subseteq S_{w'} \quad \text{if and only if} \quad \Delta_w \leq \Delta_{w'}.$$

For (i), suppose that $S_w = S_{w'}$. Suppose that $s = (T_{S_1}, \ldots, T_{S_k}) \in \Delta_w$. So $S_1, \ldots, S_k$ are distinct elements of Γ and for each m, nature can force the successive, distinct answers $T_{S_1}, \ldots, T_{S_k}$ from an arbitrary, convergent method M starting from $w|m$. Hence, $S_{w'} = S_w \subseteq S_1 \subset \ldots \subset S_k$. So for each m, nature can force $S_1 \subset \ldots \subset S_k$ from M starting with $w'|m$, so $s \in \Delta_{w'}$. Thus, $\Delta_w \subseteq \Delta_{w'}$. For the converse inclusion, reverse the roles of w and w'. For the converse implication, suppose that $\Delta_w = \Delta_{w'}$. Suppose that $s = (T_{S_1}, \ldots, T_{S_k}) \in \Delta_w$. Then for each m, nature can force M to produce s wtarting from $w|m$. Since M is a convergent solution, there exists $m' \geq n$ such that $M(w|m') = T_{S_w}$. Nature can still force M to produce $T_{S_w} * s$ starting from $w|m'$. Hence, (a) for each $s \in \Delta_w$, for each m, $T_{S_w} * s$ is forcible by nature starting from $w|m$. By a similar argument, (a) also holds as well for w'. Call that statement (a'). Since for each m, nature can force (T_{S_w}) given $w|m$, $(T_{S_w}) \in \Delta_w$. Suppose, for reductio, that $S_w \neq S_{w'}$. Then by (a'), $(T_{S_{w'}}, T_{S_w}) \in \Delta_{w'}$ and by (a), $(T_{S_w}, T_{S_{w'}}, T_{S_w}) \in \Delta_w$ and, hence, is forcible. So $S_w \subseteq S_{w'} \subseteq S_w$, so $S_w = S_{w'}$. Contradiction.

For (ii), suppose that $S_w \subseteq S_{w'}$. Suppose that $s \in \Delta_w$. Then for each m, sequence $s = (T_{S_1}, \ldots, T_{S_k})$ is forcible starting with $w'|m$. Let $\Delta_e = \Delta_w$. Recall from case (i) that $(T_{S_w}) \in \Delta_w$, so T_{S_w} is forcible starting with e and, hence, $S_e \subseteq S_w$. Since $S_w \subseteq S_{w'}$, choose $e' \geq e$ such that $S_{e'} = S_{w'}$. Since forcibility in $\mathcal{P}_\Gamma$ depends only on presentation of effects, $\Delta_{e'} = \Delta_{w'}$. Hence, $\Delta_e \leq \Delta_{e'}$. Conversely, suppose that $S_w \not\subseteq S_{w'}$, so let effect $a \in S_w \backslash S_{w'}$. Choose e such that $S_e = S_w$, since S_w is finite. Since forcibility in $\mathcal{P}_\Gamma$ depends only on presentation of effects, $\Delta_e = \Delta_w$. Recall from part (i) that $(T_{S_{w'}}) \in \Delta_{w'}$. But $a \in S_e \setminus S_{e'} = S_{w'}$, so Nature cannot force $(T_{S_{w'}})$ at arbitrary $e' \geq e$ such that $\Delta_{e'} = \Delta_{w'}$. Hence, $\Delta_w \not\leq \Delta_{w'}$, completing the proof of (ii).

Let $e \in F$. The mapping $\phi : \Gamma|e \to \Gamma'|e$ defined by:

$$\phi(S_w) = \Delta_w$$

is well-defined, by property (i). To see that ϕ is onto, let $\Delta_w \in \Gamma'|e$. Then $w \in K_\Gamma|e$, so $S_w \in \Gamma|e$, so $\phi(S_w) = \Delta_w$. To see that ϕ is injective, let $S_w \neq S_{w'}$. Without loss of generality, let $a \in S_w \setminus S_{w'}$. Then $(T_{S_w}) \in \Delta_w \setminus \Delta_{w'}$, so $\phi(S_w) = \Delta_w \neq \Delta_{w'} = \phi(S_{w'})$. Finally, ϕ is order-preserving by (ii), so ϕ is the required order-isomorphism. It follows immediately that $c(w, e) = c'(w, e)$ and $c(P, e) = c'(P, e)$. ■

LEMMA 11 (Lower cost bound for arbitrary solutions). *If M is a convergent solution given $e \in F_\Gamma$ and $w \in C_e(n)$, and $b > 0$, then there exists $w' \in C_e(n)$ and there exists path $(S_0, \ldots, S_n) \in \pi_e(*, S_{w'})$ such that*

1. *$S_0 \in (\Gamma|e)_{min}$;*
2. *$S_{w'} = S_w$;*
3. *M produces T_{S_i} at least b times in immediate succession after the end of e (if $n = 0$) or after the end of e_{i-1} (if $n > 0$) in w';*

Proof. Let $e \in F_\Gamma$ and $w \in C_e(n)$. Then there exists a path $p = (S_0, \ldots, S_n) \in \pi_e(*, S_w)$ of length $n+1$ whose length is maximal among paths in $\pi_e(*, S_w)$. Property (1) follows from Lemma 13. Let $w' = (e_n * \emptyset^\infty)$. For property (2), note that by part 2 of Lemma 14, $S_{e_n} = S_n$. By construction, $S_{w'} = S_{e_n}$. Since $p = (S_0, \ldots, S_n) \in \pi_e(*, S_w)$, $S_n = S_w$. So $S_w = S_{w'}$. Property (3) is part 3 of Lemma 14. ■

LEMMA 12 (Upper cost bound for normal Ockham strategies). *Suppose that $e \in F_\Gamma$ and $e \in K_\Gamma|e$ and M is normally Ockham from e onward, where length(e) $= k$. Then for each $n \geq 0$:*

1. *if $c(w, e) \leq n$, then M retracts at most n times in w from stage $k+1$ onward.*
2. *if $c(w, e) = 0$, then M commits no error in w from stage k onward.*

Proof. For statement (1), suppose that M retracts $> n$ times along $w \in K_\Gamma|e$ from stage $k+1$ onward, say at stages $j_0 + 1 < \ldots < j_n + 1$, where $k \leq j_0$. Let index i range between 0 and n. Then there exists $(S_0, \ldots, S_n)$ such that $M(w|j_i) = T_{S_i}$ and $M(w|j_i + 1) \neq T_{S_i}$. Since M is a normal Ockham method, Proposition 3 implies that

$$\begin{array}{lrcl} \text{i.} & (\Gamma|(w|j_i)))_{\min} & = & \{S_i\}; \\ \text{ii.} & (\Gamma|(w|(j_i+1)))_{\min} & \neq & \{S_i\}. \end{array}$$

Also, by part 1 of Lemma 20, there exists $j > j_{n+1}$ such that $(\Gamma|(w|j))_{\min} = \{S_w\}$. Then, by (i) and Lemma 18, $S_0 \subseteq \ldots \subseteq S_n \subseteq S_w$. So by (ii) and Lemma 19, $S_0 \subset \ldots \subset S_n \subset S_w$, so $p = (S_0, \ldots, S_n, S_w) \in \pi_e(*, S_w)$. Observe that length($p$) $= n+2$. So $c(w, e) > n$.
For statement (2), suppose that $c(w, e) = 0$. Then by part 1 of Proposition 2, $S_w \in (\Gamma|e)_{\min}$. Let $k' \geq k$. By part 2 of Lemma 20, $S_w \in (\Gamma|(w|k'))_{\min}$. So by Proposition 3 and the fact that M is Ockham from $w|k$ onward, either $M(w|k) = T_{S_w}$ or $M(w|k) =$ '?', neither of which is an error. ■

LEMMA 13 (Minimal beginning of maximal path). *Suppose $e \in F_\Gamma$ and $S \in \Gamma|e$ and $p \in \pi_e(*, S)$ has maximal length in $\pi_e(*, S)$. Then $p(0) \in (\Gamma|e)_{min}$.*

Proof. Suppose that $p(0) \notin (\Gamma|e)_{\min}$. Since $p \in \pi_e(*, S)$, $p(0) \in \Gamma|e$, so there exists $S' \subset p(0)$ such that $S' \in \Gamma|e$. Then $S' * p \in \pi_e(*, S)$, so p does not have maximal length in $\pi_e(*, S)$. ■

LEMMA 14 (Optimal strategy for nature). *If M is a convergent solution given $e \in F_\Gamma$ and if $w \in C_e(n)$, and $p = (S_0, \ldots, S_n) \in \pi_e(*, S_w)$ and $b > 0$, then there exists sequence $(e_0, \ldots, e_n)$ of elements of $F_\Gamma|e$ such that for each i such that $0 \leq i \leq n$ and for each j such that $0 \leq j < n$:*

1. *$e < e_j < e_{j+1}$;*
2. *$S_{e_i} = S_i$;*
3. *M produces T_{S_i} at least b times in immediate succession in e_i after the end of e (if $n = 0$) or after the end of e_{i-1} along w (if $n > 0$);*
4. *$(e_n * \emptyset^\infty) \in K_\Gamma \cap C_e(n)$.*

Proof by induction on length(p). Base case: $p = ()$. Then the lemma is trivially true, because $() \notin \Pi_e(*, S_w)$. Induction: let $p = (S_0, \ldots, S_n) \in \Pi_e(*, S_w)$. Let $e_{-1} = e$ and let $S_{-1} = S_e$. Let $R_n = S_n \setminus S_{e_{n-1}}$. Let e_n be the least initial segment of $w_n = (e_{n-1} * R_n * \emptyset^\infty)$ extending $e_{n-1} * R_n$ by which M selects theory T_{S_n} at least b times without interruption after the end of e_{n-1}. There exists such an e_n, since M is a convergent solution and $S_{w_n} = S_n \in \Gamma|e$, so $w_n \in K_\Gamma|e$. Properties (1-3) are immediate by construction and by the induction hypothesis. For property (4), observe that $(e_n * \emptyset^\infty) = (e_{n-1} * R_n * \emptyset^\infty) = w_n$. By the induction hypothesis, $S_{w_n} = S_{e_n} = S_{e_{n-1}} \cup R_n = S_{n-1} \cup (S_n \setminus S_{n-1}) = S_n$. So since $(S_0, \ldots, S_n)$ is maximal in $\pi_e(*, S_w) = \pi_e(*, S_n)$, $w_n \in K_\Gamma \cap C_e(n)$. ■

LEMMA 15 (Non-triviality). *Let $e \in F_\Gamma$. Then $(\Gamma|e)_{min} \neq \emptyset$.*

Proof. Since $e \in F_\Gamma$, there exists $w \in K_\Gamma|e$ such that $e < w$. Hence, $S_w \in \Gamma|e$. Since each member of $\Gamma|e$ is finite and $\Gamma|e \neq \emptyset$, let $S' \in \Gamma|e$ be $\subseteq$-minimal in $\Gamma|e$, so $S' \in (\Gamma|e)_{\text{min}}$. ■

LEMMA 16 (Simple alternative world). *Suppose that $e \in F_\Gamma$ and $(\Gamma|e)_{min} \neq \{S\}$. Then there exists $w \in C_e(0)$ such that $S_w \neq S$.*

Proof. Since $(\Gamma|e)_{\text{min}} \neq \{S\}$, Lemma 15 implies that there exists $S' \in (\Gamma|e)_{\text{min}}$ such that $S' \neq S$. Let $w = (e * (S' \setminus S_e) * \emptyset^\infty)$. By construction, $S_w = S' \in (\Gamma|e)_{\text{min}}$ and $w \in K_\Gamma|e$, so $w \in C_e(0)$, by Proposition 2. ■

LEMMA 17 (Simple world existence). *Let $e \in F_\Gamma$. Then there exists $w \in K_\Gamma|e$ such that $c(w, e) = 0$.*

Proof. Let $S \in (\Gamma|e)_{\text{min}}$, by Lemma 15. Let $w' = (e * (S \setminus S_e) * \emptyset^\infty)$. Then $S_w = S \in (\Gamma|e)_{\text{min}}$. So by Proposition 2, $c(w', e) = 0$. ■

LEMMA 18 (Monotonicity). *Suppose that $e, e' \in F_\Gamma$. Then:*

if $e \leq e'$ and $(\Gamma|e)_{min} = \{S\}$ and $(\Gamma|e')_{min} = \{S'\}$, then $S \subseteq S'$.

Proof. Since $S' \in \Gamma|e'$ and $e \leq e'$, $S' \in \Gamma|e$. Since $(\Gamma|e)_{\min} = \{S\}$, S is minimal in $\Gamma|e$ by Proposition 3, so $S \subseteq S'$. ■

LEMMA 19 (Down and out). *Suppose that* $e, e', e'' \in F_\Gamma$. *Then:*

$$\text{if } e < e' \leq e'' \text{ and } (\Gamma|e)_{min} = \{S\} \text{ and } (\Gamma|e')_{min} \neq \{S\}, \text{ then } S \notin \Gamma|e''.$$

Proof. Suppose that $e, e', e'' \in F_\Gamma$ and $e < e' \leq e''$ and $(\Gamma|e)_{\min} = \{S\}$ and $S \in \Gamma|e''$. It suffices to show that $(\Gamma|e')_{\min} = \{S\}$. Since $S \in \Gamma|e''$ and $e < e'$, $S \in \Gamma|e'$. Suppose $S' \in \Gamma|e'$. Then $S' \in \Gamma|e$, since $e < e'$. Since $(\Gamma|e)_{\min} = \{S\}$, Proposition 3 yields that $S \subseteq S'$. So $S' \not\subset S$, so $S \in (\Gamma|e')_{\min}$. Now, suppose that $R \in (\Gamma|e')_{\min}$. Then $R \in \Gamma|e$, since $e < e'$, so by Lemma 3, $S \subseteq R$. But since $R \in (\Gamma|e')_{\min}$, $S \not\subset R$. So $S = R$. So $(\Gamma|e')_{\min} = \{S\}$. ■

LEMMA 20 (Stable arrival). *Suppose that* $w \in K_\Gamma$. *Then*

1. *if* $k \geq \mu(w)$, *then* $(\Gamma|(w|k))_{min} = \{S_w\}$;
2. *if* $S_w \in (\Gamma|(w|i))_{min}$ *and* $i \leq i'$, *then* $S_w \in (\Gamma|(w|i'))_{min}$.

Proof. For (1), let $k \geq \mu(w)$. Then $S_w = S_{w|k}$, so $S_w \in \Gamma|(w|k)$ and for each $R \in \Gamma|(w|k)$, $S_w = S_{w|k} \subseteq R$, so $S_w \in (\Gamma|(w|k))_{\min} = \{S_w\}$. For (2), suppose that $i \leq i'$ and $S_w \in (\Gamma|(w|i))_{\min}$. Note that $S_w \in \Gamma|(w|i')$. Suppose that there exists $R \in \Gamma(w|i')$ such that $R \subset S_w$. Then $R \in \Gamma(w|i)$, so $S_w \notin (\Gamma(w|i))_{\min}$, which is a contradiction. So $S_w \in (\Gamma(w|i'))_{\min}$. ■

EPISTEMIC LOGIC AND INFORMATION UPDATE

Alexandru Baltag, Hans P. van Ditmarsch, and Lawrence S. Moss

1 PROLOGUE

Epistemic logic investigates what agents know or believe about certain factual descriptions of the world, and about each other. It builds on a model of what information is (statically) available in a given system, and isolates general principles concerning knowledge and belief. The information in a system may well change as a result of various changes: events from the outside, observations by the agents, communication between the agents, etc. This requires *information updates.* These have been investigated in computer science via *interpreted systems*; in philosophy and in artificial intelligence their study leads to the area of *belief revision.* A more recent development is called *dynamic epistemic logic.* Dynamic epistemic logic is an extension of epistemic logic with dynamic modal operators for belief change (i.e., information update). It is the focus of our contribution, but its relation to other ways to model dynamics will also be discussed in some detail.

Situating the chapter This chapter works under the assumption that *knowledge* is a *variety of true justifiable belief.* The suggestion that knowledge *is nothing but* true *justified* belief is very old in philosophy, going back to Plato if not further. The picture is that we are faced with alternative "worlds", including perhaps our own world but in addition other worlds. To know something is to observe that it is true of the worlds considered possible. Reasoners adjust their stock of possible worlds to respond to changes internal or external to them, to their reasoning or to facts coming from outside them.

The *identity* of knowledge with true justified (or justifiable) belief has been known to be problematic in light of the Gettier examples (see also our discussion in Section 3.3). Being very short again, the point is that this simple picture ignores the *reasons* that one would change the collection of possibilities considered, and in particular it has nothing to say about an agent who made a good change for bad reasons and thereby ends up "knowing" something in a counter-intuitive way.

However, the work presented in this chapter is in no way dependent on this mistaken identity: while all the forms of knowledge presented here are forms of true justified belief, the converse does not necessarily hold. On the contrary, in all the logics in this chapter that are expressive enough to include *both* knowledge and belief operators, the above-mentioned identity is *provably wrong.*

We have already mentioned that we are interested in the broader concept of *justifiable belief.* This is broader in the sense that we consider an even more ideal agent, someone who reasons perfectly and effortlessly. (Technically, this means that we are going to ignore the fact that the agents we want to model are not *logically omniscient.* So justifiable belief can be regarded as a modeling of logically omniscient agents immune from Gettier-type problems.)

In addition, justifiable belief for us diverges from knowledge in the sense that it need not imply truth. As with Gettier examples, if one accepts and uses misinformation, then whatever conclusions are drawn are in some sense "justified." We postpone a fuller discussion of this point until later, but we wanted to alert the reader who expects us to write justifiable *true* belief for what we study.

Since the topic of the chapter is "epistemic logic", and since we were quick to point out that it is mistaken to identify knowledge with (true) justifiable belief, the reader may well wonder: why are they writing about it?

We claim that the study of justifiable belief is in itself illuminating with respect to the nature of knowledge, even if it fails as a satisfactory proposal for knowledge. This is the main reason why people have worked in the area. The main contributions are technical tools that allow one to make reliable predictions about complicated *epistemic scenarios*, stories about groups of agents which deal with who knows what about whom, etc. We shall go into more detail on what this means in Section 2 just below, and then in Section 5 we shall see how it works in detail.

Getting back to our study overall, one could think of it as a first approximation to the more difficult studies that would be desirable in a formal epistemology. It is like the study of motion on a frictionless plane: it is much easier to study the frictionless case than that of real motion on a real surface, but at the same time the easier work is extremely useful in many situations. We also point out that our subject has two other things going for it. First, it is a completely formalized subject with precise definitions, examples, and results. We know that not all readers will take this to be a virtue, and we have tried hard to introduce the subject in a way that will be friendly to those for whom logic is a foreign language. But we take the formalized nature of the subject to be an attraction, and so we aim to convey its nice results. Second, in recent years the subject has concentrated on two phenomena that are clearly of interest to the project of epistemology and which for the most part are peripheral in older treatments of epistemic logic. These are the *social* and *dynamic* sides of knowledge. The modeling that we present puts these aspects in the center.

At the time of this writing, it seems fair to say that the subject matter of the first part of our chapter, modeling based on justifiable belief, is fairly advanced. There are some open issues to be sure, and also outstanding technical questions. Many of the people involved in the area have gone on to adjacent areas where the insights and technical machinery may be put to use. Two of these are *combinations of logic and game theory* and *belief revision theory.* We are not going to discuss logic and game theory in this chapter, but the last section of our chapter does present proposals on the extension of dynamic epistemic logic to the area of belief revision.

In addition, as the subject moves closer to belief revision, it is able to question the notion of justifiable belief, to develop a more general notion of *conditional belief*, and also to meaningfully distinguish between *various types of "knowledge"* and characterize them in terms of conditional belief. These and similar formal notions should appeal to the epistemologist as well.

Overview Our overarching goal is that this chapter make the case for its subject to the uninitiated. We begin work in Section 2 with discussion of a series of epistemic scenarios. These discussions illustrate the subject of the chapter by example, rather than by a direct discussion. They also are a form of "on-the-job-training" in the kinds of logical languages and representations that will be found later. This leads to a collection of issue areas for the subject that we present briefly in Section 3. Following that, we have some background in logic in Section 4. Even there, we are not only offering a catalog of a million logical systems: we attempt to say *why* the philosophically-minded reader might come to care about technical results on those systems. Dynamic epistemic logic (**DEL**, for short) is our next topic. This is the part of the chapter with the most sustained technical discussion. After this, we end the chapter with our look at belief revision theory. The proposals there are foreshadowed in Section 2, and a reader mainly interested in belief revision probably could read only Sections 2 and 7. It goes without saying that such readers should also read Hans Rott's Chapter 4c on the subject in this handbook. This is also true for readers whose main interest is in epistemology.

All readers would do well to consult Johan van Benthem and Maricarmen Martinez' Chapter 3b in order to situate our work even more broadly in studies of information modeling, especially the contrast and blending of the correlational and proof theoretic stances on the topic. In their terminology, however, the work in this chapter is squarely in the "information as range" paradigm.

The material in this chapter alternates between its main thrust, an examination of the philosophical and conceptual sides of the subject, and the technical aspects. We have chosen to emphasize the philosophical material because it is the subject of this handbook; also, there already are several introductions to the technical material. Historical pointers are mainly to be found at the ends of the sections.

2 INTRODUCTION: LOGICAL LANGUAGES AND REPRESENTATIONS

As with many works in the area, we begin with an *epistemic scenario.* The one here is probably the simplest possible such scenario, an agent ignorant of which of two exclusive alternatives holds.

A person named Amina enters a room and is then shown a closed box on a table. She has been told that box contains a coin, and that the coin lies flat in the box. What she does not know is whether the coin lies heads up or tails up.

Our tasks as modelers are (1) to provide an adequate representation of this scenario; (2) to use the representation as part of a formal account of "knowledge" and related terms; (3) to see where the representation and formal account run into problems; (4) to then "scale up" all of the previous points by considering more complicated scenarios, models, and accounts, with the same goals in (1)–(3).

The most natural representation is simply as a set of two alternatives. In pictures, we have

The two circles are intended as abstract representations of the two states of the coin. There is no significance to the symbols H (for heads) and T (for tails, but please do not confuse it with *truth*). There is also no significance to the fact that the representation has heads on the left and tails on the right. There is a very real significance to the fact that each circle has exactly one symbol. There is *some* significance to the absolutely symmetric treatment of the two alternatives. Perhaps the most important aspect of the representation is that it leaves out everything to do with Amina's state of mind: why she thinks that heads and tails are the only ones possible, her prior experience with similar situations, her emotions, etc. For the most part, the formal work of this chapter will not help with proposals on any of these important matters precisely because the representations abstract from them.

We regard the symbols H and T as *atomic propositions* (We also call them *atomic sentences*, using the two terms interchangeably.) It is problematic at this point to speak of these as true or false in our scenario: since the story was silent on the matter of whether the coin was, in fact, lying heads up or tails up, it is debatable whether there is a "fact of the matter" here or not. No matter how one feels on where there is, in fact, a fact or not, it is less controversial to hold that either the coin lies heads up or tails up, and not both. (Recall that this is part of what Amina has been told at the outset.) It is natural to use standard propositional logic, and to therefore write $\mathsf{H} \leftrightarrow \neg\mathsf{T}$. (We may read this as "heads holds just in case tails fails".) We would like this sentence $\mathsf{H} \leftrightarrow \neg\mathsf{T}$ to come out true on our representation, and so clearly we need a semantics for sentences of this type.

Even before that, we need a formal language. We take propositional logic built over the atomic propositions H and T. So we have examples such as the one we just saw, H and T, and also $\neg\neg\mathsf{H}$, $\mathsf{H} \to (\mathsf{H} \to \mathsf{T})$, $\mathsf{H} \vee \mathsf{T}$, etc. The language is built by recursion in the way all formal languages are. We'll use letters like φ for propositions of this and other languages.

In order to give the semantics, it will be very useful to change the representation a little. We had used H and T inside the circles, but this will get in the way; also, as we shall see many times in this chapter, the states in our representations are not individuated by the atomic facts that they come with. So let us change our representation to

with the additional information that H holds at s and not at t, and T holds at t and not at s. We take this extra piece of information to be part of our representation. So we have a set $\{s,t\}$ of two *(abstract) states* and some extra information about them. The set $\{s,t\}$ has four subsets: $\emptyset$, $\{s\}$, $\{t\}$, and $\{s,t\}$ itself. We also have the usual set theoretic operations of the union of two subsets $(x \cup y)$, intersection $(x \cap y)$, and relative complement $(\overline{x})$. To spell out the details of relative complement in this example: $\overline{\emptyset} = \{x,y\}$, $\overline{\{s\}} = \{t\}$, $\overline{\{t\}} = \{s\}$, and $\overline{\{s,t\}} = \emptyset$.

Now it makes sense to formally interpret our language, assigning a set of states $[\![\varphi]\!]$ to a sentence φ as follows:

$$\begin{array}{lcl@{\qquad}lcl}
[\![\mathsf{H}]\!] & = & \{s\} & [\![\varphi \wedge \psi]\!] & = & [\![\varphi]\!] \cap [\![\psi]\!] \\
[\![\mathsf{T}]\!] & = & \{t\} & [\![\varphi \vee \psi]\!] & = & [\![\varphi]\!] \cup [\![\psi]\!] \\
[\![\neg\varphi]\!] & = & \overline{[\![\varphi]\!]} & [\![\varphi \rightarrow \psi]\!] & = & \overline{[\![\varphi]\!]} \cup [\![\psi]\!] \\
 & & & [\![\varphi \leftrightarrow \psi]\!] & = & ([\![\varphi]\!] \cap [\![\psi]\!]) \cup (\overline{[\![\varphi]\!]} \cap \overline{[\![\psi]\!]})
\end{array}$$

The reader will know that we could have given only a few of these, leaving the rest to re-appear as derived properties rather than the official definition. The choice is immaterial. What counts is that we have a precise definition, and we can verify important properties such as $[\![\mathsf{H} \leftrightarrow \neg\mathsf{T}]\!] = \{s,t\}$. The reason is that

$$([\![\mathsf{H}]\!] \cap [\![\neg\mathsf{T}]\!]) \cup (\overline{[\![\mathsf{H}]\!]} \cap \overline{[\![\neg\mathsf{T}]\!]}) \quad = \quad (\{s\} \cap \overline{\{t\}}) \cup (\overline{\{s\}} \cap \overline{\overline{\{t\}}}) \quad = \quad \{s,t\}.$$

We'll use S to refer to our set of states, both in this discussion and in later ones. And we shall say that φ is *valid in a model* if its semantic interpretation $[\![\varphi]\!]$ is the full set S of states, not merely a proper subset.

We have reliable and consistent intuitions concerning *knowledge*. Surely one feels that upon walking into the room, Amina does not know whether the coin lies heads or tails up: she was informed that there is a coin in the box, but so without further information to the contrary, she should not know which alternative holds. We expand our language by adding a knowledge operator K as a sentence-forming operation, making sentences from sentences the way $\neg$ does. We thus have sentences like $K\mathsf{H}$, $K\neg K\mathsf{T}$, etc. The semantics is then given by

$$[\![K\varphi]\!] \quad = \quad \begin{cases} S & \text{if } [\![\varphi]\!] = S \\ \emptyset & \text{if } [\![\varphi]\!] \neq S \end{cases}$$

Notice that this modeling makes knowledge an "all-or-nothing" affair. One can check that $[\![K\mathsf{H}]\!] = \emptyset$, matching the intuitions that Amina does not know that the coin lies heads up. But also $[\![K\neg\mathsf{H}]\!] = \emptyset$. In contrast, $K(\mathsf{H} \vee \neg\mathsf{H})$ is valid on this semantics: its interpretation is the entire state set.

"Knowing that" and "knowing whether" Up until now, all of our modeling of knowledge is at the level of *knowing that* a given proposition, say φ, is true or false. We have no way of saying *knowing whether* φ holds or not. The easiest way to do this in our setting is to identify *knowing whether* φ with the disjunction *knowing that* φ or *knowing that* $\neg\varphi$. It will turn out that in this example and all of our other ones, this "or" is automatically an exclusive disjunction. That is, our modeling will arrange that no agents know both a sentence and its negation.

Iterated knowledge One should note that our formal semantics gives a determinate truth value to sentences with *iterated knowledge* assertions. For example, $K\neg K\mathsf{H}$ comes out true. (The reason: we saw above that $[\![K\mathsf{H}]\!] = \emptyset$. Therefore $[\![\neg K\mathsf{H}]\!] = \{s,t\}$, and so also $[\![K\neg K\mathsf{H}]\!] = \{s,t\}$.) Translating this back to our original scenario, this means that we are predicting that Amina knows that she doesn't know that the coin lies heads up. For a real person, this *introspectivity* is clearly false in general: though sometimes people can and do introspect about their knowledge, it seems that only a tiny amount of what we talk about people knowing is even susceptible to introspection. However, given our take on knowledge as justifiable belief (and with justifications modeled as surveys of all relevant possibilities), it fits. The upshot is that in the case of iterated knowledge assertions, the kind of modeling that we are doing gives predictions which are at odds with what real people do (though they are the acid test) but seem to work for ideal agents.

Note, however, that justifiable knowledge is a kind of *potential knowledge*: we would not feel that a reasoner who exhibited it was making a mistake. We would be more likely to commend them. Thus, the modeling that uses it is of value in *adversarial situations* of the kind found in game theory. When you reason about your opponent in a game (or war), you should not assume him to be stupid, but on the contrary: the safe option is to assume that he already knows everything that he could possibly know; i.e, to model him as a logically omniscient, fully introspective ideal agent. This is because you want to make sure your strategy works no matter how smart or how resourceful your opponent happens to be. (On the contrary, when reasoning about your own, or your allies', knowledge, it is safer not to idealize it, but to take into account possible failures of introspection.)

2.1 Learning

Suppose next that Amina opens the box and sees that the coin lies heads up. It is natural to assume that after she looks, she *knows* that the coin lies heads up. Furthermore, and in support of this, consider the model

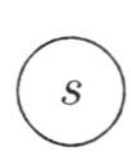

along with the information that s is a state where H is true. This model reflects the intuition that she considers only one state to be possible. Recall from Section 1 that our work here is mainly about knowledge as justifiable belief. It takes knowledge to result from a survey of the possible. The model above reflects this choice: it is a survey of the justifiable possible. (But a one-point model is so simple that it reflects other intuitions as well.)

What interests us most is that we have a *change of model*. In this case, the change was to throw away one state. Taking seriously the idea of change leads to *dynamics*, a key point of our study. Sometimes people with a little exposure to logic, or even a great deal of it, feel that logic is the study of eternal certainties. This is not the case at all. In the kinds of settings we are interested in here, we move from single models to *sequences* of them, or structures of some other kind. The particular kind of structure used would reflect intuitions about time, causality and the like. These matters are largely orthogonal to the epistemic modeling. But the important point for us now is that we can "add a dimension" to our models to reflect *epistemic actions*.

We would like to indicate the whole story as

We think of this as the two representations from before (along with the information that the picture suppresses, that s is a state where the coin lies heads up, and t the same for tails) connected by the dotted arrow. Amina discards the state t because it does not reflect what she learned. But s persists, and the dotted arrow indicates this. As it happens, it will again be confusing to use the same letter s for both the "before" and "after" states. This is not strictly needed here, but later in the paper it will be needed. So we would prefer to illustrate the story as

Again, we would supplement the picture with a description of which states have which coin faces showing: H is true at s and u, while T is true at t. Note that we *can no longer use the "all-or-nothing" notion of knowledge* from the previous section: in the original state s, Amina knows the state cannot be u (since she knows that she doesn't yet know the face of the coin); while in the new state u, Amina knows the state is neither s or t anymore. In other words, Amina *cannot distinguish* between the initial states s and t, but *can* distinguish between them and the new state u. We illustrate this by using lines to represent the agent's *indifference* between two (indistinguishable) possibilities:

The way to read an assertion like "there is a line between s and t" is as follows: Amina is indifferent between the world being as described in s and being described as in t. The agent is unable to tell apart these two descriptions: for all she knows, either of them can be a correct description of the real world. So in s, Amina thinks that the world might be t, or again it might be s itself. (This is not shown in the picture, but it is tacitly assumed.) In u, Amina thinks that u itself is the only possible state, and so she knows there that the coin lies heads up.

But what are these "states"? Since we have been using the word "state" quite a bit, a word is in order on this usage. Our states are the same as *possible worlds* in explanations of modality. That is, one should regard them as theoretical primitives that have an overall use in the modeling. They are abstract objects that we as outsiders use to discuss the examples and to further build a theory. Using them does not involve a commitment to their ontological or psychological reality. There is also a tradition of possible worlds as *(maximal consistent) sets of propositions*, and we also think of Carnap's state descriptions. But we do not follow either of these in our modeling, preferring to keep states as the primitive. It would be possible to change what we do to render states as maximal consistent sets, however, if one took the underlying language to be the full modal language which we are describing, not simply propositional logic. For our states are not individuated by the propositional facts holding in them: as we shall see shortly in (6) below, to separate states one needs the information contained in the arrows.

Knowledge The heart of the matter is the proposal for the semantics of knowledge. Building on our explanation of what the worlds and the lines represent, the idea is that Amina "knows" (in our sense) φ in a world x just in case the following holds: φ is true in all worlds that she cannot tell apart from x. In symbols, the semantics is given by:

$$[\![K\varphi]\!] \quad = \quad \{s : \text{whenever Amina is indifferent between } s \text{ and } t,\ t \in [\![\varphi]\!]\} \tag{1}$$

In other words: we relativize the previous "all-or-nothing" definition to the set of worlds that are indistinguishable from the real one. Using this modified definition, we can see that, in the initial state s of the above model, Amina doesn't know the face of the coin is H, while in the new state u she knows the face is H.

Comparison with probabilistic conditioning It will be useful at this point to note a similarity between the kind of updating that models Amina's learning in this very simple setting and what happens all the time in *probability*. Suppose that we have a *probability space*. This is a set S of *simple events*, together with a *probability* $p : S \to [0,1]$ with the property that $\sum_{s \in S} p_s = 1$. The probability space as a whole is $\mathcal{S} = (S, p)$; that is, the set S together with the function p.

The subsets of S are called *events*. Every event X gets a probability $p(X)$, defined by $p(X) = \sum_{s \in X} p_s$. We should think of S as the states of some background system $\mathcal{S}$. An event X is then like a property of the system, and p_X as the probability that a randomly observation of $\mathcal{S}$ will have the property X. Not only this, but every event X whose probability is nonzero gives a probability space on its own. This time, the space is X, and the probability $p|X$ is given by $(p|X)(s) = p(s)/p(X)$. This formula reflects the re-normalization of the probability p. We'll call this new space $S|X = (X, p|X)$. It is a *subspace of the original* S *called* S *conditioned on* X. The idea is that if we again start with a system $\mathcal{S}$ whose set of states is modeled by the probability space S, and if we obtain additional information to the effect that the background system definitely has the property X, then we should revise our modeling, and instead take use $S|X$:

$$(S,p) \xrightarrow{\text{learning that } X} (X, p|X) \ .$$

The sentences in our formal language are notations for extensional properties (subsets) of the overall state space. Adding new information, say by a direct observation, corresponds to *moving to a subspace*, to changing the representation.

2.2 Another agent enters

Let us go back to our first scenario, where Amina walks into the room in ignorance of whether the coin lies heads or tails up. (We therefore set aside Section 2.1 right above.) Being alone in a room with a closed box is not much fun. Sometime later, her friend Bao walks over. The door is open, and someone tells Bao the state of the coin and does it in a way that makes it clear to Amina that Bao now knows it. At the same time, Bao is in the dark about Amina's knowledge.

The natural representation here uses four states.

$$\boxed{u{:}\mathsf{H}} \overset{b}{\text{——}} \boxed{v{:}\mathsf{H}} \overset{a}{\text{——}} \boxed{w{:}\mathsf{T}} \overset{b}{\text{——}} \boxed{x{:}\mathsf{T}} \qquad (2)$$

Note that now we have *labeled* the indifference lines with the names of the agents. In this case, we have four worlds called u, v, w, and x. The atomic information is also shown, and we have made the picture more compact by eliminating redundant information. (So we intend u to be a world where H is true and T is false, even though the second statement is not explicit.)

As before, the way to read an assertion like "there is a line labeled b between u to v" is as follows: Bao is indifferent in u between the world being as described in u and being described as in v. So in v, Amina thinks that the world might be w, or again it might be v itself. In u, Amina thinks that u itself is the only possible state, and so she knows there that the coin lies heads up.

The world u is the *real world*, and so once we have a formal semantics, we check our intuitions against the formal semantics at u.

We begin, as we did in Section 2, with the propositional language built from symbols H and T. In our current model, we have semantics for them:

$$[\![\mathsf{H}]\!] = \{u, v\} \qquad\qquad [\![\mathsf{T}]\!] = \{w, x\}.$$

We clearly need now two knowledge operators, one for a (Amina) and one for b. We shall use K_a and K_b for those, and we define by taking for each agent the appropriate indifference lines in the definition (1):

$$\begin{array}{lcl} [\![K_a\varphi]\!] & = & \{s : \text{whenever Amina is indifferent between } s \text{ and } t,\ t \in [\![\varphi]\!]\} \\ [\![K_b\varphi]\!] & = & \{s : \text{whenever Bao is indifferent between } s \text{ and } t,\ t \in [\![\varphi]\!]\} \end{array} \quad (3)$$

We can check whether our formal semantics matches our intuitions about our model. The way we do this is by translating a sentence A of English into a sentence φ in our formal language, and evaluating the semantics. We want to be sure that $x \in [\![\varphi]\!]$, where x is the "real" or "actual" world in the model at hand.

Here are some examples:

English	Formal rendering	Semantics
the coin shows heads	H	$\{u, v\}$
a knows the coin shows heads	$K_a\mathsf{H}$	$\emptyset$
a knows whether the coin shows heads	$K_a\mathsf{H} \vee K_a\neg\mathsf{H}$	$\emptyset$
b knows that the coin shows heads	$K_b\mathsf{H}$	$\{u, v\}$
b knows whether the coin shows heads	$K_b\mathsf{H} \vee K_b\neg\mathsf{H}$	$\{u, v, w, x\}$
a knows that b knows whether the coin shows heads	$K_a(K_b\mathsf{H} \vee K_b\neg\mathsf{H})$	$\{u, v, w, x\}$

In our model, u is the "real world". In all of the examples above, the intuitions match the formal work.

But does there have to be a real world? Our representation in (2) and our semantics in (3) did not use a designated real world. So mention of a real world could be dropped. However, doing so would mean that we have less of a way to check our intuitions against the formalism (since our intuitions would be less sharp). But one who doesn't like the idea of a "real world" would then look at all the worlds in a representation, take intuitive stories for them, and then check intuitions in all cases.

Announcements We have already begun to discuss dynamics in connection with these simple scenarios. Here are ways to continue this one. Suppose that Amina and Bao go up and open the box. We again would like to say that the resulting model has a single world, and in that world both agents consider that world to be the only possible one. In a word (picture), we expect

$$\boxed{u{:}\mathsf{H}} \quad (4)$$

However, this is not quite what we get by the world-elimination definition which we have already seen. What we rather get is

$$\boxed{u{:}\mathsf{H}} \overset{b}{\text{———}} \boxed{v{:}\mathsf{H}} \quad (5)$$

(We have dropped the worlds w and x from the picture in (2), and all the lines pertaining to them.)

So we have a question at this point: can we say that the two models, (4) and (5) are equivalent, and can we do so in a principled way? We return to this point at the end of Section 4.4.

As an alternative, suppose someone outside simply shouts out "The coin lies Heads up." Again, on the modeling so far, we have the same state at the end. We thus conclude *our representations cannot distinguish sources of information.*

2.3 Another agent enters, take II

At this point, we want an alternative to the story in Section 2.2. Amina is again in the room in ignorance of the state of the coin. Bao walks over, but this time, the door is shut. Outside the door, some trusted third party says to him, "I'll tell *you* that the coin lies heads up." Then Bao walks in.

We naturally have some intuitions about what is going on. Bao should know that the coin lies heads up, and he should also know that Amina doesn't know whether the coin lies heads up or tails up. What about Amina? What should she think about Bao's knowledge? We don't know enough about her to say for sure, but to be definite, let us assume that she (falsely) believes that Bao is as ignorant as she. Moreover, let us assume that Bao believes *this* about Amina.

Belief Since we have mentioned belief, a comment is in order. We continue to deal with "knowledge" as justifiable belief. This is the only notion at play at the moment. We have used *belief* in the previous paragraph only to emphasize that the proposition believed is actually false.

At this point, we return to our scenario and to a model of these intuitions. The model we have in mind is

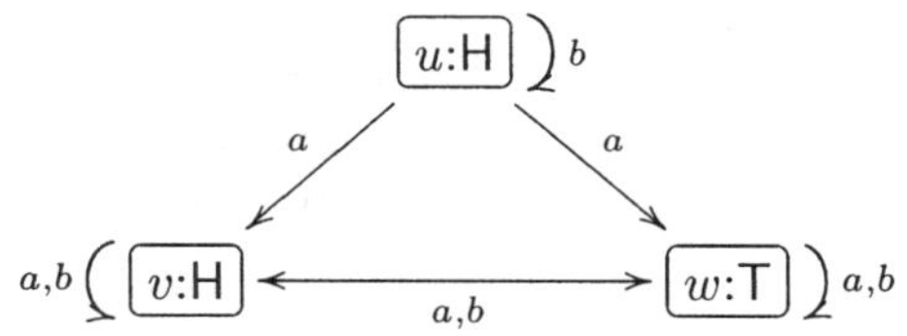

(6)

We shall shortly go into details about why this model works. Before that, a comment: If one tries to come up with a model by hand which reflects correctly the intuitions that we spelled out above, he or she will see that it is not a straightforward task. It is hard to argue that something is not easy to do, so we encourage readers to try it. This will also provide experience in the important task of examining putative models with the eye towards seeing whether they match intuitions or not.

This model generalizes the two-directional lines that we saw above to one-directional arrows. The way to read an assertion like "there is an arrow labeled a

from u to v" is as follows: if the situation were modeled by u, then Amina would be justified in considering v a possibility.

In our example, u is the "real world". In that world, Bao countenances no others. Amina, on the other hand, thinks that the world is either v or w. (But she does not think that the world u itself is even possible. This reflects the intuition that Amina doesn't think it possible that Bao knows the state of the coin.) The worlds she thinks are possible are ones pretty much like the two we saw earlier for her alone, except that we have chosen to put in arrows for the two agents.

Note that u and v in (6) have the same atomic information. However, they are very different because what counts is not just the information "inside" but also the arrows. Now given our explanation of what the epistemic arrows are intended to mean, we can see that there is some circularity here. This is not a pernicious circularity, and the use of the logical language makes this evident. Once we take the representation as merely a *site for the evaluation of the logical language*, the problematic features of the models lose their force. We turn to that evaluation now.

Building on our explanation of what the worlds now represent, we say that a *believes* φ in a world x just in case the following holds: φ is true in all worlds that she would think are possible, if x were the actual world. Formally:

$$\begin{array}{lcl} [\![B_a\varphi]\!] & = & \{s : \text{whenever } s \xrightarrow{a} t,\, t \in [\![\varphi]\!]\} \\ [\![B_b\varphi]\!] & = & \{s : \text{whenever } s \xrightarrow{b} t,\, t \in [\![\varphi]\!]\} \end{array} \tag{7}$$

We again check that our semantics and model are sensible, going via examples.

English	Formal rendering	Semantics
the coin shows heads	H	$\{u, v\}$
a knows (believes) the coin shows heads	$B_a\mathsf{H}$	$\emptyset$
a believes the coin shows tails	$B_a\mathsf{T}$	$\emptyset$
b believes the coin shows heads	$B_b\mathsf{H}$	$\{u\}$
b believes that a doesn't know (believe) it's heads	$B_b\neg B_a\mathsf{H}$	$\{u, v, w\}$
b believes that a believes that b doesn't know (believe) it's heads	$B_bB_a\neg B_b\mathsf{H}$	$\{u, v, w\}$

In our model, u is the "real world". In all of the examples above, the intuitions match the formal work.

Knowledge and Belief What does Amina actually *know* in the scenario above? She believes that Bao doesn't know the face of the coin, but this belief is not true. Is Amina aware of the possibility that her belief is false? Let us assume so: in other words, although she *believes* that Bao does not know the face, she at least countenances the possibility that he does. Note that the states u, v and w are *indistinguishable* for her: she "sees" the same things, and has the same information and the same beliefs in all these states. But then these states also are indistinguishable for her from a *fourth* possibility, namely the one in which Bob knows the face but the coin shows tails. All the information that Amina has is consistent with this fourth possibility, so she cannot exclude it either.

To distinguish between belief and (truthful) knowledge, we supplement what we so far have, in a few ways. First, we need to add to the state space a fourth state, representing the fourth possibility just mentioned. Second, we consider *two models* on the same state set. The one on the left below is intended to model *knowledge*, while the one on the right is intended for *belief*:

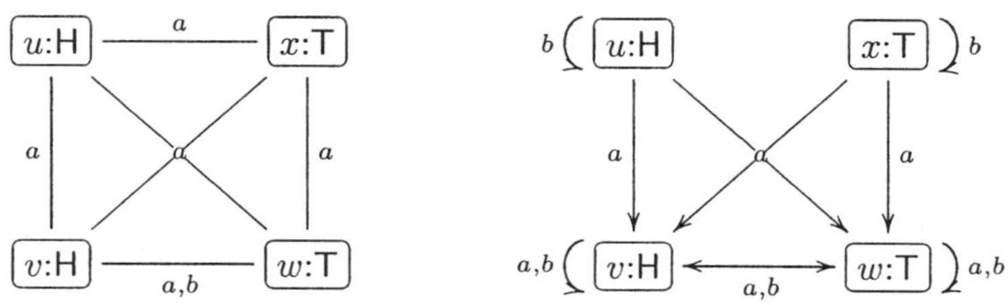

(8)

The real world is u. On the side of knowledge, Amina is indifferent between all the states. The difference between the belief model and the one in (6) is that we have added a fourth state, x. This state is *inaccessible* from u in the belief model, and so it will turn out to be irrelevant from the point of view of the agent's beliefs in the actual world; however, x will be crucial in dealing with knowledge and *conditional belief* in the real world. (Concerning knowledge, leaving x off would mean that Amina in the real world knows there is no world in which Bao knows that the coin is tails up. This would give her knowledge beyond what is justified by our story.) Recall that the lines (as on the left) are the same as two-way arrows. So we can see that all of the arrows in the right diagram (for belief) are also present in the left diagram (for knowledge). This is good: it means that everything the agents know in a model will also be believed by them. To make this precise, we of course need a formal semantics. Let us agree to write $\overset{a}{\sim}$ for the knowledge arrows (indifference lines), and $\overset{a}{\to}$ for the belief ones. In fact, it is natural to consider loops as being implicitly present in the knowledge model, so we put $s \overset{a}{\sim} t$ iff either $s = t$ or there is an indifference line between them. The relevant definitions (stated only for Amina) will then be

$$\begin{array}{lcl} [\![K_a\varphi]\!] & = & \{s : \text{whenever } s \overset{a}{\sim} t,\ t \in [\![\varphi]\!]\} \\ [\![B_a\varphi]\!] & = & \{s : \text{whenever } s \overset{a}{\to} t,\ t \in [\![\varphi]\!]\} \end{array} \tag{9}$$

On this semantics, then, Amina will *believe* that Bao does not know the state of the coin, but also, she does not *know* this.

Observe now that the two models for this situation are not independent: *the belief model contains all the information about the knowledge model.* Indeed, we can recover the knowledge model from the belief model by *closing the belief arrows under reflexivity, symmetry and transitivity.* Visually, this amounts to replacing in the model on the right all the one-way arrows by lines, adding loops everywhere and adding lines between any two states that are connected by a chain of lines. A simpler alternative procedure is to connect any two states by lines if and only if the same states are reachable from both via (one-step) belief arrows. This gives us the knowledge model on the left.

We know that there are issues involved in the translation that we are ignoring. One point worth mentioning is that translating beliefs regarding *conditionals* is problematic (and this is why none of the sentences in the table are conditionals). The reason is that the formal language suggests the material conditional, and so the mismatches between any natural language conditional and the material conditional are highlighted by the process we are suggesting.

2.4 Conditional beliefs

Consider again the belief-knowledge model (8). Where this model goes wrong is in dealing with conditional assertions that are *counterfactual with respect to the agents' beliefs*. Consider the following statements:

1. If Bao knows the state of the coin, then the coin lies either heads up or tails up.
2. If Bao knows the state of the coin, then the coin lies heads up.
3. If Bao knows the state of the coin, then Amina does, too.

We are interested in whether *Amina believes any of these statements*. As such, these are *conditional belief* assertions. Intuitively, she should believe the first statement, but not the second and third. Yet, they are all true on the definition of belief in (9), if we interpret conditional beliefs as beliefs in the conditional and we interpret the conditionals as material conditionals.

In fact, the problem is not simply the use of the material conditional: no other "belief-free" notion of conditional would do either! As argued in e.g. Leitgeb [2007], it is not possible to separate a conditional belief into a doxastic part (the "belief") and a belief-free part (the "conditional" that forms the content of the "belief"). Gärdenfors' Impossibility Theorem[1] can be understood as showing that, under reasonable assumptions, *conditional beliefs are not equivalent to beliefs in conditionals, for any belief-free notion of conditional.* As a consequence, we have to treat conditional belief as one indivisible operator $B_a^\alpha\varphi$ instead of a composed expression $B_a(\alpha \Rightarrow \varphi)$.[2] But the above models are not adequate to give a semantics to this operator. On a technical level, the problem is that to make the sentence *Amina believes that Bao doesn't know* come out true we need a belief model (such as the one above (8)) in which u and x are *not* to be accessible for Amina from u; but at the same time, for evaluating hypothetical statements like the conditionals above, we need to use a different belief model, one in which u and x become accessible from u.

---

[1]See Section 7 and Hans Rott's Chapter 4c in this handbook for.

[2]In Chapter 3b, this expression would be written $B_a(\varphi|\alpha)$.

There are several ways of getting an appropriate model, but all of them involve going beyond simple belief models. We are going to present one common modeling, essentially derived from work of Lewis, Grove, and others. We supplement our model with a *Grove system of spheres*, exactly as discussed and pictured in Section 2.3 of Chapter 4c. We need one for each agent at each state. We'll only spell out what Amina's system looks like at u. It would have v and w in the center, since she is most committed to them. In the next ring, we put u and (crucially) x. As with all such systems of spheres, the definition gives us notions of which worlds are *at least as plausible* as others, and *strictly more plausible* as others. According to the above system of spheres, states v and w are equally plausible for Amina, and they are strictly more plausible than the other two states u, x, which are also themselves equally plausible.

If we draw arrows from any given state to all the states that are at least as plausible as it, we obtain the following diagrammatic representation:

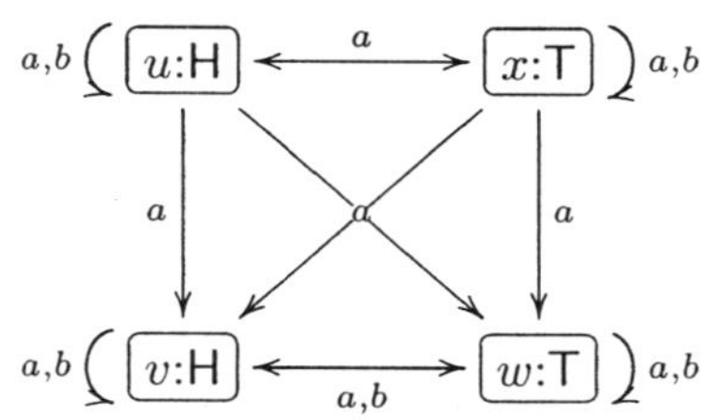

(10)

This is a *plausibility model*: it doesn't directly capture knowledge or beliefs, but only doxastic plausibility. However, *this plausibility model contains all the information about the belief and knowledge models.* Indeed, we can recover the belief model by looking at the *most plausible states* (for each agent); i.e., the states which can be reached via some (one-step) plausibility arrow from any other state that is reachable from them (via a one-step arrow). To obtain the belief model in (8), we keep for each agent only the arrows pointing to that agent's most plausible states. We can then recover the knowledge model from the belief model as in the previous section. Or we can directly obtain the knowledge model in (8) from the above plausibility model, by simply replacing all the arrows by lines (and deleting the loops).

We now *rework the definition of conditional belief.* Actually, we keep track of the antecedent of the conditional in a special way, and write $B_a^\alpha \chi$ to symbolize *Amina believes that were α to be true, χ would have been true as well.* The formal definition is:

$$
\begin{aligned}
[\![B_a^\alpha \chi]\!] \quad = \quad & \{s : t \in [\![\chi]\!], \text{ for all } t \in [\![\alpha]\!] \text{ such that } s \precsim t \\
& \quad \text{and such that there is no } u \in [\![\alpha]\!], \text{ such that } s \precsim u \\
& \quad \text{and } u \text{ is strictly more plausible than } t \text{ for Amina}\}
\end{aligned}
\tag{11}
$$

(And similarly for $B_b^\alpha \chi$, of course.)

The idea is that in evaluating a conditional whose antecedent α contradicts the current beliefs, one has to discard the most plausible worlds, and instead to "fall back" to the worlds that are most plausible given α.

Let us see how this works in our example. For Amina, x is at least as plausible as the real world u. So she should use this world in evaluating conditionals, along with others. This easily shows why sentences 2 and 3 in the beginning of this subsection come out false.

Incidentally, one desirable property of this kind of modeling is that an agent's *knowledge* (as opposed to belief) should not be overridden, even in hypothetical contexts. (This will not be suitable to modeling conditionals which are *counterfactual with respect to knowledge.*) To arrange this, we should require that the union of all spheres for a given agent in a given state coincides with the $\sim$-equivalence class of the agent there.

Modern epistemic logic started to flourish after modal logic (with its roots in Aristotle) was formalized and given a possible world semantics. It is hard to track down the exact origins of this semantics, but it is widely known as Kripke semantics, after Kripke, who devoted a number of early papers to the semantics of modal logic [Kripke, 1959]. A contemporary and thorough reference for modal logic is the monograph [Blackburn *et al.*, 2001].

Observations on how epistemic states change as a result of new information have been around since the Hintikka founded the field of epistemic logic in his 1962 book *Knowledge and Belief* [Hintikka, 1962] (republished in 2005 by College Publications, London). Hintikka is broadly acknowledged as the father of modern epistemic logic, and his book is cited as the principal historical reference. Hintikka himself thinks that von Wright [1951] deserves these credits.

From the late 1970s, epistemic logic became subject of study or applied in the areas of artificial intelligence (as in R.C. Moore's early work [1977] on reasoning about actions and knowledge), philosophy (as in Hintikka's [1986]), and game theory (e.g. Aumann [1976]). In the 1980s, computer scientists became interested in epistemic logic. In fact, the field matured a lot by a large stream of publications by Fagin, Halpern, Moses and Vardi. Their important textbook *Reasoning about Knowledge* [Fagin *et al.*, 1995] which appeared in 1995, contains the contents of many papers co-authored by (subsets of) them over a period of more than ten years. Another important textbook in both 'pure' and 'applied' epistemic logic is Meyer and van der Hoek [1995]. These both should be consulted in connection with Sections 1–4 of this chapter. The work from Section 5 onward (on dynamic epistemic logic and its extensions) is mainly newer than those books. A brief, but very good, introduction to the history, the philosophical importance and some of the technical aspects of epistemic logic is the chapter "Epistemic Logic", by Gochet and Gribomont, in the *Handbook of History of Logic* [2006]. It also gives a very brief look at some of the older work in dynamic epistemic logic.

At the same time as computer scientists became interested in the topic, linguistic semanticists were also discovering many of the basic issues, such as effects of public announcements and the problem of belief revision. Of special mention here is the long work of Robert Stalnaker, whose longstanding involvements with knowledge and belief revision theory include publications such as [Stalnaker, 1968; Stalnaker, 1996; Stalnaker, 2006].

3 FURTHER ISSUES AND AREAS IN EPISTEMIC LOGIC

At this point, we have said a little about the subject matter of the chapter. Before we go further, we mention a few issues, problems, puzzles, and discussion topics in the area. We especially comment on how they relate to the examples in the previous sections.

3.1 The Muddy children

Perhaps the most common of epistemic puzzles is one known in various guises and under names like *the muddy children, the wise men* or *the unfaithful spouses* [Gamow and Stern, 1958; Moses *et al.*, 1986]. Here is one version of it. A number of children have been playing outside. After some time, some of them might have mud on their foreheads; however, they don't discuss this with one another. But along comes one of their fathers, and says:

"At least one of you has mud on his/her forehead. Do you know if you are muddy?"

Let n be the number of who have muddy foreheads. If $n = 1$, the one muddy one sees the clean heads of the others and deduce that she herself is muddy.

Otherwise, all reply (in unison) "No, I don't know." At this point, the father again asks the same question. If $n = 2$, the two muddy ones would know see each other and know that $n \geq 1$, simply because the other did not answer Yes the first time. So they would know on the second questioning.

The story continues in this way. The mathematical upshot is that if there are n muddy children to start, then after the father asks his question n times, the muddy ones will know their status; and before the nth time, nobody will know it. The essential feature for us of the story is that it illustrates that *statements about ignorance can lead to knowledge.*

Comparing to the content of this chapter, it is not hard to draw the representations for this problem and for variations, and to watch the process of announcement (as we have seen it in Section 2.2). Indeed, we have found these to be excellent sources of exercises in modal logic. We shall see the formal logical systems related scenarios like this.

Incidentally, we saw above that statements can change an agent's knowledge. It is even possible to find a setting where an agent can believe something at the outset, then someone else's statement causes them to lose this belief, and then a third statement to regain it. We present an example in Section 6.1.

3.2 Logical omniscience

Logical omniscience is the phenomenon whereby an agent's beliefs or knowledge are modeled in such a way that they are closed under logical deduction. So the agent knows (or believes) all the consequences of their knowledge, and in particular knows infinitely many sentences, theorems whose length or complexity are absurdly great, etc. Logical omniscience is thus a *complaint* against all of the kinds of models we are considering in this chapter. To avoid the complaint, one must adopt much more fine-grained models. A number of different such models have been proposed: a logic of *awareness* ([Levesque, 1984], further extended by Fagin and Halpern), multi-valued epistemic logic (A. Wisniewski [1998]), doxastic linear logic (D'Agostino, Gabbay and Russo [1997]), resource-bounded belief revision (R. Wassermann [1999; 2000]) etc. A solution using a new type of dynamic-epistemic logic was proposed by Ho Ngoc Duc [2001].

3.3 The Gettier challenge

Gettier [1963] pointed out examples that effectively jettisoned the justified true belief analysis of knowledge. The ensuing discussions are central to modern epistemology. For an overview of the area, see, e.g., Steup [Spring 2006].

Perhaps the easiest related example in epistemic logic is the following. Consider a muddy children scenario with two children, say A and B. A is muddy and B clean. A parent announces that at least one is muddy, asks if the two know their state. Usually, A would announce affirmatively and B negatively, but this time let A lie and say that she does not; B of course truthfully replies that he doesn't know. Then on second round, both announce that they do know. The point is, that B's announcement is truthful: taking knowledge to be justifiable true belief, he will have some knowledge of his state after hearing A once, no matter what she says. B's announcement is also justified, being based on A's first declaration. At that point, B has a justified true belief that he knows his state. But we would not judge B to actually know whether he is dirty or not. This would mean either knowing that he is dirty, or knowing that he is clean: he thinks he knows the former and denies he knows the latter.

3.4 Other notions of knowledge

The Gettier examples have been used, among other things, to deny the validity of the Negative Introspection axiom for knowledge: in the example in Section 3.3, *B thinks that he knows* his state, but intuitively speaking we can't agree that he actually knows it. So agents may not know something, while believing that they know it.

Various authors proposed dropping the characteristic $S5$ axiom (Negative Introspection), and sticking with the system $S4$ instead. For instance, the $S4$ principles were as far as Hintikka [1962] was willing to go. This may also be appropriate in an *intuitionistic context*, and also fit well with a *topological interpretation* of knowledge. For other work on appropriate axioms, see Lenzen [1978; 2003].

The defeasibility analysis of knowledge We only look here at one of the alternative proposals for a knowledge concept, that fits well with our discussion of conditional beliefs in Section 2.4. This is the "defeasibility strategy", followed by many of those who attempted to respond to Gettier's challenge, see e.g Lehrer and

Paxson [1968], Swain [1974], Stalnaker [1996; 2006]. To quote Stalnaker [2006], "the idea was that the fourth condition (to be added to justified true belief) should be a requirement that there would be no 'defeater' - no true proposition that, if the knower learned that it was true, would lead her to give up the belief, or to be no longer justified in holding it". One way to do this is to add to the semantics of belief a theory of *belief revision*, and then define knowledge as belief that is stable under any potential revision by a true piece of information. But as we shall see, *conditional beliefs* and plausibility models, introduced in Section 2.4, give us a semantics for belief revision. So it is not surprising that defeasible knowledge was formalized using a logic of conditional beliefs, as in [Board, 2004] and [Baltag and Smets, 2006c], or a logic of conditionals [Stalnaker, 1996].

Knowledge and "safe belief" However, the notion of knowledge defined on plausibility models in Section 2.4 is *stronger* than the one of (true, justifiable) defeasible belief. As we shall see, it corresponds to a belief that is "absolutely unrevisable": it cannot even be defeated by revising with *false* information. Since we followed the common usage in Computer Science and called "knowledge" this strong, absolute notion, we shall follow Baltag and Smets [2006b; 2006c] and call *safe belief* the weaker notion resulting from the defeasibility analysis. In [Stalnaker, 1996; Baltag and Smets, 2006b; Baltag and Smets, 2006c], this concept is applied to reasoning about solution concepts in Game Theory.

3.5 Moore sentences

By a *Moore sentence* we mean one of the form 'p is true and I don't believe that', or 'p is true and I don't know that'. Moore's "paradox" is that such a sentence may well happen to be *true*, but it can never be *truthfully asserted*: a person uttering this sentence *cannot believe it*. As this example crops up in very different settings, and as it is so crucial for a proper understanding of dynamic epistemics, we discuss its origin in some detail, as a proper historical primer to the subject area. In this discussion, $B\varphi$ means "I believe φ " and $K\varphi$ means "I know φ ".

Moore writes that if I assert a proposition φ, I *express* or *imply* that I *think* or *know* φ, in other words I express $B\varphi$ or $K\varphi$. But φ cannot be said to *mean* $B\varphi$ [Moore, 1912, p.77] as this would cause, by substitution, an infinite sequence $BB\varphi$, $BBB\varphi$, ad infinitum. "But thus to believe that somebody believes, that somebody believes, that somebody believes ... quite indefinitely, without *ever* coming to anything which is what is believed, is to believe nothing at all" [Moore, 1912, p.77]. Moore does not state in [Moore, 1912] (to our knowledge) that $\varphi \wedge \neg B\varphi$ cannot be believed. In Moore's "A reply to my critics", a chapter in the 'Library of Living Philosophers' volume dedicated to him, he writes " 'I went to the pictures last Tuesday, but I don't believe that I did' is a perfectly absurd thing to say, although *what* is asserted is something which is perfectly possibly logically" [Moore, 1942, p.543]. The absurdity follows from the implicature 'asserting φ implies $B\varphi$' pointed out in [Moore, 1912]. In other words, $B(p \wedge \neg Bp)$ is 'absurd' for the example of factual information p. As far as we know, this is the first full-blown occurrence of a Moore-sentence. Then in [Moore, 1944, p.204] Moore writes " 'I believe he has gone out, but he has not' is absurd. This, though absurd, is not self-contradictory; for it may quite well be true."

Hintikka [1962] mentions the so-called 'Moore'-problem about the *inadequacy of information updates with such sentences.* This leads us to an interesting further development of this notion, due to Gerbrandy [1999], van Benthem [2004] and others. This development, addressed in our contribution, firstly puts Moore-sentences in a *multi-agent* perspective of announcements of the form 'I say to you that: p is true and that *you* don't believe that', and, secondly, puts Moore-sentences in a *dynamic* perspective of announcements that cannot be believed after being announced. This analysis goes beyond Moore and makes essential use of the tools of dynamic epistemic logic. The dynamic point of view asks how an agent can possibly *come to believe* (or know) that a Moore sentence φ is true. The only way to achieve this seems to be by *learning* φ, or by learning some other sentence that implies φ. But one can easily see that, when φ is a Moore sentence, *the action of learning it changes its truth value*: the sentence becomes false after being learned, though it may have been true before the learning! The same applies to any sentence that implies φ. In terms of [Gerbrandy, 1999], an update with a Moore sentence can never be "successful": indeed, in Section 5.2, a *successful formula* is defined as one that is *always* true after being announced. Observe that Moore sentences have the opposite property: they are "strongly un-successful", in the sense that they are *always* false after being announced. As a consequence, they are *known* to be un-successful: once their truth is announced, their negation is known to be true. Van Benthem [2004] calls such sentences *self-refuting.*

There is nothing inherently paradoxical about these properties of Moore sentences: the "world" that a Moore sentence is talking about is not simply the world of facts, but a "world" that comprises the agent's own beliefs and knowledge. In *this* sense, *the world is always changed by our changes of belief.* Far from being paradoxical, these phenomena can in fact be formalized within a consistent logic, using e.g. the logic of public announcements in Section 5.1: using the notation introduced there, $!\varphi$ is the action of learning (or being announced) φ. If φ is a Moore sentence of the form $p \wedge \neg Kp$, it is easy to check the validity of the dynamic logic formulas $[!\varphi]\neg\varphi$ and $[!\varphi]K\neg\varphi$. The first says that Moore sentences are strongly un-successful; the second says that Moore sentences are self-refuting. As argued in [van Benthem, 2004], self-refuting sentences are essentially *un-learnable.* This explains why a Moore sentence can never be known or believed: because it can never be learned.

A similar analysis applies to the doxastic versions $p \wedge \neg Bp$ of Moore sentences. But, in the case of belief, this phenomenon has even more far-reaching consequences: as pointed out by van Ditmarsch [2005] and others, the un-successfulness of Moore sentences shows that the standard **AGM** postulates for belief revision (in particular, the "Success" postulate) cannot accommodate higher-order beliefs. This observation leads to the distinction, made in the dynamic-epistemic literature [van Benthem, 2006; Baltag and Smets, 2006a; Baltag and Smets, 2007b], between "static" and "dynamic" belief revision. As shown in Section 7.2, in the presence of higher-order beliefs the **AGM** postulates (even in their multi-agent and "knowledge-friendly" versions) apply only to *static* belief revision.

3.6 The Knower paradox

Related to the phenomenon of Moore-sentences is what comes under the name of 'paradox of the knower', also known as Fitch's paradox [Brogaard and Salerno, 2004]. The general verification thesis states that *everything that is true can be known* to an agent; formally, if we introduce a modal possibility $\Diamond\varphi$ to express the fact that something *can* be achieved (by an agent), this says that the implication $\varphi \to \Diamond K\varphi$ is true (in our world), for all formulas φ. The following argument, due to Fitch, appears to provide a refutation of verificationism on purely logical grounds. Take a true Moore sentence φ, having the form $\psi \wedge \neg K\psi$. By the "verificationist" implication above, $\Diamond K\varphi$ must be true. But then $K\varphi$ must be true at some possible world (or some possible future "stage", achievable by the agent). But, as we have already seen in the previous subsection, this is impossible for Moore sentences: $K(\psi \wedge \neg K\psi)$ is inconsistent, according to the usual laws of epistemic logic. The only possible way out is to conclude that *there are no true Moore sentences*; in other words, the implication $\psi \to K\psi$ holds for all formulas. This simply trivializes the verificationist principle, by collapsing the distinction between truth and knowledge: all truths are already known!

Numerous solutions for this paradox have been proposed; see [Wansing, 2002; Tennant, 2002], for example. In particular, Tennant [2002] argues persuasively that the verificationist principle should be weakened, by restricting its intended applications only to those sentences φ for which $K\varphi$ is consistent. In other words: if φ is true and if it is logically consistent to know φ, then φ can be known. This excludes the principle's application to Moore sentences of the usual type.

An interesting take on this matter is proposed by van Benthem in [2004]: one can interpret the modal possibility operator $\Diamond\varphi$ above in a dynamic sense, namely as the 'ability' to achieve φ by performing some learning action, e.g. an announcement in the technical sense of Section 5, to follow. In other words, 'φ is knowable' is identified with 'a true announcement can be made after which the agent knows φ.' In this interpretation, the above-mentioned verificationist thesis reads "*what is true may come to be known* (after some learning)", while its Tennant version restricts this to sentences φ such that $K\varphi$ is consistent. The Moore-sentences are obviously unknowable (by the agent to whose knowledge they refer). But van Benthem [2004] shows that this interpretation is also incompatible with Tennant's weakened verificationist principle: in other words, there are sentences φ such that $K\varphi$ is consistent but still, $\varphi \to \Diamond K\varphi$ does not hold. A counterexample is the formula $p \wedge \neg Kq$. The dynamic epistemic logic of the 'ability' modality $\Diamond$ is completely axiomatized and thoroughly studied in [Balbiani *et al.*, 2007].

3.7 The Hangman paradox

The Hangman paradox, also known as the Surprise Examination paradox, has a relatively short history of about sixty years. Apparently the Swedish mathematician Lennart Ekbom heard a message on the radio during the second world war announcing a civil defense exercise, which was to take place in the next week. It was also announced that this exercise would be a surprise. Then he noticed that there was something paradoxical about this announcement. [Kvanvig, 1998; Sorensen, 1988, pp.253]. The paradox was first published by O'Conner in 1948.

> Consider the following case. The military commander of a certain camp announces on a Saturday evening that during the following week there will be a "Class A blackout". The date and time of the exercise are not prescribed because a "Class A blackout" is defined in the announcement as an exercise which the participants cannot know is going to take place prior to 6.00 p.m. on the evening in which it occurs. It is easy to see that it follows that the exercise cannot take place at all. It cannot take place on Saturday because if it has not occurred on the first six days of the week it must occur on the last. And the fact that the participants can know this violates the condition which defines it. Similarly, because it cannot take place on Saturday, it cannot take place on Friday either, because when Saturday is eliminated Friday is the last available day and is, therefore, invalidated for the same reason as Saturday. And by similar arguments, Thursday, Wednesday, etc., back to Sunday are eliminated in turn, so that the exercise cannot take place at all. [O'Connor, 1948]

Many authors proposed various solutions to this paradox. Williamson [2000] analyzes it as an epistemic variety of the Sorites paradox. The first analysis that uses dynamic epistemic logic was presented in [Gerbrandy, 1999], and found its final form in [van Ditmarsch and Kooi, 2006] and [Gerbrandy, 2007]. According to Gerbrandy, the commander's statement is ambiguous between two possible readings of what "Class A" means: the first reads "You will not know (before 6:00 on the evening of the blackout) when the blackout will take place, *given* (the information you have in) *the situation as it is at the present moment*", while the second reads "You will not know (before 6:00 of that evening) when it will take place, *even after you hear my announcement.*" Gerbrandy chooses the first reading, and shows that in fact there is no paradox in this interpretation, but only a Moore-type sentence: in this reading, the property of being "Class A" cannot remain true after the announcement. Unlike the previous puzzles however, there is also a more complex temporal aspect that needs to be modeled by sequences of such announcements. As for the second reading, Gerbrandy considers it to be genuinely paradoxical, similar to more standard self-referential statements, such as the Liar Paradox.

3.8 Common knowledge

One of the important concepts involved in the study of social knowledge is that of *common knowledge*. The idea is that common knowledge of a fact by a group of people is more than just the individual knowledge of the group members. This would be called *mutual knowledge*. Common knowledge is something more – what, exactly, is an issue, as is how to model it in the kinds of models we are dealing with.

Countries differ as to which side of the road one drives a car; the matter is one of social and legal convention. As it happens, at the time of this writing all three co-authors are living in countries in which people drive on the left. Suppose that in one of those, the government decides to change the driving side. But suppose that the change is made in a quiet way, so that only one person in the country, say Silvanos, finds out about it. After this, what should Silvanos do? From the point of view of safety, it is clear that he should not obey the law: since others will be disobeying it, he puts his life at risk. Suppose further that the next day the government decides to make an announcement to the press that the law was changed. What should happen now? The streets are more dangerous and more unsure this day, because many people will still not know about the change. Even the ones that have heard about it will be hesitant to change, since they do not know whether the other drivers know or not. Eventually, after further announcements, we reach a state where:

The law says *drive on the right* and everyone knows (12). (12)

Note that (12) is a circular statement. The key point is not that everyone know what the law says, but that they in addition know *this very fact*, the content of the sentence you are reading.

This is an intuitive conception of common knowledge. Obviously it builds on the notion of knowledge, and since there are differing accounts of knowledge there will be alternative presentations of common knowledge. When we turn to the logical formalism that we'll use in the rest of this chapter, the alternative presentations mostly collapse. The key here is to unwind a sentence like (12) into an infinite list:

α_0: The law says *drive on the right.*

α_1: α_0, and everyone knows α_0.

α_2: α_1, and everyone knows α_1.

...

Each of these sentences uses knowledge rather than common knowledge. Each also implies its predecessors. Taking the *infinite conjunction*

$$\alpha_0 \wedge \alpha_1 \wedge \alpha_2 \wedge \cdots \tag{13}$$

we arrive at a different proposal for common knowledge. As it happens, given the kind of modeling that we are doing in this chapter, the *fixed point account* in (12) and the *infinite iteration account* in (13) agree.

History An early paper on common knowledge is Friedell [1969]. This paper contains many insights, both mathematical and social. In view of this, it is even more noteworthy that the paper is not common knowledge for people in the field. Probably the first commonly-read source in the area is David Lewis' 'Convention' [Lewis, 1969]. Heal's 1978 paper [Heal, 1978] came a decade later and is still a good source of examples. In the area of game theory, Aumann's [1976] gives one of the first formalizations of common knowledge. McCarthy formalizes common knowledge in a rather off-hand way when solving a well-known epistemic riddle, the Sum and Product-riddle [McCarthy, 1990] (although at the time it was unknown to him that this riddle originated with the Dutch topologist Freudenthal [1969]) as an abstract means towards solving the Sum and Product-riddle. McCarthy's work dates from the seventies but was only published later in a collection of his work that appeared in 1990.

Concerning the formalizations, here is how matters stand in two textbooks on epistemic logic: Fagin et al. [1995] defines common knowledge by transitive closure, whereas Meyer and van der Hoek [1995] define it by reflexive transitive closure. There is a resurgence of interest in variants of the notion, e.g., Artemov's evidence-based common knowledge, also known as justified common knowledge [Artemov, 2004]. Another interesting variant is *relativized (or conditional) common knowledge*, which came to play an important role in some recent developments of dynamic epistemic logic [van Benthem *et al.*, 2006b; Kooi, 2007].

4 EPISTEMIC LOGIC: WHAT AND WHY?

We have introduced logical systems along with our representations in Section 2. We have presented a set of logical languages and some semantics for them. One of the first things one wants to do with a language and a semantics is to propose one or another notion of *valid* sentences; informally, these are the sentences true in all intended models. In all the settings of this paper, this information on validity includes the even more useful information of which sentences semantically imply which others. Then one wants to present a *proof system* for the valid sentences. There are several reasons why one would want an axiomatic system in the first place. One might wish to compare alternative presentations of the same system. One might also want to "boil a system down" to its essentials, and this, too, is accomplished by the study of axiomatic systems. Finally, one might study a system in order to see whether it would be feasible (or even possible) for a computer to use the system. We are not going to pursue this last point in our chapter, but we instead emphasize the "boiling down" aspect of *logical completeness results.*

4.1 Logic for ignorance of two alternatives

We return to our earliest scenario of Section 2, one person in a room with a concealed coin. We have a language including a knowledge operator K and a semantics for it using just one model. We write $\models \varphi$ to say that φ is true in that

all sentential validities	
$\mathsf{H} \leftrightarrow \neg\mathsf{T}$	exclusivity
$\neg K\mathsf{H}, \neg K\mathsf{T}$	basic ignorance axioms
$K\neg\varphi \rightarrow \neg K\varphi$	consistency of knowledge
$K(\varphi \rightarrow \psi) \rightarrow (K\varphi \rightarrow K\psi)$	distribution
$K\varphi \rightarrow \varphi$	veracity
$K\varphi \rightarrow KK\varphi$	positive introspection
$\neg K\varphi \rightarrow K\neg K\varphi$	negative introspection
From φ and $\varphi \rightarrow \psi$, infer ψ	modus ponens
From φ, infer $K\varphi$	necessitation

Figure 1. A logical system for valid sentences concerning two exclusive alternatives and a very simple semantics of K. The axioms are on top, the rules of inference below.

model. The object of our logical system is to give an alternative characterization of the true sentences.

Figure 1 contains our logical system truth. This paper is not the place to learn about logical systems in a detailed and deep way, but in the interests of keeping the interest of philosophers who may not know or remember the basics of logic, we do hope to provide a refresher course.

We say that φ is provable in our system if there is a sequence of sentences each of which is either an axiom or follows from previous sentences in the sequence by using one of the two rules of inference, and which ends in φ. In this case, one would often write $\vdash \varphi$, but to keep things simple in this chapter we are not going to use this notation.

Here is a simple deduction in the logic, showing that $K\neg K\neg\mathsf{T}$ is derivable.

1.	$\mathsf{H} \leftrightarrow \neg\mathsf{T}$	7.	$\neg K\mathsf{H}$
2.	$(\mathsf{H} \leftrightarrow \neg\mathsf{T}) \rightarrow (\neg\mathsf{T} \rightarrow \mathsf{H})$	8.	$\neg K\mathsf{H} \rightarrow ((K\neg\mathsf{T} \rightarrow K\mathsf{H}) \rightarrow \neg K\neg\mathsf{T})$
3.	$\neg\mathsf{T} \rightarrow \mathsf{H}$	9.	$(K\neg\mathsf{T} \rightarrow K\mathsf{H}) \rightarrow \neg K\neg\mathsf{T}$
4.	$K(\neg\mathsf{T} \rightarrow \mathsf{H})$	10.	$\neg K\neg\mathsf{T}$
5.	$K(\neg\mathsf{T} \rightarrow \mathsf{H}) \rightarrow (K\neg\mathsf{T} \rightarrow K\mathsf{H})$	11.	$\neg K\neg\mathsf{T} \rightarrow K\neg K\neg\mathsf{T}$
6.	$K\neg\mathsf{T} \rightarrow K\mathsf{H}$	12.	$K\neg K\neg\mathsf{T}$

Line 1 is our exclusivity axiom and line 7 the basic ignorance axiom. A distribution axiom is found in line 5, and negative introspection in 11. This deduction uses propositional tautologies in lines 2 and 8, modus ponens in 3, 6, 9, 10, and 12, and necessitation in 4.

We mentioned before that there are several different reasons why one would construct a logical system to go along with a particular semantics. The first, perhaps, is that by *formulating sound principles, one uncovers (or highlights) hidden assumptions.* In this case, we can see exactly what the assumptions are in this example: they are the principles in the figure. The interesting point is that these assumptions are *all there is*: if one reasons with the system as above, then they will obtain *all* the truths.

PROPOSITION 1. *The logical system above is* sound and complete*:* $\vdash \varphi$ *iff* φ *is true in the model.*

The point of this completeness theorem is that we have isolated all the assumptions in the scenario.

One remark which is of only minor significance in this discussion is that the veracity axioms $K\varphi \to \varphi$ may be dropped from the system. That is, all instances of them are provable anyway from the other axioms. The reason for including them is that they will be needed in all of the future systems.

Recall that the system here is based on our discussion at the beginning of Section 2. We then went on in Section 2.1 to the situation after Amina looks. It is not hard to re-work the logical system from Figure 1 to handle this second situation. We need only discard the basic ignorance axioms $\neg K\mathsf{H}$ and $\neg K\mathsf{T}$, and instead take $K\mathsf{H}$ so that we also get H. In particular, all of the sound principles that we noted for the earlier situation continue to be sound in the new one.

4.2 Logic can change the world

There is another intuition about knowledge pertinent to the simple scenario that we have been dealing with. It is that *what Amina knows about a coin in a closed box is the same as what she would know if the box were flipped over.* In this section, we show what this means.

We consider the same language as before, except we add an operator *flip* to indicate the flipping the box over. For the semantics, let us begin with two models on the same state set.

1. M, a model with two states s and t, with the information that H is true at s and false at t, and T is true at t and false at s.

2. N, a model with two states s and t, with the information that H is true at t and false at s, and T is true at s and false at t.

Then we define $M, u \models \varphi$ and $N, u \models \varphi$ in tandem, the main points being that

$$\begin{array}{lcl} M, u \models \mathit{flip}\,\varphi & \text{iff} & N, u \models \varphi \\ N, u \models \mathit{flip}\,\varphi & \text{iff} & M, u \models \varphi \end{array}$$

Finally, we say that φ is *valid* if it holds at both states in both models. (This turns out to be the same as φ holding in any one state in either model.)

We present a logical system for validity in Figure 2. Perhaps the first exercise on it would be to prove the other flipping property: $\mathit{flip}\,\mathsf{T} \leftrightarrow \mathsf{H}$. We did not include in the figure the general principles of knowledge that we have already seen, but we intend them as part of the system. (These are the consistency, distribution, veracity, and introspection axioms; and the necessitation rule.) Note that the

$\mathit{flip}\,\mathsf{H} \leftrightarrow \mathsf{T}$	flipping
$\varphi \leftrightarrow \mathit{flip}\,\mathit{flip}\,\varphi$	involution
$\mathit{flip}\,\neg\varphi \leftrightarrow \neg\mathit{flip}\,\varphi$	determinacy
$\mathit{flip}\,(\varphi \to \psi) \to (\mathit{flip}\,\varphi \to \mathit{flip}\,\psi)$	normality
$K\varphi \leftrightarrow \mathit{flip}\,K\varphi$	invariance
From φ, infer $\mathit{flip}\,\varphi$	necessitation

Figure 2. The logical system for knowledge and flipping. We also need everything from Figure 1, except the basic ignorance axioms (these are derivable).

invariance axiom $K\varphi \leftrightarrow \mathit{flip}\,K\varphi$ is exactly the opening of our discussion. We refrain from presenting the completeness proof for this system, but it does hold. One thing to note is that the invariance axiom of this system makes the earlier ignorance axioms $\neg K\mathsf{H}$ and $\neg K\mathsf{T}$ unnecessary: they are derivable in this system.

4.3 Modal logics of single-agent knowledge or belief

At this point, we review the general topic of logics of knowledge. The basic language begins with a set P of atomic propositions. From these sets, a language $\mathcal{L}$ is built from the atomic propositions using the connectives of classical logic and also the knowledge operator K. We get a language which is basically the same as what we saw in Section 2, except that our set of atomic propositions is taken to be arbitrary, not just $\{\mathsf{H}, \mathsf{T}\}$.

Semantics We interpret this language on *relational models*. These are tuples $M = \langle S, R, V \rangle$ consisting of a *domain* S of *states* (or 'worlds'), an *accessibility relation* $R \subseteq S \times S$, and a *valuation* (function) $V : P \to \mathcal{P}(S)$. We usually write V_p instead of $V(p)$. We also write $s \to t$ instead of $R(s,t)$. We call these semantic objects *relational models*, but they are more often called *Kripke models*. We call a tuple of the form (M, s), where M is a model and s is a state in it, is an *epistemic state*.

We then define the interpretation of each sentence φ on an epistemic state (suppressing the name of the underlying model):

$$\begin{array}{lcl} [\![p]\!] & = & V_p \\ [\![\neg\varphi]\!] & = & \overline{[\![\varphi]\!]} \\ [\![\varphi \wedge \psi]\!] & = & [\![\varphi]\!] \cap [\![\psi]\!] \\ [\![K\varphi]\!] & = & \{s : \text{whenever } s \to t,\ t \in [\![\varphi]\!]\} \end{array}$$

It is more common to find this kind of definition "re-packaged" to give the interpretation of a sentence *at a given point*. This rephrasing would be presented as follows:

$$\begin{array}{lll} s \models p & \text{iff} & s \in V_p \\ s \models \neg\varphi & \text{iff} & s \not\models \varphi \\ s \models \varphi \wedge \psi & \text{iff} & s \models \varphi \text{ and } s \models \psi \\ s \models K\varphi & \text{iff} & \text{for all } \ t \in S : s \to t \text{ implies } t \models \varphi \end{array}$$

The two formulations are completely equivalent. At the same time, using one notation over another might lead to different insights or problems.

Further semantic definitions A sentence φ is *valid* on a model M, notation $M \models \varphi$, if and only if for all states s in the domain of M: $s \models \varphi$. A formula φ is *valid*, notation $\models \varphi$, if and only if for all models M (of the class of models for the given parameters of A and P): $M \models \varphi$. That is, φ holds on all epistemic states.

Logic and variations on it The logical system for validity is a sub-system of one which we already have seen. Look back at Figure 1, and take only the propositional tautologies, modus ponens, the distribution axiom, and the necessitation rule. This logical system is called K. Every book on modal logic will prove its completeness: a sentence is valid in the semantics just in case it can be proved from the axioms using the rules.

What's the point? For the modeling of knowledge, all we have so far is a spare definition and logic: An agent lives in a world and can see others. What it knows in a given world is just what is true in the worlds it sees. This seems a far cry from a full-bodied analysis of knowledge. The logical completeness result underscores the point. Any agent who "knew" things in the manner of this semantics would exemplify properties indicated by the logic. In particular, it would act as if the distribution axiom and necessitation rule held. The former, turned around a bit, says that the agent would be a *perfect reasoner*: if it knows φ and also knows $\varphi \to \psi$, then it automatically and effortlessly knows ψ. Necessitation says that it also knows all the general features of this logic. Thus, the agent is *logically omniscient*. And one would be hard-pressed to maintain that such an agent "knew" in the first place. For it is even possible that in some situations (models) the agent would "know" things which are false: this might well happen if the agent lived in a world which was not among those it considered possible.

This leads to our next point. If one wants to model agents with certain epistemically-desirable properties (see below for these), one can impose mathematical conditions on the accessibility relation of models under consideration. Then one changes the definition of *valid* from *true in all epistemic states* to *true in all epistemic states meeting such-and-such a condition.*

To see how this works, we need a definition. A *frame* is the same kind of structure as what we are calling a *model*, but it lacks the valuation of atomic sentences. So it is just a pair $F = \langle S, R \rangle$, with R a relation on S. (In other terms, a frame is a *graph*.) Given a sentence φ in our logic, we say that $F \models \varphi$ if for all valuations $V : P \to \mathcal{P}(S)$, every $s \in S$ satisfies φ in the model $\langle S, R, V \rangle$.

Ax	formal statement	property of R	interpretation
K	$K(\varphi \to \psi) \to (K\varphi \to K\psi)$	(none)	closed under modus ponens
T	$K\varphi \to \varphi$	reflexive	veracity
D	$K\varphi \to \neg K \neg \varphi$	serial	consistency
4	$K\varphi \to KK\varphi$	transitive	positive introspection
5	$\neg K\varphi \to K\neg K\varphi$	Euclidean	negative introspection

Figure 3. Axiom schemes of modal logic with their relational correspondents and epistemic interpretations.

Figure 3 presents well-known *correspondences* between conditions on a frame and properties of the logic. One example: a frame F satisfies each instance of D (say) iff F meets the condition listed that every point in it has a successor. is related by R to *some* point or other. Reflexivity means that every point is related to itself. Transitivity means that if x is related to y, and y to z, then x is related to z. The Euclidean condition mentioned in connection with the 5 axioms is

$$(\forall x)(\forall y)(\forall z)((xRy \wedge xRz) \to yRz).$$

There is a further aspect of the correspondence. If one wants to study the sentences which are valid on, say, transitive models, then one need only add the corresponding axiom (in this case $K\varphi \to KK\varphi$) to the basic logic that we mentioned above. On a conceptual level, we prefer to turn things around. If one wants to model agents which are positively introspective in the sense that if the know something, then they know that they know it, then one way is to assume, or argue, that they work with transitive models. The same goes for the other properties, and for combinations of them.

There are many modal systems indeed, but we wish to mention only a few here. We already have mentioned K. If we add the axioms called T and 4, we get a system called S4. It is complete for models which are reflexive and transitive, and intuitively it models agents who only know true things and which are positively introspective. If one adds the negative introspection axioms 5, one gets a system called S5. The easiest way to guarantee the S5 properties is to work with relations which are reflexive, symmetric, and transitive (*equivalence relations*), for these are also Euclidean.

Turning from knowledge to belief, the T axiom is highly undesirable. So logics appropriate for belief will not have T. But they often have D, the seriality axioms. These may be interpreted as saying that if an agent believes φ, then it does not at the same time believe $\neg\varphi$. Probably the most common logic of belief is KD45, obtained by adding to K the other axioms in its name. KD45 is complete with respect to models which have the properties listed above. When studying belief, one usually changes the name of the modality from K to B, for obvious reasons.

We wish to emphasize that all of the intuitive properties of knowledge and belief discussed in this section, and indeed in this chapter as a whole, are highly contestable. Using logic does not commit one to any of those properties. But logic can help to clarify the commitments in a given form of modeling. For example, any modeling of knowledge using relational semantics will always suffer from problems having to do with logical omniscience, as we have seen.

4.4 Multi-agent epistemic logic

The formal move from the basic modal logic of the last section to its multi-agent generalization is very easy. One begins with a set A of *agents* in addition to the set of atomic propositions. Then the syntax adds operators K_a, and we read $K_a\varphi$ as "a knows φ." The semantics then moves from the models of the last section to what we shall call *epistemic models*. These are tuples $\langle S, R, V\rangle$ as before, except now R is an *accessibility* (function) $R : A \to \mathcal{P}(S \times S)$. That is, it is a family of accessibility relations, one for each agent. We usually write R_a instead of $R(a)$. Further, we write $s \xrightarrow{a} t$ instead of $(s,t) \in R_a$.

EXAMPLE 2. We are going to present an example which hints at the applicability of our subject to the modeling of games.

Consider three players Amina, Bao, and Chandra (a, b, c). They sit in front of a deck consisting of exactly three cards, called clubs, hearts, and spades. Each is dealt a card, they look, and nobody sees anyone else's card. We want to reason about knowledge in this situation. It makes sense to take as atoms the nine elements below

$$\{Clubs_a, Clubs_b, \ldots, Spades_b, Spades_c\}.$$

The natural model for our situation has six states. It is shown in a slightly re-packaged form in Figure 4. We call the model *Hexa*. The states are named according to who has what. For example, ♣♡♠ is the state where Amina has clubs, Bao hearts, and Chandra spades. The lines between the states have labels, and these indicate the accessibility relations. So Bao, for example, cannot tell the difference between ♡♣♠ and♠♣♡: if the real deal were one of those, he would think that it could be that same deal, or the other one (but no others).

We mentioned that the picture of *Hexa* differs slightly in its form from what the official definition calls for. That version is mathematically more elegant but does not immediately lend itself to a picture. It would have, for example,

$$\begin{array}{lcl} V(Clubs_a) & = & \{♣♡♠, ♣♠♡\} \\ & \ldots & \\ V(Spades_c) & = & \{♣♡♠, ♡♣♠\} \\ R_a & = & \{(♣♡♠, ♣♠♡), (♡♣♠, ♡♠♣), (♠♣♡, ♠♡♣)\} \\ & \ldots & \end{array}$$

We can then evaluate sentences in English by translating into the formal language and using the semantics. Here are some examples.

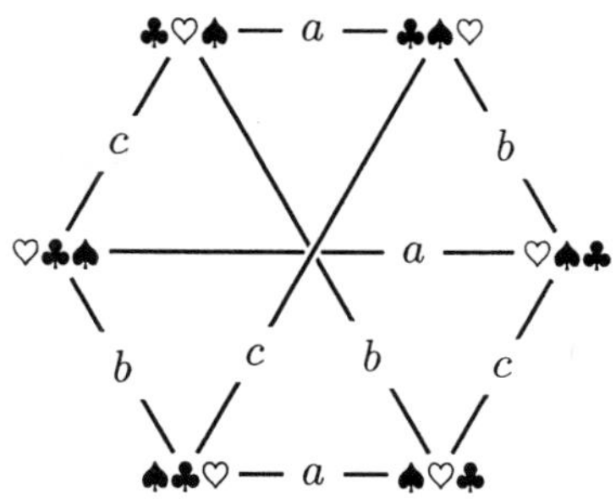

Figure 4. The model *Hexa*, with the accessibility relations pictured as edges between the states. For example, Amina cannot distinguish ♣♡♠ from ♣♠♡ as she holds clubs in both deals. But Bao cannot distinguish ♣♡♠ from ♠♡♣ as he holds hearts in both.

Amina knows she has the heart card. We translate to $K_a \mathit{Hearts}_a$. The semantics in *Hexa* is $[\![K_a \mathit{Hearts}_a]\!]$ = {♡♣♠, ♡♠♣}. (In more detail: in each of the two worlds ♡♣♠ and ♡♠♣, every world that Amina thinks is possible belongs to $V(\mathit{Hearts}_a)$. And if s is one of the four other worlds, there is some world accessible from s for Amina which does not belong to $V(\mathit{Hearts}_a)$. For example, in s = ♣♡♠, the world s itself is accessible for Amina, and she does not have hearts there.) That is, our sentence is true exactly at ♣♡♠ and ♠♡♣. Note that this is precisely the set of worlds where Amina indeed has hearts. So the sentence *Amina has hearts if and only if she knows she has hearts* comes out true at all states.

If Bao has spades, Chandra has clubs. This is $\mathit{Spades}_b \rightarrow \mathit{Clubs}_c$. The semantics is

{♣♡♠, ♡♣♠, ♡♠♣, ♠♣♡, ♠♡♣}.

If Amina has hearts, then she knows that if Bao has spades, Chandra has clubs. The translation is

$$\mathit{Hearts}_a \rightarrow K_a(\mathit{Spades}_b \rightarrow \mathit{Clubs}_c).$$

This true at all states.

Bao considers it possible that Amina has spades but actually Amina has clubs. We translate "consider it possible that φ" by "does not know that φ is false." So our sentence here is $\neg K_b \neg \mathit{Spades}_a \wedge \mathit{Clubs}_a$. Usually one prefers to avoid negation by introducing an abbreviation. So if we say that $\hat{K}_b\varphi$ abbreviates $\neg K_b \neg\varphi$, then we may read this as *Bao considers* φ *possible* and translate our sentence as above. Its semantics is {♣♡♠}.

The last sentence shows the *duality* between "consider possible" and "know". This phenomenon of dual definitions is something we shall see later as well.

Logic We define *validity* exactly as in the last section, but using the generalized language and semantics. Given a language and a semantics, one task for logic is to determine the set of sentences which are valid. In particular, we might ask for a nice logical system for the valid sentences, since this may well give insight into the underlying assumptions in the model. Now in what we have been discussing in this section, we might seek at least three logical systems:

1. A system for validity on the model *Hexa*.

2. A system for validity on all models whatsoever.

3. A system for validity on all models that are "basically similar" to *Hexa*.

For the first question, we modify the system of Figure 1 for ignorance of two alternatives. The axioms that actually pertain to H and T must be replaced, of course. Instead of exclusivity, we take an axiom that says, informally, that for exactly one of the six states s of *Hexa*, all atoms true in s hold, and all not true in s do not hold. We also add an axiom saying that *If Amina has clubs, then she knows it*, and similarly for the other players and cards, and also that *Amina does not know which card any other player holds.* All of the rest of the axioms are valid in this model, as are the rules. Each statement of the old system would be replaced by three, one for each player. So one of the introspection axioms would be $K_b\varphi \to K_bK_b\varphi$. In particular, the introspectivity and necessitation principles are valid.

In passing, we note that this logical system has no *interaction properties* between the knowledge of different agents. That is, none of the axioms mix K_a and K_b for different a and b. Mathematically, this means that the generalization of single-agent knowledge to the multi-agent case will be almost trivial. But there are two caveats: first, the phenomenon of *common knowledge* does involve discussions of different agents' knowledge, and so it turns out to be harder to study. And second, it really is possible to have situations where interaction properties make sense. For example, suppose one wants to model situations where *everything Bao knows Chandra also knows.* In the semantics, one would want $R_c \subseteq R_b$. And then in the logic one could add $K_b\varphi \to K_c\varphi$.

For the second question above, the fact that we are dealing with a larger class of models means that fewer logical principles are sound. The only sound principles would be the propositional tautologies and the distribution axioms, and the rules of modus ponens and necessitation.

The last question is obviously not precise. The point of raising it is that one can make precise the sense in which games are, or are not, similar by using the logical principles that hold. For example, one often simplifies matters by assuming that adversaries are perfect reasoners, and in this setting it is natural to *assume* the introspectivity principles in the modeling. That is, one works only with models where those principles turn out to be sound. The easiest way to arrange this is to look at the chart in Figure 3. If each accessibility relation $\xrightarrow{a}$ is reflexive, transitive, and euclidean, then the model will satisfy both introspectivity assertions. (This holds no matter what the valuation V happens to do.) It turns out that a relation with these properties is automatically *symmetric* and hence an equivalence relation. Moreover, an equivalence relation on a set is equivalently formulated as a *partition* of the set. So one often finds the tacit assumption in much of the game theory/economics literature that the models used are *partition models* consisting of a set S and a *partition* of S for each player.

Identity conditions on models We first looked at announcements in Section 2.2. In discussing (4) and (5), we noted the need for principled identity conditions on relational models. Fortunately, the general study of relational models and their logical systems gives us a usable condition, called *bisimulation*. This condition is coarser than *isomorphism*, does what we want (in particular, it tells us that (4) and (5) ought to be identified), and has an intuitive resonance as well. For the formal definition and much more, see any text on modal logic, for example Blackburn et al. [2001].

4.5 Common knowledge

We have discussed the idea of common knowledge in Section 3.8. We turn now to its formalization on top of what we saw in Section 4.4 above. We present a generalization of our previous concept, however. For each group $B \subseteq A$, we want notions of *group knowledge for the set B* and *common knowledge for the set B*. This last notion has the same intuitive basis as common knowledge itself. For example, it is common knowledge among Englishmen that one drives on the left, but this common knowledge does not hold for the entire world.

For the syntax of group knowledge, we add operators E_B to the language of multi-agent epistemic logic. The semantics is given by

$$[\![E_B\varphi]\!] = \bigcap_{a \in B} [\![K_a\varphi]\!].$$

This means that we (straightforwardly) translate $E_B\varphi$ as *Everyone in group B knows* φ. In the case of finitely many agents (the assumption in practically all papers on the topic), $E_B\varphi$ may be regarded as an abbreviation.

EXAMPLE 3. We return to the model *Hexa* from Section 4.4 (see Figure 4). We have

$$\clubsuit\heartsuit\spadesuit \models E_{\{a,b\}}\neg(\mathit{Spades}_a \wedge \mathit{Clubs}_b \wedge \mathit{Hearts}_c).$$

That is, both Amina and Bao know that the deal of cards is *not* ♠♣♡. However, despite this, each does not know that the other knows this fact.

We next turn to common knowledge. The syntax adds operators $C_B\varphi$ to the language, exactly as with group knowledge. The semantics is more problematic, and indeed there are differing proposals in the literature. We follow the most common treatment. One essentially takes the unwinding of the fixed point that we saw in (13) in Section 3.8 as the definition, and then the fixed point property becomes a semantic consequence later on.

For the semantics, for each group B we pool all the accessibility relations for the members of B together, and then take the *reflexive-transitive closure*:

$$R_B^* \equiv (\bigcup_{a \in B} R_a)^*.$$

(see below for an explanation). Then we interpret via

$$s \models C_B\varphi \quad \text{iff} \quad \text{for all } \ t \in S : R_B^*(s,t) \text{ implies } t \models \varphi$$

Alternatively said, $C_B\varphi$ is true in s if φ is true in any state s_m that can be reached by a (finite) path of zero or more states $s_1, \ldots, s_m$ such that, for not necessarily different agents $a, b, c \in B$: $R_a(s_1, s_2)$, $R_b(s_2, s_3)$, and ..., and $R_c(s_{m-1}, s_m)$. A path of zero states is just a single state alone. Hence if $C_B\varphi$ is true at s, then automatically φ is true at s as well.

As an example of both the formal and informal concepts, we consider an n-person muddy children scenario (see Section 3.1), *before* any announcement that at least one agent is muddy. It is easy to describe the model: it has 2^n states with the rest of the structure determined in the obvious way. Then it is common knowledge at all states that no agents know their own state. More interesting is the comment that in this model, if s is a state and every agent knows φ at s, then φ is already common knowledge at all states.

The logic of common knowledge adds two principles to the basic multi-agent epistemic logic. Those are the *Mix Axiom*:

$$C_B\varphi \to \varphi \wedge E_B C_B\varphi$$

(so-called because it deals with the interactions of the the two operators of this section) and the *induction rule*:

$$\text{from } \chi \to \psi \text{ and } \chi \to K_a\chi \text{ for all } a \in B, \text{ infer } \chi \to C_B\psi.$$

Using this logic, one can prove the important properties of common knowledge. For example, it is *idempotent*:

$$C_B\varphi \leftrightarrow C_B C_B\varphi.$$

The interesting direction here says that if φ is common knowledge in a group, then the fact of its being common knowledge is itself common knowledge in the group.

4.6 Belief-knowledge logic

Intuitively, *belief* and *knowledge* are related but different. However, we have heretofore conflated the two notions. We present here the simplest possible system which can sensibly separate the two by incorporating both at the same time with different semantics. It and the logical axioms are taken from Meyer and van der Hoek [1995].

We fix sets A of agents and P of atoms. To separate the two notions, we need a language with different operators K and B.

A *knowledge-belief model* (KB-model) is a Kripke model of the form $\langle S, R_a^K, R_a^B, V\rangle_{a\in A}$, where S is set of states, R_a^K and R_a^B are binary accessibility relations in $\mathcal{P}(S \times S)$, and V is a function from P to $\mathcal{P}(S)$. We write $s \overset{a}{\sim} t$ instead of $(s,t) \in R_a^K$, and $s \overset{a}{\to} t$ instead of $(s,t) \in R_a^B$. As the letters K and B indicate, the first relation $\overset{a}{\sim}$ is meant to capture the *knowledge* of agent a, while the second $\overset{a}{\to}$ captures the agent's *beliefs*.

A KB model is required to satisfy the following conditions: $\overset{a}{\sim}$ is an equivalence relation; $\overset{a}{\to}$ is serial; if $s \overset{a}{\sim} t$ and $s \overset{a}{\to} w$, then $t \overset{a}{\to} w$; and finally, $\overset{a}{\to}$ is included in $\overset{a}{\sim}$. So the modeling reflects the following intuitions: the *truthfulness and introspection of knowledge*, full belief introspection (agents *know their own beliefs*), *beliefs are consistent*, and *knowledge implies belief*. It is not necessary to assume that $\overset{a}{\to}$ is transitive and Euclidean, since these properties immediately follow from the above conditions. So we also have for free that belief is introspective, in the usual sense.

Notice also that, as observed on the example in Section 2.3, the knowledge relation $\overset{a}{\sim}$ is recoverable from the belief relation $\overset{a}{\to}$, via the following rule:

$$s \overset{a}{\sim} t \quad \text{iff} \quad (\forall w)(s \overset{a}{\to} w \text{ iff } t \overset{a}{\to} w). \tag{14}$$

To see this, first assume that $s \overset{a}{\sim} t$. Then also $t \overset{a}{\sim} s$. From this and one of the conditions in a KB model, we get the right-hand side of (14). And if the right-hand side holds, we show that $s \overset{a}{\sim} t$. First, by seriality, there must be some w so that $s \overset{a}{\to} w$. For this w we also have $t \overset{a}{\to} w$. And then using the fact that $\overset{a}{\sim}$ is an equivalence relation including $\overset{a}{\to}$, we see that $s \overset{a}{\sim} t$.

So, in fact, one could present KB-models simply as *belief models* $\langle S, \overset{a}{\to}, V\rangle$, where $\overset{a}{\to}$ is transitive, serial and Euclidean, and one can take the knowledge relation as a defined notion, given by the rule (14) above. We interpret the logical system in KB-models via

$$\begin{array}{lcl} [\![K_a\varphi]\!] & = & \{s \in S : t \in [\![\varphi]\!], \text{ for all } t \text{ such that } s \overset{a}{\sim} t\} \\ [\![B_a\varphi]\!] & = & \{s \in S : t \in [\![\varphi]\!], \text{ for all } t \text{ such that } s \overset{a}{\to} t\} \end{array} \tag{15}$$

EXAMPLE 4. The easiest example is a two-state model for ignorance of two alternatives, say heads and tails, together with a belief in one of them, say heads. Formally, we have one agent, so we drop her from the notation. There are two states s and t, for H and T. The relation $\to$ is $\mathsf{H} \to \mathsf{H}$ and $\mathsf{T} \to \mathsf{H}$. The relation $\sim$ therefore relates all four pairs of states. Then at both states $B\mathsf{H} \wedge \neg K\mathsf{H}$. In particular, the agent believes heads at the tails state t. Hence we have our first example of a *false belief*. But agents in KB-models are not so gullible that they believe absolutely *anything*: $\neg B(\mathsf{H} \wedge \mathsf{T})$, for example. And indeed, the seriality requirement prohibits an agent from believing a logical inconsistency.

The logic is then axiomatized by the S5 system for knowledge, the KD45 system for belief, and two connection properties: First, $B_a\varphi \to K_a B_a\varphi$. So an agent may introspect on her own beliefs. (It also follows in this logic that $\neg B_a\varphi \to K_a\neg B_a\varphi$.) We should mention that *introspection about beliefs* is less controversial than *introspection about knowledge*. If we take knowledge to be a relation between an agent and an external reality, then it is as problematic to account for an agent's knowledge of their own knowledge as it is to account for any other type of knowledge. But to the extent that belief is an "internal" relation, it seems easier to say that fully-aware agents should have access to their own beliefs.

The second logical axiom connecting knowledge and belief is $K_a\varphi \to B_a\varphi$. This reiterates an early point: we are thinking of knowledge as a strengthening of belief. It is sound due to the requirement that $\xrightarrow{a}$ be included in $\overset{a}{\sim}$.

Variations There are some variations on the soundness and completeness result which we have just seen. Suppose one takes an arbitrary relation $\xrightarrow{a}$, then defines $\overset{a}{\sim}$ from it using (14), and then interprets our language on the resulting structures by (15). Then $\overset{a}{\sim}$ is automatically an equivalence relation, and so the S5 axioms for knowledge will be sound. Further, the two connection axioms automatically hold, as does negative introspection. Continuing, we can add impose conditions on $\xrightarrow{a}$ (such as seriality, transitivity, the Euclidean property, or subsets of these), and then study validity on that class. In the logic, one would take the corresponding axioms, as we have listed them in Figure 3. In all of the cases, we have completeness results.

4.7 Conditional doxastic logic

We now re-work the logical system of Section 4.6, so that it can handle conditionals in the way that we did in Section 2.4. The logical system is based on Board [2004], and Baltag and Smets [2006a; 2006b; 2006c]. Following the latter, we'll call it *conditional doxastic logic* (**CDL**).

Its syntax adds to propositional logic statements of the form $B_a^\alpha\varphi$, where a is an agent and α and φ are again sentences in the logical system. This should be read as "If a were presented with evidence of the assumption α in some world, then she should believe φ describes the world (as it was before the presentation)."

The simplest semantics for this logic uses *plausibility models*. These are also special cases of the *belief revision structures* used in Board [2004]. (We shall see the general notion at the end of this section.) Plausibility frames are Kripke structures of the form $(S, \leq_a)_{a\in A}$, consisting of a set S endowed with a family of *locally pre-wellordered* relations $\leq_a$, one for each agent a. When S is *finite*, a locally pre-wellordered relation on S is just one that is *reflexive, transitive and weakly connected both forwards and backwards*[3], i.e. $s \leq_a t$ and $s \leq_a w$ implies that either $t \leq_a w$ or $w \leq_a t$, and also $t \leq_a s$ and $w \leq_a t$ implies that either $t \leq_a w$ or $w \leq_a t$. Equivalently, we have a *Grove system of spheres*, just as in Section

[3]This last property is also known as *no branching to the left or to the right.*

2.3, consisting of a number of (disjoint) "smallest spheres" (listing the worlds "in the center"), then surrounding them the next smallest spheres (containing worlds a little less plausible than these central ones), then the next ones, having worlds a little less plausible than these, etc. To match the notion of local pre-wellorder above (again, in the finite case), we need to assume that every world belongs to some sphere, and that if two spheres intersect or are both included in a larger sphere, then one is included in the other.

As for Kripke models in general, a *plausibility model* is a tuple $M = (S, \leq, V)$, where $(S, \leq)$ is a plausibility frame and V is a valuation on it. A *doxastic state* is a tuple of the form (M, s), where M is a plausibility model and $s \in M$.

REMARK 5. For readers of van Benthem and Martinez' Chapter 3b, we mention that our orderings go the other way from theirs: for them, "more plausible" is "upward" in the ordering, and for us it is "in the center" or "lower down".

As in the example in Section 2.3, we define a *knowledge (indifference) relation* by putting

$$s \stackrel{a}{\sim} t \text{ iff either } s \leq_a t \text{ or } t \leq_a s.$$

Plausibility models for only one agent have been used as models for *AGM* belief revision in [Gärdenfors, 1988; Segerberg, 1998; Spohn, 1988]. The additional indifference relations turn out to be useful in modeling, as we indicate shortly.

Similarly, we define a *belief relation* by putting:

$$s \xrightarrow{a} t \quad \text{iff} \quad s \stackrel{a}{\sim} t \text{ and } (\forall u)(s \stackrel{a}{\sim} u \rightarrow t \leq_a u) \ .$$

It is easy to see that $\stackrel{a}{\sim}$ and $\xrightarrow{a}$ satisfy all the postulates of a *KB* model. More generally, we obtain a *conditional belief* relation by putting, for any set $X \subseteq S$ of states:

$$s \xrightarrow{a,X} t \quad \text{iff} \quad s \stackrel{a}{\sim} t, t \in X, \text{ and } (\forall u)(s \stackrel{a}{\sim} u \ \& \ u \in X \rightarrow t \leq_a u) \ .$$

In other words, $s \xrightarrow{a,X} t$ if a considers t to be possible in s, if $t \in X$, and if t is among the most plausible worlds for a with these two conditions. So here we see the basic idea: reasoning about a's hypothetical beliefs assuming α involves looking at the relation $\xrightarrow{a,X}$ where $X = [\![\alpha]\!]$.

Observe that the belief relation $\xrightarrow{a}$ is the same as the conditional belief relation $\xrightarrow{a,S}$, where S is the set of all states. As a passing note, for each $X \subseteq S$ we can use the relations $\xrightarrow{a,X}$ to make a structure which is *almost* a KB-model in the sense of Section 4.6. Take S for the set of worlds, $\xrightarrow{a,X}$ for the belief relation for each agent a, take the knowledge relation $\stackrel{a}{\sim}$ for a as above, and use the same valuation as in S. The only property of KB-models lacking is that the belief relations might fail to be serial: a state s has no $\xrightarrow{a,X}$-successors if X is disjoint from the $\stackrel{a}{\sim}$-equivalence class of s. This makes sense: seriality corresponds to consistency of belief; but X being disjoint from the $\stackrel{a}{\sim}$-equivalence of s corresponds to conditionalizing on a condition X that is *known* (with absolute certainty) to be false: the resulting set of conditional beliefs should be inconsistent, since it contradicts (and at the same time it preserves) the agent's knowledge.

For the semantics, we say

$$[\![B_a^\alpha \varphi]\!] \quad = \quad \{s \in S : t \in [\![\varphi]\!], \text{ for all } t \text{ such that } s \xrightarrow{a,X} t, \text{ where } X = [\![\alpha]\!]\} \tag{16}$$

In other words, to evaluate a doxastic conditional, a looks at which of her possible worlds are the most plausible given the antecedent α, and then evaluates the conclusion on all of those worlds. If they all satisfy the conclusion φ of the conditional, then a believes φ conditional on α. It is important that the evaluation take place in the original model.

EXAMPLE 6. The model S from Section 2.4 had four worlds u, v, w, x. Amina's plausibility relation $\leq_a$ is essentially the ordered partition: $\{v, w\} < \{u, x\}$. Bao's plausibility relation is the reflexive closure of the relation $\{(v, w), (w, v)\}$. We are interested in sentences $B_a^\alpha \varphi$, where $\alpha = K_b\mathsf{H} \vee K_b\neg\mathsf{H}$. Let $X = [\![\alpha]\!] = \{u, x\}$. Then $\xrightarrow{a,X}$ relates all four worlds to u and x. Our semantics in (16) is equivalent to what we used in (11). Note as well that the right-hand model in (8) shows $\xrightarrow{a,S}$ and $\xrightarrow{b,S}$.

Knowledge in plausibility models There are two equivalent ways to define knowledge in plausibility models. One can use the definition (15) applied directly to the $\stackrel{a}{\sim}$ relations introduced above, saying that

$$s \models K_a\varphi \text{ iff } s \stackrel{a}{\sim} t \text{ implies } t \models \varphi.$$

Alternatively, one can use the following observation to get an intuitively appealing reformulation.

PROPOSITION 7. *Let S be a plausibility model and $s \in S$. The following are equivalent:*

1. $s \models K_a\varphi$.
2. $s \models B_a^{\neg\varphi}\bot$, *where $\bot$ is a contradiction such as $p \wedge \neg p$.*
3. $s \models B_a^{\neg\varphi}\varphi$.
4. $s \models B_a^\alpha\varphi$, *for* every *sentence α.*

Here is the reasoning: If (1) holds, then s has no $\xrightarrow{a,X}$ -successors at all, where $X = [\![\neg\varphi]\!]$. This means that the next two assertions hold vacuously. Also, notice that, for *every* set X, $s \xrightarrow{a,X} t$ implies $s \stackrel{a}{\sim} t$. Hence, if (1) holds then, for any sentence α, φ is true at all $\xrightarrow{a,X}$ -successors of s, where $X = [\![\alpha]\!]$. Therefore (4) holds as well. Conversely, (4) clearly implies (3); also, if either of (2) or (3) hold, then the semantics tells us that s has no $\xrightarrow{a,X}$ -successors, so every $t \stackrel{a}{\sim} s$ must satisfy φ; i.e., (1) must hold.

A proposal going back to Stalnaker [1968] *defines* "necessity"[4] in terms of conditionals, via the clause (3) above. This contains an idea concerning our notion of

[4]This is denoted in [Stalnaker, 1968] by $\Box\varphi$. Note that it corresponds in our notation to $K\varphi$, and should not be confused with our notation for "safe belief" $\Box\varphi$ in the next subsection.

Syntax	$\varphi ::= p \mid \neg\varphi \mid \varphi \wedge \psi \mid B_a^\alpha \varphi$	
Definition	$K_a\varphi := B_a^{\neg\varphi}\varphi$	(knowledge)
Main Axioms	$B_a^\alpha \alpha$	hypothetical acceptance
	$K_a\varphi \to \varphi$	veracity
	$K_a\varphi \to B_a^\alpha\varphi$	persistence of knowledge
	$B_a^\alpha\varphi \to K_a B_a^\alpha\varphi$	positive belief introspection
	$\neg B_a^\alpha\varphi \to K_a \neg B_a^\alpha\varphi$	negative belief introspection
	$\neg B_a^\alpha \neg\varphi \to (B_a^{\alpha\wedge\varphi}\theta \leftrightarrow B_a^\alpha(\varphi \to \theta))$	minimality of revision

Figure 5. Syntax and axioms for conditional doxastic logic. We also assume Modus Ponens, as well as Necessitation and the (K) axiom for B_a^α.

"knowledge": what it means to know φ (in this strong sense) is that one would still believe φ even when hypothetically assuming $\neg\varphi$.

Finally, (4) can be related to the "defeasibility analysis" discussed in Section 3.4. Indeed, what is says is that our notion of knowledge satisfies a *stronger* version of this analysis than the original one: our knowledge is the same as "*absolutely unrevisable*" belief. One "knows" φ (in this absolute sense) if giving up one's belief in φ would *never* be justified, under *any* conditions (even when receiving false information).

The logic We list a complete axiomatization in Figure 5. Note that in the syntax, we take conditional belief as the only basic operator, and define knowledge via (3). Verifying the soundness of most of the axioms is easy, and we discuss only the principle of minimality of revision. Let $X = [\![\alpha]\!]$. Let $Y = [\![\alpha \wedge \varphi]\!]$, so $Y \subseteq X$. Suppose that $s \models \neg B_a^\alpha \neg\varphi$. So there is some t so that $s \xrightarrow{a,X} t$ and $t \models \varphi$. This means that, for all w, we have $s \xrightarrow{a,X} w$ iff $s \xrightarrow{a,Y} w$. We first show that if $s \models B_a^\alpha(\varphi \to \theta)$, then $s \models B_a^{\alpha\wedge\varphi}\theta$. To see this, let w be such that $s \xrightarrow{a,Y} w$. Then $s \xrightarrow{a,X} w$, and since $w \models \varphi$, we have $w \models \theta$. For the second half, one checks directly that $B_a^{\alpha\wedge\varphi}\theta$ implies $s \models B_a^\alpha(\varphi \to \theta)$.

Incidentally, the axioms of the logic have interpretations in terms of *belief revision*, as we shall see. In particular, the last axiom ("minimality of revision") corresponds to the conjunction of the **AGM** principles of Subexpansion and Superexpansion (principles (7) and (8) in Section 7).

Adding common knowledge and belief It is also possible to expand the language with operators Cb_B^α and Ck_B^α, reflecting *conditional versions of common belief and knowledge* in a group B. For the proof systems and more on these systems, cf. Board [2004], and Baltag and Smets [2006a; 2006b; 2006c]. Board's paper also contains interesting variations on the semantics, and additional axioms. Baltag and Smets offer a generalization of the notion of a plausibility model to that of a *conditional doxastic model.* Both authors considers applications to modeling in games.

Belief revision structures The plausibility models that we have been concerned with in this section may be generalized in a number of ways. First of all, the pre-wellorders for each agent might be *world-dependent.* This would be important to model agents with incorrect beliefs about their own beliefs, for example.

In this way, we arrive at what Board [2004] calls *belief revision structures.* For more on this logic, including completeness results, see Board [2004].

4.8 The logic of knowledge and safe belief

As we saw, Stalnaker's defeasibility analysis of knowledge asks a *weaker* requirement than the one satisfied by our notion of "absolutely unrevisable knowledge": namely, it states that φ is known (in the weak, defeasible sense) if there exists no *true* piece of information X such that the agent would no longer be justified to believe φ after learning X. Following Baltag and Smets [2006b; 2006c], we call *safe belief* this weak notion of defeasible "knowledge", and we use the notation $\Box_a\varphi$ to express the fact that a safely believes φ. We can immediately formalize this notion in terms of conditional belief arrows:

$$s \models \Box_a\varphi \text{ iff for all } t \in S, \text{ and all } s \in X \subseteq S : s \xrightarrow{a,X} t \text{ implies } t \models \varphi.$$

To our knowledge, the first formalization of safe belief (under the name of "knowledge") was due to Stalnaker [1996], and used the above clause as a definition. Observe that it uses quantification over propositions (sets of states). It was only recently observed, in [Baltag and Smets, 2006b; Baltag and Smets, 2006c; Stalnaker, 2006], that this second-order definition is equivalent to a simpler one, which takes safe belief as the Kripke modality associated to the relation "at most as plausible as":

$$[\![\Box_a\varphi]\!] \quad = \quad \{s \in S : t \in [\![\varphi]\!], \text{ for all } t \leq_a s\} \tag{17}$$

This last condition was adopted by Baltag and Smets [2007b] as the definition of safe belief. The same notion was earlier defined, in this last form, by van Benthem and Liu [2004], under the name of "preference modality".

EXAMPLE 8. The situation described in Section 2.4 provides us with examples of safe and unsafe beliefs. In the model 10, Amina believes (though she doesn't know) that Bao doesn't know that the face of the coin is tails. If the real state is v, then this belief is true, and moreover it is *safe*: this is easy to see, since it is true at both v and w, which are the only two states that are at least as plausible for Amina as v. This gives us an example of a *safe belief which is not knowledge.* If instead the real state is u, then the above belief is still true (though is not known).

But it is *not safe* anymore: formally, this is because at state x (which for Amina is at least as plausible as u), Bao does know the face is tails. To witness this unsafety, consider the sentence $\alpha := B_b\mathsf{H} \vee B_b\mathsf{T}$, saying that Bao knows the face of the coin. At the real world u, the sentence α is true; but, at the same world u, Amina does not believe that, if α were true then Bao wouldn't know that the face was tails: $u \models \alpha \wedge \neg B_a^{\alpha} \neg B_b\mathsf{T}$. This shows that Amina's belief, though true, can be defeated at state u by learning the true sentence α.

Stalnaker [2006] observes that *belief can be defined in terms of safe belief*, via the logical equivalence: $B_a\varphi \leftrightarrow \neg\Box_a\neg\Box_a\varphi$, and that *the complete logic of the safe belief modality* $\Box_a$ *is the modal logic* $S4.3$.[5] Baltag and Smets [2006b; 2006c] observe that by combining safe belief $\Box_a\varphi$ with the "absolute" notion of knowledge $K_a\varphi$ introduced in the previous section, one can define conditional belief via the equivalence[6]:

$$B_a^{\alpha}\varphi \quad\leftrightarrow\quad (\neg K_a\neg\alpha \to \neg K_a\neg(\alpha \wedge \Box_a(\alpha \to \varphi))).$$

The logic of knowledge and safe belief is then axiomatized by the S5 system for knowledge, the S4 system for safe belief, and two connection properties: First, $K_a\varphi \to \Box_a\varphi$. This reiterates the earlier observation: knowledge (in our absolute sense) is a strengthening of safe belief. The second axiom says that the plausibility relation $\leq_a$ is connected within each $\overset{a}{\sim}$-equivalence class:

$$K_a(\varphi \vee \Box_a\psi) \wedge K_a(\psi \vee \Box_a\varphi) \to K_a\varphi \vee K_a\psi.$$

Belief and conditional belief are derived notions in this logic, defined via the above logical equivalences.

4.9 Propositional Dynamic Logic

This section of our chapter mainly consists of brief presentations of logical systems which are intended to model notions of importance for epistemic or doxastic logic. The current subsection is an exception: *propositional dynamic logic* (**PDL**) is a system whose original motivations and main uses come from a different area, semantic studies of programming languages. We are not concerned with this here, and we are presenting **PDL** in a minimum of detail, so that the reader who has not seen it will be able to see what it is, and how it is used in systems which we shall see later in the chapter.

The syntax of **PDL** begins with atomic sentences p, q, ..., (also called atomic propositions), and also *atomic programs* a, b, ... (sometimes called *actions*). From these atomic sentences and programs, we build a language with two types of syntactic objects, called sentences and programs. The syntax is set out in Figure 6.

[5] $S4.3$ is the logic of reflexive transitive frames with no branching to the right.

[6] Van Benthem and Liu [2004] use another logical equivalence to similarly define conditional belief.

Syntax	**Sentences** φ	p_i	$\neg\varphi$	$\varphi \wedge \psi$	$[\pi]\varphi$	
	Programs π	a	$?\varphi$	$\pi;\sigma$	$\pi \cup \sigma$	π^*
Semantics	**Main Clauses**	$[\![[\pi]\varphi]\!] = \{s : \text{if } s[\![\pi]\!]t, \text{ then } t \in [\![\varphi]\!]\}$				
		$[\![?\varphi]\!] = \{(s,s) : s \in [\![\varphi]\!]\}$				
		$[\![\pi;\sigma]\!] = [\![\pi]\!] \; ; \; [\![\sigma]\!]$				
		$[\![\pi \cup \sigma]\!] = [\![\pi]\!] \cup [\![\sigma]\!]$				
		$[\![\pi^*]\!] = ([\![\pi]\!])^*$				

Figure 6. The language of Propositional Dynamic Logic (**PDL**)

The sentence-building operations include those of standard logical systems. In addition, if φ is a sentence and π a program, then $[\pi]\varphi$ is again a sentence. The intended meaning is "no matter how we run π, after we do so, φ holds." This formulation hints that programs are going to be non-deterministic, and so one of the syntactic formation rules does allow us to take the *union* (or *non-deterministic choice*) of π and σ to form $\pi \cup \sigma$. The other formation rules include composition (;), testing whether a sentence is true or not ($?\varphi$), and iteration (π^*).

The basic idea in the semantics is that we have state set S to start, and programs are interpreted in the most extensional way possible, as relations over S. So we are identifying the program with its input-output behavior; since we are thinking of non-deterministic programs, this behavior is a relation rather than a function. Atomic programs are interpreted as relations which are given as part of a model, and the rest of the programs and sentences are interpreted by a simultaneous inductive definition given in Figure 6. With this interpretation, each program π turns into a sentence-forming operation $[\pi]$; these then behave exactly as in standard relational modal systems. The clause for program composition uses composition of relations, and the one for iteration uses the reflexive-transitive closure operation.

PDL turns out to be decidable and to have a nice axiom system. The system resembles modal logic, and indeed one takes the basic axioms and rules for the operators $[\pi]$ given by programs. The other main axioms and rules are

(Test)	$[?\varphi]\psi \leftrightarrow (\varphi \rightarrow \psi)$
(Composition)	$[\pi;\sigma]\varphi \leftrightarrow [\pi][\sigma]\varphi$
(Choice)	$[\pi \cup \sigma]\varphi \leftrightarrow ([\pi]\varphi \wedge [\sigma]\varphi)$
(Mix)	$[\pi^*]\varphi \rightarrow \varphi \wedge [\pi][\pi^*]\varphi$

One also has an Induction Rule:

$$\text{From } \chi \rightarrow \psi \text{ and } \psi \rightarrow [\pi]\chi, \text{ infer } \chi \rightarrow [\pi^*]\psi.$$

The treatment of iteration is related to what we saw for common knowledge in Section 4.5; there is a common set of mathematical principles at work. For more on **PDL**, see, e.g., Harel, Kozen, and Tiuryn [2000].

PDL and epistemic updates One important observation that links **PDL** with epistemic logic is that the changes in agents' accessibility relations as a result of an epistemic action of some sort or other are often given as *relation-changing* programs. We want to spell this out in detail, because it will be important in Section 7.2.

The semantics of a **PDL** sentence φ in a given model M may be taken to be a subset $[\![\varphi]\!] \subseteq M$, namely the set of worlds making φ true. Similarly, the semantics of a **PDL** program π may be taken to be a relation $[\![\pi]\!]$ on M. Now given a **PDL** program $\pi(r)$ with a *relation variable* r, we can interpret $\pi(r)$ by a function $[\![\pi(r)]\!]$ from relations on M to relations on M (a *relation transformer*): for each $R \subseteq M \times M$, we use R for the semantics of r and the rest of the semantics as above.

We shall see a simple example of this shortly, in Example 10 of Section 5.1.

5 DYNAMIC EPISTEMIC LOGIC

We now move on to a different form of dynamics related to the topic of our chapter. Starting from the perspective of epistemic logic, knowledge and belief change can also be modeled by expanding the logic with dynamic modal operators to express such changes. The result is known as *Dynamic Epistemic Logic(s)*, or **DEL** for short. The first and the simplest form of dynamics is that associated with *public announcements*. It is simple from the perspective of change, but not particularly simple seen as an extension of epistemic logic. Public announcement logic is discussed in Section 5.1, and some related technical results of philosophical interest are presented in Section 5.2. Next, we move on to various forms of *private announcements* and to the associated dynamic logics, presented in Section 5.3. Even more complex types of dynamics, induced by various types of *epistemic actions*, are treated in Section 5.4. Finally, in Section 5.5 we briefly introduce logical languages and axioms for epistemic actions.

5.1 Public announcements

We first saw public announcements in Section 2.1. The example there was very simple indeed, and so to illustrate the phenomenon further, it will be useful to have a more complicated scenario. This discussion in this section is based on Example 2 in Section 4.4. Assume for the moment that the card deal is described by ♣♡♠: Amina holds clubs, Bao hearts, and Chandra spades. Amina now says ('announces') that she does not have the hearts card. Therefore she makes public to all three players that all deals where $Hearts_a$ is true can be eliminated from consideration: everybody knows that everybody else eliminates those deals, etc. They can therefore be *publicly* eliminated. This results in a restriction of the model *Hexa* as depicted in Figure 7.

At this point, we only consider announcements like this in states where the announcement is true. We view the public announcement "I do not have hearts" as

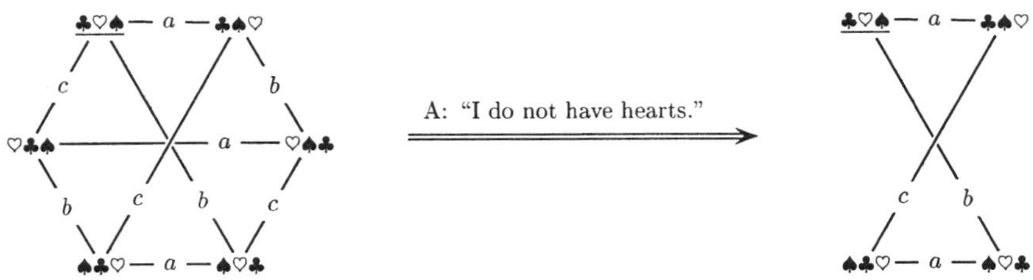

Figure 7. In the epistemic state (*Hexa*, ♣♡♠) Amina announces that she does not have hearts.

an 'epistemic program'. We interpret it as an *state transformer* just as flipping the box was so interpreted in Section 4.2. This program is interpreted as an 'epistemic state transformer' of the original epistemic state, exactly as we saw in Section 4.9 for **PDL**. We want $[!\varphi]\psi$ to mean that after (every) truthful announcement of φ, the sentence ψ holds. Continuing to borrow terminology from dynamic logic, state transformers come with *preconditions.* In this case, we want the precondition to be $\neg Hearts_a$, so that we set aside the matter of false announcements.

The effect of such a public announcement of φ is the restriction of the epistemic state to all worlds where φ holds. So, 'announce φ' can indeed be seen as an epistemic state transformer, with a corresponding dynamic modal operator $[!\varphi]$.

We *appear* to be moving away slightly from the standard paradigm of modal logic. So far, the accessibility relations were between states in a given model underlying an epistemic state. But all of a sudden, we are confronted with an accessibility relation *between* epistemic states as well. "I do not have hearts" induces a(n) (epistemic) state transition such that the pair of epistemic states in Figure 7 is in that relation. The epistemic states take the role of the points or worlds in a seemingly underspecified domain of 'all possible epistemic states'. By lifting accessibility between points in the original epistemic state to accessibility between epistemic states, we can get the dynamic and epistemic accessibility relations 'on the same level' again, and see this as an 'ordinary structure' on which to interpret a perfectly ordinary multimodal logic. (There is also a clear relation here with interpreted systems, which will be discussed in Subsection 6.3, later.) A crucial point is that this 'higher-order structure' is induced by the initial epistemic state and the actions that can be executed there, and not the other way round. So it is standard modal logic after all.

Amina's announcement "I do not have hearts" is a simple epistemic action in various respects. *It is public.* A 'private' event would be when she learns that Bao has hearts without Bao or Chandra noticing anything. This requires a more complex action description. *It is truthful.* She could also have said "I do not have clubs." She would then be lying, but, e.g., may have reason to expect that

Bao and Chandra believe her. This would also require a more complex action description. *It is deterministic.* In other words, it is a state transformer. A non-deterministic action would be that Amina whispers into Bao's ear a card she does not hold, on Bao's request for that information. This action would have two different executions: "I do not have hearts", and "I do not have spades." Such more complex actions can be modeled in the action model logic presented in Section 5.4.

Language and semantics Add an inductive clause $[!\varphi]\psi$ to the definition of the language. For the semantics, add the clause:

$$M, s \models [!\varphi]\psi \quad \text{iff} \quad M, s \models \varphi \text{ implies } M|\varphi, s \models \psi$$

where $M|\varphi = \langle S', R', V' \rangle$ is defined as

$$\begin{array}{lcl} S' & \equiv & \{s \in S \mid M, s \models \varphi\} \\ R'_a & \equiv & \{(s,t) \in S' \times S' : (s,t) \in R_a\} \\ V'_p & \equiv & \{s \in S : s \in V_p\} \end{array}$$

In other words: the model $M|\varphi$ is the model M restricted to all the states where φ holds, including access between states (a submodel restriction in the standard meaning of that term). It might be useful to look back at Section 2.1 for a discussion of the parallel case of probabilistic conditioning.

The language described above is called the language of public announcement, or *public announcement logic* (**PAL**).

EXAMPLE 9. After Amina's announcement that she does not have hearts, Chandra knows that Amina has clubs (see Figure 7). We can verify this with a semantic computation as follows: In order to check that $Hexa, \clubsuit\heartsuit\spadesuit \models [!\neg Hearts_a]K_c Clubs_a$, we have to show that $Hexa, \clubsuit\heartsuit\spadesuit \models \neg Hearts_a$ implies $Hexa|\neg Hearts_a, \clubsuit\heartsuit\spadesuit \models K_c Clubs_a$. The antecedent of this conditional being true, it remains to show that $Hexa|\neg Hearts_a, \clubsuit\heartsuit\spadesuit \models K_c Clubs_a$. The state $Hexa|\neg Hearts_a, \clubsuit\heartsuit\spadesuit$ is shown in Figure 7. Clearly, at the world $\clubsuit\heartsuit\spadesuit$ in it, $K_c Clubs_a$.

EXAMPLE 10. We mentioned at the end of Section 4.9 that program terms in **PDL** with variables may be used to specify the actions of epistemic actions. Here is how this works for public announcement of a sentence φ. For each agent a, the program $\pi(r)$ we want is $?\varphi; r; ?\varphi$. As previously explained, this defines a relation transformer $[\![\pi(r)]\!]$ on the underlying model. Then if R_a is agent a's accessibility relation before the announcement, $[\![\pi(r)]\!](R_a)$ is her accessibility relation afterwards. In more detail,

$$\begin{array}{lcl} [\![?\varphi; r; ?\varphi]\!](R_a) & = & \{(w,w) : w \in [\![\varphi]\!]\}; \{(u,v) : u\, R_a, v\}; \{(w,w) : w \in [\![\varphi]\!]\} \\ & = & \{(u,v) : u, v \in [\![\varphi]\!] \text{ and } u\, R_a, v\} \end{array}$$

The dual operators $\langle !\varphi \rangle$ Most people prefer to consider the *dual* $\langle !\varphi \rangle$ of $[!\varphi]$. That is, we take $\langle !\varphi \rangle \psi$ to an abbreviation for $\neg[!\ \varphi]\neg\psi$. This is equivalent to saying that $M, s \models \langle !\varphi \rangle \psi$ if and only if $M, s \models \varphi$ *and* $M|\varphi, s \models \psi$.

The point is that statements of the form $[!\varphi]\psi$ are conditionals and therefore are taken to be true when their antecedents are false; the duals are conjunctions. To see the difference, $\langle !Hearts_a \rangle K_c Clubs_a$ is *false* at $(Hexa, \clubsuit\heartsuit\spadesuit)$.

Announcement and knowledge In general, $[!\varphi]K_a\psi$ is not equivalent to $K_a[!\varphi]\psi$. The easiest way to see this in our running example is to note that

$$Hexa, \clubsuit\heartsuit\spadesuit \not\models K_c[!\neg Hearts_a]\, Clubs_a.$$

The correct equivalence in general requires that we make the truth of $[!\varphi]K_a\psi$ conditional on the truth of the announcement. So we get the following:

$$[!\varphi]K_a\psi \text{ is equivalent to } \varphi \to K_a[!\varphi]\psi.$$

Announcement and common knowledge Incidentally, the principle describing the interaction between common knowledge and announcement is rather involved. It turns out to be an *inference rule* rather than an axiom scheme. (One may compare it to the Induction Rule of **PDL** which we saw in Section 4.9; the rule here generalizes that one.) We therefore turn next to the proof system for validity in this logic.

A logical system See Figure 8 for a a proof system for this logic, essentially taken from [Baltag *et al.*, 1998]. It has precursors (namely completeness results for the logic *with* announcements but *without* common knowledge) in [Plaza, 1989] and [Gerbrandy and Groeneveld, 1997]; technically, this works out easier because the rules of the logic allow one to rewrite all sentences in a way that eliminates announcements altogether, and in this situation we may appeal to the known completeness result for epistemic logic. Thus the main point in the axiomatic work of [Baltag *et al.*, 1998] was the formulation of the Announcement Rule relating announcements and common knowledge, and the resulting completeness theorem.

Announcements are functional If an announcement can be executed, there is only one way to do it. So the partial functionality axiom in Figure 8 is sound. It is also convenient to write this as

$$\langle !\varphi \rangle \psi \to [!\varphi]\psi.$$

This is a simple consequence of the functionality of the state transition semantics for announcement. One might also say (from a programming perspective) that announcements are *deterministic*.

$[!\varphi]p \leftrightarrow (\varphi \rightarrow p)$	atomic permanence
$[!\varphi]\neg\psi \leftrightarrow (\varphi \rightarrow \neg[!\varphi]\psi)$	partial functionality
$[!\varphi]K_a\psi \leftrightarrow (\varphi \rightarrow K_a[!\varphi]\psi)$	announcement-knowledge

From $\chi \rightarrow [!\varphi]\psi$ and $\chi \wedge \varphi \rightarrow K_a\chi$ for all $a \in B$, infer $\chi \rightarrow [!\varphi]C_B\psi$ (the Announcement Rule)

Figure 8. The main points of the logic of public announcements. We have omitted the usual modal apparatus for modalities $[!\varphi]$.

Sequence of announcements A sequence of two announcements can always be replaced by a single, more complex announcement. Instead of first saying 'φ' and then saying 'ψ' you may as well have said for the first time 'φ, and after saying that ψ'. This is expressed in

$$[!\langle!\varphi\rangle\psi]\chi \leftrightarrow [!\varphi][!\psi]\chi.$$

This is useful when analyzing announcements that are made with specific intentions; or, more generally, conversational implicatures à la Grice. Intentions can be postconditions ψ that should hold after the announcement. So the (truthful) announcement of φ with the intention of achieving ψ corresponds to the announcement $\langle!\varphi\rangle\psi$.

If a sentence is common knowledge to all agents, there is no point in announcing it It will not change anyone's knowledge state:

$$C_A\varphi \rightarrow (K_a\psi \leftrightarrow [!\varphi]K_a\psi).$$

Here A is our set of all agents.

What can be achieved by public announcements? An interesting question, related to van Benthem's [2004] discussion of Fitch's paradox presented in Section 3.6, is to characterize the sentences that *can come to be true through some (sequence of) public announcement(s)*, at a given state (in a given model). One answer is offered by the "ability" modality $\Diamond\varphi$, informally introduced in Section 3.6. Now we can formally define the semantics of $\Diamond\varphi$, by saying it is true at a state iff there exists some epistemic sentence ψ such that $\langle!\psi\rangle\varphi$ is true at that state. So the "ability" modality can be obtained by quantifying over public announcements (for all epistemic sentences). The above observation on the fact that a sequence of public announcements can be simulated by a single announcement shows that we do not have to iterate the defining clause of $\Diamond\varphi$: it already captures what can be achieved by any iteration. It also means that this modality satisfies the axioms of the system S4. Balbiani et al [2007] call $\Diamond\varphi$ the "arbitrary announcement" modality, and give a complete axiomatization of the logic of arbitrary announcements, as well as studying its expressivity.

Alternative semantics There are some alternative semantics for public announcements. Gerbrandy [1999; 2007] and Kooi [2007] propose a different semantics for announcements in a setting possibly more suitable for 'belief'. The execution of such announcements is not conditional to the truth of the announced formula.

Yet another semantics has recently been proposed by Steiner [2006], to solve the problem of inconsistent beliefs that may be induced by public announcements. The semantics presented in this section has the disadvantage that updates induced by announcements do not necessarily preserve the seriality property (axiom D above): agents who have wrong (but consistent) beliefs may acquire inconsistent beliefs after a truthful public announcement. Steiner's alternative semantics solves this by proposing a modified semantics in which the new information is rejected if not consistent with prior beliefs. Yet another possible solution would be to incorporate some mechanism for belief revision, along the lines we discuss in Section 7.

Relativized common knowledge Recent developments in the area use a different modal notion, 'relativized common knowledge', of which standard common knowledge can be seen as a special case [van Benthem *et al.*, 2006b; Kooi, 2007]. Here is the idea. Add to the syntax an operation $C_B(\varphi, \psi)$. The semantics is

$w \models C_B(\varphi, \psi)$ iff every path from w using $\bigcup_{a\in B} \xrightarrow{a}$ consisting entirely of worlds where φ holds ends in a world where ψ holds

This results in more a expressive logic. At the same time, the relation between announcements and relativized common knowledge turns into an axiom:

$$(E_B(\varphi \to \psi) \to C_B(\varphi, \psi \to E_B(\varphi \to \psi))) \to C_B(\varphi, \psi).$$

van Benthem, van Eijck and Kooi [2006b] contains the completeness proofs for this logic and others, and also various expressivity results.

Iteration The language of **PDL** has an iteration operator on actions, but this has not been reflected in any of our example scenarios. However, there are scenarios and protocols whose natural description uses action iteration. One example is the general form of the Muddy Children-type scenario, as we described it in Section 3.1. We discuss this in connection with the sentences in Figure 9. These are based on sentences in Gerbrandy and Groeneveld [1997]. In them, d_a is an atomic sentence asserting that child a is dirty, and similarly for the other children. Informally, the sentence VISION says that every child a can see and therefore knows the status of

VISION	$\bigwedge_{a\in A}\bigwedge_{b\neq a}((\mathsf{d}_b \to K_a\mathsf{d}_b) \wedge (\neg\mathsf{d}_b \to K_a\neg\mathsf{d}_b))$
AT LEAST ONE	$\bigvee_{a\in A}\mathsf{d}_a$
BACKGROUND	$C_A(\text{VISION} \wedge \text{AT LEAST ONE})$
NOBODY KNOWS	$\bigwedge_{a\in A}(\neg K_a\mathsf{d}_a \wedge \neg K_a\neg\mathsf{d}_a)$
SOMEBODY KNOWS	$\neg$NOBODY KNOWS

Figure 9. Abbreviations in the discussion of the Muddy Children scenario, following [Gerbrandy and Groeneveld, 1997].

all other children. Note that VISION is a (finite) sentence since the set A of agents (the children here) is finite. BACKGROUND says that it is common knowledge that VISION and AT LEAST ONE hold. The intuition is that this is the background that the children have after the adult's announcement that at least one of them is dirty.

The sentence BACKGROUND is much weaker than what one would usually take to be the formalization of the overall background assumptions in the Muddy Children scenario. However, it is enough for the following result. Let

$$\varphi_A \quad \equiv \quad \text{BACKGROUND} \to \langle\text{NOBODY KNOWS}^*\rangle\text{SOMEBODY KNOWS}. \qquad (18)$$

Note the $*$ in (18). The formal semantics would make this equivalent to the infinitary sentence

$$\text{BACKGROUND} \to \bigvee_n \langle\text{NOBODY KNOWS}^n\rangle\text{SOMEBODY KNOWS}.$$

Either way, φ_A says that given the background assumption, some finite number of public announcements of everyone's ignorance will eventually result in the opposite: someone knowing their status.

PROPOSITION 11. *For each finite set A of children,* $\models \varphi_A$.

For a proof, see Miller and Moss [2005]. The point of Proposition 11 is that the statements φ_A are natural *logical validities.* So it makes sense to ask for a logical system in which such validities coincide with the provable sentences. The basic logic of announcements and common knowledge is known to be decidable, and indeed we have seen the axiomatization in Figure 8. However, it was shown in [Miller and Moss, 2005] that adding the *iterated announcement* construct that gives us the $\langle$NOBODY KNOWS$^*\rangle$ operation results in logical systems whose satisfiable sentences are not decidable. The upshot is that (unfortunately) there is no hope of a finitely axiomatized logical system for the validities in a language which includes sentences like (18).

Notes The logic of multi-agent epistemic logic with public announcements and without common knowledge has been formulated and axiomatized by Plaza [1989]. For the somewhat more general case of introspective agents, this was done by Gerbrandy and Groeneveld [1997]; they were not aware of Plaza's work at the time. In [Plaza, 1989], public announcement is seen as a binary operation $+$, such that $\varphi + \psi$ is equivalent to $\langle !\varphi \rangle \psi$. The logic of public announcements *with* common knowledge was axiomatized by Baltag, Moss, and Solecki [1998], see also [Baltag *et al.*, 1999; Baltag, 2002; Baltag and Moss, 2004], in a more general setting that will be discussed in Section 5.4: the completeness of their proof system is a special case of the completeness of their more general logic of action models. A concise introduction into public announcement logic (and also some of the more complex logics presented later) is found in [van Ditmarsch *et al.*, 2005]. A textbook presentation of the logic is [van Ditmarsch *et al.*, 2007]. This also contains a more succinct completeness proof than found in the original references. Results on complexity of the logic are presented by Lutz in [2006].

There are a fair number of precursors of these results. One prior line of research is in dynamic modal approaches to semantics, not necessarily also epistemic: 'update semantics'. Another prior line of research is in meta-level descriptions of epistemic change, not necessarily on the object level as in dynamic modal approaches. This relates to the temporal epistemics and interpreted systems approach for which we therefore refer to the summary discussion in the previous section.

The 'dynamic semantics' or 'update semantics' was followed in van Emde Boas, Groenendijk, and Stokhof [1984], Landman [1986], Groeneveld [1995], and Veltman [1996]. However, there are important philosophical and technical differences between dynamic semantics and dynamic epistemic logic as we present it here. The main one is that update semantics interprets meaning (in natural language) as a relation between states, and so it departs from standard accounts. Nevertheless, the "dynamic" feature is common to both. Work taking propositional dynamic logic (**PDL**) in the direction of natural language semantics and related areas was initiated by van Benthem [1989] and followed up in de Rijke [1994] and Jaspars [1994]. As background literature to various dynamic features introduced in the 1980s and 1990s we recommend van Benthem [1989; 1996; 1994]. More motivated by runs in interpreted systems is van Linder, van der Hoek, and Meyer [1995]. All these approaches use dynamic modal operators for information change, but (1) typically not (except [van Linder *et al.*, 1995]) in a multi-modal language that also has epistemic operators, (2) typically not for more than one agent, and (3) not necessarily such that the effects of announcements or updates are defined given the update formula and the current information state: the **PDL**-related and interpreted system related approaches *presuppose* a transition relation between information states, such as for atomic actions in **PDL**. We outline, somewhat arbitrarily, some features of these approaches. Groeneveld's approach [Groeneveld, 1995] is typical for dynamic semantics in that is has formulas $[!\varphi]_a \psi$ to express that after an update of agent a's information with φ, ψ is true. His work was later merged with that of Gerbrandy, resulting in the seminal [Gerbrandy and Groeneveld, 1997]. Gerbrandy's semantics of public announcements is given in [Gerbrandy, 1999], in terms of the *universe* V_{AFA} *of non-wellfounded sets*: this is a kind of "universal Kripke model"; i.e., a class in which every Kripke model can be embedded in a unique manner (up to bisimilarity; see the end of Section 4.4).

In this way, one can avoid changing the initial model (by eliminating states and arrows) after a public announcement: instead, one just moves to another state in the same huge, all-encompassing Kripke super-model V_{AFA}. It was later observed in [Moss, 1999] that one can do with ordinary models.

De Rijke [1994] defines theory change operators $[+\varphi]$ and $[*\varphi]$ with a dynamic interpretation that link an enriched dynamic modal language to AGM-type theory revision [Alchourrón *et al.*, 1985] (see also Section 7 addressing dynamic epistemics for belief revision). In functionality, it is not dissimilar to Jaspars' [1994] φ-addition (i.e., expansion) operators $[!\varphi]_u$ and φ-retraction (i.e., contraction) operators $[!\varphi]_d$, called updates and downdates by Jaspars. Van Linder, van der Hoek, and Meyer [1995] use a setting that combines dynamic effects with knowledge and belief, but to interpret various action operators they assume an explicit transition relation as part of the Kripke structure interpreting such descriptions.

As somewhat parallel developments to [Gerbrandy, 1999], we also mention Lomuscio and Ryan [1999]. They do not define dynamic modal operators in the language, but they define epistemic state transformers that clearly correspond to the interpretation of such operators: $M * \varphi$ is the result of refining epistemic model M with a formula φ, etc. Their semantics for updates is only an *approximation* of public announcement logic, as the operation is only defined for *finite* (approximations of) models.

5.2 Sentences true after being announced

Moore sentences, revisited Recall that Moore sentences are *strongly unsuccessful*: they are always false after being announced. In terms of our public announcement logic, we can define strongly unsuccessful formulas φ as the ones such that the formula $[!\varphi]\neg\varphi$ is valid. An interesting open problem is to give a syntactic characterization of strongly unsuccessful sentences.

Successful formulas A more interesting and natural question is to characterize syntactically the *successful* formulas: those φ such that $[!\varphi]\varphi$ is valid. That is, whenever φ holds and is announced, then φ holds after the announcement. In our setting, it is easy to see that a successful formula has also the property that $[!\varphi]C_A\varphi$ is valid.

For example, the atomic sentences p are successful, as are their boolean combinations and also the sentences Kp. *Logically inconsistent formulas* are also trivially successful: they can never be truthfully announced, so after their truthful announcement everything is true (including themselves). *Public knowledge formulas* are also successful: $[!C_A\varphi]C_A\varphi$ is valid. This follows from bisimulation invariance under point-generated submodel constructions. On the negative side, even when both φ and ψ are successful, $\neg\varphi$ may be unsuccessful (for $\varphi = \neg p \vee Kp$), $\varphi \wedge \psi$ may be unsuccessful (for $\varphi = p$ and $\psi = \neg Kp$), and as well $[!\varphi]\psi$ and $\varphi \to \psi$ may be unsuccessful.

In its general form, the question of syntactically characterizing successful sentences remains *open*. But we present now two results on this problem.

Preserved formulas One successful fragment form the *preserved formulas* (introduced for the language without announcements by van Benthem in [2002]) that are inductively defined as

$$\varphi ::= p \mid \neg p \mid \varphi \wedge \psi \mid \varphi \vee \psi \mid K_a\varphi \mid C_B\varphi \mid [!\neg\varphi]\psi$$

(where $B \subseteq A$). From $\varphi \to [!\psi]\varphi$ for arbitrary ψ, follows $\varphi \to [!\varphi]\varphi$ which is equivalent to $[!\varphi]\varphi$; therefore preserved formulas are successful formulas. The inductive case $[!\neg\varphi]\psi$ in the 'preserved formulas' may possibly puzzle the reader. Its proof [van Ditmarsch and Kooi, 2006] is quite elementary (and proceeds by induction on formula structure) and shows that the puzzling *negation* in the announcement clause is directly related to the truth of the announcement as a *condition*:

Let $M, s \models [!\neg\varphi]\psi$, and $M' \subseteq M$ such that $s \in M'$. Assume $M', s \models \neg\varphi$. Then $M, s \models \neg\varphi$ by contraposition of the inductive hypothesis for φ. From that and $M, s \models [!\neg\varphi]\psi$ follows $M|\neg\varphi, s \models \psi$. From the inductive hypothesis for ψ follows $M'|\neg\varphi, s \models \psi$. Therefore $M', s \models [!\neg\varphi]\psi$ by definition.

Universal formulas A different guess would be that φ is successful iff φ is equivalent to a sentence in the *universal* fragment of modal logic, the fragment built from atomic sentences and their negations using K, $\wedge$, and $\vee$. However, this is not to be. We discuss work on single agent models whose accessibility relation is an *equivalence relation* in this discussion. It remains open to weaken this assumption and obtain similar results.

Suppose that φ and ψ are non-modal sentences (that is, boolean combinations of atomic sentences). Suppose that $\models \psi \to \varphi$. Consider $\varphi \vee \hat{K}\psi$. (Again, $\hat{K}$ is the dual of K, the "possibility" operator.) This is clearly not in general equivalent to a sentence in our fragment. Yet we claim that

$$\models [!(\varphi \vee \hat{K}\psi)](\varphi \vee \hat{K}\psi).$$

To see this, fix a state model M and some state s in it. If $s \in [\![\varphi]\!]$ in M, then since φ is non-modal, s "survives the announcement" and satisfies $\varphi \vee \hat{K}\psi$ in the new model. On the other hand, suppose that $s \in [\![\hat{K}\psi]\!]$ in M. Let $s \to t$ with $t \in [\![\psi]\!]$ in M. Then again, t survives and satisfies ψ and even $\varphi \vee \hat{K}\psi$. Hence, s satisfies $\varphi \vee \hat{K}\psi$ in the new model.

This example is due to Lei Qian. He also found a hypothesis under which the "first guess" above indeed holds. Here is his result. Let T_0 be the set of non-modal sentences. Let

$$T_1 \quad = \quad T_0 \cup \{K\varphi : \varphi \in T_0\} \cup \{\hat{K}\varphi : \varphi \in T_0\}.$$

Finally, let T_2 be the closure of T_1 under $\wedge$ and $\vee$.

THEOREM 12 (Qian [2002]). *Let $\varphi \in T_2$ have the property that $\models [!\varphi]\varphi$. Then there is some ψ in the universal fragment of modal logic such that $\models \varphi \leftrightarrow \psi$.*

5.3 Varieties of privacy

As a warm up before meeting the general notion of "epistemic actions" in the next section, we present here two generalizations of public announcements: the first, called *fully private announcements*, is essentially due (modulo minor differences[7]) to Gerbrandy [1997; 1999], while the second, which we call *fair-game announcements*, is due to van Ditmarsch [2000; 2002]. Both can be regarded as forms of private announcements: some information is broadcast to an agent, or a group of agents, while being withheld from the outsiders. But there are important differences: a fully private announcement is so secret that the outsiders do not even suspect it is happening; while a fair-game announcement is known by outsiders to be possible, among other possible announcements.

Fully Private Announcements to Subgroups For each subgroup B of agents, $!_B\varphi$ is the action of secretely broadcasting φ to all the agents in the group B, in a way that is completely oblivious to all outsiders $a \notin B$: they do not even suspect the announcement is taking place. An example of fully private announcement was encountered in Section 2.3: Bao is informed that the coin lies Heads up, but in such a way that Amina does not suspect that this is happening. The announcement is *truthful* (as in the previous section) but completely private: after that, Amina still believes that Bao doesn't know the state of the coin.

Assuming that before Bao entered, it was common knowledge that nobody knew that state of the coin, the belief/knowledge model *before the announcement* is a multi-agent version of the model

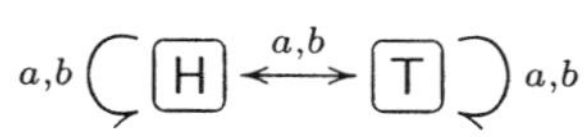

(19)

The situation after the fully private announcement (by which Bao is secretely informed that the coin lies Heads up) is given by the model (6) from Section 2.3. To recall, this was:

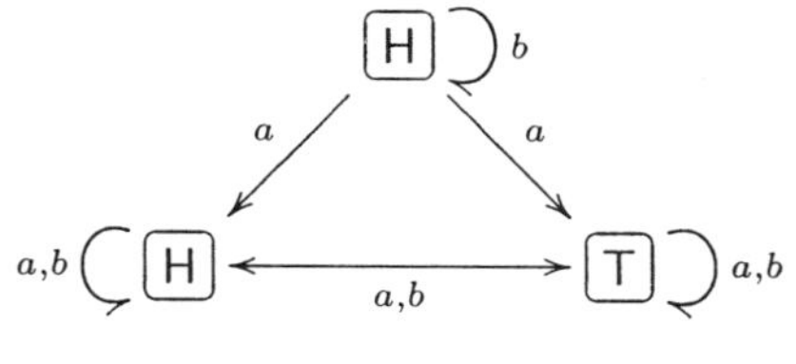

(20)

We can see that unlike the case of public announcements, the number of states *increases* after a fully private announcement. In fact, one can think of the model in the above picture as being obtained by putting together the initial model (19) and the model obtained from it by doing a public announcement, with the outsider

[7]As for public announcements, Gerbrandy's private announcements are not necessarily truthful. We present here a slightly modified version, that assumes truthfulness, in order to be able to subsume public announcements (as presented in Section 5.1) as a special case.

(Amina) having doxastic arrows between the two submodels. In other words, the state transformer for a fully private announcement combines features of the original model with the one given by the state transformer for a public announcement.

Language and semantics Add an inductive clause $[!_B\varphi]\psi$ to the definition of the language. For the semantics, add the clause:

$$M, s \models [!_B\varphi]\psi \quad \text{iff} \quad M, s \models \varphi \text{ implies } M!_B\varphi, s \models \psi$$

where $M!_B\varphi = \langle S' \cup S, R', V' \rangle$ is defined as

$$\begin{array}{lcl} S' & \equiv & \{s \in S \mid M, s \models \varphi\} \\ R'_a & \equiv & R_a \cup \{(s,t) \in S' \times S' : (s,t) \in R_a\} \\ V'_p & \equiv & V(p) \cup \{s \in S' : s \in V(p)\} \end{array}$$

The language described above is called the logic of fully private announcements to subgroups. The axioms and rules are just as in the logic of public announcements, with a few changes. We must of course consider the relativized operators $[!_B\varphi]$ instead of their simpler counterparts $[!\varphi]$. The most substantive change which we need to make in Figure 8 concerns the Action-Knowledge Axiom. It splits into two axioms, noted below:

$$\begin{array}{ll} [!_B\varphi]K_a\psi \leftrightarrow (\varphi \rightarrow K_a[!_B\varphi]\psi) & \text{for } a \in B \\ [!_B\varphi]K_a\psi \leftrightarrow (\varphi \rightarrow K_a\psi) & \text{for } a \notin B \end{array}$$

The last equivalence says: assuming that φ is true, then after a private announcement of φ to the members of B, an outsider knows ψ just in case she knew ψ before the announcement.

Fair-game Announcements In a fair-game announcement, some information is privately learned by an agent or a group of agents, but the outsiders are aware of this possibility: it is publicly known that the announcement is one of a given list of possible alternatives, although only the insiders will known which one.

We illustrate fair-game announcements with two examples. Let us reconsider the epistemic state (*Hexa*, ♣♡♠) wherein Amina holds clubs, Bao holds hearts, and Chandra holds spades. It is shown in Figure 4. Consider the following scenario:

> *Amina shows (only) Bao her clubs card. Chandra cannot see the face of the shown card, but notices that a card is being shown.*

It is assumed that it is publicly known what the players can and cannot see or hear. Call the action we are discussing showclubs. The epistemic state transition induced by this action is depicted in Figure 10. Unlike after public announcements, in the showclubs action we cannot eliminate any state. Instead, all b-links between states have now been severed: whatever was the actual deal of cards,

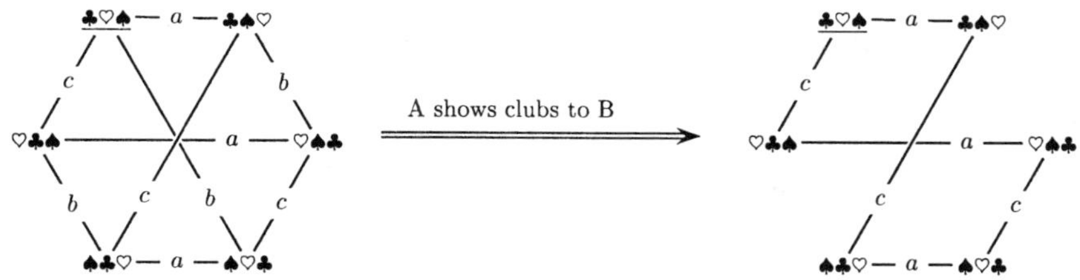

Figure 10. On the left, the Kripke model for three players each holding one card. On the right, the effect of Amina showing her clubs card to Bao.

Bao now knows that card deal and cannot imagine any alternatives. We hope to demonstrate the intuitive acceptability of the resulting epistemic state. After the action showclubs, Amina considers it possible that Chandra considers it possible that Amina has clubs. That much is obvious, as Amina has clubs anyway. But Amina also considers it possible that Chandra considers it possible that Amina has hearts, because Amina considers it possible that Chandra has spades, and so does not know whether Amina has shown clubs or hearts. It is even the case that Amina considers it possible that Chandra considers it possible that Amina has spades, because Amina considers it possible that Chandra does not have spades but hearts, in which case Chandra would not have known whether Amina has shown clubs or spades. And in all those cases where Amina shows her card, Bao obviously would have learned the deal of cards. Note that, even though for Chandra there are only two possible actions—showing clubs or showing hearts—none of the *three* possible actions can be eliminated from public consideration.

But it can become even more complex. Imagine the following action, rather similar to the showclubs action:

> *Amina whispers into Bao's ear that she does not have the spades card, given a (public) request from Bao to whisper into his ear one of the cards that she does not have.*

This is the action whispernospades. Given that Amina has clubs, she *could* have whispered "no hearts" or "no spades". And whatever the actual card deal was, she could always have chosen between two such options. We obtain a model that reflects all possible choices, and therefore consists of $6 \times 2 = 12$ different states. It is depicted in Figure 11 (wherein we assume transitivity of the accessibility relation for c). There is a method of calculating complex representations like this one, and we shall discuss this particular model in Example 17 in the next section. But for now, the reader may look at the model itself to ascertain that the desirable postconditions of the action whispernospades indeed hold. For example, given that Bao holds hearts, Bao will now have learned from Amina what Amina's

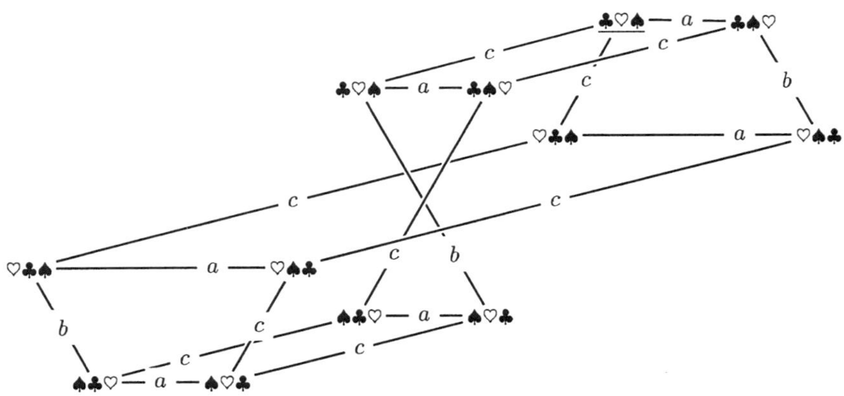

Figure 11. After Amina whispered into Bao's ear that she does not have the spades card, given a (public) request from Bao to whisper into his ear one of the cards that she does not have. Assume transitivity of the accessibility relation for Chandra.

card is, and thus the entire deal of cards. So there should be no alternatives for Bao in the actual state (the underlined state ♣♡♠ 'at the back' of the figure—for convenience, different states for the same card deal have been given the same name). But Chandra does not *know* that Bao knows the card deal, as Chandra considers it possible that Amina actually whispered "no hearts" instead. That would have been something that Bao already knew, as he holds hearts himself—so from that action he would not have learned very much. Except that Chandra could then have imagined him to know the card deal ... Note that in Figure 11, there is also another state named ♣♡♠, 'in the middle', so to speak, that is accessible for Chandra from the state ♣♡♠ 'at the back', and that witnesses that Bao doesn't know that Amina has clubs.

Notes The logic of fully private announcements has been first formulated and axiomatized by Gerbrandy [Gerbrandy and Groeneveld, 1997; Gerbrandy, 1999], in a slightly different version: as for public announcements, Gerbrandy's private announcements are not necessarily truthful. Also, Gerbrandy's semantics of fully private announcements, as the one of public announcements, is given in terms of non-wellfounded sets, rather than Kripke models. The version presented here (which assumes truthfulness and uses Kripke semantics) was formulated in Baltag and Moss [2004], as a special case of a "logic of epistemic programs". Gerbrandy [1999] considers more general actions: fully private announcements are only a special case of his operation of *(private) updating with an epistemic program.*

The logic of fair-game announcements is a special case of the work by van Ditmarsch [2000; 2002] and by Baltag, Moss, and Solecki [1998; 2004]; the latter call it "the logic of common knowledge of alternatives".

5.4 *Epistemic actions and the product update*

As we saw in the previous section, some epistemic actions are more complex than public announcements, where the effect of the action is always a restriction on the epistemic model. As in the previous examples, the model may grow in complex and surprising ways, depending on the specific epistemic features of the action. Instead of computing by hand the appropriate state transformer for each action, it would be useful to have a general setting, in which one could input the specific features of any desired action and compute the corresponding state transformer in an automatic way.

Action models We present a formal way to model such actions, and a large class of similar events, via the use of 'action models', originating in [Baltag *et al.*, 1998]. The basic idea is that the agents' uncertainty about actions can profitably be modeled by putting them in relation to other possible actions, in a way similar to how the agents' uncertainty about states was captured in a Kripke model by relating them to other possible states. When Amina shows her clubs card to Bao, this is indistinguishable for Chandra from Amina showing her hearts card to Bao— if she were to have that card. And, as Amina considers it possible that Chandra holds hearts instead of spades, Amina also considers it possible that Chandra interprets her card showing action as yet a third option, namely showing spades. These three different card showing actions are therefore, from a public perspective, all indistinguishable for Chandra, but, again from a public perspective, all different for Amina and Bao.

We can therefore visualize the 'epistemic action' of Amina showing clubs to Bao as some kind of Kripke structure, namely with a domain of three 'action points' standing for 'showing clubs', 'showing hearts', and 'showing spades', and accessibility relations for the three players corresponding to the observations above. We now have what is called an *action model*. What else do we need? To relate such 'action models' to the preconditions for their execution, we associate to each action point in such a model a formula in a logical language: the precondition of that action point.

To execute an epistemic action, we compute what is known as the *restricted modal product* of the current epistemic state and the epistemic action. The result is 'the next epistemic state'. It is a *product* because the domain of the next epistemic state is a subset of the cartesian product of the domain of the current epistemic state and the domain of the action model. It is *restricted* because we restrict that full product to those (state, action) pairs such that the precondition for the action of the pair is satisfied in the state of the pair. Two states in the new epistemic state are indistinguishable (accessible), if and only if the states in the previous epistemic state from which they evolved were already indistinguishable (accessible), and if the two different actions executed there were also indistinguishable. For example, Chandra cannot distinguish the result of Amina showing clubs in state ♣♡♠ from Amina showing hearts in state ♡♣♠, because in the first place she could not distinguish those two card deals, and in the second place she cannot distinguish Amina showing clubs from Amina showing hearts.

REMARK 13. This is perhaps a good point to make a comment on the terminology. What we are calling "action models" involve "actions" in an abstract sense, and so some of the important features of real actions are missing. For example, there is no notion of *agency* here: events like public announcement are modeled without reference to any agent(s) whatsoever as their source. Further, they may well be complex (many-step) actions, and for this reason they are also called *programs* in work such as [Baltag and Moss, 2004]. So other authors have called our "action models" *event models*. We maintain the older terminology mainly because this is how it has appeared in the literature.

We now formally define action models and their execution, for any given logical language. We leave for later the problem of finding a good such language for describing epistemic actions and their effects. As usual, we assume background parameters in the form of a set of agents A and a set of propositional variables P.

DEFINITION 14 (Action model). Let $\mathcal{L}$ be a logical language. An *action model over* $\mathcal{L}$ is a structure $\mathsf{U} = \langle \mathsf{S}, \mathsf{R}, \mathsf{pre} \rangle$ such that S is a domain of *action points*, such that for each $a \in A$, R_a is an accessibility relation on S, and such that $\mathsf{pre} : \mathsf{S} \to \mathcal{L}$ is a precondition function that assigns a *precondition* $\mathsf{pre}(\alpha) \in \mathcal{L}$ to each $\alpha \in \mathsf{S}$. An *epistemic action* is a pointed action model (U, α), with $\alpha \in \mathsf{S}$.

EXAMPLE 15. The *public announcement* of φ is modeled by a singleton action model, consisting of only one action point, accessible to all agents, and having φ as its precondition. We call this action model *Pub* φ, and denote by $!\varphi$ the (action corresponding to the) unique point of this model. A more concrete example is the action $!\neg Hearts_a$ in Section 5.1 in which Amina publicly announces that she does not have the hearts card: the action model is $Pub\neg Hearts_a$.

A *fully private announcement* of φ to a subgroup B is modeled by a two-point action model, one point having precondition φ (corresponding to the private announcement) and the other point having precondition $\top := p \vee \neg p$ (corresponding to the case in which no announcement is made):

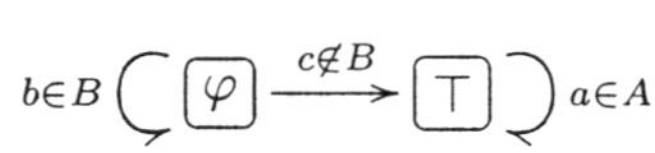

We call this action model $Pri_B\varphi$. Again, the action point on the left represents the fully private announcement of φ. This action will be denoted by $!_B\varphi$. The action point on the right has as precondition some tautology $\top$, and represents the alternative action in which *no announcement* is made: essentially nothing is happening. This action will be denoted by τ.

A more concrete example of a fully private announcement model is the action considered in Section 2.3, in which Bao was secretely informed that the coin lay heads up, without Amina suspecting this to be happening. This corresponds to the right-hand point in the action model Pri_b H:

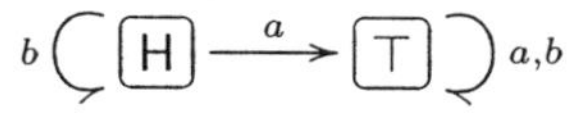

Fair-game announcements with n commonly-known alternatives can be modeled using an action model with n points, having the corresponding announcements as preconditions. For the "insiders", the accessibility relation is the identity relation, while the accessibility for the "outsiders" is the universal relation (linking every two points).

As a concrete example of fair-game announcement, the action model U for the showclubs action in the previous section has three action points α, β, and γ, with preconditions $\mathsf{pre}(\alpha) = Spades_a$, $\mathsf{pre}(\beta) = Hearts_a$, and $\mathsf{pre}(\gamma) = Clubs_a$. The epistemic structure of this model is:

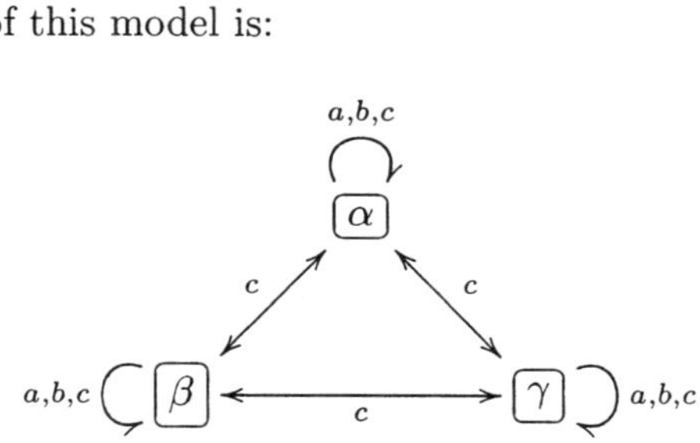

(21)

The pointed action model of interest is (U, γ). In it, the action γ which really happened is one where Amina and Bao come to share the knowledge that she has clubs: no other options are available to the two of them. Chandra, on the other hand, is in the dark about which of three announcements is taking place, but she does know the three possible messages: $Clubs_a, Spades_a, Hearts_a$.

Similarly, the action model U' for the action whispernospades in the previous section has the same structure as the model U above, except that we take: $\mathsf{pre}(\alpha) = \neg Clubs_a$, $\mathsf{pre}(\beta) = \neg Hearts_a$, $\mathsf{pre}(\gamma) = \neg Spades_a$. The pointed action model of interest is (U', γ).

DEFINITION 16 (Execution, Product Update). Consider an epistemic state (M, s) with $M = \langle S, R, V\rangle$ and an epistemic action (U, α) with $\mathsf{U} = \langle \mathsf{S}, \mathsf{R}, \mathsf{pre}\rangle$. The result of *executing* (U, α) *in* (M, s) is only defined when $M, s \models \mathsf{pre}(\alpha)$. In this case, it is the epistemic state $((M \otimes \mathsf{U}), (s, \alpha))$ where $(M \otimes \mathsf{U}) = \langle S', R', V'\rangle$ is a restricted modal product of M and U defined by

$$
\begin{array}{lll}
S' & \equiv & \{(s,\alpha) \mid s \in S, \alpha \in \mathsf{S}, \text{ and } M, s \models \mathsf{pre}(\alpha)\} \\
R'_a((s,\alpha),(t,\beta)) & \text{iff} & R_a(s,t) \text{ and } \mathsf{R}_a(\alpha,\beta) \\
(s,\alpha) \in V'_p & \text{iff} & s \in V_p
\end{array}
$$

This restricted product construction has become known in the **DEL** literature as "Product Update". Here, we simply call it *action execution*. The intuition is that *indistinguishable actions performed on indistinguishable input-states yield indistinguishable output-states*: if when the real state is s, agent a thinks it is possible that the state might be s', and if when action α is happening, agent a thinks it is possible that action α' might be happening, then after this, when the real state is (s, α), agent a thinks it is possible that the state might be (s', α').

EXAMPLE 17. At this point we can go back and justify all our previous state transformers in a uniform manner. The model in Figure 7 can be obtained by calculating $\mathit{Hexa} \otimes !\neg \mathit{Hearts}_a$, where *Hexa* is the model shown in Figure 4, and the action model $\mathit{Pub}\neg\mathit{Hearts}_a$ was described above. The model (6) from Section 2.3 can be computed by calculating the restricted modal product of the model (19) from Section 5.3 and the action model $\mathit{Pub}_b\mathsf{H}$ above. The pointed model shown in Figure 10 is obtained by calculating

$$(\mathit{Hexa} \otimes \mathsf{U}, (\clubsuit\heartsuit\spadesuit, \gamma)),$$

where U is from (21) above.

Finally, we justify the pointed model in in Figure 11, by calculating

$$(\mathit{Hexa} \otimes \mathsf{U}', (\clubsuit\heartsuit\spadesuit, \gamma)),$$

where the action model U' is as above. Let us look at this last calculation in more detail: the restricted product itself contains the twelve pairs

$$(\clubsuit\heartsuit\spadesuit, \beta), (\clubsuit\heartsuit\spadesuit, \gamma), (\clubsuit\spadesuit\heartsuit, \beta), (\clubsuit\spadesuit\heartsuit, \gamma), (\heartsuit\clubsuit\spadesuit, \alpha), (\heartsuit\clubsuit\spadesuit, \gamma),$$
$$(\heartsuit\spadesuit\clubsuit, \alpha), (\heartsuit\spadesuit\clubsuit, \gamma), (\spadesuit\clubsuit\heartsuit, \alpha), (\spadesuit\clubsuit\heartsuit, \beta), (\spadesuit\heartsuit\clubsuit, \alpha), (\spadesuit\heartsuit\clubsuit, \beta)$$

The valuation only looks at the first components. For example $(\heartsuit\spadesuit\clubsuit, \alpha) \models \mathit{Hearts}_a \wedge \mathit{Spades}_b \wedge \mathit{Clubs}_c$. The epistemic relations are determined in the usual way of products. For example, $R'_c((\spadesuit\heartsuit\clubsuit, \beta), (\heartsuit\spadesuit\clubsuit, \alpha))$ because Chandra cannot tell the difference between $\spadesuit\heartsuit\clubsuit$ and $\heartsuit\spadesuit\clubsuit$ in *Hexa*, and she also cannot tell the difference between β and α in U.

5.5 Logics for epistemic actions

There is only one more step to make: to give a logical language with an inductive construct for action models. The task of finding a natural general syntax for epistemic actions is not an easy problem. A number of different such languages have been proposed, see e.g., [Gerbrandy, 1999; Baltag, 2002; Baltag and Moss, 2004; van Benthem *et al.*, 2006b; van Ditmarsch, 2000; van Ditmarsch, 2002; van Ditmarsch *et al.*, 2003]. We follow [Baltag and Moss, 2004], presenting here only one type of syntax, based on the notion of *signature.*

DEFINITION 18 (Signature). An *action signature* is a *finite* Kripke frame Σ, together with a *list* $(\sigma_1, \ldots, \sigma_n)$ enumerating some of the elements of Σ without repetitions. The elements of Σ are called *action types.*

EXAMPLE 19. The *public announcement signature Pub* is a singleton frame, consisting of an action type !, accessible to all agents, and the list (!).

The signature Pri_B of *fully private announcements to a subgroup B* is a two-point Kripke frame, consisting of an action type $!_B$ (corresponding to fully private announcements) and an action type τ (for the case in which no announcement is made). The list is $(!_B)$, and the structure is given by:

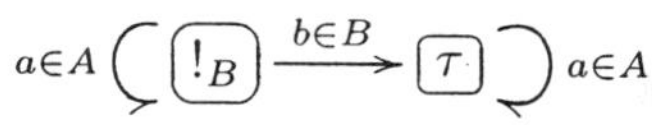

The signature of *fair-game announcements* (to a given group of insiders, and with common knowledge of a given finite set of alternatives) can be similarly formalized. For instance, the signature $Show_{a,b}$ for the logic of the actions showclubs and whispernospades (with a, b as insiders and c as outsider) is a frame with three action types listed as (σ, ρ, μ). The structure is:

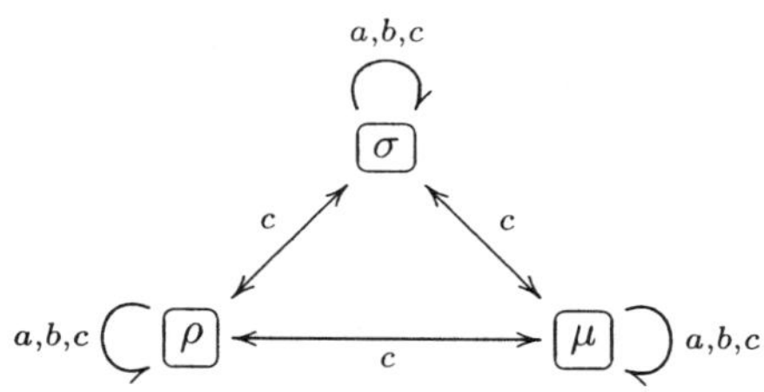

(22)

DEFINITION 20 (Language). For a given action signature Σ with a listing of some of its action types $(\sigma_1, \ldots, \sigma_n)$, the language $\mathcal{L}_\Sigma$ of the logic of Σ-actions is the union of the *formulas* φ and the *epistemic actions*[8] α defined by

$$\begin{array}{rcl}\varphi & ::= & p \mid \neg\varphi \mid \varphi \wedge \psi \mid K_a\varphi \mid C_B\varphi \mid [\alpha]\varphi \\ \alpha & ::= & \sigma\varphi_1 \ldots \varphi_n \mid \alpha \cup \beta\end{array}$$

where $p \in P$, $a \in A$, $B \subseteq A$, $\sigma \in \Sigma$, and $\sigma\varphi_1 \ldots \varphi_n$ above is an expression consisting of an action type σ followed by a string of n formulas, where n is taken from the listing in Σ.

The expressions of the form $\sigma\vec{\varphi}$ are called *basic epistemic actions.* In addition, we have included in the language $\mathcal{L}_\Sigma$ an operation of *non-deterministic choice* on the actions, mainly to show the reader familiar with dynamic logic, process algebra, and the like that it is possible to add such operations. One can also add *sequential composition, iteration* (Kleene star $*$), etc.

DEFINITION 21 (Action model induced by a signature). Given an action signature Σ with its list $(\sigma_1, \ldots, \sigma_n)$ of action types, and given a list $\vec{\varphi} = (\varphi_1, \ldots, \varphi_n)$ of n formulas in $\mathcal{L}_\Sigma$, the action model $\Sigma\vec{\varphi}$ is obtained by endowing the Kripke frame Σ with the following precondition map: if $\sigma = \sigma_i$ is in the given list $\vec{\alpha}$, we take $\mathsf{pre}(\sigma_i) := \varphi_i$; while, if σ is not in the given list $(\sigma_1, \ldots, \sigma_n)$, $\mathsf{pre}(\sigma)$ is taken

[8] We are using the letter α here for both action points in an action model and also for "epistemic actions" as syntactic expressions in our language. This ambiguity should not cause problems, but we wish to alert the careful reader of it.

to be some *tautology* $p \vee \neg p$. When seen as an *action point* in the action model $\Sigma\vec{\varphi}$, the type $\sigma \in \Sigma$ is denoted by $\sigma\vec{\varphi}$. Since the frame is the same as Σ, having the relation $\sigma\vec{\varphi} \xrightarrow{a} \sigma'\vec{\varphi}$ in the action model $\Sigma\vec{\varphi}$ is the same as having the relation $\sigma \xrightarrow{a} \sigma'$ in the frame Σ.

EXAMPLE 22. The action model *Pub* φ from the previous section is induced by the signature *Pri* above, in the obvious way. The action model $Pri_B\varphi$ from the previous section is induced by the signature *Pri* above. The action model U for the showclubs action in the previous section is induced by the signature $Show_{a,b}$, since it coincides with the model $Show_{a,b}Spades_a\,Clubs_a\,Hearts_a$. (This is an action signature followed by three propositions.) The model U′ for the whispernospades action is induced by the same signature, since it can be written as $Show_{a,b}\neg Clubs_a\neg Hearts_a\neg Spades_a$.

DEFINITION 23 (Semantics).

$$\begin{array}{lll} M,s \models [\sigma\vec{\varphi}]\psi & \text{iff} & M,s \models \mathsf{pre}(\sigma\vec{\varphi}) \text{ implies } (M \otimes \Sigma\vec{\varphi}),(s,\sigma\vec{\varphi}) \models \psi \\ M,s \models [\alpha \cup \beta]\varphi & \text{iff} & M,s \models [\alpha]\varphi \text{ and } M,s \models [\beta]\varphi \end{array}$$

Note that the preconditions in an action model are arbitrary sentences in the language, since we want to talk about announcements concerning announcements and similar things. In fact, to avoid vicious circles, the definition of the semantics of $\mathcal{L}_\Sigma$ and the definition of action execution (as in the previous section) for action model over $\mathcal{L}_\Sigma$ should be taken to form one single definition (by simultaneous double induction) of both concepts. As usual, $\langle\alpha\rangle\varphi$ is defined by duality as $\neg[\alpha]\neg\varphi$.

It is easy to see that the logic of public announcements (**PAL**) from Section 5.1, the logic of fully private announcements and the logic of fair-game announcements are examples of signature-based logics. The only syntactic difference is the presence of modalities $[\tau\varphi]\psi$ in the signature-based language for the signature *Pri*; but it is easy to see that $[\tau\varphi]\psi$ is logically equivalent to ψ, and so this language reduces to the logic of fully private announcements.

The logical system for this language is a generalization of what we have seen for the public announcement logic earlier. A statement of it may be found in Baltag and Moss [2004], and the completeness in the final version of Baltag, Moss, and Solecki [1998]. For each of the operators of basic epistemic logic, one has a *Reduction Axiom* which allows one to push the dynamic (action) modalities past that operators, given a certain context. But the main difficulty comes in the combination of the action modality with common knowledge statements. An alternative system which uses the relativized common knowledge operators may be found in van Benthem, van Eijck and Kooi [2006b]. Since we are not going to need any of these systems or any of their interesting fragments, we leave matters at that. The only exception is the generalization of the Announcement-Knowledge Axiom, which we deem worth explaining in some detail.

The Action-Knowledge Axiom The Reduction Axiom for the K operator will be a generalization of of the Announcement-Knowledge Axiom, which we call the *Action-Knowledge Axiom*: for every basic action α, we have

$$[\alpha]K_a\varphi \leftrightarrow \left(\mathsf{pre}(\alpha) \rightarrow \bigwedge_{\alpha \xrightarrow{a} \alpha'} K_a[\alpha']\varphi\right).$$

To state it in a more transparent form, we need the notion of *appearance of an action to an agent*: for each basic action α of our language and for each agent a, the *appearance of α to a* is the action

$$\alpha_a := \bigcup_{\alpha \xrightarrow{a} \alpha'} \alpha',$$

where $\bigcup$ is the non-deterministic choice of a (finite) set of actions. The action α_a describes the way action α *appears* to agent a: when α is happening, agent a thinks that (one of the deterministic actions subsumed by) α_a is happening. With this notation, the Action-Knowledge Axiom says that, for every basic action α, we have:

$$[\alpha]K_a\varphi \leftrightarrow (\mathsf{pre}(\alpha) \rightarrow K_a[\alpha_a]\varphi).$$

In other words: knowledge commutes with action modalities, modulo the satisfaction of the action's precondition and modulo the replacement of the real action with its appearance. One can regard this as a fundamental law governing the dynamics of knowledge, a law that may be used to compute or predict future knowledge states from past ones, given the actions that *appear* to happen in the meantime. The law embodies one of the important insights that dynamic-epistemic logic brings to the philosophical understanding of information change.

Notes The action model framework has been developed by Baltag, Solecki, and Moss, and has appeared in various forms [1998; 1999; 2002; 2004]. The signature-based languages are introduced in Baltag and Moss [2004]. A final publication on the completeness and expressivity results is still in preparation. A different but also rather expressive way to model epistemic actions was suggested by Gerbrandy in [1999]; this generalizes the results by Gerbrandy and Groeneveld in [1997]. Gerbrandy's action language can be seen as defined by relational composition, interpreted on non-wellfounded set theoretical structures corresponding to bisimilarity classes of pointed Kripke models. Van Ditmarsch explored another relational action language—but based on standard Kripke semantics—[van Ditmarsch, 2000; van Ditmarsch, 2002] and was influenced by both Gerbrandy and Baltag et al. His semantics is restricted to $S5$ model transformations. Van Ditmarsch et al. later proposed *concurrent epistemic actions* in [van Ditmarsch *et al.*, 2003]. How the expressivity of these different action logics compares is unclear. Recent developments include a proposal by Economou in [2005]. *Algebraic axiomatizations* of a logic of epistemic actions may be found in [Baltag *et al.*, 2005] and [Baltag *et al.*,

2007, to appear], while a *coalgebraic* approach is in [Cîrstea and Sadrzadeh, 2007]. A logic that extends the logic of epistemic actions by allowing for *factual change* and by closing epistemic modalities under regular operations is axiomatized in [van Benthem *et al.*, 2006b]. A *probabilistic version of the action model framework* is presented by van Benthem, Gerbrandy and Kooi in [2006a]. For a more extensive and up-to-date presentation of dynamic epistemic logic (apart from the present contribution), see the textbook '*Dynamic Epistemic Logic*' by van Ditmarsch, van der Hoek, and Kooi [2007].

6 TEMPORAL REASONING AND DYNAMIC EPISTEMIC LOGIC

It is very natural in a conversation about knowledge to refer to the past knowledge of oneself or others: *I didn't know that, but now I do.* We have already mentioned briefly the "Mr. Sum and Mr. Product" puzzle, illustrating that agents' comments on the past ignorance and knowledge of others can lead to further knowledge. In addition, all treatments of the Hangman paradox mentioned in Section 3.7 must also revolve around the issue of temporal reasoning concerning the future.

We begin with a scenario in which agent's knowledge and ignorance reverses itself more than once. We present an example, due to Sack [2007], because the natural summation of it involves statements about past knowledge.

Our three players Amina, Bao, and Chandra are joined by a fourth, Diego. They have a deck with two indistinguishable $\spadesuit$ cards, one $\diamondsuit$ and one $\clubsuit$. The cards are dealt, and in the obvious notation, the deal is $(\spadesuit, \spadesuit, \diamondsuit, \clubsuit)$. We assume that the following are common knowledge: the distribution of cards in the deck, the fact that each player knows which card was dealt to them, and that they do not initially know any other player's card. Then the following conversation takes place:

i. Amina: "I do not have $\diamondsuit$."

ii. Diego: "I do not have $\spadesuit$."

iii. Chandra: "Before (i), I knew φ: Bao doesn't know Amina's card. After (i), I did not know φ. And then after (ii), I again knew φ."

All three statements are intuitively correct. After Amina's statement, Chandra considers it possible that the world is $w = (\spadesuit, \clubsuit, \diamondsuit, \spadesuit)$. In w after the announcement, Bao does know that Amina holds $\spadesuit$, so φ is false. But Amina no longer reckons this world w to be possible after Diego's announcement. Indeed, she only considers possible $v = (\spadesuit, \spadesuit, \diamondsuit, \clubsuit)$. And in v after both announcements, Bao thinks that $(\clubsuit, \spadesuit, \spadesuit, \diamondsuit)$ is possible. Hence φ holds, and Chandra knows that it does.

Our first order of business is to extend the kind of modeling we have been doing to be able to say the sentence in (iii), and also to prove it in a logical system.

6.1 *Adding a 'yesterday' operator to the logic of public announcements*

To get started, we present here the simplest temporal extension of the simplest dynamic epistemic logic, the logic of public announcements from Section 5.1. We think of a multi-agent epistemic model M_0 subject to a sequence of public announcements of sentence $\varphi_1, \varphi_2, \ldots, \varphi_n$. These determine models M_i: M_0 is given, and for $i < n$, M_{i+1} is given by taking submodels via $M_{i+1} = M_i|\varphi_{i+1}$. We add a single operation Y to the language, with the intended semantics that $Y\varphi$ means that φ was true before the last announcement. Formally, we would set

$$M_i, w \models Y\varphi \quad \text{iff} \quad M_{i-1}, w \models \varphi \tag{23}$$

There are two problems here, one minor and one more significant. The slight problem: what to do about sentence $Y\varphi$ in the original model M_0? The choice is not critical, and to keep our operators $\Box$-like, we'll say that all sentences $Y\varphi$ are automatically true in M_0, w.

The larger problem has to do with the semantics of public announcement sentences $[!\psi]\chi$. We know how to deal with announcements of one of the φ sentences with which we started, since these figure into the definition of the models M_i. But for announcements of other sentences, those models are of no help. One solution is to think in terms of *histories*

$$H \quad = \quad (M_0, \varphi_1, M_1, \varphi_2, \ldots, M_{n-1}, \varphi_n, M_n) \tag{24}$$

Again, we require that the models and sentences be related by $M_{i+1} = M_i|\varphi_{i+1}$. We recast (23) as a relation involving a history H as in (24) and a world $w \in M_n$:

$$\begin{array}{lll} H, w \models Y\varphi & \text{iff} & i = n, \text{ or } w \in M_{n-1} \text{ implies } (M_0, \varphi_1, \ldots, M_{n-1}), w \models \varphi \\ H, w \models [!\psi]\chi & \text{iff} & H, w \models \psi \text{ implies } (M_0, \varphi_1, \ldots, M_n, \psi, M_n|\psi), w \models \chi \end{array}$$

So the effect of public announcements is to extend histories.

We turn to the logical principles that are reflected in the semantics. The decision to have Y be $\Box$-like means that the distribution axiom and the rule of necessitation formulated with Y are going to be sound for the logic. Here are the additional logical principles that are sound for this semantics (true in all worlds in all models in all histories):

$$\begin{array}{ll} (p \to Yp) \wedge (\neg p \to Y\neg p) & \text{atomic permanence} \\ \neg Y\bot \to (Y\neg\varphi \to \neg Y\varphi) & \text{determinacy} \\ (Y\bot \to K_aY\bot) \wedge (\neg Y\bot \to K_a\neg Y\bot) & \text{initial time} \\ (\varphi \to \psi) \leftrightarrow [!\varphi]Y\psi & \text{action-yesterday} \\ YK_a\varphi \to K_aY\varphi & \text{memory} \end{array}$$

These are due to Yap [2006] and Sack [2007]. In these, p must be atomic. And $\bot$ is a contradiction, so $Y\bot$ is only true at the runs of length 1. Most of the axioms are similar to what we have seen in other systems, except that one must be careful to consider those runs of length 1. The initial time axiom implies that it is common knowledge whether the current history is of length 1 or not The memory axiom is named for obvious reasons. Notice that the converse is false.

Sack's dissertation [Sack, 2007] also contains the completeness proof for this logic, with common knowledge operators added. In fact, his work also includes operators for the complete past (not just the one step 'yesterday'), the future, and also arbitrary epistemic actions formulated in the same language. This means that one can model private announcements concerning the past knowledge of other agents, to name just one example. His language also contains *nominals* to allow reference to particular states (we do not discuss these here) and also allows agents to, in effect, know what epistemic action they think just took place.

Returning to the flip-flop of knowledge In the previous section we presented a scenario that involved statements of previous knowledge and ignorance. Here is how this is formalized. Let φ be $\neg(K_b Spades_a \vee K_b Diamonds_a \vee K_b Clubs_a)$. Then a formalized statement of the entire conversation would be

$$\langle !\neg Diamonds_a\rangle\langle !\neg Spades_d\rangle(YYK_a\varphi \wedge Y\neg K_a\varphi \wedge K_a\varphi).$$

All of the background information about the scenario and the initial deal can be written as a sentence ψ in the language, assuming that we have common knowledge operators. Then the fact that we have a completeness result means that $\psi \vdash \varphi$ in the proof system. The logic is moreover decidable. As a result, it would be possible to have a computer program find a formal proof for us.

6.2 The future

Adding temporal operators for the future is more challenging, both conceptually and technically. To see this, let us return to the modeling of *private announcements* which we developed in Example 15 in Section 5.4. The way we modeled things, private announcements to groups seem to come from nowhere, or from outside the system as a whole. Let us enrich this notion just a bit, to see a simple setting in which temporal reasoning might be profitably modeled.

Consider a setting where each individual agent a might send a message m to some set B of agents, with the following extra assumptions: (0) m is a sentence in whatever language we are describing; (1) the names of the recipient agents B are written into m; (2) sending m takes arbitrarily long, but eventually each agent in B will receive m; (3) all agents in B receive m at the same time; (4) the sending and receipt of messages is completely private; (5) at each moment, at most one message is sent or received; (6) messages are delivered in the order sent. We make these assumptions only to clarify our discussion, not because they are the most realistic or useful.

One might like to have temporal operators in the language so that agents can "say" sentences like *at some future point, all agents in B will receive the message*

I just sent, resulting in the common knowledge for this group of φ, or *I sent* m_1 *and then* m_2 *to* b, *and the message I received just now from* b *shows me that it was sent between the time* b *received* m_1 *and the time he received* m_2.

At the time of this writing, there are no formalized systems which include knowledge and temporal operators, epistemic actions as we have been presenting them in this chapter, and also have temporally extended events such as the *asynchronous message passing* we have just mentioned. There is a separate tradition from the computer science literature which incorporates temporally extended events, knowledge, and temporal assertions along different lines than this chapter. We are going to present the ideas behind one of those approaches, that of *interpreted systems*.

Before that, we want to mention a different approach, the *history-based semantics* of messages due to Parikh and Ramanujam [2003]. This work has a different flavor than interpreted systems. (But the two are equivalent in the sense the semantic objects in them may be translated back and forth preserving truth in the most natural formal language used to talk about them. See Pacuit [to appear] for this comparison.) The only reason we present the interpreted systems work instead is that it has a larger literature.

6.3 Interpreted systems and temporal epistemic logic

A general framework involving information change as a feature of *interpreted systems* was developed by Halpern and collaborators in the 1990s [Fagin *et al.*, 1995]. There are a few basic notions.

We start with a collection of agents or processors, each of which has a *local state* (such as 'holding clubs' for agent Amina), a *global state* is a list of all the local states of the agents involved in the system, plus a state of the environment. The last represents actions, observations, and communications, possibly outside the sphere of influence of the agents. An example global state is $(\clubsuit\heartsuit\spadesuit, \emptyset)$ wherein Amina has local state $\clubsuit$, i.e., she holds clubs, Bao local state $\heartsuit$, and Chandra local state $\spadesuit$, and where 'nothing happened so far in the environment,' represented by a value $\emptyset$. It is assumed that agents know their local state but cannot distinguish global states from one another when those states have the same local state. This induces an equivalence relation among global states which will play the role of an accessibility relation. Another crucial concept in interpreted systems is that of a *run*: a run is a (typically infinite) sequence of global states. For example, when Amina says that she does not have hearts, this corresponds to a transition from global state $(\clubsuit\heartsuit\spadesuit, \emptyset)$ to global state $(\clubsuit\heartsuit\spadesuit, \mathsf{nohearts})$. Atomic propositions may also be introduced to describe facts. For example, not surprisingly, one may require an atom $Hearts_a$ to be false in both global state $(\clubsuit\heartsuit\spadesuit, \emptyset)$ and in global state $(\clubsuit\heartsuit\spadesuit, \mathsf{nohearts})$.

Formally, a *global state* $g \in \mathcal{G}$ is a tuple consisting of local states g_a for each agent and a state g_ϵ of the environment. A *run* $r \in \mathcal{R}$ is a sequence of global states. The m-th global state occurring in a run r is referred to as $r(m)$, and the local state for agent a in a global state $r(m)$ is written as $r_a(m)$. An *interpreted*

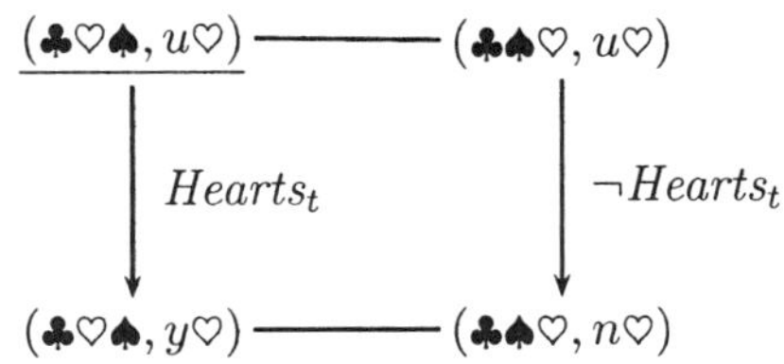

Figure 12. Amina holds clubs, hearts is on top of spades on the two-card stack on the table, and Amina does not know (in the underlined, actual global state) if it is. The two visualized runs reveal which card is on top.

system $\mathcal{I}$ is a pair $(\mathcal{G}, \mathcal{R})$ consisting of a set of global states $\mathcal{G}$ and a set of runs $\mathcal{R}$ relating those states.

A *point* (r, m) is a pair consisting of a run and a point in time m—this is the proper abstract domain object when defining epistemic models for interpreted systems. In an interpreted system, agents cannot distinguish global states from one another iff they have the same local state in both, which induces the relation shown below:

$$(r, m) \stackrel{a}{\sim} (r', m') \text{ iff } r(m) \stackrel{a}{\sim} r'(m') \text{ iff } r_a(m) = r'_a(m')$$

(For an indistinguishability relation that is an equivalence, we usually write $\sim$ instead of R.) With the obvious valuation for local and environmental state values, that defines an epistemic model. For convenience we keep writing $\mathcal{I}$ for that. Given a choice of a *real* (or *actual*) point (r', m'), we thus get an epistemic state $(\mathcal{I}, (r', m'))$. Epistemic and temporal (next) operators have the interpretation

$$\begin{array}{lll} \mathcal{I}, (r, m) \models X\varphi & \text{iff} & \mathcal{I}, (r, m+1) \models \varphi \\ \mathcal{I}, (r, m) \models K_a\varphi & \text{iff} & \text{for all } (r', m') : (r, m) \stackrel{a}{\sim} (r', m') \text{ implies } \mathcal{I}, (r', m') \models \varphi \end{array}$$

It will be clear that subject to some proper translation (see e.g. [Lomuscio, 1999]) interpreted systems correspond to some subclass of the $S5$ models: all relations are equivalence relations, but the interaction between agents is even more than that. The relation between Kripke models and interpreted systems is not entirely trivial, partly because worlds or states in Kripke models are abstract entities that may represent the same set of local states. The main difference between the treatment of dynamics in interpreted systems and that in dynamic epistemics is that in the former this is encoded in the state of the environment, whereas in the latter it emerges from the relation of a state (i.e., an abstract state in a Kripke model) to other states.

Example For a simple example, consider the case of our three players as usual. Suppose that Amina holds clubs, and the hearts card is on top of the spades card (both facedown) on the table. She may now be informed about the card on top of the stack. This is represented by the interpreted system depicted in Figure 12. It consists of four global states. The card Amina holds represents her local state. The other cards are (in this case, unlike in the three-agent card deal) part of the environment. The state of the environment is represented by which of the two cards is on top, and by an 'observation' state variable *obs* that can have three values $u\heartsuit$, $y\heartsuit$, and $n\heartsuit$, corresponding to the state before the announcement which card is on top, the state resulting from the announcement that hearts is on top, and the other state resulting from the announcement that it is at the bottom. The valuation V is now such that $V(Clubs_a) = \{(\clubsuit\heartsuit\spadesuit, u\heartsuit), (\clubsuit\spadesuit\heartsuit, u\heartsuit), (\clubsuit\heartsuit\spadesuit, y\heartsuit), (\clubsuit\spadesuit\heartsuit, n\heartsuit)\}$, and $V(Hearts_t) = \{(\clubsuit\heartsuit\spadesuit, u\heartsuit), (\clubsuit\heartsuit\spadesuit, y\heartsuit)\}$. The system consists of two runs, one from $(\clubsuit\heartsuit\spadesuit, u\heartsuit)$ to $(\clubsuit\heartsuit\spadesuit, y\heartsuit)$ (optionally extended with an infinite number of idle transitions), and the other run from $(\clubsuit\spadesuit\heartsuit, u\heartsuit)$ to $(\clubsuit\spadesuit\heartsuit, n\heartsuit)$. One can now compute that in the actual state $(\clubsuit\heartsuit\spadesuit, u\heartsuit)$ it is true that $\neg K_a Hearts_t$, but in state $(\clubsuit\heartsuit\spadesuit, y\heartsuit)$ she has learned that hearts is on top: $K_a \neg Hearts_t$ is now true. For another example: in the actual state $XK_a Hearts_t$. How the treatment of announcements in interpreted systems relates to public announcement logic will be made precise at the end of the following section.

Interpreted systems have been highly successful as an abstract architecture for multi-agent systems, where agents are either human operators or computer processors, and where the assumption that an agent 'knows its own state' is a realistic simplification. For that reason they can be said to model interaction between *ideal agents.* This assumption is also implicitly applied when modeling perfectly rational agents as in game theory and economics. Also, given that all the dynamics is *explicitly* specified in the runs through the system, it combines well with temporal epistemic logics wherein dynamics is *implicitly* specified by referring to an underlying structure wherein such a change makes information sense. Temporal epistemic logics have been fairly successful. Their computational properties are well-known and proof tools have been developed. See, for example, [van der Meyden, 1998; Dixon *et al.*, 1998; Halpern *et al.*, 2004]. The work of Fagin et al. [1995] also generated lots of complexity results on knowledge and time, we also mention the work of van der Meyden in this respect, e.g. [van der Meyden, 1994; van der Meyden, 1998].

There are two rather pointed formal differences between the temporal epistemic approach and the dynamic epistemic approach.

Closed versus open systems First, the temporal epistemic description takes as models systems together with their whole (deterministic) history and future development, in the shape of 'runs'. As such, it can be easily applied to 'closed' systems, in which all the possible developments are fixed in advance, where there are no accidents, surprises or new interactions with the outside world, and thus the future is fully determined. Moreover, in practice the approach is more applicable to closed systems having a *small* number of possible moves: that's the only ones for which it is feasible to work explicitly with the transition graph of the full history.

The dynamic epistemic approach is better suited to 'open' systems. This is for example the case with epistemic protocols which can be modified or adapted at any future time according to new needs, or which can interact with an unpredictable environment. But it is also applicable to closed systems in which the number of possible different changes is large or indefinite.

There are two analogies here. The first is with open-versus-closed-system paradigms in programming. People in concurrency are usually interested in open systems. The program might be run in many different contexts, in the presence of many other programs, etc. More recently (in the context of mobile computation), people have looked at approaches that allow programs to be changed at any time inside the same logical frame. The temporal logic approach is not fit for this, since it assumes the full current program to be fixed and given as 'the background model'. That is why people in this area have used totally different kinds of formalisms, mainly process algebraic, such as the π-calculus. In contrast to that, dynamic epistemic logics are interesting in that, although based on a modal logic, which is not an algebraic kind of formalism, they are able to express changes in an open system through the semantic trick of changing the models themselves, via 'epistemic updates'.

The second analogy is with *game theory*. The temporal approach is like the description of a game through explicitly giving its full extensive form: the graph of all possible plays. For instance, chess (in this approach) is defined as the set of all possible chess plays. But there is another way to describe a game: by giving only the 'rules of the game' (which type of actions are allowed in which type of situations), together maybe with an 'initial state' (or set of states) and some 'winning rules'. This is a much more economical and way to describe a game, and it is more common as well. Of course, once this description is given, one could draw the game in extensive form as the set of all plays, if one is given enough computational power... If we neglect the aspects of the game that deal with who wins (and what), the dynamic epistemic approach can naturally describe epistemic games in precisely this way: one gives an epistemic Kripke model of 'initial states' and also an epistemic Kripke model or other semantically precise description of possible 'epistemic actions', including preconditions that tell us on which type of states a given action may be applied. Then one can play the game by repeatedly updating the state model with the action model. A 'full play' or 'run' of the game is obtained when we reach a state (at the end of many updates) on which no action (in our given action model) can be applied.

Information change description The second difference between the interpreted systems and the dynamic epistemics approach simply concerns the ability to model and classify various 'types' or 'patterns' of information change, or information exchange, such as public announcements, private announcements, game announcements etc. The dynamic epistemic approach obviously has this in-built ability, while the temporal approach doesn't have it, at least not in a direct, usable manner. There is nothing like an "announcement". All of the structure is encoded in the set of runs that serves as a model. Even the semantics of knowledge uses this set of runs, and so if one wants to use this as a model of real knowledge, it means that the agents must have implicit access to the overall model. To put it differently, in the temporal approach, one can only say what is true 'before' and 'after' a given action, and thus only implicitly get some information about the type of the action itself, through its input-output behavior. Moreover, this information is *not* enough to isolate the type of the action, since it only gives us the *local* input-output behavior of a given action; and different actions may behave identically in one local context, but differ in general. For instance in the two players and two cards case, in an epistemic state in which the fact that the card deal is ♣♡♠ is common knowledge, a public announcement of that fact will have the same input-output description as a 'skip' action corresponding to 'nothing happens'. But in the epistemic state where the cards were dealt but not seen, or the subsequent one where all players only know their own card, this fact was not common knowledge and its public announcement will in that case induce an informative (i.e. non-skip) transition. For the same reason, actions like private announcements, announcements with suspicion, etc., are harder to model in the interpreted systems approach.

A number of people are investigating the relation between dynamic epistemic logic and either interpreted systems or history-based models. One should see, for example, van Benthem and Pacuit [2006] for hints in this direction and also for related work on temporal epistemic reasoning.

7 BELIEF CHANGE AND DYNAMIC EPISTEMIC LOGIC

Our final section is concerned with the interaction of **DEL** with the topic of belief revision. The material of this section is very new and still in a state of flux, so our discussion here cannot claim in any way to represent the definitive word on the matter.

Here is our plan for the section: First, we briefly present the classical **AGM** theory of belief revision. We then briefly mention some dynamic (but non-epistemic) versions of **AGM**. Finally, we present some of the recent work that incorporates belief revision into the **DEL** framework, in an attempt to overcome the above-mentioned classical problems: the work of van Benthem on the dynamic logic of belief upgrades, the action plausibility models of Aucher and van Ditmarsch, and the action-priority update of Baltag and Smets. As before, we follow a "logical" rather than a historical order, leaving the history for the Notes at the end of the section.

Classical AGM theory A *belief set* (or *theory*) is a set K of sentences in some language. We at first take the language to be propositional logic, but we are keen to extend this to various modal languages, where the modalities are either one of the knowledge or belief modalities which we have already seen, or an operation coming from the field of Belief Revision itself.

The notion of a belief set is intended to model the set of sentences believed by some agent. So to incorporate the reasoning of the agent, one usually works on top of some logical system or other and then requires belief sets to be closed under deduction in the system; they need not be consistent, however. They certainly need not be complete either: we might have $\varphi \notin K$ and $\neg\varphi \notin K$ as well. The **AGM** theory of belief revision deals with changes to an agent's belief set when presented with a new sentence φ. The main point is that φ might conflict with what the agent believes, and so the theory is exactly about this issue. The theory is named for its founding paper, the celebrated Alchourrón, Gärdenfors, and Makinson [1985]. Overview publications include Gärdenfors [1988] and most notably for us, Chapter 4c by Hans Rott.

The **AGM** theory employs three basic operations and presents postulates concerning them. Since belief sets are *sets*, the overall theory is second-order. Moreover, it is an interesting issue to then construct and study semantic models of the **AGM** postulates, or of related ones.

The first operation is called *expansion*. Intuitively, this is what happens when the agent takes K as a given and simply adds φ as a new belief. We write $K + \varphi$ for the result. The postulates for expansion as follows:

(1)	Closure	$K + \varphi$ is a belief set.
(2)	Success	$\varphi \in K + \varphi$
(3)	Inclusion	$K \subseteq K + \varphi$
(4)	Vacuity	If $\varphi \in K$, then $K = K + \varphi$.
(5)	Monotonicity	If $J \subseteq K$, then $J + \varphi \subseteq K + \varphi$.
(6)	Minimality	$K + \varphi$ is the minimal set with (1) - (5).

It is easy to check that these postulates exactly capture the operation of taking the consequences of $K \cup \{\varphi\}$ in the underlying logical system.

More interesting are the other two operations, *contraction* and *revision*. Intuitively, the contraction of K by φ models the result of the agent's giving up the belief in φ and doing this without giving up too much. The revision of K by φ models the agent's minimally changing her beliefs to incorporate φ. There are postulates for both operations, and we are only going to spell out those for revision. The reason is that on top of the postulates for expansion, those of revision determine the contraction operation (and vice-versa). We write the revision operation as $K * \varphi$. The postulates are:

(1)	Closure	$K * \varphi$ is a belief set.
(2)	Success	$\varphi \in K * \varphi$
(3)	Inclusion	$K * \varphi \subseteq K + \varphi$
(4)	Preservation	If $\neg\varphi \notin K$, then $K + \varphi \subseteq K * \varphi$.
(5)	Vacuity	$K * \varphi$ is inconsistent iff $\neg\varphi$ is provable.
(6)	Extensionality	If φ and ψ are equivalent, then $K * \varphi = K * \psi$
(7)	Subexpansion	$K * (\varphi \wedge \psi) \subseteq (K * \varphi) + \psi$
(8)	Superexpansion	If $\neg\psi \notin K * \varphi$, then $K * (\varphi \wedge \psi) \supseteq (K * \varphi) + \psi$.

The result of a contraction operation $K - \varphi$ satisfying some postulates which we did not list turns out to be the same as $K \cap (K * \neg\varphi)$; this is called the *Harper identity.* And given a contraction operation satisfying the postulates, one can define revision by the *Levi identity* $K * \varphi = (K - \neg\varphi) + \varphi$; this operation will then satisfy the eight postulates above.

One important issue in the area is the relation between belief revision and the older topic of conditional logics which began with Lewis' book [1969]. To see what this is about, assume that we are working over a logical system with a symbol $\Rightarrow$ that we want to use in the modeling of some natural language conditional, say the subjunctive one. Then a belief set K might well contain sentences $\varphi \Rightarrow \psi$ and $\neg\varphi$. So in this context, we would like or even expect to have $\psi \in K * \varphi$. In other words, we ask about the condition

$$\varphi \Rightarrow \psi \in K \quad \text{iff} \quad \psi \in K * \varphi.$$

This is called the *Ramsey test.* A key result in the subject is Gärdenfors' *Impossibility Theorem*: there is no operation of revision on belief sets which both satisfies the postulates of $*$ and also the Ramsey test. (The result itself depends on some non-triviality condition which we ignore here.)

Although the literature on belief revision may be read as a discussion of changes in belief, it may also be read as an extended discussion about the correspondence between various axiom systems and types of semantic structures. These include structures akin to what we have seen. In particular, the *sphere systems* of Grove (based on earlier work of Lewis) come from belief revision theory.

7.1 Dynamic versions of revision theory

Moving now to a more *semantical* setting, we show how some of the operations which we have already seen can be interpreted as belief change operators. We then present some dynamic versions of **AGM**: the Katsuno-Mendelzon theory **KM** of belief update, de Rijke's dynamic modal logic **DML** and Segerberg's dynamic doxastic logic **DDL**. These are all dynamic in some sense, but some of them are not "epistemic" in that knowledge is not modeled via the Kripke (relational) semantics, or any other semantics for that matter. We mention some of the difficulties and problems encountered by classical belief revision theory.

Examples of belief change via dynamic epistemic logic Consider expressing and changing uncertainty about the truth of a single fact p, and assume an information state where the agent (whose beliefs are interpreted by the unlabeled accessibility relation depicted) may be uncertain about p and where p is actually false (indicated by 'designating' the actual state by underlining it). Figure 13 lists all conceivable sorts of belief change.

In the top structure, uncertainty about the fact p (i.e., absence of belief in p and absence of belief in $\neg p$) is changed into belief in $\neg p$. On the left, $\neg Bp$ is true, and on the right $B\neg p$. In the second from above, belief in p is weakened to uncertainty

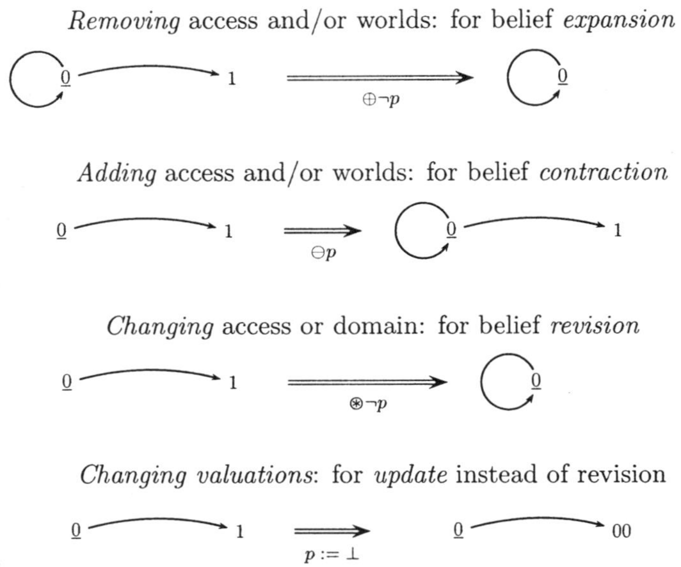

Figure 13. Possible changes of belief mirrored in Kripke structure transitions

about p, and in the third from above we change from Bp to $B\neg p$. Note that also in this semantic setting of Kripke-structure transformation, belief revision can again be seen as a contraction followed by an expansion, so we may in principle consider semantic alternatives for the Levi-identity. The last information state transition in Figure 13 depicts factual change. The state with changed valuation has suggestively been renamed from 1 to 00, although formally, of course, it is only the valuation of a named state that changes. The 'assignment' or substitution $p := \bot$ indicates that the valuation of atom p is revised into the valuation of the assigned formula. As this is $\bot$, the new valuation of p (seen as a subset of the domain) is now the empty set of states.

Updates and the KM theory A topic in traditional belief revision comes under the name of 'update'. An *update*—unfortunately a clash cannot be avoided with the more general meaning of that term in dynamic epistemic logic, where it incorporates belief revision as well—is a *factual* change, as opposed to a belief change in the three previously distinguished notions. The latter merely express a different agent stance towards a non-changing world, but in an 'update' the world itself changes. The standard reference for updates in belief revision is Katsuno and Mendelzon [1991]. Recent investigations on updates (factual change) in a dynamic epistemic setting are [van der Meyden, 1998; van Benthem, 1996; Kooi, 2007]. These ideas also deserve to be properly applied to the belief revision arena.

We mention one of the motivating examples, mainly to contrast with the **AGM** postulates. It is taken directly from [Katsuno and Mendelzon, 1991].

Consider a belief set with two atomic propositions, b and m, standing for "there is a book on the floor" and "there is a magazine on the floor". Suppose that

$$K = \uparrow \{b \wedge \neg m, m \wedge \neg b\},$$

where the arrow denotes the deductive closure in propositional logic. This models a situation in which an agent believes that exactly one item is on the floor, but does not have a specific belief of which it is. Suppose we wish to change the world by instructing a robot (as they have it) to put the book on the floor. So we wish to consider $K * b$ or $K + b$. Now $K + b = \uparrow \{b \wedge \neg m\}$. As for $K * b$, since $\neg b \notin K$, we see from the Inclusion and (especially the) Preservation postulates that $K * b = K + b$ anyways. In particular, $\neg m \in K * b$. This seems like an unintuitive result: given that we want to model a change in the world resulting from putting the book on the floor, why should we believe afterwards that the magazine is not on the floor? The **KM** theory addresses this by proposing **AGM**-like postulates on the matter of *update*, the phenomenon illustrated in this example.

Belief change with dynamic non-epistemic logic The three 'theory change operators' $\oplus$, $\ominus$, and $\circledast$ can be reinterpreted as dynamic modal operators. A straightforward way to model these operators would be a logic in which $[\circledast\varphi]\psi$ expresses that after revision with φ, (the agent believes) ψ. This approach was suggested by van Benthem in [1994][9] and further developed by de Rijke in [1994]. They propose a semantical counterpart of a total order on theories, in the form of 'updating' and 'downdating' relations between states or worlds, standing for theories, and interpret the modal operator as a transition in such a structure according to these relations. 'Updating' models expansion: it relates the current state to states that result from expansion. 'Downdating' models contraction. It relates states that result from contraction to the current state. Revision is indeed downdating followed by updating. In this overview we focus on approaches that extend epistemic logics, therefore we do not give more details on this non-epistemic approach.

Dynamic Doxastic Logic (DDL) In the approach by Segerberg and collaborators [Lindström and Rabinowicz, 1999a; Segerberg, 1999b; Segerberg, 1999a; Lindström and Rabinowicz, 1999b], beliefs are represented explicitly. We now identify a theory K with the believed formulas (or some subset of the believed formulas) in an epistemic state:

$$K = \{\psi \mid M, s \models B\psi\}.$$

[9] It is only one of many topics covered in that publication, namely Section 6, pages 714–715, 'Cognitive procedures over information patterns'. Note this work is similar to a 1991 technical report.

As in [de Rijke, 1994] , **DDL** expresses belief change with dynamic modal operators $[\oplus\varphi]$, $[\ominus\varphi]$, and $[\circledast\varphi]$. In a typical revision where we have that $\neg\varphi \in K$, $\varphi \in K \circledast \varphi$, and $\neg\varphi \notin K \circledast \varphi$, we now get

- $M, s \models B\neg\varphi$
- $M, s \models [\circledast\varphi]B\varphi$
- $M, s \models [\circledast\varphi]\neg B\neg\varphi$

For contraction, we want that in case $M, s \models B\varphi$, after contraction φ is no longer believed, i.e., $M, s \models [\ominus\varphi]\neg B\varphi$. Similarly, for expansion we aim to achieve $M, s \models [\oplus\varphi]B\varphi$.

This approach is known as *dynamic doxastic logic* or **DDL**. Similar to [de Rijke, 1994] it presumes a transition relation between states representing theories, but this is now differently realized, namely using what is known as a Segerberg-style semantics wherein factual and epistemic information— called the *world component* and *doxastic component* —are strictly separated. A dynamic operator is interpreted as a transition along the lines of minimal theory change set out by this given structure, with the additional restriction that the transitions describe epistemic (doxastic) change only, and not factual change. This restriction is enforced by not allowing the 'world component' to change in the transition relation but only the doxastic component [Lindström and Rabinowicz, 1999a, p.18].

There are now two options: either we restrict ourselves to beliefs in *objective* (boolean, non-epistemic) formulas, and we get what is known as basic **DDL**, as in [Lindström and Rabinowicz, 1999a; Segerberg, 1999b]. Or we allow higher-order beliefs, as in the dynamic epistemics described in previous sections of our chapter. We thus get 'full' or 'unlimited' **DDL**, also discussed in [Lindström and Rabinowicz, 1999a] but mainly in [Lindström and Rabinowicz, 1999b].

Incidentally, the semantic models of **DDL** are rather different from those in this chapter, at least on the surface. They are more similar to neighborhood models, or topological models for modal logics. These are too different from what we have seen to allow us to present them in this chapter. Getting back to **DDL** and the systems we have presented, we know of no publications offering detailed comparisons; surely this is because the work of this section is so new. For this, and for discussion of recent work on **DDL**, see Leitgeb and Segerberg [2007].

Problems of the classical theory Classical belief revision, and its dynamic versions presented in the previous section, encounter a number of problems: the difficulties in extending them to *iterated belief revision*; the *multiplicity of belief revision policies*; difficulties in dealing with *multi-agent beliefs* and even more with *higher-order beliefs*, that is beliefs about other beliefs. We refer for details to Chapter 4c on the subject in this handbook. We are mainly interested in higher-order beliefs and iteration. Our discussion amounts to a suggestion that the work of this chapter can be useful in dealing with these matters.

If one drops the restriction to belief in objective formulas and allows higher-order beliefs, then the standard **AGM** postulates lead to paradoxes. In particular, *the Success postulate for revision is problematic for sentences that involve doxastic modalities*: we have already noted in Section 5.2 that Moore sentences $p \wedge \neg K_a p$ are not successful. Similar examples apply to formalisms which have the syntactic means to specify semantic properties of evident interest. We now continue with the in-depth treatment of recent dynamic epistemic approaches to belief revision. The hallmark of the approach is that the transition that interprets the dynamic operators is *constructed* (as a state transformer) from the (specific action of announcing the) revision formula, instead of *assuming* as given such a transition relation.

7.2 The dynamic logic of belief change

This section is mainly based on the work of J. van Benthem [2006; 2004] on the dynamic logic of belief change and "preference upgrade" (with some additional input from Baltag and Smets [2006a]). Essentially, this work uses the **DEL** paradigm to develop a logic for belief change that completely solves the problems posed to belief revision by multi-agent beliefs and higher-order beliefs, iterated revision, as well as partially addressing the problem of the multiplicity of belief revision policies.

Static versus dynamic belief revision The first fundamental distinction underlying this work is the one between *static* and *dynamic* belief revision: the first has to do with *conditional beliefs*, while the second has to do with the *beliefs acquired after a belief-changing action.* The distinction is only significant when dealing with higher-order beliefs: in the case of factual beliefs, the two types of revision coincide. Static belief revision captures the agent's changing beliefs about an unchanging world. But, if we take the "world" as incorporating all the agents' higher-order beliefs, then *the world is in fact always changed by our changes of belief* (as shown above, using examples involving Moore sentences). As a consequence, the best way to understand static belief revision with a proposition P is as expressing *the agent's revised beliefs, after learning* P, about what *was the case, before the learning.* In contrast, dynamic belief revision captures the agent's revised beliefs about the world *as it is after revision.*

Static revision as conditional belief Classical **AGM** theory deals with changing beliefs about an unchanging reality. In our terminology, it is static belief revision. In a modal logic setting, it is natural to formalize static revision as hypothetical belief change, using *conditional belief* operators[10] $B_a^\alpha \varphi$, as in Section 4.7:

[10] It may seem that the failure of Ramsey's test for **AGM** revision would conflict with a conditional belief interpretation of **AGM**. But this is not the case. In the conditional belief setting, Gardenfors' impossibility result simply shows that "a conditional belief" is not the same as "a belief in a conditional"; more precisely, there doesn't exists any non-epistemic, non-doxastic

if K is agent a's current belief set at state s and $*_a$ is her belief revision operator, then writing $s \models B_a^{\varphi}\psi$ is just a way of saying that $\psi \in K *_a \varphi$. Based on the above discussion, we can thus read a doxastic conditional $B_a^{\varphi}\psi$ as follows: *if learning φ, agent a would come to believe that ψ was the case (before the learning).* The semantics is given by plausibility models (or systems of Grove spheres, see Section 2.4), as in Section 4.7, with a conditional belief $B_a^{\varphi}\psi$ defined via the the most plausible states (satisfying φ) and being epistemically indistinguishable from the current state. If we translate the **AGM** postulates into the language of conditional beliefs, while taking into account the concept of "(fully introspective) knowledge" and the limitations that it poses to belief revision, we obtain the axioms of conditional doxastic logic **CDL** from Section 4.7.[11] In particular, observe that the **AGM** Success postulate *is* valid for static belief revision, even if we allow doxastic modalities (and thus higher-order beliefs) in our language: it is always true (even for Moore sentences φ) that, after learning that φ is the case, agents come to believe that φ *was* the case (before the learning).

In contrast, a statement $[!\varphi]B_a\psi$ involving a dynamic modality says that *after learning φ, agent a would come to believe that ψ is the case (in the world after the learning).* Due to Moore-type sentences, dynamic belief revision will *not* satisfy the **AGM** postulates (and in particular, the Success postulate will fail).

Triggers of information change As van Benthem [2006] observes, in order to understand and formalize dynamic belief revision, it is essential to take into account the actual "learning event" that "triggered" the belief change. For example, our beliefs about *the current situation after* hearing a *public* announcement are different from our beliefs after receiving a *fully private* announcement. In the public case, we may come to believe that the content of the announcement is now *common knowledge* (or at least common belief); in the private case, we may come to believe the opposite: that the content of the announcement forms now our *secret knowledge.* In contrast, our beliefs about the triggering action are irrelevant as far our static revision is concerned: our conditional beliefs about the current situation given some hypothetical information do not depend on the way this information might be acquired. This explains a fact observed in [van Benthem, 2006], namely that by and large, the standard literature on belief revision (or belief update) does not mention in any way the explicit triggers (the actual doxastic events) that cause the belief changes (dealing instead only with types of abstract operations on beliefs, such as update, revision and contraction etc). The reason for this lies in the static character of **AGM** revision, as well as its restriction (shared with the **KM** updates and basic **DDL**) to one-agent, first-level, factual beliefs.

This is where the **DEL** paradigm can help: as already seen in this chapter, **DEL** explicitly analyzes the triggers for information change, from simple announcements of facts to individual agents to complex information-carrying events, involving many agents and their different perspectives on the learning event.

---

notion of conditional that would validate this equivalence. For more on this, cf. Leitgeb [2007].

[11]The translation is carried out in detail in [Baltag and Smets, 2006a].

Revision as relation change If we model static beliefs (including conditional beliefs) using plausibility relations, then dynamic belief revision corresponds to *relation change*: the model is modified by changing the plausibility arrows. Different types of dynamic belief revision (induced by different triggering events) will correspond to different such changes. In other words, we can *model the triggers of belief revision as relation transformers*, similarly to the way we previously modeled knowledge updates as epistemic state transformers.

Hard versus soft information The second fundamental distinction made in [van Benthem, 2006] is between learning *'hard facts'* and acquiring *'soft information'*. Unlike in the classical belief-revision setting, in an epistemic-logic setting we need to distinguish between the announcements that lead to *knowledge* (in the absolute, un-revisable sense) of some hard fact and the ones that only affect our *beliefs*. The first type is exemplified by the "truthful public announcement" actions $!P$, that we have already seen. The second type will correspond to "soft" informational actions, of the kind that is more standard in belief revision. One can also have more complex *mixtures of hard and soft information*, giving rise to more complex belief-revision policies.

Defining, classifying, and axiomatizing belief-revision policies We mentioned **PDL** in Section 4.9, and one reason for doing so is the work here. The natural language to define *relation changes* is the set of programs of **PDL**: one can then redefine the plausibility relations R_a (corresponding to the "at least as plausible as" relations $\leq_a$ in Section 4.7), via a clause of the form

$$R_a := \pi(R_a),$$

where $\pi(R_a)$ is any **PDL** program built using tests $?\varphi$, the universal relation $\top$, the old plausibility relations R_a and regular operations on relations: union $\cup$, composition ; and iteration *. In other words, one can use a *relation transformer* $\pi(r)$ as in Section 4.9, and redefine the plausibility relations via $R_a := [\![\pi(r)]\!](R_a)$. In their analysis of revision policies, van Benthem and Liu [2004] propose as a natural class of relation transformers the ones that are definable in **PDL** *without iteration*, showing that these are particularly well-behaved. In particular, one can read off the relational definition a set of *reduction laws* for each such relation transformer, automatically providing a *complete axiomatization* of the corresponding dynamic logic. It is important to note that the reduction laws that are immediately obtainable through this method are for the *safe belief*[12] and *knowledge* modalities, not for conditional belief. But one *can* derive reduction laws for conditional belief in many instances, using the observation made in Section 4.8 concerning the definability of conditional belief in terms of knowledge and safe belief.

[12]This is called the "preference modality" by van Benthem and Liu [2004].

Examples of definable revision policies and their reduction laws We give here only three examples of multi-agent belief-revision policies: truthful public announcements of hard facts, lexicographic upgrades and conservative upgrades. They were all introduced by van Benthem in [2006] as dynamic multi-agent versions of revision operators previously considered by Rott [2004] and other authors. In each case, we give here only one example of a reduction law, namely the analogue for belief of the **DEL** Action-Knowledge Axiom which we mentioned in Section 5.5.

1. Belief change under hard information Truthful public announcements $!\varphi$ of hard facts can be considered as a (limit-)case of belief revision policy. Instead of defining it by world elimination as before, we can equivalently define it using the relation transformer

$$\pi(r) \quad = \quad ?\varphi; r; ?\varphi.$$

So the new accessibility relations are $R_a := [\![\pi(r)]\!](R_a)$, where R_a are the old relations. (See Example 10 in Section 5.1.) The corresponding Reduction Axiom for belief is

$$[!\varphi]B_a\theta \leftrightarrow (\varphi \to B_a^{\varphi}[!\varphi]\theta),$$

which generalizes the Announcement-Knowledge Axiom from Section 5.1 to the case of beliefs. There also exists a more general reduction law for *conditional* belief.

2. Public Announcements of Soft Facts: The "Lexicographic Upgrade" To allow for soft belief revision, an operation $\Uparrow \varphi$ was introduced in [van Benthem, 2006], essentially adapting to public announcements the 'lexicographic' policy for belief revision described in [Veltman, 1996] and [Rott, 2004]. This operation, called *lexicographic upgrade* consists of changing the current plausibility order on any given state model as follows: all φ-worlds become more plausible than all $\neg\varphi$-worlds, and within the two zones, the old ordering remains. Following what we did at the end of Section 4.9 and in Example 10 in Section 5.1, we are using the **PDL** relation transformer

$$\pi(r) \quad = \quad (?\varphi; \top; ?\neg\varphi) \cup (?\neg\varphi; r; ?\neg\varphi) \cup (?\varphi; r; ?\varphi)$$

where $\top$ is the universal relation. As before, the accessibility relations R_a change to $[\![\pi(r)]\!](R_a)$. The important new step in verifying that this does what we want is

$$\begin{array}{lcl} [\![?\varphi; \top; ?\neg\varphi]\!] & = & \{(w,w) : w \in [\![\varphi]\!]\}; \{(u,v) : u, v \in W\}; \{(w,w) : w \in [\![\neg\varphi]\!]\} \\ & = & \{(u,v) : u \in [\![\varphi]\!] \text{ and } v \in [\![\neg\varphi]\!]\}. \end{array}$$

The corresponding Reduction Axiom for belief is

$$[\Uparrow \varphi]B_a\theta \leftrightarrow (\hat{K}_a\varphi \wedge B_a^{\varphi}[\Uparrow \varphi]\theta) \vee (\neg\hat{K}_a\varphi \wedge B_a[\Uparrow \varphi]\theta)$$

where again $\hat{K}$ is the "epistemic possibility" operator (the $\Diamond$-like dual of the K operator). As in the case of hard announcements, there also exists a more general reduction law for conditional belief.

3. Public Announcements of Soft Facts: The Conservative Upgrade. The operation $\uparrow \varphi$ of *conservative upgrade*, as defined in [van Benthem, 2006], changes any model as follows: *the best φ-worlds come on top* (i.e., the most plausible φ-states become the most plausible overall), *and apart from that, the old order remains.* The reduction law for belief is the same as in the previous case. The difference can be only be seen by looking at the reduction law for *conditional* belief. See [van Benthem, 2006] for details.

7.3 Logics for doxastic actions: the action-priority update

The work of Aucher [2005a; 2005b], Baltag and Smets [2006b; 2006c; 2007b], and van Ditmarsch and Labuschagne [2003; 2005; 2007] can be considered as an attempt to extend to dynamic belief revision the unified **DEL** setting based on *action models.* We focus on the approach by Baltag and Smets and specifically on their 'action-priority update'. This gives currently the most convincing picture given the relational approach. It also goes some way towards addressing the problem of the multiplicity of belief revision policies: as we will see below, the action-priority update unifies and subsumes many different policies and types of revision, which come to be seen as the result of applying the same update operation to different triggers given by specific learning events. Indeed, in this interpretation, the triggering events for belief revision are *doxastic actions*, modeled using *action plausibility models*, in a similar way to the way epistemic actions were modeled using epistemic action models. The actions' "preconditions" encode the *information* carried by each action. The plausibility relations between actions are meant to represent *the agent's (conditional) beliefs about the learning event at the moment of its happening.*

We assume here the setting of plausibility models and the conditional doxastic logic **CDL** from Section 4.7. The following definitions give the natural plausibility analogue of the action models from Section 5.4, by incorporating the main intuition underlying the **AGM** belief revision: that *new information has priority over old beliefs* .

DEFINITION 24 (Action plausibility model). (Aucher) Let $\mathcal{L}$ be a logical language which extends the language of **CDL**. An *action plausibility model over $\mathcal{L}$* is a structure $\mathsf{U} = \langle \mathsf{S}, \leq, \mathsf{pre} \rangle$ such that $\langle \mathsf{S}, \leq \rangle$ is a plausibility frame and $\mathsf{pre} : \mathsf{S} \to \mathcal{L}$ is a precondition function that assigns a *precondition* $\mathsf{pre}(\alpha) \in \mathcal{L}$ to each $\alpha \in \mathsf{S}$. As in Section 5.4, the elements of S are called *action points*, and for each $a \in A$, $\leq_a$ is a plausibility relation on S. For $\alpha, \beta \in \mathsf{S}$, we read $\alpha \leq_a \beta$ as follows: agent a considers action α as being at least as plausible as action β. A *doxastic action* is a pointed action plausibility model (U, α), with $\alpha \in \mathsf{S}$.

EXAMPLE 25. The *truthful public announcement of a hard fact* φ is modeled by a singleton action model consisting of an action point α, with identity as the plausibility relation for every agent, and with precondition $\mathsf{pre}(\alpha) = \varphi$. As in Section 5.4, we call this action model *Pub* φ, and denote by $! \ \varphi$ the action corresponding to the point α.[13]

Fully private announcements and *fair-game announcements* can be similarly modeled, essentially by reading the arrows in their epistemic action models from Section 5.4 as plausibility arrows.

Announcements of Soft Information. We can simulate public announcements of soft facts, as described in the previous section, using action plausibility models. For instance, the lexicographic upgrade $\Uparrow \varphi$ has the following action model:

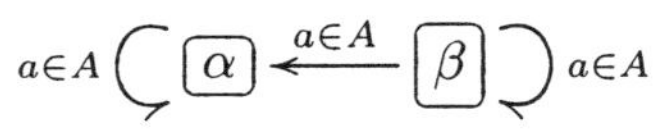

with $\mathsf{pre}(\alpha) = \varphi$ and $\mathsf{pre}(\beta) = \neg\varphi$. The action point on the left corresponds to the case that the announcement happens *true*; the action point on the right corresponds to the case that the announcement is *false*.

The conservative upgrade $\uparrow \varphi$ can be similarly encoded, using a more complicated action model.

Successful Lying. The action of an agent b's "lying in a publically successful manner" by can be modeled as follows: given a sentence φ, the model consists of two action points α and β, the first being the action in which agent b publicly lies that (he knows that) φ (while in fact he doesn't know it), and the second being the action in which b makes a truthful public announcement that (he knows that) φ. The preconditions are $\mathsf{pre}(\alpha) = \neg K_b\varphi$ and $\mathsf{pre}(\beta) = K_b\varphi$. Agent b's plausibility relation is simply the identity: she knows whether she's lying or not. The relation for any hearer $c \neq b$ should make it more plausible to him that b is telling the truth rather than lying: $\alpha <_c \beta$. This reflects the fact that we are modeling "typically successful lying": by default, in such an action, the hearer trusts the speaker, so he is inclined to believe the lie.

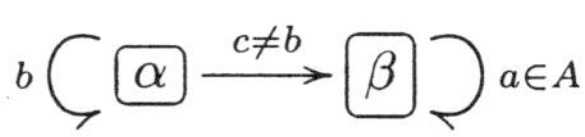

We call this model $\mathit{Lie}_b\ \varphi$, and also denote the action corresponding to the point α by $\mathit{Lie}_b\ \varphi$, and the action corresponding to the point β by $\mathit{True}_b\ \varphi$.

DEFINITION 26 (Execution, Action-Priority Update). Given a doxastic state (M, s) with $M = (S, \leq, V)$, and a doxastic action (U, α) with $\mathsf{U} = (\mathsf{S}, \leq, \mathsf{pre})$, the result of executing (U, α) in (M, s) is only defined when $M, s \models \mathsf{pre}(\alpha)$. In this case, it is the doxastic state $(M \otimes_{\leq} \mathsf{U}), (s, \alpha))$ where $M \otimes_{\leq} \mathsf{U} = \langle S', \leq', V' \rangle$ is a *restricted anti-lexicographic product* of the structures M and U, defined by

$$\begin{array}{lll} S' & \equiv & \{(s,\alpha) \mid s \in S, \alpha \in \mathsf{S}, \text{ and } M, s \models \mathsf{pre}(\alpha)\} \\ (s,\alpha) \leq'_a (t,\beta) & \text{iff} & \alpha <_a \beta \text{ and } s \overset{a}{\sim} t\text{; or else } \alpha \cong_a \beta \text{ and } s \leq_a t \\ (s,\alpha) \in V'_p & \text{iff} & s \in V_p \end{array}$$

[13]Technically, we should distinguish between this plausibility model and the corresponding action model in Section 5.4, but we choose to use the same notation, relying on the context for deciding when to interpret it as a plausibility model. The same applies to all the other plausibility action models in this section having the same notation as an epistemic model.

where $\stackrel{a}{\sim}$ is the *epistemic indifference (uncertainty)* relation[14] on states, and $\cong_a$ is the *equi-plausibility relation* on actions (defined by $\alpha \cong_a \beta$ iff both $\alpha \leq_a \beta$ and $\beta \leq_a \alpha$).

This is a generalization of one of the belief-revision policies encountered in the literature (essentially incorporating the so-called "*maximal-Spohn revision*" into plausibility action models), as well as being a natural plausibility analogue of the product update from Section 5.4. The new order is simply the *anti-lexicographic order* on (epistemically indistinguishable) pairs. The name comes from [Baltag and Smets, 2006b; Baltag and Smets, 2006c; Baltag and Smets, 2007b]. Van Benthem calls it "Action Priority Update": indeed, this construction gives priority to the *action* plausibility relation. This is not an arbitrary choice, but is motivated by a specific interpretation of action models as encoding *belief changes*. In other words, the (strict) order on actions encodes *changes of order* on states. The definition of execution is a consequence of this interpretation: it just says that a strict plausibility order $\alpha <_a \beta$ on actions corresponds indeed to a change of ordering on states, (from whatever the ordering was) between the original (indistinguishable) input-states $s \stackrel{a}{\sim} t$, to the order $(s, \alpha) <_a (t, \beta)$ between output-states; while equally plausible actions $\alpha \cong_a \beta$ will leave the initial ordering unchanged: $(s, \alpha) \leq_a (t, \beta)$ iff $s \leq_a t$. Giving priority to action plausibility does not in any way mean that the agent's belief in actions is stronger than her belief in states; it just captures the fact that, at the time of updating with a given action, *the belief about the action is what is actual, it is the current belief about what is going on, while the beliefs about the input-states are in the past.*[15]

In a nutshell: *the doxastic action is the one that changes the initial doxastic state, and not vice-versa.* The belief update induced by a given action is nothing but an update with the (presently) believed action. If the believed action α requires the agent to revise some past beliefs, then so be it: this is the whole point of believing α, namely to use it to revise or update one's past beliefs. For example, in a successful lying, the action plausibility relation makes the hearer believe that the speaker is telling the truth; so she'll accept this message (unless contradicted by her knowledge), and change her past beliefs appropriately: this is what makes the lying successful.

EXAMPLE 27. Consider the situation in Section 2.3, in which Bao was told the face of the coin, without Amina suspecting this. Assume moreover the coin lies heads up. This was represented in Section 2.4 by the plausibility model (10):

[14] Recall that, in a plausibility model, the epistemic uncertainty relation is defined by: $s \stackrel{a}{\sim} t$ iff either $s \leq_a t$ or $t \leq_a s$.

[15] Of course, *at a later moment*, the above-mentioned belief about action (*now* belonging to the past) might be itself revised. But this is another, *future update*.

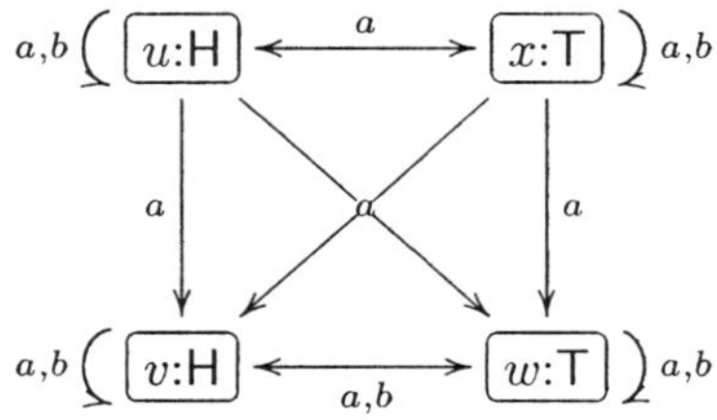

where the real state is the upper-left one.

Next, suppose Bao tells Amina: "I know the face of the coin". Let us first assume this is an evidently truthful statement, coming with a warranty of veracity of some sort or other. Then Amina takes Bao's statement as an announcement of a hard fact. So this action is represented by *Pub* ($K_b\mathsf{H} \vee K_b\mathsf{T}$), with the one-point action model described above (for truthful public announcements of hard information); the action point will be called α. Execute now this doxastic action on the doxastic state given by (upper-left point in) the model (10) above. We identify the old states u and x with the pairs (u, α) and (x, α), respectively, and then we picture the result as

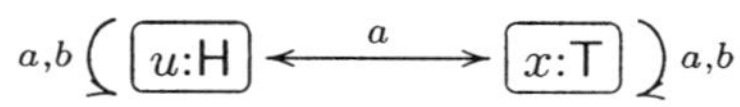

which fits our intuition about the agent's beliefs: it is now common knowledge that Bao knows the face of the coin.

What if Bao's announcement was not evidently truthful? Amina may still *believe* it, but she *doesn't know* that it is true. We model using an announcement of a soft fact rather than a hard one, corresponding to the lexicographic upgrade $\Uparrow (K_b\mathsf{H} \vee K_b\mathsf{T})$. Using the two-point action plausibility model for lexicographic update which we saw above, and computing the execution of this doxastic action on the original doxastic state given by model (10) above, we obtain:

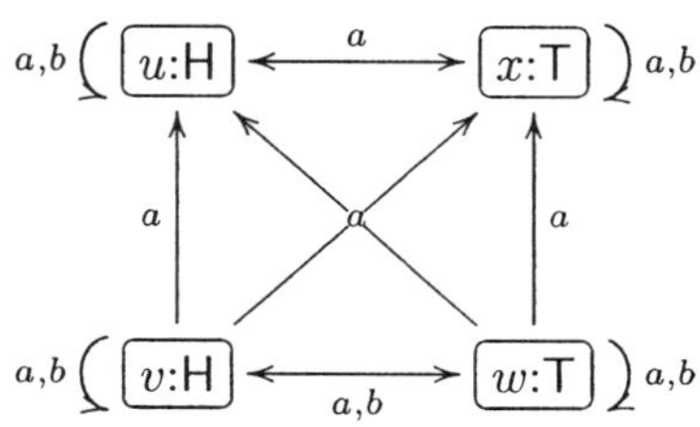

Here and just below, we are identifying the old states with certain pairs to simplify the representation.

What if, instead of making a truthful announcement, Bao chooses to *lie*? For instance (in the initial situation, after he was secretely told that the coin lies heads up), suppose he tells Amina: "I know the coin is lying tails up". This is a lie. Let us assume it is a successful one, so that Amina trusts Bao completely and therefore believes he is telling the truth. We can represent this action using, as described above, the two-point action model $Lie_b\ (K_bH)$ for successful lying. If we now we execute this doxastic action on the original doxastic state from the model (10) above, we obtain

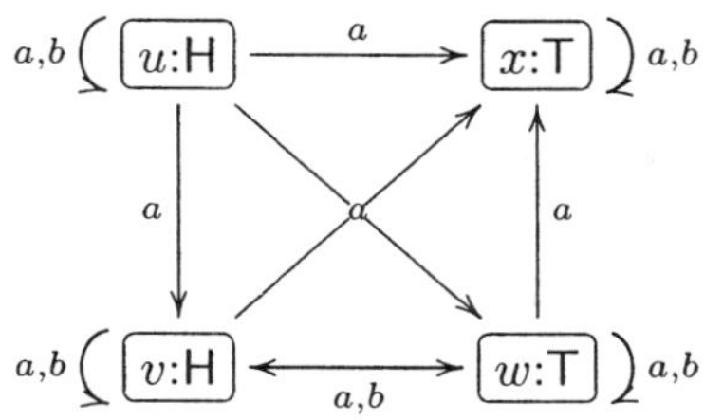

in which the upper left-hand state is the real one. Again, this fits our doxastic intuitions: Amina is deceived and believes the upper right-hand state to be the real one. However, this false belief is *revisable*: a new public announcement $Pub\ K_b\mathsf{H}$ (in effect, saying that Bao has lied and that in fact he knows the coin lies heads up) would correct Amina's wrong belief, making her know that the real state is the left-hand one.

Action-Priority Update Generalizes Product Update Recall the definition of the epistemic indistinguishability relation $\stackrel{a}{\sim}$ in a plausibility model: $s \stackrel{a}{\sim} t$ iff either $s \leq_a t$ or $t \leq_a s$. It follows that Action Priority Update implies the Product Update rule from Section 5.4:

$$(s, \alpha) \stackrel{a}{\sim} (t, \beta) \text{ iff } s \stackrel{a}{\sim} t \text{ and } \alpha \stackrel{a}{\sim} \beta.$$

The logic of doxastic actions As in Section 5.5, we can consider a signature-based language, where a *doxastic signature* is a *finite (fixed) plausibility frame* $\mathbf{\Sigma}$, together with an ordered list without repetitions $(\sigma_1, \ldots, \sigma_n)$ of some of the elements of $\mathbf{\Sigma}$. As in Section 5.5, each signature induces a syntactic action model, and it gives rise to a dynamic-doxastic logic $L(\mathbf{\Sigma})$. The language is obtained by augmenting either the language of conditional doxastic logic **CDL** from Section 4.7, or the language of the logic of knowledge and safe belief from Section 4.8, with dynamic modalities for (signature-based) doxastic actions. The semantics can be given in a similar way to the one in Section 5.5. We skip here the details, referring to [Baltag and Smets, 2006b; Baltag and Smets, 2006c; Baltag and Smets, 2007b]. Just as in **DEL**, and similarly to the approach in the previous subsection, one can automatically read off a set of *Reduction Axioms for knowledge and safe belief*, thus obtaining a complete proof system. But Baltag and Smets also derive in [Baltag and Smets, 2006c] general (though very complex) Reduction laws for conditional belief.

The Action-Safe-Belief Axiom As for **DEL**, we only present here the most important reduction axiom, namely the appropriate generalization of the Action-Knowledge Axiom to the logic of doxastic actions. In fact, there are two such laws:

one for knowledge, the other for safe belief. But the reduction law for knowledge is essentially the same as the Action-Knowledge Axiom in Section 5.5. So we only state here an "Action-Safe-Belief Axiom", saying that, for every basic action α, we have:

$$[\alpha]\Box_a\varphi \leftrightarrow \left(\mathsf{pre}(\alpha) \rightarrow \bigwedge_{\alpha' <_a \alpha} K_a[\alpha']\varphi \wedge \bigwedge_{\alpha'' \cong_a \alpha} \Box_a[\alpha'']\varphi\right)$$

where $<_a$ is the strict plausibility order on (syntactic) actions (in the action model induced by the signature) and similarly $\cong_a$ is the *equi-plausibility relation* on (syntactic) actions.

This axiom could be thought of as the "fundamental law of dynamic belief revision": it allows us to compute or predict safe beliefs after a learning event in terms of knowledge and safe beliefs before the event. In plain words, it says that: a sentence φ will be safely believed after a doxastic event iff, whenever the action can take place, it is known that φ will become true after all more plausible events and at the same time it is safely believed that φ will become true after all equi-plausible events.

Unifying Diverse Belief-Revision Policies As seen in the examples above, the Action-Priority Update can simulate the various belief revision policies considered in the previous section. More generally, the power of the action model approach is reflected in the fact that many different revision policies can be recovered, in a uniform manner, as instances of the same type of update operation. In this sense, the **DEL** approach can be seen as a change of perspective: the multiplicity of possible revision policies considered in the Belief Revision literature is replaced by the multiplicity of possible action models; the differences are now viewed as *differences in input, rather than having different "programs" for revision.* For a computer scientist, this resembles *currying* in the lambda-calculus: if every "operation" is encoded as an input-term, then *one operation* (functional application) *can simulate all operations.*[16] In a sense, this is nothing but the idea of Turing's universal machine, from the theory of computation. Note that, by incorporating the Product Update from Section 5.4, the Action-Priority Update gains all its dynamic features and its advantages: in addition to simulating a range of individual belief-revision policies, it can deal with an even wider range of complex types of multi-agent learning and communication actions. It may thus be realistic to expect that, *within its own natural limits*, Action Priority Update could play the role of a "universal qualitative machine" for dynamic interactive belief-revision. The problem of finding these natural limits remains open.

[16]Note that, as in untyped lambda-calculus, the input-term encoding the operation (i.e, the action model) and the static input-term to be operated upon (the state mode) are essentially of the same type: epistemic plausibility models for the same language (and for the same set of agents).

Notes The action plausibility models were first introduced by Aucher [2005a; 2005b], as an adaptation of the **DEL** framework of Baltag, Solecki and Moss to the case of dynamic belief revision. Aucher used an equivalent definition, inspired from the work of Spohn [1988], describing plausibility models in terms of ordinal plausibility functions, interpreted as "*degrees of belief*". This lead Aucher, and then van Ditmarsch and Labuschagne [van Ditmarsch and Labuschagne, 2003; van Ditmarsch, 2005; van Ditmarsch and Labuschagne, 2007], to propose and study various types of product update, of a different, more "quantitative" flavor than the Action-Priority update presented above; these proposals are based on using various binary operations on ordinals to compute the degree of belief of an updated state in terms of the corresponding degrees of belief of the input-state and of the action. None of these specific proposals seem to correspond to the Action-Priority update (although it is easy to see that this type of update *can* be computed via a special ordinal function, so in a sense it is *subsumed* by the general "quantitative" approach). Aucher introduced a doxastic logic, with operators $B_a^n \varphi$ for each ordinal degree of belief n, and completely axiomatized the dynamic logic corresponding to his proposal of product update. This work was generalized by van Ditmarsch [2005], who also gave a good presentation of the various proposals in the literature, as well as of the various problems encountered. A recent breakthrough in the field was the work of van Benthem [2006] on the relational approach to belief "upgrades", partially based on previous work by van Benthem and Liu [2004] on preference upgrades. At the same time, Baltag and Smets [2006a; 2006b; 2006c; 2007b] developed their own relational approach to dynamic belief revision, introducing the Action-Priority Update and the Action-Safe-Belief Axiom. Both van Benthem, and Baltag and Smets, used a qualitative logical language (either based on conditional belief operators, or on knowledge and safe belief operators) rather than one based on degrees of belief. Baltag and Sadrzadeh [2006] gave an *algebraic axiomatization* of a type of dynamic belief revision. In more recent work (still to appear), Baltag and Smets [2007a] develop a *probabilistic version of dynamic belief revision*, based on combining their previous work on safe belief and the Action-Priority update with the work of van Fraasen [1995], Boutilier [1995] and Parikh [2005] on using Popper's counterfactual probability functions to deal with belief revision.

8 CONCLUSION

As we end this chapter, we step back to try to understand what makes this particular subject of epistemic logic and information update what it is. We especially want to compare what is going on here to what is discussed in other chapters, especially Chapters 4c and 3b.

In a sense, our treatment of epistemic phenomena is *ultra-semantic*. Beginning in Section 2, we depicted representations and treated them as abstract semantic objects. Even before this, we stated openly that our modeling was slanted towards justifiable belief. This stance implicitly allowed us to ignore *reasons to believe* and instead focus on models of the phenomena of interest. All throughout our section on examples, we emphasized that one must test models and semantic definitions against intuitions, that the proof of the pudding is in the eating. Indeed, our subject is not a single pudding at all but rather a whole buffet of delectable semantic desserts. We also made it clear that the chefs used an artificial sweetener, relational models, and so those allergic to logical omniscience might prefer the fresh fruit. But except for this, the models work extremely well: the predictions of the logical languages match the intuitions. And one can use the formal tools as a real aid in building representations.

At the same time, our work is *unexpectedly syntactic*. We saw a series of logical languages crafted to exploit the key semantic features of the models. Whenever one hears about "encoding" in this subject, it is this: the semantic objects quickly become the sites for semantic evaluation in languages which are richer than one might have at first expected. The easiest example is the relational (Kripke) semantics itself. Having a set of Carnapian state descriptions living alone is at this time fairly mundane. Even adding one or more accessibility relation and calling things "possible worlds" does not go far in relating the worlds to one another. But once one has languages with modal operators, statements evaluated at one world in general must refer to other worlds. Thus the worlds *really* are related: since the logical language has iterated modal sentences, what is true here is in general influenced by what is true far away.

The models in this chapter also incorporate dynamics, social features such as common knowledge, and conditional operators. In each case, the languages are taken to be immediately higher-order: we have knowledge about knowledge, belief about beliefs, announcements about announcements, etc. What makes the subject work is that the formal semantics of the languages refer to the structure of the models, and at the same time the intuitive concepts of interest correspond most closely to statements in the formal languages. One aspect of our work which might be unexpected is the emphasis on particular logical systems for specialized phenomena. We presented a logic of public announcements in Section 5, but this is just the tip of the iceberg. One can formulate specialized logics for other epistemic actions. The point again is that we have semantic objects corresponding to these actions (this seems to be an innovation coming from this subject) and then the resulting logical systems take on an interest of their own, qua syntactic systems. And on the opposite pole from the specialized logics are the very general ones which incorporate arbitrary actions in some sense: these logical languages are unexpectedly syntactic in the sense that their very formulation is trickier than usual, as is their semantics. But the arrows inside of relational models are the same kind of thing as the arrows between the models, and this is *why* dynamic epistemic logic works.

One should compare the situation with the belief revision literature surveyed in Chapter 4c. The AGM postulates deal with several operations, most notably revision. These came first, and then later people were concerned with concrete models of them, with representations of theory-change operations, and the like.

There is much less of an emphasis on matching the predictions of models to intuitions, mainly because the intuitions are often not as clear, and also because notions like a *theory change operation* are more abstract than a *completely private announcement* in our sense. It also took longer for the matter of iterated revision to become central. So the subject developed in a different way from ours. At the same time, there are interesting similarities: as Table 1 in Chapter 4c shows, the history of work in belief revision might be organized according to the particular kinds of prior and posterior belief states discussed. In our subject, the parallel is the extension of the ideas from "hard" semantic updates to "soft" ones (in the terminology of Chapter 3b). Belief revision theory is a much more active field than dynamic epistemic logic, and so one would expect to see a further push towards varied semantic models. But overall, these parallels could be taken to indicate hidden traces of functionalism in what we are doing, though clearly the emphasis on models and languages here is the most prominent difference.

All of this could be said about other closely related topics, especially work on history-based epistemic systems, interpreted systems, and related models which we surveyed in our temporal reasoning Section 6.

Another difference between the main thrust of belief revision work and recent trends in epistemic logic is the "social" aspect of the latter area. This is not true of the earliest work in the subject, partly because philosophers have tended to look only at public information. But as one can see from our chapter, the subject is now about *public and private* types of information change: how they compare and contrast, and how they are integrated in larger theories. This is clearly of interest in mathematical areas of the social sciences, but we feel it is also of interest to philosophy. To be a person is to relate to others, and so to understand knowledge we should pay special attention to multi-agent phenomena.

What connects the ultra-semantic and unexpectedly syntactic are the results on the logical systems themselves. Details of representations often conceal significant conceptual decisions, and results on logical languages and systems can help in the evaluation of different representations. By formulating sound principles, one uncovers (or highlights) hidden assumptions. The matching completeness theorems indicate the right kind of "harmony" (see Chapter 3b). Even more indicative is the fact that those logical systems typically have axiomatic presentations that make intuitive sense. There is no mathematical reason why the axioms behind logical systems should in any way be "nice." Frequently they are not. But we would like to regard the happy coincidences of axioms and intuitions in our subject as signposts which indicate that we are on the right track and point the way ahead.

ACKNOWLEDGMENTS

We thank Anthony Gillies, Joshua Sack, and Ignacio Viglizzo for their very useful comments on this chapter at various stages.

BIBLIOGRAPHY

[Alchourrón *et al.*, 1985] C.E. Alchourrón, P. Gärdenfors, and D. Makinson. On the logic of theory change: partial meet contraction and revision functions. *Journal of Symbolic Logic*, 50:510–530, 1985.

[Arlo-Costa and Parikh, 2005] H. Arlo-Costa and R. Parikh. Conditional probability and defeasible inference. *Journal of Philosophical Logic*, 34:97–119, 2005.

[Artemov, 2004] S. Artemov. Evidence-based common knowledge. Technical report, CUNY, 2004. Ph.D. Program in Computer Science Technical Report TR-2004018.

[Aucher, 2005a] G. Aucher. A combined system for update logic and belief revision. In M.W. Barley and N. Kasabov, editors, *Intelligent Agents and Multi-Agent Systems – 7th Pacific Rim International Workshop on Multi-Agents (PRIMA 2004)*, pages 1–17. Springer, 2005. LNAI 3371.

[Aucher, 2005b] G. Aucher. How our beliefs contribute to interpret actions. In M. Pechoucek, P. Petta, and L.Z. Varga, editors, *Proceedings of CEEMAS 2005: Multi-Agent Systems and Applications IV*, volume 3690 of *Lecture Notes in Computer Science*, pages 276–285. Springer, 2005.

[Aumann, 1976] R.J. Aumann. Agreeing to disagree. *Annals of Statistics*, 4(6):1236–1239, 1976.

[Balbiani *et al.*, 2007] P. Balbiani, A. Baltag, H.P. van Ditmarsch, A. Herzig, T. Hoshi, and T. De Lima. What can we achieve by arbitrary announcements? A dynamic take on Fitch's knowability. To appear in the proceedings of TARK XI, 2007.

[Baltag and Moss, 2004] A. Baltag and L.S. Moss. Logics for epistemic programs. *Synthese*, 139:165–224, 2004. Knowledge, Rationality & Action 1–60.

[Baltag and Sadrzadeh, 2006] A. Baltag and M. Sadrzadeh. The algebra of multi-agent dynamic belief revision. *Electronic Notes in Theoretical Computer Science*, 157(4):37–56, 2006.

[Baltag and Smets, 2006a] A. Baltag and S. Smets. Conditional doxastic models: a qualitative approach to dynamic belief revision. In *Proceedings of WOLLIC'06*, Electronic Notes in Theoretical Computer Science, 2006.

[Baltag and Smets, 2006b] A. Baltag and S. Smets. Dynamic belief revision over multi-agent plausibility models. Proceedings of LOFT 2006 (7th Conference on Logic and the Foundations of Game and Decision Theory), 2006.

[Baltag and Smets, 2006c] A. Baltag and S. Smets. The logic of conditional doxastic actions: a theory of dynamic multi-agent belief revision. Proceedings of ESSLLI Workshop on Rationality and Knowledge, 2006.

[Baltag and Smets, 2007a] A. Baltag and S. Smets. From conditional probability to the logic of belief-revising actions. To appear in the proceedings of TARK IX, 2007.

[Baltag and Smets, 2007b] A. Baltag and S. Smets. A qualitative theory of dynamic interactive belief revision. To appear in G. Bonanno, W. van der Hoek, M. Wooldridge (eds). Selected Papers from LOFT'06, *Texts In Logic and Games*, Amsterdam University Press, 2007.

[Baltag *et al.*, 1998] A. Baltag, L.S. Moss, and S. Solecki. The logic of common knowledge, public announcements, and private suspicions. In I. Gilboa, editor, *Proceedings of the 7th Conference on Theoretical Aspects of Rationality and Knowledge (TARK 98)*, pages 43–56, 1998.

[Baltag *et al.*, 1999] A. Baltag, L.S. Moss, and S. Solecki. The logic of public announcements, common knowledge, and private suspicions. Technical report, Centrum voor Wiskunde en Informatica, Amsterdam, 1999. CWI Report SEN-R9922.

[Baltag *et al.*, 2005] A. Baltag, B. Coecke, and M. Sadrzadeh. Algebra and sequent calculus for epistemic actions. *Electronic Notes in Theoretical Computer Science*, 126:27–52, 2005.

[Baltag *et al.*, 2007, to appear] A. Baltag, B. Cooke, and M. Sadrzadeh. Epistemic actions as resources. *Journal of Logic and Computation*, 2007, to appear. Also in LiCS 2004 Proceedings of Logics for Resources, Programs, Processes (LRPP).

[Baltag, 2002] A. Baltag. A logic for suspicious players: epistemic actions and belief updates in games. *Bulletin Of Economic Research*, 54(1):1–46, 2002.

[Blackburn *et al.*, 2001] P. Blackburn, M. de Rijke, and Y. Venema. *Modal Logic*. Cambridge University Press, Cambridge, 2001. Cambridge Tracts in Theoretical Computer Science 53.

[Board, 2004] O. Board. Dynamic interactive epistemology. *Games and Economic Behaviour*, 49:49–80, 2004.

[Boutilier, 1995] C. Boutilier. On the revision of probabilistic belief states. *Notre Dame Journal of Formal Logic*, 36(1):158–183, 1995.

[Brogaard and Salerno, 2004] B. Brogaard and J. Salerno. Fitch's paradox of knowability. In E.N. Zalta, editor, *The Stanford Encyclopedia of Philosophy*. 2004. `http://plato.stanford.edu/archives/sum2004/entries/fitch-paradox/`.

[Cîrstea and Sadrzadeh, 2007] C. Cîrstea and M. Sadrzadeh. Coalgebraic epistemic update without change of model. In T. Mossakowski, editor, *Proceedings of 2nd Conference on Algebra and Coalgebra in Computer Science*, needed. needed, 2007.

[D'Agostino *et al.*, 1997] M. D'Agostino, D.M. Gabbay, and A. Russo. Grafting modalities onto substructural implication systems. *Studia Logica*, VII:1–40, 1997.

[de Rijke, 1994] M. de Rijke. Meeting some neighbours. In J. van Eijck and A. Visser, editors, *Logic and information flow*, pages 170–195, Cambridge MA, 1994. MIT Press.

[Dixon *et al.*, 1998] C. Dixon, M. Fisher, and M. Wooldridge. Resolution for temporal logics of knowledge. *Journal of Logic and Computation*, 8(3):345–372, 1998.

[Duc, 2001] Ho Ngoc Duc. *Resource-Bounded Reasoning About Knowledge*. PhD thesis, University of Leipzig, 2001.

[Economou, 2005] P. Economou. Sharing beliefs about actions: A parallel composition operator for epistemic programs. In *Proceedings of the ESSLLI 2005 workshop Belief Revision and Dynamic Logic*, 2005. Available on `http://www.irit.fr/~Andreas.Herzig/Esslli05/`.

[Fagin *et al.*, 1995] R. Fagin, J.Y. Halpern, Y. Moses, and M.Y. Vardi. *Reasoning about Knowledge*. MIT Press, Cambridge MA, 1995.

[Fraassen, 1995] B. C. Van Fraassen. Fine-grained opinion, probability, and the logic of full belief. *Journal of Philosophical Logic*, 24:349–377, 1995.

[Freudenthal, 1969] H. Freudenthal. (formulation of the sum-and-product problem). *Nieuw Archief voor Wiskunde*, 3(17):152, 1969.

[Friedell, 1969] M. F. Friedell. On the structure of shared awareness. *Behavioral Science*, 14(1):28–39, 1969.

[Gamow and Stern, 1958] G. Gamow and M. Stern. *Puzzle-Math*. Macmillan, London, 1958.

[Gärdenfors, 1988] P. Gärdenfors. *Knowledge in Flux: Modeling the Dynamics of Epistemic States*. Bradford Books, MIT Press, Cambridge, MA, 1988.

[Gerbrandy and Groeneveld, 1997] J.D. Gerbrandy and W. Groeneveld. Reasoning about information change. *Journal of Logic, Language, and Information*, 6:147–169, 1997.

[Gerbrandy, 1999] J.D. Gerbrandy. *Bisimulations on Planet Kripke*. PhD thesis, University of Amsterdam, 1999. ILLC Dissertation Series DS-1999-01.

[Gerbrandy, 2007] J.D. Gerbrandy. The surprise examination in dynamic epistemic logic. *Synthese*, 155(1):21–33, 2007.

[Gettier, 1963] E. Gettier. Is justified true belief knowledge? *Analysis*, 23:121–123, 1963.

[Gochet and Gribomont, 2006] P. Gochet and P. Gribomont. Epistemic logic. In D. M. Gabbay and J. Woods, editors, *Handbook of the History of Logic*, volume 7, pages 99–195. Elsevier, 2006.

[Groeneveld, 1995] W. Groeneveld. *Logical investigations into dynamic semantics*. PhD thesis, University of Amsterdam, 1995. ILLC Dissertation Series DS-1995-18.

[Halpern *et al.*, 2004] J.Y. Halpern, R. van der Meyden, and M.Y. Vardi. Complete axiomatizations for reasoning about knowledge and time. *SIAM Journal on Computing*, 33(3):674–703, 2004.

[Harel *et al.*, 2000] D. Harel, D. Kozen, and J. Tiuryn. *Dynamic Logic*. MIT Press, Cambridge MA, 2000. Foundations of Computing Series.

[Heal, 1978] J. Heal. Common knowledge. *Philosophical Quarterly*, 28:116–131, 1978.

[Hintikka, 1962] J. Hintikka. *Knowledge and Belief*. Cornell University Press, Ithaca, NY, 1962.

[Hintikka, 1986] J. Hintikka. Reasoning about knowledge in philosophy. In J.Y. Halpern, editor, *Proceedings of the 1986 Conference on Theoretical Aspects of Reasoning About Knowledge*, pages 63–80, San Francisco, 1986. Morgan Kaufmann Publishers.

[Jaspars, 1994] J. Jaspars. *Calculi for Constructive Communication*. PhD thesis, University of Tilburg, 1994. ILLC Dissertation Series DS-1994-4, ITK Dissertation Series 1994-1.

[Katsuno and Mendelzon, 1991] H. Katsuno and A. Mendelzon. On the difference between updating a knowledge base and revising it. In *Proceedings of the Second International Conference on Principles of Knowledge Representation and Reasoning*, pages 387–394, 1991.

[Kooi, 2007] B.P. Kooi. Expressivity and completeness for public update logics via reduction axioms. *Journal of Applied Non-Classical Logics*, 2007. To appear.

[Kripke, 1959] S. Kripke. A completeness theorem in modal logic. *Journal of Symbolic Logic*, 24:1–14, 1959.

[Kvanvig, 1998] J. L. Kvanvig. Paradoxes, epistemic. In E. Craig, editor, *Routledge Encyclopedia of Philosophy*, volume 7, pages 211–214. Routledge, London, 1998.

[Landman, 1986] F. Landman. *Towards a Theory of Information.* PhD thesis, University of Amsterdam, 1986.

[Lehrer and Paxson, 1968] K. Lehrer and T. Paxson. Knowledge: undefeated justified true belief. *The Journal of Philosophy*, 66:225–237, 1968.

[Leitgeb and Segerberg, 2007] H. Leitgeb and K. Segerberg. Dynamic doxastic logic: Why, how, and where to? *Synthese*, 155(2):167–190, 2007.

[Leitgeb, 2007] H. Leitgeb. Beliefs in conditionals vs. conditional beliefs. *Topoi*, 2007.

[Lenzen, 1978] W. Lenzen. Recent work in epistemic logic. *Acta Philosophica Fennica*, 30:1–219, 1978.

[Lenzen, 2003] W. Lenzen. Knowledge, belief, and subjective probability: outlines of a unified system of epistemic/doxastic logic. In V.F. Hendricks, K.F. Jorgensen, and S.A. Pedersen, editors, *Knowledge Contributors*, pages 17–31, Dordrecht, 2003. Kluwer Academic Publishers. Synthese Library Volume 322.

[Levesque, 1984] H. J. Levesque. A logic of implicit and explicit beliefs. In *Proceedings of the National Conference on Artificial Intelligence*, pages 198–202, Austin, Texas, 1984.

[Lewis, 1969] D.K. Lewis. *Convention, a Philosophical Study.* Harvard University Press, Cambridge (MA), 1969.

[Lindström and Rabinowicz, 1999a] S. Lindström and W. Rabinowicz. Belief change for introspective agents, 1999. `http://www.lucs.lu.se/spinning/`.

[Lindström and Rabinowicz, 1999b] S. Lindström and W. Rabinowicz. DDL unlimited: dynamic doxastic logic for introspective agents. *Erkenntnis*, 50:353–385, 1999.

[Lomuscio and Ryan, 1999] A.R. Lomuscio and M.D. Ryan. An algorithmic approach to knowledge evolution. *Artificial Intelligence for Engineering Design, Analysis and Manufacturing (AIEDAM)*, 13(2), 1999. Special issue on Temporal Logic in Engineering.

[Lomuscio, 1999] A.R. Lomuscio. *Knowledge Sharing among Ideal Agents.* PhD thesis, University of Birmingham, Birmingham, UK, 1999.

[Lutz, 2006] C. Lutz. Complexity and succinctness of public announcement logic. To appear in the proceedings of the Fifth International Joint Conference on Autonomous Agents and Multi-Agent Systems (AAMAS 06), 2006.

[McCarthy, 1990] J. McCarthy. Formalization of two puzzles involving knowledge. In Vladimir Lifschitz, editor, *Formalizing Common Sense: Papers by John McCarthy*, Ablex Series in Artificial Intelligence. Ablex Publishing Corporation, Norwood, N.J., 1990. original manuscript dated 1978–1981.

[Meyer and van der Hoek, 1995] J.-J.Ch. Meyer and W. van der Hoek. *Epistemic Logic for AI and Computer Science.* Cambridge Tracts in Theoretical Computer Science 41. Cambridge University Press, Cambridge, 1995.

[Miller and Moss, 2005] J.S. Miller and L.S. Moss. The undecidability of iterated modal relativization. *Studia Logica*, 79(3):373–407, 2005.

[Moore, 1912] G.E. Moore. *Ethics.* Oxford University Press, 1912. Consulted edition: The Home University Library of Modern Knowledge, volume 54, OUP, 1947.

[Moore, 1942] G.E. Moore. A reply to my critics. In P.A. Schilpp, editor, *The Philosophy of G.E. Moore*, pages 535–677. Northwestern University, Evanston IL, 1942. The Library of Living Philosophers (volume 4).

[Moore, 1944] G.E. Moore. Russell's "theory of descriptions". In P.A. Schilpp, editor, *The Philosophy of Bertrand Russell*, pages 175–225. Northwestern University, Evanston IL, 1944. The Library of Living Philosophers (volume 5).

[Moore, 1977] R.C. Moore. Reasoning about knowledge and action. In *Proceedings of the Fifth International Joint Conference on Artificial Intelligence (IJCAI-77)*, Cambridge, Massachusetts, 1977.

[Moses *et al.*, 1986] Y.O. Moses, D. Dolev, and J.Y. Halpern. Cheating husbands and other stories: a case study in knowledge, action, and communication. *Distributed Computing*, 1(3):167–176, 1986.

[Moss, 1999] L.S. Moss. From hypersets to Kripke models in logics of announcements. In J. Gerbrandy, M. Marx, M. de Rijke, and Y. Venema, editors, *JFAK. Essays Dedicated to Johan van Benthem on the Occasion of his 50th Birthday*, Vossiuspers. Amsterdam University Press, 1999.

[O'Connor, 1948] D.J. O'Connor. Pragmatic paradoxes. *Mind*, 57:358–359, 1948.

[Pacuit, to appear] E. Pacuit. Some comments on history based structures. *Journal of Applied Logic*, to appear.

[Parikh and Ramanujam, 2003] R. Parikh and R. Ramanujam. A knowledge-based semantics of messages. *Journal of Logic, Language, and Information*, 12:453–467, 2003.

[Plaza, 1989] J.A. Plaza. Logics of public communications. In M.L. Emrich, M.S. Pfeifer, M. Hadzikadic, and Z.W. Ras, editors, *Proceedings of the 4th International Symposium on Methodologies for Intelligent Systems*, pages 201–216, 1989.

[Qian, 2002] L. Qian. Sentences true after being announced. In *Proceedings of 'Student Conference of the 2002 North American Summer School in Logic, Language, and Information'*. Stanford University, 2002.

[Rott, 2004] H. Rott. Adjusting priorities: Simple representations for 27 iterated theory change operators. Manuscript, 2004.

[Sack, 2007] J. Sack. *Adding Temporal Logic to Dynamic Epistemic Logic*. PhD thesis, Indiana University, Bloomington, 2007.

[Segerberg, 1998] K. Segerberg. Irrevocable belief revision in dynamic doxastic logic. *Notre Dame Journal of Formal Logic*, 39(3):287–306, 1998.

[Segerberg, 1999a] K. Segerberg. Default logic as dynamic doxastic logic. *Erkenntnis*, 50:333–352, 1999.

[Segerberg, 1999b] K. Segerberg. Two traditions in the logic of belief: bringing them together. In H.J. Ohlbach and U. Reyle, editors, *Logic, Language, and Reasoning*, pages 135–147, Dordrecht, 1999. Kluwer Academic Publishers.

[Sorensen, 1988] R.A. Sorensen. *Blindspots*. Clarendon Press, Oxford, 1988.

[Spohn, 1988] W. Spohn. Ordinal conditional functions: a dynamic theory of epistemic states. In W.L. Harper and B. Skyrms, editors, *Causation in Decision, Belief Change, and Statistics*, volume II, pages 105–134, 1988.

[Stalnaker, 1968] R. Stalnaker. A theory of conditionals. In N. Rescher, editor, *Studies in Logical Theory*, APQ Monograph No2. Blackwell, 1968.

[Stalnaker, 1996] R. Stalnaker. Knowledge, belief and counterfactual reasoning in games. *Economics and Philosophy*, 12:133–163, 1996.

[Stalnaker, 2006] R. Stalnaker. On logics of knowledge and belief. *Philosophical Studies*, 128:169–199, 2006.

[Steiner, 2006] D. Steiner. A system for consistency preserving belief change. In *Proceedings of the ESSLLI Workshop on Rationality and Knowledge*, pages 133–144, 2006.

[Steup, Spring 2006] M. Steup. The analysis of knowledge. In E.N. Zalta, editor, *The Stanford Encyclopedia of Philosophy*. Spring 2006. `http://plato.stanford.edu/archives/spr2006/entries/knowledge-analysis/`.

[Swain, 1974] M. Swain. Epistemic defeasibility. *The American Philosophical Quarterly*, 11:15–25, 1974.

[Tennant, 2002] N. Tennant. Victor vanguished. *Analysis*, 62:135–142, 2002.

[van Benthem and Liu, 2004] J.F.A.K. van Benthem and F. Liu. Diversity of logical agents in games. *Philosophia Scientiae*, 8(2):163–178, 2004.

[van Benthem and Pacuit, 2006] J.F.A.K. van Benthem and E. Pacuit. The tree of knowledge in action: Towards a common perspective. Technical Report, ILLC, University of Amsterdam, 2006.

[van Benthem *et al.*, 2006a] J.F.A.K. van Benthem, J. Gerbrandy, and B.P. Kooi. Dynamic update with probabilities. In W. van der Hoek and M. Wooldridge, editors, *Proceedings of LOFT'06*. Liverpool, 2006.

[van Benthem *et al.*, 2006b] J.F.A.K. van Benthem, J. van Eijck, and B.P. Kooi. Logics of communication and change. *Information and Computation*, 204(11):1620–1662, 2006.

[van Benthem, 1989] J.F.A.K. van Benthem. Semantic parallels in natural language and computation. In *Logic Colloquium '87*, Amsterdam, 1989. North-Holland.

[van Benthem, 1994] J.F.A.K. van Benthem. Logic and the flow of information. In *Proceedings of the 9th International Congress of Logic, Methodology and Philosophy of Science (1991)*. Elsevier Science B.V., 1994. Also available as: Report LP-91-10, ILLC, University of Amsterdam.

[van Benthem, 1996] J.F.A.K. van Benthem. *Exploring logical dynamics.* CSLI Publications, 1996.

[van Benthem, 2002] J.F.A.K. van Benthem. One is a lonely number: on the logic of communication. Technical report, University of Amsterdam, 2002. ILLC Research Report PP-2002-27 (material presented at the Logic Colloquium 2002).

[van Benthem, 2004] J.F.A.K. van Benthem. What one may come to know. *Analysis*, 64(2):95–105, 2004.

[van Benthem, 2006] J.F.A.K. van Benthem. Dynamic logic for belief change. *Journal of Applied Non-Classical Logics*, 2006. To appear.

[van der Meyden, 1994] R. van der Meyden. Axioms for knowledge and time in distributed systems with perfect recall. In *Proceedings of the Ninth Annual IEEE Symposium on Logic in Computer Science (LICS-94)*, pages 448–457, Paris, July 1994.

[van der Meyden, 1998] R. van der Meyden. Common knowledge and update in finite environments. *Information and Computation*, 140(2):115–157, 1998.

[van Ditmarsch and Kooi, 2006] H.P. van Ditmarsch and B.P. Kooi. The secret of my success. *Synthese*, 151:201–232, 2006.

[van Ditmarsch and Labuschagne, 2003] H.P. van Ditmarsch and W.A. Labuschagne. A multimodal language for revising defeasible beliefs. In E. Álvarez, R. Bosch, and L. Villamil, editors, *Proceedings of the 12th International Congress of Logic, Methodology, and Philosophy of Science (LMPS)*, pages 140–141. Oviedo University Press, 2003.

[van Ditmarsch and Labuschagne, 2007] H.P. van Ditmarsch and W.A. Labuschagne. My preferences about your preferences – a case study in theory of mind and epistemic logic. *Knowledge, Rationality & Action (Synthese)*, 155:191–209, 2007.

[van Ditmarsch *et al.*, 2003] H.P. van Ditmarsch, W. van der Hoek, and B.P. Kooi. Concurrent dynamic epistemic logic. In V.F. Hendricks, K.F. Jørgensen, and S.A. Pedersen, editors, *Knowledge Contributors*, pages 45–82, Dordrecht, 2003. Kluwer Academic Publishers. Synthese Library Volume 322.

[van Ditmarsch *et al.*, 2005] H.P. van Ditmarsch, W. van der Hoek, and B.P. Kooi. Playing cards with Hintikka: An introduction to dynamic epistemic logic. *The Australasian Journal of Logic*, 3:108–134, 2005.

[van Ditmarsch *et al.*, 2007] H.P. van Ditmarsch, W. van der Hoek, and B.P. Kooi. *Dynamic Epistemic Logic*, volume 337 of *Synthese Library*. Springer, 2007.

[van Ditmarsch, 2000] H.P. van Ditmarsch. *Knowledge Games.* PhD thesis, University of Groningen, 2000. ILLC Dissertation Series DS-2000-06.

[van Ditmarsch, 2002] H.P. van Ditmarsch. Descriptions of game actions. *Journal of Logic, Language and Information*, 11:349–365, 2002.

[van Ditmarsch, 2005] H.P. van Ditmarsch. Prolegomena to dynamic logic for belief revision. *Synthese (Knowledge, Rationality & Action)*, 147:229–275, 2005.

[van Emde Boas *et al.*, 1984] P. van Emde Boas, J. Groenendijk, and M. Stokhof. The conway paradox: Its solution in an epistemic framework. In J. Groenendijk, T. M. V. Janssen, and M. Stokhof, editors, *Truth, Interpretation and Information: Selected Papers from the Third Amsterdam Colloquium*, pages 159–182. Foris Publications, Dordrecht, 1984.

[van Linder *et al.*, 1995] B. van Linder, W. van der Hoek, and J.-J.Ch. Meyer. Actions that make you change your mind. In A. Laux and H. Wansing, editors, *Knowledge and Belief in Philosophy and Artificial Intelligence*, pages 103–146, Berlin, 1995. Akademie Verlag.

[Veltman, 1996] F. Veltman. Defaults in update semantics. *Journal of Philosophical Logic*, 25:221–261, 1996.

[von Wright, 1951] G.H. von Wright. *An Essay in Modal Logic.* North Holland, Amsterdam, 1951.

[Wansing, 2002] H. Wansing. Diamond's are a philosopher's best friends. the knowability paradox and modal epistemic relevance logic. *Journal of Philosophical Logic*, 31:591–612, 2002.

[Wassermann, 1999] R. Wassermann. Resource bounded belief revision. *Erkenntnis*, 50(2–3):429–446, 1999.

[Wassermann, 2000] R. Wassermann. *Resource Bounded Belief Revision.* PhD thesis, University of Utrecht, 2000.

[Williamson, 2000] T. Williamson. *Knowledge and Its Limits.* Oxford University Press, Oxford, 2000.

[Wisniewski, 1998] A. Wisniewski. Two logics of occurrent belief. *Acta Universitatis Wratislavensis*, 2023(18):115–121, 1998.

[Yap, 2006] A. Yap. Product update and looking backward. Technical report, University of Amsterdam, 2006. ILLC Research Report PP-2006-39.

INFORMATION STRUCTURES IN BELIEF REVISION

Hans Rott

1 INTRODUCTION

Belief revision theories address the problem of rationally integrating new pieces of information into an agent's belief state. Belief states themselves are represented by certain kinds of information structures. The most closely studied, and arguably the most interesting type of belief revision is the one in which the new information is inconsistent with the agent's current beliefs. The main question of the present chapter is: What kinds of information structures have figured prominently in belief revision theories as they have been developed over the past 30 years?

My focus will be on the philosophy behind 'classical belief revision' of the AGM paradigm (so-called after its founders Alchourrón, Gärdenfors and Makinson and their [1985] paper) and its generalizations to iterations of belief change in the 1990s and to operations of belief fusion or merging in the present century.

I am going to presuppose that information comes in symbolic form and will be silent about non-codified information formats such as signals, symptoms, pictures, etc. I shall say very little about approaches that make essential use of numerical, quantitative information. This of course is not to suggest that probabilistic approaches, evidence theory (Shafer [1976]) or ranking functions (Spohn [1988], Goldszmidt and Pearl [1992]) are not interesting and useful, but qualitative approaches have formed a research program in belief change that merits a chapter of its own.

The present chapter should be read in conjunction with the contribution by Baltag, van Ditmarsch and Moss [2008]. In several respects, their approach is wider than the one presented here. However, the particular emphasis of belief revision theory as understood in this chapter is on the case where new information conflicts logically with the information previously accepted (in the terminology to be introduced later: with the old 'data base'). It can thus be seen as providing a module that can be combined with the framework of Baltag et al. The following presentation, however, remains in the style of the more traditional, classical work in the area.[1]

[1]For important attempts to transfer carefully belief revision theories into the framework of modal logic, see Fuhrmann [1991], Cantwell [1997], Lindström and Rabinowicz [1999], Segerberg [2001] and van Benthem [2007]. For an enlightening discussion of the merits of the somewhat non-standard attitude towards logic in the AGM program, see Makinson [2003].

Handbook of the Philosophy of Science. Volume 8: Philosophy of Information
Volume editors: Pieter Adriaans and Johan van Benthem. General editors: Dov M. Gabbay, Paul Thagard and John Woods.

The plan of this chapter is as follows. Section 2 contains some preliminary remarks on information and truth, and on belief change as embedded in a functionalist philosophy of mind. Section 3 presents the problem of belief change as being a compound of processes of reflection and processes of revision. Depending on which of these processes are given center stage, we can distinguish foundationalist and coherentist approaches to belief change (in roughly the sense that is known from contemporary epistemology). Section 4 gives three simple examples of such approaches. There are approaches that look coherentist but can be reconstructed as generated through hidden foundationalist recipes. Reconstructions of such a flavour are fairly typical of belief revision theories. Section 5 traces the idea that the static picture of belief as being represented by an information structure may encode much of the dynamics of belief. I shall interpret the history of belief revision theory of the last three decades as a story of finding an appropriate notion of a belief state, and its interpretation as providing a framework for analyzing both static and dynamic aspects of belief.

2 PRELIMINARY REMARKS ON INFORMATION, TRUTH AND MIND

2.1 Remarks on information and truth

There are countless explications of the term 'information.' I would like to propose the following informal and very general definition:

> *Information* is some structure realized in the physical world that is suitable to be interpreted or exploited by some receiver in a reasonable way.

According to this definition, the pages of a book and the platters of a hard disk, the trunk of a tree and the DNA strands in the cell of an animal, the remains of a burnt house and the sounds of a recited poem all carry information that can be interpreted or exploited by a suitable 'receiver'. By *interpretation* or *exploitation*, I mean some kind of causal interaction between two entities, the input and the receiver. The receiver need not be human or living. Usually there is a certain asymmetry between input and receiver, but this is not necessary. In DNA replication, reading molecules and read molecules are quite similar, and we shall see that current belief revision research has inputs and receiving states of the same format. Interpretation in this sense is extremely wide (too wide perhaps), it may possibly, but need not necessarily make use of cognitive or linguistic means. A *reasonable* way is one that 'makes sense' of the information structure, and leads to successful behaviour or action of the receiver.

Often information, or rather pieces of information, are taken to represent states of affairs or objects. Some kinds of information (*signals*) are invariably *truthful* (or *veridical*), but they may still be *deceptive* in the sense that they give rise to an inadequate interpretation, e.g., to false conclusions or unsuccessful behaviour. If there is no natural or necessary link between a piece of information and what it is supposed to represent, i.e., if it is *symbolic* and has a conventional meaning, then the information carrier may itself be called *false* (or *misinformation* or *misrepresentation*). However, it may be difficult to say whether the 'falsity' of the link between information and what it represents is due to the carrier of information itself or the receiver interpreting this unit of information.

What is information as it figures in theories of belief revision? *Information* is something that may enter some belief state and change or transform it into another belief state. Belief revision theories usually do not care about the truth of beliefs, nor do they address the question of the beliefs' justification or reliability.[2] For this reason, terms and phrases like 'knowledge base', 'knowledge representation' and 'knowledge in flux', though widely used, are often misnomers.

The idea, also emphasized in dynamic semantics or update semantics,[3] is to characterize (the content of) a *piece of information* by the transformations of the receivers' internal states that it has the potential of bringing about. This may best be represented as a function turning prior states into posterior states:

$$\text{prior-state} \longmapsto \text{posterior-state}$$

In the sense just introduced, such a transformation captures the interpretation or exploitation of a piece of information. The posterior state plays the role of the receiver, and as we shall see, the reasonableness of the interpretation is captured by a list of constraints. It is not required that the input be true.

Information in belief revision theories typically is

- *syntactic* in the sense that it is representable by sentences of some appropriate systematic language, and that it can combined in the typical way sentences can,
- *semantic* in the sense that
  - it is not sensitive to transformations into logical equivalents,[4] and
  - as long as the new information is consistent with the current belief state, it simply rules out possibilities.[5]

2.2 Some clues from the philosophy of mind

The main interest of belief revision theory lies in information for receivers with a mind, i.e., humans.[6] This suggests to have a look at the philosophy of mind, a philosophical sub-discipline concerned with the relation between body and mind (for the following, cf. Kim [2006, chapters 2–6]). How can it be that a person, that

[2] But see Kelly, Schulte and Hendricks [1997] and Kelly [1999].

[3] See Stalnaker [1984], Gärdenfors [1988, chapter 6], Groenendijk and Stokhof [1991], and Veltman [1996]

[4] Compare, e.g., axioms (AGM6) and (DP6) below.

[5] Compare, e.g., axioms (AGM3), (AGM4), (DP3) and (DP4) below.

[6] Or information for computing machinery, but we will see that this difference is not important for our purposes.

is first of all a biological organism, exhibits thoughts, desires, feelings, etc? How can these mental states, states that seem to be perfectly accessible to the persons that are in them but not to any third persons, be the objects of scientific studies? Let us call the mental life of a person his or her *psychology*.

According to behaviourism, the leading school in the study of mind until the middle of the 20th century, *a person's psychology* can be identified with (characterized by, reduced to) the function

$$\text{input} \longmapsto \text{output}$$

where inputs and outputs, physical stimuli and behavioural responses, are observable entities. Behaviourism turned out to be too simplistic. An alternative approach to objectify the human mind was provided by the physicalist or materialist identity theory of mind, according to which a person's psychology can be identified with (characterized by, reduced to) its physical or material state. But this idea would rule out that beings with a different physiology like non-human animals or Martians or machines can have psychological states like humans, something one would at least want to leave conceptual room for.

So a third paradigm, functionalism, appeared on the scene which in a way combined the best of the previous approaches. Functionalism is behaviorism plus internal states, or — approaching it from the other side — functionalism is materialism plus multiple realizability.[7] Alan Turing [1950] and Hilary Putnam [1960; 1967] promoted the idea that human thinking could be likened to the calculations that go on in a computer (a Turing machine). The *computer metaphor* became popular which says that mind is to brain as software is to hardware. A piece of software is a program that can be described abstractly by a (finite) set of transitions of the following type:

$$\langle \text{prior-state, input} \rangle \longmapsto \langle \text{posterior-state, output} \rangle$$

Here the prior and posterior states are *internal states* or *psychological states* of the person or the computer. The set of all such transitions was called a 'machine table' by the early functionalists, and it fully specifies *a person's psychology* or a *computer program.*

2.3 *Functionalism as applied to belief revision*

Belief is an internal affair. Any possible announcement of one's beliefsor any other action produced by a belief may be disregarded for the purposes of this paper. Thus, we do not need to deal with any manifest output.[8] This suggests that for

[7]The thesis of multiple realizability says that a mental state can be 'realized' or 'implemented' by different physical states. Beings with different physical constitutions can thus be in the same mental state.

[8]Alternatively, one could say that belief revision's outputs are the belief states themselves. I shall avoid this terminological move.

belief revision, the functionalist format can be reduced to a simpler mapping of the following format:

$$\langle\text{prior-state, input}\rangle \longmapsto \text{posterior-state}$$

The *input* to a belief state is a piece of information. *Belief states* in turn are the results of an — often long — history of information processing episodes. It is plausible to require that the representation of the prior state and the representation of the posterior state should be of the same format.[9]

The format of the input varies in theories of belief revision. The literature has studied pieces of input of the following kinds:

- propositions
- propositions coming with a specification of their relative position in a (total) pre-order
- ranked propositions (i.e., propositions coming with numerical ranks)
- ordered pairs of propositions (indicating their comparative retractability)
- propositions with a specification of their source
- full preferential structures

Beliefs are usually taken to be represented by propositions — either linguistically, as sentences or sentence-meanings, or abstractly, as sets of possibilities. I shall, somewhat sloppily, not distinguish between these two variants and will identify a sentence, i.e., an expression of a given language, with what is said by such a sentence. By a *belief set* we shall understand a set of beliefs that is closed under logical consequences (obtained by a background operation Cn). A belief set is what logicians are used to calling 'theory'.

A *belief state* determines the set of beliefs held by the agent, but may (and usually does) encode much more information than that. Individual beliefs are derivative of belief states. For instance, a belief state may be represented with the help of Grovean systems of spheres (Grove [1988], cf. Figure 1).

A *system of spheres* is a set of nested sets of possible worlds, or more precisely, a set of nested sets of models of the language. The smallest set in the center (labelled by '1') is the set of possible worlds which the agent believes to contain the actual world, i.e., the worlds that are possible according to the agent's beliefs. If she receives evidence that the actual world is not contained in this smallest set, she falls back on the next larger set of possible worlds. Thus the first shell[10] around the center (labelled by '2') contains the worlds considered second most plausible by the reasoner. And again, should it turn out that the actual world is not to be

[9] This requirement was called *the Principle of categorial matching* in Gärdenfors and Rott [1995].

[10] A *shell* is the difference set between two neighbouring spheres. Spheres are nested, shells are disjoint. The shells are numbered, but the numbers are not supposed to have any meaning beyond the indication of the ordering of spheres.

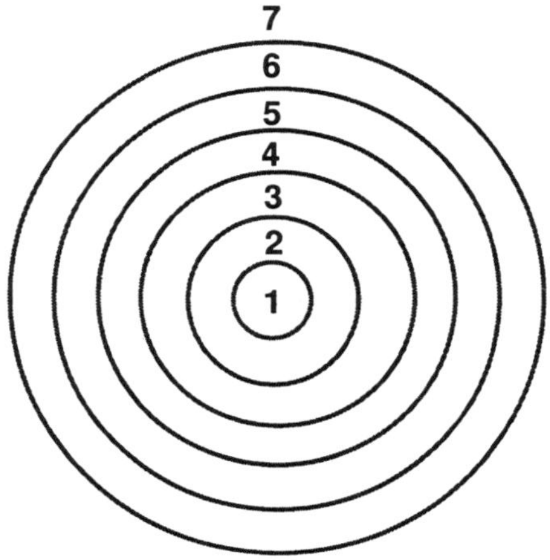

Fig. 1. A Grovean system of spheres

found in this set (the union of the sets labelled '1' and '2') either, the reasoner is prepared to fall back on her next larger set of possible worlds (the union of the sets labelled '1', '2' and '3'). And so on. A system of spheres is equivalent to a total pre-ordering of possible worlds, using the definition that $w \preceq w'$ (read 'w is at most as far-fetched as w'') if and only if w is contained in every sphere in which w' is contained.

A system of spheres at the same time determines a pre-ordering of propositions, often called *entrenchment ordering*. If the set of models of A *covers* each sphere that is covered by the set of models of B, the proposition A is at least as entrenched in the agent's belief state as the proposition B (in symbols, $B \leq A$). Conversely, an entrenchment pre-ordering $\leq$ generates a system of spheres \$ if we collect in \$ the sets of models of $\{B : A < B\}$, for all A. These two ways of linking sphere systems with entrenchments fit together and represent, in a natural sense, the same belief state. Belief states in these two senses may be regarded as *non-propositional information structures.*

The beliefs of the agent can be retrieved from her belief state. If a belief state is represented by a system of spheres \$ (or the corresponding total pre-ordering $\preceq$) of possible worlds, the beliefs *Bel* (\$) (or respectively, *Bel* ($\preceq$)) are those propositions that are true in each of the possible worlds contained in the innermost sphere (or respectively, in each of the possible worlds that are minimal under $\preceq$). If a belief state of an agent is represented by an entrenchment ordering $\leq$ over the propositions of a language, her beliefs *Bel* ($\leq$) are those propositions that are non-minimal under this ordering.

2.4 Filling in the parameters

We can now give a first overview of some of the more important stages in the development of belief revision theory. We just need to fill in the parameters into the scheme just specified in various ways (see Table 1).

	input	**prior belief state**	**posterior belief state**
AGM [1978]ff	proposition	set of beliefs (logically closed; plus preference structure on beliefs or sets of beliefs)	set of beliefs (logically closed)
Grove [1988], Katsuno-Mendelzon [1991]	proposition	preference structure on possible worlds	set of possible worlds
Veltman [1976], Kratzer [1981], Nebel [1989], Hansson [1989; 1999]	proposition	set of beliefs (syntactically structured, i.e. not logically closed)	set of beliefs (logically closed?)
Spohn [1988], Goldszmidt-Pearl [1992]	proposition plus plausibility index	ranking function (a kind of preference structure)	ranking function
Darwiche-Pearl [1994; 1997]	proposition (often plus plausibility index)	general format, with beliefs derivative	general format, with beliefs derivative
Cantwell [1997], Fermé-Rott [2004]	pair of sentences	preference structure	preference structure
Nayak [1994]ff, merging — fusion	preference structure	preference structure	preference structure

Table 1. Filling in the parameters for a functionalist account of belief change

An important turning point of the development of belief revision theory was the recognition in the 1990s that a belief state must not be identified with a (logically closed) set of beliefs. The study of the problem of iterated belief revision made it clear that AGM's belief set had to be replaced by a belief state with a selection functions or a preferential structure in order to encode a full belief state. Alternatively, Darwiche and Pearl [1994; 1997] suggested that it is best to take 'belief state' as a primitive concept. More on this in section 6 below. After the turn of the millennium, much research has focussed on the problem of fusing or merging belief states. The old idea that a belief state (perhaps carrying information about the learning history of the agent) gets revised by a single piece of information is no longer valid as a description of the standard problem. The question addressed in fusion or merging is how to merge two or more rather general information structures into a single one.

3 BELIEF CHANGE = REVISION + REFLECTION

How do belief states change? I suggest to decompose the process of belief change into two different processes. There are then two fundamentally different perspectives, depending on which of the two processes is being highlighted (see Fig. 2).

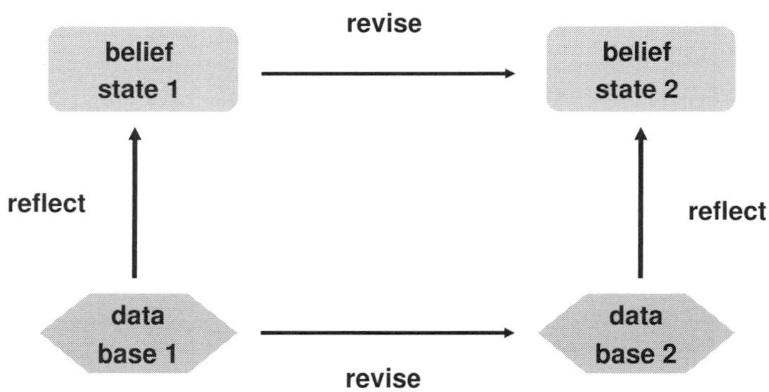

Fig. 2. Reflection and revision

The process of *revision* is that of changing the current data base or belief state in response to the receiving of some new piece of information. Models of belief change emphasizing the process of revision while employing rather a plain method of transforming data bases into belief states take the *horizontal perspective.*

The process of *reflection* is that of finding an equilibrium state by processing, or drawing inferences from, the currently available information. Models of belief change emphasizing the process of reflection while employing a rather straightforward method of revision take the *vertical perspective.*[11]

We shall consider three approaches.

- Foundationalism in vertical perspective

[11]The pair 'reflection' and 'revision' also plays an important role in Harman's seminal book *Change in View* [1986, Chapter 1], but Harman's usage of the terms is different from ours.

- Foundationalism in horizontal perspective
- Coherentism in horizontal perspective

Foundationalism assumes that there is a distinguished class of basic beliefs which are somehow given (by perception or by intuition, in any case independently from any other beliefs) and from which all other beliefs can be derived with the help of some inference operation.[12] *Coherentism*, on the other hand denies that the foundationalists' distinction between basic and derived beliefs has any clear significance. For the coherentist there is no set of propositions that enjoy the privilege of serving as a foundation for the other beliefs. Coherentists are not interested in the origin of belief states. The reflection component does not even appear in the picture, the aim of reaching (or rather remaining in) an equilibrium state is part and parcel of the revision process.

3.1 Foundationalism

Assume that inputs have come in repeatedly. A *data base* is the result of collecting the inputs and putting them together. It is important that data bases in our sense need not obey coherence constraints of a logical or any other nature. A data base will be considered as a rough and ready collection of pieces of information. It is the *basic information structure* figuring in belief revision, directly interpreting or exploiting, as it were, the structure of the inputs.

For instance, data bases can be

- sets of propositions
- totally pre-ordered sets of propositions
- ranked sets of propositions

The difference between totally pre-ordered and ranked sets is that the latter, but not the former, have numbers attached to the propositions signifying their degree of belief. These numbers represent distances, and it makes sense to perform arithmetical operations on them. We assume that a data base may grow or shrink in response to incoming input, through insertions or deletions at certain positions in the pre-ordered or ranked structure.

The process of *reflection* is that of finding an equilibrium state on the basis of a given data base. The data base is processed and thereby transformed into a belief state in equilibrium. Reflection is *static* in the sense that no new input is being dealt with. It can be thought of as an act of *information processing*. Reflection may distinguished from the equally static process of *drawing inferences* which does not yield a belief state, but only a belief set, i.e., a well-balanced set of propositions that can be inferred from, or are supported by, the agent's data base.[13] We can

[12]Note that nothing in this formulation of the foundationalist idea presumes that the elements of the data base are immune to revision!

[13]It is plausible to assume two mappings here, one taking data bases to belief states, and another one taking belief states to belief sets. Neither of these mappings is injective. The latter

thus distinguish two kinds of static operations on data bases, with information processing being more complex than the drawing inferences. The latter achieves less than the former unless a belief state is identified with a belief set (theory) to begin with. In any case a belief state uniquely determines a belief set, but in general a single belief set can be determined by many different belief states.

The process of *revision* is a response to incoming input.[14] Obviously, the nature of a change operation depends on the nature of the entities on which the change operation operates. Revision may consist in a relatively trivial change operation on the data base level, or alternatively, in a relatively sophisticated change operation on the belief state level. On the base level, the changes need not be sophisticated since data bases are not required to be coherent (Fig. 3).

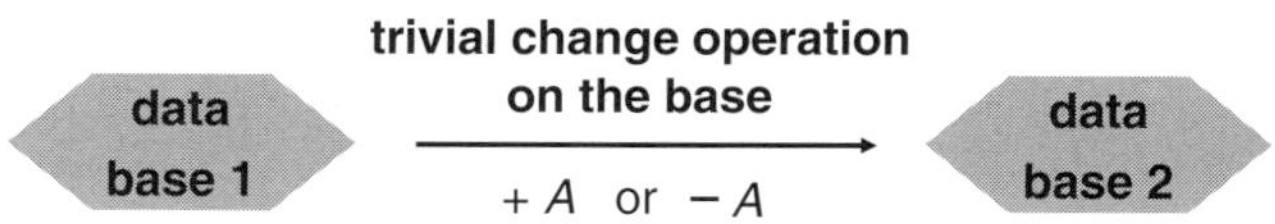

Fig. 3. Changes on the data base level

A change operating on the level of data bases may be thought of as inducing, if combined with an inference operation, a change operation on the level of belief states. The latter changes are then only derivative. The change operation on the data base level typically applies simple and straightforward, unrestrained *insertions* or *deletions* of a proposition from a set of propositions. If that set is ordered, addition of a piece of information[15] on top of the ordered list may be suitable, or the new information may come labelled with some 'rank' or 'reference sentence' specifying its new position in the existing ranking. Free and simple insertions will usually cause a violation of deductive closure and will sometimes cause a violation of consistency of the information the agent has been provided with. This calls for a sophisticated reflection operation that restores consistency and ultimately produces logical closure. Fig. 4 illustrates the foundationalist idea of the vertical perspective.

3.2 Coherentism

Alternatively, one can think the of the revision process as operating directly on the *belief state level*. Then the lower level of the data bases is not in the picture any more, all deliberation that takes place is part and parcel of the change

mapping has been called retrieval above.

[14]Here I am using the term 'revision' in the wider sense of 'change', covering both revisions (in the narrower sense) and contractions of belief states. 'Revision' and 'contraction' have been used as technical terms since AGM. I trust that no confusion arises from my double use of the term 'revision'.

[15]Or of a 'phantom belief', see footnote 22 below.

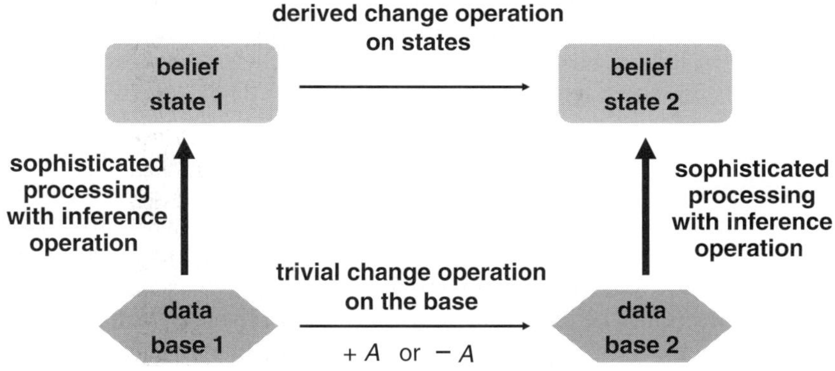

Fig. 4. Foundationalism as taking the vertical perspective

from one equilibrium state to another. Incoming information means a disturbance of an existing equilibrium, and processing information means equilibrating belief states. This idea necessarily emphasizes the revision process, it takes the horizontal perspective. Since no separate reflection process is there to eliminate inconsistencies or incoherencies, the change process itself has to be sophisticated (Fig. 5). It is important to note that the input alone does not bring about the transition from one belief state to another. In order to resolve contradictions, or to avoid unwanted implications, it is necessary to make choices which beliefs to remove. For non-trivial revision operations, the agent will therefore need some *selection structure* (for instance, a preference relation) to guide these choices, and also some *rule of application* specifying how exactly to apply the selection structure when accommodating the input.[16] This is what makes the change operation sophisticated.

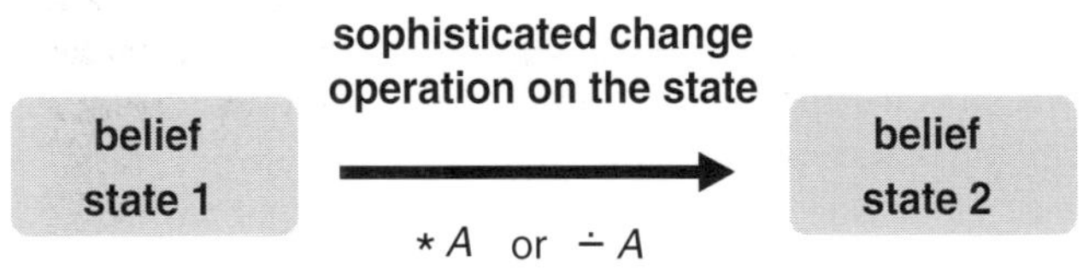

Fig. 5. Changes on the belief state level —
Coherentism as taking the horizontal perspective

[16]For instance, the belief set revised by A can be identified with the set of all sentences that are true in the least far-fetched worlds that satisfy A, or with the set of sentences B for which $A \rightarrow B$ is more entrenched than $\neg A$. I cannot deal with concrete rules of application in this paper and have to refer the reader to Gärdenfors and Rott [1995] and Hansson [1999].

Coherentism may be hidden foundationalism. In fact, researchers in belief change have often aimed at bringing in the data base level again and provided *representation results* of the following sort. Given a revision operation on the belief state level, are there data bases, a revision operation on the data base level and a reflection operation such that the revision operation on the belief state level is induced by the latter three items? We will give a few examples indicating that the answer is positive in some interesting cases.

'Processing information' or, equivalently, 'processing inputs' may thus mean two different kinds of things. Either the agent adds the new piece of information to the current data base (i.e., to some existing totally pre-ordered set of propositions) and subsequently draws inferences from the revised data base. Or the agent accommodates her belief state to the input, without recourse to any basic beliefs.[17]

4 INFERENCE OPERATIONS FOR SIMPLE CHANGE OPERATIONS: THREE EXAMPLES

Let us now have a look at the vertical perspective. Which methods of reflection are suitable for which methods of change? Here a method is called 'suitable' if it results in a reasonable change operation at the belief state level.

4.1 Example 1: Flat data bases

First, let the *data base* be a plain set Γ of propositions. The most straightforward method of *drawing inferences* from such a data base is to take the logical closure $Cn(\Gamma)$ of Γ, where the logic Cn used is some broadly classical or Tarskian logic.[18]

The use of some such standard logic, however, will frequently create *problems of interaction* with simple change operations if the latter are applied to data bases. If a new proposition is simply added to Γ, this may easily generate an inconsistency that cannot be processed by the logic Cn. This is the consistency problem for belief revisions. If a previously accepted proposition is simply eliminated from Γ, this may fail to efficiently remove Γ. The reason is that applying logical closure to the remaining propositions may easily restore the eliminated belief. This is the closure problem for belief contractions.

While it is not clear how any remotely standard logic can solve the consistency problem for belief revision, a paraconsistent logic might help.[19] Belief change theorists, however, have usually preferred to take another route and insisted on consistence (for revisions) and effective removal (for contractions) already on the

[17] For more on the perspectives described in this section, as well as a number of concrete suggestions to flesh them out, see Rott [2001].

[18] A Tarskian logic is required to be reflexive, monotonic, idempotent, and to satisfy the deduction theorem.

[19] For paraconsistent logic in general compare Priest, Routley and Norman [1989], and for its application to belief change, see Tanaka [2005].

data base level. Many authors have argued that it is better to install change operations that are not as simple-minded as set-theoretic additions and subtractions from Γ (this idea has been championed by Sven Ove Hanssson, see in particular Hansson [1999]). Instead of adding or subtracting just a single item, additional items have to be removed from the data base in order to avoid lapsing into inconsistency or to avoid that the sentence to be eliminated can be rederived from the remaining elements of the data base. A choice mechanism is necessary for determining which additional items to remove. This approach combines a foundationalist outlook with an emphasis on the revision process rather than on the reflection process. It changes from the vertical to the horizontal perspective (Fig. 6).

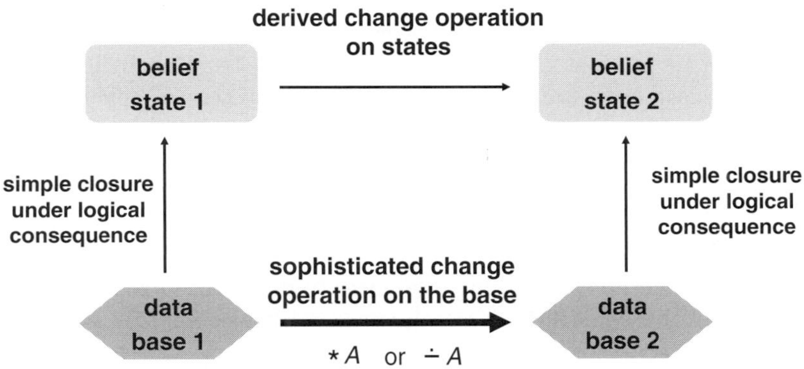

Fig. 6. Foundationalism combined with the horizontal perspective

4.2 Example 2: Up-sets in prioritized data bases

For the second and third examples, we look at two methods of drawing inferences from a *totally pre-ordered data base* $\langle\Gamma, \prec\rangle$, also written as

$$\Gamma_1 \prec \Gamma_2 \prec \ldots \prec \Gamma_n$$

Here the Γ_i's are non-empty subsets of a finite set Γ of propositions such that $\Gamma = \Gamma_1 \cup \ldots \cup \Gamma_n$.[20] We call $\prec$ a priority relation and $\langle\Gamma, \prec\rangle$ a *prioritized data base*. Intuitively, the elements in Γ_i are less certain than the elements in Γ_j if $i < j$. The most interesting case is the one in which Γ is inconsistent.

What inferences can one draw from $\langle\Gamma, \prec\rangle$? A simple way of exploiting the pre-ordering is to take the consequences under a Tarskian Cn of $\Gamma_i \cup \ldots \cup \Gamma_n$,

[20] I assume finiteness for simplicity. If the Γ_i's were allowed to be empty, then we would in effect use ordinal numbers for the description of a data base expressing quantitative differences in degrees of belief. In other words, we would have a ranked data base in the sense of Spohn [1988] and Goldszmidt and Pearl [1992].

where $i \geq 1$ is minimal such that $\Gamma_i \cup \ldots \cup \Gamma_n$ is consistent. That is, the belief set supported by a prioritized data base $\langle \Gamma, \prec \rangle$ is everything that follows from its maximally consistent upper part. Clearly, the data base Γ is not in general included in this belief set. In fact, the method in a way loses more than is intuitively needed. Sentences with low ranks are apt to be suppressed even though they do not in any way contribute to, or are not in any way relevant for, a contradiction within Γ.

This method of drawing inferences allows us to construct a full belief state (rather than a mere belief set) out of the data base $\langle \Gamma, \prec \rangle$. Let us use both the system of spheres and the entrenchment representation. A Grovean system $\$$ of spheres of possible worlds can be generated by collecting in $\$$ all the sets of models of $\Gamma_i \cup \ldots \cup \Gamma_{n-1} \cup \Gamma_n$, for $i = 1, \ldots, n$.[21] Alternatively, we can generate a total entrenchment pre-ordering of the propositions of the language by defining $A \leq B$ if and only if for all $i = 1, \ldots, n$, whenever A follows logically from $\Gamma_i \cup \ldots \cup \Gamma_n$, so does B. Given the translations between systems of spheres and entrenchment pre-orderings (see section 2.3), the two ways of constructing belief states from a pre-ordered data base are equivalent. Both the system of spheres $\$$ and the ordering $\leq$ thus constructed correspond in a natural way to the prioritized belief base $\langle \Gamma, \prec \rangle$.

4.3 *Example 3: Maximal-consistent sets in prioritized data bases*

A more powerful method of drawing inferences from a prioritized data base $\langle \Gamma, \prec \rangle$ can be obtained by constructing all maximal consistent subsets X of Γ, where maximization is subject to consistency from the top to the bottom. The method is specified as follows: First take a maximally consistent subset of Γ_n; then add in a maximally consistent way elements of Γ_{n-1}, and keep this extended set; then add in a maximally consistent way elements of Γ_{n-2}, and keep the newly extended set; and so on, until you reach Γ_1. Call the whole set obtained by this procedure X. Of course, there are many such X's, since in general there are multiple maximally consistent extending subsets on each level i from n to 1. In the face of the multiplicity of the X's, there are two inference strategies. A *bold method* of drawing inferences is to pick *one* such X and close it under the standard logic Cn. Alternatively, a *cautious method* of drawing inferences takes *all* those X's, closes each of them under the standard logic Cn and forms the intersection of the resulting theories. There are thus two different notions of what one can infer from the prioritized data base, i.e., two different ways of defining the belief set supported by $\langle \Gamma, \prec \rangle$.

Notice that even the cautious method is much bolder than the method of section 4.2. Still this boldness does not generate a consistency problem by interaction with simple change operations. One may simply add any item to a prioritized data base, at any level one likes, and still be sure that the inferences drawn from the enlarged data base remain consistent. The ordering of the data base induces as it were

[21]Equivalently, we can generate a total pre-ordering of possible worlds by defining $w \leq w'$ if and only if for all i, w satisfies $\Gamma_i \cup \ldots \cup \Gamma_n$ whenever w' does.

some sort of paraconsistent logic. But notice that Γ is not in general included in the belief set supported by $\langle \Gamma, \prec \rangle$.[22]

Does this more sophisticated method of drawing inferences allow us to construct a full belief state? It can be shown that there is a corresponding entrenchment relation which refines the entrenchment relation defined at the end of section 4.2. But this relation is not total, so that the graphic systems of spheres representation cannot be used (cf. Rott [2000]).

Let us sum up this section. We started by remarking that it does not make sense to combine simple methods of revision (just adding a piece of information to the existing beliefs) with a simple methods of inference (as represented by standard monotonic logics). Simple revision methods will soon generate inconsistencies, and simple inference operations will turn inconsistencies into 'epistemic hell', due to the classical rule of *ex falso quodlibet*. Standard logics can only be employed for foundationalist ideas in the horizontal perspective as represented by Fig. 6. The processing mechanisms of both sections 4.2 and 4.3, on the other hand, do not create problems of consistency or closure when combined with simple change operations. They are suitable for implementing foundationalist ideas in the vertical perspective as indicated in Fig. 4.

5 REPRESENTING COHERENTIST BELIEF CHANGE BY OPERATIONS ON PRIORITIZED DATA BASES

We return to the idea that some coherentist approaches may be interpreted as forms of hidden foundationalism. More precisely, sophisticated change operations on the belief state level may be representable as resulting from simple operations on the base level combined with a sophisticated inference operation. In the following we present two applications of the method of turning prioritized data bases into Grovean systems of spheres that was sketched in section 4.2. Changes on the belief state level will be represented by changes of systems of spheres, and we shall specify syntactic operations on prioritized belief bases that correspond, through the inference operation mentioned, to these changes. In this way a coherentist approach instantiating the scheme of Fig. 5 is matched by an approach exemplifying Fig. 4. We will use a slightly different format of belief bases now, in which the conjunctions $G_i = \bigwedge \Gamma_i$ replace the finite sets Γ_i above. The material of this section is put into a much wider perspective in Rott [to appear].

For the rest of this section, let the agent's prior belief state be represented by a system of spheres of possible worlds as depicted in Fig. 1, and let $G_1 \prec G_2 \prec \ldots \prec G_n$ a prioritized data base that generates this system of spheres as described in section 4.2. Intuitively, these are two representations of the same belief state. We suppose that $\neg A$ is believed in this state, and that the input is A.

[22]The closure problem after simple *eliminations* of an element A of the data base can be avoided, if this belief is eliminated by adding the 'phantom belief' $\neg A$ at the top of the data base. As a phantom belief, $\neg A$ is counted for the consistency check from the top to the bottom, but it is not used in the closure $Cn(X)$. See Rott [2001, Chapter 5].

5.1 *Moderate revision*

The idea of moderate revision (Nayak [1994], Rott [2003]), also known as lexicographic revision, is as indicated in Figure 7.

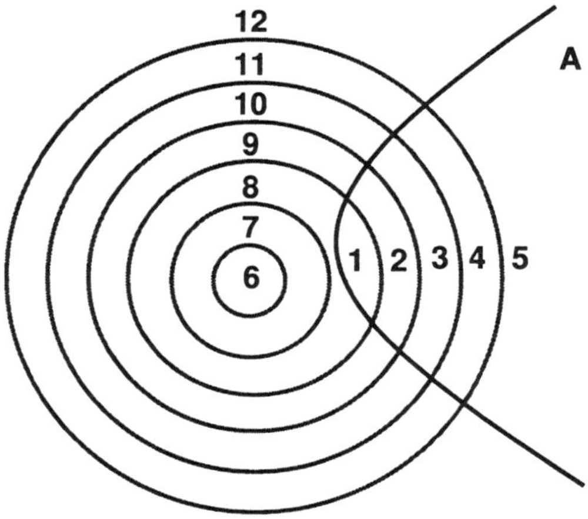

Fig. 7. Revision by input A

This is a semantic operation on systems of spheres of possible worlds. It can be shown that the corresponding syntactic operation on the prioritized belief base turns it into

$$G_1 \prec G_2 \prec \ldots \prec G_n \prec A \prec G_1 \vee A \prec G_2 \vee A \prec \ldots \prec G_n \vee A$$

Thus the coherentist moderate method of changing belief states turns out to be a hidden form of foundationalism. It is induced by a simple method of changing data bases that generate the relevant system of spheres.

5.2 *Revision by comparison*

Let $\leq$ be the entrenchment relation generated from the agent's system of spheres as in section 2.3. Suppose that B is believed in this state, and that the input is A with the proviso that A should be accepted as firmly as the reference sentence B. Alternatively, the input may be thought of as coming in the form $B \leq A$. Then the idea of revision by , also known as raising (Cantwell [1997]), is given by the operation on the system of spheres indicated in Figure 8.[23]

Again there is a syntactic operation on the prioritized belief base $G_1 \prec \ldots \prec G_n$ that corresponds precisely to this semantic operation on systems of spheres. The base gets changed into

$$G_1 \prec G_2 \prec \ldots \prec G_{i-1} \prec G_i \wedge A \prec G_{i+1} \prec G_{i+2} \prec \ldots \prec G_n$$

[23]The representation of the idea in terms of changes of entrenchments is rather complicated, see Fermé and Rott [2004].

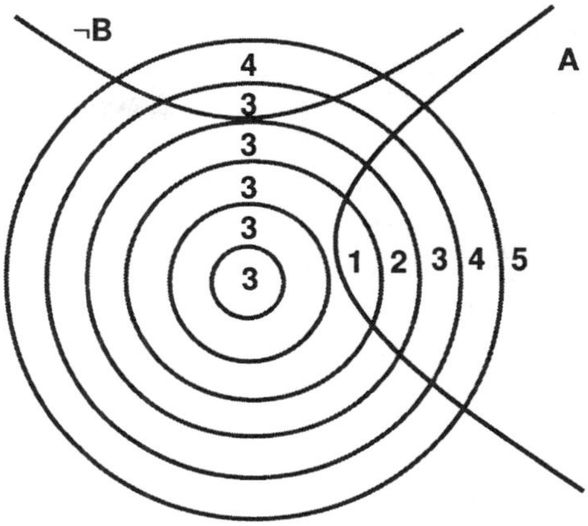

Fig. 8. Revision by A, accepting A at least as strongly as B

where i is chosen such that B follows logically from $G_i \wedge \ldots \wedge G_n$, but not from $G_{i+1} \wedge \ldots \wedge G_n$.

The binary operation of revision by comparison has features of unary revision functions with respect to the input sentence A, and at the same time features of unary belief contraction functions with respect to the reference sentence B. The prioritized base representation makes it very transparent why the belief B is lost if the negation of A is at least as entrenched as B.

6 A VERY BRIEF HISTORY OF BELIEF REVISION

Information *as structure* is a static phenomenon, but information *as being interpreted* transfers an agent's belief state and is in this sense dynamic. We may suspect that the static picture determines the dynamic one. Do structural features fully determine the evolution of belief states through time? A look at the history of belief change research will help us answer this problem.

From now on, let Φ denote a belief state. The only condition we place on this concept is that it is possible to retrieve the beliefs of an agent from Φ. Let us write $Bel(\Phi)$ for belief set supported by Φ.[24]

Let $\Phi * A$ denote the revised belief state, if Φ is the prior belief state and A is the input (piece of new information).

6.1 The 1980s: AGM's classical model

The original models of Alchourrón, Gärdenfors and Makinson [1985] were somewhat ambiguous about the notion of an epistemic state. Officially, an agent's belief state was represented by a belief set $\Phi = Bel(\Phi)$, that is, by the agent's set

[24] Usually, we suppose that $Bel(\Phi)$ is closed under some broadly classical logic.

of beliefs phrased in sentences of some regimented language. Belief sets were assumed to be closed under the (broadly classical) logic governing this language. As discussed in section 3, however, the agent has to make use of a selection structure or preference structure in order to revise her belief set in a reasonable way. Given such a structure associated with a belief set Φ, AGM's method (algorithm, recipe, ...) indeed uniquely determines the new beliefs. The resulting belief change function $*$ specifies, for a given belief set Φ, the posterior belief state $\Phi * A$ for any potential input sentence A. It satisfies the following (by now classic) set of *AGM postulates*:[25]

(AGM1) $\Phi * A$ is logically closed.

(AGM2) $\Phi * A$ implies A

(AGM3) $\Phi * A$ is a subset of $Cn(\Phi \cup \{A\})$

(AGM4) If A is consistent with Φ, then Φ is a subset of $\Phi * A$

(AGM5) If A is consistent, then $\Phi * A$ is consistent

(AGM6) If A is equivalent with B, then $\Phi * A = \Phi * B$

(AGM7) $\Phi * (A \wedge B)$ is a subset of $Cn((\Phi * A) \cup \{B\})$

(AGM8) If B is consistent with $\Phi * A$, then $\Phi * A$ is a subset of $\Phi * (A \wedge B)$

(AGM1) and (AGM) state that the posterior belief set is closed and consistent if possible. (AGM5) requires that the input be accepted, and (AGM6) says that it is the content of the input that matters, not its syntactic surface structure. The expansion postulates (AGM3) and (AGM4) jointly say that in the case where the input A is consistent with the prior belief set Φ, the revised set includes Φ and does not include more than the logical consequences of A taken together with Φ. The conjunction postulates (AGM7) and (AGM8) compare the revision by a conjunction $A \wedge B$ with the revision by the conjunct A. They state that if B is consistent with $\Phi * A$, then $\Phi * (A \wedge B)$ includes $\Phi * A$ and is included by the set of logical consequences of B taken together with $\Phi * A$.

One problem with the original AGM approach is that it did not provide for revisions of selection structures in response to new information. For this reason, iterated revisions of belief states were largely impossible, and belief revision theory was not fully dynamic.[26]

Another problem is that AGM left it open where the selection structures are supposed to come from. They were just assumed to be somewhere in the background, waiting to be exploited in belief change processes. In my view, it is hard to imagine an objective measure with which to gauge changes of beliefs. It is much

[25]Given here in a slightly adapted form, weakening (AGM4) and (AGM8) in order to make them 'purer'.

[26]A slight adaptation of the AGM definitions, however, allows the same selection structure to be used in the context of arbitrary belief sets. For various ways of implementing this idea, see Alchourrón and Makinson [1985], Areces and Becher [2001] and Rott [2003].

more plausible to assume that the structure guiding the change of an agent's beliefs is part his or her own mental state, and indeed I suggest that it is part of the agent's *belief state*. As such, it will itself be subject to changes in response to new information. This more dynamic way of thinking about the evolution of belief states came to dominate belief revision theory during the 1990s.

6.2 The 1990s: Iteration

Two main ways of extending the AGM framework have been suggested so as to make belief states rich enough to support iterated changes in response to sequences of new information. In both models, change functions do not operate on the agent's beliefs, but directly on his or her belief states. Arbitrary iterations of belief changes can be modelled, and a fully developed dynamics becomes feasible.

First, many researchers have suggested to identify belief states with *selection* or *preference structures*, of the kind that have proven suitable for one-shot AGM belief change. While such a structure is sufficient to uniquely determine the set of current beliefs as well as the AGM revisions of this belief set, it is not sufficient to determine its own revision. A method or rule how to change the selection structure if a new piece of information comes in has to be specified. Three of the most simple and plausible ideas are surveyed under the names 'radical', 'conservative' and 'moderate' revision in Rott [2003]. For instance, the moderate method introduced in section 5.1 can be fully characterized by adding a single axiom to the AGM postulates that takes care of iterations of belief set revision:[27]

$$\text{(Mod)} \quad Bel\,(\Phi * A * B) = \begin{cases} Bel\,(\Phi * (A \wedge B)) & \text{if } B \text{ is consistent with } A \\ Bel\,(\Phi * B) & \text{otherwise.} \end{cases}$$

Now it looks as if *the statics fully encodes the dynamics of belief.* Each belief state contains all information for all its future revisions. But this is not quite true: In order to perform a change of belief, one needs to specify a certain method, a 'rule of application', like, e.g., the rule for moderate revision.

Let us now turn to the second way of extending the AGM models. In the important paper of Darwiche and Pearl [1997], a belief state is introduced as a *primitive notion.* A belief state is not identical with a (logically closed) belief set, but the latter is assumed to be retrievable from the belief state with the help of the *Bel* function. The notation of the postulates then needs to be adapted accordingly. Here is the set proposed by Darwiche and Pearl:[28]

(DP1) $Bel\,(\Phi * A)$ is logically closed.

(DP2) $Bel\,(\Phi * A)$ implies A

(DP3) $Bel\,(\Phi * A)$ is a subset of $Cn\,(Bel\,(\Phi) \cup \{A\})$

---

[27] Compare Nayak [1994] and Nayak, Pagnucco and Peppas [2003].

[28] (DP4) and (DP8) are given in a slightly modified form, in order to facilitate the comparison with my presentation of AGM. Cf. footnote 25 above.

(DP4) If A is consistent with $Bel(\Phi)$, then $Bel(\Phi)$ is a subset of $Bel(\Phi * A)$

(DP5) If A is consistent, then $Bel(\Phi * A)$ is consistent

(DP6) If A is equivalent with B, then $Bel(\Phi * A) = Bel(\Phi * B)$

(DP7) $Bel(\Phi * (A \wedge B))$ is a subset of $Cn(Bel(\Phi * A) \cup \{B\})$

(DP8) If B is consistent with $Bel(\Phi * A)$, then $Bel(\Phi * A)$ is a subset of $Bel(\Phi * (A \wedge B))$

The only difference with standard AGM belief revision postulates resides in Darwiche-Pearl's sixth condition. It is much weaker than AGM's sixth postulate which must in the new notation be written as follows:[29]

(AGM6′) If $Bel(\Phi_1) = Bel(\Phi_2)$ and A is equivalent with B, then $Bel(\Phi_1 * A) = Bel(\Phi_2 * B)$

Darwiche and Pearl [1997] added four postulates for the iterated revision of belief states.

(DP9) If A is implied by B, then $Bel((\Phi * A) * B) = Bel(\Phi * B)$

(DP10) If A is inconsistent with B, then $Bel((\Phi * A) * B) = Bel(\Phi * B)$

(DP11) If A is implied by $Bel(\Phi * B)$, then it is implied by $Bel((\Phi * A) * B)$

(DP12) If A is consistent with $Bel(\Phi * B)$, then it is consistent with $Bel((\Phi * A) * B)$

These postulates have a very convincing semantic motivation. When revising by A, the agent is required not to mess up the ordering of worlds within the A-area (DP9), nor within the $\neg A$-area (DP10), and not to let any $\neg A$-worlds 'overtake' A-worlds (DP11 and DP12). But like the AGM postulates, the DP postulates do not determine a unique posterior belief state resulting from the change. The result of a revision is doubly relative now. Only *given* a doxastic preference structure and *given* a specific rule of application (like, e.g., that of moderate revision or revision by comparison), is the posterior belief state determinately fixed. While for belief set revision in the 1980s, AGM's way of using preference structures was essentially without rivals, many different rules of application have been proposed for belief state revision since the 1990s. The Darwiche-Pearl postulates give the reasoner ample leeway, but there are methods of iterated belief change that do not satisfy them. While moderate revision satisfies them, revision by comparison does not (cf. sections 5.1 and 5.2).

Iterated revision by means of Spohn's [1988] conditionalization also satisfies the Darwiche-Pearl postulates. Hild and Spohn [2008] show how to strengthen corresponding postulates for iterated belief contraction in order to ensure that a contraction function satisfying the strengthened postulates can be generated by

[29] It is the fourth postulate in Darwiche and Pearl's original numbering, see Darwiche-Pearl [1997], p. 7.

an underlying ranking function and Spohn's method of conditionalization.[30]

We said above that it is plausible to regard preference structures as parts of the agent's belief state. But where do rules of application come from? Preferences are non-propositional, but they may still be taken to represent something in the agent's mind, something that might be called the modal or nomological structure of the world. Rules of application, on the other hand, are not 'declarative' in any sense. They constitute 'procedural' information about how to apply preference structures in the process of rebuilding one's belief state. Should rules of application be thought of as belonging to the agent's belief state, too? Are they subject to changes in response to new evidence? These questions have remained unanswered so far.

6.3 The 2000s: Merging

From the end of the 1990s on, research on belief change has become more and more focused on the merging of belief states. In traditional theories of belief change, the input was usually treated as a single piece of information. In belief merging, the 'input' is one or more data bases or belief states of other agents. Earlier there had been a clear asymmetry between the input on the one hand, and the data base or the belief state on the other hand. The former was called 'new information', the latter was some representation of the result of the previous information that an agent had received and/or processed.[31] In belief merging, no such asymmetry is assumed, although it may of course be stipulated as a special constraint on a problem of merging. Belief revision in the customary style may thus be viewed as a special case of belief merging. With the turn to belief merging, the area of belief revision has left the restrictions of the single agent environment and moved into a genuine multi-agent setting. Now multiple belief states can be dealt with.

Today this new field is extremely active, and it would be presumptuous to try to survey here the diversity of paths followed in belief merging. I rather try to convey the flavour of any such undertaking by presenting the axiomatic characterization of the account of Konieczny and Pino-Pérez's [1998]. Their terminology is different from the one used so far, and we will now reproduce the terminology of the much refined paper Konieczny and Pino-Pérez [2002]. By a *belief base*, Konieczny and Pino-Pérez mean just a single proposition A representing (the conjunction of) a person's beliefs. By a *belief set*, they denote a multiset $\mathbb{A} = [A_1, \ldots, A_n]$ of propositions, where A_i is the belief base of the ith person. $\Delta(\mathbb{A})$ denotes the belief base that results from merging the elements of $\mathbb{A}$. Here are the postulates for *merging* suggested by Konieczny and Pino-Pérez:

(KP1) $\Delta(\mathbb{A})$ is a consistent proposition

(KP2) If the belief sets $A_1, \ldots, A_n$ in $\mathbb{A}$ are jointly consistent, then $\Delta(\mathbb{A}) = A_1 \wedge \ldots \wedge A_n$

[30]Spohn conditionalization achieves a contraction by A by setting the rank of $\neg A$ to zero.

[31]Perhaps together with whatever constitutes the agent's a priori beliefs.

(KP3) If $\mathbb{A}_1$ is element-wise equivalent with $\mathbb{A}_2$, then $\Delta(\mathbb{A}_1)$ is equivalent with $\Delta(\mathbb{A}_2)$

(KP4) If A_1 is inconsistent with A_2, then $\Delta([A_1] \sqcup [A_2]) \nvdash A_1$

(KP5) $\Delta(\mathbb{A}_1) \wedge \Delta(\mathbb{A}_2)$ implies $\Delta(\mathbb{A}_1 \sqcup \mathbb{A}_2)$

(KP6) If $\Delta(\mathbb{A}_1)$ is consistent with $\Delta(\mathbb{A}_2)$, then $\Delta(\mathbb{A}_1 \sqcup \mathbb{A}_2)$ implies $\Delta(\mathbb{A}_1) \wedge \Delta(\mathbb{A}_2)$

Here $\sqcup$ denotes multiset union. For the motivation and semantics of these postulates, we have to refer the reader to Konieczny and Pino-Pérez's original papers.

Given the fact that (essential aspects of) belief states can be identified with preference structures that are suitable for the resolution of potential conflicts between different units of information, it is not surprising that many tasks involved in belief merging present themselves as problems of amalgamating or aggregating preference relations. These are very general problems that can be considered in abstraction from the specific problems pertaining to information processing. The information structures used for belief merging are not fundamentally different from the ones used for the single-agent setting. It is only the rules of application that become more complex, corresponding to the increased complexity of the problems that are to be solved.

The variety of operations similar to merging includes the fusion, combination, integration, arbitration of beliefs, as well as judgement aggregation. Natural links are established with social choice theory, game theory, negotiation theory, etc. Among the many relevant papers are Baral, Kraus, Minker and Subrahmanian [1992], Revesz [1993], Nayak [1994], Liberatore and Scherf [1998], Benferhat, Dubois, Prade and Williams [1999], Meyer [2000], List and Pettit [2002], Andreka, Ryan and Schobbens [2002] and Liau [2005].

7 CONCLUSION

We have seen different *kinds* of information structures at work in belief revision. There is propositional information: beliefs and inputs. There is non-propositional information that still seems, in some sense, to represent something: the preference orderings that we have identified with belief states. And there is non-propositional, non-representing, purely procedural information: rules of application specifying how to use the preference orderings in the process of belief revision.

We have further seen different models of *changes* of such information structures, models that are supposed to characterize rational changes. In the classical models of the 1980s, beliefs were determined by preferences and rules of application for the use of these preferences. In the 1990s, preferences themselves were determined by prior preferences and rules of application for the change of these preferences. The question concerning the choice or change of these rules of application, however, has remained unanswered. Ultimately, this brings us to the question whether we believers are free to use information as we like. Do we possess 'informational freedom' in this sense? Or is, in the picture afforded by the literature on belief revision, everything about the development of our beliefs determined?

The philosopher Galen Strawson [1986; 1994] has put forward an argument to the effect that there can be no free, responsible action. Strawson's argument, which has gained some notoriety, goes as follows. Any free action of an agent a, that is, according to Strawson, any action for which a is responsible or any action performed by a for a reason, is a consequence of (among other things) 'the way a is, mentally speaking' or her 'mental nature' or 'character.' Thus agent a is responsible for her action only if she is responsible for her character. She is responsible for the latter only if she has intentionally chosen it. She can intentionally choose her character only if she is equipped with 'principles of choice, "P1" — preferences, values, pro-attitudes, ideals — in the light of which [she] chooses how to be.' ([1994], p. 6) Thus agent a is responsible for her character only if she is responsible for her principles of choice P1. She is responsible for the latter only if she has intentionally chosen them, which in turn is possible only if she is equipped with second-order principles of choice P2 for the choice of her first-order principles of choice P1. And so on, *ad infinitum.* Strawson concludes: 'True self-determination is impossible because it requires the actual completion of an infinite series of choices of principles of choice.' ([1994], p. 7) Therefore, there can be neither true freedom nor true responsibility.

What is striking about this argument from our point of view is that the initial steps of Strawson's a priori argument appear to describe quite exactly what has in fact happened in the historical development of belief revision theory. The agent's changes of beliefs are rational if and only if they are determined by higher-order information structures (preferences and rules of application). Such structures are naturally viewed as parts of the agent's mind. It turned out that what helps to solve the problem of rational belief change at one level is itself subject to revision. On the next level, the changes of the preference-orderings are themselves rational if and only if they are determined by some principled method of preference change. But such methods are themselves rational only to the extent they are determined in a principled way. And one is tempted to go on and reiterate the same argument on each new level. But so far no mechanisms for the rational choice of rules how to revise doxastic preferences have been proposed. And assuming that we had such choice mechanisms, how would *they* be rationally selected? What is the right interpretation of higher-order preferences? Should we reckon to find, at some higher level, deep a priori principles decreeing what is ultimately rational? Or should we assume that the agent enjoys an unrestricted personal freedom of choice at a certain point in the hierarchy of preferences and rules of application? Or are we to expect that at some level, the agent turns out to be just a slave of the mental nature or character that she happens to possess, so that questions of rationality fade away into plain matters of fact? These are not questions that haunt computer scientists and information technologists in their daily work, but they seem to be important and hard to dismiss from a philosophical point of view.

ACKNOWLEDGMENTS

I would like to thank Horacio Arló-Costa, Johan van Benthem and Cristiano Castelfranchi for stimulating comments on a previous version of this chapter. It is only restrictions of time that have unfortunately prevented me from taking up more of their valuable suggestions than I actually managed to do.

BIBLIOGRAPHY

[Alchourrón and Makinson, 1985] Carlos Alchourrón and David Makinson. On the logic of theory change: Safe contraction. *Studia Logica*, 44:405–422, 1985.

[Alchourrón *et al.*, 1985] Carlos Alchourrón, Peter Gärdenfors, and David Makinson. On the logic of theory change: Partial meet contraction and revision functions. *Journal of Symbolic Logic*, 50:510–530, 1985.

[Andréka *et al.*, 2002] Hajnal Andréka, Mark Ryan, and Pierre-Yves Schobbens. Operators and laws for combining preference relations. *Journal of Logic and Computation*, 12:13–53, 2002.

[Areces and Becher, 2001] Carlos Areces and Verónica Becher. Iterable AGM functions. In Mary-Anne Williams and Hans Rott, eds., *Frontiers in Belief Revision*, pages 261–277. Kluwer, 2001.

[Baltag *et al.*, 2008] Alexandru Baltag, Hans van Ditmarsch, and Larry Moss. Epistemic logic and information update. In Pieter Adriaans and Johan van Benthem, eds., *Handbook of the Philosophy of Information*. Elsevier, 2008.

[Baral *et al.*, 1992] Chitta Baral, Sarit Kraus, Jack Minker, and V.S. Subrahmanian. Combining multiple knowledge bases consisting of first order theories. *Computational Intelligence*, 8:45–71, 1992.

[Benferhat *et al.*, 1999] Salem Benferhat, Didier Dubois, Henri Prade, and Mary-Anne Williams. A practical approach to fusing prioritized knowledge bases. In Pedro Barahona and José Júlio Alferes, eds., *Progress in Artificial Intelligence. 9th Portuguese Conference on Artificial Intelligence, EPIA '99, Évora, Portugal, September 21-24, 1999*, volume 1695 of *LNAI*, pages 222–236. Springer, 1999.

[Cantwell, 1997] John Cantwell. On the logic of small changes in hypertheories. *Theoria*, 63:54–89, 1997.

[Darwiche and Pearl, 1994] Adnan Darwiche and Judea Pearl. On the logic of iterated belief revision. In R. Fagin, ed., *TARK'94 — Proceedings of the Fifth Conference on Theoretical Aspects of Reasoning About Knowledge*, pages 5–23, San Mateo, CA, 1994. Morgan Kaufmann.

[Darwiche and Pearl, 1997] Adnan Darwiche and Judea Pearl. On the logic of iterated belief revision. *Artificial Intelligence*, 89:1–29, 1997.

[Fermé and Rott, 2004] Eduardo Fermé and Hans Rott. Revision by comparison. *Artificial Intelligence*, 157:5–47, 2004.

[Fuhrmann, 1991] André Fuhrmann. On the modal logic of theory change. In André Fuhrmann and Michael Morreau, eds., *The Logic of Theory Change*, LNCS 465, pages 259–281. Springer, Berlin, 1991.

[Gärdenfors and Rott, 1995] Peter Gärdenfors and Hans Rott. Belief revision. In Dov M. Gabbay, Christopher J. Hogger, and John A. Robinson, eds., *Handbook of Logic in Artificial Intelligence and Logic Programming*, volume Volume IV: Epistemic and Temporal Reasoning, pages 35–132. Oxford University Press, Oxford, 1995.

[Gärdenfors, 1978] Peter Gärdenfors. Conditionals and changes of belief. In Illkka Niiniluoto and Raimo Tuomela, eds., *The Logic and Epistemology of Scientific Change*, volume 30, nos. 2–4 of *Acta Philosophica Fennica*, pages 381–404. North Holland, Amsterdam, 1978.

[Gärdenfors, 1988] Peter Gärdenfors. *Knowledge in Flux: Modeling the Dynamics of Epistemic States*. Bradford Books, MIT Press, Cambridge, Mass., 1988.

[Goldszmidt and Pearl, 1992] Moises Goldszmidt and Judea Pearl. Rank-based systems: A simple approach to belief revision, belief update, and reasoning about evidence and actions. In *Proceedings of the Third International Conference on Principles of Knowledge Representation and Reasoning (KR'92)*, pages 661–672, Cambridge, MA, 1992. Morgan Kaufmann.

[Groenendijk and Stokhof, 1991] Jeroen Groenendijk and Martin Stokhof. Dynamic predicate logic. *Linguistics and Philosophy*, 14:39–101, 1991.

[Grove, 1988] Adam Grove. Two modellings for theory change. *Journal of Philosophical Logic*, 17:157–170, 1988.

[Hansson, 1989] Sven Ove Hansson. New operators for theory change. *Theoria*, 55:114–132, 1989.

[Hansson, 1999] Sven Ove Hansson. *A Textbook on Belief Dynamics*. Kluwer, Dordrecht, 1999.

[Harman, 1986] Gilbert Harman. *Change in View*. Bradford Books, MIT Press, Cambridge, Mass, 1986.

[Hild and Spohn, 2008] Matthias Hild and Wolfgang Spohn. The measurement of ranks and the laws of iterated contraction. To appera in *Artificial Intelligence*, 2008.

[Katsuno and Mendelzon, 1991] Hirofumi Katsuno and Alberto O. Mendelzon. Propositional knowledge base revision and minimal change. *Artificial Intelligence*, 52:263–294, 1991.

[Kelly *et al.*, 1997] Kevin Kelly, Oliver Schulte, and Vincent Hendricks. Reliable belief revision. In Maria Luisa Dalla Chiara, Kees Doets, Daniele Mundici, and Johan van Benthem, eds., *Logic and Scientific Methods — Proceedings of the 10th International Congress of Logic, Methodology and Philosophy of Science*, pages 383–398. Kluwer, Dordrecht, 1997.

[Kelly, 1999] Kevin Kelly. Iterated belief revision, reliability, and inductive amnesia. *Erkenntnis*, 50:11–58, 1999.

[Kim, 2006] Jaegwon Kim. *Philosophy of Mind*. Westview Press, Boulder, Colorado, second edition, 2006.

[Konieczny and Pino-Pérez, 1998] Sébastien Konieczny and Ramón Pino-Pérez. On the logic of merging. In A. G. Cohn, L. Schubert, and S. C. Shapiro, eds., *Principles of Knowledge Representation and Reasoning: Proceedings of the Sixth International Conference (KR'98)*, pages 488–498, San Francisco, CA, 1998. Morgan Kaufmann.

[Konieczny and Pino Pérez, 2002] Sébastien Konieczny and Ramón Pino Pérez. Merging information under constraints: A qualitative framework. *Journal of Logic and Computation*, 12:773–808, 2002.

[Kratzer, 1981] Angelika Kratzer. Partition and revision: The semantics of counterfactuals. *Journal of Philosophical Logic*, 10:201–216, 1981.

[Liau, 2005] Churn-Jung Liau. A modal logic framework for multi-agent belief fusion. *ACM Transactions on Computational Logic (TOCL)*, 6:124–174, 2005.

[Liberatore and Scherf, 1998] Paolo Liberatore and Marco Scherf. Arbitration (or how to merge knowledge bases). *IEEE Transactions Knowledge and Engineering*, 10:76–90, 1998.

[Lindström and Rabinowicz, 1999] Sten Lindström and Wlodzimierz Rabinowicz. DDL unlimited: Dynamic doxastic logic for introspective agents. *Erkenntnis*, 50:353–385, 1999.

[List and Pettit, 2002] Christian List and Philip Pettit. Aggregating sets of judgments: An impossibility result. *Economics and Philosophy*, 18:89–110, 2002.

[Makinson, 2003] David Makinson. Ways of doing logic: What was different about AGM 1985? *Journal of Logic and Computation*, 13:3–13, 2003.

[Meyer, 2000] Thomas Meyer. Merging epistemic states. In Riichiro Mizoguchi and John Slaney, eds., *PRICAI 2000: Topics in Artificial Intelligence*, volume 1886 of *LNAI*, pages 286–296. Springer-Verlag, Berlin, 2000.

[Nayak *et al.*, 2003] Abhaya Nayak, Maurice Pagnucco, and Pavlos Peppas. Dynamic belief revision operators. *Artificial Intelligence*, 146:193–228, 2003.

[Nayak, 1994] Abhaya Nayak. Iterated belief change based on epistemic entrenchment. *Erkenntnis*, 41:353–390, 1994.

[Nebel, 1989] Bernhard Nebel. A knowledge level analysis of belief revision. In Ronald Brachman, Hector Levesque, and Ray Reiter, eds., *Proceedings of the 1st International Conference on Principles of Knowledge Representation and Reasoning*, pages 301–311, San Mateo, CA, 1989. Morgan Kaufmann.

[Priest *et al.*, 1989] Grahm Priest, Richard Routley, and Jean Norman, eds. *Paraconsistent Logic. Essays on the Inconsistent.* Philosophia Verlag, München, 1989.

[Putnam, 1960] Hilary Putnam. Minds and machines. In Sidney Hook, ed., *Dimensions of Mind*, pages 138–164. State University of New York Press, Albany, N.Y., 1960. Reprinted in Hilary Putnam, *Mind, Language and Reality*, Cambridge 1975, pp. 362–385.

[Putnam, 1967] Hilary Putnam. Psychological predicates. In W.H. Captain and D.D. Merrill, edd., *Art, Mind and Religion*, pages 37–48. University of Pittsburgh Press, 1967. Reprinted as 'The Nature of Mental States', in Hilary Putnam, *Mind, Language and Reality*, Cambridge, 1975, pp. 429–440.

[Revesz, 1993] Peter Z. Revesz. On the semantics of theory change: Arbitration between old and new information. In *Proceedings PODS'93: 12th ACM SIGACT SIGMOD SIGART Symposium on the Principles of Database Systems*, pages 71–82, 1993.

[Rott, 2000] Hans Rott. "Just because": Taking belief bases seriously. In Samuel R. Buss, Petr Hájek, and Pavel Pudlák, editors, *Logic Colloquium '98 — Proceedings of the Annual European Summer Meeting of the Association for Symbolic Logic held in Prague*, volume 13 of *Lecture Notes in Logic*, pages 387–408. Association for Symbolic Logic, Urbana, Ill., 2000.

[Rott, 2001] Hans Rott. *Change, Choice and Inference: A Study in Belief Revision and Nonmonotonic Reasoning.* Oxford University Press, Oxford, 2001.

[Rott, 2003] Hans Rott. Coherence and conservatism in the dynamics of belief. Part II: Iterated belief change without dispositional coherence. *Journal of Logic and Computation*, 13:111–145, 2003.

[Rott, to appear] Hans Rott. Shifting priorities: Simple representations for twenty-seven iterated theory change operators. In David Makinson, Jacek Malinowski, and Heinrich Wansing, eds., *Towards Mathematical Philosophy*, Trends in Logic. Springer Verlag, Berlin, to appear.

[Segerberg, 2001] Krister Segerberg. The basic dynamic doxastic logic of AGM. In Mary-Anne Williams and Hans Rott, eds., *Frontiers in Belief Revision*, pages 57–84. Kluwer, Dordrecht, 2001.

[Shafer, 1976] Glenn Shafer. *A Mathematical Theory of Evidence.* Princeton University Press, Princeton, New Jersey, 1976.

[Spohn, 1988] Wolfgang Spohn. Ordinal conditional functions: A dynamic theory of epistemic states. In W.L. Harper and B. Skyrms, eds., *Causation in Decision, Belief Change, and Statistics*, pages 105–134. Kluwer, Dordrecht, 1988.

[Stalnaker, 1984] Robert C. Stalnaker. *Inquiry.* Bradford Books, MIT Press, Cambridge, MA, 1984.

[Strawson, 1986] Galen Strawson. *Freedom and Belief.* Clarendon Press, Oxford, 1986. Reprinted with corrections 1991.

[Strawson, 1994] Galen Strawson. The impossibility of moral responsibility. *Philosophical Studies*, 75:5–24, 1994. (Reprinted in Gary Watson, ed., *Free Will*, Oxford University Press, 2nd edition 2002, pp. 212–228).

[Tanaka, 2005] Koji Tanaka. The AGM theory and inconsistent belief change. *Logique et Analyse*, 48 (189–192):113–150, 2005.

[Turing, 1950] Alan Turing. Computing machinery and intelligence. *Mind*, 59:433–460, 1950.

[van Benthem, 2007] Johan van Benthem. Dynamic logic for belief revision. *Journal of Applied Non-Classical Logics*, 17:129–155, 2007.

[Veltman, 1976] Frank Veltman. Prejudices, presuppositions and the theory of counterfactuals. In Jeroen Groenendijk and Martin Stokhof, eds., *Amsterdam Papers of Formal Grammar*, volume 1, pages 248–281, 1976.

[Veltman, 1996] Frank Veltman. Defaults in update semantics. *Journal of Philosophical Logic*, 25:221–261, 1996.

INFORMATION, PROCESSES AND GAMES

Samson Abramsky

1 PRELUDE: SOME BASIC PUZZLES

Before attempting a conventional introduction to this article, we shall formulate some basic puzzles which may serve as motivation for, and an indication of, some of the themes we shall address.

1.1 Does Information Increase in Computation?

Let us begin with a simple-minded question:

Why do we compute?

The natural answer is: *to gain information* (which we did not previously have)! But how is this possible?[1]

Problem 1: Isn't the output *implied* by the input?

Problem 2: Doesn't this contradict the second law of thermodynamics?

A logical form of Problem 1 This problem lies adjacent to another one at the roots of logic. If we extract logical consequences of axioms, then surely the answer was already there implicitly in the axioms; what has been added by the derivation? Since computation can itself, via the Curry-Howard isomorphism [Curry and Feys, 1958; Howard, 1980; Girard *et al.*, 1989], be modelled as performing *Cut elimination* on proofs, or *normalization* of terms, the same question can be asked of computation. A normal form which is presented as the result of a computation is logically *equal* to the term we started with:

$$M \longrightarrow^* N \quad \Longrightarrow \quad [\![M]\!] = [\![N]\!].$$

so what has been added by computing it?

[1]Indeed, I was once challenged on this point by an eminent physicist (now knighted), who demanded to know how I could speak of information increasing in computation when Shannon Information theory tells us that it cannot! My failure to answer this point very convincingly at the time led me to continue to ponder the issue, and eventually gave rise to this discussion.

Handbook of the Philosophy of Science. Volume 8: Philosophy of Information
Volume editors: Pieter Adriaans and Johan van Benthem. General editors: Dov M. Gabbay, Paul Thagard and John Woods.

The same issue can be formulated in terms of the logic programming paradigm, or of querying a relational database [Ceri *et al.*, 1990]: in both cases, the result of the query is a logical consequence of the data- or knowledge-base.

The challenge here is to build a useful theory which provides convincing and helpful answers to these questions. We simply make some preliminary observations. Note that normal forms are in general *unmanagably big* [Vorobyov, 1997]. Useful output has two aspects:

- Making information explicit—*i.e.* extracting the normal form.
- Data reduction—getting rid of a lot of the information in the input.

(Note that it is *deletion of data* which creates thermodynamic cost in computation [Landauer, 1961]). Thus we can say that much (or all?) of the actual usefulness of computation lies in getting rid of the hay-stack, leaving only the needle.

Problem 2: Discussion While information is presumably conserved in the *total* system, there can be information flow between, and information increase in, *subsystems.* (A body can gain heat from its environment). More precisely, while the entropy of an isolated (total) system cannot decrease, a sub-system *can* decrease its entropy by consuming energy from its environment.

Thus if we wish to speak of information flow and increase, this must be done relative to subsystems. Indeed, the fundamental objects of study should be *open systems*, whose behaviour must be understood in relation to an external environment. Subsystems which can *observe* incoming information from their environment, and *act* to send information to their environment, have the capabilities of *agents.*

Observer-dependence of information increase Yorick Wilks (personal communication) has suggested the following additional twist. Consider an equation such as

$$3 \times 5 = 15.$$

The forward direction $3 \times 5 \rightarrow 15$ is obviously a natural direction of computation, where we perform a multiplication. But the reverse direction $15 \rightarrow 3 \times 5$ is also of interest — finding the prime factors of a number! So the *direction of possible information increase* must be understood as relative to the observer or user of the computation.[2]

Moral: Agents and their interactions are intrinsic to the study of information flow and increase in computation. The classical theories of information *do not reflect this adequately.*

[2]Formally, this can be understood in terms of different choices of normal forms. For a general perspective on rewriting as a computational paradigm, see [Baader and Nipkow, 1999; Terese, 2003].

1.2 What Function Does the Internet Compute?

Our second puzzle reflects the changing conception of computation which has been developing within Computer Science over the past three decades. The traditional conception of computation is that we compute an output as a function of an input, by an algorithmic process. This is the basic setting for the entire field of algorithms and complexity, for example. So *what* we are computing is clear — it is a function.[3] But the reality of modern computing: distributed, global, mobile, interactive, multi-media, embedded, autonomous, virtual, pervasive, ...[4] — forces us to confront the limitations of this viewpoint.

Traditionally, the *dynamics* of computing systems — their unfolding behaviour in space and time — has been a mere means to the end of computing the function which specifies the algorithmic problem which the system is solving.[5] In much of contemporary computing, the situation is reversed: the *purpose* of the computing system is to exhibit certain behaviour. The *implementation* of this required behaviour will seek to reduce various aspects of the specification to the solution of standard algorithmic problems.

What does the Internet compute?

Surely not a mathematical function ...

Moral: We need a theory of the dynamics of informatic processes, of *interaction*, and *information flow*, as a basis for answering such fundamental questions as :

- *What* is computed?
- What *is* a process?
- What are the analogues to Turing-completeness and universality when we are concerned with processes and their behaviours, rather than the functions which they compute?

2 INTRODUCTION: MATTER AND METHOD

Philosophers of science are concerned with explaining various aspects of science, and often, moreover, with viewing science as a kind of gold-mine of philosophical opportunity. The direction in both cases is *philosophy from science*. For a

[3]We may, if we are willing to countenance non-deterministic or probabilistic computation, be willing to stretch this functional paradigm to accomodate relations or stochastic relations of some kind. These are minor variations, compared to the shift to a fully-fledged dynamical perspective.

[4]See e.g. [Milner, 2006a; Milner, 2006b].

[5]Insofar as the dynamics has been of interest, it has been in quantitative terms, counting the resources which the algorithmic process consumes — leading of course to the notions of algorithmic complexity. Is it too fanciful to speculate that the lack of an adequate structural theory of processes may have been an impediment to fundamental progress in complexity theory?

theoretical or mathematical scientist, the primary inclination is often to see conceptual analysis as a preliminary to a more technical investigation, which may lead to a new theoretical development. In short: *science from philosophy.* This article is written mainly in the latter spirit, from the stand-point of Theoretical Computer Science, or perhaps more broadly "Theoretical Informatics": a — still largely putative — general science of information. That being said, we hope that our conceptual discussions may also provide some useful grist to the philosopher's mill.

2.1 Towards Information Dynamics

The best-known existing mathematical theories of information are (largely) *static* in nature. That is, they do not explicitly describe informatic processes and information flow, but rather certain *invariants* of these processes and flows. There is by now ample experience from Computer Science which indicates that it is fruitful, and eventually necessary, to develop fully-fledged dynamical theories. We shall try to map some steps in this direction.

We begin by reviewing some of the theories developed in Computer Science which form the background for our discussion. Then we consider another important issue in theories of information: the distinction between *qualitative* and *quantitative* theories, and how they can be reconciled — or, more positively, combined. Our discussion here will still be at the level of *static* theories. We then go on to consider dynamic theories proper.

This article is well outside the author's usual remit as a researcher. While it is clearly not a contribution to philosophy, it cannot be said to be the usual kind of conceptually-oriented overview of a scientific field which one might find in such a Handbook (and of which there are some fine examples in the present volume) either; not least for the reason that the scientific field we are attempting to overview does not exist yet, in a fully realized form at any rate. Rather, the main purpose of this article is to play some small part in helping this field to come into being.

What, then, is this nascent field? We would like to use the term *Information Dynamics*, which was proposed some time ago by Robin Milner, to suggest how the area of Theoretical Computer Science usually known as "Semantics" might emancipate itself from its traditional focus on interpreting the syntax of pre-existing programming languages, and become a more autonomous study of the fundamental structures of Informatics.[6] The development of such a field would transform our scientific vision of Information, and give us a whole new set of tools for thinking about it. Hence its relevance for any attempt to develop a Philosophy of Information.

Rather than a developed field of Information Dynamics, with some consensus as to what its fundamental notions and methods are, what we have at present

[6]Robin Milner has also written several articles in the same general spirit as this one, notably [Milner, 1996].

are some *partial exemplifications*; some theories which have been shown to work well over certain ranges of applications, and which exhibit both conceptual and mathematical depth. Our approach to conveying the current state of the art, and indicating the major objectives visible from where we stand now, is necessarily largely based on describing (some of) these current theories. The obvious danger with this approach is that this article will appear to be a disjointed series of descriptions of various formalisms. We have probably not succeeded in avoiding this completely—despite the author's best efforts. But we regard the expository aspect of this article as important in itself. The theories we shall expound deserve to be known in wider circles than they presently are. And our discussions of Domain Theory, Game semantics and Geometry of Interaction delve more into conceptual issues, while minimizing the level of technical detail, than other accounts of which we are aware.

2.2 Some Themes

To assist the reader in keeping their bearings, we mention some of the main themes which will thread through our discussion:

Information Increase in Computation We compute in order to gain information: but how is this possible, logically or thermodynamically? How can it be reconciled with the point of view of Information Theory? How does information increase appear in the various extant theories? This will be an important explicit theme in our discussion of background theories in Section 3, and particularly in Section 4. Obtaining a good account in the context of dynamic theories, as exemplified by those presented in Sections 5 and 6, is a key desideratum for future work.

Unifying Quantitative and Qualitative Theories of Information We mainly discuss this explicitly in Section 4, where we describe some striking recent progress which has been achieved by Keye Martin and Bob Coecke, in the setting of current static theories of information (Scott Domain Theory and Shannon Information Theory). A similar development in the setting of the dynamic theories described in Sections 5 and 6 is a major objective for future research.

Information Dynamics: Logic and Geometry We introduce Game Semantics and Geometry of Interaction in Sections 5 and 6 as substantial partial exemplications of Information Dynamics. They have strong connections to both Logic and Geometry, and form a promising new bridge between these two fields. While we shall not be able to do full justice to these topics, we hope at least to raise the reader's awareness of these developments, and to provide pointers into the literature.

The Power of Copying, and Logical Emergence This is mainly developed in Section 6, in the context of Geometry of Interaction-type models. The theme here is to look at how logically complex behaviour can emerge from very simple "copy-cat processes", showing the power of interaction. The links between the interactive and geometric points of view become very clear at this basic level.

One theme which we have, regretfully, omitted is that of the emerging connections with Physics, in particular with **Quantum Information and Computation**. Here there is already much to say (see e.g. [Abramsky and Coecke, 2002; Abramsky and Coecke, 2004; Abramsky and Coecke, 2005]). We have not included this material simply for lack of the appropriate physical resources of space, time and energy.

3 SOME BACKGROUND THEORIES

Following our previous discussion, we can classify theories of information along two axes: as static or dynamic, and as qualitative or quantitative. We list some examples in the following table.

	Static	Dynamic
Qualitative	Domain Theory, Dynamic Logic	Process Algebra
Quantitative	Shannon Information theory	

It may seem strange to list Dynamic Logic as a static theory — and indeed, not everyone would agree with this classification! We regard it as static because it considers input-output relations only, and not the structure of the processes which realize these relations. The distinction we have in mind will become clearer when we go on to discuss Process Algebra.

Shannon Information theory is discussed in detail in another Chapter of this Handbook. In this Section, we shall give brief overviews of the other three theories listed above, which have all been developed within Computer Science—Domain Theory and Dynamic Logic originating in the 1970s, and Process Algebra in the 1980s.

It may be useful to give a timeline for some of the seminal publications:

1948	Claude Shannon	*A Mathematical Theory of Communication*	Information Theory
1963	Saul Kripke	*Semantical Considerations on Modal Logic*	Kripke Structures
1969	Dana Scott	*Outline of a Mathematical Theory of Computation*	Domain Theory
	Tony Hoare	*An Axiomatic Basis for Computer Programming*	Hoare Logic
1976	Vaughan Pratt	*Semantical Considerations on Floyd-Hoare Logic*	Dynamic Logic
	Johan van Benthem	*Modal Correspondence Theory*	Bisimulation
1980	Robin Milner	*A Calculus of Communicating Systems*	Process Algebra

The work on Game Semantics and Geometry of Interaction to be covered in Sections 5 and 6 comes from the 1990's. As always, a full intellectual history is complex, and we shall not attempt this here.

We shall devote rather more space to Domain Theory than to the other two theories, for the following reasons:

- Domain Theory is more intrinsically and explicitly a theory of information than Dynamic Logic or Process Algebra, and will figure significantly in our subsequent discussions.
- The other theories will receive some coverage elsewhere in this Handbook, notably in the Chapter by Baltag and Moss.

3.1 Domain Theory

Domain Theory was introduced by Dana Scott *c.* 1970 [Scott, 1970] as a mathematical foundation for the denotational semantics of programming languages which had been pioneered by Christopher Strachey. A *domain* is a partially ordered structure $(D, \sqsubseteq)$. The best intuitive reading of elements of D is as *information states.* We pass immediately to some illustrative examples.

Examples of Domains

Flat Domains Given a set X, we can form a domain $X_\bot$ by adjoining an element $\bot \notin X$, and defining an order by

$$x \sqsubseteq y \iff x = \bot \vee x = y.$$

Frequently used examples : $\mathbb{N}_\bot$, $\mathbb{B}_\bot$, $\mathbb{O} = \mathbf{1}_\bot$. Here $\mathbb{N} = \{0, 1, 2, \ldots\}$, the set of *natural numbers*; $\mathbb{B} = \{\mathsf{tt}, \mathsf{ff}\}$, the set of *booleans*; and $\mathbf{1} = \{*\}$, an (arbitrary) one-element set.

We can use such flat domains to model computations in terms of very simple *processes of information increase.* Thus a (possibly non-terminating) natural number computation can be modelled in $\mathbb{N}_\bot$ in the following sense. Initially, no output has been produced. This "zero information state" is represented by the bottom element $\bot$. If the computation terminates, a natural number n is produced. Thus we obtain the "process"

$$\bot \sqsubseteq n.$$

The case where no output is ever produced is captured by the "stationary process" $\bot$, which we can view more "dynamically" as

$$\bot \sqsubseteq \bot \sqsubseteq \cdots$$

Streams Now consider the scenario where we have an unbounded or potentially infinite tape (much as for the *output tape* of a Turing machine), on successive squares of which symbols from some finite alphabet Σ can be printed. This computational scenario is naturally modelled by the domain Σ^∞, the set of finite and infinite sequences of elements of Σ. This is ordered by *prefix*: $x \sqsubseteq y$ if $x = y$, or x is finite, and for some (finite or infinite) sequence z, $xz = y$. Example:

$$\langle 0\rangle \sqsubseteq \langle 0,0\rangle \sqsubseteq \langle 0,0,0\rangle \sqsubseteq \cdots \sqsubseteq 0^\omega$$

where 0^ω is the infinite sequence of 0's.

This example shows the ability of domain theory to model infinite computations as *limits* of processes of information increase, where at each stage in the process the information state is finite.

It is important to distinguish a finite stream in this domain from a finite list as a standard programming data structure, e.g. in LISP. A finite list in standard usage is a *complete*, informationally perfect object, just like a natural number in our previous example. A finite stream, by contrast, has a "sting in the tail"; a potentially infinite computation to determine what the remaining elements to be printed on the output tape will be. Thus a finite stream in the above domain is an informationally incomplete object, which can be extended to a more defined stream, which it then approximates.

The Interval Domain Now suppose our computational scenario is that we are computing a real number in the unit interval $[0,1]$. Clearly we can only compute to finite precision in finite time (and with finite resources), so we are *forced* to consider a scenario of approximation. The appropriate domain here is $\mathbb{I}[0,1]$, consisting of all closed non-empty intervals $[a,b]$ where $0 \leq a \leq b \leq 1$. We read an interval $[a,b]$ as expressing our current state of information about the real $r \in [0,1]$ we are computing, namely that $a \leq r \leq b$. The ordering is by reverse inclusion of intervals, or equivalently by

$$[a,b] \sqsubseteq [c,d] \iff a \leq c \wedge d \leq b.$$

This corresponds to *refinement* of our information state to a more accurate determination of the location of the ideal element r. Note that the case $[r, r]$ is allowed, for any $r \in [0, 1]$. In fact, this embeds the unit interval into the interval domain as the set of *maximal elements* of $\mathbb{I}[0, 1]$. Note that for any real number $r \in [0, 1]$, there is a process of information increase

$$[0, 1] \sqsubseteq [a_1, b_1] \sqsubseteq [a_2, b_2] \sqsubseteq \cdots$$

where $a_{n+1} = a_n$ and $b_{n+1} = (a_n + b_n)/2$ if r is in the left half-interval of $[a_n, b_n]$, and $a_{n+1} = (a_n + b_n)/2$ and $b_{n+1} = b_n$ if r is in the right half-interval. Clearly r is the supremum of the a_n and the infimum of the b_n. Thus every real can be computed as the limit of a process of information increase where at each finite stage of the process the interval has rational end-points, and hence is a finitely representable information state.[7]

Partial Functions A somewhat more abstract example is provided by the set $\mathbf{Pfn}(X, Y)$ of partial functions from X to Y, ordered by inclusion. To see how this can be used in computational modelling, consider the recursive definition of the factorial function:

$$\mathsf{fact}(n) = n! = n \cdot (n-1) \cdot (n-2) \cdots 2 \cdot 1.$$

$$\mathsf{fact}(n) = \mathbf{if}\ n = 0\ \mathbf{then}\ 1\ \mathbf{else}\ n \times \mathsf{fact}(n-1).$$

We can understand this recursive definition as specifying a process of information increase over the domain $\mathbf{Pfn}(\mathbb{N}, \mathbb{N})$. Initially, we are at the zero information state (least element of the domain) $\varnothing$; we know nothing about which ordered pairs are in the graph of the function being defined recursively. Inspection of the base case of the recursion (where $n = 0$) allows us to deduce that the pair $(0, 1)$ is in the graph of the function. Once we know this, we can infer that in the case $n = 1$,

$$\mathsf{fact}(1) = 1 \times \mathsf{fact}(0) = 1 \times 1 = 1.$$

Thus the process of information increase proceeds as follows:

$$\varnothing \subseteq \{(0, 1)\} \subseteq \{(0, 1), (1, 1)\} \subseteq \cdots$$

We can see inductively that the n'th term in this sequence will give the values of factorial on the arguments from 0 to $n - 1$; and the *least upper bound* of this increasing sequence, given simply by its union, will be the factorial function.

[7] We are glossing over some technical subtleties here. The interval domain is a basic example of a *continuous domain*—the only one we shall encounter in this brief sketch of domain theory. This means that "finiteness" does not have the same "absolute" status in this case that it does in our other examples.(Formally, intervals with rational end-points are not *compact*.) Nevertheless, these finitely representable intervals do play a natural role in the *effective presentation* of the domain, and the example is an important one for conveying the basic intuitions of Domain Theory. See [Abramsky and Jung, 1994; Gierz *et al.*, 2003] for extensive coverage of continuous domains.

Technical Issues

These examples serve to motivate a number of additional *axioms for domains*. There is in fact no unique axiom system for domains. We shall mention the most fundamental forms of such axioms.

Completeness As we have seen, an essential point of Domain Theory is to allow the description of infinite computations or computational objects as *limits* of processes of information increase. A corresponding property of completeness of domains is required, to ensure that a well-defined unique limit exists for every such process. Such limits are expressed as *least upper bounds* in order-theoretic terms. The idea is that for a process

$$d_0 \sqsubseteq d_1 \sqsubseteq d_2 \sqsubseteq \cdots$$

the limit should contain *all* the information produced at any stage of the process; and *only* the information produced by some stage of the process. The first point implies that the limit should be an upper bound; the second, that it should be the *least* upper bound.

Which class of increasing sets should be regarded as processes of information increase? The most basic class, which has figured in all our examples to date, is that of *increasing sequences*, or "ω-chains" in the usual technical parlance. The axiom requiring completeness for all such chains, which picks out the class of "ω-complete partial orders", or ω-cpos for short, is often used in Domain Theory. We shall henceforth assume that all domains we consider are ω-cpos. Sometimes completeness for a larger class of sets, the *directed sets*, is used. This reflects technical issues akin to the distinction in Topology between sequential completeness and completeness for nets or ultrafilters, and we shall not pursue this here.

Least Elements All our examples have had a least element: $\bot$ for flat domains, the empty stream for Σ^∞, the unit interval $[0,1]$ for $\mathbb{I}[0,1]$, and the empty set for $\mathbf{Pfn}(X,Y)$. This provides a zero information point, and hence a canonical starting point for processes of information increase. Mathematically, least elements are essential for the least fixed point theorem which we shall encounter shortly. There are schemes for Domain Theory in which domains (or "pre-domains") are not required to have least elements in general, but they always enter the theory at crucial points, sometimes through a general operation of adjoining a least element to a predomain to form a domain ("lifting").

Approximation The intuition developed through our examples for how general elements of the domain can be approximated by others, which may in particular be of finite character, is captured formally by requiring domains to be *algebraic* or *continuous*. We shall not develop these notions here, but will simply note for our examples:

- For flat domains such as $\mathbb{N}_\perp$, we can regard *all* elements as of finite character.
- Every stream in Σ^∞ can be realized as the least upper bound of an increasing sequence of *finite* streams.
- Every real in $[0, 1]$, and more generally every interval in $\mathbb{I}[0, 1]$, can be realized as the least upper bound of an increasing sequence of intervals with *rational* end-points.
- Every partial function in $\mathbf{Pfn}(X, Y)$, and in particular every total function from X to Y, where X and Y are countable, can be realized as the least upper bound of an increasing sequence of *finite* partial functions. (The case where X is uncountable is a typical example where we would naturally resort to general directed sets rather than sequences.)

Conceptual Issues

Why Partial Orders? Having developed some examples and intuitions, we now re-examine the basic concept of domains as partial orders $(D, \sqsubseteq)$. If we think of the elements of D as information states, the way we articulate this structure is *qualitative* in character. That is, we don't ask *how much* information a given state contains, but rather a relational question: does one state convey *more* information than another? We read $d \sqsubseteq e$ as "e conveys at least as much information as d". If we consider the partial order axioms with this reading:

Reflexivity	$x \sqsubseteq x$	
Transitivity	$x \sqsubseteq y \ \wedge \ y \sqsubseteq z$	$\Longrightarrow \quad x \sqsubseteq z$
Anti-Symmetry	$x \sqsubseteq y \ \wedge \ y \sqsubseteq x$	$\Longrightarrow \quad x = y.$

then Reflexivity is clear; and Transitivity also very natural. Anti-Symmetry can be seen as embodying an important *Principle of Extensionality*: if two states convey the *same* information, they are regarded as *equal*.

States of What? We have been using the term "information state" to convey the intuition for what the elements of a domain represent. In fact, there is a certain creative ambiguity lurking here, between two interpretations of what these are states *of*:

- We may think of states of a *computational system* in itself, characterized in terms of the information they contain, as an "intrinsic" or "objective" property of the system, independently of any observer.
- We may implicitly introduce an *observer* of the system, and understand the information content of a system in terms of the observer's state of information about it.

In the first reading, we think of the partial elements of the domain in an ontological way, as necessary extensions to our universe of discourse to represent the range of possible outputs of computational systems which may run for ever, and may fail to terminate or to produce information beyond some finite stage of the computation. In the second reading, we are thinking epistemologically: what information can the observer gain about the computation.

In fact, both readings are useful—and are widely used. It is very common to slip without explicit mention from one to the other—nor, for the technical purposes of the theory, does this seem to do any harm. Mathematically, this distinction can be related to the duality between *points* and *properties*, in the sense of Stone-type dualities: the duality between the points of a topological space, and its basic "observable properties"—the open sets [Johnstone, 1982]. The particular feature of domains which allows this creative ambiguity between points and properties to be used so freely without incurring any significant conceptual confusions or overheads is that *basic points and basic properties (or observations) are essentially the same things.* We explain this in terms of an example. Consider a finite steam s in Σ^∞. On the one hand, this can be viewed as a point, *i.e.* as an element of the domain — which may be produced by some system which computes the elements of s in finite time, and then continues to run forever without producing any more output. On the other hand, we may view this finite stream s as a property: the property satisfied by any system with output stream t such that $s \sqsubseteq t$. It is a *finitely observable property*, since we can tell whether a system satisfies it after only a finite time spent observing the system. Whether we take Σ^∞ as the space of points X generated as limits of increasing sequences of finite streams, or as the "logic" (or open-set lattice) L of properties generated by the basic observations given by finite streams, we get the same thing: the topology of X will be L, and the space of points generated (as completely prime filters) over L will be X. This is Stone duality. An extensive development of Stone duality for Domain Theory has been given in [Abramsky, 1991]; see also [Abramsky and Jung, 1994; Zhang, 1991; Bonsangue and Kok, 1999].

In fact, we would argue that it is hard to avoid the epistemic stance entirely. For example, the plausibility of something as basic as the Anti-Symmetry axiom is much greater if we think in terms of an observer. Much of the conceptual power of Domain Theory comes from the idea that it articulates how we can approximate infinite ideal objects by processes which use only finite resources at each finite stage.

Static or Dynamic? Another subtle underlying issue which is not usually made explicit is that Domain Theory is *a static theory resting on dynamic intuitions.* Indeed, we have motivated the theory in terms of certain *processes of information increase.* Processes happen in time; thus time is present implicitly in Domain Theory. This underlying temporality can be developed more explicitly within the Domain Theoretic framework:

- One can add axioms to the basic ones for domains to pick out those domains which are *concrete* [Kahn and Plotkin, 1978], in the sense that we can understand information increase in terms of a temporal flow of events. Now the ordering is not simply one of information content, but involves an idea of *causality*, so that some events *must* temporally precede others. This leads to notions of *event structures* [Nielsen *et al.*, 1981], which have been applied to the study of concurrent processes. Very similar structures have shown up recently in Theoretical Physics, in the Causal Sets approach to quantum gravity [Sorkin, online].

- In some remarkable recent work, Domain Theoretic tools are used to characterize globally hyperbolic space-time manifolds in terms of their causal orderings [Martin and Panangaden, 2006].

However, it should be said that most of the applications of Domain Theory in denotational semantics are carried out at a much higher level of abstraction, where temporality appears only in the most residual form. This arises from the fact that computations or programs are modelled in the Domain-Theoretic denotational framework essentially as *functions* from inputs to outputs.

Continuous Functions

We now consider the appropriate notion of function between domains. Let D, E be ω-complete partial orders. A function $f : D \to E$ is *monotonic* if, for all $x, y \in D$:

$$x \sqsubseteq y \implies f(x) \sqsubseteq f(y).$$

It is *continuous* if it is monotonic, and for all ω-chains $(x_n)_{n\in\omega}$ in D:

$$f(\bigsqcup_{n\in\omega} x_n) = \bigsqcup_{n\in\omega} f(x_n).$$

Examples We consider a number of examples of functions $f : \Sigma^\infty \to \mathbb{B}_\bot$, where $\Sigma = \{0, 1\}$.

1. $f(x) = \mathsf{tt}$ if x contains a 1, $f(x) = \bot$ otherwise.

2. $f(x) = \mathsf{tt}$ if x contains a 1, $f(0^\infty) = \mathsf{ff}$, $f(x) = \bot$ otherwise.

3. $f(x) = \mathsf{tt}$ if x contains a 1, $f(x) = \mathsf{ff}$ otherwise.

Of these: (1) is continuous, (2) is monotonic but not continuous, and (3) is not monotonic.

As these examples indicate, the conceptual basis for monotonicity is that *the information in Domain Theory is positive; negative information is not regarded as stable observable information.* That is, if we are at some information state s, then for all we know, s may still increase to t, where $s \sqsubseteq t$. This means that if we decide to produce information $f(s)$ at s, then we must produce all this information, and possibly more, at t, yielding $f(s) \sqsubseteq f(t)$. Thus we can only make decisions at a given information state which are stable under every possible information increase from that state. This idea is very much akin to the use of partial orders in Kripke semantics for Intuitionistic Logic, in particular in connection with the interpretation of negation in that semantics. The continuity condition, on the other hand, reflects the fact that a computational process will only have access to a finite amount of information at each finite stage of the computation. If we are provided with an infinite input, then any information we produce as output at any finite stage can only depend on some finite observation we have made of the input. This is reflected in one of the inequations corresponding to continuity:

$$f(\bigsqcup_{n\in\omega} x_n) \sqsubseteq \bigsqcup_{n\in\omega} f(x_n)$$

which says that the information produced at the limit of an infinite process of information increase is no more than what can be obtained as the limit of the information produced at the finite stages of the process. Note that the "other half" of continuity

$$\bigsqcup_{n\in\omega} f(x_n) \sqsubseteq f(\bigsqcup_{n\in\omega} x_n)$$

follows from monotonicity.

Note by the way how this discussion is permeated with the epistemic stance. Continuous functions produce *points* as outputs on the basis of *observations* they make of their inputs. Thus the duality between these two points of view plays a basic rôle in our very *understanding* of continuous functions.[8] This can be (and often is) glossed over in Domain Theory, by virtue of the coincidence of finite points and finite properties which we have already discussed.

The Fixpoint Theorem

We now consider a simple but powerful and very widely applicable theorem, which is one of the main pillars of Domain Theory, since by virtue of this result it provides a general setting in which recursive definitions can be understood.[9]

THEOREM 1 (The Fixpoint Theorem). *Let D be an ω-cpo with a least element, and $f : D \to D$ a continuous function. Then f has a least fixed point $\mathsf{lfp}(f)$. Moreover, $\mathsf{lfp}(f)$ is defined explicitly by:*

(1) $\mathsf{lfp}(f) = \bigsqcup_{n\in\omega} f^n(\bot)$.

[8]Mathematically, this duality appears in the guise of the compact-open topology for function spaces. We can think of open sets in functions spaces as observations which can be made on functions viewed as black boxes. Dually to the point of view of the function, which *observes* an input and *produces* an output, a *function environment* must produce an input (a point, or in more general topological situations, a compact set), and observe the corresponding output.

[9]A *fixed point* of a function $f : X \longrightarrow X$ is an element $x \in X$ such that $f(x) = x$. "Fixpoint" is (standard) jargon for fixed point. For some historical information on this theorem and its variations, see [Lassez *et al.*, 1982].

We give the proof, since it is elementary, and exhibits very nicely how the basic axiomatic structure of Domains is used.

Proof. Note that $f^n(\bot)$ is defined inductively by:

$$f^0(\bot) = \bot, \qquad f^{k+1}(\bot) = f(f^k(\bot)).$$

We show firstly that this sequence is indeed an ω-chain . More precisely, we show for all $k \in \mathbb{N}$ that $f^k(\bot) \sqsubseteq f^{k+1}(\bot)$. For $k = 0$, this is just $\bot \sqsubseteq f(\bot)$. For the inductive case, assume that $f^k(\bot) \sqsubseteq f^{k+1}(\bot)$. Then by monotonicity of f, $f(f^k(\bot)) \sqsubseteq f(f^{k+1}(\bot))$, *i.e.* $f^{k+1}(\bot) \sqsubseteq f^{k+2}(\bot)$, as required.

Next we show that (1) does yield a fixpoint. This is a simple calculation using the continuity of f:

$$f(\bigsqcup_{n\in\omega} f^n(\bot)) = \bigsqcup_{n\in\omega} f^{n+1}(\bot) = \bigsqcup_{n\in\omega} f^n(\bot).$$

The last step uses the (easily verified) fact that removing the first element of an ω-chain does not change its least upper bound.

Finally, suppose that a is a fixpoint of f. Then we show by induction that, for all k, $f^k(\bot) \sqsubseteq a$. The basis is just $\bot \sqsubseteq a$. For the inductive step, assume $f^k(\bot) \sqsubseteq a$. Then by monotonicity of f,

$$f^{k+1}(\bot) = f(f^k(\bot)) \sqsubseteq f(a) = a.$$

Thus a is an upper bound of $(f^n(\bot) \mid n \in \omega)$, and hence $\bigsqcup_{n\in\omega} f^n(\bot) \sqsubseteq a$. ■

Factorial revisited We now reconstrue the definition of the factorial function we considered previously, as a function on *domains*:

$$F : \mathbf{Pfn}(\mathbb{N}, \mathbb{N}) \longrightarrow \mathbf{Pfn}(\mathbb{N}, \mathbb{N}),$$

defined by

$$F(f)(n) = \textbf{if } n = 0 \textbf{ then } 1 \textbf{ else } n \times f(n-1).$$

We can check that F *is continuous.* Hence we can apply the fixpoint theorem to F, and conclude that it has a least fixpoint $\mathsf{lfp}(f)$, defined explicitly by (1). Now we can make the (explicit, non-circular) definition:

$$\mathsf{fact} = \mathsf{lfp}(F).$$

One can check that this definition yields exactly the expected definition of factorial. In fact, the increasing sequence constructed in forming the least fixpoint according to (1) is exactly the one we described concretely in our previous discussion of the factorial.

Thus in particular the processes of information increase we have been emphasizing are involved directly in the construction underpinning the Fixpoint Theorem.

Further Developments in Domain Theory

This is of course just the beginning of an extensive subject. We mention a few principal further features of Domain Theory:

Function Spaces A key point of the theory is that, given domains D and E, the set of continuous functions from D to E, written as $[D \longrightarrow E]$, will again be a domain, with the following *pointwise ordering*:

$$f \sqsubseteq g \iff \forall x \in D.\, f(x) \sqsubseteq_E g(x).$$

Moreover, operations such as function application and currying or lambda-abstraction are continuous. This means that we can form models of typed λ-calculi and higher-order computation within Domain Theory, which is of central importance for the denotational semantics of programming languages. Of course, such domains of higher-order functions are very "abstract"—they are in fact the prime examples of domains which are *not* concrete in the sense of [Kahn and Plotkin, 1978]—and notions of temporality are left quite far behind. (There have attempts to capture more of these notions by varying the definition of the order on function spaces, but these have not been completely successful—and in some cases, provably cannot be).

Recursive Types Remarkably, the idea of the Fixpoint Theorem, and its use to give meaning to recursive definitions of elements of domains, can be lifted to the level of domains themselves, to give meaning to *recursive definitions of types.* This even extends to the free use of function spaces in recursive definitions of domains, leading to the construction of domains D whose continuous function spaces $[D \longrightarrow D]$ are isomorphic to D or to a subspace of D. This allows models of the type-free λ-calculus, and of various strongly impredicative type theories, to be given within Domain Theory.

Powerdomains There are also a number of *powerdomain* constructions $P(D)$, which build a domain of subsets of D. This allows various forms of non-deterministic and concurrent computation to be described. There is also a *probabilistic powerdomain* construction, which provides semantics for probabilistic computation.

Some suggestions for further reading on Domain Theory The text [Davey and Priestley, 2002] gives a fairly gentle introduction to partial orders and lattices, with some material on domains. The Handbook article [Abramsky and Jung, 1994] is a comprehensive technical survey of domain theory. The monograph [Gierz *et al.*, 2003] focusses on the connections to topology and lattice theory. Gordon Plotkin's classic lecture notes [Plotkin, online] are available on-line. The texts [Winskel, 1993; Amadio and Curien, 1998] show how domain theory is used in the semantics of programming languages.

3.2 Dynamic Logic

Dynamic Logic originates at the confluence of two sources: modal logic and its Kripke semantics [Kripke, 1963; Blackburn *et al.*, 2001]; and Hoare logic of programs [Hoare, 1969].

Modal Logic Modal Logic adds to a standard background logic (say classical propositional calculus) the propositional operators $\Box$ and $\Diamond$, expressing ideas of "necessity" and "possibility". This was transformed from a philosophical curiosity to a vibrant and highly applicable branch of mathematical logic by the introduction of Kripke semantics [Kripke, 1963]. This is based on Kripke structures (W, R, V), where W is a set of worlds, $R \subseteq W \times W$ is an "accessibility relation", and $V : \mathbb{P} \to \mathcal{P}(W)$ is a valuation which for each propositional atom in $\mathbb{P}$ assigns the set of worlds in which it is true. This valuation is then extended to one on formulas, with the key clauses:

$$w \models \Box\phi \equiv \forall w'.\, wRw' \Rightarrow w' \models \phi$$

$$w \models \Diamond\phi \equiv \exists w'.\, wRw' \wedge w' \models \phi.$$

The importance of the Kripke semantics is that it gives modal logic a clear mathematical purpose: it is a logical language for talking about such structures, which strikes a good balance between expressive power and tractability. Computer Science provides a wealth of situations where such structures arise naturally, and where there is a clear need for the verification of their logical properties. The dominant interpretation of Kripke structures in Computer Science replaces metaphysical talk of "possible worlds" by the more prosaic terminology of *states*. Here we think of states of a system, which are generally characterized by the information we have about them. In a Kripke structure, the *direct* information we have about a state is which atomic propositions are true in that state. However, while we seem again to be speaking about information states, as in our discussion of Domain Theory, there is an important difference. In Domain Theory, (as in Kripke semantics for Intuitionistic Logic), information is in general *partial*, but also *persistent*. Information can only *increase* along a computation. We may never reach total information, but we will never lose what we had—just as we can never (in current Physics) change the past. (Indeed, the two are intimately related. In the implicit temporality of Domain Theory, the current information state summarizes all the information produced in the computation up till now; whatever happens in the future cannot change that). By contrast, Kripke structures for modal logics correspond to a less stable world. We may have perfect knowledge of the current state, but the dynamics of the system, as described by the accessibility relation, allow in general for arbitrary state change. A basic Computer Science model for this scenario is provided by taking the states to be memory states of a computer. At some instant of time we may have a complete snap-shot of the memory. But our repertoire of actions allow us to assign an arbitrary new value into any memory cell, so we can go from any given state to any other (possibly by a sequence of basic actions). In particular, the key feature of computer memory, the fact that we can destructively over-write the previous contents of a memory cell, (a feature which is not, apparently, available for our own memories!), ensures that the past is not in general carried forward.

Hoare Logic Hoare Logic [Hoare, 1969; de Bakker, 1980] provides a compositional proof theory for reasoning about imperative programs. It is a two-sorted system. We have a syntax for *programs* P, and one for *formulas* ϕ, which are generally taken to be formulas of predicate calculus. Such formulas can be used to express properties of program states (*i.e.* memory state snap-shots as in our previous discussion, or more formally assignments of values to the variables appearing in the program), by a **variable pun** by which the individual variables used in formulas are identified with the program variables. The basic assertions of the system are taken to be *Hoare triples* $\phi\{P\}\psi$. Such a triple is said to be valid if, in any initial state satisfying the formula ϕ (the *precondition*), execution of the program P will, if it terminates, result in a final state satisfying the formula ψ (the *post-condition*).

The **variable pun** is put to use in the axiom for assignment statements:

$$\phi[e/x]\{x := e\}\phi$$

which says that ϕ is true after executing the assignment statement $x := e$ if ϕ with e substituted for x was true before.

The key rules of the system allow for compositional derivation of assertions about complex programs from assertions about their immediate sub-programs.

$$\frac{\phi\{P\}\psi \quad \psi\{Q\}\theta}{\phi\{P;Q\}\theta} \qquad \frac{\phi \wedge B\{P\}\psi \quad \phi \wedge \neg B\{Q\}\psi}{\phi\{\textbf{if } B \textbf{ then } P \textbf{ else } Q\}\psi} \qquad \frac{\phi \wedge B\{P\}\phi}{\phi\{\textbf{while } B \textbf{ do } P\}\phi \wedge \neg B}$$

Here $P;Q$ is the *sequential composition* which firstly performs P, then Q; **if** B **then** P **else** Q is the *conditional* which evaluates B in the current state; if it is **true** then P is performed, while if it is **false**, Q is performed. Finally, **while** B **do** P evaluates B; if it is **true**, then P is performed, after which the whole statement is repeated; while if it is **false**, the statement terminates immediately.

Dynamic Logic Dynamic Logic [Pratt, 1976] arises by combining salient features of these two systems. Note that we are reasoning about programs in terms of the *input-output relations on states* which they define. If the program is deterministic, this relation will actually be a partial function, but there is no need to insist on this. We can thus view each program P as defining a relation $R \subseteq S \times S$, where S is the set of states. Thus for each individual program, we obtain a Kripke structure (S, R, V), where V is the valuation which assigns truth conditions on states for some repertoire of state predicates. The key point of contact between the two systems is that validity of the Hoare triple $\phi\{P\}\psi$ corresponds exactly to the validity of the modal formula

$$\phi \to \Box\psi$$

in the Kripke structure (S, R, V), where by validity we mean that

$$s \models \phi \to \Box\psi$$

for every $s \in S$.

As a first extension, we can consider multiple programs, each defining an accessibility relation R. To keep track of which program we are talking about at any given point, we replace $\Box$ by $[R]$, so that the formula corresponding to the Hoare triple now reads as

$$\phi \to [R]\psi.$$

Just as $[R]$ replaces $\Box$, so $\langle R\rangle$ replaces $\Diamond$. Thinking of R as the input-output relation defined by a program, we can read $[R]\phi$ as holding in all states (worlds) s such that any output state obtained by executing R starting in s will satisfy ϕ. Similarly, $\langle R\rangle$ will be true in any state s such that there is some output state than can be obtained by running R starting in s which satisfies ϕ.

This is just *multi-modal logic*, with mutiple accessibility relations, each with its own modalities. Note that it is now completely meaningful to consider modal formulas which make assertions about programs which go well beyond Hoare triples, e.g.

$$[R]\langle S\rangle\phi \to \langle S\rangle[R]\phi.$$

However, at this point we lack the compositional analysis of programs offered by Hoare Logic.

The final step to (propositional) Dynamic Logic comes by considering a two-sorted system with a mutually recursive syntax. We have a set $\mathbb{P}$ of propositional atoms as before, and also a set Rel of basic relations. The syntax of formulas is given by

$$\phi \quad ::= \quad p \in \mathbb{P} \mid \neg\phi \mid \phi \wedge \psi \mid [R]\phi$$

while the syntax of relations R is given by

$$R \quad ::= \quad r \in \mathsf{Rel} \mid R; S \mid R \cup S \mid R^* \mid \phi?$$

We have not included the modal operator $\langle R\rangle$ as primitive syntax, since we can define

$$\langle R\rangle\phi \quad \equiv \quad \neg[R]\neg\phi.$$

In this syntax, any program is allowed to appear as a modal operator on formulas, while in addition to the usual regular operations of relational algebra (composition, union, and reflexive transitive closure), any formula is allowed to appeared as a program test (we may call this the **formula pun**). In general, this is too strong, and only a restricted class of tests should be allowed. Tests are interpreted as sub-identity relations—so $\phi?$ is the set of all (s, s) such that ϕ is true in s.

Note that the usual imperative program constructs can be recovered from these relational constructs. Sequential composition is provided directly, while

$$\textbf{if } b \textbf{ then } R \textbf{ else } S \;\equiv\; b; R \cup \neg b; S \qquad \textbf{while } b \textbf{ do } R \;\equiv\; (b; R)^*; \neg b.$$

The Hoare Logic axioms can now be derived from the following modal axioms:

$$\begin{array}{rcl} [R;S]\phi & \leftrightarrow & [R][S]\phi \\ [R \cup S]\phi & \leftrightarrow & [R]\phi \wedge [S]\phi \\ [\psi?]\phi & \leftrightarrow & \psi \rightarrow \phi \end{array}$$

and the rule

$$\frac{\phi \rightarrow [R]\phi}{\phi \rightarrow [R^*]\phi}.$$

Discussion

While Hoare Logic is specifically tailored to the needs of conventional imperative programming languages, Dynamic Logic is much more generic in style; and indeed, it has been applied in a range of contexts, including Natural Language and Quantum Logic. In the Chapter in this Handbook by Baltag and Moss, a version of Dynamic Logic is described in which the states are *information states of agents*, and the actions are *epistemic actions* by these agents, such as public announcements.

As a general formalism, though, Dynamic Logic offers only a limited analysis of information dynamics. Indeed, despite its name, it is not really very dynamic, as it is limited to speaking of the input-output behaviour of relations. This is confirmed by the simple translation it admits into first-order logic (augmented with fixpoints to account for the reflexive transitive closure operation on relations). We briefly sketch this. To each relation term R, we associate a formula $B_R(x,y)$ in two free variables, and to each modal formula ϕ we associate a formula $A_\phi(x)$ in one free variable. The main clauses in the definition of B_R are as follows:

$$\begin{array}{rcl} B_{R\cup S}(x,y) & \equiv & B_R(x,y) \vee B_S(x,y) \\ B_{R;S}(x,y) & \equiv & \exists z.\, B_R(x,z) \wedge B_S(z,y) \\ B_{\phi?}(x,y) & \equiv & A_\phi(x) \wedge x = y \\ B_{R^*}(x,y) & \equiv & \mu S.\, [(x=y) \vee (\exists z.\, B_R(x,z) \wedge S(z,y))] \end{array}$$

The clauses for modal formulas are standard. The one for the modality is:

$$A_{[R]\phi}(x) \equiv \forall y.\, B_R(x,y) \rightarrow A_\phi(y).$$

Suggestion for further reading The book [Harel *et al.*, 2000] is a comprehensive technical reference, while [van Benthem, 1988] is a wide-ranging study. Applications to Natural Language appear in [Groenendijk and Stockhof, 1991; van Eijck and Stokhof, 2006], and to Quantum Logic in [Baltag and Smets, 2006].

3.3 Process Algebra

Background

One of the major areas of activity in Theoretical Computer Science over the past three decades has been Concurrency Theory, and in particular Process Algebra.

Whereas modelling sequential computation in terms of input-output functions or relations essentially uses off-the-shelf tools from Discrete Mathematics and Logic, albeit in novel combinations and with new technical twists, and even Domain Theory can be seen as an off-shoot of General Topology and Lattice Theory, Concurrency Theory has really opened up some new territory. In Concurrency Theory, the computational processes themselves become the objects of study; concurrent systems are executed for the behaviour they produce, rather than to compute some pre-specified function. In this setting, even such corner-stones of computation as Turing's analysis of computability do not provide all the answers. For all its conceptual depth, Turing's analysis of computability was still calibrated using familiar mathematical objects: which *functions* or *numbers* are computable? When we enter the vast range of possibilities for the behaviour of computational systems in general, the whole issue of what it means for a concurrent formalism to be *expressively complete* must be re-examined. There is in fact no generally accepted form of Church-Turing thesis for concurrency; and no widely accepted candidate for a universally expressive formalism. Instead, there are a huge range of concurrency formalisms, embodying a host of computational features.

Another question which ramifies alarmingly in this context is what is the right notion of *behavioural equivalence* of processes. Again, a large number of candidates have arisen. Experts use what seems most appropriate for their purpose; it is not even plausible that a single notion will gain general acceptance as "the right one".

In fact, a great deal of progress has been achieved, and the situation is much more positive than might appear from these remarks. There is a great diversity of particular formalisms and definitions in Concurrency Theory; but underpinning these are a much smaller number of underlying paradigms and technical tool-kits, which do provide effective intellectual instruments, both for fundamental research and applications.

Examples include:

- labelled transition systems and bisimulation
- naming and scope restriction and extrusion
- the automata-theoretic paradigm for model-checking

These tool-kits are the real fruits of these theories. They may be compared to the traditional tool-kits of physics and engineering: Differential Equations, Laplace and Fourier Transforms, Numerical Linear Algebra, etc. They can be applied to a wide range of situations, going well beyond those originally envisaged, e.g. Security, Computational Biology, and Quantum Computation.

Some Basics of Process Algebra

We now turn to a brief description of a few basic notions, in a subject on which there is a vast literature. We begin with the key semantic structure, namely *labelled transition systems.* A labelled transition system is a structure (S, Act, T), where S is a set of states, Act is a set of actions, and $T \subseteq S \times \mathsf{Act} \times S$ is the transition relation. We write $s \xrightarrow{a} t$ for $(s, a, t) \in T$. Note how close this is to the notion of Kripke structure we have already encountered. However, that notion is tuned to a *state-based* view of computation, in which we focus on assertions which are true in given states. The transition relation plays an indirect rôle, in controlling the behaviour of the modal operators. By contrast, the point of view in labelled transition systems is that states are not directly observable, and hence do not have properties directly attributable to them. Rather, it is the actions which are the basic observables, and we infer information about states indirectly from their potential for observable behaviour. Thus the point of view here is closer to automata theory. A key difference from classical automata theory, however, is that we look beyond the classical notion of behaviour in terms of the words or traces (sequences of actions) accepted or generated by the system, and also encompass *branching behaviour*. The classical example which illustrates this is the following [Milner, 1980]:

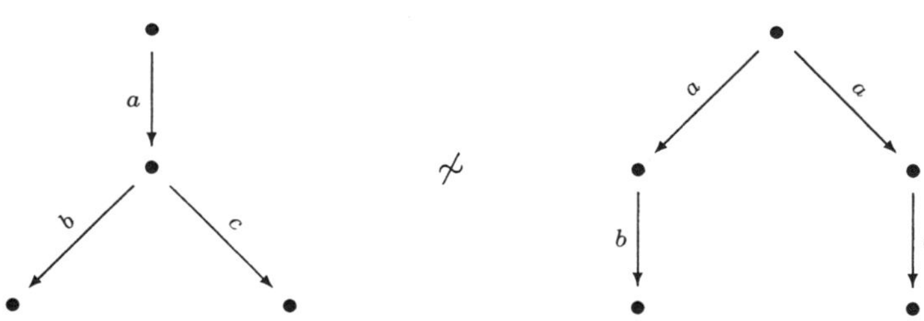

These systems have the same linear traces $\{ab, ac\}$. However, if we think of a scenario where we can perform experiments by pressing buttons labelled with the various actions, and observe if the experiments succeed, *i.e.* whether the system performs the corresponding action, then after observing an a in the first system, it is clear that whether we press the b button or the c button, we will succeed; whereas in the second system, one button will succeed and the other won't. A fundamental notion of process equivalence which enforces this distinction is *bisimulation*. We define a *bisimulation* [van Benthem, 1976; Hennessy and Milner, 1980; Park, 1981; Milner, 1989; Sangiorgi, 2004] on a labelled transition system (S, Act, T) to be a relation $R \subseteq S \times S$ such that:

$$\begin{array}{lcl} sRt \ \wedge \ s \xrightarrow{a} s' & \Rightarrow & \exists t'.\, t \xrightarrow{a} t' \ \wedge \ s'Rt' \\ \wedge & & \\ sRt \ \wedge \ t \xrightarrow{a} t' & \Rightarrow & \exists s'.\, s \xrightarrow{a} s' \ \wedge \ s'Rt' \end{array}$$

We write $s \sim t$ if there is a bisimulation R such that sRt. We can see that indeed the root states of the two trees in the above example are not bisimilar, since the first has an action a to a state in which both the actions b and c are possible, while the second has no a-move to a matching state.

We now turn to a suitable modal logic for labelled transition systems. The basic form for such a logic is Hennessy-Milner Logic. This has modal operators

$[a]$, $\langle a \rangle$ for each action a. In general, this logic does not have (or require) any propositional atoms; just constants tt (true) and ff (false). The semantic clauses are as expected for a multi-modal logic, where we view the transition relation as an Act-indexed family of relations $\{T_a\}_{a \in \mathsf{Act}}$, where $T_a \subseteq S \times S$ is defined by

$$T_a = \{(s,t) \mid (s,a,t) \in T\}.$$

Thus we have the clauses

$$s \models [a]\phi \;\equiv\; \forall t.\, s \xrightarrow{a} t \;\Rightarrow\; t \models \phi$$

$$s \models \langle a \rangle\phi \;\equiv\; \exists t.\, s \xrightarrow{a} t \;\wedge\; t \models \phi.$$

The basic result here is that, under suitable hypotheses, two states in a labelled transition system are bisimilar if and only if they satisfy the same formulas in this modal logic. Thus in our example above, the first system satisfies the formula $\langle a \rangle(\langle b \rangle \mathtt{tt} \;\wedge\; \langle c \rangle \mathtt{tt})$, while the second does not.

We now turn, finally, to the *algebraic* aspect of process algebra. Just as we structured the programs in Dynamic Logic using relational algebra, so we seek an algebraic structure to generate a wide class of process behaviours. As we have already discussed, there is no one universally adopted set of process combinators, but we shall consider a standard set of operations, essentially a fragment of Milner's CCS [Milner, 1980; Milner, 1989]. The syntax of process terms P is defined, assuming a set Act of actions, as follows:

$$P \;::=\; a.P \;\; (a \in \mathsf{Act}) \mid P + Q \mid \; 0 \; \mid P \parallel Q.$$

Here $a.P$ is *action prefixing*; first do a, then behave as P. $P + Q$ is *non-deterministic choice* between P and A, while 0 is *inaction*; the process which can do nothing. Finally, $P \parallel Q$ is *parallel composition*, which we take here in a simple form, not involving any interaction between P and Q.

We formalize these intuitions as a labelled transition system in which the states are the process terms, while the transition relation is defined by structural induction on the syntax of terms—the Structural Operational Semantics paradigm [Plotkin, 2004].

The transition relation is specified as follows.

$$\frac{}{a.P \xrightarrow{a} P} \qquad \frac{P \xrightarrow{a} P'}{P + Q \xrightarrow{a} P'} \qquad \frac{Q \xrightarrow{a} Q'}{P + Q \xrightarrow{a} Q'}$$

$$\frac{P \xrightarrow{a} P'}{P \parallel Q \xrightarrow{a} P' \parallel Q} \qquad \frac{Q \xrightarrow{a} Q'}{P \parallel Q \xrightarrow{a} P \parallel Q'}$$

This labelled transition system gives rise to a notion of bisimulation, which is an equivalence relation, and in fact a congruence for the process algebra. The corresponding equational theory for the algebra can be axiomatized as follows:

$$\begin{array}{rcl} P+P & = & P \\ P+0 & = & 0 \\ P+Q & = & Q+P \\ P+(Q+R) & = & (P+Q)+R \end{array}$$

together with the following equational scheme. If $P \equiv \sum_{i \in I} a_i.P_i$ and $Q \equiv \sum_{j \in J} b_j.Q_j$, then:

$$P \parallel Q \;=\; \sum_{i \in I} a_i.(P_i \parallel Q) + \sum_{j \in J} b_j.(P \parallel Q_j).$$

This is an infinite family of equations. In fact, the equational theory of bisimulation on process terms is not finitely axiomatizable [Moller, 1990b]; however, with the aid of an auxiliary operator (the "left merge"), a finite axiomatization can be achieved [Moller, 1990a].

Communication and Interaction in Process Algebra

We shall take a brief glimpse at this large topic. For illustration, we shall describe the CCS approach [Milner, 1980]. However, it should be emphasized that there is a huge diversity of approaches in the process algebra literature, with none having a clear claim to being considered canonical. (For further remarks on this issue of non-canonicity, see the final section of this article, and [Abramsky, 2006]).

We assume some structure on the set A: a fixed-point free involution $a \mapsto \bar{a}$, so that we have $a \neq \bar{a}$ and $\bar{\bar{a}} = a$. The idea is that a and $\bar{a}$ will be complementary partners to an interaction or synchronized action. We also introduce a special action τ, which is intended to be a "silent action", unobservable to the external environment.

We can now introduce a parallel composition $P \,|\, Q$ which does allow for interaction, in the form of synchronization between P and Q. Its dynamics are given by the following rules:

$$\frac{P \xrightarrow{a} P'}{P \,|\, Q \xrightarrow{a} P' \,|\, Q} \qquad \frac{P \xrightarrow{a} P' \quad Q \xrightarrow{\bar{a}} Q'}{P \,|\, Q \xrightarrow{\tau} P' \,|\, Q'} \; (a, \bar{a} \neq \tau) \qquad \frac{Q \xrightarrow{a} Q'}{P \,|\, Q \xrightarrow{a} P \,|\, Q'}$$

The new ingredient is the middle rule, which allows for synchronization between P and Q. Note that a and $\bar{a}$ "complete" each other into the action τ which is now an internal step of the system, and hence unobservable to the external environment.

To take proper account of the unobservable character of τ, we introduce the observable transition relation for each $a \neq \tau$:

$$\stackrel{a}{\Longrightarrow} \;=\; \xrightarrow{\tau}{}^{*} \xrightarrow{a} \xrightarrow{\tau}{}^{*}$$

We can then define *weak bisimulation* with respect to these observable transitions. However, a new complication arises: this weak bisimulation is not a congruence with respect to the operations of the process algebra. It is necessary to take the largest congruence compatible with weak bisimulation, finally yielding the notion of *observational congruence.* This notion can be equationally axiomatized, but it is considerably more complex and less intuitive than the "strong bisimulation" we encountered previously.

Discussion

Process Algebra can be used as a *vehicle* for discussions of information flow and information dynamics, e.g. [Lowe, 2002]. It does not in itself offer a fully fledged theory of these notions.

Process Algebra is a qualitative theory of process behaviours. It is our first example of a *dynamic* theory, since it makes temporality and the flow of events explicit.

Suggestions for further reading Introductory textbooks include [Hoare, 1985; Milner, 1989; Baeten and Weijland, 1990; Milner, 1999]. The Handbook of Process Algebra [Bergstra and Ponse, 2000] provides wide technical coverage of the field.

4 COMBINING QUALITATIVE AND QUANTITATIVE THEORIES OF INFORMATION

4.1 Scott domain theory and Shannon information theory

Two important theories of information give contrasting views on the question of information increase, which we discussed in Section 1. Information theory *à la* Shannon is a quantitative theory which considers how *given* information can be transmitted losslessly on noisy channels. In this process, information may only be lost, never increased. Domain Theory *à la* Scott is, as we have seen, a qualitative theory in which the key notion is the partial order $x \sqsubseteq y$, which can be interpreted as: "y has more information content than x". This theory is able to model a wide range of computational phenomena. To take a classical example, consider the interval bisection methods for finding the root of a function. We start with an interval in which the root is known to lie. At each step, we halve the length of the interval being considered. This represents an increase in our information about the location of the root, in an entirely natural sense. In the limit, this nested sequence of intervals contains a single point, the root – we have perfect information about the solution.

More generally, in Domain Theory recursion (and thereby control mechanisms such as iteration) is modelled as the least fixed point of a monotonic and (order-)continuous function:

$$\bot \sqsubseteq f\bot \sqsubseteq f^2\bot \sqsubseteq \cdots \bigsqcup_k f^k\bot$$

since $f(\bigsqcup_k f^k \bot) = \bigsqcup_k f^{k+1} \bot = \bigsqcup_k f^k \bot$. Thus a basic tenet of this theory is that information *does* increase during computation, and in particular this is how the meaning of recursive definitions is given.

It is intriguing to consider that the different viewpoints taken by Information theory and Domain Theory may have been influenced by their technological roots. Information theory was summoned forth by the needs of the telecommunications industry, whose task is to transmit the customer's data with the highest possible fidelity. Domain Theory arose as a mathematical theory of *computation*; the task of computation is to "add value" to the customer's data.[10]

How can these views be reconciled? Information theory is a thermodynamic theory; Shannon information is negative entropy. From this viewpoint, the *total information* of a system can only decrease; however, information can flow from one subsystem into another, just as a body can be warmed by transferring heat from its environment.

The Domain Theory view, we suggest, arises most naturally if we think of adding an observer to a system. It is the observer's information which increases during a computation. This reading has a precise mathematical analogue in the view of Domain Theory as a "logic of observable properties" [Abramsky, 1991]. Information increase is always, necessarily it seems, *relative to a sub-system.* Moreover, this is a subsystem which can *observe* its environment, and which may, symmetrically, act itself to direct information to the environment. It is then a small step to viewing such sub-systems as *agents.*

It is worth adding that Shannon Information Theory also relies on such a view for its guiding intuitions. One of the standard ways of motivating Shannon information is in terms of "twenty questions": the number of bits of information in a message is how many yes/no questions we would need to have answered in order to know the contents of the message. Again, implicit here is some interaction between agents. And of course, the purpose of communication itself is to transfer information from one agent to another.

We need a quantitative theory to deal with essentially quantitative issues such as complexity, information content, rate of information flow etc. However, the *weakness* of a purely quantitative theory is that numbers are always comparable, so that some more subtle issues are obscured, such as, crucially, distinguishing different *directions* of information increase. Beyond this, by combining quantitative and qualitative aspects, e.g. in formulating conditions on "informatic processes", a unified theory can be more than the sum of its parts.

---

[10]Which of course raises our question of how this can be possible thermodynamically. The answer is, again, that it is the *customer's data* which is having value added to it; just as buying energy from the National Grid does not violate the Second Law.

4.2 Domains with measurements: connecting the quantitative and qualitative views

An important step towards unifying the qualitative and the quantitative points of view was taken in Keye Martin's Ph.D. thesis [Martin, 2002] and subsequent publications [Martin, 2001a; Martin, 2001b; Martin *et al.*, 2002]. Martin introduced a simple notion of *measurement* on domains. In its most concrete form, a measurement assigns real numbers to domain elements, which can be said to measure the degree of undefinedness or uncertainty of the element. Thus the maximal elements, which can be regarded as having perfect information, will have measurement 0. The axioms for measurements, while quite simple and intuitive, tie the quantitative notion in with the qualitative domain structure in a very rich way. Just to mention some of the highlights:

- There is a rich theory of fixpoints which applies to increasing, *but not necessarily monotonic*, functions on domains. This is already a remarkable departure from 'classical' domain theory, in which monotonicity is always assumed. However, Martin shows that there are compelling natural examples, such as interval bisection, which require this broader framework. Not only are there existence and uniqueness theorems for fixpoints in this frameworks, but also novel induction principles.

- As the previous point suggests, there is a move away from the use of domain theory to model purely extensional aspects of computation, and towards using it to capture important features of *computational processes.* This leads to a notion of 'informatic derivative' which can be used to gain information about the *rate of convergence* of a computational process.

- A notable aspect of this development is the unified basis on which it puts the study of both discrete and continuous (e.g. real-number) computation.

It is also important that there are many natural examples of measurement covering most of the domains standardly arising as data-types for computation, including the domain of intervals, for which the natural measurement is the *length* of the interval; finite lists and other standard finite data-structures; streams; partial functions on the natural numbers; and both non-deterministic and probabilistic powerdomains.

However, the example which has really revealed the possibilities of this framework has only appeared recently, and is a major development in its own right.

4.3 Combining Scott Information and Shannon Information

Recently, Bob Coecke and Keye Martin have produced a very interesting construction which can be seen as a first step towards a unification of these two theories of information [Coecke and Martin, 2002]. The problem which they attacked can be formulated as follows. Consider the set of probability distributions on a finite set. For an n-element set, these are the "classical n-states" of Physics:

$$\Delta^n := \{x \in [0,1]^n : \sum_{i=1}^{n} x_i = 1\}.$$

This is the setting in which Shannon entropy, the fundamental quantitative notion in classical Information Theory, is defined. It assigns a number, the "expected information", to each classical state. The question is: can we place a *partial order* on Δ^n such that:

1. This partial order forms a domain.

2. Shannon entropy is a measurement with respect to this domain.

3. The order extends to quantum states (density operators).

These are highly non-trivial requirements to satisfy. Note that the set of probability distributions on a 3-element set, seen as a subset of Euclidean space, form a (solid) triangle, and in general those on a n-element set form an n-simplex. The distribution corresponding to maximum uncertainty is the uniform distribution, with each point assigned probability $1/n$ — geometrically, the barycenter of the simplex; while the maximal elements $\max(\Delta^n)$, corresponding to perfect information, are the *pure states* assigning probability 1 to one element, and 0 to all others — geometrically, the vertices of the simplex. This geometrical aspect brings a rich mathematical structure to this example which seems different to anything previously encountered in Domain Theory.

Note also the contrast with previous work on the probabilistic powerdomain [Jones and Plotkin, 1989]. Classical probability distributions are *maximal elements* in the probabilistic powerdomain; non-standard elements (valuations) are introduced which provide approximations to measures, but the order restricted to the measures themselves is discrete. By contrast, we are seeking a rich informatic structure on the standard objects of probability (distributions) and quantum mechanics (density operators) *themselves*, without introducing any non-standard elements. It is by no means *a priori* obvious that this can be done at all; once we see that it can, many new possibilities will unfold.

A classical state $x \in \Delta^n$ is *pure* when $x_i = 1$ for some $i \in \{1, \ldots, n\}$; we denote such a state by e_i. Pure states $\{e_i\}_i$ are the actual states a system can be in, while general mixed states x and y are epistemic entities.

If we know $x \in \Delta^{n+1}$ and by some means determine that outcome i is not possible, our knowledge improves to

$$p_i(x) = \frac{1}{1-x_i}(x_1, \ldots, \hat{x_i}, \ldots, x_{n+1}) \in \Delta^n,$$

where $p_i(x)$ is obtained by first removing x_i from x and then renormalizing. The partial mappings which result,

$$p_i : \Delta^{n+1} \rightharpoonup \Delta^n$$

with $\mathrm{dom}(p_i) = \Delta^{n+1} \setminus \{e_i\}$, are called the *Bayesian projections* and lead one directly to the following inductively defined relation on classical states.

DEFINITION 2. For $x, y \in \Delta^2$:

(2) $x \sqsubseteq y \equiv (y_1 \leq x_1 \leq 1/2)$ or $(1/2 \leq x_1 \leq y_1)$.

For $n \geq 2$, and $x, y \in \Delta^{n+1}$:

(3) $x \sqsubseteq y \equiv (\forall i)(x, y \in dom(p_i) \Rightarrow p_i(x) \sqsubseteq p_i(y))$.

The relation $\sqsubseteq$ on Δ^n is called the Bayesian order.

See [Coecke and Martin, 2002] for motivation, and results showing that the order on Δ^2 is uniquely determined under minimal assumptions.

The key result is:

THEOREM 3. $(\Delta^n, \sqsubseteq)$ *is a domain with maximal elements*

$$\max(\Delta^n) = \{e_i : 1 \leq i \leq n\}$$

and least element $\perp := (1/n, \ldots, 1/n)$. *Moreover, Shannon entropy*

$$\mu(x) = -\sum_{i=1}^{n} x_i \log x_i$$

is a measurement of type $\Delta^n \to [0, \infty)^*$.

The Bayesian order can also be described in a more direct manner, the *symmetric characterization*. Let $S(n)$ denote the group of permutations on $\{1, \ldots, n\}$, and

$$\Lambda^n := \{x \in \Delta^n : (\forall i < n)\, x_i \geq x_{i+1}\}$$

the collection of *monotone* classical states.

THEOREM 4. *For $x, y \in \Delta^n$, we have $x \sqsubseteq y$ iff there is a permutation $\sigma \in S(n)$ such that $x \cdot \sigma, y \cdot \sigma \in \Lambda^n$ and*

$$(x \cdot \sigma)_i (y \cdot \sigma)_{i+1} \leq (x \cdot \sigma)_{i+1} (y \cdot \sigma)_i$$

for all i with $1 \leq i < n$.

In words, this result says that the Bayesian order holds between states x and y if we can find a permutation σ which rearranges them both as monotone states, and such that x falls less rapidly than y as we proceed through the ordered list of component probabilities.

Thus, the Bayesian order is order isomorphic to $n!$ many copies of Λ^n identified along their common boundaries. This fact, together with the pictures of $\uparrow x := \{y \in \Delta^n \mid x \sqsubseteq y\}$ at representative states x in Figure 1, will give the reader a good feel for the geometric nature of the Bayesian order.

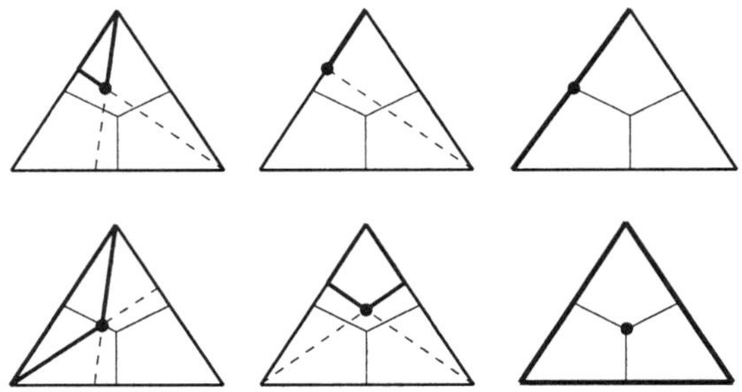

Figure 1. Pictures of $\uparrow x$ for $x \in \Delta^3$.

4.4 The Quantum Case

The real force of the construction for classical states becomes apparent in the further development in [Coecke and Martin, 2002], to show that it can be lifted to analogous constructions for *quantum states.* Here, rather than probability distributions on finite sets, one is looking at *mixed states on finite-dimensional Hilbert spaces.* Let $\mathcal{H}^n$ denote an n-dimensional complex Hilbert space. A *quantum state* is a density operator $\rho : \mathcal{H}^n \to \mathcal{H}^n$, i.e., a self-adjoint, positive, linear operator with $\mathrm{tr}(\rho) = 1$. The quantum states on $\mathcal{H}^n$ are denoted Ω^n. A quantum state ρ on $\mathcal{H}^n$ is *pure* if

$$\mathrm{spec}(\rho) \subseteq \{0, 1\}.$$

The set of pure states is denoted Σ^n. They are in bijective correspondence with the one-dimensional subspaces of $\mathcal{H}^n$. Classical states are distributions on the set of pure states $\max(\Delta^n)$. By Gleason's theorem [Gleason, 1957], an analogous result holds for quantum states: Density operators encode distributions on Σ^n.[11]

If our knowledge about the state of a system is represented by density operator ρ, then quantum mechanics predicts the probability that a measurement of observable e yields the value $\lambda \in \mathrm{spec}(e)$. It is

$$\mathrm{pr}(\rho \to e_\lambda) := \mathrm{tr}(p_e^\lambda \cdot \rho),$$

where p_e^λ is the projection corresponding to eigenvalue λ and e_λ is its associated eigenspace in the *spectral representation* of e.

Let e be an observable on $\mathcal{H}^n$ with $\mathrm{spec}(e) = \{1, \ldots, n\}$. For a quantum state ρ in Ω^n,

$$\mathrm{spec}(\rho|e) := (\mathrm{pr}(\rho \to e_1), \ldots, \mathrm{pr}(\rho \to e_n)) \in \Delta^n.$$

So what does it mean to say that we have more information about the system when we have $\sigma \in \Omega^n$ than when we have $\rho \in \Omega^n$? It means that there is an observable e such that (a) the meaurement of e serves as a physical realization of the knowledge each state imparts to us, and (b) we have a better chance of

[11] Of course, Gleason's theorem also applies to separable infinite-dimensional spaces.

predicting the result of the measurement of e in state σ than we do in state ρ. Formally, (a) means that $\text{spec}(\rho) = \text{Im}(\text{spec}(\rho|e))$ and $\text{spec}(\sigma) = \text{Im}(\text{spec}(\sigma|e))$ (where the image Im simply converts a list to the underlying set), which is equivalent to requiring $[\rho, e] = 0$ and $[\sigma, e] = 0$, where $[a, b] = ab - ba$ is the commutator of operators.

DEFINITION 5. *Let $n \geq 2$. For quantum states $\rho, \sigma \in \Omega^n$, we have $\rho \sqsubseteq \sigma$ iff there is an observable $e : \mathcal{H}^n \to \mathcal{H}^n$ such that $[\rho, e] = [\sigma, e] = 0$ and $\text{spec}(\rho|e) \sqsubseteq \text{spec}(\sigma|e)$ in Δ^n.*

Taking this definition together with our reading of the Bayesian order on classical states, we capture the idea of being able to predict the result of an experiment more confidently on σ than on ρ in terms of the less rapid falling off of the values of $\text{spec}(\rho|e)_i$ than of $\text{spec}(\sigma|e)_i$.

THEOREM 6. *$(\Omega^n, \sqsubseteq)$ is a domain with maximal elements*

$$\max(\Omega^n) = \Sigma^n$$

and least element $\bot = I/n$, where I is the identity matrix. Moreover, von Neumann entropy

$$S(\rho) = -\text{tr}(\rho \log \rho)$$

is a measurement of type $\Omega^n \to [0, \infty)^*$.

This order can be characterized in a similar fashion to the Bayesian order on Δ^n, in terms of symmetries and projections. In its symmetric formulation, *unitary operators* on $\mathcal{H}^n$ take the place of permutations on $\{1, \ldots, n\}$, while the projective formulation of $(\Omega^n, \sqsubseteq)$ shows that each classical projection $p_i : \Delta^{n+1} \rightharpoonup \Delta^n$ is actually the restriction of a special quantum projection $\Omega^{n+1} \rightharpoonup \Omega^n$.

4.5 The Logics of Birkhoff and von Neumann

Quantum Logic in the sense of Birkhoff and von Neumann [Birkhoff and von Neumann, 1936] consists of the propositions one can make about a physical system. Each proposition takes the form "The value of observable e is contained in $E \subseteq \text{spec}(e)$." For classical systems, the logic is $\mathcal{P}\{1, \ldots, n\}$, while for quantum systems it is $\mathbb{L}^n$, the lattice of (closed) subspaces of $\mathcal{H}^n$. In each case, implication of propositions is captured by inclusion, and a fundamental distinction between classical and quantum — that there are pairs of quantum observables whose exact values cannot be simultaneously measured at a single moment in time — finds lattice theoretic expression: $\mathcal{P}\{1, \ldots, n\}$ is distributive; $\mathbb{L}^n$ is not.

The classical and quantum logics can be *derived* from the Bayesian and spectral orders using the *same* order theoretic construction.

DEFINITION 7. *An element x of a dcpo D is* irreducible *when*

$$\bigwedge(\uparrow x \cap \max(D)) = x$$

The set of irreducible elements in D is written Ir(D).

The order dual of a poset $(D, \sqsubseteq_D)$ is written D^*; its order is $x \sqsubseteq y \Leftrightarrow y \sqsubseteq_D x$. The following result is proved in [Coecke, 2003].

THEOREM 8. *For $n \geq 2$, the classical lattices arise as*

$$\mathrm{Ir}(\Delta^n)^* \simeq \mathcal{P}\{1,\ldots,n\} \setminus \{\emptyset\},$$

and the quantum lattices arise as

$$\mathrm{Ir}(\Omega^n)^* \simeq \mathbb{L}^n \setminus \{0\}.$$

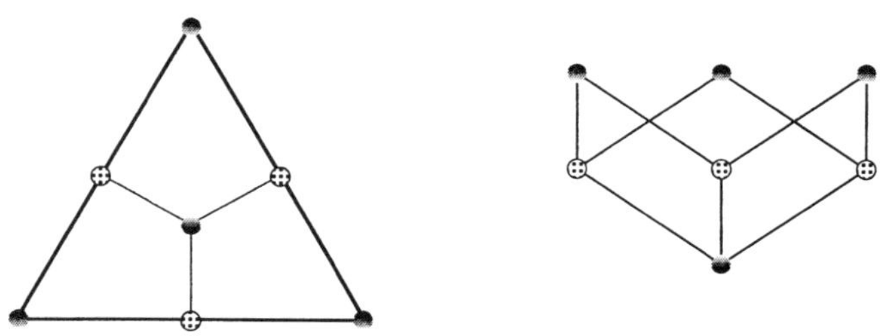

Figure 2. The irreducibles of Δ^3 with the corresponding Hasse diagram.

4.6 Discussion

The foregoing development has been quite technical, but the underlying programme which these ideas illustrate has a clear conceptual interest. The broad agenda of developing a unified quantitative/qualitative theory of information, applicable to a wide range of situations in logic and computation, is highly attractive, and likely to lead to new perspectives on information in general.

Our discussion thus far has largely been couched in terms of *static* theories, although we have already hinted at the importance of agents and explicit dynamics. We now turn to *interactive models* of logic and computation.

5 GAMES, LOGICAL EQUILIBRIA AND CONSERVATION OF INFORMATION FLOW

In this Section and the next, we shall discuss some dynamical theories of computation which are explicitly based on *interaction* between agents, and which expose a structure of information flow which is both *geometrical* and *logical* in character. These theories, which go under the names of *Game Semantics* and *Geometry of Interaction*, have played a considerable rôle in recent work on the semantics both of programming languages, and of logical proofs.

Figure 3. Computing "in the isolation ward".

5.1 Changing Views of Computation

To set the scene, we begin by recalling how perspectives on computation have changed since the first computers appeared. The early practice of computing can be pictured as in Figure 3. This is the era of stand-alone machines and programs: computers are served by an elite priesthood, and have only a narrow input-output interface with the rest of the world.

First-generation models of computation Given this limited vision of computing, there is a very natural abstraction of computation, in which programs are seen as computing *functions* or *relations* from inputs to outputs.[12]

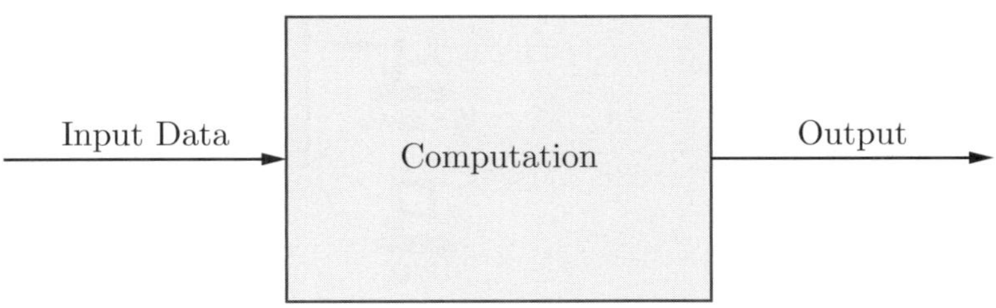

These models live on the existing intellectual inheritance from discrete mathematics and logic. *Time* and *processes* lurk in the background, but are largely suppressed.

Computation in the Age of the Internet As we know, the technology has changed dramatically. Even a conventional Distributed Systems picture, as illustrated in Figure 4, which has been common-place for the last 20 years, tells a very different story. We have witnessed the progression

[12]This is the exactly the point of view on which, as we have seen, program logics such as Hoare Logic and Dynamic Logic are based.

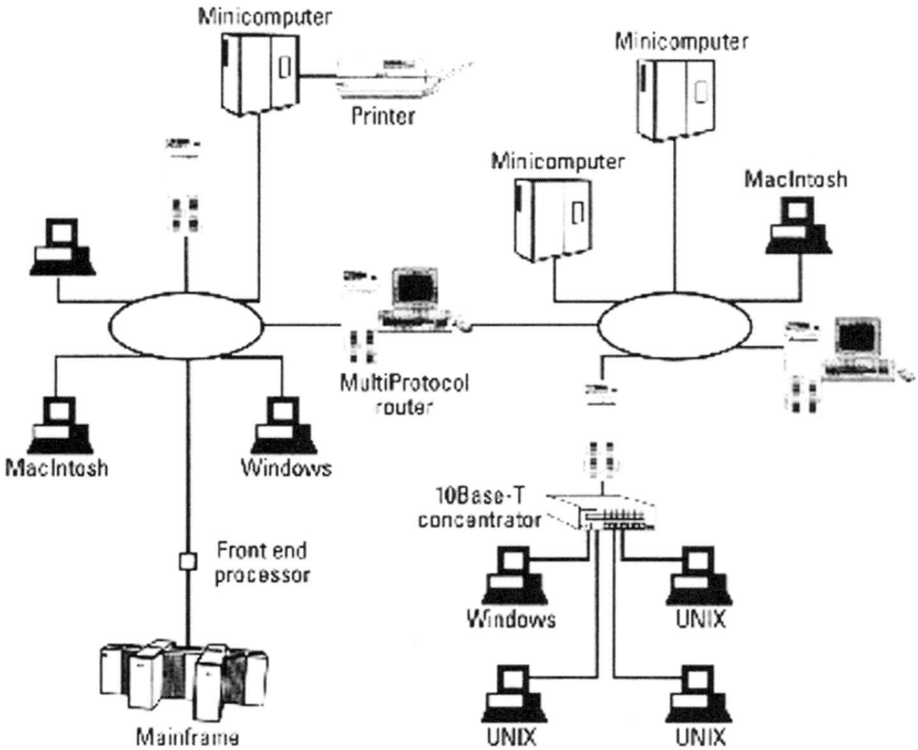

Figure 4. Distributed Computing

multitasking → distributed systems → Internet → "mobile" and "global" computing

Key features of this unfolding new computational universe include: *agents interacting* with each other, and *information flowing* around the system.

The insufficiency of the first-generation models of computation for this new computational environment is evident. The old concepts fail to match the modern world of computing and its concerns:

Robustness in the presence of failures.

Atomicity of transactions.

Security of information flows.

Quality of user interface.

Quantitative aspects.

Processes vs. Products We see a shift in emphasis and importance between *How* we compute *vs.* *What* we compute. Processes were in the background, but now come to the fore: the "how" *becomes* the new "what".

This leads ineluctably to the need for **Second-generation models** of computation, and in particular *Process Models* such as Petri nets, Process Algebra, etc. Whereas 1st-generation models lived off the intellectual inheritance from mathematics and logic, there is no adequate pre-existing theory of *processes* or *agents*, *interaction*, and *information flow*, as we see by considering the following questions (which have already been mentioned in Section 1):

- *What* is computed?
- What *is* a process?
- What are the analogues to Turing-completeness, universality?

There are indeed a plethora of models, but no definitive conceptual analysis, comparable to Turing's analysis of computation in its "classical" sense: not least, perhaps, because it is indeed *a harder problem*!

5.2 Some New Perspectives

Instead of isolated systems, with rudimentary interactions with their environment, the standard unit of description or design becomes a *process* or *agent*, the essence of whose behaviour is *how it interacts* with its environment.

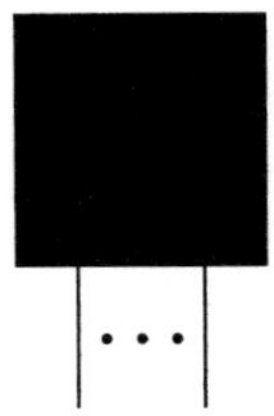

Who is the System? Who is the Environment? This depends on point of view. We may designate some agent or group of agents as the System currently under consideration, with everything else as the Environment; but it is always possible to contemplate a rôle interchange, in which the Environment becomes the System and vice versa. (This is, of course, one of the great devices, and imaginative functions, of creative literature). This *symmetry* between System and Environment carries a first clue that there is some structure here; it will lead us to a key *duality*, and a deep connection to logic.

5.3 Interaction

Complex behaviour arises as the global effect of a *system* of *interacting agents* (or processes).

The key building block is the agent. The key operation is *interaction* – plugging agents together so that they interact with each other

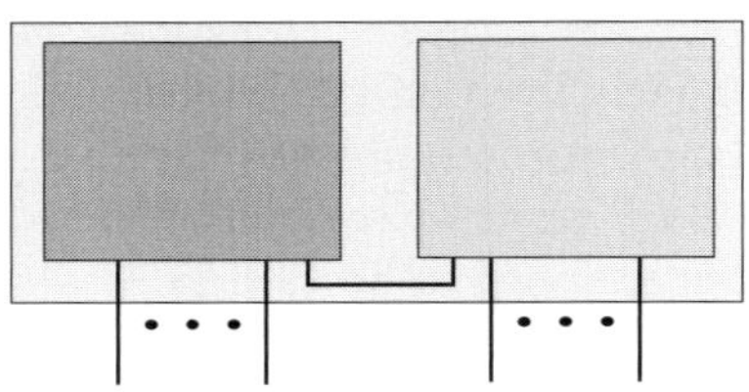

This conceptual model works at all "scales" :

- Macro-scale: processes in operating systems, software agents on the Internet, transactions.
- Micro-scale: how programs are implemented (subroutine call-return protocols, register transfer) all the way down into hardware.

It is applicable both to *design* (synthesis) and to *description* (analysis); to *artificial* and to *natural* information-processing systems.

There are of course large issues lurking here, e.g. in the realm of "Complex Systems": *emergent behaviour* and even *intelligence*. Is is helpful, or even feasible, to understand this complexity *compositionally*? We need new conceptual tools, new theories, to help us analyze and synthesize these systems, to help us to *understand* and to *build*.

5.4 *Towards a "Logic of Interaction"*

Specifying and reasoning about the behaviour of computer programs takes us into the realm of logic. For the first-generation models, logic could be taken "as it was"—static and timeless. For our second-generation models, getting an adequate account—a genuine "logic of interaction"—may require a fundamental reconceptualization of logic itself. This radical revision of our view of logic is happening anyway—prompted partly by the applications, and partly by ideas arising within logic.

The Static Conception of Logic

We provide an unfair caricature of the standard logical idea of tautology to make our point. The usual "static" notion of tautology is as "a statement which is vacuously true because it is compatible with all states of affairs".

$$A \vee \neg A$$

"It is raining *or* it is not raining"—truth-functional semantics. This is illustrated (subversively) in Figure 5.

Figure 5. Tertium non datur?

But what could a *dynamic notion of tautology* look like?

The Copy-Cat Strategy

We begin with a little fable, illustrated by Figure 6:

How to beat an International Chess Grandmaster by the power of pure logic

Since we are relying on logic, rather than on any talent at Chess, we proceed as follows. We arrange to play two games of Chess with the grandmaster, say Gary Kasparov, once as White and once as Black. Moreover, we so arrange matters that we start with the game in which we play as Black. Kasparov makes his opening move; we respond by playing the *same* move in the *other* game—this makes sense,

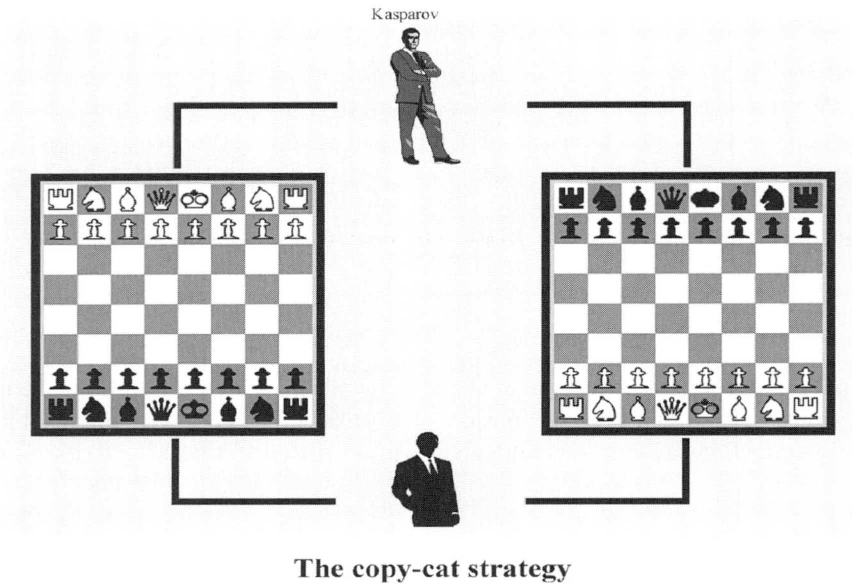

Figure 6. How to beat a Grandmaster

since we are playing as White there. Now Kasparov responds (as Black) to our move in that game; and we copy that response back in the first game. We simply proceed in this fashion, copying the moves that our opponent makes in one board to the other board. The net effect is that *we play the same game twice—once as White, and once as Black.* (We have essentially made Kasparov play against himself). Thus, whoever wins that game, we can claim a win in one of our games against Kasparov! (Even if the game results in a stalemate, we have done as well as Kasparov over the two games—surely still a good result!)[13]

Of course, this idea has nothing particularly to do with Chess. It can be applied to any two-person game of a very general form. We shall continue to use Chess-boards to illustrate our discussion, but this underlying generality should be kept in mind.

What are the salient features which can be extracted from this example?

A dynamic tautology There is a sense (which will shortly be made more precise) in which the copy-cat strategy can be seen as a *dynamic version* of the

[13]Our fable is actually recorded as having happened at least once in the chronicles of Chess. Two players conspired to play this copy-cat strategy against Alekhine in the 1920's. Alekhine realized what was happening, and made a tempting offer of a sacrifice to one of his opponents. That opponent was not able to resist such a coup against the great Alekhine, and departed from the copy-cat strategy to swallow the bait. Then the symmetry was broken, and Alekhine was able to win easily in both games. Thus we are reminded of the familiar truth, that logic rarely prevails over psychology in "real life".

tautology $A \vee \neg A$. Note, indeed, that an essential condition for being able to play the copy-cat is that the rôles of the two players are inter-changed on one board as compared to the other. Note also the disjunctive quality of the argument that we must win in one or other of the two games. But the copy-cat strategy is a *dynamic process*: a two-way channel which maintains the correlation between the plays in the two games.

Conservation of information flow The copy-cat strategy does not *create* any information; it reacts to the environment in such a way that information is conserved. It ensures that exactly the same information flows out to the environment as flows in from it. Thus one gets a sense of logic appearing in the form of *conservation laws for information dynamics.*

The power of copying Another theme which appears here, and which we will see more of later, concerns the surprising power of simple processes of copying information from one place to another. Indeed, as we shall eventually see, such processes are *computationally universal.*

The geometry of information flow From a dynamical point of view, the copy-cat strategy realizes a channel between the two game boards, by performing the *actions* of copying moves. But there is also some implicit *geometry* here. Indeed, the very idea of two boards laid out side by side appeals to some basic underlying spatial structure. In these terms, the copy-cat channel can also be understood geometrically, as creating a graphical link between these two spatial locations.These two points of view are complementary, and link the logical perspective to powerful ideas arising in modern geometry and mathematical physics.

To provide further evidence that the copy-cat strategy embodies more substantial ideas than might at first be apparent, we consider varying the scenario. Consider now the case where we play against Kasparov on *three boards*; one as Black, two as White.

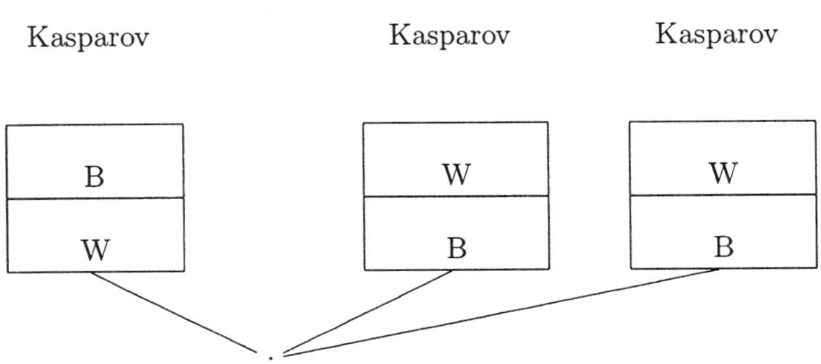

Does the Copy-Cat strategy still work here? In fact, we can easily see that it does *not.* Suppose Kasparov makes an opening move m_1 in the left-hand board where he plays as White; we copy it to the board where we play as White; he responds with m_2; and we copy m_2 back to the board where Kasparov opened. So far, all has proceeded as in our original scenario. But now Kasparov has the option of playing a *different* opening move, m_3 say, in the rightmost board. We have no idea how to respond to this move; nor can we copy it anywhere, since the board where we play as White is already "in use". This shows that these simple ideas already lead us naturally to the setting of a *resource-sensitive* logic, in which in particular the Contraction Rule, which can be expressed as $A \rightarrow A \wedge A$ (or equivalently as $\neg A \vee (A \wedge A)$) cannot be assumed to be valid.

What about the other obvious variation, where we play on two boards as White, and one as Black?

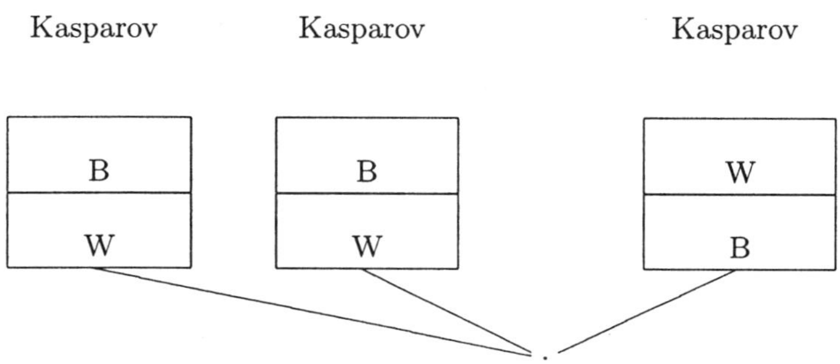

It seems that the copy-cat strategy *does* still work here, since we can simply ignore one of the boards where we play as White. However, a geometrical property of the original copy-cat strategy has been lost, namely a *connectedness* property, that information flows to every part of the system. This at least calls the corresponding logical principle of Weakening, which can be expressed as $A \wedge A \rightarrow A$, (or equivalently as $\neg A \vee \neg A \vee A$) into question.

We see from these remarks that we are close to the realm of Linear Logic and its variants; and, mathematically, to the world of monoidal (rather than cartesian) categories.

Game Semantics

These ideas find formal expression in *Game Semantics*. Games play the role of:

- Interface types for computation modules
- Propositions with dynamic content.

In particular, 2-person games capture the duality of:

- Player *vs.* Opponent
- System *vs.* Environment.

Agents are strategies In this setting, we model our agents or processes as *strategies* for playing the game. These strategies *interact* by playing against each other. We obtain a notion of correctness which is *logical* in character in terms of the idea of *winning* strategy—one which is guaranteed to reach a successful outcome however the environment behaves. This in a sense replaces (or better, *refines*) the logical notion of "truth": winning strategies are our dynamic version of tautologies (more accurately, of *proofs*).

Building complex systems by combining games We shall now see how games can be combined to produce more complex behaviours while retaining control over the interface. This provides a basis for the *compositional* understanding of our systems of interacting agents—understanding the behaviour of a complex system in terms of the behaviour of its parts. This is crucial for both analysis and synthesis, *i.e.* for both description and design. These operations for building games can be seen as (dynamic forms of) "type constructors" or "logical connectives". (The underlying logic here will in fact be Linear Logic).

Duality—"Linear Negation" $A^{\perp}$ — interchange rôles of Player and Opponent (reflecting the symmetry of interaction).

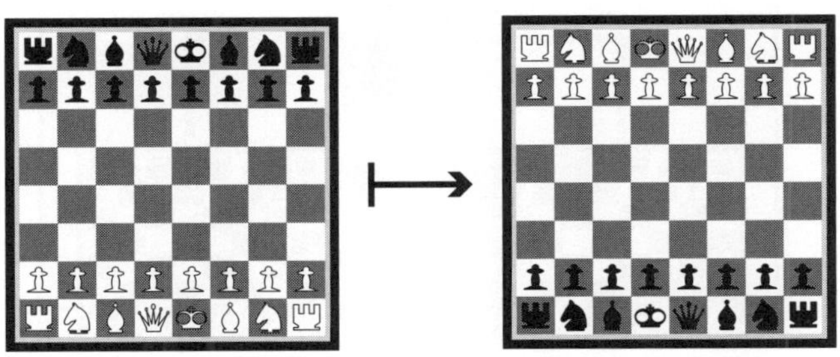

Note that, with this interpretation, negation is involutive:

$$A^{\perp\perp} = A.$$

Tensor — "Linear conjunction"

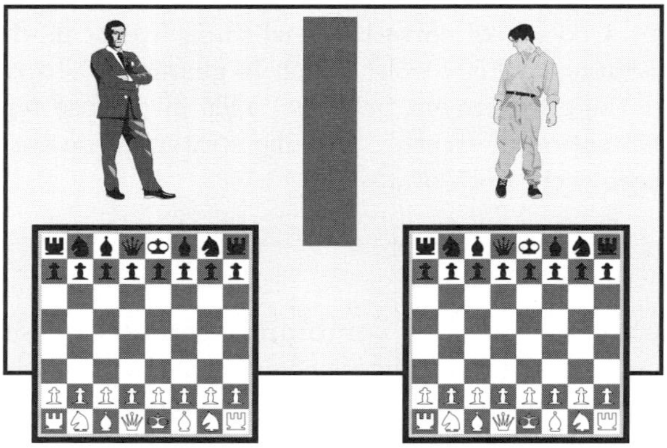

$A \otimes B$

The idea here is that we combine the two game boards into one system, *without any information flow between the two sub-systems.* (This is the significance of the "wall" separating our two players, who we shall refer to as Gary (Kasparov) and Nigel (Short)). This connective has a conjunctive quality, since we must independently be able to play (and to win) in each conjunct. Note however, that there is no constraint on information flow for the environment, as it plays against this compound system.

Par — "Linear disjunction"

$A \parr B$

In this case, we have two boards, but one player (who we refer to as the Copy-Cat), indicating that we *do* allow information flow for this player between the two game boards. This for example allows information revealed in one game board by the Opponent to be used against him on the other game board—as exemplified by the copy-cat strategy. However, note that the wall appears on the environment's side now, indicating that the environment is constrained to play separately on the two boards, with no communication between them.

Thus we have a De Morgan duality between these two connectives, mediated by the Linear negation:

$$\begin{aligned}(A \otimes B)^{\perp} &= A^{\perp} \parr B^{\perp} \\ (A \parr B)^{\perp} &= A^{\perp} \otimes B^{\perp}\end{aligned}$$

The idea is that on one side of the mirror of duality (Player/System for the Tensor, Opponent/Environment for the Par), we have the constraint of no information flow, while on the other side, we do have information flow.

We can now reconstrue the Copy-Cat strategy in logical terms:

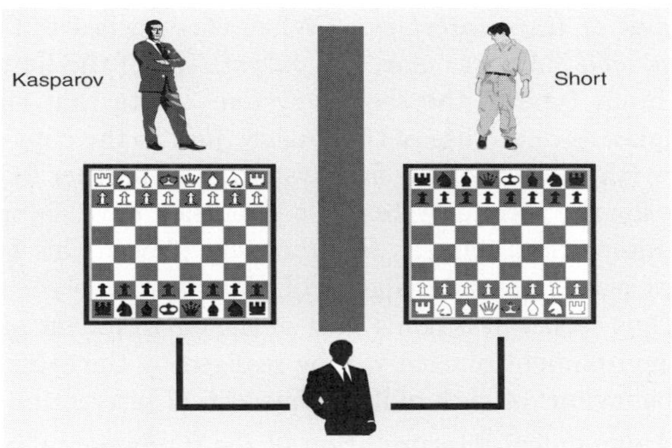

We see that it is indeed a winning strategy for $A^{\perp} \parr A$. Moreover, we can define $A \multimap B$ ("Linear implication") by

$$A \multimap B \equiv A^{\perp} \parr B$$

(*cf.* $A \supset B \equiv \neg A \vee B$.) Then the copy-cat strategy becomes the canonical proof of the most basic tautology of all: $A \multimap A$.

The information flow possibilities of Par receive a more familiar logical interpretation in terms of the Linear implication; namely, that we can use information about the antecedent in proving the consequent (and conversely with respect to their negations, if we think of proof by contraposition).

Thus an entire "linearized" logical structure opens up before us, with a natural interpretation in terms of the dynamics of information flow.

Interaction

We now turn to a key step in the development: the modelling of *interaction* itself. Constructors create "potentials" for interaction; the operation of plugging modules together so that they can communicate with each other *releases* this potential into *actual computation.*

$A^{\perp} \parr B \; \equiv \; A \multimap B$

$B^{\perp} \parr C \; \equiv \; B \multimap C$

Here we see two separate sub-systems, each with a compound structure, expressed by the *logical types of their interfaces.* What these types tell us is that these systems are *composable*; in particular, the *output type* of the first system, namely B, matches the input type of the second system. Note that this "logical plug-compatibility" makes essential use of the duality, just as the copy-cat strategy did. What makes Gary (the player for the first system) a fit partner for interaction with Nigel (the player for the second system), is that they have *complementary views* of their locus of interaction, namely B. Gary will play in this type "positively", as Player (he sees it as B), while Nigel will play "negatively", as Opponent (he sees it as $B^{\perp}$). Thus each will become part of the environment of the other—part of the potential environment of each will be realized by the other, and hence part of the *potential* behaviour of each will become *actual* interaction.

This leads to a dynamical interpretation of the fundamental operation of *composition*, in mathematical terms:

$$\frac{A \xrightarrow{\text{Gary}} B \xrightarrow{\text{Nigel}} C}{A \xrightarrow{\text{Gary;Nigel}} C}$$

or of the *Cut rule*, in logical terms:

$$\text{Cut:} \quad \frac{\vdash \Gamma, A \quad \vdash A^{\perp}, \Delta}{\Gamma, \Delta}$$

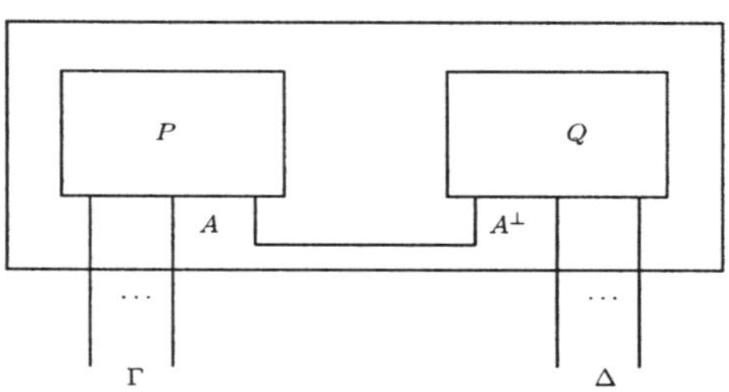

Composition as Interaction

The Interaction Game

The picture here shows the new system formed by plugging together the two subsystems. The "external interface" to the environment now shows just the left hand board A as input, and the right hand board C as output. The Cut formula B is hidden from the environment, and becomes the locus of interaction inside the black box of the system. Suppose that the Environment makes some move m in C. This is visible only to Nigel, who as a strategy for $B \multimap C$ has a response. Suppose this response m_1 is in B. This is a move by Nigel as Player in $B^{\perp}$, hence appears to Gary as a move by Opponent in B. Gary as a strategy for $A \multimap B$ has a response m_2 to this move. If this response is again in B, Nigel sees it as a response by the environment to his move, and will have a response again; and so on. We thus have a sequence of moves $m_1, \ldots, m_k$ in B, ping-ponging back and forth between Nigel and Gary. If, eventually, Nigel responds to Gary's last move by playing in C, or Gary responds to Nigel's last move by playing in A, then we have the response of the *composed strategy* Gary; Nigel to the original move m. Indeed, all that is visible to the Environment is that it played m, and eventually some response appeared, in A or C.

Moreover, if both Nigel and Gary are winning strategies, then so is the composed strategy; and the composed strategy will not get stuck forever in the internal ping-pong in B. To see this, suppose for a contradiction that it did in fact get stuck in B. Then we would have an infinite play in B following the winning strategy Gary for *Player* in B, and the *same* infinite play following the winning strategy Nigel for Player in $B^{\perp}$, hence for *Opponent* in B. Hence the same play would count as a win for both Player and Opponent. This yields the desired contradiction.

5.5 Discussion

Game Semantics in the sense discussed in this section has had an extensive development over the past decade and a half, with a wealth of applications to the semantics of programming languages, type theories and logics [Abramsky and Jagadeesan, 1994b; Abramsky *et al.*, 2000; Abramsky and McCusker, 1997; Abramsky and McCusker, 1999a; Abramksy and McCusker, 1998; Abramsky and McCusker, 1999b; Abramsky and Mellies, 1999; Hyland and Ong, 2000]. More recently, there has been an algorithmic turn, and some striking applications to verification and program analysis [Ghica and McCusker, 2000; Abramsky, 2002; Abramsky *et al.*, 2004a; Murawski *et al.*, 2005].

From the point of view of the general analysis of Information, we see the following promising lines of development:

- Game semantics provides a promising arena for exploring the combination of quantitative and qualitative theories of information, as discussed in Section 4, but now in a dynamic setting. In particular, it provides a setting for quantifying information flow between agents. We would like to ask quantitative questions about *rate of information flow* through a strategy (representing a program, or a proof); how can a system gain *maximum* information from its environment while providing *minimal* information in return; robustness in the presence of *noise*, etc.

- As we saw in our discussion of the copy-cat strategy, there is an intuition of logical principles arising as *conservation laws for information flow.* (And indeed, in the case of Multiplicative Linear Logic, the proofs correspond exactly to "generalized copy-cat strategies"). Can we develop this intuition into a fully-fledged theory? Can we *characterize* logical principles as those expressing the conservation principles of this information flow dynamics?

- There is also the hope that the more structured setting of game semantics will usefully constrain the exuberant variety of possibilities offered by process algebra, and allow a sharper exploration of the logical space of possibilities for information dynamics[14]. This has already been borne out in part, by the success of game semantics in exploring the space of programming language semantics. It has been possible to give crisp characterizations of the "shapes" of computations carried out within certain *programming disciplines*: including purely functional programming [Abramsky *et al.*, 2000; Hyland and Ong, 2000], stateful programming [Abramsky and McCusker, 1997; Abramsky and McCusker, 1999a], general references [Abramsky *et*

[14]It should be said that the exuberant variety of process algebras has been directly motivated by the vast range of real-word informatic processes which we need to model. The whole area of information dynamics is in a dynamic tension between the need on the one hand for descriptive adequacy, and on the other for mathematical structure and tractability [Milner, 2006a; Milner, 2006b]. Process algebra, game semantics, and other approaches are making valuable inroads into this territory. We need to combine the strengths of all these ideas!

al., 1998], programming with non-local jumps and exceptions [Laird, 1997a; Laird, 1997b], non-determinism [Harmer and McCusker, 1999], probability [Danos, 2002], concurrency [Ghica andMurawski, 2004; Ghica and Murawski, 2006], names [Abramsky *et al.*, 2004b], polymorphism [Hughes, 2000; Abramsky and Jagadeesan, 2005] and more. See [Abramsky and McCusker, 1999b] for an overview (now rather out of date).

There has also been a parallel line of development of giving *full completeness* results for a range of logics and type theories, characterizing the "space of proofs" for a logic in terms of informatic or geometric constraints which pick out those processes which are proofs for that logic [Abramsky and Jagadeesan, 1994b; Abramsky and Mellies, 1999; Loader, 1994; Blute, 1998; Devarajan *et al.*, 1999; Blute *et al.*, 2005]. This allows a new look at such issues as the boundaries between classical and constructive logic, or the fine structure of polymorphism and second-order quantification.

- This also gives some grounds for optimism that we can capture—in a "machine-independent", and moreover "geometrical", non-inductive way—what *computational processes* are, *without* referring back to Turing machines or any other explicit machine model.

- In the same spirit as for computability, can we characterize *polynomial-time computation* and other complexity classes in such terms?

6 EMERGENT LOGIC: THE GEOMETRY OF INFORMATION FLOW

Game Semantics carries many vivid intuitions arising from our experiences of game-playing as a human activity. We were able to take advantage of this in the previous section to explain some key ideas without resorting to any explicit formalization. We now turn to a related but somewhat different development of interactive models for logic and computation, known loosely as "Geometry of Interaction particle-style models".[15] We will use this setting to carry forward our discussion of dynamic models for information flow, with particular emphasis on the following themes:

- Firstly, the model or family of models we shall discuss is technically simpler to formalize mathematically than Game Semantics, although also less cloaked in familiar intuitions. Thus we can introduce some more precision into our discussion without unduly taxing the reader.

- Secondly, the simple yet expressive nature of these models is itself of conceptual interest. They show how logic and computation can be understood

[15]See [Girard, 1989; Girard, 1990; Girard, 1995; Malacaria and Reginer, 1991; Danos and Reginer, 1993; Danos and Reginer, 1996], and [Abramsky and Jagadeesan, 1994a; Abramsky and Jagadeesan, 1994b; Abramsky, 1996; Abramsky, 1997; Abramsky and Lenisa, 2005; Abramsky *et al.*, 2002].

in terms of simple processes of copying information from one "place" to another, generalizing what we have already seen of the copy-cat strategy. In fact, we shall see that *mere copying is computationally universal.* Moreover, models of logics and type theories arise from these models; because of the simplicity of the models, we may reasonably speak of *emergent logic*—where, as discussed in the previous section, we may think of the logical character of certain principles as arising from the fact that they express conservation laws of information flow.

- We will also be able to make visible how geometrical structure unfolds in these models, in a striking and unexpected fashion. This part of the development can be carried much further than we can describe here; there is a thread of ideas linking logical processes of cut-elimination to diagram algebras, knot theory and topological quantum field theory [Abramsky, 2007].
- We shall also begin to see the beginnings of links between *Logic* and *Physics.* The processes we shall describe will be *reversible* in a very strong sense. This link can in fact be carried much further, and the same kind of structures we are discussing here can be used to axiomatize Quantum Mechanics, and to give an incisive analysis of quantum entanglement and information flow [Abramsky and Coecke, 2002; Abramsky and Coecke, 2004; Abramsky and Coecke, 2005; Abramsky, 2007].

6.1 Background: Combinatory Logic

It will be convenient to work in the setting of Combinatory Logic [Curry and Feys, 1958; Hindley and Seldin, 1986], which provides one of the simplest of all the formulations of computability—and moreover one which is purely algebraic. Combinatory Logic is also the basis for realizability constructions, which provide powerful methods for building extensional models of strong impredicative type theories and higher-order logics.

We recall that combinatory logic is the algebraic theory **CL** given by the signature with one binary operation (application) written as an infix $\_ \cdot \_$, and two constants **S** and **K**, subject to the equations

$$\begin{array}{lcl} \mathbf{K} \cdot x \cdot y & = & x \\ \mathbf{S} \cdot x \cdot y \cdot z & = & x \cdot z \cdot (y \cdot z) \end{array}$$

(application associates to the left, so $x \cdot y \cdot z = (x \cdot y) \cdot z$). Note that we can define $\mathbf{I} \equiv \mathbf{S} \cdot \mathbf{K} \cdot \mathbf{K}$, and verify that $\mathbf{I} \cdot x = x$.

The key fact about the combinators is that they are *functionally complete*, i.e. they can simulate the effect of λ-abstraction. Specifically, we can define bracket abstraction on combinatory terms built using a set of variables X:

$$\begin{array}{lcl} \lambda^* x.\, M & = & \mathbf{K} \cdot M \quad (x \notin \mathsf{FV}(M)) \\ \lambda^* x.\, x & = & \mathbf{I} \\ \lambda^* x.\, M \cdot N & = & \mathbf{S} \cdot (\lambda^* x.\, M) \cdot (\lambda^* x.\, N) \end{array}$$

Moreover (Theorem 2.15 in [Hindley and Seldin, 1986]):

$$\mathbf{CL} \vdash (\lambda^* x. M) \cdot N = M[N/x].$$

The $\mathbf{B}$ combinator can be defined by bracket abstraction from its defining equation:

$$\mathbf{B} \cdot x \cdot y \cdot z = x \cdot (y \cdot z).$$

The combinatory *Church numerals* are then defined by

$$\bar{n} \equiv (\mathbf{S} \cdot \mathbf{B})^n \cdot (\mathbf{K} \cdot \mathbf{I})$$

where we define

$$a^n \cdot b = a \cdot (a \cdots (a \cdot b) \cdots).$$

A partial function $\phi : \mathbb{N} \rightharpoonup \mathbb{N}$ is *numeralwise represented* by a combinatory term M if for all $n \in \mathbb{N}$, if $\phi(n)$ is defined and equal to m, then

$$\mathbf{CL} \vdash M \cdot \bar{n} = \bar{m}$$

and if $\phi(n)$ is undefined, then $M \cdot \bar{n}$ has no normal form.
The basic result on computational universality of $\mathbf{CL}$ is then the following (Theorem 4.18 in [Hindley and Seldin, 1986]):

THEOREM 9. *The partial functions numeralwise representable in* **CL** *are exactly the partial recursive functions.*

Principal Types of Combinators The functional behaviour of combinatory terms can be described using *types*. The type expression $T \to U$ denotes the set of terms which, when applied to an argument of type T, produce a result of type U. By convention, $\to$ associates to the right, so we write $T_1 \to T_2 \to \cdots T_k \to U$ as short-hand for $T_1 \to (T_2 \to \cdots (T_k \to U) \cdots)$.

Now consider the combinator $\mathbf{K}$. The equation $\mathbf{K} \cdot x \cdot y = x$ tells us that this combinator expects to receive an argument x, say of type α, then an argument y, say of type β, and then returns a result, namely x, of type α. Thus its type has the form

$$\mathbf{K} : \alpha \to (\beta \to \alpha).$$

In fact, if we take α and β to be type variables, this is the *principal*, *i.e.* the most general, type of this combinator. A similar but more complicated argument establishes that the principal type of the $\mathbf{S}$ combinator is

$$\mathbf{S} : (\alpha \to \beta \to \gamma) \to (\alpha \to \beta) \to (\alpha \to \gamma).$$

These principal types can in fact be computed by the Hindley-Milner algorithm [Hindley, 1997] from the defining equations for the combinators. (This algorithm is nowadays routinely used to perform "type-checking" for modern programming languages with polymorphic types.)

Curry observed [Curry and Feys, 1958] that the principal types of the combinators correspond to *axiom schemes* for a Hilbert-style proof system for Intuitionistic implicational logic—with the application operation corresponding to *Modus Ponens*. This is the "Curry" part of the Curry-Howard isomorphism. Thus combinators are to Hilbert-style systems as λ-calculus is to Natural Deduction.

The Curry Combinators Curry's original set of combinators was not the Schönfinkel combinators **S** and **K**, but rather the combinators **B**, **C**, **K**, and **W**:

$$\begin{array}{lcl} \mathbf{B} \cdot x \cdot y \cdot z & = & x \cdot (y \cdot z) \\ \mathbf{C} \cdot x \cdot y \cdot z & = & x \cdot z \cdot y \\ \mathbf{W} \cdot x \cdot y & = & x \cdot y \cdot y \end{array}$$

These combinators are equivalent to the Schönfinkel combinators, in the sense that the two sets are inter-definable [Barendregt, 1984; Hindley and Seldin, 1986]. In particular, **S** can be defined from **B**, **C**, **I** and **W**. They have the following principal types:

$$\begin{array}{lcll} \mathbf{I} & : & \alpha \to \alpha & \text{Axiom} \\ \mathbf{B} & : & (\beta \to \gamma) \to (\alpha \to \beta) \to \alpha \to \gamma & \text{Cut} \\ \mathbf{C} & : & (\alpha \to \beta \to \gamma) \to \beta \to \alpha \to \gamma & \text{Exchange} \\ \mathbf{K} & : & \alpha \to \beta \to \alpha & \text{Weakening} \\ \mathbf{W} & : & (\alpha \to \alpha \to \beta) \to \alpha \to \beta & \text{Contraction} \end{array}$$

Thus we see that in logical terms, **B** expresses the transitivity of implication, or the Cut rule; **C** is the Exchange rule; **W** is Contraction; and **K** is Weakening. Curry's analysis of *substitution* is close to Gentzen's analysis of *proofs*.

6.2 Linear Combinatory Logic

We shall now present another system of combinatory logic: *Linear Combinatory Logic* [Abramsky, 1997; Abramsky *et al.*, 2002; Abramsky and Lenisa, 2005]. This can be seen as a finer-grained system into which standard combinatory logic, as presented in the previous section, can be interpreted. By exposing some finer structure, Linear Combinatory Logic offers a more accessible and insightful path towards our goal of mapping universal functional computation into a simple model of computation as copying.

Linear Combinatory Logic can be seen as the combinatory analogue of Linear Logic [Girard, 1987]; the interpretation of standard Combinatory Logic into Linear Combinatory Logic corresponds to the interpretation of Intuitionistic Logic into Linear Logic. Note, however, that the combinatory systems we are considering are type-free and "logic-free" (*i.e.* purely equational).

DEFINITION 10. A *Linear Combinatory Algebra* $(A, \cdot, !)$ consists of the following data:

- An applicative structure $(A, \cdot)$

- A unary operator $! : A \to A$
- Distinguished elements **B**, **C**, **I**, **K**, **D**, δ, **F**, **W** of A

satisfying the following identities (we associate $\cdot$ to the left and write $x \cdot !y$ for $x \cdot (\,!(y))$, etc.) for all variables x, y, z ranging over A):

1.	$\mathbf{B} \cdot x \cdot y \cdot z$	$=$	$x \cdot (y \cdot z)$	Composition/Cut
2.	$\mathbf{C} \cdot x \cdot y \cdot z$	$=$	$(x \cdot z) \cdot y$	Exchange
3.	$\mathbf{I} \cdot x$	$=$	x	Identity
4.	$\mathbf{K} \cdot x \cdot !y$	$=$	x	Weakening
5.	$\mathbf{D} \cdot !x$	$=$	x	Dereliction
6.	$\delta \cdot !x$	$=$	$!!x$	Comultiplication
7.	$\mathbf{F} \cdot !x \cdot !y$	$=$	$!(x \cdot y)$	Monoidal Functoriality
8.	$\mathbf{W} \cdot x \cdot !y$	$=$	$x \cdot !y \cdot !y$	Contraction

The notion of LCA corresponds to a Hilbert style axiomatization of the $\{!, \multimap\}$ fragment of linear logic [Abramsky, 1997; Avron, 1988; Troelstra, 1992]. The *principal types* of the combinators correspond to the axiom schemes which they name. They can be computed by a Hindley-Milner style algorithm [Hindley, 1997] from the above equations:

1.	**B**	:	$(\beta \multimap \gamma) \multimap (\alpha \multimap \beta) \multimap \alpha \multimap \gamma$
2.	**C**	:	$(\alpha \multimap \beta \multimap \gamma) \multimap (\beta \multimap \alpha \multimap \gamma)$
3.	**I**	:	$\alpha \multimap \alpha$
4.	**K**	:	$\alpha \multimap !\beta \multimap \alpha$
5.	**D**	:	$!\alpha \multimap \alpha$
6.	δ	:	$!\alpha \multimap !!\alpha$
7.	**F**	:	$!(\alpha \multimap \beta) \multimap !\alpha \multimap !\beta$
8.	**W**	:	$(!\alpha \multimap !\alpha \multimap \beta) \multimap !\alpha \multimap \beta$

Here $\multimap$ is a *linear function type* (linearity means that the argument is used exactly once), and $!\alpha$ allows arbitrary copying of an object of type α.

A *Standard Combinatory Algebra* consists of a pair $(A, \cdot_s)$ where A is a nonempty set and $\cdot_s$ is a binary operation on A, together with distinguished elements $\mathbf{B}_s, \mathbf{C}_s, \mathbf{I}_s$, $\mathbf{K}_s$, and $\mathbf{W}_s$ of A, satisfying the following identities for all x, y, z ranging over A:

1.	$\mathbf{B}_s \cdot_s x \cdot_s y \cdot_s z$	$=$	$x \cdot_s (y \cdot_s z)$
2.	$\mathbf{C}_s \cdot_s x \cdot_s y \cdot_s z$	$=$	$(x \cdot_s z) \cdot_s y$
3.	$\mathbf{I}_s \cdot_s x$	$=$	x
4.	$\mathbf{K}_s \cdot_s x \cdot_s y$	$=$	x
5.	$\mathbf{W}_s \cdot_s x \cdot_s y$	$=$	$x \cdot_s y \cdot_s y$

This is just a combinatory algebra with interpretations of the Curry combinators. Note that this is equivalent to the more familiar definition of **SK**-combinatory algebra as discussed in the previous sub-section.

Let $(A, \cdot, !)$ be a linear combinatory algebra. We define a binary operation $\cdot_s$ on A as follows: for $a, b \in A$, $a \cdot_s b \equiv a \cdot !b$. We define $\mathbf{D}' \equiv \mathbf{C} \cdot (\mathbf{B} \cdot \mathbf{B} \cdot \mathbf{I}) \cdot (\mathbf{B} \cdot \mathbf{D} \cdot \mathbf{I})$. Note that

$$\mathbf{D}' \cdot x \cdot !y = x \cdot y.$$

Now consider the following elements of A.

1. $\mathbf{B}_s \equiv \mathbf{C} \cdot (\mathbf{B} \cdot (\mathbf{B} \cdot \mathbf{B} \cdot \mathbf{B}) \cdot (\mathbf{D}' \cdot \mathbf{I})) \cdot (\mathbf{C} \cdot ((\mathbf{B} \cdot \mathbf{B}) \cdot \mathbf{F}) \cdot \delta)$
2. $\mathbf{C}_s \equiv \mathbf{D}' \cdot \mathbf{C}$
3. $\mathbf{I}_s \equiv \mathbf{D}' \cdot \mathbf{I}$
4. $\mathbf{K}_s \equiv \mathbf{D}' \cdot \mathbf{K}$
5. $\mathbf{W}_s \equiv \mathbf{D}' \cdot \mathbf{W}$

THEOREM 11. *Let $(A, \cdot, !)$ be a linear combinatory algebra. Then $(A, \cdot_s)$ with $\cdot_s$ and the elements $\mathbf{B}_s, \mathbf{C}_s, \mathbf{I}_s, \mathbf{K}_s, \mathbf{W}_s$ as defined above is a standard combinatory algebra.*

Finally, we mention a special case which will arise in our model. An *Affine Combinatory Algebra* is a Linear Combinatory Algebra such that the $\mathbf{K}$ combinator satisfies the stronger equation

$$\mathbf{K} \cdot x \cdot y = x.$$

Note that in this case we can *define* the identity combinator: $\mathbf{I} \equiv \mathbf{C} \cdot \mathbf{K} \cdot \mathbf{K}$.

6.3 Universal Computation by Copy-Cats

Our aim is to describe an interactive model for logic and computation, which can be understood in two complementary ways:

- A model built from simple dynamical processes of copying information from one place to another.
- A model built from simple geometrical constructions, in which computation is interpreted as geometric simplification—tracing paths through tangles, and yanking them straight.

We begin with the dynamical interpretation. Here we think of an informatic *token* or *particle* traversing a path through logical (discrete) "space" and "time". For this purpose, we assume a set Pos of *positions* or *places* in "logical space". For the purposes of obtaining a type-free universal model of computation, it is important that Pos is (countably) infinite. (So we could just take it to be the set $\mathbb{N}$ of natural numbers). The only significant property of the instantaneous state of the particle is its current position $p \in \mathsf{Pos}$.

The *processes* we shall consider will be very simple, "history-free" or "time-independent" reversible dynamics, which we represent as *partial injective functions*

$$f : \mathsf{Pos} \rightharpoonup \mathsf{Pos}.$$

Such a process maps a particle in position p at any time t to the position $f(p)$ at time $t+1$; or may be undefined. In fact, we will have no need to make time explicit, since discrete time will be modelled adequately by *function composition.*[16]Thus the path traced by the particle starting from position p_0 under the dynamics f will be

$$p_0, p_1, p_2, \ldots, p_n, \ldots$$

where $p_{i+1} = f(p_i)$. This dynamics is clearly reversible. Since f is a partial injective map, its inverse f^{-1} (*i.e.* the relational converse of f) is also a partial injective function on Pos, and $p_i = f^{-1}(p_{i+1})$, so we can trace the reverse path using the inverse dynamics.

In fact, it will be possible to restrict ourselves to an even simpler class of dynamics: namely the *fixed-point free partial involutions*, *i.e.* those partial injective functions $f : \mathsf{Pos} \rightharpoonup \mathsf{Pos}$ satisfying

$$f = f^{-1}, \qquad\qquad f \cap 1_{\mathsf{Pos}} = \varnothing.$$

Thus such a map satisfies:

$$f(x) = y \iff x = f(y), \qquad\qquad f(x) \neq x.$$

A partial involution on a set X is equivalently described as a partial partition of X into 2-element subsets:

$$X \supseteq \bigcup E, \qquad\qquad \text{where } E = \{\{x, y\} \mid f(x) = y\}.$$

This defines an undirected graph $G_f = (X, E)$. Clearly each vertex in this graph has at most one incident edge. Conversely, every graph $G = (X, E)$ with this property determines a unique partial involution f on X, with $G_f = G$.

Partial involutions will be our model at this basic level of "copy-cat processes"; they simply copy information back and forth between pairs of "locations". It is somewhat remarkable that such simple maps can form a universal computational model.

Function Application as Interaction

Our next and key step is to model *functional application* by *interaction* of these simple dynamical processes. This will in fact be a bare-bones version of the game-theoretic model of composition as interaction which we gave in the previous section. We shall view a "functional process" which can be applied to other processes as a two-input two-output function

[16]The non-trivial dynamics we shall actually consider, which will arise when we model function application, will in fact come from the interaction between *two* very simple functions — the "ping-ponging" back and forth between them, in the terms of our informal discussion of interaction in Section 5. Note that *all* elements of our combinatory algebra, whether they appear as "functions" or "arguments" in the context of a given application, will be represented by functions on positions, corresponding to processes or strategies.

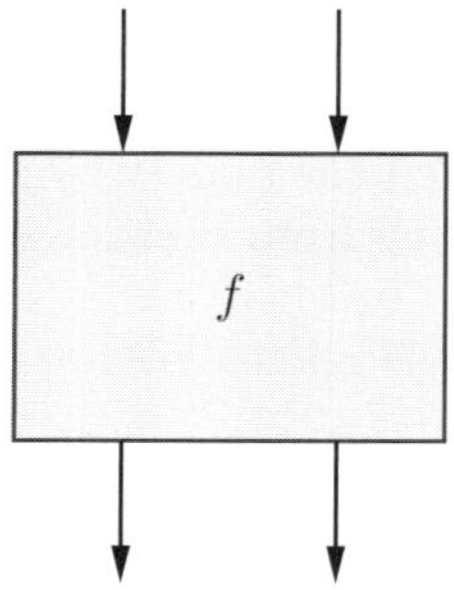

The interpretation of these two pairs of input-output lines is that the first will be used to connect the functional process to its argument, and the second to connect it to its external environment or context—which will interact with the function to consume its output. Formally, this is a function

$$f : \mathsf{Pos} + \mathsf{Pos} \longrightarrow \mathsf{Pos} + \mathsf{Pos}.$$

Note that we have the used the disjoint union (two copies of Pos) rather than the cartesian product $\mathsf{Pos} \times \mathsf{Pos}$ (infinitely many copies of Pos). This is because a particle coming in as input must *either* be on the first input line, *or* (in the exclusive sense) on the second input line; and similarly for the outputs.

However, since we want to make a type-free universal model of computation, we must reduce *all* our processes to one-input one-output functions. This is where our assumption that Pos is infinite becomes important. It allows us to define a *splitting function* s:

$$\mathsf{s} : \mathsf{Pos} + \mathsf{Pos} \xrightarrow{\cong} \mathsf{Pos}.$$

We can think of this as splitting logical space into two disjoint "address spaces". This allows us to transform any one-input/one-output function into a two-input/two-output function, or conversely, by conjugation. Thus if $f : \mathsf{Pos} \rightarrow \mathsf{Pos}$ is any process, we can view it as a two-input, two-output functional process, namely

$$\mathsf{s}^{-1} \circ f \circ \mathsf{s} : \mathsf{Pos} + \mathsf{Pos} \longrightarrow \mathsf{Pos} + \mathsf{Pos}.$$

Geometrical Representation of Application

Suppose we wish to apply f, *qua* functional process, to g, where both f and g are partial involutions on Pos. The application operation $f \bullet g$ is indicated pictorially as follows.

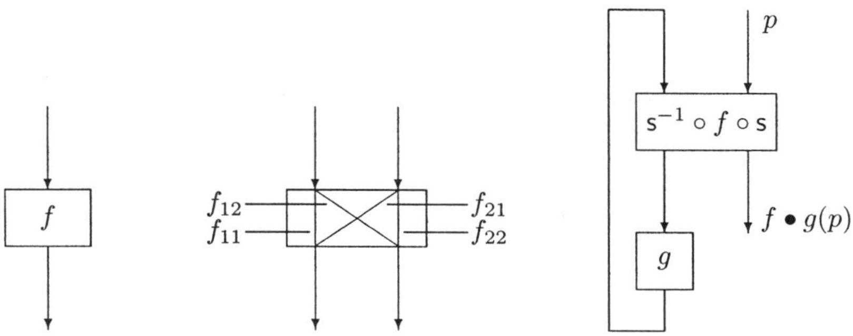

As already explained, we conjugate f by s to turn it into something with the right shape to be a functional process. Then we connect it to its argument, g, by a feedback loop using the first input and output lines of $f' = s^{-1} \circ f \circ s$. The residual behaviour by which the process resulting from the application communicates with its environment uses the second input and output lines. The full geometric significance of how this notion of application works will become apparent when we discuss the interpretation of the combinators in this setting. But we can give the dynamical interpretation of application immediately. Suppose the process $f \bullet g$ receives a token p on its input. The function f' may immediately dispatch this to its second output line as p'—in which case, that will be the response of $f \bullet g$. This would correspond to the behaviour of a constant function, which knows its output without consulting its input. Otherwise, f' may dispatch p to its *first* output line, as p_1. This is then fed as input to g. Thus this corresponds to the function represented by f' interrogating its argument. If $g(p_1) = p_2$, then this is fed back around the loop as input to f' (now on its first input line). We may continue in this fashion, ping-ponging between f' (on its first input/output lines) and g around the feedback loop. Eventually, f' may have seen enough, and decide to despatch the token on its second output line, as p'. We then say that $f \bullet g(p) = p'$. In other words, to the external environment, the whole interaction between f' and g has been hidden inside the black box of the application $f \bullet g$; it only sees the final response p' to the initial entry of the token at p.

All of this should seem very familiar. It follows exactly the same general lines as the game-semantical interpretation of composition which we presented in the previous section. We note the following points of difference:

- The notion of composition we discussed in the previous section was fully symmetrical between the two agents involved, reflecting the classical nature of the underlying logic. Here, we are discussing *functional computation*, and our description of application reflects the asymmetry between function and argument.

- Since we are dealing with a type-free universal model of computation, we must allow some partiality in our model. The token may get trapped in the feedback loop for ever, for example, so the involutions giving the dynamics must be partial in general. This is unavoidable, for well-known metamathematical reasons.

- We are also considering a very restricted, simple notion of dynamics here. Certainly in the game semantics context, we would not want in general to limit ourselves to such a restricted class of strategies.

Algebraic Description of Application

We now give a formal definition of the application operation. Firstly, consider the map $f' = \mathsf{s}^{-1} \circ f \circ \mathsf{s} : \mathsf{Pos} + \mathsf{Pos} \longrightarrow \mathsf{Pos} + \mathsf{Pos}$. Each input lies in *either* the first component of the disjoint union, *or* (exclusive or) the second, and similarly for the corresponding output. This leads to a decomposition of f' into four *disjoint partial maps* f_{ij}, $i, j \in \{1, 2\}$, where f_{ij} maps the i'th input summand to the j'th output summand. Note that f' can be recovered as the union of these four maps. Since f' is a partial involution, these maps will also be partial involutions. The decomposition is indicated pictorially in the preceding diagram. Now we can define

$$f \bullet g = f_{22} \cup f_{21}; g; (f_{11}; g)^*; f_{12}\,,$$

where we use relational algebra (union $R \cup S$, relational composition $R; S$ and reflexive transitive closure R^*) to write down formally exactly the information flow we described in our informal explanation of application above. It is a nice exercise to show that partial involutions are closed under application; that is, that $f \bullet g$ is again a partial involution.

6.4 Combinators as Copy-Cats

At this point, we have defined our applicative structure $(A, \bullet)$, where A is the set of partial involutions. We must now show that we can define combinators as partial involutions such that this structure will indeed form a Linear Combinatory Algebra. From now on, we shall mainly resort to drawing pictures, rather than writing algebraic expressions.

The Identity Combinator

Our first example is the simplest, and yet already shows the essence of the matter. The identity combinator **I** is represented by the *twist map*, which copies any token on its first input line to the second output line, and vice versa. This is depicted as follows.

I

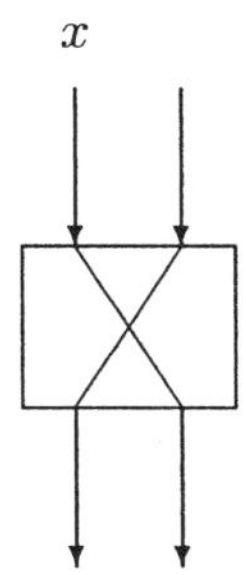

What is surprising, and striking, is the geometric picture of *why* this works: that is, why the equation $\mathbf{I} \bullet x = x$ holds:

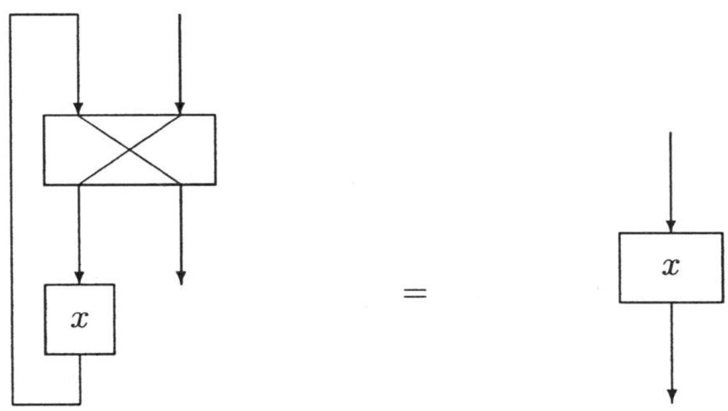

$$\mathbf{I} \bullet x = x$$

We see that geometrically, this is a matter of *yanking the string straight*; while dynamically, we picture the token flowing once around the feedback loop, and exiting exactly according to x.

Once again, we can recognize this combinator as a new description of an old friend from the previous section. *This is exactly the copy-cat strategy!* Reduced to its essence, it simply copies "tokens" or "moves" from one place to another, and *vice versa*; the logical requirement is that one of these places should be positive (or output); while the other should be negative (or input).

The Composition Combinator

We now consider the composition combinator **B**. We interpret it as a *combination of copy-cats.* That is, it plays copy-cat between three pairs of input and output lines. (Thus, in particular, it is a partial involution).

B

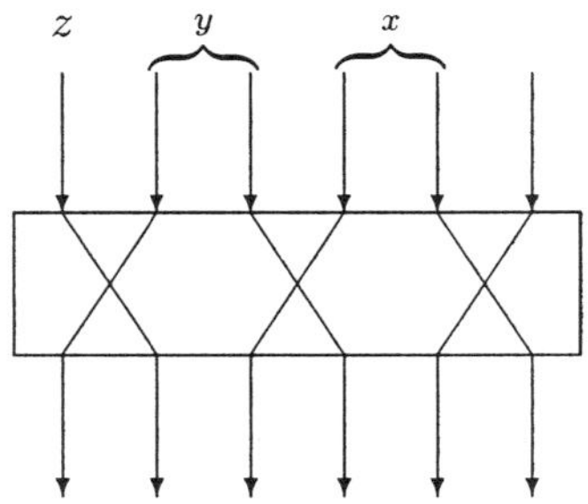

Note that we can regard this combinator as having six inputs and six outputs, as shown in the diagram, simply by iterating the trick of conjugating it by the splitting map s. Our reason for giving it this many inputs and outputs is based on the *functionality* of **B**, *i.e.* its principal type. It expects to get three arguments, the first two of which will themselves be applied to arguments, and hence should each have two inputs and two outputs.

Once again, the real insight as to how this combinator works will come from the geometry, or equivalently the particle dynamics. We let the picture speak for itself.

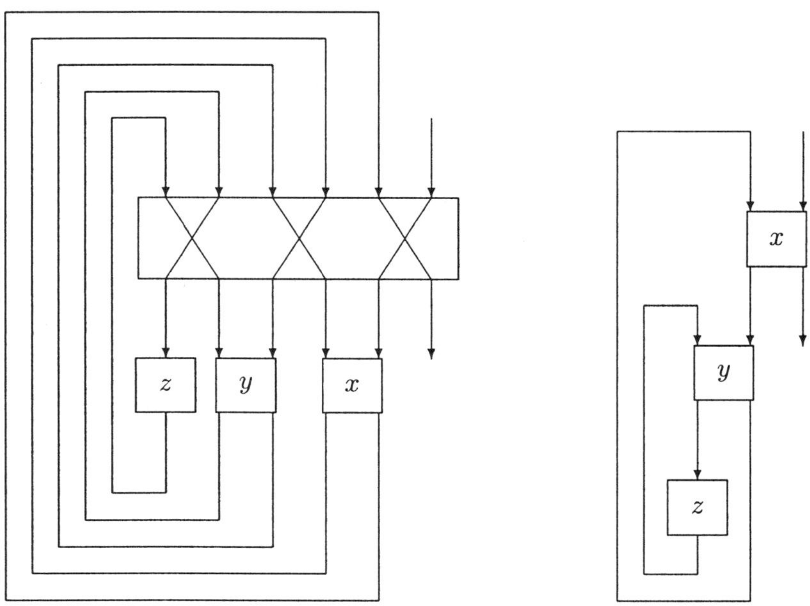

$$\mathbf{B} \cdot x \cdot y \cdot z = x \cdot (y \cdot z)$$

Other Affine Combinators

The remaining Linear Combinators can be described in similar style. We simply show the definition for **C**.

C

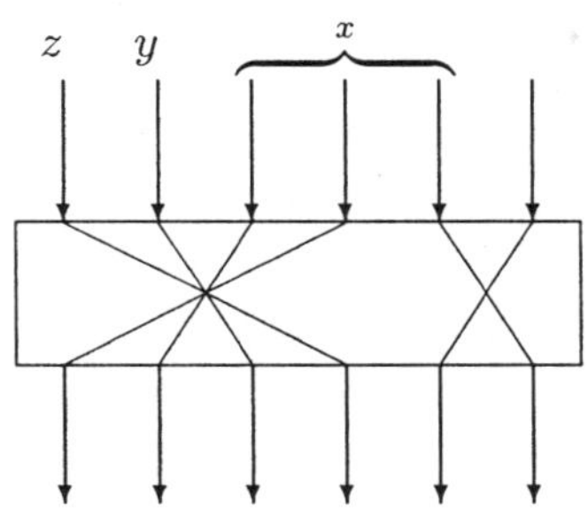

$$\mathbf{C} \cdot x \cdot y \cdot z = x \cdot z \cdot y$$

We note that geometrically, this is our first example of a *non-planar* combinator.[17] This gives a hint of the geometrical possibilities lurking just below the surface. We shall not pursue this fascinating theme here for lack of space, but see [Abramsky, 2007].

In fact, the algebra is naturally *affine*. We can define a **K** combinator:

K

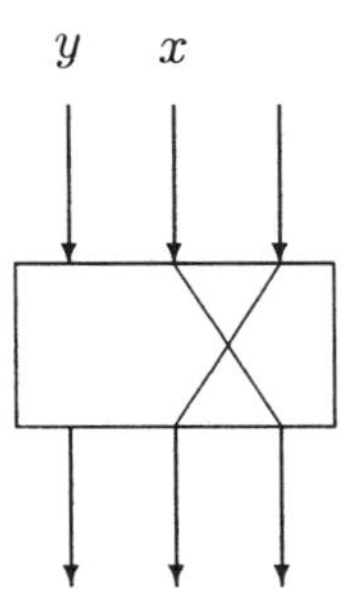

$$\mathbf{K} \cdot x \cdot y = x$$

However, note that another topological property is violated here; the first input and output lines are *disconnected* from the information flow. (Recall our discussion of the second variation of the Chess copy-cat scenario).

[17]Our diagrammatic conventions obscure this point, since all our diagrams involve over-crossing lines. For an explicit discussion of planarity and an alternative diagrammatics, see [Abramsky, 2007].

Duplication

We shall conclude our discussion of the algebra by sketching how explicit duplication of arguments is handled. This is needed for full expressive power.

We define another auxiliary function

$$\mathsf{p} : \mathbb{N} \times \mathsf{Pos} \xrightarrow{\cong} \mathsf{Pos}$$

which splits logical space into countably many disjoint copies. Again, this requires the assumption that Pos is infinite. Using this, we can define an operation $!f$ which is intended to produce *infinitely many copies of f*. These are obtained by simply tagging each copy with a natural number, i.e. we define:

$$!f = \mathsf{p} \circ (1_{\mathbb{N}} \times f) \circ \mathsf{p}^{-1}.$$

We can then define **W** satisfying

$$\mathbf{W} \cdot x \cdot !y = x \cdot !y \cdot !y.$$

The W combinator

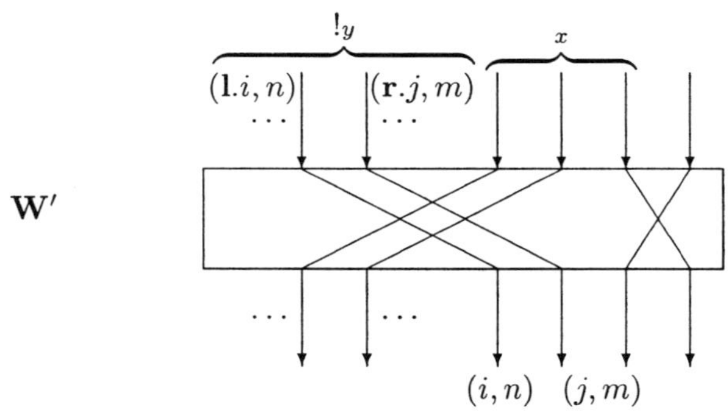

This combinator can be understood as effecting a "translation between dialects":

- x sees two arguments, each in many copies.
- $!y$ provides one argument, in as many copies as needed.

The combinator in effect decomposes into infinitely many copy-cat strategies, using a suitable splitting function to split the "address space" of the countably many copies of $!y$ into two infinite, disjoint parts, and copying between each of these and the corresponding argument position of x.

6.5 *Putting the Pieces Together*

We can round out the descriptions of the combinators as partial involutions to obtain a Linear Combinatory Algebra. By Theorem 11, this yields a standard Combinatory Algebra, and hence by Theorem 9 a universal model of computation. Moreover, realizability constructions over this Combinatory Algebra provide models for higher-order logics and type theories. Thus we have fulfilled our programme for this Section, of exhibiting the power of copying, leading to emergent models of logic and computation.

6.6 *Discussion*

Our gentle description of the partial involutions model in this section has merely indicated some first steps in this topic. We list some further directions:

- There is a general axiomatic formulation of this construction in terms of *traced monoidal categories* [Joyal *et al.*, 1996], with instances for deterministic, non-deterministic, probabilistic and quantum interaction [Abramsky, 1996; Abramsky *et al.*, 2002].
- The connections with reversible computation have been mentioned; this topic is carried further in [Abramsky, 2005].
- These models have some striking applications to the analysis of proofs, and of definability in various type theories, via *Full Completeness theorems* for models arising by realizability constructions over the basic Geometry of Interaction models [Abramsky and Lenisa, 2005].
- Current work is showing that the suggestive connections with geometry can be carried much further. In particular, there are connections with diagram algebras such as the Temperley-Lieb algebra, and hence with the Jones polynomial and ensuing developments [Abramsky, 2007].
- Finally, as already mentioned, there are strong connections with Quantum Information and Computation, which deserve a proper account of their own. Some references are [Abramsky and Coecke, 2002; Abramsky and Coecke, 2004; Abramsky and Coecke, 2005].

7 CONCLUDING REMARKS

The underlying project we have tried to illuminate in this article, via some partial exemplifications, has been that of developing a free-standing, syntax-independent Information Dynamics, worthy of the name.

In our view, this project is significant not just for Computer Science, but for Applied Logic, and for the theory and philosophy of information in general.

7.1 Combining Static and Dynamic

We have already emphasized the importance of combining qualitative and quantitative aspects of information, in the context of both static and dynamic theories. We conclude by making the point that it can be fruitful to combine static and dynamic aspects as well.[18]

We can set this in a wider context. One can distinguish two views on how Logic relates to Structure:

1. **The Descriptive View.** Logic is used to *talk about* structure. This is the view taken in Model Theory, and in most of the uses of Logic (Temporal logics, MSO etc.) in Verification in Computer Science. It is by far the more prevalent and widely-understood view.

2. **The Intrinsic View.** Logic is taken to *embody* structure. This is, implicitly or explicitly, the view taken in the Curry-Howard isomorphism, and more generally in Structural Proof Theory, and in (much of) Categorical Logic. In the Curry-Howard isomorphism, one is not using logic to *talk about* functional programming; rather, logic (in this aspect) *is* functional programming.

The descriptive view is well exemplified by Dynamic Logic and other modal logics. Indeed, one can use modal logics to *talk about* games and strategies, while on the other hand these can be used as a manifestation of the intrinsic view, modelling proofs as certain interactive processes. In some sense the intrinsic view is *global*, giving the structure of a type theory or semantic category; while the descriptive view is *local*, exploring the structure of particular games (objects) or strategies (morphisms). There is no reason why these two views cannot be brought fruitfully together, e.g. using a suitable modal logic to verify the logical soundness properties of strategies such as the copy-cat strategies we have discussed.

This further attempt to draw some of the strands we have examined in this article together is one of many promising directions for future work.

7.2 The Fundamental Challenge

The most fundamental challenge faced by the project of an Informatic Dynamics is in our view this: how to expose what is really robust and intrinsic structure, a bedrock on which we can build, as opposed to what is more or less arbitrarily chosen.[19] This problem is all the more acute, given the ever-increasing range of concrete informatic phenomena which we are continually being challenged to model by the rapidly-moving world of Information Technology.

Without under-estimating these difficulties, we find numerous, if "local", grounds for optimism in the theories we have surveyed, in the insights and structures which

---

[18]This point was emphasized by Johan van Benthem (personal communication).

[19]There is a name in Computer Science for the syndrome of a profusion of choices, none canonical: the "next 700 syndrome". It comes from Peter Landin's classic paper (from 1966!), "The Next 700 Programming Languages" [Landin, 1966]. For further discussion of this syndrome, and where we might find inspiration in addressing it, see [Abramsky, 2006].

they have brought to light. We venture to believe that real and exciting progress will continue to be made, and that a fundamental and widely applicable scientific theory of Information, incorporating qualitative and structural as well as quantitative features, is in the making.

ACKNOWLEDGEMENTS

A number of people kindly agreed to read a draft of this article, and provided very perceptive and helpful comments: Adam Brandenburger, Jeremy Butterfield, Robin Milner and Yorick Wilks. I would like to express my warmest appreciation for their input. My thanks also to Jan van Eijck, who commented on the paper on behalf of the Editors, and produced two rounds of useful comments. Special thanks to Johan van Benthem, who asked me to write the article in the first place, and whose encouragement, suggestions and gentle reminders have kept me on track.

BIBLIOGRAPHY

[Abramsky, 1991] S. Abramsky. Domain theory in logical form. *Annals of Pure and Applied Logic* 51, 1–77, 1991.

[Abramsky, 1996] S. Abramsky. Retracing some paths in process algebra. In *Proceedings of CONCUR 96*, Lectures Notes in Computer Science Volume 1119, pp. 1–17, Springer-Verlag, 1996.

[Abramsky, 1997] S. Abramsky, *Interaction, Combinators and Complexity*. Lecture Notes, Siena, Italy, 1997.

[Abramsky, 2002] S. Abramsky. Algorithmic game semantics: a tutorial introduction. In: *Proof and System-Reliability*, Kluwer, 2002.

[Abramsky, 2005] S. Abramsky. A structural approach to reversible computation. *Theoretical Computer Science* vol. 347(3), 441–464, 2005.

[Abramsky, 2006] S. Abramsky. What are the fundamental structures of concurrency? We still don't know!, in Proceedings of the Workshop "Essays on Algebraic Process Calculi" (APC 25). Electronic Notes in Theoretical Computer Science, Volume 162:37–41, 2006.

[Abramsky, 2007] S. Abramsky. Temperley-Lieb algebra: from knot theory to logic and computation via quantum mechanics. In *Mathematics of Quantum Computation and Quantum Technology*, G. Chen, L. Kauffman and S. Lomonaco, eds., Taylor and Francis, pages 515–558, 2007.

[Abramsky and Coecke, 2002] S. Abramsky and B. Coecke. Physical traces: classical vs. quantum information processing. In Electronic Notes in Theoretical Computer Science, vol. 69, 2003, 1–26.

[Abramsky and Coecke, 2004] S. Abramsky and B. Coecke, A Categorical Semantics of Quantum Protocols, in Proceedings of the 19th Annual IEEE Symposium on Logic in Computer Science: LICS 2004, IEEE Computer Society, 415–425, 2004.

[Abramsky and Coecke, 2005] S. Abramsky and B. Coecke, Abstract Physical Traces, in Theory and Applications of Categories, vol 14, 111–124, 2005.

[Abramsky and Jagadeesan, 1994a] S. Abramsky and R. Jagadeesan. New foundations for the geometry of interaction. *Information and Computation* **111**, 53–119, 1994.

[Abramsky and Jagadeesan, 1994b] S. Abramsky and R. Jagadeesan. Games and full completeness for multiplicative linear logic. *Journal of Symbolic Logic* **59**, 543–574, 1994.

[Abramsky and Jagadeesan, 2005] S. Abramsky and R. Jagadeesan. A game semantics for generic polymorphism. In *Annals of Pure and Applied Logic*, 133: 3-37, 2005.

[Abramsky and Jung, 1994] S. Abramsky and A. Jung. Domain theory. In: *Handbook of Logic in Computer Science* Volume III, S. Abramsky, D. Gabbay and T. S. E. Maibaum, eds., 1–168. Oxford University Press, 1994.

[Abramsky and Lenisa, 2005] S. Abramsky and M. Lenisa. Linear realizability and full completeness for typed lambda-calculi, in *Annals of Pure and Applied Logic*, vol 134, 122–168, 2005.

[Abramsky and McCusker, 1997] S. Abramsky and G. McCusker. Linearity, sharing and state. In: P. O'Hearn and R. D. Tennent, eds. *Algol-like languages*, pp. 317–348. Birkhauser, 1997.

[Abramksy and McCusker, 1998] S. Abramsky and G. McCusker. Call-by-value games. In Proceedings of the Eleventh International Workshop on Computer Science Logic, M. Nielsen and W. Thomas, eds., Springer Lecture Notes in Computer Science Vol. 1414, pp. 1–17, Springer-Verlag, 1998.

[Abramsky and McCusker, 1999a] S. Abramsky and G. McCusker. Full abstraction for idealized Algol with passive expressions. *Theoretical Computer Science*, vol. 227, pp. 3–42, 1999.

[Abramsky and McCusker, 1999b] S. Abramsky and G. McCusker. Game semantics. In: *Computational Logic: Proceedings of the 1997 Marktoberdorf Summer School*, pp.1–56. Springer-Verlag, 1999.

[Abramsky and Mellies, 1999] S. Abramsky and P.-A. Mellies. Concurrent games and full completeness. In Proceedings of the Fourteenth International Symposium on Logic in Computer Science, (Computer Society Press of the IEEE), pp. 431–442, 1999.

[Abramsky *et al.*, 1998] S. Abramsky and K. Honda and G. McCusker. A fully abstract game semantics for general references. In Proceedings of the Thirteenth International Symposium on Logic in Computer Science, (Computer Society Press of the IEEE) 1998, 334–344.

[Abramsky *et al.*, 2000] S. Abramsky, R. Jagadeesan and P. Malacaria. Full abstraction for PCF. *Information and Computation* **163**, 409–470, 2000.

[Abramsky *et al.*, 2002] S. Abramsky and E. Haghverdi and P. J. Scott. Geometry of Interaction and Linear Combinatory Algebras. *Mathematical Structures in Computer Science* 12:625–665, 2002.

[Abramsky *et al.*, 2004a] S. Abramsky, D. R. Ghica, A. S. Murawski, and C.-H. L. Ong. Applying game semantics to compositional software modelling and verification. In *Proc. TACAS'04*, LNCS 2988, pages 421–435, 2004.

[Abramsky *et al.*, 2004b] S. Abramsky, D. R. Ghica, A. S. Murawski, I. D. B. Stark and C.-H. L. Ong. Nominal games and full abstraction for the nu-calculus. In *Proc. LICS'04*, pages 150–159, 2004. IEEE Computer Society Press.

[Alberti and Uhlmann, 1982] P. M. Alberti and A. Uhlmann. *Stochasticity and Partial Order: Doubly Stochastic Maps and Unitary Mixing.* Math. Monographien 18. VEB Deutscher Verlag der Wissenschaften, 1982.

[Amadio and Curien, 1998] R. Amadio and P.-L. Curien. *Domains and Lambda Calculi.* Cambridge University Press 1998.

[Avron, 1988] A. Avron, The semantics and proof theory of linear logic. *Theoretical Computer Science* 57:161–184, 1988.

[Baader and Nipkow, 1999] F. Baader and T. Nipkow, *Term Rewriting and All That.* Cambridge University Press, 1999.

[Baeten and Weijland, 1990] J.C.M. Baeten and W.P. Weijland, *Process algebra*, Cambridge Tracts in Theoretical Computer Science 18, Cambridge University Press, 1990.

[de Bakker, 1980] J. W. de Bakker. *Mathematical Theory of Program Correctness.* Prentice Hall 1980.

[Baltag and Smets, 2006] A. Baltag and S. Smets. LQP: The Dynamic Logic of Quantum Information, submitted to Mathematical Structures in Computer Science, special issue on Quantum Programming Languages.

[Barendregt, 1984] H. P. Barendregt *The Lambda Calculus.* Studies in Logic, Vol. 103, North-Holland, 1984.

[Barwise and Seligman, 1997] J. Barwise and J. Seligman. *Information Flow: The Logic of Distributed Systems.* Cambridge University Press, 1997.

[van Benthem, 1976] Johan van Benthem. *Modal Correspondence Theory.* Ph.D. thesis, University of Amsterdam, 1976.

[van Benthem, 1988] J. van Benthem. *Exploring Logical Dynamics.* CSLI Publications, 1998.

[Bergstra and Ponse, 2000] J. Bergstra and A. Ponse, eds. *Handbook of Process Algebra.* Elsevier 2000.

[Birkhoff and von Neumann, 1936] G. Birkhoff and J. von Neumann. The logic of quantum mechanics. *Annals of Mathematics* **37**, 823–843, 1936.

[Blute, 1998] R. Blute and P. J. Scott. The Shuffle Hopf Algebra and Noncommutative Full Completeness. *J. Symb. Log* **63**, 1413–1436, 1998.

[Blute *et al.*, 2005] R. Blute, M. Hamano, and P. J. Scott. Softness of hypercoherences and MALL full completeness. *Ann. Pure Appl. Logic*, **131**, 1–63, 2005.

[Blackburn *et al.*, 2001] P. Blackburn, M. de Rijke and Y. Venema. *Modal Logic.* Cambridge University Press, 2001.

[Bonsangue and Kok, 1999] M.M. Bonsangue and J.N. Kok Towards an infinitary logic of domains: Abramsky logic for transition systems. In Information and Computation 155:170–201, 1999.

[Ceri *et al.*, 1990] S Ceri, G Gottlob and L Tanca. *Logic programming and databases.* Springer-Verlag 1990.

[Coecke, 2003] B. Coecke. Entropic geometry from logic. In: *MFPS XIX.* 2003. `arXiv: quant-ph/0212065`

[Coecke and Martin, 2002] B. Coecke and K. Martin. A partial order on classical and quantum states. Research Report PRG-RR-02-07 OUCL. `web.comlab.ox.ac.uk/oucl/publications/tr/rr-02-07.html`

[Curry and Feys, 1958] H. B. Curry and R. Feys. *Combinatory Logic* North Holland, 1958.

[Danos and Reginer, 1993] V. Danos and L. Regnier. Local and asynchronous beta-reduction. In *Proceedings of the Eighth International Symposium on Logic in Computer Science*, IEEE Press, 296–306, 1993.

[Danos and Reginer, 1996] V. Danos and L. Regnier, Reversible, Irreversible and Optimal λ-machines, in *Electronic Notes in Theoretical Computer Science*, 1996.

[Danos, 2002] V.Danos. Russell Harmer: Probabilistic game semantics. *ACM Trans. Comput. Log.*, **3**, 359-382, 2002.

[Davey and Priestley, 2002] B. Davey and H. A. Priestley. *Introduction to Lattices and Order*, 2nd edition. Cambridge University Press, 2002.

[Devarajan *et al.*, 1999] H. Devarajan, D. Hughes, G. Plotkin, and V. Pratt. Full completeness of the multiplicative linear logic of Chu spaces. *Proceedings of the 14th Annual IEEE Symposium on Logic in Computer Science*, 234–242, 1999.

[van Eijck and Visser, 2994] J. Van Eijck and A. Visser, eds. *Logic and Information Flow.* MIT Press, 1994.

[van Eijck and Stokhof, 2006] J. Van Eijck and M. Stokhof. The Gamut of Dynamic Logics. In *The Handbook of the History of Logic*, D. M. Gabbay and J. Woods, eds. Volume 7: Logic and the Modalities in the Twentieth Century, pages 499–600. Elsevier, 2006.

[Ghica and McCusker, 2000] D. R. Ghica and G. McCusker. Reasoning about idealized algol using regular languages. In *Proc. ICALP'00*, LNCS 1853, pp. 103–116. 2000.

[Ghica andMurawski, 2004] D. R. Ghica and A. S. Murawski. Angelic semantics of fine-grained concurrency. In *Proc. FOSSACS'04*, pp. 211–225. 2004. LNCS 2987.

[Ghica and Murawski, 2006] D. R. Ghica and A. S. Murawski. Compositional model extraction for higher-order concurrent programs. In *Proc. TACAS'06*, LNCS, 2006.

[Gierz *et al.*, 2003] G. Gierz, K. H. Hofmann, K. Keimel. J. Lawson, M. Mislove and D. S. Scott. *Continuous Lattices and Domains.* Cambridge University Press, 2003.

[Girard, 1987] J.-Y. Girard, Linear Logic. *Theoretical Computer Science*, **50**, 1-102, 1987.

[Girard, 1989] J.-Y. Girard, Geometry of Interaction I: Interpretation of System F, in: *Logic Colloquium '88*, ed. R. Ferro, et al. North-Holland, pp. 221-260, 1989.

[Girard, 1990] J.-Y. Girard, Geometry of Interaction II: Deadlock- free Algorithms. In *Proceedings of COLOG-88* (P. Martin-Lof, G. Mints, eds.) Springer LNCS Vol. 417, pp. 76-93, 1990.

[Girard, 1995] J.-Y. Girard, Geometry of Interaction III: accomodating the additives. In [Girard *et al.*, 1995], 329–389.

[Girard *et al.*, 1989] J.-Y. Girard, Y. Lafont, and P. Taylor. *Proof and Types.* Cambridge Tracts in Theoretical Computer Science, 1989.

[Girard *et al.*, 1995] J.-Y. Girard, Y. Lafont, L. Regnier, eds. *Advances in Linear Logic*, London Math. Soc. Series 222, Camb. Univ. Press, 1995.

[Gleason, 1957] A. M. Gleason. Measures on the closed subspaces of a Hilbert space. *Journal of Mathematics and Mechanics* **6**, 885–893, 1957.

[Groenendijk and Stockhof, 1991] J. R. Groenendijk and M. R. Stokhof. Dynamic Predicate Logic. *Linguistics and Philosophy*, 1991.

[Harel *et al.*, 2000] D. Harel, D. Kozen and J. Tiuryn. *Dynamic Logic.* MIT Press, 2000.

[Harmer and McCusker, 1999] R. Harmer and G. McCusker. A fully abstract game semantics for finite nondeterminism. In Proceedings, Fourteenth Annual IEEE Symposium on Logic in Computer Science, IEEE Computer Society Press, 1999.

[Hennessy and Milner, 1980] M. Hennessy and R. Milner. On Observing Nondeterminism and Concurrency. Proceedings ICALP 1980: 299-309, 1980.

[Hindley, 1997] R. Hindley. *Basic Simple Type Theory*, Cambridge Tracts in Theoretical Computer Science, no. 42, Cambridge Univ. Press, 1997.

[Hindley and Seldin, 1986] J. R. Hindley and J. P Seldin. *Introduction to Combinators and the λ-calculus.* Cambridge University Press, 1986.

[Hoare, 1969] C. A. R. Hoare. An Axiomatic Basis for Computer Programming. *Commun. ACM*, **12**, 576-580, 1969.

[Hoare, 1985] C. A. R. Hoare. *Communicating Sequential Processes.* Prentice Hall, 1985.

[Howard, 1980] W. A. Howard. The formulae-as-types notion of construction. In *To HB Curry: Essays on Combinatory Logic and Lambda Calculus*, 1980.

[Hughes, 2000] D. Hughes. *Hypergame Semantics: Full Completeness for System F.* D.Phil. Mathematical Sciences, Oxford University, 2000.

[Hyland and Ong, 2000] J. M. E. Hyland and C.-H. L. Ong. On full abstraction for PCF: i. models, observables and the full abstraction problem, ii. dialogue games and innocent strategies, iii. a fully abstract and universal game model. *Information and Computation* **163**, 285–408, 2000.

[Johnstone, 1982] P. T. Johnstone. *Stone Spaces.* Cambridge University Press 1982.

[Jones and Plotkin, 1989] C. Jones and G. Plotkin. A probabilistic powerdomain of valuations. In: *LiCS'89.*

[Joyal *et al.*, 1996] Andre Joyal, Ross Street, and Dominic Verity. Traced monoidal categories. *Mathematical Proceedings of the Cambridge Philosophical Society*, 119(3):447–468, 1996.

[Kahn and Plotkin, 1978] G. Kahn and G. Plotkin. Concrete Domains. *Theoretical Computer Science*, 121:187–277, 1993. Appeared as TR IRIA-Laboria 336 in 1978.

[Kripke, 1963] S. Kripke. Semantical Considerations on Modal Logic. *Acta Philosophica Fennica*, 1963.

[Laird, 1997a] J. Laird. A fully abstract games semantics of local exceptions. Extended abstract, in the *Proceedings of the 16th Annual Symposium on Logic in Computer Science*, LICS '01, 2001.

[Laird, 1997b] J. Laird. Full abstraction for functional languages with control. Extended abstract, in the *Proceedings of the 12th Annual Symposium on Logic in Computer Science*, LICS '97, 1997.

[Laird, 1998] J. Laird. *A semantic analysis of control.* Ph.D. thesis, University of Edinburgh, 1998.

[Landauer, 1961] R. Landauer. Irreversibility and heat generation in the computing process. *IBM Journal of Research and Development*, **5**, 183–191, 1961.

[Landin, 1966] P. J. Landin. The next 700 programming languages. *Communications of the ACM*, **9**, 157–166, 1966.

[Lassez *et al.*, 1982] J.-L. Lassez, V. L. Nguyen, and L.Sonenberg. Fixed Point Theorems and Semantics: A Folk Tale. *Inf. Process. Lett.*, **14**, 112-116, 1982.

[Loader, 1994] R. Loader. *Models of Lambda Calculi and Linear Logic.* D.Phil. thesis, Oxford University, 1994.

[Lowe, 2002] G. Lowe. Quantifying Information Flow. In *Proceedings of the 15th IEEE Computer Security Foundations Workshop*, 2002.

[Lowe, 2004] G. Lowe. Semantic Models for Information Flow. In *Theoretical Computer Science*, Volume 315, pages 209-256, 2004.

[Malacaria and Reginer, 1991] P. Malacaria and L. Regnier. Some results on the interpretation of λ-calculus in Operator Algebras. In *Proceedings of the Sixth International Symposium on Logic in Computer Science*, IEEE Press, 63–72, 1991.

[Martin, 2001a] K. Martin. A principle of induction. Lecture Notes in Computer Science Volume 2142, Springer Verlag, 2001.

[Martin, 2001b] K. Martin. Unique fixed points in domain theory. Proceedings of MFPS XVII. *Electronic Notes in Theoretical Computer Science*, Volume 45, Elsevier, 2001.

[Martin, 2002] K. Martin. *A Foundation for Computation.* Ph.D. Thesis, Department of Mathematics, Tulane University, 2000.

[Martin, 2004] K. Martin. Entropy as a fixed point. ICALP 2004. Springer Lecture Notes in Computer Science Volume 3142, 2004.

[Martin and Panangaden, 2006] K. Martin and P. Panangaden. A Domain of spacetime intervals for general relativity Appeared online (Aug 15 2006) in *Communications of Mathematical Physics.*

[Martin *et al.*, 2002] K. Martin, M. Mislove and J. Worrell. Measuring the probabilistic powerdomain. Lecture Notes in Computer Science Volume 2380, Springer Verlag 2002.

[McLean, 1990] J. McLean. Security models and information flow. In: *1990 IEEE Symposium on Security and Privacy*, 180–187, 1990.

[Milner, 1980] R. Milner. *A Calculus of Communicating Systems*, Springer Verlag, 1980.

[Milner, 1989] R. Milner. *Communication and Concurrency.* Prentice Hall, 1989.

[Milner, 1996] R. Milner. Semantic Ideas in Computing. In *Computing Tomorrow: Future Research Directions in Computer Science.* Cambridge University Press, 1996.

[Milner, 1999] R. Milner. *Communicating and Mobile Systems: The Pi Calculus.* Cambridge University Press, 1999.

[Milner, 2006a] R. Milner. Ubiquitous Computing: Shall we Understand It? *Comput. J.*, **49**, 383-389, 2006.

[Milner, 2006b] R. Milner. Scientific Foundation for Global Computing. *T. Comp. Sys. Biology*, 1-13, 2006.

[Moller, 1990a] F. Moller. The Importance of the Left Merge Operator in Process Algebras. *ICALP 1990*, pp. 752-764, 1990.

[Moller, 1990b] F. Moller. The Nonexistence of Finite Axiomatisations for CCS Congruences. *LICS 1990*, pp. 142-153, 1990.

[Murawski *et al.*, 2005] A. S. Murawski, C.-H. L. Ong, and I. Walukiewicz. Idealized Algol with ground recursion and DPDA equivalence. In *Proc. ICALP'05*, LNCS 3580, pp. 917–929. 2005.

[Nielsen *et al.*, 1981] M. Nielsen, G. D. Plotkin, and G. Winskel. Petri Nets, Event Structures and Domains, Part I. *Theor. Comput. Sci.*, **13**, 85-108, 1981.

[Park, 1981] D. Park. Concurrency and Automata on Infinite Sequences. *Theoretical Computer Science*, 167-183, 1981.

[Plotkin, online] G. Plotkin. Pisa Notes on Domain Theory. Available at: `http://homepages.inf.ed.ac.uk/gdp/publications/`.

[Plotkin, 2004] G. D. Plotkin: A structural approach to operational semantics. *J. Log. Algebr. Program.*, **60-61**, 17-139, 2004. A reprint of Technical Report FN-19, Computer Science Department, Aarhus University, 1981.

[Pratt, 1976] V. R. Pratt: Semantical Considerations on Floyd-Hoare Logic. *Proceedings FOCS 1976*, pp. 109-121, 1976.

[Sangiorgi, 2004] D. Sangiorgi. Bisimulation: From The Origins to Today. *Proceedings LICS 2004*, pp. 298-302, 2004.

[Scott, 1970] D. S. Scott. *Outline of a Mathematical Theory of Computation.* Technical Monograph PRG-2 OUCL, 1970.

[Scott, 1982] D. S. Scott. Domains for Denotational Semantics. *ICALP 1982*, pp. 577-613, 1982.

[Shannon, 1948] C. E. Shannon. A mathematical theory of communication. *Bell Systems Technical Journal*, **27**, 379–423 and 623–656, 1948.

[Sorkin, online] R. Sorkin. First Steps in Causal Sets. Available at `http://physics.syr.edu/$\sim$sorkin/some.papers/`.

[Terese, 2003] Terese. *Term Rewriting Systems.* Cambridge Tracts in Theoretical Computer Science, Vol. 55, Cambridge University Press, 2003.

[Troelstra, 1992] A. S. Troelstra, *Lectures on Linear Logic.* Center for the Study of Language and Information Lecture Notes No. 29, 1992.

[Vorobyov, 1997] S. G. Vorobyov. The "Hardest" Natural Decidable Theory. *Proceedings LICS 1997*, pp. 294-305, 1997.

[Winskel, 1993] G. Winskel. *The Formal Semantics of Programming Languages*, MIT Press, 1993.

[Zhang, 1991] G. Q. Zhang. *Logic of Domains.* Birkhauser, 1991.

INFORMATION AND BELIEFS IN GAME THEORY

Bernard Walliser

Game theory is devoted to the study of strategic interactions established between several rational players. It is concerned by information since players have some uncertainty about their environment and compensate for it by direct observation or by communication with other players. It is concerned by beliefs since players form expectations about their environment by relying on representations about it as well as representations about others' representations. Information and beliefs are strongly linked since acquired information modifies the previous beliefs and revised beliefs shape new information. Hence, information and beliefs became together a central topic of game theory and even appear as the main device allowing the players to coordinate. Moreover, since game theory grounds economic theory, their study leads respectively to the 'economics of information' [Macho-Stadler *et al.*, 2001] and to the 'economics of knowledge' [Foray, 2004].

The aim of this paper is not to summarize the huge technical literature concerning information and beliefs in game theory. Many chapters of game theory textbooks are devoted to the study of information, with a lot of illustrative examples [Osborne-Rubinstein, 1994; Aumann-Hart, 1992; 1994; 2002]. Many proceedings of specialized conferences such as LOFT (1994 to 2006) and TARK (1986 to 2005) are devoted to the treatment of beliefs in game theory, even if a synthesis is not already available. The aim of the article is rather to proceed to a rational reconstruction of information and beliefs in game theory. It shows in what terms the problems were initially conceptualized, how some formal tools were imported and considered to be adequate, and in what ways these problems were then considered as being solved. Problems and solutions will be presented in a rather informal way, but various toy examples (used at length in game theory) will be mentioned.

First section recalls how information and beliefs were historically introduced in game theory and induced an evolution of the usual equilibrium notions. Second section presents what type of information is needed and gathered by a player and describes the formal tools used by the modeller in order to formalize it. Third section considers in which way such information is treated by a player, and especially how it contributes to the revision of its basic or crossed beliefs. Fourth section examines how information is captured by a specific equilibrium notion and how it is assessed by a player by means of an 'information value'. Fifth section analyzes information as a strategic item that a player has interest or not to diffuse, and examines if an equilibrium state reveals or not the players' information. Sixth section is devoted to the way information is diffused among the players and examines

Handbook of the Philosophy of Science. Volume 8: Philosophy of Information
Volume editors: Pieter Adriaans and Johan van Benthem. General editors: Dov M. Gabbay, Paul Thagard and John Woods.

if it leads or not that information to become common knowledge. Seventh section considers the learning process of players endowed with bounded rationality and examines if it converges or not towards some usual equilibrium state.

1 ONTOLOGY OF GAME THEORY

Game theory is intended to describe formally how rational players engage in strategic interactions. Interactions are strategic in the sense that the consequences of a player's action depend not only on his action, but on others' action. Hence, each player follows a decision process taking into account his expectations about the others' behaviour. Such expectations are grounded on prior beliefs of the player as well as on information progressively gathered by him. Hence, in the history of game theory, information and beliefs became progressively more and more explicit. More precisely, game theory developed its own formalism even if it appeared later to be similar to the one developed in epistemic logics.

1.1 Classical game theory

Any game is defined by two kinds of actors, a set of players (acting as genuine decision-makers) and nature (acting mechanically) which are jointly involved in strategic interactions. Each player is assumed to choose an 'action', independently from others, by a rational deliberation process. More precisely, he is endowed with three 'determinants': opportunities (delimitating his action set), beliefs (representing his view of his material and social environment) and preferences (evaluating the expected consequences of his actions). He combines these three determinants into a 'choice rule' which reflects two distinct forms of rationality (Walliser, 1989). Instrumental rationality refers to the adequacy of pursued objectives to available means and cognitive rationality refers to the adequacy of designed beliefs to available information. Nature, which summarizes the common material environment, takes a state by some passive process, characterized by a 'generation law'. It is assumed to be neutral with regard to the players, taking its states in a deterministic way independently of players' actions.

Originally [von Neumann and Morgenstern, 1944; Nash, 1962], the players were considered to play simultaneously. Their joint actions (constituting a 'profile of actions') lead to common consequences which are evaluated differently by the players. Such a one-shot game is structurally expressed by the modeller under the 'normal form' which makes the players' determinants more precise. A game matrix is obtained by considering every combination of an action for each player and of a state of nature. In each cell of the matrix, the payoff (or the utility) obtained by each player is indicated (nature is passive and has no payoffs). According to the payoff structure, different types of games are distinguished, well illustrated for two-player games without nature. In 'twin games', the players have the very same interests, hence their payoffs are identical in each cell. In 'zero-sum games', the players have contradictory interests, hence their payoffs are opposite in each cell.

In 'symmetric games', the players play analogous roles in the sense that they have the same set of actions and, when they exchange their actions, they exchange their payoffs too.

In a normal form game, the two basic notions of 'dominance' and 'best response' give rise to three main equilibrium concepts. First, for some player, an action dominates another if it gives a higher payoff whatever the other's action. Hence, an action of a player is dominant if it dominates all other actions of that player and an action is dominated if some other action of that player dominates it. A 'dominant equilibrium' state is just a profile of dominant actions. A 'sophisticated equilibrium' state is a profile of actions which survive 'iterated elimination of dominated actions'. In that procedure, the dominated actions of each player are first deleted, the same is done on the remaining game and so on. A dominant equilibrium seldom exists, while sophisticated equilibria are often numerous. Second, an action of some player is a best response to the other's one if no other action gives him a higher payoff to it. A 'Nash equilibrium' state is just a profile of actions for which each action of some player is a best response to the other's action. A Nash equilibrium (in pure actions) may not exist, be unique or be multiple. A dominant equilibrium is a Nash equilibrium, and a Nash equilibrium is a sophisticated equilibrium.

1.2 Introduction of time and information

In the seventies [Selten, 1975], a time dimension was introduced by enlarging the game theory framework. Players were considered to play sequential actions which induce either at each move or globally at the end of the game some common consequences. Such a game was structurally expressed by the modeller under the 'extensive form' which again makes the players' determinants more precise. A game tree is obtained by considering at each node the possible moves for the player or nature playing at this node. Hence, one move by one actor conditions the further moves of the other ones. For each path in the tree formed of successive actions and states and leading to a terminal node, the utility obtained by each player is indicated. A special case obtains when the players play a (finite or infinite) sequence of the same basic game. Each player receives then a payoff at each period, these payoffs being aggregated thanks to a discount rate (which depreciates the payoff of each period with regard to the preceding one).

At the same period [Harsanyi, 1967], uncertainty was first made explicit in the normal form of a game. Initially, uncertainty was related to nature since a player does not precisely know, when he plays, what state nature adopts. Usually, the modeller assumes that nature takes its states randomly, according to some prior probability law, moreover assumed to be known by the players from the beginning of the game. Moreover, uncertainty was related to the determinants of a player which are badly known by another one, and eventually even by him. In that case, the modeller assumes that the determinants of a player can be summarized in the 'type' of the player, that this type can be treated as a state of nature

(nature attributes a type to each player at the beginning of the game), and that a probability distribution exists on the types, again known by the players. Finally, uncertainty was introduced in the extensive form of a game. It is essentially related with past actions of a player which are not perfectly observed by another one and even by him. The modeller assumes that the nodes of a player associated to the non discriminated actions of another player are gathered in some 'information set' associated to the last.

The equilibrium notions were extended in order to integrate time and uncertainty. As far as time is concerned, it can first be observed that, by introducing the notion of a 'strategy' (defined as the action played by a player at each relevant node), the extensive form game (on actions) can be expressed under a normal form (on strategies), hence a Nash equilibrium can be defined. In other respects, a 'subgame perfect equilibrium' is defined as an action path which is a Nash equilibrium in the game and all its subgames. When the game is finite, such an equilibrium is obtained by the 'backward induction procedure'. This procedure states that a player at a terminal node chooses his best action, a player at a penultimate node chooses his best action, knowing what the next player will do, and so on till the root node. A subgame perfect equilibrium is a refinement of a Nash equilibrium, but it always exists and is generically unique. As far as uncertainty is concerned, a 'Bayesian equilibrium' state is defined as a Nash equilibrium of the game where the players are decomposed in as much agents as possible types and choose their best action, in average on others' types. A Bayesian equilibrium is a Nash equilibrium of an extended game, and it frequently exists and is even multiple.

1.3 Introduction of beliefs and learning

In the eighties [Aumann, 1976; 1999], beliefs were introduced explicitly in the game theoretical framework. The usual framework formalizes essentially the opportunities and the preferences of the players, respectively as action sets and payoffs. But beliefs do not appear in a formal way, even if observations and even expectations are considered as basic beliefs. Beliefs are representations made by a player about his environment, essentially nature and other agents. In fact, the players have crossed beliefs since each player holds beliefs about others' beliefs. Moreover, in dynamic games, the players have prior beliefs that they modify when receiving new messages. Beliefs were represented semantically in a specific framework which revealed further to be identical to that of epistemic logics, more precisely a semantical (possible world) framework. Moreover, strong assumptions were implicitly introduced: truth (what a player knows is true), positive introspection (a player knows what he knows), negative introspection (a player knows what he does not know). Hence, beliefs were expressed as 'information partitions' in a possible world space. When some world is the actual one, a player ignores in what world of the corresponding partition cell he is.

In the same period, learning rules were introduced and form the basis of a new research program labelled as 'evolutionist[1] game theory'. The classical program considers that the players are strongly rational and optimize their behaviour. The new program integrates the fact that the players are boundedly rational and just react to their environment in a utility improving direction. First, the models were based on biological analogies where the players are exclusively directed by mutation and selection mechanisms. The players consequently abandon all their reasoning faculties and adopt zero rationality. Later, psychological features were reintroduced since the players react to new observations by revising their beliefs or adapting their behaviour rules. Moreover, random elements were added in the description of players' encounters, observations, expectations and decisions. In these models, the work of crossed beliefs is replaced by the work of repeated experience. Technically, since the players have a purely reactive behaviour with randomness, their sequential actions follows some stochastic process.

The equilibrium notions are completely revisited by these original views. The main problem about a usual equilibrium notion is that it is not constructively defined. An equilibrium state is a stationary state in the sense that if the players find themselves in it, they have no (cognitive or preferential) incentive to deviate from it. But the modeller exhibits no concrete process by which the players come to such an equilibrium state. Especially, he does not make precise how an equilibrium state is selected in case of multiplicity. At the opposite, when introducing beliefs, it becomes possible to study under what conditions hyper-intelligent players come to an equilibrium state by their sole reasoning. These conditions constitute the 'epistemic justifications' of the usual equilibrium notions (see § 6.3.). Likely, when introducing learning, it becomes possible to study under what conditions a dynamic process followed by boundedly rational players asymptotically leads to an equilibrium state. They constitute the 'evolutionist justifications' of the usual equilibrium notions (see § 7.3.).

2 INDIVIDUAL GATHERING OF INFORMATION

In game theory, players are endowed with beliefs conceived as mental representations of their surrounding environment. These beliefs are considered as imperfect as well as incomplete with regard to the modeller's model acting as a reference. They are formalized in the same terms than the model itself, but with truncated relations between partial variables. Information is then conceived as some message received by the player from outside either by direct observation or mere communication. It gives further details about some actual phenomenon or about some law followed by the system. It is formalized in a very simple way, as a more precise specification of a variable or as a signal correlated to the actual value of the variable.

[1]Evolutionist refers to a general dynamic process while evolutionary refers to a biology inspired dynamic process.

2.1 *Sources of uncertainty*

Game theory assumes that the modeller, like God, has a complete and perfect model of the system under study, including of course the players' beliefs. This model is represented by a set of generic relations between observable or non observable variables and defines alternative paths of the system. Besides, game theory assumes that each player retains beliefs according to the same general ontology than the modeller, as concerns the basic entities and their interactions. But his private model is more imperfect and incomplete than the modeller's one, and even possibly false. Uncertainty is associated to some event or to some structural feature of the system. It concerns the value of some variable or the specification of some relation, especially some parameter. To compensate for uncertainty, information makes more precise some fuzzy element of the overall system. It concerns some statement about the actual value of a variable or the true specification of a relation.

According to the modeller's view, each player faces nine basic types of uncertainty obtained by crossing two independent criteria about the uncertain characteristics of the system. On the one hand, considering the concerned entities, a player may be uncertain about nature (physical uncertainty), about the other players (actorial uncertainty) or even about himself (personal uncertainty). On the other hand, concerning the entities' attributes, he may be uncertain about past events (factual uncertainty), about atemporal structural features (structural uncertainty) or about future events (strategic uncertainty). More precisely, the considered (past or future) events are the nature's states or the players' actions. The considered (atemporal) structural features are the nature's generation law (explaining the genesis of states) and the consequence law (linking states to actions and consequences) as well as the players' determinants. The last are generally summarized in players' types having their own generation law.

Likely, each player is able to receive information from various sources before or during the play of the game. The player has some prior information about the game structure, especially about the other's determinants. He becomes naturally informed about some past events, since he observes realized states of nature, watches implemented actions of another player or feels his own utility obtained by a past action. Information received by a player is believed by him with some degree of credibility according to its source. But information is distinguished from belief in the following way. Information is considered as a flux coming from the environment of the player by observation or communication (hence is explicit). Belief is a stock anchored in the player himself and transformed by outside information or by inside restructuring (hence may be tacit). Moreover, information is generally an elementary mental state while a belief is more structured.

2.2 *Structure of uncertainty*

Uncertainty affected to some (discrete or continuous) relevant variable can be expressed semantically under two main forms. In the probabilistic form, it is expressed by a probability distribution defined over its values. In the set-theoretic form, it is expressed by a subset of its values to which the actual value belongs. The two cases cannot be easily compared, even if certainty appears as a limit case of both (probability 1 on some value or unique value). Probabilities are objective when the modeller considers the phenomenon as really stochastic and when it informs the player of the actual probabilities, the last endorsing them. Such probabilities are frequencies (in a sequence of similar events) or proportions (in a population of objects). Probabilities are subjective when the player forms a personal probability about the phenomenon, whether or not actually stochastic. Such probabilities are logical (expressing a relation between variables) or decisional (expressing the willingness to bet for a player). The last ones may be revealed by the modeller from player's actions, under strong conditions.

Uncertainty may also be expressed under a hierarchical form, meaning that a player is uncertain either about his own or about other's uncertainty. On the one hand, a player may have some second order uncertainty on his first order uncertainty. The first order uncertainty is generally objective (basic uncertainty) while the second order uncertainty is always subjective (ambiguity). Each level may be expressed in a probabilistic or a set-theoretic form, leading to original two-level belief structures: a probability distribution on probability distributions, a probability distribution on sets defining Dempster-Shafer belief functions [Shafer, 1976], a set of probability distributions defining multi-prior belief functions [Gilboa and Schmeidler, 1989]. On the other hand, a player may have a second order uncertainty on the first order uncertainty of another player. Here again, the second order uncertainty is a subjective one while the first is of any kind. The levels are then generally expressed either both in a set-theoretic form or both in a probabilistic form.

In game theory, uncertainty is expressed in a way which is precisely adapted to the badly known element as illustrated by some examples. Firstly, uncertainty of a player about nature's states can be expressed in a more or less precise way [Knight, 1921]. Bernoullian uncertainty happens when the player forms a probability distribution over the states. Knightian uncertainty happens when the player just knows the set of possible states without weighting their respective occurrences. Radical uncertainty happens when a player does not even know the set of possible states, due to 'unexpected contingencies'. Secondly, uncertainty of a player about the other's player type (and even about his own type) is usually expressed in a probabilistic form (the players' types are in fact treated as initial states of nature). Thirdly, uncertainty of a player about another player's past actions is expressed in a set-theoretic way, by 'information sets' gathering all nodes he cannot discriminate

2.3 Structure of information

Information is generally represented by a 'message' received by a player. Such a message concerns directly a given variable or parameter or indirectly some other variable correlated to it and called a 'signal'. For instance, when checking the existence of gas in the soil, a gas company may search for salt when assuming that the presence of salt is correlated with the presence of oil. Hence, information is modular in the sense that it appears as a 'psychical quantum' understandable independently of another piece of information or of prior beliefs. Information is endowed by the modeller with a truth value since a direct observation is correct or not while a communicated item is true or not. Information is moreover unambiguously interpreted by players since it does not depend on its material support or its language. More precisely, it is univocally interpreted by each player in the sense that he knows what variable is concerned. Likely, it is interpreted in the same way by all players since they agree on a structural representation of the system.

Information is expressed in different forms, but it depends now on the actual value of the relevant variable. A message is set-theoretic when it indicates, for each actual value, a subset of possible values. A message is probabilistic when it indicates, for each actual value, a probability distribution over the possible values. In the last case, the player receives, in a set of possible signals, some specific signal and, knowing the probability of that signal conditional on the actual value, he computes a probability distribution on the real value. When usual properties are attributed to the message (truth, positive introspection, negative introspection), it is defined either as a partition on the set of values or as a unique probability distribution on the values. In many applications, a player receives both public information characterized by a prior objective probability distribution and private information characterized by a subjective partition, specific to each player.

With regard to information, two extreme types of action are considered for any player. An 'operational action' aims at producing some material consequences. An 'informational action' aims at gathering some additional information. In fact, a same action may simultaneously provide information and produce a material impact. More precisely, information can be obtained in three ways, corresponding to different forms of 'experimentation'. Firstly, in 'exogenous experimentation', a player buys some information at some cost from a specialized instance. Secondly, in 'passive experimentation', a player obtains information as a by-product of an operational action, generally for free. Thirdly, in 'active experimentation', a player obtains information by deviating voluntary from some efficient operational action. He looses some utility (or incurs some cost) at short term, but compensates by the advantage brought by information in enlightening operational actions at long term.

3 INDIVIDUAL TREATMENT OF INFORMATION

A player is willing to acquire information since it may help to modify his beliefs in the direction of truth. The fundamental mental operation is then belief revision, which can be implemented for any message, true or false. Belief revision is a prototypical reasoning operation which is realized according to some rules independently of choice. Individual belief revision concerns only beliefs about the material environment while collective belief revision is extended to crossed beliefs between agents. Moreover, information is helpful when a player 'knows more' with his final belief than with his initial one. In order to check this, the modeller has to formalize the fact that some belief is more 'accurate' than another.

3.1 Individual belief revision

Consider first a single decision-maker who holds an initial belief over his material environment (summarized by states of nature). The decision-maker receives a message on the actual state, which may be compatible or contradict his initial belief. Two main revision contexts are generally considered. In a 'revising context', the player just receives some additional information about a fixed environment. In an 'updating context', the player receives information about how an evolving environment has changed. A third context, the 'focusing context', happens when a specific instance of the environment (a state of nature) is sorted out and information is given about it. Formally, the focusing context can be reduced to a revising context if a 'projection principle' is satisfied. For instance, a merchant may sell different products according to future meteorological conditions (tempest, rain, clouds, little sun, high sun) crudely assessed. In a revising context, he learns that it never rains in the region. In an updating context, he learns that a depression affects globally the meteorological conditions. In a focusing context, he learns that it does not rain this day.

The revision process, conveniently represented by a revision rule in semantics, is sustained by an axiom system in syntax. For set-theoretic beliefs, in a revising context, the AGM axiom system [Alchourron *et al.*, 1985] designs a 'conditioning rule'. In the possible worlds space, nested coronas more and more distant from the first one constituted by the initial belief are defined. The final belief is just the intersection of the message with the first corona intersecting it. Especially, when the initial belief and the message are compatible, the final belief resumes to their intersection. In an updating context, the KM axiom system [Katsuno and Mendelzon, 1992] designs an 'imaging rule' in a similar way. When dealing with probabilistic beliefs, the usual revision rule used in game theory is Bayes rule, restricted to the case where initial belief and message are compatible. The posterior probability distribution is obtained by renormalizing homothetically the prior distribution to the worlds not excluded by the message. In fact, Bayes rule is relevant only in a revising context and can only be justified by very strong axioms [Walliser and Zwirn, 2002].

Belief revision is strongly related to other reasoning modes attributed to a player in game theory in order to solve specific problems. It appears as his central reasoning mode from which the other modes are variants. A correspondence can be established in syntax between their respective axiom systems and in semantics between their corresponding rules. For instance, nonmonotonic reasoning is isomorphic to belief revision in a revising context [Kraus *et al.*, 1990]. Likely, abductive reasoning (used in game theory for an explanation of other's player behaviour) is isomorphic to some reverse belief revision process, always in a revising context [Walliser *et al.*, 2005]. Finally, counterfactual reasoning (used in game theory for prediction of other's player behaviour) is isomorphic to belief revision, but in an updating context [Stalnaker, 1968]. Other reasoning modes attributed to players remain unrelated to belief revision, for instance analogical reasoning (used in game theory for case-based reasoning) or taxonomical reasoning (used in game theory for game categorization of the game structure).

3.2 Collective belief revision

In a game, each player treats his available information in order to reduce sequentially the different types of uncertainty he faces. First, he uses his acquired information in order to reduce factual uncertainty by implementing directly some belief revision process. For instance, when he gets a message about the actual state of nature, he revises his prior belief about the actual state accordingly. Second, he uses his factual beliefs in order to reduce structural uncertainty by implementing some abductive process. For instance, when observing some actions of another player assumed to be rational, he tries to reveal the other's preferences (and/or beliefs) by making some case about him. Third, he uses structural beliefs in order to reduce strategic uncertainty by implementing some counterfactual reasoning. For instance, when knowing the determinants of another player assumed to be rational, he infers more or less precisely what actions he will choose in a game tree, even in nodes which may not be reached by the equilibrium path.

Consider now that each player is endowed with a hierarchical belief structure in a set-theoretic form. Such a structure expresses the crossed beliefs of the player about the states of nature (I know that you know... that p). The player then receives a message, defined both by its content and its status. The 'content' of the message expresses as usual what information is given to the players. For instance, one defines a material message (about the state of nature) or an epistemic message (about a player's belief on the state of nature). The 'status' of the message indicates to whom the message is diffused and what the players know about that diffusion. For instance, one defines a public message (the message is sent to all players and this is common belief), a private message (the message is sent to one player, the other knowing that the first received a message, but not its content, and all this is common belief) or a secret message (one player receives a message, the other being unaware of this). But a lot of other types of messages are conceivable (a private message believed public, a quasi secret message). Such a two-dimensional message can conveniently be expressed by an auxiliary belief structure, called the message structure.

Moreover, a 'specification message' is a message which does not contradict the initial belief, hence does not surprise the player. However, the initial belief and the final belief can include errors, as illustrated by a secret message which turns a true belief into a false one. In the framework of dynamic logics, a multi-agent belief revision rule combines in a precise way the initial belief structure and the message structure in order to obtain a final belief structure [Baltag and Moss, 2004]. The syntactic counterpart of the revision rule can be expressed by a few axioms, the main one being some kind of *modus ponens* [Billot *et al.*, 2006]. It assesses that a player believes a proposition in the final belief when he learns a message and believes initially that the message entails that proposition (or conversely for message and initial belief). The case of a 'rectification message', which contradicts the initial belief, is harder to deal with and needs again to introduce an order between possible worlds.

3.3 Accuracy orders on belief structures

Two belief structures can be compared by defining 'accuracy orders', which express that one structure is more informative than another. For set-theoretic structures, in semantics, stronger and stronger accuracy orders can be defined [Billot *et al.*, 2006]. In semantics, a relation is defined between corresponding worlds and conditions are expressed on the accessibility relations in these worlds. For instance, a belief structure is collectively more accurate than another if, in two corresponding worlds, the accessibility domain of the first is always included in the accessibility domain of the second. This just means that, in any world, each player considers fewer worlds as accessible in the second than in the first. The less accurate structure considers all worlds as accessible in each world and the most accurate structure considers only itself as accessible in each world. These accuracy orders receive again well defined syntactical counterparts. For probabilistic structures, an accuracy order is less obvious to define. However, a first probability distribution was defined as less accurate than a second one if it is a 'mixture' of it.

A specific order on belief structures can be constructed when these structures are defined in a group of players and concern some given material proposition. In syntax, the proposition is 'distributed belief' when the players may (jointly) deduce it by gathering their beliefs. The proposition is 'individual belief' when one player at least believes it. The proposition is 'shared belief' when all players believe it. The proposition is shared belief at order k when all players believe it, believe that the others believe it and so on till level k. The proposition is 'common belief' when it is shared belief at any order [Lewis, 1969]. This hierarchical definition of common belief is weaker than a circular (fixed point) definition [Barwise, 1988]. The last states that a proposition is common belief if everybody believes that it is true and common belief. All these operators have well defined semantic counterparts. Especially, in semantics where players have information partitions, the 'common belief partition' is the finest partition of the coarsened mutual partitions.

Coming back to belief revision, the various types of messages can be partially ordered according to their relative accuracy. For instance, a public message is collectively more accurate than any other message, for instance a private message or a null message. A fundamental result links accuracy orders to belief revision. It states that the accuracy order is preserved when carried from the message to the final belief [Billot *et al.*, 2006]. More precisely, for a given initial belief structure, if some message is more accurate than another (in any sense), the corresponding final structure is itself more accurate (in the same sense) than the other final structure. For instance, the final belief obtained by a public message is collectively more accurate (when compared to the null message) than the initial belief. Such a condition, satisfied by a specification message, precisely states that the (true) message has improved the player's knowledge.

4 COLLECTIVE IMPACT OF INFORMATION

Player's information acts not directly on his selected actions, but only indirectly through his revised beliefs. Since player's beliefs are affected by uncertainty, the choice rules and equilibrium notions are adapted to it. Information gets evaluated no more essentially with regard to its contribution to truth, but to its utility in decision. The value of information is precisely introduced in order to check if more accurate information is also more efficient. As expected, the answer is positive when considering individual decision-making against probabilistic uncertainty. But surprisingly, it may be negative in a game context since information has a complex impact on crossed beliefs.

4.1 Choice under uncertainty

When a single decision-maker is confronted to nature, cognitive rationality resumes to belief revision and instrumental rationality to utility maximization. For a static choice against nature, the relevant choice rule is the maximization of expected utility. When the law of nature is probabilistic and the decision-maker is aware of these objective probabilities, he computes the choice rule according to these probabilities [von Neumann and Morgenstern, 1944]. When the law of nature is unknown to him, he applies the same choice rule, but according to subjective probabilities [Savage, 1954]. For a dynamic choice against nature, the same choice rule is again assumed to be relevant, but associated with a new principle, the 'backward induction principle'. The decision-maker proceeds in the 'decision tree' (the game tree reduced to one player and nature) from the terminal nodes to the root node. At a nature's node, he computes the expected utility over states endowed with progressively revised probabilities. At a decision-maker's node, he retains the action which maximizes expected utility.

The choice rules receive cognitive justifications under the form of an axiom system relying on preferences about strategies. This axiom system was progressively extended from Bernoullian uncertainty to Knightian uncertainty, from static choice to dynamic choice. The choice rules receive moreover pragmatic justifications, especially the Dutch book argument stating that if a decision-maker does not maximize his expected utility, he can be confronted to a sequence of choices at the end of which he always looses. The choice rules receive finally evolutionist justifications, stating that in competition with others, a decision-maker who does not maximize his expected utility will be eliminated. In other respects, two main interpretations are generally given to the choice rules. The instrumental interpretation considers that the decision-maker behaves as if he maximized expected utility (like a billiard player). The realist interpretation considers that the decision-maker consciously maximizes his expected utility (like a poker player).

When several players are involved in an uncertain and dynamic game, a specific equilibrium notion called 'perfect Bayesian equilibrium' is stated. It extends simultaneously a subgame perfect equilibrium and a Bayesian equilibrium. At each information set of the game tree, the decision-maker associates a probability distribution to the constituting nodes (expressing his beliefs about the past moves of an opponent). The two usual rationality principles are then applied. Instrumental rationality states that, at each node, the player chooses the action which maximizes his expected utility, according to the fact that the future nodes were already optimized (backward induction procedure). Cognitive rationality states that the probability distribution, at each node, is adjusted along Bayes rule, with respect to the information gathered about past moves (Bayes conditioning procedure). Hence, uncertainty about another player's action is treated in the same way that uncertainty about the nature's state, even if the first is endogenous and the second exogenous.

4.2 Information value in individual decision-making

Consider a decision-maker who, before choosing an operational action, proceeds first to an exogenous experimentation. More precisely, he is endowed with a prior probability distribution about the states of nature and may buy in some agency a partitional message. As for any item, he chooses to buy the message if its (exogenous) cost is smaller than its value. By definition, the information value brought by the message is the difference of actor's expected utility before receiving the message and after receiving it. In a fundamental theorem, Blackwell [1951] proves that the information value of a message is always positive for a strongly rational decision-maker. The decision-maker cannot be worse after receiving the message than before receiving it. However, the information value may nevertheless be negative in two unusual cases: the decision-maker adopts another choice rule than expected utility maximization, the message is not a partitional one. Similar results are obtained with probabilistic messages.

Consider now that the decision-maker is involved in a sequential choice in which he acquires progressively some information by active experimentation. More precisely, he faces a repeated choice process with an infinite horizon, receives a payoff at each period and aggregates them thanks to a discount factor. He faces then a typical trade-off between exploration and exploitation. Exploration consists in gathering as much information as possible, exploitation consists in using at best the available information. In fact, this trade-off is automatically solved by computing the dynamic optimal choice rule. Intuitively, the decision-maker will do much exploration at the beginning of the process and much exploitation at its end. Moreover, an increase in the discount rate induces more exploration at the beginning of the process. When the discount rate tends to 1, he proceeds only to exploration in the first periods till acquiring the information he wants, then he shifts to pure exploitation.

For instance, consider that the decision-maker is confronted in a casino to a two-armed bandit. He may use at successive periods one of two levers with a random effect. Each lever gives him a payoff of 1 with a given probability and a payoff of 0 with the complementary probability. He faces structural uncertainty since the probability of winning with each lever is unknown to him. He is only endowed with a second-order probability distribution about the probability assigned to a positive payoff (for each lever). The optimal behaviour can be proved to be a deterministic index rule [Gittins, 1989]. In all periods, the decision-maker associates an index to each lever, which depends on the prior probability and on the discount rate. In simple cases, the index aggregates an exploration value and an exploitation value of each lever. Within a period, the decision-maker chooses the lever with greatest index, observes the payoff he obtains and adapts the index consequently. After some time, the decision-maker always uses the same lever, even if he has a (small) probability of using the wrong one.

4.3 Information value in games

In a one-shot game involving nature, the players receive from outside a set-theoretic message of any content and status about its actual state. The information value of the message for some player is again the differential utility he gets at some Bayesian equilibrium state computed before and after receiving the message. In fact, different notions of information value, more and more averaged, can be defined and assessed. The actual value is the utility differential really improved by the player, but it is only known by the modeller. The ex post value is the utility differential measured with the final beliefs, and it is computable by the player after he got the message. The ex ante value is the utility differential measured with the initial beliefs, and it is computable by the player before receiving the message, hence allows him to decide to acquire it or not.

Contrary to decision-making against nature, the ex ante information value in games may well be negative for any player [Kamien *et al.*, 1990]. More precisely, when one of two players receives a message, all combinations of signs of information value may be realized : both may become better, both may become worse, the receiver may become better and the other worse, the receiver may become worse and the other better. However, under technical assumptions, the information value is always positive for the receiver of a message for some types of messages and some classes of games. The first case corresponds to a secret message in any game [Neyman, 1991]. The player is then in a similar position than a single decision-maker. The second case corresponds to a private message in a zero-sum game (Gossner-Mertens, 2001). The player receives a message such that its impact is opposite to the impact of the message received by the other player. The third case corresponds to a public message in a pure coordination game. The players act as a team and both can only be beneficial from the message.

In a repeated game involving nature, the players are again confronted to the exploration-exploitation dilemma. The last may concern the nature's law, but also the distribution of players' types. The dilemma is far trickier than for an individual decision-maker and receives no general solution, but it can be solved in specific situations. For instance, consider an investor in some product, who is confronted to a high or low demand with some prior probability. His investment can be realized in one step or decomposed in two successive steps, the second option involving an additional cost with regard to the first option, but allowing him to observe the demand after the first step. Hence, at the first period, he can adopt an irreversible option (invest completely) or a flexible option (invest partially, observe the demand and complete the investment if and only if demand is high). It can be shown that, in the choice process, the flexible option has to be given some 'bonus', which is precisely equal to the information value given by the message about the demand.

5 INDIVIDUAL PROVIDING OF INFORMATION

A player not only receives information from other players, he also provides information to other players. Transmission of information is again a voluntary action, but is realized either directly or through a material action. Information becomes then a strategic item which is delivered only if it is in the player's interest. Such a phenomenon is precisely studied in specific classes of games with asymmetric information such as 'signalling games'. The sender, provided with some information about the context, may transmit a message to the receiver who acts materially according to it. The results obtained are based on equilibrium concepts which assume that the players simulate each other, and even form self-fulfilling expectations.

5.1 Asymmetry of information

Many concrete situations involve the existence of prior asymmetric information between players. Especially, a player knows generally more about himself than about the others. Such an asymmetry may be reduced if a player provides his information to a second one by direct transmission or through an action (which reflects more or less some information he holds). However, depending on how the second player is expected to use such information, hence on the utility differential it will induce on him, the first player has an interest or not to provide it. When information is directly communicated, he may provide it, abstain to provide it or even distort it. When information transits through an action, he may implement the intended action, render the action fuzzy or even try to transmit biased information. Two situations are usually distinguished, according to the item concerned by information. 'Moral hazard' happens when a player has no interest to publicize the action he implements. 'Adverse selection' happens when a player has no interest to diffuse a state of nature he privately knows (for instance his type).

Reflecting the moral hazard situation, the 'agency game' considers two players (principal and agent) facing nature [Grossman and Hart, 1983]. The agent first performs some action of interest for the principal. Nature provides then a message — related both to that action and to its actual state — to the principal. The principal finally observes the message (and knows its dependence on the action), but not the agent's action itself, and gives a retribution to the agent according to the signal. The payoff of both players depends on the agent's action and on the principal's retribution, and eventually on the message itself. The problem faced by the principal is to induce the agent to act in his own interest by a 'contract' fixing an adequate retribution. In a perfect Bayesian equilibrium state, the agent takes an action which departs more or less from the action he would have token if this action were observable by the principal.

For instance, on a car insurance market, the insured may or not implement some self-protection action, nature indicates the occurrence of an accident (which includes a random element) and the insurer just observes the accidents and pays a reimbursement linked to the premium. However, the insurer may adapt the premium and the reimbursement to the number of past accidents or even to variables more correlated to the self-protection actions (driver's age). Likely, in a firm, the employee is able to modulate his effort rate at work, nature indicates the production of the firm (which depends on other random factors) and the employer just observes the production and gives a wage to the employee related to the production. Here again, the employer may adjust the wage not only to its final profit, but to variables more correlated with the effort rate.

5.2 Signalling games

Reflecting the adverse selection situation, a 'signalling game' considers two players (sender and receiver) facing nature [Rasmusen, 1989]. Nature first defines a state according to some probability distribution which is common belief. The sender observes the actual state and sends one of two signals to the receiver about that state, eventually a mixed one (probability distribution over the pure signals). The receiver observes the signal, but not the state of nature, and implements an action, eventually a mixed one (probability distribution over the pure actions). The payoff of both players depends in general on the state, on the signal as well as on the action. However, in 'cheap talk' for instance, the players' payoff does not depend on the signal, hence one player may talk freely to the other by expressing what is in its interest. Finally, the receiver revises his beliefs (the probability of the state conditional to the signal) according to public information.

Two contrasted types of perfect Bayesian equilibrium states may appear. In a 'separating equilibrium', a different signal is transmitted by the sender in each state of nature, hence the receiver is able to reveal from the signal the actual state. The sender has an interest to transmit his private information and the receiver learns it perfectly. In a 'pooling equilibrium', the same signal is transmitted by the sender for all states of nature, hence the receiver is not able to reveal anything. The sender has no interest to transmit his information and the receiver learns nothing. Some hybrid equilibrium states may also be available for which the message transmitted by the sender is a mixed one. The equilibrium state which actually occurs is conditioned on the main parameters of the problem (probability of states, cost of actions). For given values of the parameters, one or more equilibrium states may happen simultaneously.

For instance, on a health insurance market, nature defines the actual risk of illness of the insured, the insured gives a signal to the insurer in form of the degree of insurance he wants to buy and the insurer fixes the insurance premium according to the cover degree. In a separating equilibrium, the insurer is able to differentiate high risk insured (asking for full cover) and low risk insured (asking for partial cover). Again, the insurer may condition the treatment reimbursement to variables correlated to the health (age, gender). Likely, on a car market [Akerlof, 1976], nature fixes the quality of the car (with a probability commonly known), the sender knows the quality of the car and proposes some price, the buyer observes only the price and accepts or not to transact. In many cases, only a pooling equilibrium happens, the one where nobody transacts. However, when the price is exogenously fixed, a transaction always takes place with private information, but fails when information about the quality of the car becomes public. This is a specific case of negative value of information for both players.

5.3 Self-fulfilling expectations

The previous examples show that the players make some expectations about the others' behaviour and these expectations are realized at the equilibrium state. In fact, such self-fulfilling expectations can happen at two levels. At the first level, they concern directly some action of another player. By definition of an equilibrium, these expectations have to be fulfilled at the equilibrium state. At the second level, they concern a relation between some structural variable (especially the other's type) and an action. This relation has again to be satisfied at the equilibrium state, but it may not hold out of equilibrium. However, in both cases, no process is described showing how the expectations are computed by the players, hence how the equilibrium state is concretely achieved. Moreover, if many self-fulfilling expectations are available, hence if many equilibrium states are possible, no process describes how one is selected.

A self-fulfilling expectation, in its structural form, is made explicit in the 'job market game' [Spence, 1973] showing adverse selection. Nature attributes to an employee a strong or weak (exogenous) ability, acting as his type. The employee may acquire a high or low education level at a cost which is inversely proportional to its ability. The employer gives a wage to the employee according to his assessment concerning his type. It is fixed with regard to a prior belief relating causally the ability of the employee to his education level. More precisely, the employer believes that a highly educated employee acquires a strong ability and conversely.

In fact, a reversed causality holds since, for the modeller, education follows from ability. If the equilibrium is separating, such a belief becomes self-fulfilled. The belief considered to be true by the employer contributes to realize what it asserts (except for the direction of causality).

Two more illustrations can be given in a dynamic setting. Firstly, the 'simplified poker game' exemplifies the notion of 'bluff'. Nature distributes a high or low card, the first player stakes or not, the second player asks to see or not (if the first stakes). Of the two forms of bluff theoretically possible for the first player (to stake with a low card, not to stake with a high card), only the first appears at equilibrium. Secondly, the 'repeated entry game' exemplifies the notion of 'reputation' astutely formalized. Nature defines the hard or soft type of a monopolist, the incumbent enters or not, the monopolist is aggressive or pacific (if the incumbent enters). In a one-shot game, a soft monopolist is always pacific while a hard one is always aggressive. In a repeated game, even a soft monopolist may be aggressive in order to acquire a reputation of being hard. At equilibrium, the monopolist is aggressive in the first periods and the incumbent keeps out, then the monopolist stays aggressive only with a given probability, and the first time he is pacific, he looses his reputation and the incumbent enters for ever.

6 COLLECTIVE DIFFUSION OF INFORMATION

At a collective level, communication of information is the main device able to ensure an efficient coordination between the players. An equilibrium state appears no longer as an equilibrium in actions, but becomes an equilibrium in beliefs. A first question is whether some private information becomes shared belief and even common belief after their interactions took place. It is answered in classical puzzles which were developed outside game theory, but are easily reinterpreted in game theory. A related question is whether hyper-intelligent players can coordinate on some equilibrium state by their sole reasoning. Unexpectedly, if some equilibrium notions can easily be justified, this is not the case for Nash equilibrium.

6.1 Communication between players

The fundamental question is how information diffuses among players, according to their more or less convergent interests. Assume that the information of players concerns only states of nature. As usual, each player is endowed both with public information (prior probability distribution on states) and private information (information partition, signal correlated to the actual state). He combines these two sources of information in a Bayesian way. Moreover, the players exchange sequentially some information either by direct communication or through their actions. A first problem is to examine if they achieve asymptotically a shared belief which gathers in some sense all their private beliefs (homogeneisation of beliefs). A second problem is to examine if this shared belief becomes even a common belief (homogeneisation of crossed beliefs). However, when considering similar agents, their actions may become identical even if their beliefs do not.

A general result about communication is the 'not agreeing to disagree' theorem [Aumann, 1976]. In a dynamic version, two actors share a common prior probability over a set of states, receive initially some private information about the nature's state and announce sequentially and publicly their posterior probability about some specific event. The players have no utility associated to their announcement, which means that their announcements have no strategic component. It can be shown that their beliefs converge in a finite number of steps to a common posterior probability of the event. The result follows from the pre-coordination of the players by their common prior probability, reflecting some common culture. To be sure, the result no longer holds when the players have different priors due for instance to different past experiences.

An application of the preceding theorem is the 'no-trade' result [Milgrom and Stokey, 1982]. Consider two risk-averse agents who are able to trade together, an exchange contingent on the actual state of nature. They share a common prior probability and receive partitional private information over the states of nature. The players have opposite interests and the game is moroever submitted to two more restrictive conditions. At a Bayesian equilibrium state, it appears that no trade will actually take place. The reason is that any player thinks that if the other is willing to trade, he holds some private information working in his own favour. Hence, this information is in his disfavour and he will abstain. However, if the players do not share a common prior (for instance, if one has an optimistic belief and the other a pessimistic one about the state of nature), trade may take place.

6.2 Usual puzzles

In the 'three hats problem', three boys have a white or red hat on their head, actually all three hats are red. Each boy observes the others' hat but not his own one. At successive periods, he has to say if he knows the colour of his hat. Before the initial period, an observer says that one hat at least is red, such information being already shared knowledge but becoming then common knowledge. Each boy gets a positive utility if he is right, a zero utility if he does not answer and a negative one if he is false; hence, his utility function depends only on his own action. In a Bayesian equilibrium state, each player gives no answer at the first two periods and answers rightly at the third (as can be shown by iteration on the number of boys). In that case, the players converge towards a common belief about the colours of their hats. Technically, the possible worlds are finite, since they are materially constituted by the possible combinations of hats and associated beliefs about them. With finite worlds, shared belief goes up one level at each period and necessarily becomes common belief.

In the 'Byzantine generals problem' [Rubinstein, 1989], two allied generals have to choose to attack or not a common enemy. One general observes the situation which may be good or bad (with some probability). If the situation is good, he sends a message to the other, but the message has a small probability of being lost. Hence, the second general sends a confirming counter-message which has the same probability of being lost and so on the messages traveling go and forth. The payoffs are such that a general gets a high disutility when attacking alone, and gets a smaller disutility when they attack together in a bad situation. In a perfect Bayesian equilibrium state, the generals never attack. In fact, even if at least one message is sent in both ways, the shared belief that the situation is good never becomes common belief. Technically, the game involves an infinite number of possible worlds, each world expressing either that the situation is bad or that the situation is good and n messages exactly arrived. Since common belief is necessary to jointly attack, this never happens, but some conventions (attack if two messages are sent) make nevertheless an attack possible.

In the 'two restaurants problem', two restaurants such that one is slightly better than the other are situated in a same street. The customers come sequentially in order to choose a restaurant. Public information consists in a common prior probability distribution over the quality of the restaurants. Private information gives to each customer a signal correlated with the relative quality of the restaurants. Additional (public) information comes from the observation of the previous choices of the customers. The payoffs just attribute a higher utility to the best restaurant, hence consider the customers as payoff-independent without externalities. In a perfect Bayesian equilibrium, after some period, all customers go to the same restaurant, with (small) probability that it is the wrong one. The fact that one restaurant is the best one becomes common belief, even if it may be the wrong one. Technically, the possible worlds reduce to two for their material parts, but the beliefs about them are here truly probabilistic.

6.3 Epistemic justifications of equilibria

An extended reasoning process followed by hyper-intelligent players leads to 'epistemic justifications' of static equilibrium notions [Walliser, 2006]. With the only assumptions of common knowledge of the game structure and of the players' rationality, the relevant equilibrium notion is 'sophisticated equilibrium', obtained by iterated elimination of dominated strategies. With the additional assumption that players play independently, the relevant notion is 'rationalizable equilibrium', where each strategy is a best response to others' strategies, considered as best responses, and so on. With the other additional assumption that the players have a common prior on the state space, the relevant notion is 'correlated equilibrium', where an outside entity, the 'correlator', chooses probabilistically an issue of the game and indicates to each player what it should do. Surprisingly, the two alternative assumptions taken together are not enough to justify a Nash equilibrium. To obtain it, it must moreover (rather heroically) be stated [Aumann and Brandenburger, 1995] that the strategies of the players become shared belief (for two players) or common belief (for more players).

Dynamic equilibrium notions are apparently easier to justify cognitively. For an extensive form game without uncertainty, a 'subgame perfect equilibrium' obtains when common knowledge of rationality applies at each node. The problem is that a player may observe a deviation from the equilibrium path during the play of the game and needs to interpret it. According to a standard result [Aumann, 1995], a deviation just cannot happen under the main assumption and the subgame perfect equilibrium is perfectly justified. However, when such a deviation is counterfactually considered as possible, a player may wonder what assumption sustaining the equilibrium is not satisfied [Reny, 1993; Binmore, 1997]. The relevant equilibrium notion depends on that precise assumption. For instance, considering the 'trembling hand' assumption (the player may deviate from the intended action with some exogenous probability) preserves the subgame perfect equilibrium [Selten, 1975]. At the opposite, lack of common belief of rationality enlarges the set of possible equilibria.

In both cases, the results were obtained by introducing more and more epistemic logics in classical game theory. In that respect, the state space of the system, which already includes the nature's state, has to be extended to the players' strategies. Conversely, the selection of some equilibrium state in case of multiplicity is treated in a more informal way. Some 'selection principles' are exhibited which are more or less homogenous with the former 'implementation principles'. A first path is to consider that some states are 'culturally' salient hence are conjointly selected as 'focal points' [Schelling, 1960]. But salience refers to cultural traits which are not included in the game structure and are essentially context-dependent. A second path is to assume that some selection rules are grounded on various properties of equilibrium states (Pareto optimality, simplicity, symmetry) and act as common 'conventions' among players. But the origin of such conventions is not made explicit and may be history-dependent.

7 INFORMATION AND LEARNING

On the one hand, a player gathers limited information since he faces search costs when he voluntary acquires it. On the other hand, a player is endowed with bounded rationality since he faces computation costs when he treats it. Hence, learning models are introduced where the players supply for the lack of cognitive capacities by repeated experience. However, if the strong rationality model is unique, there exists a large spectrum of bounded rationality models. The strategic dimension of game theory is lost since the players just react to past information without expectation loops. Nevertheless, the usual equilibrium notions can easily be justified as asymptotic states of such processes.

7.1 Bounded rationality

Players are more and more considered as endowed with bounded rationality [Rubinstein, 1994], related to limited capacities for treating information [Simon, 1982].

Bounded rationality was initially associated with instrumental rationality. A first primitive model is the 'satisficing model' [Simon, 1957] where an actor chooses the first action which entails results above some aspiration levels on partial criteria. A second primitive model is the 'stochastic choice model' [Luce, 1959] where an actor chooses an action with a probability proportional to its utility. But it is difficult to state precisely the link of these rules with limited reasoning capacities. Hence, bounded rationality is better associated with cognitive rationality. Two further models inspired by Artificial Intelligence are the 'automaton model' (an actor computes his intended action with a finite number of inner states) and the 'complexity model' (an actor has complexity constraints in his computation). More subtly, limited reasoning may be related to some violations of the cognitive rationality axioms. Especially, the actor may lack 'logical omniscience' in the sense that he is not able to deduce all consequences of what he knows.

Of course, the usual equilibrium notions may easily be extended to boundedly rational agents, keeping the idea of a fixed point of players' actions. But bounded rationality is more naturally expressed in learning processes at work in situations where the same players play sequentially the same basic game (a static or a dynamic one). Learning just means that a player behaves by adjusting his actions to what he observed in the past in order to perform better. Due to limited information and bounded rationality, he no more takes into account the strategic dimension of the game. He generally assumes that the other players are not influenced by his own action and have even a stationary strategy (even if he knows that they learn too). Moreover, he holds little prior structural information and relies essentially on factual information. His behaviour rule is fixed, only his action changes, but it now may change even if he is faced to the same situation. Of course, a second order learning may happen on the behaviour rule itself, but the two learning levels can formally be collapsed into one.

Five principles are involved in evolutionist game theory, only the first being common to traditional game theory [Fudenberg and Levine, 1998]. They all introduce some form of randomness. First, the 'utility principle' indicates that each player has a given action set and receives at each period an utility from any combination of players' actions. Second, the 'interaction principle' describes which players meet at each period, the matches being situated in some 'interaction neighbourhood' and involving some 'encounter randomness'. The 'information principle' describes the information gathered by a player, such an information being limited to some 'information neighbourhood' and implying some 'sampling randomness'. The 'evaluation principle' describes how a player treats his past information in order to build 'indices' for enlightening the future, eventually introducing some 'computation randomness'. The 'decision principle' indicates how a player uses the preceding indices in order to compute the chosen strategy, eventually introducing some 'behaviour randomness'.

7.2 *Basic learning processes*

A first learning process, 'belief-based learning' (or 'epistemic learning'), is grounded on a belief revision procedure. Each player first observes the other's past actions. He then revises his belief about the other's behaviour and forms an expectation about the other's future action. In doing so, he assumes that the other follows a stationary mixed strategy. He finally takes the action which maximizes his expected utility, hence ensuring an exploitation behaviour. In order to add an exploration behaviour, he may alternatively and with a small probability implement a random action. Especially, in the 'fictitious play model', a player considers the past frequency of the other's action, transforms it into a future probability of other's action and implements a best response to it. For instance, on the road, a driver observes if the others drive more often right or left and adopts the side followed by the majority.

A second learning process, 'reinforcement learning' (or 'behavioural learning'), is grounded on reinforcement of best actions. Each player first contents with observing the past utility obtained with his own actions. He then computes a 'performance index' for each action by aggregating its past utilities. In doing so, he assumes that the performance of each own action is stationary, even if it concretely evolves. He finally chooses an action with a probability increasing with the past performance. This rule incorporates the exploration-exploitation dilemma since the players use more and more the best performing actions without eliminating totally any other one. Especially, with the 'basic reinforcement model' (Roth-Erev, 1995), a player computes the cumulated utility obtained by each action and chooses an action with a probability which is proportional to that index. For instance, on the road, a driver observes the accidents he had when driving right or left and drives on the side with the less accidents.

A third process, an 'evolutionary process', is no more a learning process, but shares some features with it even if inspired by biology (Weibull, 1995). Each player belongs to a subpopulation of players using the same fixed (pure or mixed) strategy and interacts randomly with players from another subpopulation or from the whole population. He observes nothing (except in order to implement his strategy) and he even computes nothing. He reproduces according to his utility, assimilated to the biological notion of 'fitness', and this ensures exploitation. Moreover, some mutants in small quantity may be introduced randomly in the population in order to ensure exploration. Especially, in 'replicator dynamics', the players of some subpopulation reproduce proportionally to the average utility they get from their interactions, without mutation. For instance, on the road, a driver dies if he meets another one driving on the other side and duplicates if he meets another one driving on the same side.

7.3 *Evolutionist justifications of equilibria*

Of course, these learning or evolution rules can be mixed in different ways. Different players may use different rules, a same player may adapt the rule to the context or to the present stage of a game, the rules may be combined in hybrid ones. Moreover, it is easy to observe that the evolutionary process is isomorphic to reinforcement learning. The proportion of players playing some action is replaced by the probability that a player chooses some action. Hence, if an evolutionary process based on biological analogies was historically an important framework to produce original results, it is more and more replaced by the two more realistic learning processes. However, since these learning processes appear themselves as too passive and past oriented, learning processes tend to endow the players with more elaborated cognitive activity (categorization of the choice frame, analogical reasoning in the choice process).

The modeller may be interested by the transitory behaviour of the learning system since it is such a transitory behaviour which is essentially observable. Nevertheless, he is essentially interested by the asymptotic behaviour, characterized by various notions of convergence. The modeller looks for insights not only about the attractors of the process, but also about its speed of convergence. He studies not only the final distribution of strategies, but the emergence of some spatial or qualitative regularities (segmentation of the agents, construction of permanent links). The main problem is that the learning processes are numerous and lead to dispersed results, especially linked to the types of randomness introduced by each principle. The only result which is valid for almost all processes concerns their capacity to eliminate the (strongly) dominated strategies.

The asymptotic properties of the learning processes provide an 'evolutionist justification' to the equilibrium notions when it can be shown that the system converges to a corresponding equilibrium state. For a basic static game, many processes lead to some Nash equilibrium, either strict pure strategies, or in pure strategies or even in mixed strategies. Some processes even lead to 'refinements' of Nash equilibrium (such as a 'risk-dominant' equilibrium). For a basic dynamic game, many processes lead to a subgame perfect equilibrium. However, some learning processes may converge towards other states, for instance Pareto-optimal states which are not equilibrium states. A good news is that the selection problem is no more relevant in a dynamical view. The system always evolves and, if it converges, it converges towards a well-defined equilibrium state (at least probabilistically). But this state not only depends on the initial conditions, but also on the history of the game ('path-dependency')

Game theory has progressively internalized information and beliefs, information being treated as an item which makes the player's belief more precise with regard to the modeller's model. In order to give to these notions a more formal account, it imported some frameworks from outside, namely probability theory and epistemic logics. By making minimal and empirically naïve assumptions about them, game theory was moreover able to derive strong statements meeting empirical phenomena. Especially, it dealt with the value of information or the diffusion of information and their collective consequences. Other formal developments were ignored, for instance the Shannon theory of communication or the Kolmogorov theory of complexity. The reason is just that it was not yet possible to deduce interesting conclusions from them. However, game theory stays open to new tools, for instance able to deal with the 'meaning' of information or to the 'acceptance' of beliefs.

In its economic applications, game theory was used to explore other phenomena involving heavily information. For instance, various institutions are justified by informational principles, especially when they help to coordinate the agents facing uncertainty. Since Hayek (1973), the competitive market is seen essentially as providing prices which are a good summary of what agents have to know about scarcity and desirability of goods. Some institutions are specialized for dealing with risk, especially the insurance market. Some further institutions sustain the market when information is imperfect or incomplete, especially trust or money. Some other institutions may replace the market by providing a more local treatment of information, such as auction mechanisms. Even institutions framing the market such as language can be analyzed in a game-theoretical framework [Rubinstein, 2000; Benz *et al.*, 2005].

BIBLIOGRAPHY

[Akerlof, 1970] G. Akerlof. The market for lemons: quality uncertainty and the market mechanism, *Quarterly Journal of Economics*, 84(3), 488-500, 1970.

[Alchourron *et al.*, 1985] C. E. Alchourron, P. Gärdenfors, and D. Makinson. On the logic of theory change : partial meet contraction and revision functions, *Journal of Symbolic Logic*, 50, 510-530, 1985.

[Aumann, 1976] R. J. Aumann. Agreeing to disagree, *The Annals of Statistics*, 4(6), 1236-39, 1976.

[Aumann, 1995] R. J. Aumann. Backwards induction and common knowledge of rationality, *Games and Economic Behavior*, 17, 138-146, 1995.

[Aumann, 1999] R. J. Aumann. Interactive epistemology, *International Journal of Game Theory*, 28(3), 263-314, 1999.

[Aumann and Brandenburger, 1995] R. J. Aumann and A. Brandenburger. Epistemic conditions for Nash equilibrium, *Econometrica*, 63 (5), 1161-80, 1995.

[Aumann and Hart, 1992–2002] R. J. Aumann and S. Hart. *Handbook of game theory with economic applications*, 3 volumes, Elsevier, 1992; 1994; 2002..

[Baltag and Moss, 2004] A. Baltag and L. S. Moss. Logics for epistemic programs, *Synthese*, 139(2), 165-224, 2004.

[Barwise, 1988] J. Barwise. Three views of common knowledge, in M. Vardi ed., *Proceedings of the TARK Conference*, Morgan Kaufmann, 1988.

[Benz *et al.*, 2005] A. Benz, G. Jaeger, and R. van Rooij, eds. *Game theory and pragmatics*, Palgrave Macmillan, 2005.

[Billot *et al.*, 2006] A. Billot, J. C. Vergnaud, and B. Walliser. Multiplayer belief revision and accuracy orders, *Proceedings of the LOFT Conference*, 2006.

[Binmore, 1987] K. Binmore. Modeling rational players, *Economics and Philosophy*, 3, 9-55 ; 4, 179-214, 1987.

[Binmore, 1997] K. Binmore. Rationality and backward induction, *Journal of Economic Methodology*, 4, 23-41, 1997.

[Binmore and Brandenburger, 1990] K. Binmore and A. Brandenburger. Common knowledge and game theory, in *Essays in the Foundations of Game Theory*, Blackwell, 105-150, 1990.

[Blackwell, 1951] D. Blackwell. Comparison of experiments, *Proceedings of the second Berkeley symposium on mathematical statistics and probability*, University of California Press, 93-102, 1951.

[Foray, 2004] D. Foray. *The economics of knowledge*, MIT Press, 2004.

[Fudenberg and Levine, 1998] D. Fudenberg and D. Levine. *The theory of learning in games*, MIT Press, 1998.

[Gilboa and Schmeidler, 1989] I. Gilboa and D. Schmeidler. Maxmin expected utility with non unique prior, *Journal of Mathematical Economics*, 18, 141-153, 1989.

[Gittins, 1989] J. Gittins. *Multi-armed bandits allocation indices*, Wiley, 1989.

[Gossner and Mertens, 2001] O. Gossner and J. F. Mertens. The value of information in zero-sum games, mimeo, 2001.

[Grossman and Hart, 1983] S. Grossman and O. Hart. An analysis of the principal-agent problem, *Econometrica*, 51 (1), 42-64, 1983.

[Harsanyi, 1967] J. C. Harsanyi. Game with incomplete information played by Bayesian players, *Management Science*, 159-82, 320-34, 486-502, 1967.

[Kamien *et al.*, 1990] M. Kamien, Y. Taumann, and S. Zamir. On the value of information in a strategic conflict, *Games and Economic Behavior*, 2, 129-53, 1990.

[Katzuno and Mendelzon, 1992] A. Katzuno and A. Mendelzon. Propositional knowledge base revision and nonmonotonicity, in P. Gärdenfors ed.: *Belief revision*, Cambridge University Press, 1992.

[Knight, 1921] F. Knight. *Risk, uncertainty and profit*, Kelley, 1921.

[Kraus *et al.*, 1990] S. Kraus, D. Lehmann, and M. Magidor. Non monotonic reasoning, preferential models and cumulative logics, *Artificial Intelligence*, 44, 167-208, 1990.

[Lewis, 1969] D. Lewis. *Conventions: a philosophical study*, Harvard University Press, 1969.

[LOFT, 1994–2006] LOFT. Proceedings of the Conferences 'Logic and the Foundations of game and decision Theory', 1994; 96; 98; 2000; 02; 04; 06.

[Luce, 1959] R. A. Luce. *Individual choice behaviour: a theoretical analysis*, John Wiley, 1959.

[Macho-Stadler *et al.*, 2001] I. Macho-Stadler, D. Perez-Castillo, and R. Watt. *An introduction to the economics of information: incentives and contracts*, Oxford University Press, 2001.

[Milgrom and Stokey, 1982] P. Milgrom and N. Stokey. Information, trade and common knowledge, *Journal of Economic Theory*, 1982.

[Nash, 1950] J. Nash. Equilibrium points in N-person games, *Proceedings of the National Academy of Science (USA)*, 1950.

[Neyman, 1991] A. Neyman. The positive value of information, *Games and Economic Behavior*, 3, 350-55, 1991.

[Osborne and Rubinstein, 1994] M. J. Osborne and A. Rubinstein. *A course in game theory*, MIT Press, 1994.

[Rasmusen, 1989] E. Rasmusen. *Games and Information*, Blackwell, 1989.

[Reny, 1993] P. J. Reny. Common belief and the theory of games with imperfect information, *Journal of Economic Theory*, 59, 257-274, 1993.

[Roth and Erev, 1995] A. Roth and I. Erev. Learning in extensive form game, *Games and Economic Behavior*, 8, 164-212, 1995.

[Rubinstein, 1989] A. Rubinstein. The electronic mail game: strategic behavior under almost common knowledge, *American Economic Review*, 79(3), 385-391, 1989.

[Rubinstein, 1994] A. Rubinstein. *Models of bounded rationality*, MIT Press, 1994.

[Rubinstein, 2000] A. Rubinstein. *Economics and Language, five essays*, Cambridge University Press, 2000.

[Savage, 1954] L. J. Savage. *The foundations of statistics*, John Wiley, 1954.

[Schelling, 1960] T. Schelling. *The strategy of conflict*, Harvard University Press, 1960.

[Selten, 1975] R. Selten. Reexamination of the perfectness concept for equilibrium points in in extensive games, *International Journal of Game Theory*, 4(1), 25-55, 1975.

[Shafer, 1976] G. Shafer. *A mathematical theory of evidence*, Princeton University Press, 1976.

[Simon, 1957] H. Simon. *Models of man, social and rational*, John Wiley, 1957.

[Simon, 1982] H. Simon. *Models of bounded rationality*, MIT Press, 1982.

[Spence, 1973] A. M. Spence. Job market signalling, *Quarterly Journal of Economics*, 87(3),355-74, 1973.

[Stalnaker, 1968] R. C. Stalnaker. A theory of conditionals, in N. Rescher ed. *Studies in Logical Theory*, Blackwell, 1968.

[Stalnaker, 1996] R. C. Stalnaker. Knowledge, belief and counterfactual reasoning in games, *Economics and Philosophy*, 12, 133-163, 1996.

[TARK, 1986–2005] TARK. Proceedings of the Conferences 'Theoretical Aspects of Rationality and Knowledge', 1986; 88; 90; 92; 94; 96; 98, 2001; 03; 05.

[von Hayek, 1973] F. von Hayek. *Law, legislation and liberty*, University of Chicago Press, 1973.

[von Neumann and Morgenstern, 1944] J. von Neumann and O. Morgenstern. *Theory of Games and Economic Behaviour*, Princeton University Press, 1944.

[Walliser, 1989] B. Walliser. Instrumental rationality and cognitive rationality, *Theory and Decision*, 27, 7-36, 1989.

[Walliser, 2006] B. Walliser. Justifications of game equilibrium notions, in R. Arena and A. Festre eds. *Knowledge, beliefs and Economics*, Edward Elgar, 2006.

[Walliser and Zwirn, 2002] B. Walliser and D. Zwirn. Can Bayes rule be justified by cognitive rationality principles?, *Theory and Decision*, 53, 95-135, 2002.

[Walliser *et al.*, 2005] B. Walliser, D. Zwirn, and H. Zwirn. Abductive logics in a belief revision framework, *Journal of Logic, Language and Information*, 14, 87-117, 2005.

[Weibull, 1995] J. Weibull. *Evolutionary game theory*, MIT Press, 1995.

Part E

Information in the Humanities, Natural Sciences and Social Sciences

INFORMATION IN COMPUTER SCIENCE

J. Michael Dunn

WHAT IS INFORMATION?

Before addressing the topic of "Information in Computer Science," it is obvious that I will have to say something about what is meant by "information." I expect this to be a common preoccupation of the many authors contributing to this volume. Though perhaps not as obvious, it turns out to be equally important that we should be clear what is meant by "computer science," which we will address in the next section.

Information is of course closely linked to knowledge (and in some contexts they are even confused, as we shall discuss below). Let us start then with knowledge. In several of his dialogs, Plato considers the definition of knowledge as "true belief with an account." It is not clear whether Plato himself accepts this definition — in some dialogs he seems to, but in the *Theatetus* he seems to reject this definition. Whatever Plato intended, this definition seems to have been widely adopted in philosophy as "justified true belief" until the seemingly decisive, and in any event widely accepted objections of Gettier [1963]. These objections have led a whole industry of philosophers to add other requirements that supplement, entail, or otherwise account for justification. Just to cite one example, Goldman [1967] says that knowledge requires an appropriate causal connection between the fact which is believed and the existence of the belief, and this was subsequently refined by him so that a true belief counts as knowledge only if it is produced by a reliable process.

I like to think of information, at least as a first approximation, as what is left from knowledge when you subtract, justification, truth, belief, and any other ingredients such as reliability that relate to justification. Information is, as it were, a mere "idle thought." Oh, one other thing, I want to subtract the thinker. Anyone who has searched for information on the Web does not have to have this concept drummed home. So much of what we find on the Web has no truth or justification, and one would have to be a fool to believe it, and it is not even clear that anyone would want to claim credit for thinking it. It is something like a Fregean "thought," i.e., the "content" of a belief that is equally shared by a doubt, a concern, a wish, etc. It might be helpful to say that it is what philosophers call a "proposition," but that term itself would need explanation.

Some people believe information must be true. Floridi [2003] has claimed this, and Fetzer [2004] has responded. Floridi's point has more to do with technical

Handbook of the Philosophy of Science. Volume 8: Philosophy of Information
Volume editors: Pieter Adriaans and Johan van Benthem. General editors: Dov M. Gabbay, Paul Thagard and John Woods.

considerations than natural language considerations, most notably to deal with a semantic paradox from Carnap and Bar-Hillel, namely that on the standard technical definition of information a contradiction contains the maximum amount of information (see below).

Fetzer gives several examples from ordinary life about false information, or "misinformation," e.g., giving wrong directions to Hyde Park. He does give Floridi a conceivable defense of his position, saying that "Floridi might want to defend his position by claiming that false information is to information as artificial flowers are to flowers." I have heard a similar defense in a story of the "Information Booth" in a railway station and how it would be misnamed if it gave out false information. But note that I said "false information" in a very natural way. I think it is part of the pragmatics of the word "information" that when one asks for information, one expects to get true information, but it is not part of the semantics, the literal meaning of the term. If there is a booth in the train station advertising "food," one expects to get edible, safe food, not rotten or poisoned food. But rotten food is still food.

Another way to approach the notion of information is through the so-called "DIKW Hierarchy." It is common in some circles to make a useful distinction between Data and Information, Information and Knowledge, and even Knowledge and Wisdom.[1] It is unfortunately equally common to conflate at least two the first of these three terms. Thus a so-called "data base" might be better called an "information base," and "knowledge representation" is more accurately called "information representation." There seems to be no standard agreement about how to define the elements in the DIKW Hierarchy. Some authors talk of data as symbols or numbers, others allow pictures or sounds. The main point seems to be that data does not necessarily come with a context or interpretation.

Data is what is produced by instruments, but does not become information until it is somehow recorded in an interpreted way — typically these days in a computer. And information does not become knowledge, at least in the traditional sense much discussed by philosophers, until it at least meets Plato's three tests (believed, justified, true). Information technology has enhanced human cognitive abilities much more profoundly than perhaps anything but the invention of writing, and in some sense it is a natural extension of what began with signs and symbols and progressed to the printed book.

But information technology makes information much more immediately accessible than does a book, and it also makes it much more easily manipulable. Many of us when stuck with only a paper copy of an elaborate document have wished that we had an electronic version that we could search or transform. Think of

[1] A very good account of the history of this distinction can be found in Sharma [2005], starting with the poet T.S. Eliot, of all sources, and tracing it through Harland Cleveland, Russell Ackoff, and others. An interesting mathematical abstraction of the DIK layers is in Burgin [2004]. Bo Dahlbom and Lars Mathiassen [1993] give an interesting critique of the traditional "bottom up" DKI pyramid, and in effect invert it, starting with the notion that knowledge is situated. This is somewhat reminiscent of debates regarding logical empiricist philosophy as to whether "sense data" are completely uninterpreted or not.

a spreadsheet as a simple example. The issues of accessibility and manipulability become even more extreme with large, distributed databases, say a genomic database with tools such as BLAST to do similarity searches on the sequences — or to use the ultimate example, the Web with search engines such as Google. It seems to me that examples such as these change in some fundamental sense the old fashioned paradigm of the expert who has internalized not just the belief but also all of the justification for that belief. How many of us have much of a sense as to how BLAST or Google work? I believe that information technology, viewed as an augmentation of the human condition, raises new issues about the meaning of knowledge, or perhaps some extended/updated sense of it, and challenges the idea that information somehow has to be "believed" (internalized, mastered) by humans or other suitable agents in order to be knowledge.

Once we discuss what Computer Science is we shall return to the concept of information, and see how it has been defined in a more technical setting, perhaps the most famous of these definitions being due to Claude Shannon [1948]. In approaching the concept of information we should bear in mind the advice given by Shannon [1993, p. 180]:

> The word 'information' has been given different meanings by various writers in the general field of information theory. It is likely that at least a number of these will prove sufficiently useful in certain applications to deserve further study and permanent recognition. It is hardly to be expected that a single concept of information would satisfactorily account for the numerous possible applications of this general field.

WHAT IS "COMPUTER SCIENCE"?

It turns out that there are philosophical or at least conceptual issues arising in about just what is included under the label "Computer Science."

At various universities and other institutions the units that house at least some computer scientists have many variations in their name and structure. Variants include Computing Science, Computer Engineering, Informatics, Information Science, Information Systems, Information Technology (IT), and various combinations of these and other names, e.g., Computer and Information Science, Library and Information Science, Information and Communications Technology (ICT), Bioinformatics, Medical Informatics, Legal Informatics, etc. This keeps you on your toes in getting the names right.

The exact boundaries of "computer science" are clearly difficult to define, but I take it from the above that it can include the study of computing, not just computing machines, and that it can also include the study of information, at least in digital form. This is not just an accident of institutional nomenclature, or even of technology. As we shall see below, because of the von Neumann notion of a "stored program," the very distinction in a "computer" between computation and information becomes subtle and there is in fact a kind of duality between the two.

COMPUTERS AS INFORMATION ("DATA") PROCESSORS

"Computers" store and process digital information, which can be thought of as a series of bits (1, 0) or a series of switch settings (on, off) or a series of voltages (high, low). The future may replace the current electronics with "spintronics" where the bits can be represented on single electrons by "spin up," "spin down."

The numbers 1 and 0 are abstract, but the numerals "1" and "0," switch settings, and voltages are not. These are implementations of the abstract bits, and they are conventional in nature. Not only might there be different implementations, but they could have been reversed for example. They need to be distinguishable (discrete). In reality, a switch is not just in the two states: closed (on) or open (off). It can instead be in the process of closing (or opening). And a voltage can be somewhere in the middle of the process, say of dropping from high to low. It depends on when the measurement is taken. Computers are typically based on Boolean/2-valued logic, but there have been proposals to base them on many-valued, even continuous logics, because of the transition of voltages.

This helps emphasize the arbitrariness of picking out certain parts of the physical world as implementations of the two bits. Computers may be digital, but the world that they are a part of is not (except at the quantum level, at least when a measurement is made). In practice computers avoid the intermediate values by settling on ranges of "fault tolerance" of the strength of the voltages. Also, a clock counts the number of cycles per second and only measures the voltages at approximately their maximum and minimum. This is why we talk of computers being so-and-so numbers of cycles per second. This relates to the refresh rate of the central processing unit (CPU chip).

Information stored in a computer is, in a purely technical sense, just a string of bits. But in an intuitive/practical sense, the information coded in those bits is a derivative notion depending on the encoding (encoder?) and the program (programmer?) and the interpreter (user?). Must the meaning of those bits derive from human intelligence, or can there be "real" AI that does not depend on the parenthetical items? This is the most profound question for AI. I do not pretend to know the answer, but we shall discuss this a bit more when we talk explicitly about AI below.

COMPUTERS DO MORE THAN "COMPUTE"

Before the advent of the modern computer in the form of a machine, there were people who did complicated series of mathematical calculations, and these people were actually called computers. During WWII many of these were women doing ballistics calculations. Indeed, the first electronic computers were developed at the end of WWII to do calculations for ballistics, the atomic bomb, code-breaking, etc. Sometime in the late 1950's "computers" became information processors. This all led as we know successively to main-frame computers, mini computers, workstations, and most familiar to most of us now, personal computers.

Consider the common uses of "personal computers": word processing, e-mail, calendar, notes, address book, games, digital photos/video, CD/DVD player, etc. And we now have new applications such as VOIP (Voice Over IP).[2] Spreadsheets are the exception in "office suites" — they actually are used in computing in something like the original meaning of the word.

Other "computers" that are not used for computation in any usual sense include: digital cameras, digital thermostats, cell phones, PDA's, digital music editors, slot machines, GPS devices, trains, planes and automobiles, etc.[3] Pervasive or ubiquitous computing envisages computers, or at least network nodes, in more or less everything that we use in our daily lives. With RFID (Radio-Frequency IDentification) tags this can even include the clothes we purchase, and wear, and can lead to issues concerning security and privacy. The digital revolution is with us! Computers greatly enhance our abilities to deal with information, but they also control us to some extent. Think of the PDA (Personal Digital Assistant) as an analog of a real human assistant, who certainly serves both functions.

HISTORY OF THE CONCEPT OF INFORMATION

Early History

I defer to Pieter Adriaans and Johan van Benthem's introductory chapter in this Handbook as a kind of division of labor, and also reference an earlier work of my own on the history of the concept of information, Dunn [2001a].

The part of this last that I need to go over quickly is the development of what has been called a "UCLA Proposition." I believe the term originated with Alan Anderson, but the concept originated in the work of Boole, with his dual interpretation of what is now called "Boolean algebra." Famously, the elements of a Boolean algebra can be interpreted as either classes (operated upon by relative complement, intersection, union) or as propositions (operated upon by negation, conjunction, disjunction). Boole was well aware of this and he termed the first his primary interpretation and the second his secondary interpretation. Boole connected these: a proposition can be regarded as the set of "times" in which it is

[2] A colleague told me that he was startled when his computer "rang" when a "phone" message came in. I replied "I wish I had been there with my cell phone to take your picture." It is clear that not only are computers everywhere but that their universal character allows for devices to blend into each other.

[3] Alan Cooper [1999], the creator of Visual Basic and an advocate for goal oriented interaction design, makes this point very amusingly in asking a series of questions. Question: What do you get when you combine a camera with a computer? Answer: A computer. Question: What do you get when you combine a car with a computer? Answer: A computer. Etc. Cooper's point is that putting a computer in a device does not necessarily make it easier to use, indeed quite the opposite. Thankfully they are usually not programmable by the user, though many modern day "hot rodders" do "chip" their cars, e.g., to remove the built in speed limit.

true [Dipert, 1978]. If we take "times" metaphorically (as something like occasions, cases, states, or possible worlds) we find the beginning of a thread that weaves through Carnap, Montague, Kripke, where propositions are viewed respectively as sets of "state-descriptions," "indices," "possible worlds." Carnap and Bar-Hillel have actually two concepts: "information," i.e., the set of state-descriptions that make a proposition true, and "content" the set of state descriptions that make a proposition false. Of course classically, one is just the set-theoretic complement of the other, and so one can choose to work with either one. This may be an over-simplifying assumption as we will see later.

Carnap and Bar-Hillel also suggested a numeric measure for a UCLA proposition A that could be given by counting the number of states, and they also suggested another numeric measure:

$$Cnt\#(A) = 1 - \text{prob}(A).$$

Computer science represents information as a string of zeros and ones ("bits"). Given a set of indices I, any set $A \subseteq I$ can be understood as an indexed set of bits: 1 if $i \in A$, 0 if $i \notin A$. This is an abstract version of the Carnap and Bar-Hillel notion of information: view the members of I as being abstractions of state-descriptions, and view the indexing functions as characteristic functions.

CLASSICAL INFORMATION THEORY

Shannon Information

While logicians were busy with various variations on the theme we have labeled "UCLA propositions," Claude Shannon [1948] (see also Shannon and Weaver [1949]) was independently developing the quantitative counterpart to a UCLA proposition (actually to its complement). Shannon suggested that we measure the information in a message as roughly the inverse of probability; formally log to the base 2 ($\log_2$) of the inverse of the probability.

> Frequently the messages have meaning: that is they are referred to or correlated according to some system with certain physical or conceptual entities. These semantic aspects of communication are irrelevant to the engineering problem. The significant aspect is that the actual message is one selected from a set of possible messages.

The intuitive idea behind Shannon's measure is that the more surprising a message is, the more information it conveys. If I tell you that the sun will rise tomorrow, this is very unsurprising. But if I say that it won't, this is very surprising indeed, and in some intuitive sense more informative.

This corresponds to the quantitative measure of content proposed by Carnap and Bar-Hillel in that rather than the more probable getting the highest measure, it is the least probable. Carnap and Bar-Hillel used the arithmetic inverse of addition, which is subtraction (from 1). Alternatively, Shannon chose in effect to use the multiplicative inverse, which is division (of 1). So the multiplicative inverse of n is $1/n$. Both have the effect of inverting a high number to a low number (and vice versa) so as to make the more surprising be the more informative.

Shannon's definition corresponds elegantly with the notion of information as a string of bits:

$$\text{Info}(s) = \log_2 \text{Inv Prob}(s) = \text{Length}(s).$$

Thus the information in a binary string is just the length of the string.

Remember that log to the base 2 ($\log_2$) is the inverse of the corresponding power of 2, i.e., the following are equivalent:

$$x = 2^y,$$
$$y = \log_2 x.$$

Thus:

$$\log_2 \text{ of } 2 = 1 \text{ since } 2^1 = 2,$$
$$\log_2 \text{ of } 4 = 2 \text{ since } 2^2 = 4,$$
$$\log_2 \text{ of } 8 = 3 \text{ since } 2^3 = 8,$$
$$\log_2 \text{ of } 16 = 4 \text{ since } 2^4 = 16, \text{ etc.}$$

Formally the Shannon information measure of an event E is given by:

$$\text{Info}(E) = \log_2[1/\text{Prob}(E)].$$

Considering a few examples helps. Let us suppose that someone in the next room is tossing a fair coin. Let us decide that 1 means heads, and 0 means tails. This incidentally indicates a very important aspect of information, which Shannon emphasized. All information is relative to an initial encoding.

If they do just one toss, and I, standing in the doorway, yell out "1," I have given you a certain amount of information. How much? Well since Prob $= 1/2$, Info $= \log_2(\text{Inv } 1/2) = \log_2(2) = 1$.

What if there are two tosses, and I yell "1 0"? This time we figure that since Prob $= 1/4$, then Info $= \log_2(\text{Inv } 1/4) = \log_2 (4) = 2$.

And with three tosses, and "101"? This time we figure that since Prob $= 1/8$, then Info $=\log_2(\text{Inv } 1/8) = \log_2 (8) = 3$.

Not only is this pleasant mathematically, but as Shannon noted it has a common sense appeal as well. If I have 2 books and buy a 3^{rd}(as an idealization let's assume that all are of equal length), from an intuitive point I view I have not doubled the amount of information I possess, rather added just one more book (increased it by half). Shannon's formula above fits nicely with this intuition.[4]

[4]Imagine the economic and practical consequences of pricing books based on Shannon's formula if "$\log_2$" were not thrown in. Before Amazon sells you a book, they would first have to send out an appraiser to find out how many books you already have, or perhaps more feasibly check how many books you have bought from them.

THE "PARADOX OF THE MONKEYS"

Not everyone finds Shannon's definition intuitive. It is a kind of "paradox" that this means that the works of Shakespeare contain less information that a random rearrangement of their letters and punctuation marks. There is the well-known story, commonly attributed to Aldous Huxley,[5] that 100 monkeys, typing randomly for a sufficiently long time, would eventually (through pure chance and statistics), type all of Shakespeare's works. Most of the time, the monkey will type complete nonsense, but occasionally it will type *Hamlet.* Of course *Hamlet* would appear over and over again out of this "mist" of typing (and so would Faulkner's *Sound and the Fury,* and Miss Manners' column from last weekend's newspaper, etc.) but the vast preponderance would be totally meaningless.

The "Paradox of the Monkeys, " as I call it, is that the utter nonsense of most of this typing would contain more information than the small amount of time that *Hamlet* might appear.[6]

On one way of thinking about this, it would seem that when the monkey is typing scenes from Hamlet, the characters are more predictable than when typing total nonsense. We know that spaces occur rather frequently, that most of the strings between spaces (words) occur in a reasonably small dictionary, that the words "Hamlet" and "Ophelia" occur with some relative frequency, etc. The Monkey Paradox, put quickly, is that complete nonsense would seem to carry more information than the words of a great author.

I believe there are several answers, in stages, to this supposed "paradox" (and hence the quotes). First, let us suppose that the monkey is in fact typing completely at random. Then there is in fact no more likelihood that he will type the 18 character string "signifying nothing" than the string "gnihton gniyfingis" (the last is just a reversal of the first). This is just like the fact that if the monkey had typed 3 "a"s, this in no way increases the probability of the next character being "a" — this no more than does a (fairly) tossed coin turning up "heads" 3 times in a row increase the probability that the next toss will be "heads." In truly random sequences, patterns are in the mind of the beholder.

In a recent paper, Dalkilic *et al.*, [2006] have devised a computer program that can recognize "authentic texts" as those that fall into "the sweet spot" between total randomness and total predictability.

Of course, patterns can arise because of hidden causal influences. I am reminded of a story about Raymond Smullyan, a first-class magician as well as a first-class logician, who, while teaching some elementary probability to a class, pulled out a deck of cards and proceeded to deal a Royal Flush. Raymond told the class that

[5]Luciano Floridi has pointed out to me the *Wikipedia* article on the Infinite Monkey Theorem `http://en.wikipedia.org/wiki/Infinite_Monkey_Theorem` (accessed July 24, 2007), which challenges the Huxley attribution and at the same time talks about a number of early anticipations going back to Aristotle.

[6]The "Paradox of the Monkeys" is my name because of my way of putting it, but the point that random nonsense would on Shannon's characterization contain a high amount of information is not original with me.

surprising as this might seem, this hand was just as likely as any other hand. And of course he repeated the procedure producing one Royal Flush after another until the class broke up with laughter.

Can it be that a predictable sequence can in fact provide significant information? Consider the following, which is based on a true event.[7] During the Second World War, the Allies had broken the German's Enigma code. But the Germans used certain special code names for ships, so it didn't matter if one had deciphered the code name — it was still a code name. But the Allies had good reasons to predict that a certain code name was that of a certain ship they knew. So they sent a message with low encryption mentioning something of interest about that ship. They listened to the German traffic, and, as predicted, that information was sent on mentioning that ship using its code name. This of course confirmed that the prediction was right. Thus, even though it was say highly probable that the allies had the right code name, and hence highly probable that it would show up in the German transmission, it seems it still contained significant information when it did.

My reading of this is that what was significantly increased was not the actual information. The probability of the code name being the name of that specific ship was not significantly changed from a purely numerical point of view. Risk was reduced.

It is commonly recognized that the concept of risk involves both probability and the cost of the consequences, and the standard mathematization of "risk" used by insurance companies (and in fact anyone involved in so-called "risk assessment") involves a function in both of those variables. The standard notion is that the risk of an event is a product of the probability and the cost of the consequences, and those of us who are rational use that notion on a daily basis, with some very rough idea of both the probability and the cost of a given event.

I suggest that some similar composite notion is involved with what we might call the "significance" of a piece of information. I am not sure of how to best mathematicize it, but as at least a first approximation I suggest that we just multiply the inverse of the probability by the cost.

SOLOMONOFF-KOLMOGOROV-CHAITIN "ALGORITHMIC INFORMATION"

We mention this just briefly to indicate that there are other possible mathematical definitions of information. Suppose that one has a very long sequence:

$$11.0010010000111111011010101000100010000101101000110000100011010011$$

[7] Or at least I believe it is. I tried finding a reference to this incident, which I believe I once read about someplace, but failed. Fortunately it does not matter to the point I am making with it whether it actually happened or not. A good general reference about the Enigma Machine is Wilcox [2004].

Perhaps it can be given by an even shorter algorithm that computes this sequence. In fact it can as is given by the hint of the decimal point in the third position — this is just π in binary notation and expressed to 64 places. Roughly the algorithmic information of a sequence is the length of the shortest algorithm that generates the sequence. If that algorithm is essentially just to list the sequence, then that sequence is viewed as incompressible or "random."

The notion of "algorithmic information" was developed independently by Chaitin, Solomonoff, and Kolmogorov in the 1960s. The exact history is complicated, and the creators had somewhat different motivations (though all had something to do with information, probability, and randomness).[8] Algorithmic information is known by a number of different names, the most common likely being "Kolmogorov complexity." The general subject heading now though seems to be "Algorithmic Information Theory" (AIT).

Kolmogorov formally defined the complexity of a string s as the length of its shortest description d on a universal Turing machine U. Chaitin's definition is essentially the same. Solomonoff was interested in addressing the philosophical issues of induction in a formal way, and used the idea that the a priori probability of a finite sequence of symbols is determined by the shortest input to a universal Turing machine whose output is the sequence in question.

VON NEUMANN DUALITY

The so-called von Neumann model of a computer emphasizes that there are both static and dynamic aspects of a computer. This is the duality of programs and "data" (stored programs). A stored program is simply a series of bits (information) that can be taken as an input to a program (even itself, perhaps copying it first to make the process transparent). And in principle any series of bits can be called as a program. Yes, it is very likely that it will not execute and there will be a "syntax error" message — but this can be viewed as just an "identity program" that leaves the input unchanged.

The von Neumann model was actually anticipated by Turing's Universal Machine. Turing observed that all of what we now call Turing machines could be enumerated

$$M_0, M_1, M_2, \ldots$$

and then by some clever reasoning he constructed a universal machine U that given an ordinary input n plus the input of the appropriate index i would compute

[8]Solomonoff says he distributed "A Preliminary Report on a General Theory of Inductive Inference" at the conference "Cerebral Systems and Computers" held at the California Institute of Technology February 8-11, 1960. See the preface to the revised version at `http://world.std.com/~rjs/pubs.html` (accessed July 24, 2007). Chaitin claims he had the basic idea in 1962 and worked it out in 1965. See `http://www.cs.umaine.edu/~chaitin/60.html` (accessed July 24, 2007). I do not know when Kolmogorov might first have had the idea before his published paper [1965].

$$U(n,i) = M_i(n).$$

In effect the index i codes up the program that the machine M_i implements. A Turing machine M_i can even be applied in effect to itself by "diagonalization": $M_i(i)$.

One can similarly argue that the von Neumann model was also anticipated by other early models of computation, e.g., by the Schönfinkel-Curry Combinatory Logic and by Church's Lambda Calculus. In both cases these were originally regarded as untyped so that a term can function as either a function or as an argument, and a term can be applied to itself. For some time, it was puzzling how to give mathematical models of these systems because self-reference or self-application is notorious in its tendency to produce paradoxes. Think of the famous Russell paradox of the set of all sets that are not members of themselves.

But Dana Scott, working with Christopher Strachey, produced models of the Lambda Calculus and of Combinatory Logic. There were in fact two kinds of models, a model based on "continuous lattices" first introduced by Scott [1969], and an easier to understand "graph model" introduced by Plotkin [1972] and Scott [1974]. This general approach has been very fruitful in the semantics of programming languages, and is studied under the headings "domain theory" and "denotational semantics" (with slightly different connotations). See Stoy [1977].

The basic idea of the graph model P_ω is to start with the set of natural numbers ω and then to consider a binary relation R on ω. (This is why it is called the "graph model," because a (directed) graph is essentially just a binary relation.) mRn can be thought of as n is a possible output of a computation with input m. Given $A \subseteq \omega$, The R-image of A,

$$R^*(A) = \{n : \exists m \in A.mRn\}.$$

Set-theoretically, R is of course just a set of ordered pairs, and an ordered pair (m,n) can be coded up arithmetically in some standard way as just a single natural number. Scott chose to code (m,n) as $1/2(m+n)(m+n+1)+m$. So R can be regarded as just a set of natural numbers. Now given two sets $A, B \subseteq \omega$, we can view say A as implicitly a relation $R_A = \{(m,n) : \exists k \in A, k = 1/2(m+n)(m+n+1)+m\}$ and define

$$A^*(B) = [R_A]^*(B).$$

Computational magic has occurred. Objects of the same type (sets of numbers) can be applied one to another, and the output is again an object of the same type. An object can be even applied to itself: $A^*(A)$.

There is another way of approaching the von Neumann duality that stems from the Routley-Meyer semantics for relevance logic. This semantics famously uses a ternary frame (U,R), where U is a non-empty set and R is a 3-placed relation on U.[9] The trick we use is to view the ternary relation $R\alpha\beta\gamma$ as a binary relation

[9]In the actual semantics for relevance logic there may be other features on the frame, such as a distinguished world or set of worlds, and an involution * to model negation. We overlook these here.

$R_\alpha \beta\gamma$ indexed by α.[10] This means that for $A \subseteq U$, A can simultaneously be viewed as a UCLA proposition (a set of information states) and as a program, i.e., a set of actions (binary relations) on states. This allows for a semantics for combinatory logic and (with further additions) relation algebras. In the first AB is interpreted as "apply the actions coded up by the states in A to the states in B." For the second we understand both of A and B to be sets of actions (relations) and we take AB to be the relative product of these relations. Details can be found in Dunn and Meyer [1997] and Dunn [2001c] (see also Dunn [2001b]). Comparisons should also be made to "arrow logic" developed by van Benthem [1991] and his collaborators, and also the "logic of information flow" developed by Barwise and Seligman [1997].

INFORMATION RETRIEVAL, INFORMATION REPRESENTATION, NETWORKS, AND DISTRIBUTED INFORMATION

Information Retrieval (IR) has a long and distinguished history. It can be said to go back in antiquity at least to the Library of Alexandria. Unfortunately the story of the burning library underscores the importance to information retrieval the presupposition of information storage. An important reference to the history of information retrieval is Crestani, Lalmas and van Rijsbergen [1998].

There are many ways of representing information that are of direct or indirect interest to philosophers. Those that are the closest to familiar logics are the most obvious. I am thinking of the programming language PROLOG, the Rich Description Framework (RDF) for the Semantic Web, etc.

Perhaps the most remarkable recycling of a philosophical notion is "ontology," which has become very important in the area of knowledge representation and object oriented programming.[11] Recently *Communications of the ACM* devoted an entire issue to the subject of ontology (February 2002, vol. 45, no. 2). One philosophical or at least conceptual issue is just what there is to "ontology" that goes beyond more familiar and mundane classification familiar from such diverse subjects as botany, chemistry, genealogy, library science, and medicine. *Wikipedia* has an interesting discussion of ontologies in computer science and the relations to the philosophical origins of the concept: `http://en.wikipedia.org/wiki/Ontology_%28computer_science%29` (accessed July 23, 2007).

Information representation can be viewed in a quadrant. Along say the side we have Structured/Unstructured, and across the top we have Text/Multi-Media. The primary example of structured information is to be found in a traditional

---

[10] The frame models can actually be seen as generalizing the graph model. Note that in the graph model there is an implicit ternary relation on numbers $Rkmn$: $k = \frac{1}{2}(m+n)(m+n+1)+m$. The difference between the graph model and the frame model is basically one of type level: each point in the graph model codes up an ordered pair, while in the frame model each point represents a set of ordered pairs. The latter essentially reduces to the former when the set is a singleton.

[11] Incidentally knowledge representation, or KR as it is familiarly called, is a good example of the tendency to use "knowledge" when what is really meant is "information."

tabular database where one has tables say with the names of employees, their classification, date of hiring, salary, etc. Relational databases are much more flexible, but they are still relatively rigid. The primary example of unstructured information is to be found in the World Wide Web.

Once upon a time all information was in effect textual, but again the World Wide Web has brought multi-media to the fore. One can browse the Web and find pictures and music, not just text. Perhaps an example of a structured multimedia database would be the Apple i-Store, and an unstructured example would be Napster.

Unstructured information, the Web in particular, emphasizes the importance of search engines. There is an interesting, emerging distinction between memory (storage) and search, particularly when one emphasizes unstructured storage. External storage and search can be seen as genuine extensions of human abilities, much like the first writing on a cave wall or on stone tablets.

Another philosophical issue has to do with the meaning of negation. In traditional data bases one can assume the so-called "closed world assumption." Thus one does not list the non-employees in the database, and one can assume that if Joe Smith is not listed as an employee, then he is not an employee. This was made explicit in the programming language PROLOG, based on a fragment of first-order logic (Horn clauses), with an added "negation as failure."

Belnap [1977] introduced the idea of databases that might contain inconsistent information, and used this to motivate the need for a 4-valued logic to limit the effect of inconsistencies (since in the usual 2-valued logic a contradiction implies anything whatsoever). This was in the high day of structured tabular databases when inconsistent entries were difficult to imagine, and so was very prescient, as the now commonplace inconsistency of the Web demonstrates. While I have given credit to Belnap for the explicit application to databases (and by extension to WWW), the idea of using the 4-valued semantics to limit the effect of inconsistencies is implicit in my dissertation and explicit in Dunn [1976]. See also Dunn [1986] for relevant history. Another motivation for a 4-valued logic might come from a computer network, where the state of information about the network at say a given node is very partial and much of what goes on in the network is invisible — truth values are under determined. This is a concern to security since one would like to know of abnormal behavior anywhere in the network signaling an attack. It might seem harder to motivate a situation where information is contradictory — truth values are over determined. I hope I am not stretching too far by pointing out the growing prevalence of mirror sites and the serious possibility of discrepancies, especially if updating is done periodically.

There are of course attempts to make the Web more structured. "Markup languages" such as HTML and particularly XML add structure to Websites. Incidentally, the increasing importance of audio and video files on the Web and elsewhere raises the question again of just what information is. There is no obvious way to structure audio and video information, and this makes it difficult to search. This is a realm in digital libraries where "metadata" is particularly important.

There is a whole planned architecture of the Web led by the group W3C (World Wide Web Consortium). It starts with the Rich Description Framework (RDF) which is very roughly a very restricted form of positive first-order logic — no negation. RDF viewed as a sublanguage of FOL has just conjunction, existential quantification, and binary predication. The official description of RDF may be found in the W3C *RDF Primer* at `http://www.w3.org/TR/rdf-primer/` (accessed July 23, 2007).

The underlying structure of any expression in RDF is a collection of triples, each consisting of a subject, a predicate and an object. A set of such triples is called an "RDF graph." This can be illustrated by a node and directed-arc diagram, in which each triple is represented as a node-arc-node link (hence the term "graph").

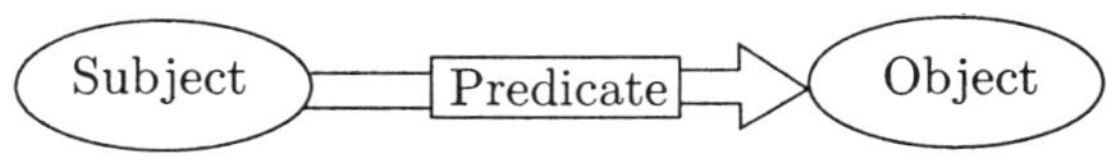

One significant difference between RDF and FOL is that RDF lacks negation. Another is that RDF has only binary predication. While many seeming ternary predications can be reduced to conjunctions of binary predications, it seems not obvious that all ternary predications can be reduced to binary predications, unless that is one adds to RDF the computationally heavy apparatus of set theory.[12] But RDF is intended to be computationally light. One wonders about adopting a "Peircean" framework in which ternary predicates are the primitive, since it seems true, as Peirce thought, that all predication can be reduced to ternary predications, and without using set theory but essentially just conjunction and existential quantification (see Dunn and Meyer [1997]).

There are many issues raised by the digital storage and transmission of information, and by parallel computation on that information. The World Wide Web is just one example in this general framework, and perhaps the most extreme in terms of being the "Wild, Wild Web." We cannot go into these many issues here, but do want to mention the philosophically important work by Dretske [1981], Barwise & Seligman [1997], Devlin [1997], etc. Some of this relates to Shannon's mathematical theory of communication, but there is much more to it.

Let us close this discussion by noting the fact that the form of representation can greatly influence the ease with which information can be manipulated. The introduction of the decimal notation to replace Roman numerals certainly made much easier the school-child arithmetic of addition, multiplication, etc. (and is another good example, in addition to our earlier example of writing, of an advanced enhancement of human cognitive abilities that predates computers). One very important example of how the form can influence the communication of information is visualization, and particularly beautiful examples come from the work of

[12]In set theory, all relations are in effect reducible to the binary relation $x \in y$. A ternary relation is construed as a set of ordered triples, and an ordered triple (a, b, c) is taken to be an ordered pair $((a, b), c)$. An ordered pair (a, b) is taken to be the set $\{\{a, b\}, \{a\}\}$. This is clearly generalizable to n-ary relations as sets of ordered n-tuples.

Edward Tufte, e.g. [Tufte, 1990].

QUANTUM COMPUTATION/INFORMATION

Our discussion here shall be brief because of limitations of space and time, and because the concepts of quantum computation and information are still somewhat speculative. Let me take the occasion to recommend an excellent book to the reader on this subject, Julian Brown [2001]. Although this is a "popular" book it is extremely informative on both the detail and history of quantum computation, and includes a foreword by one of the early proponents of quantum computing, David Deutsch. Deutsch's contributions go back to 1983, and were preceded by Richard Feynman's suggestion in 1979, about using a quantum computer to model quantum mechanics.

In the classical model of a computer the most fundamental building block, the bit, can only exist in one of two distinct states, "0" or "1." We have already indicated that with spintronics bits might be represented as say the spin up or spin down of an electron. This suggests great advantages in miniaturization and hence speed, and this is itself of great practical potential. Gordon Moore, the co-founder of Intel, suggested that computing power (actually literally the number of transistors on an integrated circuit of a given area) doubles roughly every two years, and this has become famously known as "Moore's Law." Moore's Law has been predicted to run out in a decade or so because the transistors on a chip would by then have to be down to the molecular level. But spintronics promises to go past that level and down to the atomic level. But the promise of quantum computing is much more fundamental than extending Moore's Law by some decades, important as that might be.

A "quantum bit" (qubit) can be in the classical "0" and "1" states, but it can also be in a superposition (coherent state) of both "0" and "1." This has to do with the well-known concept of "superposition" in which say an electron can be simultaneously in both the states "spin up" and "spin down" until measured, when it will then "decohere" into a single one of those states.[13] Some examples of qubits include the polarization state of a photon (i.e., parallel or perpendicular polarized to a given axis), an atomic two level system (e.g., hydrogen atom, with the electron in the ground state and the first excited state) and of course poor "Schrödinger's cat" (which is both dead and alive until a measurement is taken by looking at the cat to determine whether a radioactive atom decayed).

Consider a register of 3 classical bits: it would be possible to use this register to represent any one of the numbers from 0 to 7 at any one time. In a register of 3 qubits, the register can represent all the numbers from 0 to 7 simultaneously!

A processor that can use registers of qubits will in effect be able to perform calculations using all the possible values of the input registers simultaneously. This

---

[13]Thomas Siegfried [2000], in his somewhat journalistic but very readable book *The Bit and the Pendulum*, has a clever metaphor for a qubit: it is like a tossed coin, spinning in the air, and neither heads not tails until say you grab it and slap it on the back of your other hand.

phenomenon is sometimes called quantum parallelism. There are thus conceivably more advantages to quantum computing that just the miniaturization given by spintronics. In principle, quantum computations can involve completely new algorithms on qubits that exploit the phenomenon of quantum parallelism. Perhaps the most famous of these is "Shor's algorithm," created by Peter Shor of AT&T Bell laboratories. This algorithm factors a large number into its prime factors. This task is classically so difficult that it forms the basis of RSA public-key encryption, the standard method of public-key encryption used today.[14] This raises strongly the issues of quantum complexity theory and whether this might differ from classical complexity theory, and perhaps even that a quantum computer might be able to solve at least certain NP problems in polynomial time. But it is an open question whether factoring is classically an NP problem.

Another important quantum algorithm, though not as impressive in it seeming speed advantage is "Grover's algorithm" which can search an unsorted list of length n on average in time on the order $\sqrt{n}$ as opposed to the usual $n/2$ for the classical linear search algorithm which checks every element of a list until a match is found. Grover's algorithm starts by setting a quantum register to a superposition of all possible items in the search space. Grover's algorithm involves a sequence of simple quantum operations on the register's state. Grover describes these in terms of wave mechanics: "All the paths leading to the desired results interfere constructively, and the others ones interfere destructively and cancel each other out."

Some interpretations of quantum parallelism have used the "many worlds interpretation" (MWI) of quantum mechanics, first proposed by Hugh Everett [1957] and "popularized" by Bryce Seligman De Witt [1970], who actually gave it the exciting label "many worlds." "Many universes" is actually better terminology that "many worlds," and the term "multiverse" is sometime used for a "super universe" of all possible universes.

The basic idea of MWI, familiar to philosophers from the debate surrounding "modal realism" in the possible worlds semantics for modal logic, is that there are many possible worlds and we are in only one. Where quantum mechanics enters in is that every time a quantum experiment is performed with different possible outcomes, each of these outcomes exists in a different possible world. All are real, even though "I" will be aware of only the one containing the outcome I have seen. I put "I" in quotes because obviously there will be actually a "me" corresponding to each of the outcomes. This of course excludes quantum murder or suicide where I am in effect Schrödinger's cat and have been killed as one of the outcomes.

David Deutsch, whom I already mentioned as foundational in quantum computing, is a believer in the many worlds interpretation. In Deutsch [1985] he suggested quantum computation could take place simultaneously in many possible worlds, giving a kind of parallelism which would give "a method by which certain probabilistic tasks can be performed faster by a universal quantum computer than by any classical restriction of it."

According to Deutsch the single photon interference pattern observed in the

[14]We shall say more about public-key encryption below when we discuss "quantum encryption."

double slit experiment, can be explained by interference of photons in multiple universes. Viewed in this way, the single photon interference experiment is indistinguishable from the multiple photon interference experiment.

It is interesting that regarding the potential for quantum computation to break RSA encryption, that one can "make lemonade out of lemons" by using certain quantum encryption devices, which in fact appear to be making faster practical headway than building quantum computers — there are already several companies offering commercial quantum cryptography. What the quantum taketh away, the quantum giveth back.

As noted above, classical public-key cryptography relies on the computational difficulty of certain mathematical problems, e.g., factorization. However quantum cryptography relies on one of the two peculiarities of quantum mechanics, either Heisenberg uncertainty of quantum entanglement. To convey even a limited knowledge of how these peculiarities are invoked, we must digress and talk briefly about cryptography.

The key to cryptography is of course a "key," i.e., a mechanism for generating cypher text from plain text, and/or vice versa. We are probably all familiar from our childhood with the "alphabetical shift" key, where plain text is encrypted with "a" becoming "b," "b" becoming "c," etc., and "z" becoming "a." The cipher text is of course decrypted by the reverse. A key can actually be regarded as a positive integer, say in this case 1 (for "shift 1," as opposed to 2 for "shift 2"). This is so-called "secret key" or "symmetric" cryptography, because essentially the same key can be used to code (shift 1 left) or decode (shift 1 right). "Public key" or "asymmetric" cryptography uses a pair of keys, a public key known in principle to everyone, and a private key known only to the receiver of the message. When Alice sends a message to Bob she uses Bob's widely distributed public key, and then he decodes it using his private key.

With secret key cryptography the key must be transmitted secretly, either in person or by a transmission (in olden times, a courier). Finding the best way to do this is the so-called Key Distribution Problem. The difficulty with doing this in person is that it takes a great deal of effort, and the issue with transmissions is that they can be intercepted. An underlying problem with secret key cryptography that exacerbates the Key Distribution Problem is the need to send keys frequently. The best guarantee that a key cannot be broken by looking for patterns is to use a different key each time, making sure it is longer than the message to be sent. The great advantage of public key cryptography is that your private key need never be transmitted (you can just randomly generate it), and yet the public key can be transmitted openly and widely. But the corresponding difficulty is that it is then open to breaking, say by prime factorization.[15] Wouldn't it be nice for you to have a way of distributing a secret key that is relatively effortless and that cannot be compromised without your detection? This is the promise of quantum

[15] It is actually common practice to use a combination of secret key and public key cryptography, using public key cryptography to distribute the secret keys. The Advance Encryption Standard (AES), adopted by the U.S. government, is a private key encryption used this way.

cryptography.

Quantum cryptography originated with Stephen Wiesner [1983] (he actually drafted the paper around 1970), where he showed how to transmit a key using complementary observables. The most familiar complementary observables are position and momentum of a particle, made famous by many discussions of the Heisenberg Uncertainty Principle. But perpendicular photon polarization states of light are also complementary observables, e.g. rectilinear (vertical and horizontal) and also diagonal polarization (at 45 ° and 135 °). Just as it is impossible to determine with precision both the position and momentum of a particle (as measurements zero in on one, they blur out on the other), it is also impossible to simultaneously measure both say the vertical and horizontal polarization of light. Charles Bennett and Giles Brassard [1984] developed a protocol (called BB84) using photon polarization states to transmit the cryptographic key. We will not go into the precise protocol of BB84, but simply remark the feature that protects against compromise. Measuring the value of one complementary observable implies an uncertainty about the other. This means in particular that obtaining some information about an unknown quantum system generally causes a disturbance to the quantum state of that system. The security of quantum cryptography relies on this trade-off.

A second method of quantum cryptography, using entangled pairs of photons, was developed by Artur Ekert [1991]. We shall introduce it by way of a "wishful thinking" thought experiment. Suppose Alice and Bob each have one of a pair of "entangled coins." By "entangled" is meant that if in a given sequence Alice's coin comes up heads, Bob's coin will come up tails when he tosses it, and vice versa, and this will happen instantaneously even over large distances. Now if Alice wants to transmit a random key to Bob she tosses her coin some appropriate number of times, making note of each toss. She then sends an uncoded message to Bob, say by phone, and tells him to do the same. If Alice tosses HTT ..., then Bob will toss THH ... , and with the prior understanding that H = 1, T = 0 they know that the key is 100 ... .

Does this sound too good to be true? Well so far it is just magic. But now let us introduce quantum entanglement. Quantum entanglement is when the states of two objects cannot be described separately, and thus there are correlations between observable properties. For example, it is possible to prepare two particles in a single quantum state such that when one is observed to be spin-up, the other one will always be observed to be spin-down and vice versa. There are two important features of quantum entanglement: first, quantum entanglement can persist even though the two objects are widely separated by space, and second, quantum entanglement can exist even though it is impossible to predict which properties will be observed.

So let us now substitute a sequence of entangled pairs of particles for the two sequences of coin tosses, and let Alice measure the first particle in each pair, and Bob measure the second particle. We then have in effect the same thing as the magic coins, but now it is not magic, it is quantum mechanics![16]

How are the particles communicated? There could be a secret meeting between Alice and Bob where the sequences of particles are distributed to the two, but it is also possible that the particles could be distributed say through a fiber optic cable if they are photons, or even through free space. An eavesdropper (always named "Eve") on this cable would have to observe a photon to read the signal, and this measurement could be detected by an application of something called Bell's Theorem.

Returning to quantum computation, one philosophical question that I believe has not been sufficiently examined is the relationship between quantum computation and quantum logic. The latter was initially introduced by Birkhoff and von Neumann [1936] but had very different motivations and the concept of a "qubit" was not explicitly introduced. Dunn, Hagge, Moss, and Wang [2004] is a beginning.

Another philosophically interesting aspect of quantum computing is that it is reversible. No information is lost! This can also happen in a classical closed system, or it can be programmed with great overhead into a computation on a classical computer (one needs to keep somehow all of the previous steps). But the remarkable fact about quantum computing is that reversibility comes for free with no special attention.

Is the world ultimately digital because of quantum mechanics? The answer is yes, and no. Yes when it is measured, no when it is not. Quantum computing is a kind of hybrid between digital and analog computing. Perhaps it represents the best of both?

Rather than end on this high note, honesty compels me to mention the great practical difficulty with building a quantum computer due to the fact that coherent states are easily destroyed by small changes in their environment. For this reason it is important to develop fault tolerant quantum computing. One promising direction is "topological quantum computing," where the qubits are stored as "quantum knots." As all of us know who have tried to untangle a knot in our shoestring, knots are very resistant to even large changes in their environment. This was first proposed by Freedman, Kitaev, and Wang [2000], where the "knots" are braids in a 2-dimensional quasi-particle called an "anyon." The serious implementation issue is whether appropriate anyons can actually be found in nature. An excellent general article with references is Collins [2006]. This paradigm for quantum computing is being pursued by Microsoft Station Q in Santa Barbara. `http://stationq.ucsb.edu/research.html` (accessed July 23, 2007).

The first quantum computers were independently constructed in 1998 at Oxford University and at IBM's Almaden Research Center.[17] They were based on nuclear

[16]Einstein would likely think it is still magic, for the Ekert protocol uses the famous Einstein-Podolsky-Rosen [1935] effect, which Einstein saw as raising serious doubts about the very foundations of quantum mechanics. Entanglement though has come to be an accepted part of quantum mechanics, and there is recent experimental evidence for entanglement. See e.g., Xu *et al.* [2005].

[17]A "Timeline of Quantum Computing," can be found in *Wikipedia*, `http://en.wikipedia.org/wiki/Timeline_of_quantum_computing` (accessed July 23, 2007).

magnetic resonance (NRM) and had 2 qubits. Then 5, 6, 7, qubit computers were demonstrated, culminating in a 12 qubit NRM model in 2006. D-Wave Systems, Inc. demonstrated in February 2007 a working prototype of a commercially 16-qubit "adiabatic" quantum computer.

MODELING, SIMULATION, COMPLEX SYSTEMS, VIRTUAL REALITY

There are distinctions to be made between "modeling" and "simulation" (though they are often used interchangeably), but it seems to me difficult to get agreement on what those distinctions are. In the context of computer science, simulation of course means simulation on a computer, and modeling usually means the construction of a formal mathematical theory that is somehow implemented by the program run on the computer. I think of a model, perhaps a set of differential equations describing the motions of some bodies, as static, whereas the simulation is dynamic. Sometimes modeling is taken to be more scientifically serious than simulation, trying to (mathematically) represent the reality of the underlying causes of a phenomenon, and simulation is allowed to be mere mimicry at the phenomenal level. Sometimes simulation is taken to presuppose modeling, modeling being the (scientifically serious) structure that underlies simulation, which itself must be programmed into a machine.[18] I shall tend to use the terms "modeling" and "simulation" in this last sense, which I think is the most common in the computing community, and talk about "imitation" when I have the weaker sense of mimicry in mind. I should point out that there is even a stronger sense of modeling/simulation, as when a person tries to model the flight of a bird by building an actual machine with flapping wings. Let us call that "emulation."

Whatever the terminology, computers have become more and more used in modeling and simulation, and can be used to model complex systems in a way that often produces unexpected outcomes. In biology it has been common place for sometime to distinguish between experiments *in vitro* (in the glass, "test tube") and *in vivo* (in the living organism). Now a new type of experiment has arisen given computer modeling: "*in silico*" (*in silicon,* or in the computer). This raises certain issues in scientific methodology and statistics.

A related issue is "virtual reality," where computers can simulate the real world, or wildly imaginary worlds. The CAVE was developed at the University of Illinois in 1992. "CAVE" is an acronym for "Computer Assisted Virtual Environment," and the name was cleverly chosen as a take off on Plato's metaphor of The Cave in his *Republic.* In the *Republic* denizens of a cave see shadows reflected on a wall, and mistake them for the real things they represent. Plato of course wanted to say that the "real things" we see in everyday life are but poor reflections of their ideal forms. But there is another way to interpret this and that is that at least under certain conditions people cannot tell illusion from reality.

[18]To paraphrase Kant, simulation without modeling is empty, modeling without simulation is blind.

This reminds us of Descartes' Evil Demon who might deceive him (or us) into believing that we are sitting in front of a fire, etc. when in fact we are not. It is by now a commonplace skeptical argument, and is sometimes gotten at through talking about a "brain in a vat." The movie *The Matrix* is a more current example and is based on the premises that (most) people are in fact encased as cells in a large power-generating "matrix" and are deluded into thinking that they live ordinary lives by complicated computer programs manipulating their brains. Bostrom [2003] with his "simulation argument" is an extreme but well argued version of this. Heim [2001] is a good source of philosophical issues, both old and new, raised by so-called "virtual reality." I love his phrase: Cyberspace is Platonism as a working product. One might substitute "Computer Science" for "Cyberspace."

ARTIFICIAL INTELLIGENCE

There has also been a significant role for logic and philosophy in Artificial Intelligence (AI) since its beginnings. It is common to cite John McCarthy among the trinity of legendary founders of AI (the other two being Marvin Minsky and Herbert Simon), and McCarthy [1959], expanded in McCarthy and Hayes [1969], clearly articulates the importance of logic/philosophy to AI.

Alan Turing raised the issue as to what would count as a computer showing intelligence, and devised what has been canonized as "The Turing Test." The Turing Test famously has to do with whether having a "conversation" with a computer under suitably disguised circumstances would allow you to determine that it is a machine and not a human. Artificial Intelligence (AI) might thus be viewed as a special case of simulation (in either the strong or the weak sense), and of course this raises many interesting philosophical issues.

John Searle [1980] introduced a distinction between "strong AI" and "weak AI":

> According to weak AI, the principal value of the computer in the study of the mind is that it gives us a very powerful tool. For example, it enables us to formulate and test hypotheses in a more rigorous and precise fashion. But according to strong AI, the computer is not merely a tool in the study of the mind; rather, the appropriately programmed computer really is a mind, in the sense that computers given the right programs can be literally said to understand and have other cognitive states. In strong AI, because the programmed computer has cognitive states, the programs are not mere tools that enable us to test psychological explanations; rather, the programs are themselves the explanation. (p. 417)

There is even something else, which I dub "spineless AI." Put quickly, spineless AI might be described as the engineering approach to intelligence — do whatever is easy and effective to solve the problem in question, whether that is the way humans do it or not. For Searle, even weak AI seems to be based on models of cognitive processes. Perhaps the distinction between weak and strong AI should be placed with three distinctions: engineering (spineless) AI, cognitive science (weak) AI, and metaphysical (strong) AI. Viewed as "simulation," these correspond, at least roughly, to imitation (mimicry), simulation, and emulation.

I believe that the fundamental philosophical question regarding AI relates to the representation of information: how much "lies in the eye of the beholder?" This is meant metaphorically of course — replace the word "eye" with "I" or "mind."[19] Dretske [1985] argued that since the semantics of the symbols manipulated by machines are defined by humans and can change irrespective of the machine, there is no meaning in the machine.[20] Stevan Harnad [1990] labeled the issue of how symbols get their meaning The Symbol Grounding Problem: "How can the semantic interpretation of a formal symbol system be made intrinsic to the system, rather than just parasitic on the meanings in our heads?"[21] Dretske made some important first steps in setting out "specs" for solving this problem, and central among these was that the symbols play a causal role in determining the machine's interaction with its environment. "Embodied Embedded Cognition" has become the catch phrase for this, and "cognitive robotics" is a promising and rapidly growing field. An early anticipation of this is Clark and Grush [1999]. Andy Clark has written a number of pieces on embedded cognition. See for example Clark [1998].

It would seem to me to be obvious that for the Turing Test to reveal anything beyond imitation, the use of symbols need to be grounded. A related and perhaps more difficult question has to do with whether "intelligence" requires consciousness, and just what is meant by consciousness. It would seem logically possible, as has been pointed out by Chalmers [1997] and others, for there to be a creature with seeming intelligence in navigating its environment, but which might not have an "inner life," and certainly not self-consciousness. Chalmers defends consciousness as a primitive, non-reducible property. For quite the opposite see Dennett [1992]. For an interesting more recent attempt to understand consciousness see Hofstadter [2007].

Nick Bostrom, Ray Kurzweil, and Bill Joy have all independently been concerned with machines becoming more intelligent than humans. "Concerned" is perhaps the wrong term to use for Bostrom and Kurzweil. Bostrom co-founded (with David Pearce) the World Transhumanist Association (WTA) `http://www.transhumanism.org/` (accessed July 23, 2007), "an interdisciplinary approach to understanding and evaluating the possibilities for overcoming biological limitations through technological progress," and Kurzweil [1999] has written with striking optimism about the time when machines will outrun humankind in intelligence and perhaps even in "spirituality" — quite in opposition to the view of say the *Ter-*

[19] I shamelessly borrow here from the title of the book *The Mind's I: Fantasies and Reflections on Self and Soul* by Douglas Hofstadter and Daniel Dennett [1981].

[20] This can be nicely illustrated by an anecdote the late Australian philosopher Ian Hinkfus once told me. He was working for IBM in the design of an early computer and they ran out of nand gates in building their prototype. So they used nor gates instead and just reinterpreted the output.

[21] See Kay [2001] for an excellent review of the issues and the literature on whether artificial intelligence systems can incorporate intrinsic meaning.

minator films. Bostrom and Kurzweil in various of their writing also suggest that human brains may in effect be uploaded into powerful computers thus extending human capabilities and even experiences by a kind of virtual reality, giving to my mind the title *The Matrix Revisited* a new meaning.

ETHICAL ISSUES

By a natural transition this takes us to ethical issues relating to information and computer science. It is difficult not to be a bit of a luddite when contemplating issues such as the above. The original Luddites were English textile workers in the early 19th century who objected to the introduction of power looms, which they felt threatened their jobs. It is ironic that Charles Babbage, the inventor of the "analytical engine" which foreshadowed the modern digital computer, was heavily influenced by the use of punched cards for programming the weaving in a Jacquard loom. Needless to say there is often the perception today that information technology threatens various people's jobs. What started off as an issue about the "digital divide" between the information have's and have not's, has now often become an issue about off-shoring, finding the cheapest labor that still has the relevant IT knowledge. As it was famously put by Nandan Nilekani, CEO of Infosys Technologies, an Indian outsourcing company at the World Economic Forum in 2004: "Everything you can send down a wire is up for grabs" (reported by Drezner 2004).

Another important ethical issue has to do with the "digital divide," the growing division between those individuals (and parts of the world) that have access to the fruits of information technology and those that do not.

The familiar story of Napster and musical file sharing makes clear that there are ethical issues regarding shared information. We talk about "the information economy," but there seems to be no general consent that "information workers" should be paid for the information they produce. Intellectual Property (IP) raises a number of philosophical/conceptual/legal issues under the headings of Digital Rights Management (DRM) and Digital Content Management (DCM) — this last is newer and oriented towards music and video content.

Another set of ethical issues concerning the Internet has to do with confidentiality, privacy, identity theft, etc. It is worth pointing out that one of the Ten Commandments was "Thou shall not bear false witness against thy neighbor," and not lying (I might put it as intentionally conveying false information) is taken as a general rule in all cultures. Clearly information and communications technologies unfortunately give feasible ways of distributing false information to many millions of people. The "Nigerian scam" is perhaps the most familiar of these and it also is intended to violate another of the Commandments: "Thou shall not steal." "Phishing" as a means of identity theft is another increasingly familiar example.[22]

[22] If any reader is not familiar with these, I suggest that they "google" to find out more, and in the meantime that they not send any money or divulge any personal information to untrusted

It is left as "an exercise to the reader" to say more about the important ethical issues raised by computing and information.

CONCLUSION

This essay has been a bit rambling. I have tried to convince myself that it is not my fault — it is because of the wide range of topics. But frankly it was my choosing to broaden the definition of "computer science" so as to include the study of digital information that made the task of writing this chapter especially difficult. It made it somewhat like my having to write the chapter on the role of the library in library science in a volume on "the philosophy of libraries." Anyway, I conclude not by giving a simple summary, but trying to leave the reader with a correct understanding of the limits of this connection of information with computer science. I do not mean to suggest that all there is to the study of information, even digital information, is to be found in traditional areas of computer science and relates to mathematical and logical aspects of information, even though the topic of this chapter brings a focus on these aspects.

As an antidote to such a misconception, just consider "Information Science," in the sense in which it is used in the common academic label of "Library and Information Science." While it overlaps with computer science in areas such as databases, it typically emphasizes the social sciences, studying the human, social, and organizational aspects of information.

Also, information can not only be studied by the social sciences, it can be an important ingredient in them, say in the context of decision making. Such studies are particularly advanced in economics. The other chapters of this volume make clear that the concept of information plays a role not only in the social sciences, but also in the biological and physical sciences.

So it seems clear that the label "information science" is misleading. It should be "information sciences" (plural) or even "information studies." It should perhaps be compared to "medical sciences," where the emphasis is on health and healing, but the topics range from the most applied to the most basic. No one seems to mind much if a medical school has a biochemistry department, or even a biomedical ethicist, and medical schools increasingly include topics in the social and behavioral sciences, e.g., patient behavior, physician role and behavior, physician–patient interactions, social and cultural issues in health care, and health policy and economics. I think the real challenge for computing and the information sciences is to be transdisciplinary while keeping ties with the traditional disciplines.

---

sources who approach them over the Internet, and to be cautious even if they appear to be trusted since the whole point of phishing is to deceive on that score. ☺

ACKNOWLEDGMENTS

I want to tip my hat to Luciano Floridi for his insight in establishing the very concept of the "philosophy of information." His books Floridi [1999; 2004] and various papers have really legitimized this as a major area in philosophy and not just a minor topic. My minor disagreement with him above about whether information must be true should not be seen as undercutting my fundamental respect. I also want to thank the editors of this *Handbook on the Philosophy of Information,* Pieter Adriaans and Johan van Benthem, for their leadership in fostering this area. I wish to thank Pieter Adriaans, Johan van Benthem, Katalin Bimbó, David Hakken, Dana Scott and Luciano Floridi for their helpful comments and suggestions on various drafts of this chapter. I didn't in all cases accept them, so please blame me, not them, for any remaining infelicities.

BIBLIOGRAPHY

[Barendregt, 1984] H. P. Barendregt. *The Lambda Calculus: Its Syntax and Semantics*, North-Holland Publishing Co., Amsterdam, 1984.

[Barwise and Seligman, 1997] J. Barwise and J. Seligman. *Information Flow: the Logic of Distributed Systems*, Cambridge Tracts in Theoretical Computer Science, vol. 44, Cambridge University Press, Cambridge, 1997.

[Belnap, 1977] N. D. Belnap. "A Useful Four-valued Logic," in *Modern Uses of Multiple-valued Logic*, J. M. Dunn and G. Epstein (eds.), D. Reidel Publishing Co., Dordrecht, 1977. Reprinted as "A Useful Four-valued Logic: How a Computer Should Think," in *Entailment. The Logic of Relevance and Necessity*, vol. II, A. R. Anderson, N. D. Belnap, Jr., and J. M. Dunn, Princeton University Press, Princeton, 1992, rewritten/combined with Belnap's "How a Computer Should Think," originally published in *Contemporary Aspects of Philosophy*, G. Ryle (ed.), Oriel Press, Boston, 1977.

[Bennett and Brassard, 1984] C. H. Bennett and G. Brassard. *Proceedings of the IEEE International Conference on Computers, Systems, and Signal Processing*, IEEE Press, p. 175, 1984.

[Birkhoff and von Neumann, 1936] G. Birkhoff and J. von Neumann. "The Logic of Quantum Mechanics," *Annals of Mathematics*, 37, pp. 823-843, 1936.

[Bostrom, 1998] N. Bostrom. "How Long Before Superintelligence?," *International Journal of Futures Studies*, 2, `http://www.nickbostrom.com/superintelligence.html`, 1998, accessed July 23, 2007.

[Bostrom, 2003] N. Bostrom. "Are You Living In a Computer Simulation?," *Philosophical Quarterly*, 53, pp. 243-255, 2003.

[Brown, 2001] J. Brown. *Quest for the Quantum Computer*, foreword by D. Deutsch, Touchstone (Simon and Shuster), 2001. Originally published in 2000 as *Minds, Machines, and the Multiverse: The Quest for the Quantum Computer,* Simon and Schuster.

[Burgin, 2004] M. Burgin. "Data, Information, and Knowledge," *Information*, 7, pp. 47-57, 2004.

[Chaitin, 1966] G. J. Chaitin. "On the Length of Programs for Computing Finite Binary Sequences," *Journal of the ACM*, 13, pp. 547-569, 1966.

[Chalmers, 1997] D. Chalmers. *The Conscious Mind: In Search of a Fundamental Theory*, Oxford University Press, Oxford, 1997.

[Clark, 1998] A. Clark. "Embodiment and the Philosophy of Mind," in *Current Issues in Philosophy of Mind: Royal Institute of Philosophy Supplement 43*, A. O'Hear (ed.), Cambridge University Press, Cambridge, pp. 35-52, 1998.

[Clark and Grush, 1999] A. Clark and R. Grush. "Towards a Cognitive Robotics," *Adaptive Behavior*, 7, no.1, pp. 5-16, 1999.

[Collins, 2006] G. P. Collins. "Computing with Quantum Knots," *Scientific American,* April, pp. 57-63, 2006.

[Cooper, 1999] A. Cooper. *The Inmates are Running the Asylum: Why High Tech Products Drive us Crazy and How to Restore the Sanity,* SAMS (Macmillan Computer, Publishing) Indianapolis, 1999.

[Crestani, Lalmas and van Rijsbergen, 1998] F. Crestani, M. Lalmas and C. J. van Rijsbergen. *Advanced Models for the representation and Retrieval of Information*, Springer, 1998.

[Cuff and Vanselow, 2004] P. A. Cuff and N. Vanselow, eds. *Improving Medical Education: Enhancing the Behavioral and Social Science Content of Medical School Curricula*, National Academy of Medicine Committee on Behavioral and Social Sciences in Medical School Curricula, The National Academy Press, Washington, DC, 2004.

[Curry and Feys, 1958] H. B. Curry and R. Feys. *Combinatory Logic*, vol. 1, North-Holland Publishing Co., Amsterdam, 1958.

[Dahlbom and Mathiassen, 1993] B. Dahlbom and L. Mathiassen. *Computers in Context: The Philosophy and Practice of Systems Design.* Blackwell, Oxford, 1993.

[Dalkilic *et al.*, 2006] M. M. Dalkilic, W. T. Clark, J. C. Costello and P. Radiovojac. "Using Compression to Identify Classes of Inauthenic Texts," *Proceedings of the 2006 SIAM Conference on Data Mining*, 2007/ `http://www.siam.org/meetings/sdm06/proceedings.htm`, accessed July 23, 2007.

[Davis, 1988] M. Davis. "Influences of Mathematical Logic on Computer Science," in *The Universal Turing Machine: A Half-century Survey*, R. Herken (ed.), Oxford University Press, Oxford, pp. 315-326, 1988.

[Dennett, 1992] D. Dennett, with P. Weiner (illustrator). *Consciousness Explained*, Back Bay Books, 1992.

[Deutsch, 1985] D. Deutsch. "Quantum Theory, the Church-Turing Principle and the Universal Quantum Computer," *Proceedings of the Royal Society of London A*, 400, pp. 97-117, 1985.

[Devlin, 1997] K. Devlin. *Logic and Information*, John Wiley & Sons, 1997.

[De Witt, 1970] B. S. M. De Witt. "Quantum Mechanics and Reality," *Physics Today*, 23, pp. 30-35, 1970.

[Dipert, 1978] R. R. Dipert. *Development and Crisis in Late Boolean Logic: The Deductive Logics of Peirce, Jevons and Schröder*, Ph.D. Thesis, Indiana University, Bloomington, (UMI, Ann Arbor), 1978.

[Dretske, 1981] F. Dretske. *Knowledge and the Flow of Information*, 2nd Ed., Oxford University Press, Oxford, 1981.

[Dretske, 1985] F. Dretske. "Machines and the Mental," *Presidential Address, Eighty-third Annual Meeting of the Western Division of the American Philosophical Association, APA Proceedings*, pp. 23-33, 1985.

[Drezner, 2004] D. W. Drezner. "The Outsourcing Bogeyman," *Foreign Affairs*, May/June, 2004.

[Dunn, 1976] J. M. Dunn. "Intuitive Semantics for FirstDegree Entailments and 'Coupled Trees'," *Philosophical Studies*, vol. 29, pp. 149168, 1976.

[Dunn, 1986] J. M. Dunn. "Relevance Logic and Entailment," in *Handbook of Philosophical Logic*, vol. 3, D. Gabbay and F. Guenthner (eds.), D. Reidel, Dordrecht, pp. 117-224, 1986.

[Dunn, 2001a] J. M. Dunn. "The Concept of Information and the Development of Modern Logic," in *Non-classical Approaches in the Transition from Traditional to Modern Logic*, W. Stelzner (ed.), W. de Gruyter, Berlin, 2001.

[Dunn, 2001b] J. M. Dunn. "Ternary Relational Semantics and Beyond: Programs as Data and Programs as Instructions," *Logical Studies* (on-line journal), no. 7, Institute of Logic, Russian Academy of Sciences, Special Issue: Proceedings of the International Conference *Third Smirnov Readings* (Moscow, May 24-27, 2001), Part 2, `http://logic.ru/en/node/116`, accessed July 23, 2007.

[Dunn, 2001c] J. M. Dunn. "A Representation of Relation Algebras Using Routley-Meyer Frames," in *Logic, Meaning and Computation: Essays in Memory of Alonzo Church,* C. A. Anderson and M. Zelëny (eds.), pp. 77-108, 2001. Preliminary version in Indiana University Logic Group Preprint Series, IULG-93-28, 1993.

[Dunn *et al.*, 2004] J. M. Dunn, T. J. Hagge, L. S. Moss, and Z. Wang. "Quantum Logic as Motivated by Quantum Computing," *The Journal of Symbolic*, 70, pp. 353-359, 2004.

[Dunn and Meyer, 1997] J. M. Dunn and R. K. Meyer. "Combinators and Structurally Free Logic," *The Logic Journal of IGPL*, 5, pp. 505-537, 1997.

[Einstein *et al.*, 1935] A. Einstein, B. Podolsky and N. Rosen. "Can Quantum-Mechanical Description of Physical Reality be Considered Complete?," *Physical Review*, 47, pp. 777-780, 1935.

[Ekert, 1991] A. Ekert. "Quantum Cryptography Based on Bell's Theorem," *Physical Review Letters*, 67, p. 661, 1991.

[Everett, 1957] H. Everett III. " 'Relative State' Formulation of Quantum Mechanics," *Review of Modern Physics*, 29, p. 454, 1957.

[Fetzer, 2004] J. H. Fetzer. "Information: Does it Have to be True?," *Minds and Machines*, 14, pp. 223-229, 2004.

[Floridi, 1999] L. Floridi. *Philosophy and Computing — An Introduction*, Routledge, London, 1999.

[Floridi, 2003] L. Floridi. "Outline of a Theory of Strongly Semantic Information," *Minds and Machines,* 14, pp. 197-221, 2003.

[Floridi, 2004] L. Floridi, ed. *The Blackwell Guide to the Philosophy of Computing and Information*, Blackwell Publishing, Malden, 2004.

[Floridi, 2005] L. Floridi. "Is Information Meaningful Data?," *Philosophy and Phenomenological Research*, 70, pp. 351-370, 2005.

[Freedman *et al.*, 2000] M. H. Freedman, A. Kitaev, and Z. Wang. "Simulation of Topological Field Theories by Quantum Computers," 17th of March 2000, Physics e-Print archive, `http://arXiv.org/abs/quant-ph/0001071`, accessed July 23, 2007.

[Gettier, 1963] E. Gettier. "Is Justified True Belief Knowledge?," *Analysis*, 23, pp. 121-123, 1963.

[Goldman, 1967] A. Goldman. "A Causal Theory of Knowing," *The Journal of Philosophy,* 64, pp. 335-372, 1967.

[Grover, 1996] L. Grover. "A Fast Quantum Mechanical Algorithm for Database Search," *Proceedings of the 28th Annual ACM Symposium on Theory of Computing*, pp. 212-219, 1996.

[Halpern *et al.*, 2001] J. Y. Halpern, R. Harper, N. Immerman, P. G. Kolaitis, M. Y. Vardi and V. Vianu. "On the Unusual Effectiveness of Logic in Computer Science," *The Bulletin of Symbolic Logic*, 7, pp. 213-236, 2001.

[Harnad, 1990] S. Harnad. "The Symbol Grounding Problem," *Physica D*, 42, 335-346, 1990.

[Heim, 1994] M. Heim. *The Metaphysics of Virtual Reality*, Oxford University Press, Oxford, 1994.

[Hofstadter and Dennett, 1981] D. Hofstadter and D. Dennett. *The Mind's I: Fantasies and Reflections on Self and Soul*, Basic Books, New York, 1981.

[Hofstadter, 2007] D. Hofstadter. *I Am a Strange Loop*, Basic Books, New York, 2007.

[Kay, 2001] K. Kay. "Machines and the Mind," *The Harvard Brain*, 8, 1-12, 2001. `http://www.hcs.harvard.edu/~husn/BRAIN/vol8-spring2001/ai.htm`, accessed July 23, 2007.

[Kolmogorov, 1965] A. N. Kolmogorov. "Three Approaches to the Quantitative Definition of Information," *Problems of Information and Transmission*, 1, pp. 1-7, 1965.

[Kurzweil, 1999] R. Kurzweil. *The Age of Spiritual Machines: When Computers Exceed Human Intelligence*, Viking, New York, 1999.

[McCarthy, 1959] J. McCarthy. "Programs with Common Sense," in *Proceedings of the Teddington Conference on the Mechanization of Thought Processes*, Her Majesty's Stationary Office, London, pp. 75-91, 1959.

[McCarthy and Hayes, 1969] J. McCarthy and P. J. Hayes. "Some Philosophical Problems from the Standpoint of Artificial Intelligence," in *Machine Intelligence*, 4, B. Meltzer and D. Michie, (eds.), Edinburgh University Press, Edinburgh, pp. 463-502, 1969.

[Plotkin, 1972] G. D. Plotkin. "A Set-Theoretical Definition of Application," School of Artificial Intelligence, Memo MIP-R-95, University of Edinburgh, 1972.

[Plotkin, 1976] G. D. Plotkin. "A Power Domain Construction," *SIAM Journal of Computing*, 5, pp. 452-487, 1976.

[Popper, 1984] K. R. Popper. "Evolutionary Epistemology," in *Evolutionary Theory: Paths into the Future*, J. W. Pollard (ed.), John Wiley & Sons, 1984.

[Scott, 1969] D. Scott. "Models for the λ-Calculus," unpublished manuscript, 53 pp, 1969.

[Scott, 1972] D. Scott. "Continuous Lattices," in *Toposes, Algebraic Geometry and Logic*, F. W. Lawvere (ed.), Lecture Notes in Mathematics, vol. 274, Springer, Berlin, 1972.

[Scott, 1974] D. Scott. "The Language LAMBDA" (Abstract), *The Journal of Symbolic Logic*, 39, pp. 425-427, 1974.

[Searle, 1980] J. R. Searle. "Minds, Brains, and Programs," *Behavioral and Brain Sciences*, 3, pp. 417-457, 1980.

[Shannon, 1948] C. E. Shannon. "A Mathematical Theory of Communication," *Bell System Technical Journal,* 27, pp. 379-423 and 623-656, 1948.

[Shannon and Weaver, 1949] C. E. Shannon and W. Weaver. *The Mathematical Theory of Communication,* University of Illinois Press, Urbana. Foreword by Richard E. Blahut and Bruce Hajek, 1949; reprinted in 1998.

[Shannon, 1993] C. E. Shannon. *Collected Papers*, edited by N. J. A. Sloane and A. D. Wyner, IEEE Press, New York, 1993.

[Sharma, 2005] N. Sharma. "The Origin of the 'Data Information Knowledge Wisdom' Hierarchy," 2005. `http://www-personal.si.umich.edu/~nsharma/dikw_origin.htm`, accessed July 23, 2007.

[Shor, 1994] P. Shor. "Algorithms for Quantum Computation: Discrete Logarithms and Factoring," *Proceedings of the 35th Annual IEEE Symposium on Foundations of Computer Science*, pp. 124-134, 1994.

[Siegfried, 2000] T. Siegfried. *The Bit and the Pendulum: From Quantum Computing to M Theory — The New Physics of Information*, John Wiley & Sons, New York, 2000.

[Simon, 1969] H. A. Simon. *The Sciences of the Artificial,* 1^{st}edition, (2nd edition 1981, 3rd edition 1996), MIT Press, Cambridge, (MA), 1969.

[Solomonoff, 1964] R. J. Solomonoff. "A Formal Theory of Inductive Inference: Parts 1 and 2," *Information and Control*, 7, pp. 1-22 and 224-254, 1964.

[Stoy, 1977] J. Stoy. *Denotational Semantics: The Scott-Strachey Approach to Programming Languages,* MIT Press, Cambridge, (MA), 1977.

[Thomason, 2005] R. Thomason. "Logic and Artificial Intelligence," *The Stanford Encyclopedia of Philosophy*, E. N. Zalta (ed.), 2005. `http://plato.stanford.edu/archives/sum2005/entries/logic-ai/`, accessed July 23, 2007.

[Tufte, 1990] E. R. Tufte. *Envisioning Information*, Graphics Press, Cheshire, CT, 1990.

[Turing, 1936] A. M. Turing. "On Computable Numbers, with an Application to the Entscheidungsproblem," *Proceedings of the London Mathematical Society,* ser.2, vol. 42, pp. 230-265, 1936.

[Turing, 1950] A. M. Turing. "Computing Machinery and Intelligence," *Mind*, 59, pp. 433-460, 1950.

[van Benthem, 1991] J. van Benthem. *Language in Action. Categories, Lambdas and Dynamic Logic*, Elsevier Science Publishers (Studies in Logic, vol. 130), Amsterdam, 1991.

[Wiesner, 1983] S. Wiesner. "Conjugate Coding," *SIGACT News*, 15, pp. 78-88, 1983.

[Wigner, 1960] E. P. Wigner. "The Unreasonable Effectiveness of Mathematics in the Natural Sciences," *Communications on Pure and Applied Mathematics*, 13, pp. 1-14, 1960.

[Wilcox, 2004] J. Wilcox. *Solving the Enigma: History of the Cryptanalytic Bombe*, NSA, 2004.

[Xu *et al.*, 2005] H. Xu, F. W. Strauch, S. K. Dutta, P. R. Johnson, R.C. Ramos, A. J. Berkley, H. Paik, J. R. Anderson, A. J. Dragt, C. J. Lobb and F. C. Wellstood, "Spectroscopy of Three-Particle Entanglement in a Macroscopic Superconducting Circuit," *Physical Review Letters*, 94, 027003, pp. 1-4, 2005.

THE PHYSICS OF INFORMATION

F. Alexander Bais and J. Doyne Farmer

1 THE PHYSICS OF INFORMATION

> Why cannot we write the entire 24 volumes of the Encyclopedia Brittanica on the head of a pin?
>
> R. P. Feynman

Information is carried, stored, retrieved and processed by machines, whether they be electronic computers or living organisms. All information, which in an abstract sense one may think of as a string of zeros and ones, has to be carried by a physical substrate, be it paper, silicon chips or holograms, and the handling of this information is physical, so information is ultimately constrained by the fundamental laws of physics. It is therefore not surprising that physics and information share a rich interface.

The notion of information as used by Shannon is a generalization of the notion of entropy, which first appeared in thermodynamics. In thermodynamics entropy is an abstract quantity depending on heat and temperature whose interpretation is not obvious. This changed with the theory of statistical mechanics, which explains and generalizes thermodynamics. Statistical mechanics exploits a decomposition of a system into microscopic units such as atoms to explain macroscopic phenomena such as temperature and pressure in terms of the statistical properties of the microscopic units. Statistical mechanics makes it clear that entropy can be regarded as a measure of microscopic disorder. The entropy S can be written as $S = -\sum p_i \log p_i$, where p_i is the probability of a particular microscopic state, for example the likelihood that a given atom will have its velocity and position within a given range.

Shannon realized that entropy is useful to describe disorder in much more general settings, which might have nothing to do with atoms or physics. The entropy of a probability distribution $\{p_i\}$ is well defined as long as p_i is well defined. In this more general context he argued that measuring order and measuring disorder are essentially the same — in a situation that is highly disordered, making a measurement gives a great deal of information, and conversely, in a situation that is highly ordered, making a measurement gives little information. Thus for a system that can randomly be in one of several different states the entropy of its distribution is the same as the information gained by knowing which state i it is

Handbook of the Philosophy of Science. Volume 8: Philosophy of Information
Volume editors: Pieter Adriaans and Johan van Benthem. General editors: Dov M. Gabbay, Paul Thagard and John Woods.

in. It turns out that the concept of entropy or equivalently information is useful in many applications that have nothing to do with physics.

It also turns out that thinking in these more general terms is useful for physics. For example, Shannon's work makes it clear that entropy is in some sense more fundamental than the quantities from which it was originally derived. This led Jaynes to formulate all of statistical mechanics as a problem of maximizing entropy. In fact, all of science can be viewed as an application of the principle of maximum entropy, which provides a means of quantifying the tradeoff between simplicity and accuracy of description. If we want to understand how physical systems can be used to perform computations, or construct computer memories, it can be useful to define entropies that may not correspond to thermodynamic entropy. But if we want to understand the limits to computation it is very useful to think in thermodynamic or statistical terms. This has become particularly important in efforts to understand how to take advantage of quantum mechanics to improve computation. These considerations have given rise to a subfield of physics that is often called the physics of information.

In this chapter we attempt to explain to a non-physicist where the idea of information came from. We begin in Section 2 by describing the origin of the concept of entropy in thermodynamics, where entropy is just a macroscopic state variable related to heat flow and temperature, a rather mathematical device without a concrete physical interpretation. In Section 3 We then discuss how the microscopic theory of atoms led to statistical mechanics, which makes it possible to derive and extend thermodynamics. This led to the definition of entropy in terms of probabilities on the set of accessible microscopic states of a system and provided the inspiration for modern information theory starting with the seminal work of Shannon [Shannon, 1948]. A close examination of the foundations of statistical mechanics and the need to reconcile the probabilistic and deterministic views of the world leads us to a discussion of chaotic dynamics in Section 4, where information plays a crucial role in quantifying predictability. In Section 5 we discuss a variety of fundamental issues that emerge in defining information and how one must exercise care in discussing concepts such as order, disorder, and incomplete knowledge. We also discuss an alternative form of entropy and its possible relevance for nonequilibrium thermodynamics.

Toward the end of the chapter in Section 6 we give a brief axpose of how quantum mechanics gives rise to the concept of quantum information. Entirely new possibilities for information storage and transfer and computation are possible due to the massive parallel processing inherent in quantum mechanics. We also point out how entropy can be extended to apply to quantum mechanics to provide a useful measurement for quantum entanglement. Finally, in Section 7 we make a small excursion to the interface betweeen quantum theory and general relativity, where one is confronted with the "ultimate information paradox" posed by the physics of Black Holes. In this review we have limited ourselves; not all relevant topics that touch on physics and information have been covered.

In our quest for more and more volume and speed in storing and processing information we are naturally led to the smallest scales we can physically manipulate. We began the introduction by quoting Feynman's visionary 1959 lecture "Plenty of room at the bottom" [Feynman, February, 1959) (and http://www.zyvex.com/feynman.html] where he discusses storing and manipulating information on the atomic level. Currently commercially available processors work at scales of 60 nm (1 nm = 1 nanometer = 10^{-9} meter). In 2006, IBM announced circuitry on a 30 nm scale, which indeed makes it possible to write the Encyclopedia Britannica on the head of a pin, so Feynmann's speculative remark in 1959 is now just a marker of the current scale of computation. To make it clear how close this is to the atomic scale, a square with sides of length 30 nm contains about 1000 atoms. Under the historical pattern of Moore's law, integrated circuitry halves in size every 2 years. If we continue on the same trajectory of improvement, within about 20 years the components will be the size of individual atoms, and it is difficult to imagine that computers will be able to get any smaller. Once this occurs information at the atomic scale will be directly connected to our use of information on a macroscopic scale. There is a certain poetry to this: Once a computer has components on a quantum scale, the motion of its atoms will no longer be random, and in a certain sense will not be described by classical statistical mechanics, at the same time that it will be used to process information on a macroscopic scale.

2 THERMODYNAMICS

> The truth of the second law is, therefore, a statistical and not a mathematical truth, for it depends on the fact that the bodies we deal with consist of millions of molecules and that we never can get a hold of single molecules
>
> J.C. Maxwell

Thermodynamics is the study of macroscopic physical systems.[1] These systems contain a large number of degrees of freedom, typically of the order of Avogadro's number, i.e. $N_A \approx 10^{23}$. The three laws of thermodynamics describe processes in which systems exchange energy with each other or with their environment. For example, the system may do work, or exchange heat or mass through a diffusive process. A key idea is that of *equilibrium*, which in thermodynamics is the assumption that the exchange of energy or mass between two systems is the same in both directions; this is typically only achieved when two systems are left alone for a long period of time. A process is *quasistatic* if it always remains close to equilibrium, which also implies that it is *reversible*, i.e that the process can be undone and the system can return to its original state without any external energy inputs. We distinguish various types of processes, for example an *isothermal* process in which the system is in thermal contact with a reservoir that keeps it at

[1] Many details of this brief expose of selected items from thermodynamics and statistical mechanics can be found in standard textbooks on these subjects [Reif, 1965; Kittel, 1966; Huang, 1987; Lifschitz and Landau, 1980].

a fixed temperature. Another example is an *adiabatic* process in which the system is kept thermally isolated and the temperature is allowed to change. A system may also go from one equilibrium state to another via a nonequilibrium process, such as the free expansion of a gas or the mixing of two fluids, in which case it is not reversible. No real system is fully reversible, but it is nonetheless a very useful concept.

The remarkable property of systems in equilibrium is that the macro states can be characterized by only very few variables, such as the volume V, pressure P, temperature T, entropy S, chemical potential μ and particle number N. These state variables are in general not independent, but rather are linked by an *equation of state*, which describes the constraints imposed by physics. A familiar example is the ideal gas law $PV = N_A kT$, where k is the Boltzmann constant relating temperature to energy ($k = 1.4 \times 10^{-23}$ $joule/Kelvin$). In general the state variables come in pairs, one of which is *intensive* while the other conjugate variable is *extensive*. Intensive variables like pressure or temperature are independent of system size, while extenstive variables like volume and entropy are proportional to system size.

In this lightning review we will only highlight the essential features of thermodynamics that are most relevant in connection with information theory.

2.1 The laws

The first law of thermodynamics reads[2]

$$(1) \quad dU = đQ - đW$$

and amounts to the statement that heat is a form of energy and that energy is conserved. More precisely, the change in internal energy dU equals the amount of heat $đQ$ absorbed by the system minus the work done by the system, $đW$.

The second law introduces the concept of entropy S, which is defined as the ratio of heat flow to temperature. The law states that the entropy for a closed system (with constant energy, volume and number of particles) can never decrease. In mathematical terms

$$(2) \quad dS = \frac{đQ}{T}, \qquad \frac{dS}{dt} \geq 0.$$

By using a gas as the canonical example, we can rewrite the first law in proper differentials as

$$(3) \quad dU = TdS - PdV,$$

where PdV is the work done by changing the volume of the container, for example by compressing the gas with a piston. It follows from the relation between entropy, heat and temperature that entropy differences can be measured by measuring the

[2]The bars through the differentials indicate that the quantities following them are not state variables: the d-bars therefore refer to small quantities rather then proper differentials.

temperature with a thermometer and the change in heat with a calorimeter. This illustrates that from the point of view of thermodynamics entropy is a purely macroscopic quantity.

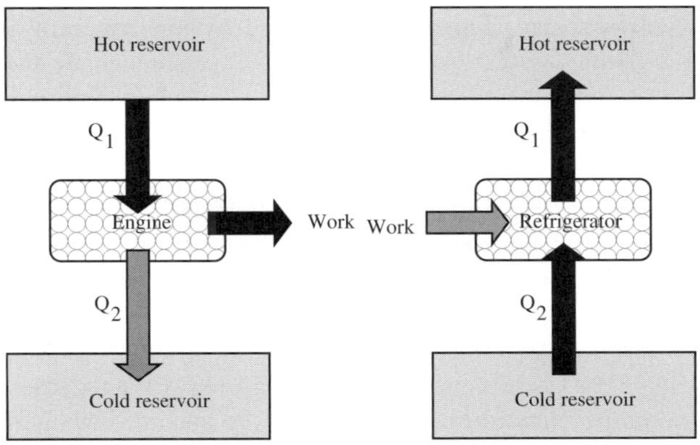

Figure 1. The relation between heat and work illustrating the two formulations of the second law of thermodynamics. On the left we have the Kelvin formulation. The ideal engine corresponds to the diagram with the black arrows only. The second law tells us that the third, grey arrow is necessarily there. The right picture with only the black arrows corresponds to the ideal refrigerator, and the third, grey arrow is again required by the second law.

There are two different formulations of the second law. The Kelvin formulation states that it is impossible to have a machine whose sole effect is to convert heat into work. We can use heat to do work, but to do so we must inevitably make other alterations, e.g. letting heat flow from hot to cold and thereby bringing the system closer to equilibrium. Clausius' formulation says that it is impossible to have a machine that only extracts heat from a reservoir at low temperature and delivers that same amount of heat to a reservoir at higher temperature. Rephrasing these formulations, Kelvin says that ideal engines cannot exist and Clausius says that ideal refrigerators can't exist. See figure 1.

The action of a heat engine or refrigerator machines can be pictured in a diagram in which the reversible sequence of states the system goes through are a closed curve, called a Carnot cycle. We give an example for the Kelvin formulation in figure 2. Imagine a piston in a chamber; our goal is to use the temperature differential between two reservoirs to do work. The cycle consists of four steps: In step $a \to b$, isothermal expansion, the system absorbs an amount Q_1 of heat from the reservoir at high temperature T_1, which causes the gas to expand and push on the piston, doing work; In step $b \to c$, adiabatic expansion, the gas continues to expand and do work, but the chamber is detached from the reservoir, so that it no longer absorbs any heat. Now as the gas expands it cools until it reaches temperature T_2. In step $c \to d$, isothermal compression, the surroundings do work on the gas, as heat flows into the cooler reservoir, giving off an amount Q_2 of heat; and in step $d \to a$, adiabatic compression, the surroundings continue to do work, as the gas is further compressed (without any heat transfer) and brought back up to the original temperature. The net work done by the machine is given by the line integral:

$$(4) \quad W = \oint_{cycle} PdV = \text{enclosed area}$$

which by the first law should also be equal to $W = Q_1 - Q_2$ because the internal energy is the same at the beginning and end of the cycle. We also can calculate the total net change in entropy of the two reservoirs as

$$(5) \quad \Delta S = \frac{-Q_1}{T_1} + \frac{Q_2}{T_2} \geq 0\,,$$

where the last inequality has to hold because of the second law. Note that the two latter equations can have solutions with positive W. The efficiency of the engine η is by definition the ratio of the work done to the heat entering the system, or

$$(6) \quad \eta = \frac{W}{Q_1} = 1 - \frac{Q_2}{Q_1} \leq 1 - \frac{T_1}{T_2}.$$

This equals one for an ideal heat engine, but is less then one for a real engine.

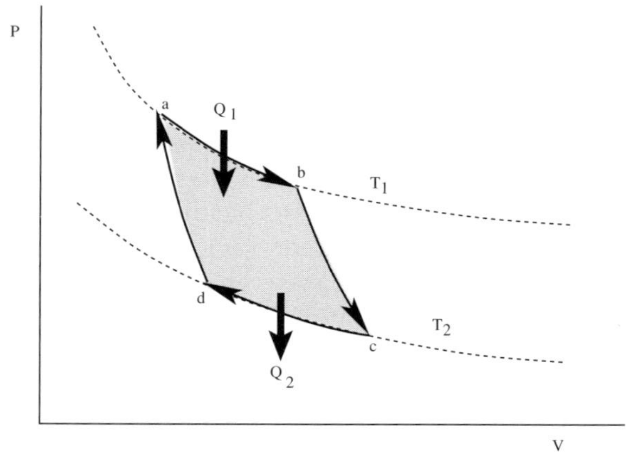

Figure 2. The Carnot cycle corresponding to the Kelvin formulation of the second law. The work done by the engine equals the line integral along the closed contour and is therefore equal to the enclosed area.

A modern formulation of the second law, which in the setting of statistical mechanics is equivalent to the statements of Kelvin and Clausius, is the *Landauer principle*, which says that there is no machine whose sole effect is the erasure of information. There is a price to forgetting: The principle states that the erasure of information (which is irreversible) is inevitably accompanied by the generation of heat. In other words, logical irreversibility necessarily involves thermodynamical irreversibility. One has to generate at least $kT \ln 2$ to get rid of one bit of information [Landauer, 1961; Landauer, 1991]. We return to the Landauer principle in the section on Statistical mechanics.

We just showed that the second law sets fundamental limits on the possible efficiency of real machines like steam engines, refrigerators and information processing devices. As everybody knows, real engines give off heat and real refrigerators and real computers need power to do their job. The second law tells us to what extent heat can be used to perform work. The increase of entropy as we go from one equilibrium situation to another is related to dissipation and the production of heat, which is intimately linked to the important notion of *irreversibility*. A given action in a closed system is irreversible if it makes it impossible for the system to return to the state it was in before the action took place without external inputs. Irreversibility is always associated with production of heat, because heat cannot be freely converted to other forms of energy (whereas any other form of energy can always be converted to heat). One can decrease the entropy of a system by doing work on it, but in doing the work one has to increase the entropy of another system (or of the system's environment) by an equal or greater amount.

The theory of thermodynamics taken by itself does not connect entropy with information. This only comes about when the results are interpreted in terms of a microscopic theory, in which case temperature can be interpreted as being related to uncertainty and incoherence in the position of particles. This requires a discussion of statistical mechanics, as done in the next section.

There is another fundamental aspect to the second law which is important from an operational as well as philosophical point of view. A profound implication of the second law is that it defines an "arrow of time", i.e., it allows us to distinguish the past from the future. This is in contrast to the fundamental microscopic laws of physics which are time reversal invariant (except for a few exotic interactions, that are only very rarely seen under normal conditions as we find them on earth). If one watches a movie of fundamental processes on the microscopic level it is impossible to tell whether it is running forwards or backwards. In contrast, if we watch a movie of macroscopic events, it is not hard to identify irreversible actions such as the curling of smoke, the spilling of a glass of water, or the mixing of bread dough, which easily allow us to determine whether we are running in forward or reverse. More formally, even if we didn't know which way time were running, we could pick out some systems at random and measure their entropy at times $t_1, t_2, \ldots$ The direction in which entropy increases is the one that is going forward in time. Note that we didn't define an a priori direction of time in formulating the second law — it establishes a time direction on its own, without any reference to atomic theory or any other laws of physics.

The second law of thermodynamics talks only about the difference between the entropy of different macrostates. The absolute scale for entropy is provided by the third law of thermodynamics. This law states that when a system approaches the absolute zero of temperature the entropy will go to zero, i.e.

$$(7) \quad T \to 0 \quad \Rightarrow \quad S \to 0.$$

When $T = 0$ the heat is zero, corresponding classically to no atomic motion, and the energy takes on its lowest possible value. In quantum theory we know that such a lowest energy "ground" state also exists, though, if the ground state of the system turns out to be degenerate the entropy will approach a nonzero constant at zero temperature. We conclude by emphasizing that the laws of thermodynamics have a wide applicability and a rich phenomenology that supports them unequivocally.

2.2 Free energy

Physicists are particularly concerned with what is called the (Helmholtz) *free energy*, denoted F. It is a very important quantity because it defines the amount of energy available to do work. As we discuss in the next section, the free energy plays a central role in establishing the relation between thermodynamics and statistical mechanics, and in particular for deriving the microscopic definition of entropy in terms of probabilities.

The free energy is defined as

$$(8) \quad F \equiv U - TS.$$

This implies that in differential form we have

$$(9) \quad dF = dU - TdS - SdT,$$

which using (3) can be written as

$$(10) \quad dF = -PdV - SdT.$$

The natural independent variables to describe the free energy of a gas are volume and temperature.

Let us briefly reflect on the meaning of the free energy. Consider a system A in thermal contact with a heat bath A' kept at a constant temperature T_0. Suppose the system A absorbs heat $đQ$ from the reservoir. We may think of the total system consisting of system plus bath as a closed system: $A^0 = A + A'$. For A^0 the second law implies that its entropy can only increase: $dS^0 = dS + dS' \geq 0$. As the temperature of the heat bath A' is constant and its absorbed heat is $-đQ$, we may write $T_0 dS' = -đQ$. From the first law applied to system A we obtain that $-đQ = -dU - đW$, so that we can substitute the expression $T_0 dS' = -dU - đW$ in $T_0 dS + T_0 dS' \geq 0$ to get $-dU + T_0 dS \geq đW$. As the system A is kept at a constant temperature the left hand side is just equal to $-dF$, demonstrating that

$$(11) \quad -dF \geq đW.$$

The maximum work that can be done by the system in contact with a heat reservoir is $(-dF)$. If we keep the system parameters fixed, i.e. $đW = 0$, we obtain that $dF \leq 0$, showing that for a system coupled to a heat bath the free energy can only decrease, and consequently in a thermal equilibrium situation the free energy reaches a minimum. This should be compared with the entropy, which reaches a maximum at equilibrium.

We can think of the second law as telling us how different kinds of energy are converted into one another: In an isolated system, work can be converted into heat, but heat cannot be converted into work. From a microscopic point of view forms of energy that are "more organized", such as light, can be converted into those that are "less organized", such as the random motion of particles, but the opposite is not possible.

From Equation (10) the pressure and entropy of a gas can be written as partial derivatives of the free energy

$$(12) \quad P = \left(\frac{\partial F}{\partial V}\right)_T, \quad S = \left(\frac{\partial F}{\partial T}\right)_V.$$

So we see that for a system in thermal equilibrium the entropy is a state variable, meaning that if we reversibly traverse a closed path we will return to the same value (in contrast to other quantities, such as heat, which do not satisfy this property). The variables P and S are dependent variables. This is evident from the Maxwell relation, obtained by equating the two second derivatives

$$(13) \quad \frac{\partial^2 F}{\partial T \partial V} = \frac{\partial^2 F}{\partial V \partial T},$$

yielding the relation

$$(14) \quad \left(\frac{\partial P}{\partial T}\right)_V = \left(\frac{\partial S}{\partial V}\right)_T.$$

3 STATISTICAL MECHANICS

In dealing with masses of matter, while we do not perceive the individual molecules, we are compelled to adopt what I have described as the statistical method of calculation, and to abandon the strict dynamical method, in which we follow every motion by the calculus.

J.C. Maxwell

We are forced to be contented with the more modest aim of deducing some of the more obvious propositions relating to the statistical branch of mechanics. Here there can be no mistake in regard to the agreement with the facts of nature.

J.W. Gibbs

Statistical mechanics is the explanation of the macroscopic behavior of physical systems using the underlying microscopic laws of physics, even though the microscopic states, such as the position and velocity of individual particles, are unknown. The key figures in the late 19th century development of statistical mechanics were Maxwell, Boltzmann and Gibbs [Maxwell, 1872; Boltzmann, 1896-1898; Gibbs, 1902]. One of the outstanding questions was to derive the laws of thermodynamics, in particular to give a microscopic definition of the notion of entropy. Another objective was the understanding of phenomena that cannot be computed from thermodynamics alone, such as transport phenomena. For our purpose of highlighting the links with information theory we will give a brief and somewhat lopsided introduction. Our main goal is to show the origin of the famous expression due to Gibbs for the entropy, $S = -\sum_i p_i \ln p_i$, which was later used by Shannon to define information.

3.1 Definitions and postulates

> Considerable semantic confusion has resulted from failure to distinguish between prediction and interpretation problems, and attempting a single formalism to do both.
>
> T.S. Jaynes

Statistical mechanics considers systems with many degrees of freedom, such as atoms in a gas or spins on a lattice. We can think in terms of the microstates of the system which are, for example, the positions and velocities of all the particles in a vessel with gas. The space of possible microstates is called the *phase space.* For a monatomic gas with N particles, the phase space is $6N$-dimensional, corresponding to the fact that under Newtonian mechanics there are three positions and three velocities that must be measured for each particle in order to determine its future evolution. A microstate of the whole system thus corresponds to a single point in phase space.

Statistical mechanics involves the assumption that, even though we know that the microstates exist, we are largely ignorant of their actual values. The only information we have about them comes from macroscopic quantities, which are bulk properties such as the total energy, the temperature, the volume, the pressure, or the magnetization. Because of our ignorance we have to treat the microstates in statistical terms. But the knowledge of the macroscopic quantities, along with the laws of physics that the microstates follow, constrain the microstates and allow us to compute relations between macroscopic variables that might otherwise not be obvious. Once the values of the macroscopic variables are fixed there is typically only a subset of microscopic states that are compatible with them, which are called the *accessible states.* The number of accessible states is usually huge, but differences in this number can be very important. In this chapter we will for simplicity assume a discrete set of microstates, but the formalism can be straightforwardly generalized to the continuous case.

The first fundamental assumption of statistical mechanics is that in equilibrium a closed system has an equal a priori probability to be in any of its accessible states. For systems that are not closed, for example because they are in thermal contact or their particle number is not constant, the set of accessible states will be different and their probabilities have to be calculated. In either case we associate an *ensemble* of systems with a characteristic probability distribution over the allowed microscopic states. Tolman [Tolman, 1938] clearly describes the notion of an ensemble:

> In using ensembles for statistical purposes, however, it is to be noted that there is no need to maintain distinctions between individual systems since we shall be interested merely in the number of systems at any time which would be found in the different states that correspond to different regions of phase space. Moreover, it is also to be noted for statistical purposes that we shall wish to use ensembles containing a large enough population of separate members so that the number of systems in such different states can be regarded as changing continuously as we pass from the states lying in one region of the phase space to those in another. Hence, for the purpose in view, it is evident that the condition of an ensemble at any time can be regarded as appropriately specified by the density r with which representative points are distributed over phase space.

The second postulate of statistical mechanics, called *ergodicity*, says that time averages correspond to ensemble averages. That is, on one hand we can take the time average by following the deterministic motion of the all the microscopic variables of all the particles making up a system. On the other hand, at a given instant in time we can take an average over all possible accessible states, weighting them by their probability of occurrence. The ergodic hypothesis says that these two averages are the same. We return to the restricted validity of this hypothesis in the section on nonlinear dynamics.

3.2 Counting microstates for a system of magnetic spins

In the following example we show how it is possible to derive the distribution of microscopic states through the assumption of equipartition and simple counting arguments. This also illustrates that the distribution over microstates becomes extremely narrow in the thermodynamic (i.e. $N \to \infty$ limit). Consider a system of N magnetic spins that can only take two values $s_j = \pm 1$, corresponding to whether the spin is pointing up or down (often called *Ising spins*). The total number of possible configurations equals 2^N. For convenience assume N is even, and that the spins do not interact. Now put these spins in an upward pointing magnetic field H and ask how many configurations of spins are consistent with each possible value of the energy. The energy of each spin is $e_j = \mp \mu H$, and because they do not interact, the total energy of the system is just the sum of the energies of each spin. For a configuration with k spins pointing up and $N-k$ spins pointing down the total energy can be written as $\varepsilon_m = 2m\mu H$ with $m \equiv (N-2k)/2$ and $-N/2 \leq m \leq N/2$. The value of ε_m is bounded : $-N\mu H \leq \varepsilon_m \leq N\mu H$ and the difference between two adjacent energy levels, corresponding to the flipping of one spin, is $\Delta\varepsilon = 2\mu H$. The number of microscopic configurations with energy ε_m equals

$$(15)\quad g(N,m) = g(N,-m) = \frac{N!}{(\frac{1}{2}N+m)!(\frac{1}{2}N-m)!}.$$

The total number of states is $\sum_m g(N,m) = 2^N$. For a thermodynamic system N is really large, so we can approximate the factorials by the Stirling formula

$$(16)\quad N! \cong \sqrt{2\pi N} N^N e^{-N+\cdots}$$

Some elementary math gives the Gaussian approximation for the binomial distribution for large N,

$$(17)\quad g(N,m) \cong 2^N \left(\frac{2}{\pi N}\right)^{\frac{1}{2}} e^{-2m^2/N}.$$

We will return to this system later on, but at this point we merely want to show that for large N the distribution is sharply peaked. Roughly speaking the width of the distribution grows with $\sqrt{N}$ while the peak height grows as 2^N, so the degeneracy of the states around $m = 0$ increases very rapidly. For example $g(50,0) = 1.264 \times 10^{14}$, but for $N \approx N_A$ one has $g(N_A,0) \cong 10^{10^{22}}$. We will return to this example in the following section to calculate the magnetization of a spin system in thermal equilibrium.

3.3 The Maxwell-Boltzmann-Gibbs distribution

Maxwell was the first to derive an expression for the probability distribution p_i for a system in thermal equilibrium, i.e. in thermal contact with a heat reservoir kept at a fixed temperature T. This result was later generalized by Boltzmann and Gibbs. An equilibrium distribution function of an ideal gas without external force applied to it should not depend on either position or time, and thus can only depend on the velocities of the individual particles. In general there are interactions between the particles that need to be taken into account. A simplifying assumption that is well justified by probabilistic calculations is that processes in which two particles interact at once are much more common than those in which three or more particles interact. If we assume that the velocities of two particles are independent before they interact we can write their joint probability to have velocities v_1 and v_2 as a product of the probability for each particle alone. This implies $p(v_1,v_2) = p(v_1)p(v_2)$. The same holds after they interact: $p(v_1',v_2') = p(v_1')p(v_2')$. In equilibrium, where nothing can depend on time, the probability has to be the same afterward, i.e. $p(v_1,v_2) = p(v_1',v_2')$. How do we connect these conditions before and after the interaction? A crucial observation is that there are conserved quantities that are preserved during the interaction and the equilibrium distribution function can therefore only depend on those. Homogeneity and isotropy of the distribution function selects the total energy of the particles as the only function on which the distribution depends. The conservation of energy in this situation boils down to the simple statement that $\frac{1}{2}mv_1^2 + \frac{1}{2}mv_2^2 = \frac{1}{2}mv_1'^2 + \frac{1}{2}mv_2'^2$. From these relations Maxwell derived the well known thermal equilibrium velocity distribution,

$$(18)\quad p_0(v) = n\left(\frac{m}{2\pi T}\right)^{3/2} e^{-mv^2/2kT}.$$

The distribution is Gaussian. As we saw, to derive it Maxwell had to make a number of assumptions which were plausible even though they couldn't be derived from the fundamental laws of physics. Boltzmann generalized the result to include the effect of an external conservative force, leading to the replacement of the kinetic energy in (18) by the total conserved energy, which includes potential as well as kinetic energy.

Boltzmann's generalization of Maxwell's result makes it clear that the probability distribution p_i for a general system in thermal equilibrium is given by

$$(19)\quad p_i = \frac{e^{-\varepsilon_i/T}}{Z}.$$

Z is a normalization factor that ensures the conservation of probability, i.e. $\sum_i p_i = 1$. This implies that

$$(20)\quad Z \equiv \sum_i e^{-\varepsilon_i/T}.$$

Z is called the *partition function.* The Boltzmann distribution describes the *canonical ensemble*, that is it applies to any situation where a system is in thermal equilibrium and exchanging energy with its environment. This is in contrast to the *microcanonical ensemble*, which applies to isolated systems where the energy is constant, or the *grand canonical ensemble*, which applies to systems that are exchanging both energy and particles with their environment[3]. To illustrate the power of the Boltzmann distribution let us briefly return to the example of the thermal distribution of Ising spins on a lattice in an external magnetic field. As we pointed out in section (3.2), the energy of a single spin is $\pm\mu H$. According to the Boltzmann distribution, the probabilities of spin up or spin down are

$$(21)\quad p_\pm = \frac{e^{\mp\mu H/T}}{Z}.$$

The spin antiparallel to the field has lowest energy and therefore is favored. This leads to an average field dependent magnetization m_H (per spin)

[3]Gibbs extended the Boltzmann result to situations where the number of particles is not fixed, leading to the introduction of the *chemical potential.* Because of its complicated history, the exponential distribution is referred to by a variety of names, including Gibbs, Boltzmann, Boltzmann-Maxwell, and Boltzmann-Gibbs.

(22) $m_H = \langle \mu \rangle = \dfrac{\mu p_+ + (-\mu) p_-}{p_+ + p_-} = \mu \tanh \dfrac{uH}{T}.$

This example shows how statistical mechanics can be used to establish relations between macroscopic variables that cannot be obtained using thermodynamics alone.

3.4 Free energy revisited

In our discussion of thermodynamics in section 2.2 we introduced the concept of the free energy F defined by equation 8, and argued that it plays a central role for systems in thermal contact with a heat bath, i.e. systems kept at a fixed temperature T. In the previous section we introduced the concept of the partition function Z defined by equation 20. Because all thermodynamic quantities can be calculated from it, the importance of the partition function Z goes well beyond its role as a normalization factor. The free energy is of particular importance, because its functional form leads directly to the definition of entropy in terms of probabilities. We can now directly link the thermodynamical quantities to the ones defined in statistical mechanics. This is done by postulating[4] the relation between the free energy and the partition function as[5]

(23) $F = -T \ln Z,$

or alternatively $Z = e^{-F/T}$. From this definition it is possible to calculate all thermodynamical quantities, for example using equations (12). We will now derive the expression for the entropy in statistical mechanics in terms of probabilities.

3.5 Gibbs entropy

The definition of the free energy in equation (8) implies that

(24) $S = \dfrac{U - F}{T}.$

From (23) and (19) it follows that

(25) $F = \varepsilon_i + T \ln p_i.$

Note that even though both the terms on the right depend on i the free energy F is independent of i. The equilibrium value for the internal energy is by definition

(26) $U = \langle \varepsilon \rangle \equiv \sum_i \varepsilon_i \, p_i \; .$

[4]Once we have identified a certain macroscopic quantity like the free energy with a microscopic expression, then of course the rest follows. Which expression is taken as the starting point for the identification is quite arbitrary. The justification is *a posteriori* in the sense that the well known thermodynamical relations should be recovered.

[5]Boltzmann's constant k relates energy to temperature. Its value in conventional units is 1.4×10^{-23} *joule/kelvin*, but we have set it equal to unity, which amounts to choosing a convenient unit for energy or temperature.

With these expressions for S, F and U, and making use of the fact that F is independent of i and $\sum_i p_i = 1$, we can rewrite the entropy in terms of the probabilities p_i and arrive at the famous expression for the entropy:

$$(27) \quad S = -\sum_i p_i \ln p_i \ .$$

This expression is usually called the Gibbs entropy[6].

In the special case where the total energy is fixed, the w different (accessible) states all have equal a priori probability $p_i = p = 1/w$. Substitution in the Gibbs formula yields the expression in terms of the number of accessible states, originally due to Boltzmann (and engraved on his tombstone):

$$(28) \quad S = \ln w.$$

We emphasize that the entropy grows logarithmically with the number of accessible states[7]. Consider a system consisting of a single particle that can be in one of two states. Assuming equipartition the entropy is $S_1 = \ln 2$. For a system with Avogadro's number of particles $N \sim 10^{23}$, there are 2^N states and if we assume independence the entropy is $S_N = \ln 2^N = NS_1$, a very large number. The tendency of a system to maximize its entropy is a probabilistic statement: The number of states with half of the particles in one state and half in the other is enormously larger than the number in which all the particles are in the same state, and when the system is left free it will relax to the most probable accessible state. The state of a gas particle depends not only on its allowed position (i.e. the volume of the vessel), but also on its allowed range of velocities: If the vessel is hot that range is larger then when the vessel is cold. So for an ideal gas one finds that the entropy increases with the logarithm of the temperature. The fact that the law is a probabilistic implies that it is not completely impossible that the system will return to a highly improbable initial state. Poincaré showed that it is bound to happen and gave an estimate of the recurrence time (which for a macroscopic system is much larger than the lifetime of the universe).

The Gibbs entropy transcends its origins in statistical mechanics. It can be used to describe any system with states $\{\psi_i\}$ and a given probability distribution $\{p_i\}$. Credit for realizing this is usually given to Shannon [Shannon, 1948], although antecedents include Szilard, Nyquist and Hartley. Shannon proposed that by analogy to the entropy S, information can be defined as

$$(29) \quad H \equiv -\sum_i p_i \log_2 p_i.$$

[6]In quantum theory this expression is replaced by $S = -Tr\ \rho \ln \rho$ where ρ is the density matrix of the system.

[7]These numbers can be overwhelmingly large. Imagine two macrostates of a system which differ by 1 millicalorie at room temperature. The difference in entropy is $\Delta S = -\Delta Q/T = 10^{-3}/293 \approx 10^{-5}$. Thus the ratio of the number of accessible states is $w_2/w_1 = \exp(\Delta S/k) \approx \exp(10^{18})$, a big number!

In information theory it is common to take logarithms in base two and drop the Boltzmann constant[8]. Base two is a natural choice of units when dealing with binary numbers and the units of entropy in this case are called *bits*; in contrast, when using the natural logarithm the units are called *nats*, with the conversion that 1 *nat =1.443* bits . For example a memory consisting of 5 bits (which is the same as a system of 5 Ising spins), has $N = 2^5$ states. Without further restrictions all of these states (messages) have equal probability i.e. $p_i = 1/N$ so that the information content is $H = -N\frac{1}{N}\log_2\frac{1}{N} = \log_2 2^5 = 5\ bits$. Similarly consider a DNA-molecule with 10 billion base pairs, each of which can be in one of four combinations (A-T,C-G,T-A,G-C). The molecule can a priori be in any of $4^{10^{10}}$ configurations so the naive information content (assuming independence) is $H = 2 \times 10^{10}\ bits$. The logarithmic nature of the definition is unavoidable if one wants the additive property of information under the addition of bits. If in the previous spin example we add another string of 3 bits then the total number of states is $N = N_1 N_2 = 2^5 \times 2^3 = 2^8$ from which it also follows that $H = H_1 + H_2 = 8$. If we add extra ab initio correlations or extra constraints we reduce the number of independent configurations and consequently H will be smaller.

As we will discuss in Section 5, this quantitative definition of information and its applications transcend the limited origin and scope of conventional thermodynamics and statistical mechanics, as well as Shannon's original purpose of describing properties of communication channels. See also [Brillouin, 1956].

4 NONLINEAR DYNAMICS

> The present state of the system of nature is evidently a consequence of what it was in the preceding moment, and if we conceive of an intelligence which at a given instant comprehends all the relations of the entities of this universe, it could state the respective position, motions, and general effects of all these entities at any time in the past or future.
>
> Pierre Simon de Laplace (1776)

> A very small cause which escapes our notice determines a considerable effect that we cannot fail to see, and then we say that the effect is due to chance.
>
> Henri Poincaré (1903).

From a naive point of view statistical mechanics seems to contradict the determinism of Newtonian mechanics. For any initial state $x(0)$ (a vector of positions and velocities) Newton's laws define a dynamical system ϕ^t (a set of differential equations) that maps $x(0)$ into its future states $x(t) = \phi^t(x(0))$. This is completely deterministic. As Laplace so famously asserted, if mechanical objects obey Newton's laws, why do we need to discuss perfect certainties in statistical terms? Laplace partially answered his own question:

---

[8]In our convention k=1, so $H = S/\ln 2$.

> ... But ignorance of the different causes involved in the production of events, as well as their complexity, taken together with the imperfection of analysis, prevent our reaching the same certainty [as in astronomy] about the vast majority of phenomena. Thus there are things that are uncertain for us, things more or less probable, and we seek to compensate for the impossibility of knowing them by determining their different degrees of likelihood. So it is that we owe to the weakness of the human mind one of the most delicate and ingenious of mathematical theories, the science of chance or probability.

Laplace clearly understood the need for statistical descriptions, but at that point in time was not fully aware of the consequences of nonlinear dynamics. As Poincaré later showed, even without human uncertainty (or quantum mechanics), when Newton's laws give rise to differential equations with chaotic dynamics, we inevitably arrive at a probabilistic description of nature. Although Poincaré discovered this in the course of studying the three body problem in celestial mechanics, the answer he found turns out to have relevance for the reconciliation of the deterministic Laplacian universe with statistical mechanics.

4.1 The ergodic hypothesis

As we mentioned in the previous section, one of the key foundations in Boltzmann's formulation of statistical mechanics is the *ergodic hypothesis*. Roughly speaking, it is the hypothesis that a given trajectory will eventually find its way through all the accessible microstates of the system, e.g. all those that are compatible with conservation of energy. At equilibrium the average length of time that a trajectory spends in a given region of the state space is proportional to the number of accessible states the region contains. If the ergodic hypothesis is true, then time averages equal ensemble averages, and equipartition is a valid assumption.

The ergodic hypothesis proved to be highly controversial for good reason: It is generally not true. The first numerical experiment ever performed on a computer took place in 1947 at Los Alamos when Fermi, Pasta, and Ulam set out to test the ergodic hypothesis. They simulated a system of masses connected by nonlinear springs. They perturbed one of the masses, expecting that the disturbance would rapidly spread to all the other masses and equilibrate, so that after a long time they would find all the masses shaking more or less randomly. Instead they were quite surprised to discover that the disturbance remained well defined — although it propagated through the system, it kept its identity, and after a relatively short period of time the system returned very close to its initial state. They had in fact rediscovered a phenomenon that has come to be called a *soliton*, a localized but very stable travelling disturbance. There are many examples of nonlinear systems that support solitons. Such systems do not have equal probability to be in all their accessible states, and so are not ergodic.

Despite these problems, there are many examples where we know that statistical mechanics works extremely well. There are even a few cases, such as the hard sphere gas, where the ergodic hypothesis can actually be proved. But more typically this is not the case. The evidence for statistical mechanics is largely empirical: we know that it works, at least to a very high degree of approximation. Subsequent work has made it clear that the typical situation is much more complicated than was originally imagined. While some trajectories may wander in more or less random fashion around much of the accessible phase space, they are blocked from entering certain regions by what are called KAM (Kolomogorov-Arnold-Moser) tori. Other initial conditions yield trajectories that make regular motions and lie on KAM tori trajectories. The KAM tori are separated from each other, and have a lower dimension than the full accessible phase space. Such KAM tori correspond to situations in which there are other conversation laws in addition to the conservation of energy, which may depend on initial conditions as well as other parameters[9]. Solitons are examples of this in which the solutions can be interpreted as a geometrically isolated pulse.

There have now been an enormous number of studies of ergodicity in nonlinear dynamics. While there are no formal theorems that definitively resolve this, the accumulated lore from these studies suggests that for nonlinear systems that do not have hidden symmetries, as the number of interacting components increases and nonlinearities become stronger, the generic behavior is that chaotic behavior becomes more and more likely — the KAM tori shrink, fewer and fewer initial conditions are trapped on them, and the regions they exclude become smaller. The ergodic hypothesis becomes an increasingly better approximation, a typical single trajectory can reach almost all accessible states, and equipartition becomes a good assumption. The problems occur in understanding when there are hidden symmetries that can support phenomena like solitons. The necessary and sufficient conditions for ergodicity to be a good assumption remains an active field of research.

4.2 Chaos and limits to prediction

The discovery of chaos makes it clear that Boltzmann's use of probability is even more justified than he realized. When motion is chaotic, two infinitesimally nearby trajectories separate at an exponential rate [Lorenz, 1963; Shaw, 1981; Crutchfield *et al.*, 1986; Strogatz, 1994]. This is a geometric property of the underlying nonlinear dynamics. From a linear point of view the dynamics are locally unstable. To make this precise, consider two N dimensional initial conditions $x(0)$ and $x'(0)$ that are initially separated by an infinitesimal vector $\delta x(0) = x(0) - x'(0)$. Providing the dynamical system is differentiable, the separation will grow as

$$(30)\quad \delta x(t) = D\phi^t(x(0))\delta x(0),$$

where $D\phi^t(x(0))$ is the derivative of the dynamical system ϕ^t evaluated at the initial condition $x(0)$. For any fixed time t and initial condition $x(0)$, $D\phi^t$ is just

[9]Dynamical systems that conserve energy and obey Newton's laws have special properties that cause the existence of KAM tori. Dissipative systems typically have *attractors*, subsets of the state space that orbits converge onto. Energy conserving systems do not have attractors, and often have chaotic orbits tightly interwoven with regular orbits.

an $N \times N$ matrix, and this is just a linear equation. If the motion is chaotic the length of the separation vector δx will grow exponentially with t in at least one direction, as shown in Figure 3. The figure shows how the divergence of nearby trajectories is the underlying reason chaos leads to unpredictability. A perfect measurement would correspond to a point in the state space, but any real measurement is inaccurate, generating a cloud of uncertainty. The true state might be anywhere inside the cloud. As shown here for the Lorenz equations (a simple system of three coupled nonlinear differential equations [Lorenz, 1963]), the uncertainty of the initial measurement is represented by 10,000 dark dots, initially so close together that they form a single dark spot ($t = 0$, top right); a single trajectory is shown for reference in light dark. As each point moves under the action of the equations, the cloud is stretched into a long, thin dark thread, which then folds over onto itself many times, until the points are mixed more or less randomly over the entire attractor. Prediction has now become impossible: the final state can be anywhere on the attractor. For a regular motion, in contrast, all the final states remain close together. We can think about this in information theoretic terms; for a chaotic motion information is initially lost at a linear rate which eventually results in all the information being lost — for a regular motion the information loss is relatively small. The numbers above the illustration are in units of 1/200 of the natural time units of the Lorenz equations. (From [Crutchfield *et al.*, 1986]).

Nonetheless, at the same time the motion can be globally stable, meaning that it remains contained inside a finite volume in the phase space. This is achieved by stretching and folding — the nonlinear dynamics knead the phase space through local stretching and global folding, just like a baker making a loaf of bread. Two trajectories that are initially nearby may later be quite far apart, and still later, may be close together again. This property is called *mixing*. More formally, the dynamics are mixing over a given set Σ and invariant measure[10] μ with support Σ such that for any subsets A and B

$$(31)\quad \lim_{t\to\infty} \mu(\phi^t B \cap A) = \mu(A)\mu(B).$$

Intuitively, this just means that B is smeared throughout Σ by the flow, so that the probability of finding a point originating in B inside of A is just the original probability of B, weighted by the probability of A. Geometrically, this happens if and only if the future trajectory of B is finely "mixed" throughout Σ by the stretching and folding action of ϕ^t.

Mixing implies ergodicity, so any dynamical system that is mixing over Σ will also be ergodic on Σ. It only satisfies the ergodic hypothesis, however, if Σ is the set of accessible states. This need not be the case. Thus, the fact that a system has orbits with chaotic dynamics doesn't mean that it necessarily satisfies the ergodic

[10] A measure is invariant over a set Σ with respect to the dynamics ϕ^t if it satisfies the condition $\mu(A) = \mu(\phi^{-t}(A))$, where A is any subset of Σ. There can be many invariant measures, but the one that we have in mind throughout is the one corresponding to time averages.

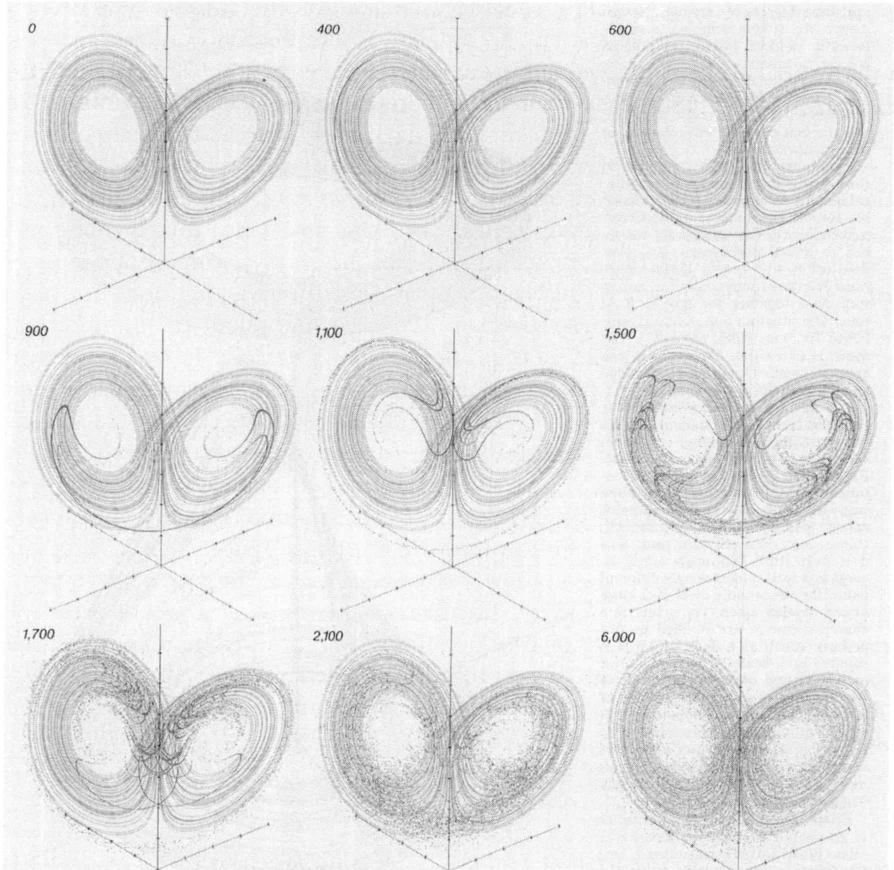

Figure 3. The divergence of nearby trajectories for the Lorenz equations. See the text for an explanation

hypothesis — there may be still be subsets of finite volume in the phase space that are stuck making regular motions, for example on KAM tori.

Nonetheless, chaotic dynamics has strong implications for statistical mechanics. If a dynamical system is ergodic but not mixing[11], by measuring the microstates it is in principle possible to make detailed long range predictions by measuring the position and velocity of all its microstates, as suggested by Laplace. In contrast, if it is mixing then even if we know the initial values of the microstates at a high (but finite) level of precision, all this information is asymptotically lost, and statistical mechanics is unavoidable[12].

4.3 Quantifying predictability

Information theory can be used to quantify predictability [Shaw, 1981]. To begin the discussion, consider a measuring instrument with a uniform scale of resolution ϵ. For a ruler, for example, ϵ is the distance between adjacent graduations. If such a measuring instrument is assigned to each of the N real variables in a dynamical system, the graduations of these instruments induce a *partition* Π of the phase space, which is a set of non-overlapping N dimensional cubes, labeled C_i, which we will call the outcomes of the measurement. A measurement determines that the state of the system is in a given cube C_i. If we let transients die out, and restrict our attention to asymptotic motions without external perturbations, let us assume the motion is confined to a set Σ (which in general depends on the initial condition). We can then compute the asymptotic probability of a given measurement by measuring its frequency of occurrence p_i, and if the motion is ergodic on Σ, then we know that there exists an invariant measure μ such that $p_i = \mu(C_i)$. To someone who knows the invariant measure μ but knows nothing else about the state of the system, the average information that will be gained in making a measurement is just the entropy

$$(32)\quad I(\epsilon) = -\sum_i p_i \log p_i.$$

We are following Shannon in calling this "information" since it represents the element of surprise in making the measurement. The information is written $I(\epsilon)$ to emphasize its dependence on the scale of resolution of the measurements. This can be used to define a dimension for μ. This is just the asymptotic rate of increase of the information with increasing resolution, i.e.

$$(33)\quad D = \lim_{\epsilon \to 0} \frac{I(\epsilon)}{|\log \epsilon|}.$$

[11] A simple example of a system that is ergodic but not mixing is a dynamical system whose solution is the sum of two sinusoids with irrationally related frequencies.

[12] An exception is that some systems display phase invariance even while they are chaotic. The orbits move around an attractor, being chaotically scrambled transverse to their direction of motion but keeping their timing for completing a circuit of the attractor [Farmer *et al.*, 1980].

This is called the *information dimension* [Farmer, 1982]. Note that this reduces to what is commonly called the fractal dimension when p_i is sufficiently smooth, i.e. when $\sum_i p_i \log p_i \approx \log n$, where n is the number of measurement outcomes with nonzero values of p_i.

This notion of dimension can be generalized by using the Rényi entropy R_α

$$(34)\quad R_\alpha = \frac{1}{1-\alpha} \log \sum_i p_i^\alpha$$

where $\alpha \geq 0$ and $\alpha \neq 1$. The value for $\alpha = 1$ is defined by taking the limit as $\alpha \to 1$, which reduces to the usual Shannon entropy. By replacing the Shannon entropy by the Rényi entropy it is possible to define a generalized dimension d_α. This contains the information dimension in the special case $\alpha = 1$. This has proved to be very useful in the study of multifractal phenomena (fractals whose scalings are irregular). We will say more about the use of such alternative entropies in the next section.

The discussion so far has concerned the amount of information gained by an observer in making a single, isolated measurement, i.e. the information gained in taking a "snapshot" of a dynamical system. We can alternatively ask how much new information is obtained per unit time by an observer who is watching a movie of a dynamical system. In other words, what is the information acquisition rate of an experimenter who makes a series of measurements to monitor the behavior of a dynamical system? For a regular dynamical system (to be defined more precisely in a moment) new measurements asymptotically provide no further information in the limit $t \to \infty$. But if the dynamical system is chaotic, new measurements are constantly required to update the knowledge of the observer in order to keep the observer's knowledge of the state of the system at the same resolution.

This can be made more precise as follows. Consider a sequence of m measurements $(x_1, x_2, \ldots, x_m) = X_m$, where each measurement corresponds to observing the system in a particular N dimensional cube. Letting $p(X_m)$ be the probability of observing the sequence X_m, the entropy of this sequence of measurements is

$$(35)\quad H_m = -\sum_i p(X_m) \log p(X_m)$$

We can then define the information acquisition rate as

$$(36)\quad h = \lim_{m \to \infty} \frac{H_m}{m \Delta t}.$$

Δt is the sampling rate for making the measurements. Providing Δt is sufficiently small and other conditions are met, h is equal to the *metric entropy*, also called the *Kolmogorov-Sinai (KS) entropy*[13]. Note that this is not really an entropy,

[13] In our discussion of metric entropy we are sweeping many important mathematical formalities under the rug. For example, to make this definition precise we need to take a supremum over all partitions and sampling rates. Also, it is not necessary to make the measurements in N dimensions — there typically exists a one dimensional projection that is sufficient, under an optimal partition.

but an entropy production rate, which (if logs are taken to base 2) has units of bits/second. If $h\rangle 0$ the motion is chaotic, and if $h = 0$ it is regular. Thus, when the system is chaotic, the entropy H_m contained in a sequence of measurements continues to increase even in the limit as the sequence becomes very long. In contrast, for a regular motion this reaches a limiting value.

Although we have so far couched the discussion in terms of probabilities, the metric entropy is determined by geometry. The average rates of expansion and contraction in a trajectory of a dynamical system can be characterized by the spectrum of Lyapunov exponents. These are defined in terms of the eigenvalues of $D\phi^t$, the derivative of the dynamical system, as defined in equation 30. For a dynamical system in N dimensions, let the N eigenvalues of the matrix $D\phi^t(x(0))$ be $\alpha_i(t)$. Because $D\phi^t$ is a positive definite matrix, the α_i are all positive. The Lyapunov exponents are defined as $\lambda_i = \lim_{t\to\infty} \log \alpha_i(t)/t$. To think about this more geometrically, imagine an infinitesimal ball that has radius $\epsilon(0)$ at time $t = 0$. As this ball evolves under the action of the dynamical system it will distort. Since the ball is infinitesimal, however, it will remain an ellipsoid as it evolves. Let the principal axes of this ellipsoid have length $\epsilon_i(t)$. The spectrum of Lyapunov exponents for a given trajectory passing through the initial ball is

$$(37)\ \lambda_i = \lim_{t\to\infty} \lim_{\epsilon(0)\to 0} \frac{1}{t} \log \frac{\epsilon_i(t)}{\epsilon(0)}.$$

For an N dimensional dynamical system there are N Lyapunov exponents. The positive Lyapunov exponents λ^+ measure the rates of exponential divergence, and the negative ones λ^- the rates of convergence. They are related to the metric entropy by *Pesin's theorem*

$$(38)\ h = \sum_i \lambda_i^+.$$

In other words, the metric entropy is the sum of the positive Lyapunov exponents, and it corresponds to the average exponential rate of expansion in the phase space.

Taken together the metric entropy and information dimension can be used to estimate the length of time that predictions remain valid. The information dimension allows an estimate to be made of the information contained in an initial measurement, and the metric entropy estimates the rate at which this information decays.

As we have already seen, for a series of measurements the metric entropy tells us the information gained with each measurement. But if each measurement is made with the same precision, the information gained must equal the information that would have been lost had the measurement not been made. Thus the metric entropy also quantifies the initial rate at which knowledge of the state of the system is lost after a measurement.

To make this more precise, let $p_{ij}(t)$ be the probability that a measurement at time t has outcome j if a measurement at time 0 has outcome i. In other words, given the state was measured in partition element C_i at time 0, what is the probability it will be in partition element C_j at time t?. By definition $p_{ij}(0) = 1$ if $i = j$ and $p_{ij}(0) = 0$ otherwise. With no initial information, the information gained from the measurement is determined solely by the asymptotic measure μ, and is $-\log \mu(C_j)$. In contrast, if C_i is known the information gained on learning outcome j is $-\log p_{ij}(t)$. The extra information using a prediction from the initial data is the difference of the two or $\log(p_{ij}(t)/\mu(C_j))$. This must be averaged over all possible measurements C_j at time t, and all possible initial measurements C_i. The measurements C_j are weighted by their probability of occurrence $p_{ij}(t)$, and the initial measurements are weighted by $\mu(C_i)$. This gives

$$(39)\quad I(t) = \sum_{i,j} \mu(C_i) p_{ij}(t) \log(\frac{p_{ij}(t)}{\mu(C_j)}).$$

It can easily be shown that in the limit where the initial measurements are made arbitrarily precise, $I(t)$ will initially decay at a linear rate, whose slope is equal to the metric entropy. For measurements with signal to noise ratio s, i.e. with $\log s \approx |\log \epsilon|$, $I(0) \approx D_I \log s$. Thus $I(t)$ can be approximated as $I(t) \approx D_I \log s - ht$, and the initial data becomes useless after a characteristic time $\tau = (D_I/h) \log s$.

To conclude, chaotic dynamics provides the link that connects deterministic dynamics with probability. While we can discuss chaotic systems in completely deterministic terms, as soon as we address problems of measurement and long-term predictability we are forced to think in probabilistic terms. The language we have developed above, of information dimension, Lyapunov exponents, and metric entropy, provide the link between the geometric and probabilistic views. Chaotic dynamics can happen even in a few dimensions, but as we move to high dimensional systems, e.g. when we discuss the interactions between many particles, probability is thrust on us for two reasons: The difficulty of keeping track of all the degrees of freedom, and the "increased likelihood" that nonlinear interactions will give rise to chaotic dynamics. "Increased likelihood" is in quotations because, despite more than a century of effort, understanding the necessary and sufficient conditions for the validity of statistical mechanics remains an open problem.

5 ABOUT ENTROPY

In this section we will discuss various aspects of entropy, its relation with information theory and the sometimes confusing connotations of order, disorder, ignorance and incomplete knowledge. This will be done by treating several well known puzzles and paradoxes related with the concept of entropy. A derivation of the second law using the procedure called *coarse graining* is presented. The extensivity or additivity of entropy is considered in some detail, also when we discuss nonstandard extensions of the definition of entropy.

5.1 *Entropy and information*

The important innovation Shannon made was to show that the relevance of the concept of entropy considered as a measure of information was not restricted to thermodynamics, but could be used in any context where probabilities can be defined. He applied it to problems in communication theory and showed that it can be used to compute a bound on the information transmission rate using an optimal code.

One of the most basic results that Shannon obtained was to show that the choice of the Gibbs form of entropy to describe uncertainty is not arbitrary, even when it is used in a very general context. Both Shannon and Khinchin [Khinchin, 1949] proved that if one wants certain conditions to be met by the entropy function then the functional form originally proposed by Gibbs is the unique choice. The fundamental conditions as specified by Khinchin are:

1. For a given n and $\sum_{i=1}^{n} p_i = 1$, the required function $H(p_1, ...p_n)$ is maximal for all $p_i = 1/n$.

2. The function should satisfy $H(p_1, ...p_n, 0) = H(p_1, ...p_n)$. The inclusion of an impossible event should not change the value of H.

3. If A and B are two finite sets of events, not necessarily independent, the entropy $H(A, B)$ for the occurrence of joint events A and B is the entropy for the set A alone plus the weighted average of the conditional entropy $H(B|A_i)$ for B given the occurrence of the ith event A_i in A,

   $$H(A, B) = H(A) + \sum_i p_i H(B|A_i) \tag{40}$$

   where event A_i occurs with probability p_i.

The important result is that given these conditions the function H given in equation (29) is the *unique* solution. Shannon's key insight was that the results of Boltzmann and Gibbs in explaining entropy in terms of statistical mechanics had unintended and profound side-effects, with a broader and more fundamental meaning that transcended their physical origin of entropy. The importance of the abstract conditions formulated by Shannon and Khinchin show the very general context in which the Gibbs-Shannon function is the unique answer. Later on we will pose the question of whether there are situations where *not* all three conditions are appropriate, leading to alternative expressions for the entropy.

5.2 *The Landauer principle*

Talking about the relation between information and entropy it may be illuminating to return briefly to the Landauer principle[Landauer, 1961; Landauer, 1991], which as we mentioned in the first section, is a particular formulation of the second law of thermodynamics well suited for the context of information theory. The principle expresses the fact that erasure of data in a system necessarily involves producing heat, and thereby increasing the entropy. We have illustrated the principle in figure 4. Consider a "gas" consisting of a single atom in a symmetric container

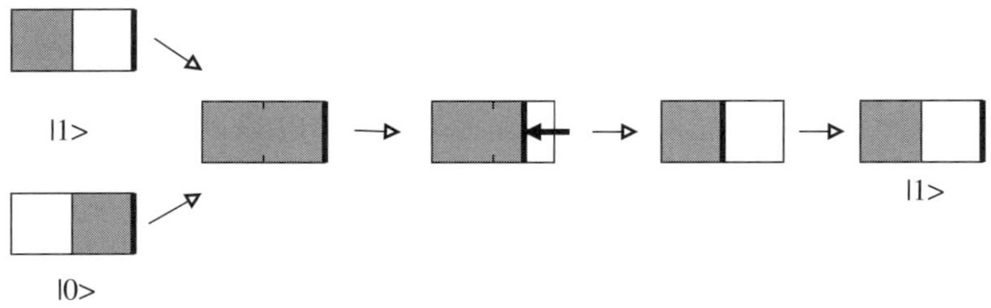

Figure 4. An illustration of the Landauer principle using a very simple thermodynamical system.

with volume 2V, in contact with a heat bath. We imagine that the position of the particle acts as a memory with one bit of information, corresponding to whether the atom is on the left or on the right. Erasing the information amounts to resetting the device to the "reference" state 1 independent of the initial state. Erasure corresponds therefore to reinitializing the system rather then making a measurement. It can be done by first opening a diaphragm in the middle, then reversibly moving the piston from the right in, and finally closing the diaphragm and moving the piston back. In the first step the gas expands freely to the double volume. The particle doesn't do any work, the energy is conserved, and therefore no heat will be absorbed from the reservoir. This is an irreversible adiabatic process by which the entropy S of the gas increases by a factor $k \ln 2V/V = k \ln 2$. (The number of states the particle can be in is just the volume; the average velocity is conserved because of the contact with the thermal bath and will not contribute to the change in entropy). In the second part of the erasure procedure we bring the system back to a state which has the same entropy as the initial state. We do this through a quasistatic (i.e. reversible) isothermal process at temperature T. During the compression the entropy decreases by $k \ln 2$. This change of entropy is nothing but the amount of heat delivered by the gas to the reservoir divided by the temperature, i.e. $\Delta S = \int dS = \int dQ/T = \Delta Q/T$. The heat produced ΔQ equals the net amount of work W that has been done in the cycle by moving the piston during the compression. The conclusion is that during the erasure of one bit of information the device had to produce at least $\Delta Q = kT \ln 2$ of heat.

We may look at the same process somewhat more abstractly, purely from the point of view of information. We map the erasure of information for the simple memory device on the sequence of diagrams depicted in figure 5. We choose this representation of the accessible (phase) space to clearly mark the differences between the situation where the particle is in the left *or* the right (A), the left *and* the right (B), and the left compartment only (C). In part A the memory

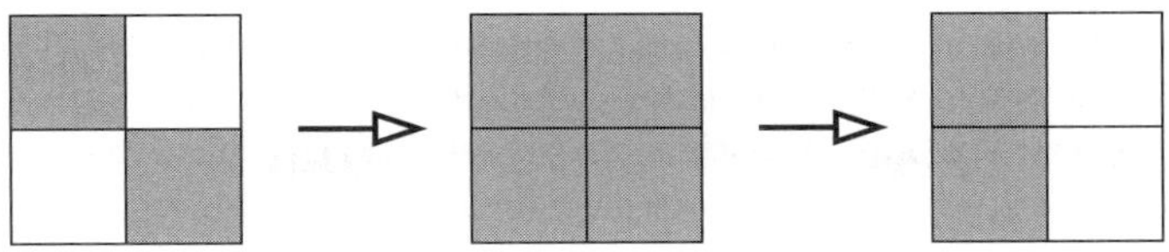

A B C

Figure 5. A phase space picture of Landauer's principle. See text for an explanation.

corresponds to the particle being either in the left or in the right compartment. In B the partition has been removed and through the free expansion the phase space has doubled and consequently the entropy increased by $\ln 2$. In C the system is brought back to the reference state, i.e. the particle is brought in the left compartment. This is done by moving a piston in from the right, inserting the partition, and moving the piston out again. It is in the compressing step that the phase space is reduced by a factor of two and hence entropy is reduced by $\ln 2$. This is possible because we did work, producing a corresponding amount of heat ($\Delta Q \geq T \ln 2$). Note that in this representation one can in principle change the sizes of the partitions along the horizontal directions and the a priori probabilities along the vertical direction to model different types or aspects of memory devices.

5.3 The entropy as a relative concept

> Irreversibility is a consequence of the explicit introduction of ignorance into the fundamental laws.
>
> M. Born

There is a surprising amount of confusion about the interpretation and meaning of the concept of entropy [Guttman, 1999; Denbigh and Denbigh, 1985]. One may wonder to what extent the "entropic principle" just is an "anthropocentric principle"? That is, does entropy depend only on our perception, or is it something more fundamental? Is it a subjective attribute in the domain of the observer or is it an intrinsic property of the physical system we study? Let us consider the common definition of entropy as a measure of disorder. This definition can be confusing unless we are careful in spelling out what we mean by order or disorder. We may for instance look at the crystallization of a supercooled liquid under conditions where it is a closed system, i.e. when no energy is exchanged with the environment. Initially the molecules of the liquid are free to randomly move about, but then (often through the addition of a small perturbation that breaks the symmetry) the liquid suddenly turns into a solid by forming a crystal in which the molecules are pinned to the sites of a regular lattice. From one point of view this a splendid example of the creation of order out of chaos. Yet from standard calculations in statistical mechanics we know that the entropy increases during crystallization. This is because what meets the eye is only part of the story. During crystallization entropy is generated in the form of latent heat, which is stored in the vibrational modes of the molecules in the lattice. Thus, even though in the crystal the individual molecules are constrained to be roughly in a particular location, they vibrate around their lattice sites more energetically than when they were free to wander. From a microscopic point of view there are more accessible states in the crystal than there were in the liquid, and thus the entropy increases. The thermodynamic entropy is indifferent to whether motions are microscopic or macroscopic — it only counts the number of accessible states and their probabilities.

In contrast, to measure the sense in which the crystal is more orderly, we must measure a different set of probabilities. To do this we need to define probabilities that depend only on the positions of the particles and not on their velocities. To make this even more clear-cut, we can also use a more macroscopic partition, large enough so that the thermal motions of a molecule around its lattice site tend to stay within the same partition element. The entropy associated with this set of probabilities, which we might call the "spatial order entropy", will behave quite differently from the thermodynamic entropy. For the liquid, when every particle is free to move anywhere in the container, the spatial order entropy will be high, essentially at its largest possible value. After the crystallization occurs, in contrast, the spatial order entropy will drop dramatically. Of course, this is *not* the thermodynamic entropy, but rather an entropy that we have designed to quantitatively capture the aspect of the crystalline order that we intuitively perceive.

As we emphasized before, Shannon's great insight was that it is possible to associate an entropy with any set of probabilities. However, the example just given illustrates that when we use entropy in the broader sense of Shannon we must be very careful to specify the context of the problem. Shannon entropy is just a function that reduces a set of probabilities to a number, reflecting how many nonzero possibilities there are as well as the extent to which the set of nonzero probabilities is uniform or concentrated. Within a fixed context, a set of probabilities that is smaller and more concentrated can be interpreted as more "orderly", in the sense that fewer numbers are needed to specify the set of possibilities. Thermodynamics dictates a particular context — we have to measure probabilities in the full state space. Thermodynamic entropy is a special case of Shannon entropy. In the more general context of Shannon, in contrast, we can define probabilities however we want, depending on what we want to do. But to avoid confusion we must always be careful to keep this context in mind, so that we know what our computation means.

5.4 Maxwell's demon

The "being" soon came to be called Maxwell's demon, because of its far-reaching subversive effects on the natural order of things. Chief among these effects would be to abolish the need for energy sources such as oil, uranium and sunlight.

C.H. Bennett

The second law of thermodynamics is statistical, deriving from the fact that the individual motions of the molecules are not observed or controlled in any way. Would things be different if we could intervene on a molecular scale? This question gives rise to an important paradox posed by Maxwell in 1872, which appeared in his *Theory of Heat* [Maxwell, 1872]. This has subsequently been discussed by generations of physicists, notably Szilard [Szilard, 1929], Brillouin [Brillouin, 1956], Landauer [Landauer, 1961], Bennett [Bennett, 1982] and others.

Maxwell described his demonic setup as follows: "Let us suppose that a vessel is divided in two portions, A and B, by a division in which there is a small hole, and that a being who can see individual molecules opens and closes this hole, so as to allow only the swifter particles to to pass from A to B, and only the slower ones to pass from B to A. He will thus, without expenditure of work, raise the temperature of B and lower that of A, in contradiction with the second law of thermodynamics." In attempts to save the second law from this demise, many aspects of the problem have been proposed for its resolution, including Brownian motion, quantum uncertainty and even Gödel's Theorem. The resolution of the paradox touches on some very fundamental issues that center on the question of how the demon might actually realize his subversive interventions.

Szilard clarified the discussion by introducing an engine (or thermodynamic cycle), which is depicted in the left half of figure 6. He and Brillouin focused on the measurement the demon has to perform in order to find out in which half of the vessel the particle is located after the partition has been put into place. For the demon to "see" the actual molecules he has to use a measurement device, such as a source of light (photons) and a photon detector. He will in principle be able to measure whether a molecule is faster or slower then the thermal average by scattering a photon off of it. Brillouin tried to argue that the entropy increase to the whole system once the measurement is included would always be larger or equal then the entropy gain achieved by the subsequent actions of the demon. However, this argument didn't hold; people were able to invent devices that got around the measurement problem, so that it appeared the demon could beat the second law.

Instead, the resolution of the paradox comes from a very different source. In 1982 Bennett gave a completely different argument to rescue the second law. The fundamental problem is that under Landauer's principle, production of heat is necessary for erasure of information (see section 5.2). Bennett showed that a reversible measurement could in principle be made, so that Brillouin's original argument was wrong — measurement does not necessarily produce any entropy. However, to truly complete the thermodynamic cycle, the demon has to erase the information he obtained about the location of the gas molecule. As we already discussed in section 5.2, erasing that information produces entropy. It turns out

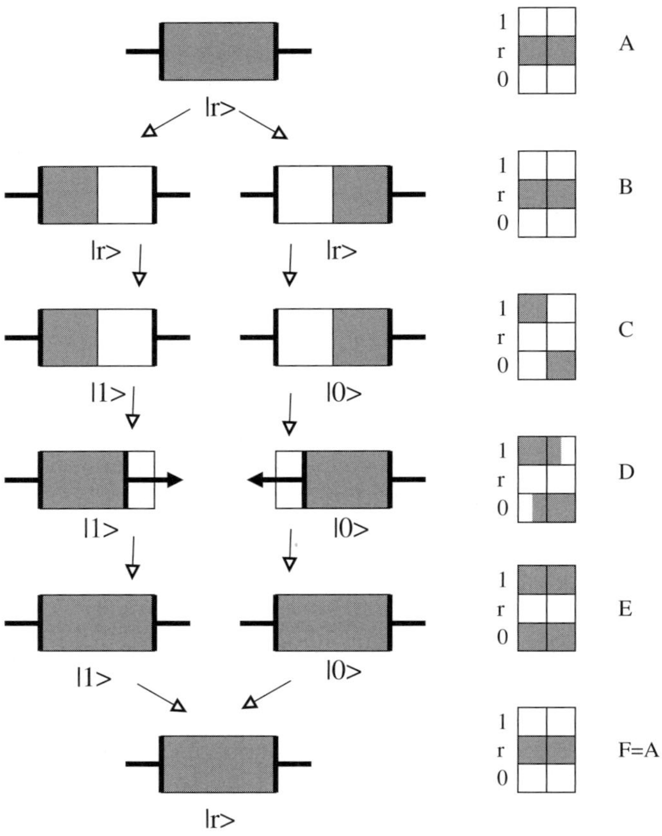

Figure 6. The one-particle Maxwell demon apparatus as envisaged by Bennett [Bennett, 1982; Bennett, 1987]. An explanation is given in the text.

that the work that has to be done to erase the demon's memory is at least as much as was originally gained.

Figure 6 illustrates the one-particle Maxwell demon apparatus as envisaged by Bennett [Bennett, 1982; Bennett, 1987], which is a generalization of the engine proposed by Szilard [Szilard, 1929]. On the left in row (A) is a gas container containing one molecule with a partition and two pistons. On the right is a schematic representation of the phase space of the system, including the demon. The state of mind of the demon can be in three different states: He can know the molecule is on the right (state 0), on the left (state 1), or he can be in the reference or blank state r, where he lacks any information and knows that he doesn't know where the particle is. In the schematic diagram of the phase space, shown on the right, the vertical direction indicates the state of memory of the demon and the horizontal direction indicates the position of the particle. In step (B) a thin partition is placed in the container, trapping the particle in either the left or right half. In step (C) the demon makes a (reversible) measurement to determine the location of the particle. This alters his state of mind as indicated — if the particle is on the right, it goes into state 0, if on the left, into state 1. In step (D), depending on the outcome of the measurement, he moves either the right or left piston in and removes the partition. In (E) the gas freely expands, moving the piston out and thereby doing work. In state (E) it appears as if the system has returned to its original state — it has the same volume, temperature and entropy — yet work has been done. What's missing? The problem is that in (E) the demon's mind has not returned to its original blank state. He needs to know that he doesn't know the position of the particle. Setting the demon's memory back into its original state requires erasing a bit of information. This is evident in the fact that to go from (E) to (F) the occupied portion of the phase space is reduced by a factor of two. This reduction in entropy has to be accompanied by production of heat as a consequence of Landauer's principle (see figure 4 and figure 5) — the work that is done to erase a bit of information is greater than or equal to the work gained by the demon. This ensures that the full cycle of the complete system respects the second law after all.

This resolution of the paradox is remarkable, because it is not the acquisition of information (the measurement) which is irreversible and thermodynamically costly, but it is the process of erasure, which is both logically and thermodynamically irreversible, that leads to the increase of entropy required by the second law. The information comes for free, but it poses a waste disposal problem which is costly. It is gratifying to see information theory come to rescue of one of the most cherished physical laws.

5.5 The Gibbs paradox

The Gibbs paradox provides another interesting chapter in the debate on the meaning of entropy. The basic question is to what extent entropy is a subjective notion. In its simplest form the paradox concerns the mixing of two ideal gases (kept at the same temperature and pressure) after removing a partition. If it has been removed the gases will mix, and if the particles of the two gases are distinguishable the entropy will increase due to this mixing. However, if the gases are *identical*, so that their particles are indistinguishable from those on the other side, there is no increase in the entropy. Maxwell imagined the situation where the gases were initially supposed to be identical, and only later recognized to be different. This reasoning led to the painful conclusion that the notion of irreversibility and entropy would depend on our knowledge of physics. He concluded that the entropy would thus depend on the state of mind of the experimenter and therefore lacked an objective ground. It was again Maxwell with a simple question who created an uncomfortable situation which caused a long debate. After the development of quantum mechanics, it became clear that particles of the same species are truly indistinguishable. There is no such thing as labeling N individual electrons, and therefore interchanging electrons doesn't change the state and this fact reduces the number of states by a relative factor of N!. Therefore the conclusion is that the entropy does not increase when the gases have the same constituent particles, and it does increase when they are different.

However, the resolution of Gibbs paradox does not really depend on quantum mechanics. Jaynes has emphasized that in the early works of Gibbs, the correct argument was already given (well before the advent of quantum mechanics) [Jaynes, 1996]. Gibbs made an operational definition, saying that if "identical" means anything, it means that there is no way an "unmixing" apparatus could determine whether a particular molecule came from a given side of the box, short of having followed its entire trajectory. Thus if the particles of the gas are identical in this sense, the entropy will not change. We conclude that the adequate definition of entropy reflects the objective physical constraints we put on the system, i.e. what measurements are possible or admissible. This has nothing to do with our lack of knowledge but rather with our choices. The 'incompleteness of our knowledge' is an exact and objective reflection of a particular set of macroscopic constraints imposed on the physical system we want to describe. The system's behavior depends on these constraints, and so does the entropy.

5.6 The maximal entropy principle of Jaynes

> The statistical practice of physicists has tended to lag about 20 years behind current developments in the field of basic probability and statistics.
>
> E.T. Jaynes (1963)

There are two equivalent sets of postulates that can be used as a foundation to derive an equilibrium distribution in statistical mechanics. One is to begin with the hypothesis that equilibrium corresponds to a minimum of the free energy, and the other is that it corresponds to a maimum of the entropy. The latter approach is a relatively modern development. Inspired by Shannon, Jaynes turned the program of statistical mechanics upside down [Jaynes, 1983]. Starting from a very general set of axioms he showed that under the assumption of equilibrium the Gibbs expression for the entropy is unique. Under Jaynes' approach, any problem in equilibrium statistical mechanics is reduced to finding the set of p_i for which the entropy is maximal, under a set of constraints that specify the macroscopic conditions, which may come from theory or may come directly from observational data [Jaynes, 1963]. This variational approach removes some of the arbitrariness that was previously present in the foundations of statistical mechanics. The principle of maximum entropy is very simple and has broad application. For example if one maximizes S only under the normalization condition $\sum_i p_i = 1$, then one finds the unique solution that $p_i = 1/N$ with N the total number of states. This is the uniform probability distribution underlying the equipartition principle. Similarly, if we now add the constraint that energy is conserved, i.e. $\sum_i \varepsilon_i p_i = U$, then the unique solution is given by the Boltzmann distribution, equation (19).

The maximum entropy principle as a starting point clearly separates the physical input and purely probabilistic arguments that enter the theory. Let us derive the Maxwell-Boltzmann distribution to illustrate the maximal entropy principle. We start with the function $L(p_i, \alpha, \beta)$ which depends on the probability distribution and two Lagrange multipliers to impose the constraints:

$$(41)\quad L(p_i, \alpha, \beta) = -\sum_{i=1}^{N} p_i \ln p_i - \alpha(\sum_{i=1}^{N} p_i - 1) - \beta(\sum_{i=1}^{N} p_i \varepsilon_i - U)$$

The maximum is determined by setting the partial derivatives of L equal zero:

$$(42)\quad \frac{\partial L}{\partial p_i} = -\ln p_i - 1 - \alpha - \beta\varepsilon_i = 0$$

$$(43)\quad \frac{\partial L}{\partial \alpha} = \sum_{i=1}^{N} p_i - 1 = 0$$

$$(44)\quad \frac{\partial L}{\partial \beta} = \sum_{i=1}^{N} p_i\varepsilon_i - U = 0$$

From the first equation we immediately obtain that:

$$(45)\quad p_i = e^{-(1+\alpha+\beta\varepsilon_i)} \Rightarrow p_i = \gamma e^{-\beta\varepsilon_i}$$

The parameters γ and β are determined by the constraint equations. If we first substitute the above solution in the normalisation constraint, and then use the defining equation for the partition sum (20), we find that $\gamma = 1/Z$. The solution for β is most easily obtained using the following argument. First substitute (45) in the definition (27) of S to obtain the relation:

$$(46)\quad S = \beta U - \mathit{const.}\,.$$

Next we use the thermodynamic relation between energy and entropy (3), from which we obtain hat $\partial U/\partial S = T$. Combining these two relations we find that $\beta = 1/T$, which yields the thermal equilibrium distribution (19).

The *maximal entropy formalism* has a much wider validity than just statistical mechanics. It is widely used for statistical inference in applications such as optimizing data transfer and statistical image improvement. In these contexts it provides a clean answer to the question, "given the constraints I know about in the problem, what is the model that is as random as possible (i.e. minimally biased) subject to these constraints?". A common application is missing data: Suppose one observes a series of points x_i at regular time intervals, but some of the observations are missing. One can make a good guess for the missing values by solving for the distribution that maximizes the entropy, subject to the constraints imposed by the know data points.

One must always bear in mind, however, that in physics the maximum entropy principle only applies to equilibrium situations, which are only a small subset of the problems in physics. For systems that are not in equilibrium one must take a different approach. Attempts to understand non-equilibrium statistical mechanics have led some researchers to explore the use of alternative notions of entropy, as discussed in Section 5.11.

5.7 Ockham's razor

> Entia non sunt multiplicanda praeter neccessitatem
> (*Entities should not be introduced except when strictly necessary*)
> William van Ockham (1285-1347)

An interesting and important application of information is to the process of modeling itself. When developing a model it is always necessary to make a tradeoff between models that are too simple and fail to explain the data properly, and models that are too complicated and fit fluctuations in the data that are really just noise. The desirability of simpler models is often called "Ockham's razor": If two models fit the existing data equally well, the simplest model is preferable, in the sense that the simpler model is more likely to make good predictions for data that has not yet been seen. While the value of using simple models seems like something we can all agree on, the tradeoff in real problems is typically not so obvious. Suppose model A fits the data a little better than model B, but has one more parameter. How does one trade off goodness of fit against number of parameters?

Using ideas from information theory Akaike [Akaike, 1974] introduced a method for making tradeoffs between goodness of fit and model complexity that can be applied in the context of simple linear models. Rissenen subsequently introduced a more general framework to think about this problem based on a principle that he called minimum description length (MDL) [Rissanen, 1978; Grunwald *et al.*, 2004; Grunwald and Vitányi,]. The basic idea is that the ability to make predictions and the ability to compress information are essentially two sides of the same coin. We can only compress data if it contains regularities, i.e. if the structure of the data is at least partially predictable. We can therefore find a good prediction model by seeking the model that gives the shortest description of the data we already have. When we do this we have to take the description length of the model into account, as well as the description length of the deviations between the model's predictions and the actual data. The deviations between the model and the data can be treated as probabilistic events. A model that gives a better fit has less deviation from the data, and hence implies a tighter probability distribution, which translates into a lower entropy for the deviations from the data. This entropy is then added to the information needed to specify the model and its parameters. The best model is the one with the lowest sum, i.e. the smallest total description length. By characterizing the goodness of fit in terms of bits, this approach puts the complexity of the model and the goodness of fit on the same footing, and gives the correct tradeoff between goodness of fit and model complexity, so that the quality of any two models can be compared, at least in principle.

This shows how at some level the concept of entropy underlies the whole scientific method, and indeed, our ability to make sense out of the world. To describe the patterns in the world, we need to make a trade-off between overfitting (fitting every bump even if it is a random variation, i.e. fitting noise) and overgeneralization (identifying events that really are different). A similar trade-off arises in assigning a causal mechanism to the occurrence of an event or explaining it as random. This problem of how to exactly make such trade-offs based on time series analysis has a rather long history but on the other hand is still an active topic of research [Kan, 2006; Crutchfield and Young, 1989; Still and Crutchfield, 2007]. Even if we do not do these trade-offs perfectly and do not think about it quantitatively, when we discover and model regularities in the world, we are implicitly relying on a model selection process of this type. Any generalization makes a judgment that trades off the information needed to specify the model and the entropy of the fit of the model to the world.

5.8 Coarse graining and irreversibility

> Our aim is not to 'explain irreversibility' but to describe and predict the observable facts. If one succeeds in doing this correctly, from first principles, we will find that philosophical questions about the 'nature of irreversibility' will either have been answered automatically or else will be seen as ill considered and irrelevant.
>
> E.T. Jaynes

The second law of thermodynamics says that for a closed system the entropy will increase until it reaches its equilibrium value. This corresponds to the irreversibility we all know from daily experience. If we put a drop of ink in a glass of water the drop will diffuse through the water and dilute until the ink is uniformly spread through the water. The increase of entropy is evident in the fact that the ink is initially in a small region, with $p_i = 0$ except for this region, leading to a probability distribution concentrated on a small region of space and hence a low entropy. The system will not return to its original configuration. Although this is not impossible in principle, it is so improbable that it will never be observed[14].

Irreversibility is hard to understand from the microscopic point of view because the microscopic laws of nature that determine the time evolution of any physical

[14] "Never say never" is a saying of unchallenged wisdom. What we mean here by "never", is inconceivably stronger then "never in a lifetime", or even "never in the lifetime of the universe". Let's make a rough estimate: consider a dilute inert gas, say helium, that fills the left half of a container of volume V. Then we release the gas into the full container and ask what the recurrence time would be, i.e. how long it would take before all particles would be in the left half again. A simple argument giving a reasonable estimate, would be as follows: At any given instant the probability for a given particle to be in the left half is 1/2, but since the particles are independent, the probability of $N \sim N_A$ particles to be in the left half is $P = (1/2)^{10^{23}} \approx 10^{(-10^{20})}$. Assuming a typical time scale for completely rearranging all the particles in the container of, say, $\tau_0 = 10^{-3}$ seconds, the typical time that will pass before such a fluctuation occurs is $\tau = \tau_0/P = 10^{10^{20}} 10^{-3} \approx 10^{10^{20}}$ *sec.*

system on the fundamental level are all symmetric under time reversal. That is, the microscopic equations of physics, such as $F = ma$, are unchanged under the substitution $t \to -t$. How can irreversibility arise on the macroscopic level if it has no counterpart on the microscopic level?

In fact, if we compute the entropy at a completely microscopic level it is conserved, which seems to violate the second law of thermodynamics. This follows from the fact that momentum is conserved, which implies that volumes in phase space are conserved. This is called Liouville's theorem. It is easy to prove that this implies that the entropy S is conserved. This doesn't depend on the use of continuous variables — it only depends on applying the laws of physics at the microscopic level. It reflects the idea of Laplace, which can be interpreted as a statement that statistical mechanics wouldn't really be necessary if we could only measure and track all the little details. The ingenious argument that Gibbs used to clarify this, and thereby to reconcile statistical mechanics with the second law of thermodynamics, was to introduce the notion of *coarse graining.* This procedure corresponds to a systematic description of what we could call "zooming out". As we have already mentioned, this zooming out involves dividing phase space up in finite regions δ according to a partition Π. Suppose, for example, that at a microscopic level the system can be described by discrete probabilities p_i for each state. Let us start with a closed system in equilibrium, with a uniform distribution over the accessible states. For the Ising system, for example, $p_i = 1/g(N,i)$ is the probability of a particular configuration of spins. Now we replace in each little region δ the values of p_i by its average value $\bar{p}_i$ over δ:

$$(47)\quad \bar{p}_i \equiv \frac{1}{\delta}\sum_{i\in\delta} p_i,$$

and consider the associated coarse grained entropy

$$(48)\quad \bar{S} \equiv -\sum_i \bar{p}_i \ln \bar{p}_i.$$

Because we start at time $t = 0$ with a uniform probability distribution, $S(0) = \bar{S}(0)$. Next we change the situation by removing a constraint of the system so that it is no longer in equilibrium. In other words, we enlarge the space of accessible states but have as an initial condition that the probabilities are zero for the new states. For the new situation we still have that $S(0) = \bar{S}(0)$, and now we can compare the evolution of the fine-grained entropy $S(t)$ and the coarse-grained entropy $\bar{S}(t)$. The evolution of $S(t)$ is governed by the reversible microscopic dynamics and therefore it stays constant, so that $S(t) = S(0)$. To study the evolution of the coarse-grained entropy we can use a few simple mathematical tricks. First, note that because $\bar{p}_i$ is constant over each region with δ elements,

$$(49)\quad \bar{S}(t) = -\sum_i \bar{p}_i \ln \bar{p}_i = -\sum_i p_i \ln \bar{p}_i.$$

Then we may write

$$(50)\quad \bar{S}(t) - \bar{S}(0) = \sum_i p_i(\ln p_i - \ln \bar{p}_i) = \sum_i p_i \ln \frac{p_i}{\bar{p}_i} = \sum_i \bar{p}_i(\frac{p_i}{\bar{p}_i} \ln \frac{p_i}{\bar{p}_i}),$$

which in information theory is called he Kullback-Leibler divergence. The mathematical inequality $x \ln x \geq (x - 1)$, with $x = p_i/\bar{p}_i$, then implies the Gibbs inequality:

$$(51)\quad \bar{S}(t) - \bar{S}(0) \geq \sum_i p_i - \sum_i \bar{p}_i = 1 - 1 = 0.$$

Equality only occurs if $p_i/\bar{p}_i = 1$ throughout, so except for the special case where this is true, this is a strict inequality and the entropy increases. We see how the second law is obtained as a consequence of coarse graining.

The second law describes mathematically the irreversibility we witness when somebody blows smoke in the air. Suppose we make a film of the developing smoke cloud. If we film the movie at an enormous magnification, so that what we see are individual particles whizzing back and forth, it will be impossible to tell which way the movie is running — from a statistical point of view it will look the same whether we run the movie forward or backward. But if we film it at a normal macroscopic scale of resolution, the answer is immediately obvious — the direction of increasing time is clear from the diffusion of the smoke from a well-defined thin stream to a diffuse cloud.

From a philosophical point of view one should ask to what extent coarse graining introduces an element of subjectivity into the theory. One could object that the way we should coarse grain is not decided upon by the physics but rather by the person who performs the calculation. The key point is that, as in so many other situations in physics, we have to use some common sense, and distinguish between observable and unobservable quantities. Entropy does not increase in the highly idealized classical world that Laplace envisioned, as long as we can observe all the microscopic degrees of freedom and there are no chaotic dynamics. However, as soon as we violate these conditions and observe the world at a finite level of resolution (no matter how accurate), chaotic dynamics ensures that we will lose information and entropy will increase. While the coarse graining may be subjective, this is not surprising — measurements are inherently subjective operations. In most systems one will have that the entropy may stabilize on plateaus corresponding to certain ranges of the fineness of the coarseness. In many applications the increase of entropy will therefore be constant (i.e. well defined) for a sensible choice for the scale of coarse graining. The increase in (equilibrium) entropy between the microscopic scale and the macroscopic scale can also be seen as the amount of information that is lost by increasing the graining scale from the microscopic to the macroscopic. A relevant remark at this point is that a system is of course never perfectly closed – there are always small perturbations from the environment that act as a stochastic perturbation of the system, thereby continuously smearing out the actual distribution in phase space and simulating the effect of coarse graining. Coarse graining correctly captures the fact that entropy is a measure of our uncertainty; the fact that this uncertainty does not exist for regular motions and perfect measurements is not relevant to most physical problems.

5.9 *Coarse graining and renormalization*

In a written natural language not all finite combinations of letters are words, not all finite combinations of words are sentences, and not all finite sequences of sentences make sense. So by identifying what we call meaningful with accessible, what we just said means that compared with arbitrary letter combinations, the entropy of a language is extremely small.

Something similar is true for the structures studied in science. We are used to thinking of the rich diversity of biological, chemical and physical structures as being enormous, yet relative to what one might imagine, the set of possibilities is highly constrained. The complete hierarchy starting from the most elementary building blocks of matter such as *leptons* and *quarks*, all the way up to living organisms, is surprisingly restricted. This has to do with the very specific nature of the interactions between these building blocks. To our knowledge at the microscopic level there are only four fundamental forces that control all interactions. At each new structural level (quarks, protons and neutrons, nuclei, atoms, molecules, etc) there is a more or less autonomous theory describing the physics at that level involving only the relevant degrees of freedom at that scale. Thus moving up a level corresponds to throwing out an enormous part of the phase space available to the fundamental degrees of freedom in the absence of interactions. For example, at the highest, most macroscopic levels of the hierarchy only the long range interactions (electromagnetism and gravity) play an important role — the structure of quantum mechanics and the details of the other two fundamental forces are more or less irrelevant.

We may call the structural hierarchy we just described "coarse graining" at large. Although this ability to leave the details of each level behind in moving up to the next is essential to science, there is no cut and dried procedure that tells us how to do this. The only exception is that in some situations it is possible to do this coarse graining exactly by a procedure called *renormalization* [Zinn-Justin, 1989]. This is done by systematically studying how a set of microscopic degrees of freedom at one level can be averaged together to describe the degrees of freedom at the next level. There are some situations, such as phase transitions, where this process can then be used repeatedly to demonstrate the existence of fixed points of the mapping from one level to the next (an example of a phase transition is the change from a liquid to a gas). This procedure has provided important insights in the nature of phase transitions, and in many cases it has been shown that some of their properties are universal, in the sense that they do not depend on the details of the microscopic interactions.

5.10 *Adding the entropy of subsystems*

Entropy is an extensive quantity. Generally speaking the extensivity of entropy means that it has to satisfy the fundamental linear scaling property

$$(52)\quad S(T, qV, qN) = qS(T, V, N), \quad 0 < q < \infty.$$

Extensivity translates into additivity of entropies: If we combine two noninteracting systems (labelled 1 and 2) with entropies S_1 and S_2, then the total number of states will just be the product of those of the individual systems. Taking the logarithm, the entropy of the total system S becomes:

$$(53)\quad S = S_1 + S_2.$$

Applying this to two spin systems without an external field, the number of states of the combined system is $w = 2^{N_1+N_2}$, i.e. $w = w_1\, w_2$. Taking the logarithm establishes the additivity of entropy.

However if we allow for a nonzero magnetic field, this result is no longer obvious. In Section 3.2 we calculated the number of configurations with a given energy $\varepsilon_k = -k\mu H$ as $g(N, k)$. If we now allow two systems to exchange energy but keep the total energy fixed, then this generates a dependence between the two systems that lowers the total entropy. We illustrate this with an example:

Let the number of spins pointing up in system 1 be k_1 and the number of particles be N_1, and similarly let this be k_2 and N_2 for system 2. The total energy $k = k_1 + k_2$ is conserved, but the energy in either subsystem (k_1 and k_2) is not conserved. The total number of spins, $N = N_1 + N_2$ is fixed, and so are the spins (N_1 and N_2) in either subsystem. Because the systems only interact when the number of up spins in one of them (and hence also the other one) changes, we can write the total number of states for the combined system as

$$(54)\quad g(N, k) = \sum_{k_1} g_1(N_1, k_1) g_2(N_2, k_2),$$

where we are taking advantage of the fact that as long as k_1 is fixed, systems one and two are independent. Taking the log of the above formula clearly does not lead to the additivity of entropies because we have to sum over k_1. This little calculation illustrates the remark made before: Since we have relaxed the constraint that each system has a fixed energy to the condition that only the sum of their energies is fixed, the number of accessible states for the total system is increased. The subsystems themselves are no longer closed and therefore the entropy will change.

The extensivity of entropy is recovered in the thermodynamic limit in the above example, i.e. when $N \to \infty$. Consider the contributions to the sum in (54) as a function of k_1, and let the value of k_1 where g reaches a maximum be $k_1 = \hat{k}_1$. We can now write the contribution in the sum in terms $\delta = k_1 - \hat{k}_1$ as

$$(55)\quad \Delta g(N, k) = g_1(N_1, \hat{k}_1 + \delta) g_2(N_2, \hat{k}_2 - \delta) = f(\delta) g_1(N_1, \hat{k}_1) g_2(N_2, \hat{k}_2)\ ,$$

where the correction factor can be calculated by expanding the g functions around their respective $\hat{k}$ values. Not surprisingly, in the limit where N is large it turns out that f is on the order of $f \sim \exp(-2\delta^2)$ so that the contributions to $g(N,k)$ of the nonmaximal terms in the sum (54) are exponentially suppressed. Thus in the limit that the number of particles goes to infinity the entropy becomes additive. This exercise shows that when a system gets large we may replace the averages of a quantity by its value in the most probable configuration, as our intuition would have suggested. From a mathematical point of view this result follows from the fact that the binomial distribution approaches a gaussian for large values of N, which becomes ever sharper as $N \to \infty$. This simple example shows that the extensivity of entropy may or may not be true, depending on the context of the physical situation and in particular on the range of the inter-particle forces.

When two subsystems interact, it is certainly possible that the entropy of one decreases at the expense of the other. This can happen, for example, because system one does work on system two, so the entropy of system one goes up while that of system two goes down. This is very important for living systems, which collect free energy from their environment and expel heat energy as waste. Nonetheless, the total entropy S of an organism plus its environment still increases, and so does the sum of the independent entropies of the non interacting subsystems. That is, if at time zero

$$(56)\quad S(0) = S_1(0) + S_2(0)\ ,$$

then at time t it may be true that

$$(57)\quad S(t) \leq S_1(t) + S_2(t)\ ,$$

This is due to the fact that only interactions with other parts of the system can lower the entropy of a given subsystem. In such a situation we are of course free to call the difference between the entropy of the individual systems and their joint entropy a *negative* correlation entropy. However, despite this apparent decrease of entropy, both the total entropy and the sum of the individual entropies can only increase, i.e.

$$(58)\quad \begin{aligned} S(t) &\geq S(0) \\ S_1(t) + S_2(t) &\geq S_1(0) + S_2(0). \end{aligned}$$

The point here is thus that equations (57) and (58) are not in conflict.

5.11 Beyond the Boltzmann, Gibbs and Shannon entropy: the Tsallis entropy

> The equation $S = k \log W + const$ appears without an elementary theory - or however one wants to say it - devoid of any meaning from a phenomenological point of view.
>
> A. Einstein (1910)

As we have already stressed, the definition of entropy as $-\sum_i p_i \log p_i$ and the associated exponential distribution of states apply only for systems in equilibrium. Similarly, the requirements for an entropy function as laid out by Shannon and Khinchin are not the only possibilities. By modifying these assumptions there are other entropies that are useful. We have already mentioned the Rényi entropy, which has proved to be valuable to describe multi-fractals.

Another context where considering an alternative definition of entropy appears to be useful concerns power laws. Power laws are ubiquitous in both natural and social systems. A power law[15] is something that behaves for large x as $f(x) \sim x^{-\alpha}$, with $\alpha > 0$. Power law probability distributions decay much more slowly for large values of x than exponentials, and as a result have very different statistical properties and are less well-behaved[16]. Power law distributions are observed in phenomena as diverse as the energy of cosmic rays, fluid turbulence, earthquakes, flood levels of rivers, the size of insurance claims, price fluctuations, the distribution of individual wealth, city size, firm size, government project cost overruns, film sales, and word usage frequencies [Newman, 2005; Farmer and Geanakoplos, 2006]. Many different models can produce power laws, but so far there is no unifying theory, and it is not yet clear whether any such unifying theory is even possible. It is clear that power laws (in energy, for instance) can't be explained by equilibrium statistical mechanics, where the resulting distributions are always exponential. A common property of all the physical systems that are known to have power laws and the models that purport to explain them is that they are in some sense nonequilibrium systems. The ubiquity of power laws suggests that there might be nonequilibrium generalizations of statistical mechanics for which they are the standard probability distribution in the same way that the exponential is the standard in equilibrium systems.

From simulations of model systems with long-range interactions (such as stars in a galaxy) or systems that remain for long periods of time at the "edge of chaos", there is mounting evidence that such systems can get stuck in nonequilibrium meta-stable states with power law probability distributions for very long periods of time before they finally relax to equilibrium. Alternatively, power laws also occur in many driven systems that are maintained in a steady state away from equilibrium. Another possible area of applications is describing the behaviour of small subsystems of finite systems.

From a purely statistical point of view it is interesting to ask what type of entropy functions are allowed. The natural assumption to alter is the last of the Khinchin postulates as discussed in Section 5.2. The question becomes which entropy functions satisfy the remaining two conditions, and some sensible alternative for the third? It turns out that there is at least one interesting class of solutions called q-entropies introduced in 1988 by Tsallis [Tsallis, 1988; Gell-Mann and

[15]It is also possible to have a power law at zero or any other limit, and to have $\alpha < 0$, but for our purposes here most of the examples of interest involve the limit $x \to \infty$ and positive α.

[16]The mth moment $\int x^m p(x)dx$ of a power law distribution $p(x) \sim x^{-\alpha}$ does not exist when $m > \alpha$.

Tsallis, 2004]. The parameter q is usually referred to as the *bias* or *correlation* parameter. For $q \neq 1$ the expression for the q-entropy S_q is

$$(59)\quad S_q[p] \equiv \frac{1 - \sum_i p_i^q}{q - 1}.$$

For $q = 1$, S_q reduces to the standard Gibbs entropy by taking the limit as $q \to 1$. Following Jaynes's approach to statistical mechanics, one can maximize this entropy function under suitable constraints to obtain distribution functions that exhibit power law behavior for $q \neq 1$. These functions are called q-exponentials and are defined as

$$(60)\quad e_q(x) \equiv \begin{cases} [1 + (1-q)x]^{1/(1-q)} & (1 + (1-q)x\rangle 0) \\ 0 & (1 + (1-q)x\langle 0). \end{cases}$$

An important property of the q-exponential function is that for $q > 1$ and $x \ll -1$ it has a power law decay. The inverse of the q-exponential is the $\ln_q(x)$ function

$$(61)\quad \ln_q \equiv \frac{x^{1-q} - 1}{1 - q}.$$

The q-exponential can also be obtained as the solution of the equation

$$(62)\quad \frac{dx}{dt} = x^q.$$

This is the typical behavior for a dynamical system at the edge of linear stability, where the first term in its Taylor series vanishes. This gives some alternative insight into one possible reason why such solutions may be prevalent. Other typical situations involve long range interactions (such as the gravitational interactions between stars in galaxy formation) or nonlinear generalizations of the central limit theorem [Umarov *et al.*, 2006] for variables with strong correlations.

At first sight a problem with q-entropies is that for $q \neq 1$ they are not additive. In fact the following equality holds:

$$(63)\quad S_q[p^{(1)}p^{(2)}] = S_q[p^{(1)}] + S_q[p^{(2)}] + (1-q)S_q[p^{(1)}]S_q[p^{(2)}]$$

with the corresponding product rule for the q-exponentials:

$$(64)\quad e_q(x)e_q(y) = e_q(x + y + (1-q)xy)$$

This is why the q-entropy is often referred to as a non-extensive entropy. However, this is in fact a blessing in disguise. If the appropriate type of scale invariant correlations between subsystems are typical, then the q-entropies for $q \neq 1$ are strictly additive. When there are sufficiently long-range interactions Shannon entropy is not extensive; Tsallis entropy provides a substitute that is additive (under the right class of long-range interactions), thereby capturing an underlying regularity with a simple description.

This alternative statistical mechanical theory involves another convenient definition which makes the whole formalism look like the "old" one. Motivated by the fact that the Tsallis entropy weights all probabilities according to p_i^q, it is possible to define an "escort" distribution $P_i^{(q)}$

$$(65)\quad P_i^{(q)} \equiv \frac{(p_i)^q}{\sum_j (p_j)q},$$

as introduced by Beck [Beck, 2001]. One can then define the corresponding expectation values of a variable A in terms of the escort distribution as

$$(66)\quad \langle A \rangle_q = \sum_i P_i^{(q)} A_i.$$

With these definitions the whole formalism runs parallel to the Boltzmann-Gibbs program.

One can of course ask what the Tsallis entropy "means". The entropy S_q is a measure of lack of information along the same lines as the Boltzmann-Gibbs-Shannon entropy is. In particular, perfect knowledge of the microscopic state of the system yields $S_q = 0$, and maximal uncertainty (i.e., all W possible microscopic states are equally probable) yields maximal entropy, $S_q = \ln_q W$. The question remains how generic such correlations are and which physical systems exhibit them, though at this point quite a lot of empirical evidence is accumulating to suggest that such functions are at least a good approximation in many situations. In addition recent results have shown that q-exponentials obey a central limit-like behavior for combining random variables with appropriate long-range correlations.

A central question is what determines q? There is a class of natural, artificial and social systems for which it is possible to choose a unique value of q such that the entropy is simultaneously extensive (i.e., $S_q(N)$ proportional to the number of elements N, $N \gg 1$) and there is finite entropy production per unit time (i.e., $S_q(t)$ proportional to time t, $t \gg 1$)[Tsallis *et al.*, 2005b; Tsallis *et al.*, 2005a]. It is possible to acquire some intuition about the nature and meaning of the index q through the following analogy: If we consider an idealized planar surface, it is only its $d = 2$ Lebesgue measure which is finite; the measure for any $d > 2$ vanishes, and that for any $d < 2$ diverges. If we have a fractal system, only the $d = d_f$ measure is finite, where d_f is the Hausdorff dimension; any $d > d_f$ measure vanishes, and any $d < d_f$ measure diverges. Analogously, only for a special value of q does the entropy S_q match the thermodynamical requirement of extensivity and the equally physical requirement of finite entropy production. The value of q reflects the geometry of the measure in phase space on which probability is concentrated.

Values of q differing from unity are consistent with the recent q-generalization of the Central Limit Theorem and the alpha-stable (Levy) distributions. Indeed, if instead of adding a large number of exactly or nearly independent random variables, we add globally correlated random variables, the attractors shift from Gaussians and Levy distributions to q-Gaussian and (q,α)-stable distributions respectively [Moyano *et al.*, 2006; Umarov *et al.*, 2006; Umarov et al., 2006].

The framework described above is still in development. It may turn out to be relevant to 'statistical mechanics' not only in nonequilibrium physics, but also in quite different arenas, such as economics.

6 QUANTUM INFORMATION

> Until recently, most people thought of quantum mechanics in terms of the uncertainty principle and unavoidable limitations on measurement. Einstein and Schrödinger understood early on the importance of entanglement, but most people failed to notice, thinking of the EPR paradox as a question for philosophers. The appreciation of the positive application of quantum effects to information processing grew slowly.
>
> Nicolas Gisin

Quantum mechanics provides a fundamentally different means of computing, and potentially makes it possible to solve problems that would be intractable on classical computers. For example, with a classical computer the typical time it takes to factor a number grows exponentially with the size of the number, but using quantum computation Shor has shown that this can be done in polynomial time [Shor, 1994]. Factorization is one of the main tools in cryptography, so this is not just a matter of academic interest. To see the huge importance of exponential vs. polynomial scaling, suppose an elementary computational step takes Δt seconds. If the number of steps increases exponentially, factorizing a number with N digits will take $\Delta t \exp(aN)$ seconds, where a is a constant that depends on the details of the algorithm. For example, if $\Delta t = 10^{-6}$ and $a = 10^{-2}$, factoring a number with $N = 10,000$ digits will take 10^{37} seconds, which is much, much longer than the lifetime of the universe (which is a mere 4.6×10^{17} seconds). In contrast, if the number of steps scales as the third power of the number of digits, the same computation takes $a'\Delta t N^3$ seconds, which with $a' = 10^{-2}$, is 10^4 seconds or a little under three hours. Of course the constants a, a' and Δt are implementation dependent, but because of the dramatic difference between exponential vs. polynomial scaling, for sufficiently large N there is always a fundamental difference in speed. In fact for the factoring problem as such, the situation is more subtle: at present the best available classical algorithm requires $\exp(O(n^{1/3} \log^{2/3} n))$ operations, whereas the best available quantum algorithm would require $O(n^2 \log n \log\log n)$ operations. Factorization is only one of several problems that could potentially benefit from quantum computing. The implications go beyond quantum computing, and include diverse applications such as quantum cryptography and quantum communication [Nielsen and Chuang, 1990; Kaye *et al.*, 2007; Mermin, 2007; Lloyd, 2008].

The possibility for such huge speed-ups comes from the intrinsically parallel nature of quantum systems. The reasons for this are sufficiently subtle that it took many decades after the discovery of quantum mechanics before anyone realized that its computational properties are fundamentally different. The huge interest in quantum computation in recent years has caused a re-examination of the concept of information in physical systems, spawning a field that is sometimes referred to as "quantum information theory".

Before entering the specifics of quantum information and computing, we give a brief introduction to the basic setting of quantum theory and contrast it with its classical counterpart. We describe the physical states of a quantum systems, the definition of quantum observables, and time evolution according to the Schrödinger equation. Then we briefly explain the measurement process, the basics of quantum teleportation and quantum computation. To connect to classical statistical physics we describe the density matrix and the von Neumann entropy. Quantum computation in practice involves sophisticated and highly specialized subfields of experimental physics which are beyond the scope of this brief review — we have tried to limit the discussion to the essential principles.

6.1 Quantum states and the definition of a qubit

In classical physics we describe the state of a system by specifying the values of dynamical variables, for example, the position and velocity of a particle at a given instant in time. The time evolution is then described by Newton's laws, and any uncertainty in its evolution is driven by the accuracy of the measurements. As we described in Section 4.2, uncertainties can be amplified by chaotic dynamics, but within classical physics there is no fundamental limit on the accuracy of measurements — by measuring more and more carefully, we can predict the time evolution of a system more and more accurately. At a fundamental level, however, all of physics behaves according to the laws of quantum mechanics, which are very different from the laws of classical physics. At the macroscopic scales of space, time and energy where classical physics is a good approximation, the predictions of classical and quantum theories have to be roughly the same, a statement that is called the *correspondence principle*. Nonetheless, understanding the emergence of classical physics from an underlying quantum description is not always easy.

The scale of the quantum regime is set by Planck's constant, which has dimensions of *energy* $\times$ *time* (or equivalently *momentum* $\times$ *length*). It is extremely small in ordinary units[17]: $\hbar = 1.05 \times 10^{-34}$ Joule-seconds. This is why quantum properties only manifest themselves at very small scales or very low temperatures. One has to keep in mind however, that radically different properties at a microscopic scale (say at the level of atomic and molecular structure) will also lead to fundamentally different collective behavior on a macroscopic scale. Most phases of condensed matter realized in nature, such as crystals, super, ordinary or semi-conductors or magnetic materials, can only be understood from the quantum mechanical perspective. The stability and structure of matter is to a large extent a consequence of the quantum behavior of its fundamental constituents.

To explain the basic ideas of quantum information theory we will restrict our attention to systems of *qubits*, which can be viewed as the basic building blocks of quantum information systems. The physical state of a quantum system is described by a wavefunction that can be thought of a vector in an abstract multidimensional space, called a *Hilbert space*. For our purposes here, this is just a finite dimensional vector space where the vectors have complex rather than real coefficients, and where the length of a vector is the usual length in such a space, i.e. the square root

[17] We are using the reduced Planck's constant, $\hbar = h/2\pi$.

of the sum of the square amplitudes of its components[18]. Hilbert space replaces the concept of phase space in classical mechanics. Orthogonal basis vectors defining the axes of the space correspond to different values of measurable quantities, also called observables, such as spin, position, or momentum.

As we will see, an important difference from classical mechanics is that many quantum mechanical quantities, such as position and momentum or spin along the x-axis and spin along the y-axis, cannot be measured simultaneously. Another essential difference from classical physics is that the dimensionality of the state space of the quantum system is huge compared to that of the classical phase space. To illustrate this drastic difference think of a particle that can move along an infinite line with an arbitrary momentum. From the classical perspective it has a phase space that is two dimensional and real (a position x and a momentum p), but from the quantum point of view it it is given by a wavefunction Ψ of one variable (typically the position x or the momentum p). This wave function corresponds to an element in an infinite dimensional Hilbert space.

We discussed the classical Ising spin in section 3.2. It is a system with only two states, denoted by $s = \pm 1$, called spin up or spin down, which can be thought of as representing a classical bit with two possible states, "0" and "1". The quantum analog of the Ising spin is a very different kind of animal. Where the Ising spin corresponds to a classical bit, the quantum spin corresponds to what is called a *qubit*. As we will make clear in a moment, the state space of a qubit is much larger then that of its classical counterpart, making it possible to store much more information. This is only true in a certain sense, as one has to take into account to what extent the state is truly observable and whether it can be precisely prepared, questions we will return to later.

Any well-defined two level quantum system can be thought of as representing a qubit. Examples of two state quantum systems are a photon, which possesses two polarization states, an electron, which possesses two possible spin states, or a particle in one of two possible energy states. In the first two examples the physical quantities in the Hilbert space are literally spins, corresponding to angular momentum, but in the last example this is not the case. This doesn't matter — even if the underlying quantities have nothing to do with angular momentum, as long as it is a two state quantum system we can refer to it as a "spin". We can arbitrarily designate one quantum state as "spin up", represented by the symbol $|1\rangle$, and the other "spin down", represented by the symbol $|0\rangle$.

The state of a qubit is described by a wavefunction or state vector $|\psi\rangle$, which can be written as

$$(67) \quad |\psi\rangle = \alpha|1\rangle + \beta|0\rangle \text{ with } |\alpha|^2 + |\beta|^2 = 1.$$

Here α and β are complex numbers[19], and thus we can think of $|\psi\rangle$ as a vector in the 2-dimensional complex vector space, denoted $\mathbf{C}^2$, and we can represent the

[18]More generally it is necessary to allow for the possibility of infinite dimensions, which introduces complications about the convergence of series that we do not need to worry about here.

[19]A complex number α has a real and imaginary part $\alpha = a_1 + ia_2$, where a_1 and a_2 are both

state as a column vector $\begin{pmatrix} \alpha \\ \beta \end{pmatrix}$. We can also define a dual vector space in $\mathbf{C}^2$ with dual vectors that can either be represented as row vectors or alternatively be written

(68) $\langle\psi| = \langle 0|\alpha^* + \langle 1|\beta^*$.

This allows us to define the inner product between two state vectors $|\psi\rangle$ and $|\phi\rangle = \gamma|1\rangle + \delta|0\rangle$ as

(69) $\langle\phi|\psi\rangle = \langle\psi|\phi\rangle^* = \gamma^*\alpha + \delta^*\beta$.

Each additional state (or configuration) in the classical system yields an additional orthogonal dimension (complex parameter) in the quantum system. Hence a finite state classical system will lead to a finite dimensional complex vector space for the corresponding quantum system.

Let us describe the geometry of the quantum configuration space of a single qubit in more detail. The constraint $|\alpha|^2 + |\beta|^2 = 1$ says that the state vector has unit length, which defines the complex unit circle in $\mathbf{C}^2$, but if we write the complex numbers in terms of their real and imaginary parts as $\alpha = a_1 + ia_2$ and $\beta = b_1 + ib_2$, then we obtain $|a_1 + a_2 i|^2 + |b_1 + b_2 i|^2 = a_1^2 + a_2^2 + b_1^2 + b_2^2 = 1$. The geometry of the space described by the latter equation is just the three dimensional unit sphere S^3 embedded in a four dimensional Euclidean space, $\mathbf{R}^4$.

To do any nontrivial quantum computation we need to consider a system with multiple qubits. Physically it is easiest to imagine a system of n particles, each with its own spin. (As before, the formalism does not depend on this, and it is possible to have examples in which the individual qubits might correspond to other physical properties). The mathematical space in which the n qubits live is the tensor product of the individual qubit spaces, which we may write as $\mathbf{C}^2 \otimes \mathbf{C}^2 \otimes ... \otimes \mathbf{C}^2 = \mathbf{C}^{2^n}$. For example, the Hilbert space for two qubits is $\mathbf{C}^2 \otimes \mathbf{C}^2$. This is a four dimensional complex vector space spanned by the vectors $|1\rangle \otimes |1\rangle$, $|0\rangle \otimes |1\rangle$, $|1\rangle \otimes |0\rangle$, and $|0\rangle \otimes |0\rangle$. For convenience we will often abbreviate the tensor product by omitting the tensor product symbols, or by simply listing the spins. For example

$$|1\rangle \otimes |0\rangle = |1\rangle|0\rangle = |10\rangle.$$

The tensor product of two qubits with wave functions $|\psi\rangle = \alpha|1\rangle + \beta|0\rangle$ and $|\phi\rangle = \gamma|1\rangle + \delta|0\rangle$ is

$$|\psi\rangle \otimes |\phi\rangle = |\psi\rangle|\phi\rangle = \alpha\gamma|11\rangle + \gamma\delta|10\rangle + \beta\gamma|01\rangle + \beta\delta|00\rangle.$$

The most important feature of the tensor product is that it is multi-linear, i.e. $(\alpha|0\rangle + \beta|1\rangle) \otimes |\psi\rangle = \alpha|0\rangle \otimes |\psi\rangle + \beta|1\rangle \otimes |\psi\rangle$. Again we emphasize that whereas the classical $n-$bit system has 2^n states, the $n-$qubit system corresponds to a vector

real, and i is the imaginary unit with the property $i^2 = -1$. Note that a complex number can therefore also be thought of as a vector in a two dimensional real space.The complex conjugate is defined as $\alpha^* = a_1 - ia_2$ and the square of the modulus, or absolute value, as $|\alpha|^2 = \alpha^*\alpha = a_1^2 + a_2^2$.

of unit length in a 2^n dimensional complex space, with twice as many degrees of freedom. For example a three-qubit can be expanded as:

$$\begin{aligned}|\psi\rangle = \alpha_1|000\rangle &+ \alpha_2|001\rangle + \alpha_3|010\rangle + \alpha_4|011\rangle \\ &+ \alpha_5|100\rangle + \alpha_6|101\rangle + \alpha_7|110\rangle + \alpha_8|111\rangle\end{aligned}$$

Sometimes it is convenient to denote the state vector by the column vector of its components $\alpha_1, \alpha_2, ..., \alpha_{2^n}$.

6.2 *Observables*

How are ordinary physical variables such as energy, position, velocity, and spin retrieved from the state vector? In the quantum formalism observables are defined as *hermitian* operators acting on the state space. In quantum mechanics an *operator* is a linear transformation that maps one state into another, which providing the state space is finite dimensional, can be represented by a matrix. A hermitian operator or matrix satisfies the condition $A = A^\dagger$, where $A^\dagger = (A^{tr})^*$ is the complex conjugate of the transpose of A. The fact that observables are represented by operators reflects the property that measurements may alter the state and that outcomes of different measurements may depend on the order in which the measurements are performed. In general observables in quantummechanics do not necessarily *commute*, by which we mean that for the product of two observables A and B one may have that $AB \neq BA$. The reason that observables have to be hermitian is because the outcome of measurements are the eigenvalues of observables, and hermitian operators are guaranteed to have real eigenvalues.

For example consider a single qubit. The physical observables are the components of the spin along the x, y or z directions, which are by convention written $s_x = \frac{1}{2}\sigma_x$, $s_y = \frac{1}{2}\sigma_y$, etc. The operators σ are the Pauli matrices

$$(70)\quad \sigma_x = \begin{pmatrix} 0 & 1 \\ 1 & 0 \end{pmatrix}, \sigma_y = \begin{pmatrix} 0 & -i \\ i & 0 \end{pmatrix}, \sigma_z = \begin{pmatrix} 1 & 0 \\ 0 & -1 \end{pmatrix},$$

which obviously do not commute. In writing the spin operators this way we have arbitrarily chosen[20] the z-axis to have a diagonal representation, so that the *eigenstates*[21] for spin along the z axis are the column vectors

$$|1\rangle = \begin{pmatrix} 1 \\ 0 \end{pmatrix}, \qquad |0\rangle = \begin{pmatrix} 0 \\ 1 \end{pmatrix}.$$

[20] We can rotate into a different representation that makes either of the other two axes diagonal, and in which the z-axis is no longer diagonal — it is only possible to make one of the three axes diagonal at a time. Experimental set-ups often have conditions that break symmetry, such as an applied magnetic fields, in which case it is most convenient to let the symmetry breaking direction be the z-axis.

[21] The eigenstates $|\chi_k\rangle$ of a linear operator A are defined by the equation $A|\chi_k\rangle = \lambda_k|\chi_k\rangle$. If A is hermitian the eigenvalue λ_k is a real number. It is generally possible to choose the eigenstates as orthonormal, so that $\langle\chi_j|\chi_k\rangle = \delta_{jk}$, where $\delta_{ij} = 1$ when $i = j$ and $\delta_{ij} = 0$ otherwise.

6.3 Quantum evolution: the Schrödinger equation

The wave function of a quantum system evolves in time according to the famous Schrödinger equation. Dynamical changes in a physical system are induced by the underlying forces acting on the system and between its constituent parts, and their effect can be represented in terms of what is called the energy or Hamiltonian operator H. For a single qubit system the operators can be represented as 2×2 matrices, for a two qubit system they are 4×4 matrices, etc. The Schrödinger equation can be written

$$(71)\quad i\hbar \frac{d|\psi(t)\rangle}{dt} = H|\psi(t)\rangle.$$

This is a linear differential equation expressing the property that the time evolution of a quantum system is generated by its energy operator. Assuming that H is constant, given an initial state $|\psi(0)\rangle$ the solution is simply

$$(72)\quad |\psi(t)\rangle = U(t)|\psi(0)\rangle \text{ with } U(t) = e^{-iHt/\hbar}.$$

The time evolution is *unitary*, meaning that the operator $U(t)$ satisfies $UU^\dagger = 1$.

$$(73)\quad U^\dagger = \exp(-iHt/\hbar)^\dagger = \exp(iH^\dagger t/\hbar) = \exp(iHt/\hbar) = U^{-1}.$$

Unitary time evolution means that the length of the state vector remains invariant, which is necessary to preserve the total probability for the system to be in any of its possible states. The unitary nature of the the time evolution operator U follows directly from the fact that H is hermitian. Any hermitean 2×2 matrix can be written

$$(74)\quad A = \begin{pmatrix} a & b+ic \\ b-ic & -a \end{pmatrix},$$

where a, b and c are real numbers[22].

For the simple example of a single qubit, suppose the initial state is

$$|\psi(0)\rangle = \sqrt{\frac{1}{2}}(|1\rangle + |0\rangle) \equiv \sqrt{\frac{1}{2}}\begin{pmatrix} 1 \\ 1 \end{pmatrix}.$$

On the right, for the sake of convenience, we have written the state as a column vector. Consider the energy of a spin in a magnetic field B directed along the positive z-axis[23]. In this case H is given by $H = Bs_z$. From (70)

$$(75)\quad U(t) = \exp(\frac{-iBt}{2\hbar}\sigma_z) = \begin{pmatrix} \exp(-iBt/2\hbar) & 0 \\ 0 & \exp(iBt/2\hbar) \end{pmatrix}.$$

Using (72) we obtain an oscillatory time dependence for the state, i.e.

[22]We omitted a component proportional to the unit matrix as it acts trivially on any state.

[23]Quantum spins necessarily have a magnetic moment, so in addition to carrying angular momentum they also interact with a magnetic field.

$$(76)\quad |\psi(t)\rangle = \sqrt{\frac{1}{2}} \begin{pmatrix} e^{-iBt/2\hbar} \\ e^{iBt/2\hbar} \end{pmatrix} = \sqrt{\frac{1}{2}} \left[\cos\frac{Bt}{2\hbar} \begin{pmatrix} 1 \\ 1 \end{pmatrix} + i \sin\frac{Bt}{2\hbar} \begin{pmatrix} -1 \\ 1 \end{pmatrix} \right].$$

We thus see that, in contrast to classical mechanics, time evolution in quantum mechanics is always linear. It is in this sense much simpler than classical mechanics. The complication is that when we consider more complicated examples, for example corresponding to a macroscopic object such as a planet, the dimension of the space in which the quantum dynamics takes place becomes extremely high.

6.4 Quantum measurements

Measurement in classical physics is conceptually trivial: One simply estimates the value of the classical state at finite precision and approximates the state as a real number with a finite number of digits. The accuracy of measurements is limited only by background noise and the precision of the measuring instrument. The measurement process in quantum mechanics, in contrast, is not at all trivial. One difference with classical mechanics is that in many instances the set of measurable states is discrete, with quantized values for the observables. It is this property that has given the theory of quantum mechanics its name. But perhaps an even more profound difference is that quantum measurement typically causes a radical alteration of the wavefunction. Before the measurement of an observable we can only describe the possible outcomes in terms of probabilities, whereas after the measurement the outcome is known with certainty, and the wavefunction is irrevocably altered to reflect this. In the conventional Copenhagen interpretation of quantum mechanics the wave function is said to "collapse" when a measurement is made. In spite of the fact that quantum mechanics makes spectacularly successful predictions, the fact that quantum measurements are inherently probabilistic and can "instantly" alter the state of the system has caused a great deal of controversy. In fact, one can argue that historically the field of quantum computation emerged from thinking carefully about the measurement problem [Deutsch, 1985].

In the formalism of quantum mechanics the possible outcomes of an observable quantity A are given by the eigenvalues of the matrix A. For example, the three spin operators defined in Eq. 70 all have the same two eigenvalues $\lambda_{\pm} = \pm 1/2$. This means that the possible outcomes of a measurement of the spin in any direction can only be plus or minus one half. This is completely different than a spinning object in classical physics, which can spin at any possible rate in any direction. This is why quantum mechanics is so nonintuitive!

If a quantum system is in an eigenstate then the outcome of measurements in the corresponding direction is certain. For example, imagine we have a qubit in the state with $\alpha = 1$ and $\beta = 0$ so $|\psi\rangle = |1\rangle$. It is then in the eigenstate of s_z with eigenvalue $+\frac{1}{2}$, so the measurement of s_z will always yield that value. This is reflected in the mathematical machinery of quantum mechanics by the fact that for the spin operator in the z-direction, $A = s_z$, the eigenvector with eigenvalue $\lambda_+ = +1/2$ is $|1\rangle = \begin{pmatrix} 1 \\ 0 \end{pmatrix}$ and the eigenvector with $\lambda_- = -1/2$ is

$|0\rangle = \begin{pmatrix} 0 \\ 1 \end{pmatrix}$. In contrast, if we make measurements in the orthogonal directions to the eigenstate, e.g. $A = s_x$, the outcomes become probabilistic. In the example above the eigenvectors of s_x are $|\chi_+\rangle = \sqrt{\frac{1}{2}}(|1\rangle + |0\rangle)$ and $|\chi_-\rangle = \sqrt{\frac{1}{2}}(|1\rangle - |0\rangle)$. In general the probability of finding the system in a given state in a measurement is computed by first expanding the given state $|\psi\rangle$ into the eigenstates $|\chi_k\rangle$ of the matrix A corresponding to the observable, i.e.

$$(77)\quad |\psi\rangle = \sum_k \alpha_k |\chi_k\rangle \text{ where } \alpha_k = \langle\chi_k|\psi\rangle.$$

The probability of measuring the system in the state corresponding to eigenvalue λ_k is $p_k = |\alpha_k|^2$. The predictions of quantum mechanics are therefore probabilistic but the theory is essentially different from classical probability theory. On the one hand it is clear that a given operator defines a probability measure on Hilbert space, however as the operators are non-commuting (like matrices) one is dealing with a non-commutative probability theory [Holevo, 1982]. It is the non-commutativity of observables that gives rise to the intricacies in the quantum theory of measurement.

Let us discuss an example for clarification. Consider the spin in the x-direction, $A = s_x$, and $|\psi\rangle = |1\rangle$, i.e. spin up in the z-direction. Expanding in eigenstates of σ_x we get $|\psi\rangle = |1\rangle = \sqrt{\frac{1}{2}}|\chi_+\rangle + \sqrt{\frac{1}{2}}|\chi_-\rangle$. The probability of measuring spin up along the x-direction is $|\alpha_+|^2 = 1/2$, and the probability of measuring spin down along the x-direction is $|\alpha_-|^2 = 1/2$. We see how probability enters quantum mechanics at a fundamental level. The average of an observable is its *expectation value*, which is the weighted sum

$$(78)\quad \langle\psi|A|\psi\rangle = \sum_k |\alpha_k|^2 \lambda_k = \sum_k p_k \lambda_k.$$

In the example at hand $\langle\sigma_x\rangle = 0$.

The act of measurement influences the state of the system. If we measure $s_x = +\frac{1}{2}$ and then measure it again immediately afterward, we will get the same value with certainty. Stated differently, doing the measurement somehow forces the system into the eigenstate $|\chi_+\rangle$, and once it is there, in the absence of further interactions, it stays there. This strange property of measurement, in which the wavefunction collapses onto the observed eigenstate, was originally added to the theory in an ad hoc manner, and is called the *projection postulate*. This postulate introduces a rather arbitrary element into the theory that appears to be inconsistent: The system evolves under quantum mechanics according to the Schrödinger equation until a measurement is made, at which point some kind of magic associated with the classical measurement apparatus takes place, which lies completely outside the rest of the theory.

To understand the measurement process better it is necessary to discuss the coupling of a quantum system and a classical measurement apparatus in more detail. A measurement apparatus, such as a pointer on a dial or the conditional emission of a light pulse, is also a quantum mechanical system. If we treat the measurement device quantum mechanically as well, it should be possible to regard the apparent "collapse" of the wavefunction as the outcome of the quantum evolution of the combined system of the measurement device and the original quantum system under study, without invoking the projection postulate. We return to this when we discuss decoherence in Section 6.7 .

Note that a measurement does not allow one to completely determine the state. A complete measurement of the two-qubit system yields at most two classical bits of information, whereas determining the full quantum state requires knowing seven real numbers (four complex numbers subject to a normalization condition). In this sense one cannot just say that a quantum states "contains" much more information that its classical counterpart. In fact, due to the non-commutativity of the observables, with simultaneous measurements one is able to extract less information than from the corresponding classical system.

There are two ways to talk about quantum theory: If one insists it is a theory of a single system, then one has to live with the fact that it only predicts the probability of things to happen and as such is a retrenchment from the ideal of classical physics. Alternatively one may take the view that quantum theory is a theory that only applies to ensembles of particles. To actually measure probability distributions one has to make many measurements on "identically prepared" quantum systems. From this perspective the dimensionality of Hilbert space should be compared to that of classical distributions defined over a classical phase space, which makes the difference between classical and quantum theories far less dramatic. This raises the quest for a theory underlying quantum mechanics which applies for a single system. So far nobody has succeeded in producing such a theory, and on the contrary, attempts to build such theories based on "hidden variables" have failed. The Bell inequalities suggest that such a theory is probably impossible [Omnes, 1999].

6.5 Multi qubit states and entanglement

When we have more than one qubit an important practical question is when and how measurements of a given qubit depend on measurements of other qubits. Because of the deep properties of quantum mechanics, qubits can be coupled in subtle ways that produce consequences for measurement that are very different from classical bits. Understanding this has proved to be important for the problems of computation and information transmission. To explain this we need to introduce the opposing concepts of separability and entanglement, which describe whether measurements on different qubits are statistically independent or statistically dependent.

An n-qubit state is *separable* if it can be factored into n-single qubit states[24], i.e. if it can be written as $n-1$ tensor products of sums of qubit states, with each

[24] Strictly speaking this is only true for pure states, which we define in the next section.

factor depending only on a single qubit. An example of a separable two-qubit state is

$$(79)\quad |\psi\rangle = \frac{1}{2}(|00\rangle + |01\rangle + |10\rangle + |11\rangle) = \frac{1}{2}(|0\rangle + |1\rangle) \otimes (|0\rangle + |1\rangle).$$

If an n-qubit state is separable then measurements on individual qubits are statistically independent, i.e. the probability of making a series of measurements on each qubit can be written as a product of probabilities of the measurements for each qubit.

An n-qubit state is *entangled* if it is not separable. An example of an entangled two-qubit state is

$$(80)\quad |\psi\rangle = \frac{1}{\sqrt{2}}(|00\rangle + |11\rangle),$$

which cannot be factored into a single product. For entangled states measurements on individual qubits depend on each other.

We now illustrate this for the two examples above. Suppose we do an experiment in which we measure the spin of the first qubit and then measure the spin of the second qubit. For both the separable and entangled examples, there is a 50% chance of observing either spin up or spin down on the first measurement. Suppose it gives spin up. For the separable state this transforms the wave function as

$$\frac{1}{2}(|0\rangle + |1\rangle) \otimes (|0\rangle + |1\rangle) \rightarrow \frac{1}{\sqrt{2}}(|1\rangle) \otimes (|0\rangle + |1\rangle) = \frac{1}{\sqrt{2}}(|10\rangle + |11\rangle).$$

If we now measure the spin of the second qubit, the probability of measuring spin up or spin down is still 50%. The first measurement has no effect on the second measurement.

In contrast, suppose we do a similar experiment on the entangled state of equation 80 and observe spin up in the first measurement. This transforms the wave function as

$$(81)\quad \frac{1}{\sqrt{2}}(|00\rangle + |11\rangle) \longrightarrow |11\rangle.$$

(Note the disappearance of the factor $1/\sqrt{2}$ due to the necessity that the wave function remains normalized). If we now measure the spin of the second qubit we are certain to observe spin up! Similarly, if we observe spin down in the first measurement, we will also observe it in the second. For the entangled example above the measurements are completely coupled — the outcome of the first determines the second[25]. This property of entangled states was originally pointed out by Einstein, Podolsky and Rosen [Einstein *et al.*, 1935], who expressed concern about the possible consequences of this when the qubits are widely separated in space.

[25] One may argue that a perfectly correlated classical system would exhibit similar behaviour. The difference between classical and classical system would still become manifest in the dependence of the correlation on the angle of two successive measurements with different measurement angle.

This line of thinking did not point out a fundamental problem with quantum mechanics as they perhaps originally hoped, but rather led to a deeper understanding of the quantum measurement problem and to the practical application of quantum teleportation as discussed in Section 6.9.

The degree of entanglement of a system of qubits is a reflection of their past history. By applying the right time evolution operator, i.e. by introducing appropriate interactions, we can begin with a separable state and entangle it, or begin with an entangled state and separate it. Separation can be achieved, for example, by applying the inverse of the operator that brought about the entanglement in the first place — quantum dynamics is reversible. Alternatively separation can be achieved by transferring the entanglement to something else, such as the external environment. (In the latter case there will still be entanglement, but it will be between one of the qubits and the environment, rather than between the two original qubits).

6.6 *Entanglement and entropy*

So far we have assumed that we are able to study a single particle or a few particles with perfect knowledge of the state. This is called a statistically pure state, or often more simply, a *pure state.* In experiments it can be difficult to prepare a system in a pure state. More typically there is an ensemble of particles that might be in different states, or we might have incomplete knowledge of the states. Such a situation, in which there is a nonzero probability for the particle to be in more than one state, is called a *mixed state.* As we explain below, von Neumann developed an alternative formalism for quantum mechanics in terms of what is called a density matrix, which replaces the wavefunction as the elementary level of description. The density matrix representation very simply handles mixed states, and leads to a natural way to measure the entropy of a quantum mechanical system and measure entanglement.

Consider a mixed state in which there is a probability p_i for the system to have wavefunction ψ_i and an observable characterized by operator A. The average value measured for the observable (also called its expectation value) is

$$(82)\quad \langle A\rangle = \sum_i p_i \langle \psi_i|A|\psi_i\rangle .$$

We can expand each wavefunction ψ_i in terms of a basis $|\chi_j\rangle$ in the form

$$|\psi_i\rangle = \sum_j \langle \chi_j|\psi_i\rangle |\chi_j\rangle,$$

where in our earlier notation $\langle \chi_j|\psi_i\rangle = \alpha_j^{(i)}$. Performing this expansion for the dual vector $\langle\psi_i|$ as well, substituting into (82) and interchanging the order of summation gives

$$\langle A\rangle \;=\; \sum_{j,k}\left(\sum_i p_i \langle \chi_j|\psi_i\rangle\langle\psi_i|\chi_k\rangle\right)\langle \chi_k|A|\chi_j\rangle$$

$$= \sum_{j,k} \langle \chi_j | \rho | \chi_k \rangle \langle \chi_k | A | \chi_j \rangle$$
$$= \mathrm{tr}(\rho A),$$

where

$$(83)\quad \rho = \sum_i p_i |\psi_i\rangle\langle\psi_i|$$

is called the *density matrix*[26]. Because the trace $\mathrm{tr}(\rho A)$ is independent of the representation this can be evaluated in any convenient basis, and so provides an easy way to compute expectations. Note that $\mathrm{tr}(\rho) = 1$. For a pure state $p_i = 1$ for some value of i and $p_i = 0$ otherwise. In this case the density matrix has rank one. This is obvious if we write it in a basis in which it is diagonal — there will only be one nonzero element. When there is more than one nonzero value of p_i it is a mixed state and the rank is greater than one.

To get a better feel for how this works, consider the very simple example of a single qubit, and let $\psi_1 = |1\rangle$. If this is a pure state then the density matrix is just

$$\rho = |1\rangle\langle 1| = \begin{pmatrix} 1 & 0 \\ 0 & 0 \end{pmatrix}.$$

The expectation of the spin along the z-axis is $\mathrm{tr}(\rho s_z) = 1/2$. If, however, the system is in a mixed state with 50% of the population spin up and 50% spin down, this becomes

$$\rho = \frac{1}{2}\left(|1\rangle\langle 1| + |0\rangle\langle 0|\right) = \frac{1}{2}\begin{pmatrix} 1 & 0 \\ 0 & 1 \end{pmatrix}.$$

In this case the expectation of the spin along the z-axis is $\mathrm{tr}(\rho s_z) = 0$.

This led von Neumann to define the entropy of a quantum state in analogy with the Gibbs entropy for a classical ensemble as

$$(84)\quad S(\rho) = -\mathrm{tr}\ \rho \log \rho = -\sum_i p_i \log p_i\ .$$

The entropy of a quantum state provides a quantitative measure of "how mixed" a system is. The entropy of a pure state is equal to zero, whereas the entropy of a mixed state is greater than zero.

In some situations there is a close relationship between entangled and mixed states. An entangled but pure state in a high dimensional multi-qubit space can appear to be a mixed state when viewed from the point of view of a lower dimensional state space. The view of the wavefunction from a lower dimensional factor in a tensor product space is formally taken using a *partial* trace. This is done by

[26]The density matrix provides an alternative representation for quantum mechanics — the Schrödinger equation can be rewritten in terms of the density matrix so that we never need to use wavefunctions at all.

summing over all the coordinates associated with the factors we want to ignore. This corresponds to leaving some subsystems out of consideration, for example, because we can only measure a certain qubit and can't measure the qubits on which we perform the partial trace. As an example consider the entangled state of equation (80), and trace it with respect to the second qubit. To do this we make use of the fact that $\mathrm{tr}(|\psi\rangle\langle\phi|) = \langle\psi|\phi\rangle$. Using labels A and B to keep the qubits straight, and remembering that because we are using orthogonal coordinates terms of the form $\langle 0|1\rangle = 0$, the calculation can be written

$$\begin{aligned}
\mathrm{tr}\left(|\psi_{AB}\rangle\langle\psi_{AB}|\right) &= \frac{1}{2}\mathrm{tr}\left(|1\rangle_A\langle 1|_B + |0\rangle_A\langle 0|_B\right)\left(\langle 0|_B\langle 0|_A + \langle 1|_B\langle 1|_A\right) \\
&= \frac{1}{2}\left(|1\rangle_A\langle 1|_A\langle 1|1\rangle_B + |0\rangle_A\langle 0|_A\langle 0|0\rangle_B\right) \\
&= \frac{1}{2}\left(|1\rangle_A\langle 1|_A + |0\rangle_A\langle 0|_A\right)
\end{aligned}$$

This is a mixed state with probability 1/2 to be either spin up or spin down. The corresponding entropy is also higher: In base two $S = -\log(1/2) = 1$ bit, while for the original pure state $S = \log 1 = 0$. In general if we begin with a statistically pure separable state and perform a partial trace we will still have a pure state, but if we begin with an entangled state, when we perform a partial trace we will get a mixed state. In the former case the entropy remains zero, but in the latter case it increases. Thus the von Neumann entropy yields a useful measure of entanglement.

6.7 Measurement and Decoherence

In this section we return to the measurement problem and the complications that arise if one wants to couple a classical measurement device to a quantum system. A classical system is by definition described in terms of macro-states, and one macro-state can easily correspond to 10^{40} micro-states. A classical measurement apparatus like a Geiger counter or a photo multiplier tube is prepared in a meta-stable state in which an interaction with the quantum system can produce a decay into a more stable state indicating the outcome of the measurement. For example, imagine that we want to detect the presence of an electron. We can do so by creating a detector consisting of a meta-stable atom. If the electron passes by its interaction with the meta-stable atom via its electromagnetic field can cause the decay of the meta-stable atom, and we observe the emission of a photon. If it doesn't pass by we observe nothing. There are very many possible final states for the system, corresponding to different micro-states of the electron and the photon, but we aren't interested in that — all we want to know is whether or not a photon was emitted. Thus we have to sum over all possible combined photon-electron configurations. This amounts to tracing the density matrix of the complete system consisting of the electron and the measurement apparatus over all states in which a photon is present in the final state. This leads to a reduced density matrix describing the electron after the measurement, with the electron in a mixed state, corresponding to the many possible photon states. Thus even though we started with a zero entropy pure state in the combined system of the electron and photon, we end up with a positive entropy mixed state in the space of the electron alone. The state of the electron is reduced to a classical probability distribution, and due to the huge number of microstates that are averaged over, the process of measurement is thermodynamically irreversible. Even if we do not observe the outcoming photon with our own eyes, it is clear whether or not the metastable atom decayed, and thus whether or not the electron passed by.

The description of the measurement process above is an example of *decoherence*, i.e. of a process whereby quantum mechanical systems come to behave as if they were governed by classical probabilities. A common way for this to happen is for a quantum system to interact with its environment, or for that matter any other quantum system, in such a way that the reduced density matrix for the system of interest becomes diagonal in a particular basis. The phases are randomized, so that after the measurement the system is found to be in a mixed state. According to this view, the wavefunction does not actually collapse, there is just the appearance of a collapse due to quantum decoherence. The details of how this happen remain controversial, and is a subject of active research [Zurek, 1991; Zurek, 2003; Schlosshauer, 2004; Omnes, 1999]. In Section 6.11 we will give an example of how decoherence can be generated even by interactions between simple systems.

6.8 The no-cloning theorem

We have seen that by doing a measurement we may destroy the original state. One important consequence connected to this destructive property of the act of measurement is that a quantum state cannot be cloned; one may be able to transfer a state from one register to another but one cannot make a xerox copy of a given quantum state. This is expressed by the no-cloning theorem [Wootters and Zurek, 1982; Dieks, 1982]. Worded differently, the no-cloning theorem states that for an arbitrary state $|\psi_1\rangle$ on one qubit and some particular state $|\phi\rangle$ on another, there is no quantum device $[A]$ that transforms $|\psi_1\rangle \otimes |\phi\rangle \rightarrow |\psi_1\rangle \otimes |\psi_1\rangle$, i.e. that transforms $|\phi\rangle$ into $\psi_1\rangle$. Letting U_A be the unitary operator representing A, this can be rewritten $|\psi_1\rangle|\psi_1\rangle = U_A|\psi_1\rangle|\phi\rangle$. For a true cloning device this property has to hold for any other state $|\psi_2\rangle$, i.e. we must also have $|\psi_2\rangle|\psi_2\rangle = U_A|\psi_2\rangle|\phi\rangle$. We now show the existence of such a device leads to a contradiction. Since $\langle\phi|\phi\rangle = 1$ and $U_A^\dagger U_A = 1$, and $U_A|\psi_i\rangle|\phi\rangle = U_A|\phi\rangle|\psi_i\rangle$, the existence of a device that can clone both ψ_1 and ψ_2 would imply that

$$\begin{aligned}
\langle\psi_1|\psi_2\rangle &= (\langle\psi_1|\langle\phi|)\,(|\phi\rangle|\psi_2\rangle) = (\langle\psi_1|\langle\phi|U_A^\dagger)\,(U_A|\phi\rangle|\psi_2\rangle) \\
&= (\langle\psi_1|\langle\psi_1|)(|\psi_2\rangle|\psi_2\rangle) \\
&= \langle\psi_1|\psi_2\rangle^2.
\end{aligned}$$

The property $\langle\psi_1|\psi_2\rangle = \langle\psi_1|\psi_2\rangle^2$ only holds if ψ_1 and ψ_2 are either orthogonal or equal, i.e. it does not hold for arbitrary values of ψ_1 and ψ_2, so there can be no such general purpose cloning device. In fact, in view of the uncertainty of quantum measurements, the no-cloning theorem does not come as a surprise: If it were possible to clone wavefunctions, it would be possible to circumvent the uncertainty of quantum measurements by making a very large number of copies of a wavefunction, measuring different properties of each copy, and reconstructing the exact state of the original wavefunction.

6.9 Quantum teleportation

Quantum teleportation provides a method for privately sending messages in a way that ensures that the receiver will know if anyone eavesdrops. This is possible because a quantum state is literally teleported, in the sense of StarTrek: A quantum state is destroyed in one place and recreated in another. Because of the no-cloning theorem, it is impossible to make more than one copy of this quantum state, and as a result when the new teleported state appears, the original state must be destroyed. Furthermore, it is impossible for both the intended receiver and an eavesdropper to have the state at the same time, which helps make the communication secure.

Quantum teleportation takes advantage of the correlation between entangled states as discussed in Section 6.5. Suppose Alice wants to send a secure message to Charlie at a (possibly distant) location. The process of teleportation depends on Alice and Charlie sharing different qubits of an entangled state. Alice makes a measurement of her part of the entangled state, which is coupled to the state she wants to teleport to Charlie, and sends him some classical information about the entangled state. With the classical information plus his half of the entangled state, Charlie can reconstruct the teleported state. We have indicated the process in figure 7. We follow the method proposed by Bennett *et al.* [Bennett et al., 1993], and first realized in an experimental setup by the group of Zeilinger in 1997 [Bouwmeester et al., 1997]. In realistic cases the needed qubit states are typically implemented as left and right handed polarized light quanta (i.e. photons).

The simplest example of quantum teleportation can be implemented with three qubits. The (A) qubit is the unknown state to be teleported,

$$|\psi_A\rangle = \alpha|1\rangle + \beta|0\rangle. \tag{85}$$

This state is literally teleported from one place to another. If Charlie likes, once he has the teleported state he can make a quantum measurement and extract the same information about α and β that he would have been able to extract had he made the measurement on the original state.

The teleportation of this state is enabled by an auxiliary two-qubit entangled state. We label these two qubits B and C. For technical reasons it is convenient to represent this in a special basis consisting of four states, called Bell states, which are written

$$|\Psi^{(\pm)}_{BC}\rangle \quad = \quad \sqrt{\frac{1}{2}}(|1_B\rangle|0_C\rangle \pm |0_B\rangle|1_C\rangle)$$

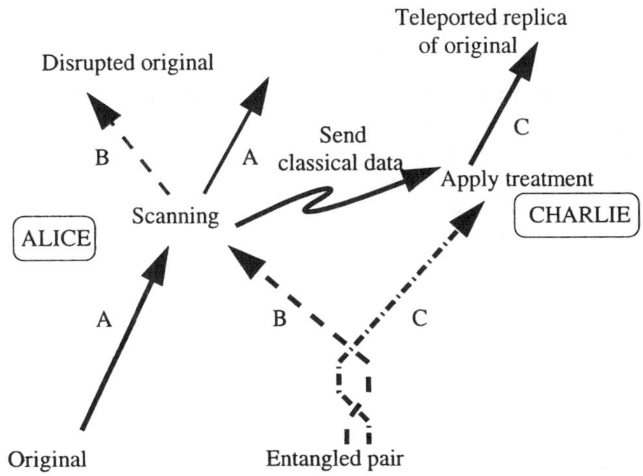

Figure 7. Quantum teleportation of a quantum state as proposed by Bennett *et al.* [Bennett et al., 1993], using an entangled pair. An explanation is given in the text.

$$(86)\quad |\Phi^{(\pm)}_{BC}\rangle \;=\; \sqrt{\frac{1}{2}}(|1_B\rangle|1_C\rangle \pm |0_B\rangle|0_C\rangle).$$

The process of teleportation can be outlined as follows (please refer to Figure 7).

1. Someone prepares an entangled two qubit state BC (the *Entangled pair* in the diagram).
2. Qubit B is sent to Alice and qubit C is sent to Charlie.
3. In the *Scanning* step, Alice measures in the Bell states basis the combined wavefunction of qubits A (the *original* in the diagram) and the entangled state B, leaving behind the *Disrupted original.*
4. Alice sends two bits of classical data to Charlie telling him the outcome of her measurements (*Send classical data*).
5. Based on the classical information received from Alice, Charlie applies one of four possible operators to qubit C (*Apply treament*), and thereby reconstructs A, getting a *teleported replica of the original.* If he likes, he can now make a measurement on A to recover the message Alice has sent him.

We now explain this process in more detail. In step (1) an entangled two qubit state ψ_{BC} such as that of (80) is prepared. In step (2) qubit B is transmitted to Alice and qubit C is transmitted to Charlie. This can be done, for example, by sending two entangled photons, one to each of them. In step (3) Alice measures the joint state of qubit A and B in the Bell states basis, getting two classical bits of information, and projecting the joint wavefunction ψ_{AB} onto one of the Bell states. The Bell states basis has the nice property that the four possible outcomes of the measurement have equal probability. To see how this works, for convenience suppose the entangled state BC was prepared in state $|\Psi^{(-)}_{BC}\rangle$. In this case the combined wavefunction of the three qubit state is

$$
\begin{aligned}
(87)\quad |\psi_{ABC}\rangle &= |\psi_A\rangle|\Psi^{(-)}_{BC}\rangle \\
&= \frac{\alpha}{\sqrt{2}}(|1_A\rangle|1_B\rangle|0_C\rangle - |1_A\rangle|0_B\rangle|1_C\rangle) + \\
&\quad \frac{\beta}{\sqrt{2}}(0_A\rangle|1_B\rangle|0_C\rangle - |0_A\rangle|0_B\rangle|1_C\rangle).
\end{aligned}
$$

If this is expanded in the Bell states basis for the pair AB, it can be written in the form

$$
\begin{aligned}
|\psi_{ABC}\rangle = &\ \tfrac{1}{2}\Big[|\Psi^{(-)}_{AB}\rangle(-\alpha|1_C\rangle - \beta|0_C\rangle) + |\Psi^{(+)}_{AB}\rangle(-\alpha|1_C\rangle + \beta|0_C\rangle) \\
(88)\qquad & |\Phi^{(-)}_{AB}\rangle(\beta|1_C\rangle + \alpha|0_C\rangle) + |\Phi^{(+)}_{AB}\rangle(-\beta|1_C\rangle + \alpha|0_C\rangle)\Big] \ .
\end{aligned}
$$

We see that the two qubit AB has equal probability to be in the four possible states $|\Psi^{(-)}_{AB}\rangle$, $|\Psi^{(+)}_{AB}\rangle$, $|\Phi^{(-)}_{AB}\rangle$ and $|\Phi^{(+)}_{AB}\rangle$.

In step (4), Alice transmits two classical bits to Charlie, telling him which of the four basis functions she observed. Charlie now makes use of the fact that in the Bell basis there are four possible states for the entangled qubit that he has, and his qubit C was entangled with Alice's qubit B before she made the measurement. In particular, let $|\phi_C\rangle$ be the state of the C qubit, which from (88) is one of the four states:

$$
(89)\quad |\phi_C\rangle = \begin{pmatrix} \alpha \\ \beta \end{pmatrix}; \begin{pmatrix} -\alpha \\ \beta \end{pmatrix}; \begin{pmatrix} \beta \\ \alpha \end{pmatrix}; \text{ and } \begin{pmatrix} -\beta \\ \alpha \end{pmatrix} .
$$

In step (5), based on the information that he receives from Alice, Charlie selects one of four possible operators F_i and applies it to the C qubit. There is one operator F_i for each of the four possible Bell states, which are respectively:

$$
(90)\quad F = -\begin{pmatrix} 1 & 0 \\ 0 & 1 \end{pmatrix}; \begin{pmatrix} -1 & 0 \\ 0 & 1 \end{pmatrix}; \begin{pmatrix} 0 & 1 \\ 1 & 0 \end{pmatrix}; \text{ and } \begin{pmatrix} 0 & -1 \\ 1 & 0 \end{pmatrix} .
$$

Providing Charlie has the correct classical information and an intact entangled state he can reconstruct the original A qubit by evolving $|\phi_C\rangle$ with the appropriate unitary operator F_i.

$$
(91)\quad |\psi_A\rangle = \alpha|1\rangle + \beta|0\rangle = F_i|\phi_C\rangle \ .
$$

By simply multiplying each of the four possibilities it is easy to verify that as long as his information is correct, he will correctly reconstruct the A qubit $\alpha|1_A\rangle+\beta|0_A\rangle$.

We stress that Charlie needs the classical measurement information from Alice. If he could do without it the teleportation process would violate causality, since information could be transferred instantaneously from Alice to Charlie. That is, when Alice measures the B qubit, naively it might seem that because the B and C qubits are entangled, this instantaneously collapses the C qubit, sending Charlie the information about Alice's measurement, no matter how far away he is. To understand why such instantaneous communication is not possible, suppose Charlie just randomly guesses the outcome and randomly selects one of the four operators F_i. Then the original state will be reconstructed as a random mixture of the four possible incoming states $|\phi_C\rangle$. This mixture does not give any information about the original state $|\psi_A\rangle$.

The same reasoning also applies to a possible eavesdropper, conveniently named Eve. If she manages to intercept qubit (C) and measures it before Charlie does, without the two bits of classical information she will not be able to recover the original state. Furthermore she will have affected that state. If Charlie somehow gets the mutilated state he will not be able to reconstruct the original state A. Security can be achieved if Alice first sends a sequence of known states which can be checked by Charlie after reconstruction. If the original and reconstructed sequence are perfectly correlated then that guarantees that Eve is not interfering. Note that the cloning theorem is satisfied, since when Alice makes her measurement she alters the state ψ_A as well as her qubit B. Once she has done that, the only hope to reconstruct the original ψ_A is for her to send her measurement to Charlie, who can apply the appropriate operator to his entangled qubit C.

The quantum security mechanism of teleportation is based on strongly correlated, highly non-local entangled states. While a strength, the non-locality of the correlations is also a weakness. Quantum correlations are extremely fragile and can be corrupted by random interactions with the environment, i.e. by decoherence. As we discussed before, this is a process in which the quantum correlations are destroyed and information gets lost. The problem of decoherence is the main stumbling block in making progress towards large scale development and application of quantum technologies. Nevertheless, in 2006 the research group of Gisin at the University of Geneva succeeded in demonstrating teleportation over a distance of 550 meters using the optical fiber network of Swisscom [Landry *et al.*, 2007].

6.10 Quantum computation

Quantum computation is performed by setting up controlled interactions with non-trivial dynamics that successively couple individual qubits together and alter the time evolution of the wavefunction in a predetermined manner. A multi-qubit system is first prepared in a known initial state, representing the input to the program. Then interactions are switched on by applying forces, such as magnetic fields, that determine the direction in which the wavefunction rotates in its state space. Thus a quantum program is just a sequence of unitary operations that are externally applied to the initial state. This is achieved in practice by a corresponding sequence of quantum gates. When the computation is done measurements are made to read out the final state.

Quantum computation is essentially a form of analog computation. A physical system is used to simulate a mathematical problem, taking advantage of the fact that they both obey the same equations. The mathematical problem is mapped onto the physical system by finding an appropriate arrangement of magnets or other fields that will generate the proper equation of motion. One then prepares the initial state, lets the system evolve, and reads out the answer. Analog computers are nothing new. For example, Leibnitz built a mechanical calculator for performing multiplication in 1694, and in the middle of the twentieth century, because of their vastly superior speed in comparison with digital computers, electronic analog computers were often used to solve differential equations.

Then why is quantum computation special? The key to its exceptional power is the massive parallelism at intermediate stages of the computation. Any operation on a given state works exactly the same on all basis vectors. The physical process that defines the quantum computation for an n qubit system thus acts in parallel on a set of 2^n complex numbers, and the phases of these numbers (which would not exist in a classical computation) are important in determining the time evolution of the state. When the measurement is made to read out the answer at the end of the computation we are left with the n-bit output and the phase information is lost.

Because quantum measurements are generically probabilistic, it is possible for the 'same' computation to yield different "answers", e.g. because the measurement process projects the system onto different eigenstates. This can require the need for error correction mechanisms, though for some problems, such as factoring large numbers, it is possible to test for correctness by simply checking the answer to be sure it works. It is also possible for quantum computers to make mistakes due to decoherence, i.e. because of essentially random interactions between the quantum state used to perform the computation and the environment. This also necessitates error correction mechanisms.

The problems caused by decoherence are perhaps *the* central difficulty in creating physical implementations of quantum computation. These can potentially be overcome by constructing systems where the quantum state is not encoded locally, but rather globally, in terms of topological properties of the system that cannot be disrupted by external (local) noise. This is called *topological quantum computing.* This interesting possibility arises in certain two-dimensional physical media which exhibit *topological order*, referring to states of matter in which the essential quantum degrees of freedom and their interactions are topological [Kitaev, 2003; DasSarma *et al.*, 2007].

6.11 Quantum gates and circuits

In the same way that classical gates are the building blocks of classical computers, quantum gates are the basic building blocks of quantum computers. A gate used for a classical computation implements binary operations on binary inputs, changing zeros into ones and vice versa. For example, the only nontrivial single bit logic operation is NOT, which takes 0 to 1 and 1 to 0. In a quantum computation the situation is quite different, because qubits can exist in superpositions of 0 and 1. The set of allowable single qubit operations consists of unitary transformations corresponding to 2×2 complex matrices U such that $U^\dagger U = 1$. The corresponding action on a single qubit is represented in a circuit as illustrated in figure 8. Some quantum gates have classical analogues, but many do not. For example, the

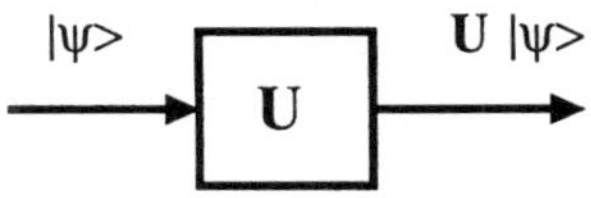

Figure 8. The diagram representing the action of a unitary matrix U corresponding to a quantum gate on a qubit in a state $|\psi\rangle$.

operator $X = \begin{pmatrix} 0 & 1 \\ 1 & 0 \end{pmatrix}$ is the quantum equivalent of the classical NOT gate, and serves the function of interchanging spin up and spin down. In contrast, the operation $\begin{pmatrix} 1 & 0 \\ 0 & -1 \end{pmatrix}$ rotates the phase of the wavefunction by 180 degrees and has no classical equivalent.

A general purpose quantum computer has to be able to transform an arbitrary n-qubit input into an n-qubit output corresponding to the result of the computation. In principle implementing such a computation might be extremely complicated, and might require constructing quantum gates of arbitrary order and complexity.

Fortunately, it is possible to prove that the transformations needed to implement a universal quantum computer can be generated by a simple — so-called universal — set of elementary quantum gates, for example involving a well chosen pair of a one-qubit and a two-qubit gate. Single qubit gates are unitary matrices with three real degrees of freedom. If we allow ourselves to work with finite precision, the set of all gates can be arbitrary well approximated by a small well chosen set. There are many possibilities — the optimal choice depends on the physical implementation of the qubits. Typical one-qubit logical gates are for example the following:

$$(92) \quad X = \begin{pmatrix} 0 & 1 \\ 1 & 0 \end{pmatrix}$$

$$(93) \quad P(\theta) = \begin{pmatrix} 1 & 0 \\ 0 & \exp^{i\theta} \end{pmatrix}$$

$$(94) \quad H = \sqrt{\frac{1}{2}} \begin{pmatrix} 1 & 1 \\ 1 & -1 \end{pmatrix}$$

X is the quantum equivalent of the classical NOT gate, serving the function of interchanging $|1\rangle$ and $|0\rangle$. The two other ones have no classical equivalent. The $P(\theta)$ operation corresponds to the phase gate, it changes the relative phase by θ degrees, typically with θ an irrational multiple of π. For the third gate we can choose the so-called Hadamard gate H which creates a superposition of the basis states, e.g. $|1\rangle \Rightarrow \frac{1}{2}(|1\rangle + |0\rangle)$.

From the perspective of experimental implementation, a convenient two-qubit gate is the $CNOT$ gate. It has been shown that for example the $CNOT$ in combination with a Hadamard and a phase gate forms a universal set [Barenco *et al.*, 1995]. The $CNOT$ gate acts as follows on the state $|A\rangle \otimes |B\rangle$:

$$(95)\quad CNOT : |A\rangle \otimes |B\rangle \Rightarrow |A\rangle \otimes |[A+B]\text{mod } 2\rangle$$

In words, the $CNOT$ gate flips the state of B if $A = 1$, and does nothing if $A = 0$. In matrix form one may write the $CNOT$ gate as

$$(96)\quad CNOT : \begin{pmatrix} 1 & 0 & 0 & 0 \\ 0 & 1 & 0 & 0 \\ 0 & 0 & 0 & 1 \\ 0 & 0 & 1 & 0 \end{pmatrix}.$$

We have fully specified its action on the basis states in figure 9.

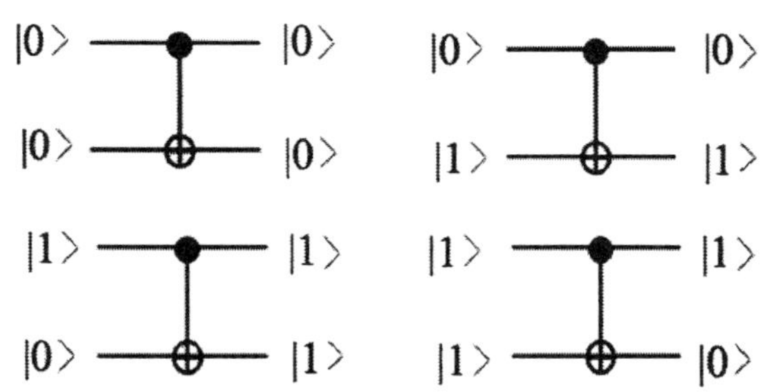

Figure 9. The circuit diagram representing the action of the $CNOT$ gate defined in (96) on the four possible two-qubit basis states. The filled dot on the upper qubit denotes the control and the cross is the symbol for the conditional one qubit NOT gate.

With the CNOT gate one can generate an entangled state from a separable one, as follows:

$$(97)\quad \text{CNOT} : \frac{1}{\sqrt{2}}(|0\rangle + |1\rangle) \otimes |0\rangle \Rightarrow \frac{1}{\sqrt{2}}(|00\rangle + |11\rangle)\,.$$

In fact, from an intuitive point of view the ability to generate substantial speed-ups using a quantum computer vs. a classical computer is related to the ability to operate on the high dimensional state space including the entangled states. To describe a separable n-qubit state with k bits of accuracy we only need to describe each of the individual qubits separately, which only requires the order of nk bits. In contrast, to describe an n-qubit entangled state we need the order of k bits for each dimension in the Hilbert space, i.e. we need the order of $k2^n$ bits. If we were to simulate the evolution of an entangled state on a classical computer we would have to process all these bits of information and the computation would be extremely slow. Quantum computation, in contrast, acts on all this information at once — a quantum computation acting on an entangled state is just as fast as one acting on a separable state. Thus, if we can find situations where the evolution of an entangled state can be mapped into a hard mathematical problem, we can sometimes get substantial speedups.

The CNOT gate can also be used to illustrate how decoherence comes about. Through the same action that allows it to generate an entangled state from a separable state, when viewed from the perspective of a single qubit, the resulting state becomes decoherent. That is, suppose we look at (97) in the density matrix representation. Looking at the first qubit only, the wavefunction of the separable state is $|\psi\rangle = 1/\sqrt{2}(|1\rangle + |0\rangle)$, or in the density matrix representation

$$\begin{aligned} |\psi\rangle\langle\psi| &= \frac{1}{2}(|1\rangle\langle 1| + |1\rangle\langle 0| + |0\rangle\langle 1| + |0\rangle\langle 0|) \\ &= \frac{1}{2}\begin{pmatrix} 1 & 1 \\ 1 & 1 \end{pmatrix}. \end{aligned}$$

Under the action of CNOT this becomes $\frac{1}{2}\begin{pmatrix} 1 & 0 \\ 0 & 1 \end{pmatrix}$, i.e. it becomes diagonal and clearly has positive entropy.

6.12 Applications

At the present point in time there are many different efforts in progress to implement quantum computing. In principle all that is needed is a simple two level quantum system that can easily be manipulated and scaled up to a large number of qubits. The first requirement is not very restrictive, and many different physical implementations of systems with a single or a few qubits have been achieved, including NMR, spin lattices, linear optics with single photons, quantum dots, Josephson junction networks, ion traps and atoms and polar molecules in optical lattices [Di Vincenzo, 2001]. The much harder problem that has so far limited progress toward practical computation is to couple the individual qubits in a controllable way and to achieve a sufficiently low level of decoherence. With the great efforts now taking place, future developments could be surprisingly fast[27]. If we had quantum computers at our disposal, what miracles would they perform? As we said in the introduction to this section, there are many problems where the intrinsic massive parallelism of quantum evolution might yield dramatic speedups.

[27] A first 16-qubit quantum computer has been announced by D-Wave Systems Inc. in California, but at the time of writing this product is not available yet.

The point is not that a classical computer would not be able to do the same computation — after all, one can always simulate a quantum computer on a classical one — but rather the time that is needed. As we mentioned already, the most spectacular speedup is the Shor algorithm (1994) for factorizing large numbers into their prime factors [Shor, 1994]. Because many security keys are based on the inability to factor large numbers into prime factors, the reduction from an exponentially hard to a polynomial hard problem has many practical applications for code breaking. Another important application is the quadratic speedup by Grover's algorithm [Grover, 1996] for problems such as the traveling salesman, in which large spaces need to be searched. Finally, an important application is the simulation of quantum systems themselves [Aspuru-Guzik *et al.*, 2005]. Having a quantum computer naturally provides an exponential speed-up, which in turn feeds back directly into the development of new quantum technologies.

Quantum computation and security are another challenging instance of the surprising and important interplay between the basic concepts of physics and information theory. If physicists and engineers succeed in mastering quantum technologies it will mark an important turning point in information science.

7 BLACK HOLES: A SPACE TIME INFORMATION PARADOX

In this section we make a modest excursion into the realm of curved space-time as described by Einstein's theory of general relativity. As was realized only in the 1970's, this theory poses an interesting and still not fully resolved information paradox for fundamental physics. In general relativity gravity is understood as a manifestation of the curvature of space-time: the curvature of space-time determines how matter and radiation propagate, while at the same time matter and radiation determine how space-time is curved. Particles follow geodesics in curved space-time to produce the curvilinear motion that we observe.

An unexpected and long-ignored prediction of general relativity was the existence of mysterious objects called *black holes* that correspond to solutions with a curvature singularity at their center. Black holes can be created when a very massive star burns all of its nuclear fuel and subsequently collapses into an ultra-compact object under its own gravitational pull. The space-time curvature at the surface of a black hole is so strong that even light cannot escape — hence the term "black hole". The fact that the escape velocity from a black hole is larger then the speed of light implies, at least classically, that no information from inside the black hole can ever reach far away observers. The physical size of a black hole of mass M is defined by its *event horizon*, which is an imaginary sphere centered on the hole with a radius (called the *Schwarzchild radius*)

$$(98)\quad R_S = \frac{2G_N M}{c^2}\,,$$

where G_N is Newton's gravitational constant and c is the velocity of light. For a black hole with the mass of the sun this yields $R_S = 3km$, and for the earth only

$R_S = 1cm$! The only measurable quantities of a black hole for an observer far away are its mass, its charge and its angular momentum.

But what about the second law of thermodynamics? If we throw an object with non-zero entropy into black hole, it naively seems that the entropy would disappear for ever and thus the total entropy of the universe would decrease, causing a blunt violation of the second law of thermodynamics. In the early 1970's, however, Bekenstein [Bekenstein, 1973] and Hawking [Bardeen *et al.*, 1973] showed that it is possible to assign an entropy to a black hole. This entropy is proportional to the area $A = 4\pi(R_S)^2$ of the event horizon,

$$(99) \quad S = \frac{Ac^3}{4G_N\hbar}.$$

A striking analogy with the laws thermodynamics became evident: The change of mass (or energy) as we throw things in leads according to classical general relativity to a change of horizon area, as the Schwarzchild radius also increases. For an electrically neutral, spherically symmetric black hole, it is possible to show that the incremental change of mass dM of the black hole is related to the change of area dA as

$$(100) \quad dM = \frac{\kappa}{2\pi} dA$$

where $\kappa = \hbar c/2R_s$ is the *surface gravity* at the horizon. One can make an analogy with thermodynamics, where dA plays the role of "entropy", dM the role of "heat", and the κ the role of "temperature". Since no energy can leave the black hole, dM is positive and therefore $dA \geq 0$, analogous to the second law of thermodynamics. At this point the correspondence between black hole dynamics and thermodynamics is a mere analogy, because we know that a classical black hole does not radiate and therefore has zero temperature. One can still argue that the information is not necessarily be lost, it is only somewhere else and unretrievable for certain observers.

What happens to this picture if we take quantum physics into account? Steven Hawking was the first to investigate the quantum behavior of black holes and his results radically changed their physical interpretation. He showed [Hawking, 1974; Hawking, 1975] that if we apply quantum theory to the spacetime region close to the horizon then black holes aren't black at all! Using arguments based on the spontaneous creation of particle-antiparticle pairs in the strong gravitational field near the horizon he showed that a black hole behaves like a stove, emitting black body thermal radiation of a characteristic temperature, called the *Hawking temperature*, given by[28]

$$(101) \quad T_H \equiv \frac{\hbar c}{4\pi R_S} = \frac{\hbar c^3}{8\pi G_N M} ,$$

fully consistent with the first law (100). We see that the black hole temperature is inversely proportional to its mass, which means that a black hole becomes hotter

[28]We recall that we adopted units where Boltzmann's constant k is equal to one.

and radiates more energy as it becomes lighter. In other words, a black hole will radiate and lose mass at an ever-increasing rate until it finally explodes[29].

We conclude that quantum mechanics indeed radically changes the picture of a black hole. Black holes will eventually evaporate, presumably leaving nothing behind except thermal radiation, which has a nonzero entropy. However, as we discussed in the previous section, if we start with a physical system in a pure state that develops into a black hole, which subsequently evaporates, then at the level of quantum mechanics the information about the wavefunction should be rigorously preserved — the quantum mechanical entropy should not change.

It may be helpful to compare the complete black hole formation and evaporation process with a similar, more familiar situation (proposed by Sidney Coleman) where we know that quantum processes conserve entropy. Imagine a piece of coal at zero temperature (where by definition $S = 0$) that gets irradiated with a given amount of high entropy radiation, which we assume gets absorbed completely. It brings the coal into an excited state of finite temperature. As a consequence the piece of coal starts radiating, but since there is no more incoming radiation, it eventually returns to the zero temperature state, with zero entropy. As the quantum process of absorbing the initial radiation and emitting the outgoing radiation is unitary, it follows that the outcoming radiation should have exactly the same entropy as the incoming radiation.

Thus, if we view the complete process of black hole formation and subsequent evaporation from a quantum mechanical point of view there should be no loss of information. So if the initial state is a pure state than a pure state should come out. But how can this be compatible with the observation that only thermal radiation comes out, independent of what we throw in? Thermal radiation is produced by entropy generating processes, is maximally uncorrelated and random, and has maximal entropy. If we throw the Encyclopedia Brittanica into the black hole and only get radiation out, its highly correlated initial state would seem to have been completely lost. This suggests that Hawking's quantum calculation is in some way incomplete. These conflicting views on the process of black hole formation and evaporation are referred to as the *black hole information paradox*. It has given rise to a fundamental debate in physics between the two principle theories of nature: the theory of relativity describing space-time and gravity on one hand and the theory of quantum mechanics describing matter and radiation on the other. Does the geometry of Einstein's theory of relativity prevail over quantum theory, or visa versa?

If quantum theory is to survive one has to explain how the incoming information gets transferred to the outgoing radiation coming from the horizon[30], so that a

[29]The type of blackholes that are most commonly considered are very massive objects like collapsed stars. The lifetime of a black hole is given by $\tau \simeq G_N^2 M^3/\hbar c^4$ which implies that the lifetime of such a massive black hole is on the order of $\tau \geq 10^{50}$ years (much larger than the lifetime of the universe $\tau_0 \simeq 10^{10}$ y). Theoretical physicists have also considered microscopic black holes, where the information paradox we are discussing leads to a problem of principle.

[30]It has been speculated by a number of authors that there is the logical possibility that the black hole does not disappear altogether, but leaves some remnant behind just in order to preserve

clever quantum detective making extremely careful measurements with very fancy equipment could recover it. If such a mechanism is not operative the incoming information is completely lost, and the laws of quantum mechanics are violated. The question is, what cherished principles must be given up?

There is a generic way to think about this problem along the lines of quantum teleportation and a so-called *final state projection* [Horowitz and Maldacena, 2004; Lloyd, 2006]. We mentioned that Hawking radiation can be considered as a consequence of virtual particle-antiparticle pair production near the horizon of the black hole. The pairs that are created and separated at the horizon are in a highly entangled state, leading to highly correlated in-falling and outgoing radiation. It is then possible, at least in principle, that the interaction between the in-falling radiation and the in-falling matter (making the black hole) would lead to a projection in a given quantum state. Knowing that final state - for example by proving that only a unique state is possible - one would instantaneously have teleported the information from the incoming mass (qubit A) to the outgoing radiation (qubit C) by using the entangled pair (qubit pair BC) in analogy with the process of teleportation we discussed in section 6.9 . The parallel with quantum teleportation is only partial, because in that situation the sender Alice (inside the black hole) has to send some classical information on the outcome of her measurements to the receiver Charlie (outside the black hole) before he is able decode the information in the outcoming radiation. But sending classical information out of a black hole is impossible. So this mechanism to rescue the information from the interior can only work if there is a projection onto an a priori known unique final state, so that it is as if Alice made a measurement yielding this state and sent the information to Charlie. But how this assumption could be justified is still a mystery.

A more ambitious way to attack this problem is to attempt to construct a quantum theory of gravity, where one assumes the existence of microscopic degrees of freedom so that the thermodynamic properties of black holes could be explained by the statistical mechanics of these underlying degrees of freedom. Giving the quantum description of these new fundamental degrees of freedom would then allow for a unitary description. Before we explain what these degrees of freedom might be, let us first consider another remarkable property of black holes. As we explained before, the entropy of systems that are not strongly coupled is an extensive property, i.e. proportional to volume. The entropy of a black hole, in contrast, is proportional to the area of the event horizon rather than the volume. This dimensional reduction of the number of degrees of freedom is highly suggestive that all the physics of a black hole takes place at its horizon, an idea introduced by 't Hooft and Susskind [Susskind and Lindesay, 2004], that is called the *holographic principle.*[31]

---

the information. The final state of the remnant should then somehow contain the information of the matter thrown in.

[31]A hologram is a two dimensional image that appears to be a three dimensional image; in a similar vein, a black hole is a massive object for which everything appears to take place on the surface.

Resolving the clash between the quantum theory of matter and general relativity of space-time is one of the main motivations for the great effort to search for a theory that overarches all of fundamental physics. At this moment the main line of attack is based on *superstring theory*, which is a quantum theory in which both matter and space-time are a manifestation of extremely tiny strings ($l = 10^{-35}m$). This theory incorporates microscopic degrees of freedom that might provide a statistical mechanical account of the entropy of black holes. In 1996 Strominger and Vafa[Strominger and Vafa, 1996] managed to calculate the Bekenstein-Hawking entropy for (extremal) black holes in terms of microscopic strings using a property of string theory called *duality*, which allowed them to count the number of accessible quantum states. The answer they found implied that for the exterior observer information is preserved on the surface of the horizon, basically realizing the holographic principle.

There are indeed situations (so-called Anti-de Sitter/Conformal Field Theory dualities or AdS/CFT models) in string theory describing space-times with a boundary where the holographic principle is realized explicitly. One hopes that in such models the complete process of formation and evaporation of a black hole can be described by the time evolution of its holographic image on the boundary, which in this case is a super-symmetric gauge theory, a well behaved quantum conformal field theory (CFT). A caveat is that in this particular Anti-de Sitter (AdS) classical setting so far only a static "eternal" black hole solution has been found, so interesting as that situation may be, it doesn't yet allow for a decisive answer to a completely realistic process of black hole formation and evaporation. Nevertheless, the communis opinion - at least for the moment - is that the principles of quantum theory have successfully passed a severe test[32] [Susskind and Lindesay, 2004].

8 CONCLUSION

The basic method of scientific investigation is to acquire information about nature by doing measurements and then to make models which optimally compress that information. Therefore information theoretic questions arise naturally at all levels of scientific enterprise: in the analysis of measurements, in performing computer simulations, and in evaluating the quality of mathematical models and theories.

The notion of entropy started in thermodynamics as a rather abstract mathematical property. With the development of statistical mechanics it emerged as a measure of disorder, though the notion of disorder referred to a very restricted context. With the passage of time the generality and the power of the notion of entropy became clearer, so that now the line of reasoning is easily reversed — fol-

[32] Indicative is that a long standing bet between Hawking and Presskil of Caltech was settled in 2004 when Hawking officially declared defeat. In doing so he recognized the fact that information is not lost when we throw something into a black hole — quantum correlations between the in-falling matter and the out-coming radiation should in principle make it possible to retrieve the original information.

lowing Jaynes, statistical mechanics is reduced to an application of the maximum entropy principle, using constraints that are determined by the physical system. Forecasting is a process whose effectiveness can be understood in terms of the information contained in measurements, and the rate at which the geometry of the underlying dynamical system, used to make the forecast, causes this information to be lost. And following Rissanen, the whole scientific enterprise is reduced to the principle of minimum description length, which essentially amounts to finding the optimal compromise between the information contained in a model and the information contained in the discrepancies between the model and the data.

Questions related to the philosophy of information have lead us naturally back to some of the profound debates in physics on the nature of the concept of entropy as it appears in the description of systems about which we have *a priori* only limited information. The Gibbs paradox, for example, centers around the question of whether entropy is subjective or objective. We have seen that while the description might have subjective components, whenever we use the concept of entropy to ask concrete physical questions, we always get objective physical answers. Similarly, when we inject intelligent actors into the story, as for Maxwell's demon, we see that the second law remains valid — it applies equally well in a universe with sentient beings.

Fundamental turning points in physics have always left important traces in information theory. A particularly interesting example is the development of quantum information theory, with its envisaged applications to quantum security, quantum teleportation and quantum computation. Another interesting example is the black hole information paradox, where the notions of entropy and information continue to be central players in our attempts to resolve some of the principal debates of modern theoretical physics. In a sense, our ability to construct a proper statistical mechanics is a good test of our theories. If we could only formulate an underlying statistical mechanics of black holes, we might be able to resolve fundamental questions about the interface between gravity and quantum mechanics.

Finally, as we enter the realm of nonequilbrium statistical mechanics, we see that the question of what information means and how it can be used remains vital. New entropies are being defined, and their usefulness and theoretical consistency are topics that are actively debated. The physics of information is an emerging field, one that is still very much in progress.

ACKNOWLEDGEMENTS

The authors would like to thank Seth Lloyd, Peter Harremoës, David Bacon, Jim Crutchfield, David Meyer, Cris Moore, Constantino Tsallis, Bill Wooters and Erik Verlinde for illuminating conversations and critical comments. Doyne Farmer appreciates support from Barclays Bank and National Science Foundation grant 0624351. Sander Bais thanks the Santa Fe Institute for its hospitality, which allowed this work to take shape. Any opinions, findings, and conclusions or recommendations expressed in this material are those of the authors and do not necessarily reflect the views of the National Science Foundation.

BIBLIOGRAPHY

[Akaike, 1974] H. Akaike. A new look at the statistical model identification. *IEEE Transactions on Automatic Control*, 19(6):716–723, 1974.

[Aspuru-Guzik *et al.*, 2005] A. Aspuru-Guzik, A. D. Dutoi, P. J. Love, and M. Head-Gordon. Simulated quantum computation of molecular energies. *Science*, 309:1704, 2005.

[Bardeen *et al.*, 1973] J.M. Bardeen, B. Carter, and S.W. Hawking. The four laws of black hole mechanics. *Commun.Math.Phys.*, 31:161, 1973.

[Barenco *et al.*, 1995] A. Barenco, C. H. Bennett, R. Cleve, D. P. DiVincenzo, N. Margolus, P. Shor, T. Sleator, J. Smolin, and H. Weinfurter. Elementary gates for quantum computation. *Quantum Physics Archive (Los Alamos National Labaratory)*, 1995.

[Beck, 2001] C. Beck. Dynamical foundations of nonextensive statistical mechanics. *Phys.Rev.Lett.*, 87:180601, 2001.

[Bekenstein, 1973] J.D. Bekenstein. Black holes and entropy. *Phys.Rev.*, D(7):2333, 1973.

[Bennett et al., 1993] C.H. Bennett et al. Teleporting an unknown quantum state via dual classical and epr channels. *Phys.Rev.Lett.*, 70:1895–1899, 1993.

[Bennett, 1982] C.H. Bennett. The thermodynamics of information - a review. *Int. Journ. of Theor. Phys*, 21:905–940, 1982.

[Bennett, 1987] C.H. Bennett. Demons, engines and the second law. *Scientific American*, November:108–116, 1987.

[Boltzmann, 1896-1898] L. Boltzmann. *Vorlesungen über Gastheorie*, volume I and II of *translated by S.G. Brush*. J.A. Barth, Leipzig, *Lectures on Gas Theory*, university of california press, berkeley, 1964 edition, 1896-1898.

[Bouwmeester et al., 1997] D. Bouwmeester et al. Experimental quantum teleportation. *Nature*, 390:575–9, 1997.

[Brillouin, 1956] L. Brillouin. *Science and Information Theory*. Academic Press, 1956.

[Crutchfield and Young, 1989] J.P. Crutchfield and K. Young. Inferring statistical complexity. *Phys.Rev.Lett.*, 63:105–108, 1989.

[Crutchfield *et al.*, 1986] J.P. Crutchfield, J.D. Farmer, N.H. Packard, and R.S. Shaw. Chaos. *Scientific American*, 254(12):46–57, 1986.

[DasSarma *et al.*, 2007] S. DasSarma, M. Freedman, C. Nayak, S. H. Simon, and A. Stern. Non-abelian anyons and topological quantum computation. `http://arXiv:0707.1889`, 2007.

[Denbigh and Denbigh, 1985] K.G. Denbigh and J.S. Denbigh. *Entropy in relation to incomplete knowledge*. Cambridge University Press, 1985.

[Deutsch, 1985] D. Deutsch. Quantum theory, the Church-Turing principle and the universal quantum computer. *Proceedings of the Royal Society of London*, A400:97, 1985.

[Di Vincenzo, 2001] D. P. Di Vincenzo. The physical implementation of quantum computation. In *Scalable Quantum Computers: Paving the Way to Realisation*, edited by S.L. Braunstein, Hoi-Kwong Lo and P. Kok. Wiley-VCH, 2001.

[Dieks, 1982] D. Dieks. Communication by epr devices. *Phys.Lett.*, A 92:271, 1982.

[Einstein *et al.*, 1935] A. Einstein, B. Podolsky, and N. Rosen. Can quantum-mechanical description of physical reality be considered complete. *Phys. Rev.*, 47, 1935.

[Farmer and Geanakoplos, 2006] J.D. Farmer and J. Geanakoplos. Power laws in economics and elsewhere. Technical report, Santa Fe Institute, 2006.

[Farmer *et al.*, 1980] J.D. Farmer, J. P. Crutchfield, H. Froehling, N. H. Packard, and R. S. Shaw. Power spectra and mixing properties of strange attractors. *Annals of the New York Academy of Science*, 375:453–472, 1980.

[Farmer, 1982] J.D. Farmer. Information dimension and the probabilistic structure of chaos. *Zeitschrift in Naturforshung*, 37A:1304–1325, 1982.

[Feynman, February, 1959)(and http://www.zyvex.com/feynman.html] R.P. Feynman. There is plenty of room at the bottom. *Engineering and Science, Caltech*, February, 1959) (and `http://www.zyvex.com/feynman.html`

[Gell-Mann and Tsallis, 2004] M. Gell-Mann and C. Tsallis. *Nonextensive Entropy: Interdisciplinary Applications*. Proceedings Volume in the Santa Fe Institute Studies in the Sciences of Complexity. Oxford University Press, 2004.

[Gibbs, 1902] J.W. Gibbs. *Elementary Principles in Statistical Physics*. Yale University Press, 1902.

[Grover, 1996] L.K. Grover. A fast quantum mechanical algorithm for database search. *Proceedings of the 28th Annual ACM Symposium on the Theory of Computing*, page 212219, 1996.

[Grunwald and Vitányi,] P.D. Grunwald and P.M.B. Vitányi. *Algorithmic informaton theory.* Chapter 3 of this book.

[Grunwald *et al.*, 2004] P.D. Grunwald, I.J. Myung, and M.A. Pitt. *Advances in Minimum Description Length: Theory and Applications.* MIT Press, 2004.

[Guttman, 1999] Y.M. Guttman. *The Concept of Probability in Statistical Physics.* Cambridge University Press, 1999.

[Hawking, 1974] S.W. Hawking. Black hole explosions. *Nature*, 248:30, 1974.

[Hawking, 1975] S.W. Hawking. Particle creation by black holes. *Commun.Math.Phys.*, 43:199, 1975.

[Holevo, 1982] A.S. Holevo. *Probabilistic and statistical aspects of Quantum Theory.* North Holland Publishing Company, Amsterdam, 1982.

[Horowitz and Maldacena, 2004] G.T. Horowitz and J. Maldacena. The black-hole final state. *JHEP*, 02:8, 2004.

[Huang, 1987] K. Huang. *Statistical Mechanics.* Wiley and Sons, 1987.

[Jaynes, 1963] E.T. Jaynes. *Information Theory and Statistical Mechanics.* in Statistical Physics (K. Ford ed.). Benjamin, New York, 1963.

[Jaynes, 1983] E.T. Jaynes. *Papers on probability, Statistics and Statistical Physics.* (R.D. Rosenkranz ed.), 1983.

[Jaynes, 1996] E.T. Jaynes. The Gibbs paradox. Technical report, `http://bayes.wustl.edu/etj/articles/gibbs.paradox.pdf`, 1996.

[Kan, 2006] *Nonlinear Time Series Analysis.* Cambridge University Press, Cambridge, UK, 2006.

[Kaye *et al.*, 2007] P.R. Kaye, R. Laflamme, and M. Mosca. *Introduction to Quantum Computing.* Oxford University Press, USA, 2007.

[Khinchin, 1949] A.I. Khinchin. *Mathematical Foundations of Statistical Mechanics.* Dover,New York, 1949.

[Kitaev, 2003] A. Kitaev. Fault tolerant quantum computation by anyons. *Annals of Physics*, 321:2–111, 2003.

[Kittel, 1966] C. Kittel. *Elementary Statistical Physics.* Wiley and Sons, 1966.

[Landauer, 1961] R. Landauer. Irreversibility and heat generation in the computing process. *IBM Journal of Research and Development*, 5:183–191, 1961.

[Landauer, 1991] R. Landauer. Information is physical. *Physics Today*, 44:23–29, 1991.

[Landry *et al.*, 2007] O. Landry et al. Quantum teleportation over the Swisscom telecommunication network. *Journ.Opt.Soc.Am. B*, 24:February 2007, 2007.

[Lifschitz and Landau, 1980] E.M. Lifschitz and L.D. Landau. *Statistical Physics*, volume 5 of *Course of Theoretical Physics.* Butterworth-Heinemann, 1980.

[Lloyd, 2006] S. Lloyd. Almost certain escape from black holes in final state projection models. *Phys.Rev.Lett.*, 96:061302, 2006.

[Lloyd, 2008] S. Lloyd. *Quantum Computers.* Wiley-International, 2008.

[Lorenz, 1963] E.N. Lorenz. Deterministic nonperiodic flow. *Journal of Atmospheric Sciences*, 20:130–141, 1963.

[Maxwell, 1872] J.C. Maxwell. *Theory of Heat.* D. Appleton & Co, New York, 1872.

[Mermin, 2007] N.D. Mermin. *Quantum Computer Science: An Introduction.* Cambridge University Press, 2007.

[Moyano *et al.*, 2006] L.G. Moyano, C. Tsallis, and M. Gell-Mann. Numerical indications of a q-generalised central limit theorem. *Europhys. Lett.*, 73:813, 2006.

[Newman, 2005] M.E.J. Newman. Power laws, Pareto distributions and Zipf's law. *Contemporary Physics*, 46:323–351, 2005.

[Nielsen and Chuang, 1990] M. Nielsen and I. Chuang. *Quantum Computation and Quantum Information.* Cambridge University Press, 1990.

[Omnes, 1999] R. Omnes. *Understanding Quantum Mechanics.* Princeton University Press, 1999.

[Reif, 1965] F. Reif. *Fundamentals of Statistical and Thermal Physics.* McGraw-Hill, 1965.

[Rissanen, 1978] J. Rissanen. Modeling by the shortest data description. *Automatica*, 14:465–471, 1978.

[Schlosshauer, 2004] M. Schlosshauer. Decoherence, the meaurement problem, and interpretations of quantum mechanics. *Reviews of Modern Physics*, 76:1267–3105, 2004.

[Shannon, 1948] C.E. Shannon. A mathematical theory of communication. *The Bell System Technical Journal*, 27:379–423,623–656, 1948.

[Shaw, 1981] R.S. Shaw. Strange attractors, chaotic behavior and information flow. *Z. Naturfor*, 36A(1):80–112, 1981.

[Shor, 1994] P. Shor. Algorithms for quantum computation: discrete logarithms and factoring. *Proceedings 35th Annual Symposium on Foundations of Computer Science, IEEE Comput. Soc. Press*, page 124134, 1994.

[Still and Crutchfield, 2007] S. Still and J.P. Crutchfield. Structure or noise. `http://arXiv:physics.gen-ph/0708.0654`, 2007.

[Strogatz, 1994] S. Strogatz. *Nonlinear Dynamics and Chaos*. Addison Wesley, 1994.

[Strominger and Vafa, 1996] A. Strominger and C. Vafa. Microscopic origin of the Bekenstein-Hawking entropy. *Phys.Lett.*, B379:99, 1996.

[Susskind and Lindesay, 2004] L. Susskind and J. Lindesay. *An Introduction To Black Holes, Information And The String Theory Revolution: The Holographic Universe*. World Scientific, 2004.

[Szilard, 1929] L. Szilard. *Z. Physik*, 53:840, 1929.

[Tolman, 1938] R.C. Tolman. *Principles of Statistical Physics*. Clarendon, Oxford, UK, 1938.

[Tsallis *et al.*, 2005a] C. Tsallis, M. Gell-Mann, and Y. Sato. *Europhysics News 36, Special Issue "Nonextensive Statistical Mechanics: New Trends, new perspectives", eds. J.P. Boon and C. Tsallis*, page 186, 2005.

[Tsallis *et al.*, 2005b] C. Tsallis, M. Gell-Mann, and Y. Sato. Asymptotically scale-invariant occupancy of phase space makes the entropy s_q extensive. *JProc. Natl. Acad. Sc. USA*, 102:15377, 2005.

[Tsallis, 1988] C. Tsallis. Possible generalization of Boltzmann-Gibbs statistics. *J.Stat.Phys.*, 52:479, 1988.

[Umarov *et al.*, 2006] S. Umarov, C. Tsallis, and S. Steinberg. A generalization of the central limit theorem consistent with nonextensive statistical mechanics. Technical report, arXiv:cond-mat/0603593, 2006.

[Umarov et al., 2006] S. Umarov et al. q-Generalization of symmetric alpha-stable distributions. *cond-mat/0606038/40*, 2006.

[Wootters and Zurek, 1982] W.K. Wootters and W.H. Zurek. A single quantum cannot be cloned. *Nature*, 299:802–803, 1982.

[Zinn-Justin, 1989] J. Zinn-Justin. *Quantum Field Theory and Critical Phenomena*. Clarendon Press, Oxford, UK, 1989.

[Zurek, 1991] W.H. Zurek. Decoherence and the transition from quantum to classical. *Physics Today*, 44:36–44, 1991.

[Zurek, 2003] W.H. Zurek. Decoherence, einselection, and the quantum origins of the classical. *Reviews of Modern Physics*, 75:715, 2003.

INFORMATION IN THE STUDY OF HUMAN INTERACTION

Keith Devlin and Duska Rosenberg

INFORMATION AS AN ANALYTIC TOOL

This chapter describes one way that information — as a conceptual entity — may be used (by an analyst, as a tool) in a study of human interaction. (Actually, most of what we say will apply to interaction in general, for instance human–machine interaction, but our examples will be taken from human interaction.) The "analyst" here may be a professional social scientist (as is the case for our main technical example), or could be an ordinary person trying to make sense of a particular interaction. When applied to such latter cases, our article also provides insight into much of the common talk about "information" that takes place in today's "information society", and in that way our essay can be viewed as an analysis of the rational structure that lies behind (and is implicit in) the modern, information-oriented view of the world.

To give a very simple example, suppose Alice (A) issues the instruction "Sit down" to Bill (B). We may view this as an attempt by A to achieve a particular action by B. A makes this attempt by herself carrying out a particular action, namely uttering certain words. The analysis could proceed by examining why A chooses the particular words she does, why B interprets those words the way he does, and what action B carries out as a result and why. Typically, this might be done by identifying social norms that describe (or prescribe) how people use language to achieve their ends. (An example we shall examine in some depth later in the paper will show just how such an analysis may proceed.)

But there is another way we could analyze the same interaction; namely as being mediated by the transmission of information from A to B. In the alternative, information-based approach, we analyze Alice and Bill's interaction in terms of the issuance of certain information by A, its reception by B, and the consequences of this transmission in terms of the actions of the two participants.

What is gained (or lost, or obscured) by the introduction of the mediating notion of information? Which (if any) approach is better (for what purpose), and why?

An analogy might help to explain the distinction between the two approaches. Suppose we want to study a wrestling match between two people. Then we would most naturally analyze the interaction in terms of the forces each exerts on the other. In contrast, if we want to examine a game of tennis between the two individuals, it is more appropriate (and surely more productive) to look at the way the ball is batted from one to the other. Why, in the second case, do we not analyze the game in terms of the forces each player exerts (through the racket) on the ball? After all, the ball is an inert object; the entire play of the game is dictated by the actions of the two players, just as it is in the wrestling match.

The reason we analyze the tennis match in terms of the motion of the ball, is precisely that the ball does indeed *mediate* between the actions of the two players in the tennis match. Mediation of a human–human interaction, even if by an inert object, changes things sufficiently that a framework appropriate for analyzing one form of interaction may be unsuitable for analyzing another. This is why newspaper accounts of tennis games typically include descriptions of the motion of the ball as well as the two players.

In the case of a human–human linguistic interaction, however (such as the "Sit down" example we just gave), we seem to have an entirely free choice between two different forms of analysis. We can adopt one of several traditional (non-information-based) approaches, focusing on the (descriptive or prescriptive) rules and protocols that describe or prescribe how interaction is done, the choice of words each participant makes, and the way each understands the words spoken by the other. This corresponds to the way we analyze the wrestling match, where we look at the various capacities each participant brings to the encounter and the manner in which those capacities result in the physical interaction that ensues. Or we may equally well consider the linguistic interaction as a transmission of information. This would correspond to our analysis of the tennis match, with the information passed from one person to the other at any one stage being the analogue of the tennis ball.

Of course, as with any analogy, it is important to recognize the limitations of the comparison. In the case of a tennis game, the same ball gets passed back and forth between one player and another; in human interaction, considered as mediated by an exchange of information, *different* information is conveyed at each stage.[1] In a typical human *linguistic* interaction (such as a conversation), for instance, there is something physical passed from one participant to another at each stage, namely the individual utterances (tokens); but these are not the information, rather they (can be said to) *carry* the information. Part of any formal account of information exchange has not only to include a definition of information, but also provide a mechanism for how tokens can in fact carry information. (The theory we make use

[1]Interaction also involves feedback — implicit information — that helps A and B coordinate their actions, which is necessary if they work together on a shared task, or perform any kind of joint action. Conversation can be viewed as a joint action whereby participant establish mutual understanding, or, in the words of Clark [Clark, 1996], "common ground". However, joint action does not always involve using language. Imagine two people carrying a plank. Each of the individual movements is felt through the movement of the plank, which can be said to carry information about the participants' moves. We can also call this "feed-through" (Dix in [Rosenberg and Hutchinson, 1994]). In this context the medium plays an important part. If, instead of the plank, the two people carry a mattress, the medium does not transmit information about their movements in the same way. The characteristics of information in interaction and joint action therefore depend significantly on the medium or the way the interaction is mediated.

of in our account, situation theory, does just that.) The purpose of the analogy is to distinguish between the conceptualization or analysis of a wrestling match as an *unmediated interaction* and a tennis match as being *mediated by a neutral object*, namely the ball.

Is one approach better than the other, and if so how? The answer is that each offers advantages the other does not. For some purposes, a descriptive analysis is better, on other occasions the information-based approach is more suitable. In some cases, carrying out *both* forms of analysis may result in greater understanding.

This distinction between the two analytic approaches is not unlike the one that arises in several different guises in physics, between "action at a distance" and the transmission of a particle. For example, do we think of gravity in terms of geometric distortion of space–time or as the transmission of gravitons? Again, is light a wave (a perturbation in the fabric of space–time) or a particle (a photon)?[2]

The distinction is not merely one of theoretical interest to the analyst; it gets at a fundamental feature of the way we conceive of and live in our current world. Today, much of our everyday thinking, writing, and talk about human activities is couched in terms of information. Yet, this way of talking about the world is relatively recent. The change was brought about largely by the development of various communication technologies — printing, the newspapers, postal services, dictionaries and reference books, radio, telephone, television, photocopiers, the Internet — that, by mediating human–human interaction, made possible (indeed encouraged) an information-based (tennis game or particle) way of thinking about communication.[3] We say a little more about the development of the modern, popular concept of information in just a moment.

INFORMATION

As indicated by some of the other articles in this collection, the word "information" has several different meanings, including a fundamental entity (closely related to entropy) that exists in the universe, a measure of order in the universe, a number of (different) mathematical concepts, and the less precise but more common, everyday (and, of particular relevance to this article, socially constructed) notion implicit

[2]Our mathematical treatment of information, described later, takes this analogy a step further by regarding information as made up of discrete items called "infons." Indeed, the invention of that word, by Devlin, was motivated entirely by that analogy.

[3]The use of a concept of information as a mediator is not restricted to communication. In human action and interaction (especially in computer-mediated communication), we are not talking only about transmission of information. We define the concept of mediation further, to include sharing of information as well. This refers in particular to the use of information to coordinate action, express communicative intent, and ultimately create trust, identity with a group or a community, and shared culture, all of which are essential features of social life. Sharing information is different from exchange and utilization; in particular, sharing is more profoundly social than transactional. When we exchange information, nothing changes unless the exchange causes some kind of change in the cognition of individuals involved. When we share information, then the information that is shared changes, because the act of sharing gives rise to new and different information.

in terms such as "information desk," "departure/arrivals information," and "Can you give me some information about renting bicycles in Amsterdam?"

In this article, we take as our concept of information the socially constructed, everyday notion mentioned last in the above list. In the more technical part of the paper, we shall make that everyday notion a little more precise by way of a mathematical definition, and use that additional precision to examine in some detail the way that information may be used to analyze human interaction (and, more generally, human action).

We make no attempt to provide a comprehensive overview of the topic. The field is far too broad for any short survey such as this to come even close to completeness. Rather we shall outline the main themes and *illustrate* the way information can play a role in an analysis of a social phenomenon.

It will be helpful to begin with a few brief (and hence simplified) remarks about the origins of the notion of information we shall focus on.

Prior to the nineteenth century, the word "information" (which first appeared in the English language in the fourteenth century) was used to refer to a knowledgable or informed individual. For example, the term "man of information" would translate into modern English as "man of learning", "well educated man", or "well informed man".

During the nineteenth century, the generally accepted conception of information shifted from something possessed by an individual (if indeed it was conceived as some-*thing* that could be "possessed") to one of a public commodity — something (and in this case definitely a *thing*) that could be shared. The cause of this shift in meaning can be traced to the growth of communication technologies, in particular the publication of mass market newspapers in the early eighteenth century and onwards. With the appearance of newspapers, and also dictionaries, encyclopedias, and general reference books and the introduction of postal services, the telegraph, and later the telephone, it was possible to identify (or conceive of) a "substance" or "commodity" that could be called "information".

That substance was largely autonomous, having an existence outside the individual human mind. It came in identifiable chunks. For instance, newspapers impose the same physical structure (a block of text within a very small range of size) on every topic reported on, be it politics, war, sport, theater, science, or whatever. Moreover, the organizations that produce newspapers, reference books, and the like provide an institutional "stamp of approval" on the information they impart, giving it the air of being neutral, free of bias and personal perspective or interpretation — "the truth."

The nineteenth century concept of information was thus an itemized one that was largely identified with its representation. It became possible to talk in terms of "amount of information." Information was also true; otherwise it would be called *mis*information.

With the rise of itemized, autonomous information, it was no longer appropriate to use the term "information" to describe personal facts. For instance, only in very special circumstances would a person today say "Alice provided me with the information that she enjoyed last night's movie." Rather one might say "Alice told me she enjoyed the movie," this fact neither being public property nor having an "institutional stamp of authority" that would grant it the status of information.

With the nineteenthc century shift in meaning, information also came to be viewed not as the *result* of a person being informed, but its *cause*.[4]

The modern everyday conception of information is different again. Whereas the nineteenth century notion was closely tied to the "containers" of information that gave rise to the notion — the books, encyclopedias, newspapers, etc. — the concept of information that arose around the middle of the twentieth century *transcends its representation*. Moroever, whereas nineteenth century information was, by definition, true, the same cannot be said for today's concept.

The modern notion of information did not fully develop until the 1970s, although the beginnings of the shift can be seen as far back as the 1940s. Like its predecessor, this new notion also developed as a result of changes in communication technologies — in this case the development of the digital computer and the growth of the many associated electrical and electronic "information and communication" media that are now part of our everyday lives.

Today, most of us think of information as a commodity that is largely independent of how it is embodied. It can be bought, sold, stolen, exchanged, shared, stored, sent along wires and through the ether, and so forth. It can also be *processed*, using *information technologies*, both concepts that would have sounded alien (and probably nonsensical) to anyone living in the nineteenth century, and even the first half of the twentieth.

The separation of information from its various representations is what made it possible for contemporary technology guru Ted Nelson to make his oft-repeated observation "Paper is just an object that information has been sprayed onto in the past."

The way present-day society conceives of information today is well captured by the following passage from *Business Week* (special issue on "The Information Revolution") in 1994:

> We can glean it [information] from the pages of a book or the morning newspaper and from the glowing phosphors of a video screen. Scientists find it stored in our genes and in the lush complexity of the rain forest. And it's always in the air where people come together, whether to work, play, or just gab.

It is the use of today's concept of (disembodied) information as a means to understand (and, when done more formally, analyze) human interaction that is the subject of this paper.

[4]Nunberg [1996], a good reference for much of the present discussion, observes that a similar shift in meaning occurred when the terms *mystery* and *horror* began to be used to describe literary genres.

HOW DOES INFORMATION ARISE?

A fundamental question to be answered at the start is, how is it possible for something in the world, say a book or a magnetic disk, to store, or represent, information? This question immediately generalizes. For, although we generally think of information as being stored (by way of representations) in things such as books and computer databases, any physical object may store information. In fact, during the course of a normal day, we acquire information from a variety of physical objects, and from the environment.

For example, if we see dark clouds in the sky, we may take an umbrella as we leave for work, the state of the sky having provided us with the information that it might rain. On Halloween night in North America, a light on in the porch provides the information that it is acceptable for children to approach the house and ask for candy; no light indicates that the householders do not want to be disturbed. In rural parts of North America, setting the flag on the mailbox in the upright position indicates to the mail carrier that there is outgoing mail to pick up.

How can an object or a collection of objects encode or represent information? How can part of the environment encode or represent information? For instance, how does smoke provide information that there is fire, and how do dark clouds provide information that it is likely to rain? Part of the explanation is that this is the way the world is: there is a systematic regularity between the existence of smoke and the existence of fire, and a systematic regularity between dark clouds in the sky and rain. Human beings and other creatures that are able to recognize those systematic regularities can use them in order to extract information. The person who sees dark clouds can take an umbrella to work, the animal that sees smoke on the horizon can take flight.

Notice that we are definitely talking about information in these examples, not what the information is about. For example, people or animals that see smoke do not necessarily see fire, but they nevertheless acquire the information that there is a fire. And the sight of dark clouds can provide the information that rain is on the way long before the first drop falls.

In general then, one way information can arise is by virtue of systematic regularities in the world. People (and certain animals) learn to recognize those regularities, either consciously or subconsciously, possibly as a result of repeated exposure to them. They may then utilize those regularities in order to obtain information from aspects of their environment.

What about the acquisition of information from books, newspapers, radio, etc., or from being spoken to by fellow humans? This too depends on systematic regularities. In this case, however, those regularities are not natural in origin like dark clouds and rain, or smoke and fire. Rather they depend on regularities created by people, the regularities of human language.

In order to acquire information from the words and sentences of English, you have to understand English — you need to know the meanings of the English words and you need a working knowledge of the rules of English grammar. In addition, in the case of written English, you need to know how to read — you need to know the conventions whereby certain sequences of symbols denote certain words. Those conventions of word meaning, grammar, and symbol representation are just that: conventions. Different countries have different conventions: different rules of grammar, different words for the same thing, different alphabets, even different directions of reading — left to right, right to left, top to bottom, or bottom to top.

At an even more local level, there are the conventional information encoding devices that communities establish on an ad hoc basis. For example, a school may designate a bell ring as providing the information that the class should end, or a factory may use a whistle to signal that the shift is over.

The fact is, anything can be used to store information. All it takes to store information by means of some object — or more generally a configuration of objects — is a convention that such a configuration represents that information. In the case of information stored by people, the conventions range from ones adopted by an entire nation (such as languages) to those adopted by a single person (such as a knotted handkerchief). For a non-human example, DNA encodes the information required to create a lifeform (in an appropriate environment).

People also have the ability to obtain information from a configuration of objects in a particular context. An example is a hotel key rack. The original purpose of the key rack is to store keys. However, because it is commonly understood that each room in a hotel has a key, the number of keys on the key rack gives information about the size of the hotel. Because the traditional key racks were also used to store passports, messages, bills, a glance at the key rack can result in obtaining information about guests who are in their rooms, who have just checked in, or who are about to leave. In this respect, an object such as a key rack can be said to carry information because of the way it is used by a community of people who share experience of hotels — hotel employees, guests, visitors and others.

For a more modern example, a management consultancy today employs ever increasing number of mobile workers. Since the consultants travel a lot, information about their whereabouts is quite important. If, for example, A's mobile phone is on the charger rack, most of his colleagues will assume he is in the office. Otherwise, the mobile phone would not be there. The phone charger carries that information for people who understand work practices in the organisation and can make reasonably accurate assumptions about the meaning of their colleagues' actions. Information in this context is often related to knowledge and understanding — the phone charger is what is often called a "common artefact" that functions as a focus of interaction. It can only fulfil this function, however, if there is shared understanding of how it is used.

This is more than convention — the result of some kind of mutual agreement by a group of people that "table" will refer to an object with a flat surface and one or more legs. Information carrying capacity of common artefacts is more dynamic, as it arises from action and interaction whose significance is understood by a given community.

To make any progress in understanding information in a precise, scientific way, we need, first, to provide a precise, representation-free[5] definition of information, and, second, to examine the regularities, conventions, etc. whereby things in the world represent information. This is what two Stanford University researchers, Jon Barwise and John Perry, set out to do in the late 1970s and early 1980s. The mathematical framework they developed to do this they named Situation Theory, initially described in their book *Situations and Attitudes* [Barwise and Perry, 1983], with a more developed version of the theory subsequently presented by Devlin in [Devlin, 1991]. We shall provide a brief summary of part of situation theory in due course.[6]

One question that arises naturally in a study such as ours is whether information really exists. Perhaps talk of information is just that: so much twentieth and twenty-first century talk. This is a fascinating question, and one that we will touch on again at the end of the chapter. For the purposes of our discussion, however, we may sidestep the issue, and remain completely agnostic as to whether information has any kind of real existence. To do this, we can adopt what we shall call the *information stance.* This refers to the information-based way of thinking about (and analyzing) human action that we shall outline. When we adopt the information stance, we agree to talk *as if* information really exists and we approach human action and interaction in terms of the creation, acquisition, storage, transmission, exchange, sharing, and utilization of information. In adopting such an approach, we are taking our lead from the philosopher Daniel Dennett [Dennett, 1989], who sidestepped many thorny questions about intentionality by viewing it as a stance ("the intentional stance") that may be adopted for various purposes.

SITUATION THEORY

In situation theory, recognition is made of the partiality of information due to the finite, *situated* nature of the agent (human, animal, or machine) with limited cognitive resources. Any agent must employ necessarily limited information extracted from the environment in order to reason and communicate effectively.

The theory takes its name from the mathematical device introduced in order to take account of that partiality. A *situation* can be thought of as a limited part of reality. Such parts may have spatio-temporal extent, or they may be more abstract, such as fictional worlds, contexts of utterance, problem domains, mathematical structures, databases, or Unix directories. The distinction between situations and individuals is that situations have a *structure* that plays a significant role in the theory whereas individuals do not. Examples of situations of particular relevance

[5] Of course, our theoretical framework will have to have its own representations. The theory we will use adopts the standard application-domain-neutral representation used in science, namely mathematics.

[6] However, since situation theory is not the focus of this paper, our description will be very partial; we introduce just those situation-theoretic concepts and tools we require for our present purposes.

to the subject matter of this paper will arise as our development proceeds.

The basic ontology of situation theory consists of entities that a finite, cognitive agent individuates and/or discriminates as it makes its way in the world: spatial locations, temporal locations, individuals, finitary relations, situations, types, and a number of other, higher-order entities.

The objects (known as *uniformities*) in this ontology include the following:

- *individuals* — objects such as tables, chairs, tetrahedra, people, hands, fingers, etc. that the agent either individuates or at least discriminates (by its behavior) as single, essentially unitary items; usually denoted in situation theory by $a, b, c, \ldots$
- *relations* — uniformities individuated or discriminated by the agent that hold of, or link together specific numbers of, certain other uniformities; denoted by $P, Q, R, \ldots$
- spatial *locations*, denoted by $l, l', l'', l_0, l_1, l_2$, etc. These are not necessarily like the points of mathematical spaces (though they may be so), but can have spatial extension.
- *temporal locations*, denoted by $t, t', t_0, \ldots$. As with spatial locations, temporal locations may be either points in time or regions of time.
- *situations* — structured parts of the world (concrete or abstract) discriminated by (or perhaps individuated by) the agent; denoted by $s, s', s'', s_0, \ldots$
- *types* — higher order uniformities discriminated (and possibly individuated) by the agent; denoted by $S, T, U, V, \ldots$
- *parameters* — indeterminates that range over objects of the various types; denoted by $\dot{a}$, $\dot{s}$, $\dot{t}$, $\dot{l}$, etc.

The intuition behind this ontology is that in a study of the activity (both physical and cognitive) of a particular agent or species of agent, we notice that there are certain regularities or *uniformities* that the agent either individuates or else discriminates in its behavior.[7]

For instance, people individuate certain parts of reality as *objects* ('individuals' in our theory), and their behavior can vary in a systematic way according to spatial location, time, and the nature of the immediate environment ('situation types' in our theory).

We note that the ontology of situation theory allows for the fact that different people may discriminate differently. For instance, Russians discriminate as two different colors what Americans classify as merely different shades of blue.

[7]This is true not only of individuals but also of groups, teams, communities. If A and B are engaged in a dialogue or a conversation, or indeed any other form of joint action, they recognize uniformities as individuals in similar ways. Socially, they negotiate the precise meanings of these, so that they can agree the exact shape of the uniformities that apply in the situation they are in.

Information is always taken to be information *about* some situation, and is taken to be in the form of discrete items known as *infons.* These are of the form

$$\ll R, a_1, \ldots, a_n, 1 \gg \;, \; \ll R, a_1, \ldots, a_n, 0 \gg$$

where R is an n-place relation and $a_1, \ldots, a_n$ are objects appropriate for R (often including spatial and/or temporal locations). These may be thought of as the informational item that objects $a_1, \ldots, a_n$ do, respectively, do not, stand in the relation R.

Infons are items of information. They are not things that in themselves are true or false. Rather a particular item of information may be true or false *about a certain part of the world* (a situation).[8]

Given a situation, s, and an infon σ, we write

$$s \models \sigma$$

to indicate that the infon σ is made factual by the situation s, or, to put it another way, that σ is an item of information that is true of s. The official name for this relation is that *s supports σ.*

It should be noted that this approach treats information as a *commodity.* Moreover a commodity that does not have to be true. Indeed, for every positive infon there is a dual negative infon that can be thought of as the opposite informational item, and both of these cannot be true (in the same situation).

Over the years, several people have misunderstood the role of infons in situation theory, and more generally have misunderstood the purpose of the situation-theoretic ontology, so it is worth making a few remarks here. A fundamental assumption underlying the situation-theoretic approach to information is that information is not intrinsic to any signal or to any object or configuration of objects in the world; rather information arises from ineractions of agents with their environment (including interactions with other agents). The individuals, relations, types, etc. of the situation-theoretic ontology are (third-party) theorist's inventions. For an agent to carry out purposeful, rational activities, however, and even more so for two or more agents to communicate effectively, there must be a substantial agreement first between the way an agent carves up the world from one moment to another, and second between the uniformities of two communicating agents. For instance, if Alice says to Bob, "My car is dirty," and if this communicative act is successful, then the words Alice utters must mean effectively the same to both individuals. In order for a successful information flow to take place, it is not necessary that Alice and Bob share exactly the same concept of "car" or of "dirty," whatever it might mean (if anything) to have or to share an

[8]One of the advantages of the framework and notation provided by situation theory is that it allows us to express partial information about complex relations. For example, the relation *eat* presupposes agent, object, instrument, place, time, but much of this information can remain implicit, as in "I'm eating." This makes it possible to choose which aspect of the structure to emphasize in a given instance of interaction. This choice of emphasis also carries information in its own right, since it is recognised and interpreted as attitude or intent.

exact concept. Rather, what is required is that their two concepts of "car" and of "dirty" overlap sufficiently. The objects in the ontology of situation theory are intended to be theorist's idealized representatives — prototypes — of the common part of the extensions of individual agent's ontologies. In consequence, the infons are theoretical constructs that enable the theorist to analyze information flow. (In terms of our tennis-ball analogue of communication, the tennis ball — the infon — is a figment of the analyst's imagination, but one that facilitates a useful and meaningful analysis of a communicative act.)

Moving on now, situation theory provides various mechanisms for defining types. The two most basic methods are type-abstraction procedures for the construction of two kinds of types: situation-types and object-types.

Situation-types. Given a *SIT*-parameter, $\dot{s}$, and a compound infon σ, there is a corresponding *situation-type*

$$[\dot{s} \mid \dot{s} \models \sigma],$$

the *type* of situation in which σ obtains.

This process of obtaining a type from a parameter, $\dot{s}$, and a compound infon, σ, is known as *(situation-) type abstraction.*

For example,

$$[SIT_1 \mid SIT_1 \models \langle\langle \text{running}, \dot{p}, LOC_1, TIM_1, 1\rangle\rangle]$$

Object-types. These include the basic types *TIM, LOC, IND,* REL^n, *SIT, INF, TYP, PAR,* and *POL*, as well as the more fine-grained uniformities described below.

Object-types are determined over some initial situation.

Let s be a given situation. If $\dot{x}$ is a parameter and σ is some compound infon (in general involving $\dot{x}$), then there is a type

$$[\dot{x} \mid s \models \sigma],$$

the *type* of all those objects x to which $\dot{x}$ may be anchored in the situation s, for which the conditions imposed by σ obtain.

This process of obtaining a type $[\dot{x} \mid s \models \sigma]$ from a parameter, $\dot{x}$, a situation, s, and a compound infon, σ, is called *(object-) type abstraction.*

The situation s is known as the *grounding* situation for the type. In many instances, the grounding situation, s, is the world or the environment we live in (generally denoted by w).

For example, the *type* of all people could be denoted by

$$[IND_1 \mid w \models \langle\langle \text{person}, IND_1, \dot{l}_w, \dot{t}_{now}, 1\rangle\rangle]$$

Again, if s denotes Jon's environment (over a suitable time span), then

$$[\dot{e} \mid s \models \langle\langle \text{sees, Jon}, \dot{e}, LOC_1, TIM_1, 1\rangle\rangle]$$

denotes the type of all those situations Jon sees (within s).

This is a case of an object-type that is a type of situation.

This example is not the same as a *situation-type*. Situation-types classify situations according to their internal structure, whereas in the type

$$[\dot{e} \mid s \models \langle\langle \text{sees, Jon, } \dot{e}, LOC_1, TIM_1, 1 \rangle\rangle]$$

the situation is typed from the outside.

Types and the type abstraction procedures provide a mechanism for capturing the fundamental process whereby a cognitive agent classifies the world. Applying the distinction between situation types and object types to interaction phenomena, we may say that we all recognise that the relationship between situation-type *fire* and the situation-type *smoke* obtains only if both are in the same place at the same time. This is then a part of the shared knowledge among members of the same group or community that is often assumed and therefore rarely articulated. Situation theory offers a mechanism for articulating these assumptions by means of defined constraints. *Constraints* provide the situation theoretic mechanism that captures the way that agents make inferences and act in a rational fashion. Constraints are linkages between situation types. They may be natural laws, conventions, logical (i.e., analytic) rules, linguistic rules, empirical, law-like correspondences, etc.

For example, humans and other agents are familiar with the constraint:

Smoke means fire.

If S is the type of situations where there is smoke present, and S' is the type of situations where there is a fire, then an agent (e.g. a person) can pick up the information that there is a fire by observing that there is smoke (a type S situation) and being aware of, or *attuned to*, the constraint that links the two types of situation.

This constraint is denoted by

$$S \Rightarrow S'$$

(This is read as "*S involves S'*.")

Another example is provided by the constraint

FIRE *means fire.*

This constraint is written

$$S'' \Rightarrow S'$$

It links situations (of type S'') where someone yells the word FIRE to situations (of type S') where there is a fire.

Awareness of the constraint

FIRE *means fire*

involves knowing the meaning of the word FIRE and being familiar with the rules that govern the use of language.

The three types that occur in the above examples may be defined as follows:

$$S = [\dot{s} \mid \dot{s} \models \langle\langle \text{smokey}, \dot{t}, 1 \rangle\rangle]$$

$$S' = [\dot{s} \mid \dot{s} \models \langle\langle \text{firey}, \dot{t}, 1 \rangle\rangle]$$

$$S'' = [\dot{u} \mid \dot{u} \models \langle\langle \text{speaking}, \dot{a}, \dot{t}, 1 \rangle\rangle \wedge \langle\langle \text{utters}, \dot{a}, \textit{fire}, \dot{t}, 1 \rangle\rangle]$$

Notice that constraints link types, not situations. However, any particular instance where a constraint is utilized to make an inference or to govern/influence behavior will involve specific situations (of the relevant types). Constraints function by capturing various regularities across actual situations.

A constraint

$$C = [S \Rightarrow S']$$

allows an agent to make a logical inference, and hence facilitates information flow, as follows. First the agent must be able to discriminate the two types S and S'. Second, the agent must be aware of, or behaviorally attuned to, the constraint. Then, when the agent finds itself in a situation s of type S, it knows that there must be a situation s' of type S'. We may depict this diagrammatically as follows:

$$\begin{array}{ccc} S & \stackrel{C}{\Longrightarrow} & S' \\ s : S \uparrow & & \uparrow s' : S' \\ s & \stackrel{\exists}{\longrightarrow} & s' \end{array}$$

For example, suppose $S \Rightarrow S'$ represents the constraint *smoke means fire.* Agent $\mathcal{A}$ sees a situation s of type S. The constraint then enables $\mathcal{A}$ to conclude correctly that there must in fact be a fire, that is, there must be a situation s' of type S'. (For this example, the constraint $S \Rightarrow S'$ is most likely reflexive, in that the situation s' will be the same as the encountered situation s.)

A particularly important feature of this analysis is that it separates clearly the two very different kinds of entity that are crucial to the creation and transmission of information: one the one hand the abstract types and the constraints that link them, and on the other hand the actual situations in the world that the agent either encounters or whose existence it infers.

For further details of situation theory, the reader should consult [Devlin, 1991].

AN EXAMPLE OF HUMAN INTERACTION

In his seminal article [Sacks, 1972], published in 1972, the sociologist Harvey Sacks sought to illustrate the role played by social knowledge in our everyday use of language. He took the following two sentences from the beginning of a child's story

The baby cried. The mommy picked it up.

and examined the way these two sentences are normally understood, paying particular attention to the role played by social knowledge in our interpretation of the story.[9]

As Sacks observes, virtually every competent speaker of English understands this story the same way. In particular, we all hear it as referring to a very small human (though the word 'baby' has other meanings in everyday speech) and to that baby's mommy (though there is no genitive in the second sentence, and it is certainly consistent for the mommy to be some other child's mother). Moreover it is the baby that the mother picks up (though the 'it' in the second sentence could refer to some object other than the baby).

To continue, we are also likely to regard the second sentence as describing an action (the mommy picking up the baby) that follows, and is caused by, the action described by the first sentence (the baby crying), though there is no general rule to the effect the sentence order corresponds to temporal order or causality of events (though it often does so).

Moreover, we may form this interpretation without knowing what baby or what mommy is being talked of.

Why do we almost certainly, and without seeming to give the matter any thought, choose this particular interpretation? Sacks asks.

Having made all of his observations, Sacks explains [Sacks, 1972, p.332]:

> My reason for having gone through the observations I have so far made was to give you some sense, right off, of the fine power of a culture. It does not, so to speak, merely fill brains in roughly the same way, it fills them so that they are alike in fine detail. The sentences we are considering are after all rather minor, and yet all of you, or many of you, hear just what I said you heard, and many of us are quite unacquainted with each other. I am, then, dealing with something real and something finely powerful.

It is worth pausing at this point to emphasize our purpose in working through Sacks' example in some detail, as we shall do momentarily. After all, as Sacks himself notes, "the sentences we are considering are . . . rather minor." Yet, from the point of view of understanding the complexities of human interaction, the example embodies many of the key issues that arise. As Sacks himself observes, almost all of us understand the two sentences the same way. We do so despite the fact the practically none of that understanding is within the sentences themselves; it depends on our experience — what Sacks calls the 'fine power of a culture'.

One way to analyze the way the sentences are (normally) understood is to explicate the social relationships that are not overtly expressed. Sacks himself studied the semantic strategies people use in communication. He showed how

[9]We first discussed Sacks' example in our research monograph [Devlin and Rosenberg, 1996]. Much of the technical material in this article is taken from that monograph.

they may draw upon their knowledge of the social systems in order to arrive at shared interpretations of the actions they observe (or imagine, as in the case of the example of the child's story). His main concern was to explain how shared social norms make such actions intelligible and interpretable (cf. [Gumpertz and Hymes, 1972, p.327]).

An alternative approach — which is the one we shall adopt here — is to identify the informational and cognitive *structures* that lead to the understanding, in particular the relational structures where relations that apply in a given situation represent the regularities the agent discriminates. The underlying structural form is indicated by the diagram on page 697. Our analysis has two main components. To identify which types S and S' are used and identify which constraints C connect those types. Paralleling Sacks' analysis in our framework, we formulate *rules* that explicate how his "fine power of a culture" leads to the choice of types used to describe or understand the event or action. We use the type structure (i.e., the information-supporting structure) to explicate how that same "fine power of a culture" guides the interpretation in a structural way.

Because the example, even though it may seem mundane, encompasses all of the main elements of human interaction, either form of analysis will result in insights and methods that have wide applicability.

The importance of such studies goes beyond the internal goals of social science. For, the better our understanding of human action and interaction, the better we will be able to design information and communication technologies. For this particular application, structural analyses are particularly well suited, of course. Descriptive analyses were created to enhance understandimg, not to design technologies. To bring that understanding closer to design, we need to be able to use a different framework, which is what we explore here.

In our analysis of the example, we shall concentrate on both speaker and listener, as we seek to describe the mechanisms they invoke to achieve successful communication. One of the advantages that is gained by including the information flow as part of our study is that we are able to pull apart the speaker and listener actions, and track the manner in which the speaker invokes mechanisms that enable the listener to correctly interpret the utterance.[10]

We should note that our analysis assumes that the speaker's perspective has been determined. That is, we shall not, at this stage, ask ourselves *why* the speaker chooses the particular form of words; she does, an issue closely related to the question why we see things in a certain way, but rather shall use the framework of situation theory to track the way the speaker and listener cooperate in order for the communicative act to be successful.

By carrying out our analysis in terms of information flow inspired by the framework of situation theory, we will be able to achieve a level of granularity that is conceptually (and intellectually) closer (compared with standard descriptive analyses) to the concept of information that is the concern of those working with

[10] Just as a description of a tennis game can be given in terms of the individual actions of the two players in a way that is simply not possible for a wrestling match.

Information Technology, Information and Communication Technology, etc.

It is important to observe that, although Sacks' example concerns a linguistic event, his analysis (and the information-mediated alternative account we subsequently present here) is not a linguist's analysis — neither he nor we are doing a syntactic or semantic analysis. (In particular, we are not doing situation semantics, the application of situation theory that motivated the original development of situation theory by Barwise and Perry.) Our focus is on the *interaction* between the speaker and the listener and between the speaker and what he or she hears. We seek to highlight how information-mediated analysis can lead to the development (or uncovering) of information structure (more precisely the information-supporting structure).[11]

AN INFORMATION-BASED ANALYSIS OF THE SACKS EXAMPLE

In order to carry out our analysis, we need to introduce some situation-theoretic structures to represent the way that information flows from the speaker to the listener.

Reference to babies and mommies is captured in our framework by means of the types:

$$\text{`baby'} = T_{baby} = [\dot{p} \mid w \models \ll \text{baby}, \dot{p}, t_{now}, 1 \gg],$$

$$\text{`mommy'} = T_{mother} = [\dot{p} \mid w \models \ll \text{mother}, \dot{p}, t_{now}, 1 \gg],$$

where $\dot{p}$ is a parameter for a person. (In these type definitions, the situation w is "the world", by which we mean any situation big enough to include everything under discussion. It is purely a convenience to think of this situation as the world, thereby providing a fixed context for the type definitions.)

We observe (as did Sacks in his original analysis) that both babies and mommies have different aspects. For instance, a baby can be thought of as a young person or as a member of a family, and a mommy can be viewed in relation to a child or to a father. These aspects, which affect the choice of words speakers make and the way listeners interpret them, are captured in our framework by the hierarchical structure on types (types of types, types of types of types, etc.).

Let:

$$T_{family} = [\dot{e} \mid w \models \ll \text{family}, \dot{e}, t_{now}, 1 \gg],$$

$$T_{stage\text{-}of\text{-}life} = [\dot{e} \mid w \models \ll \text{stage-of-life}, \dot{e}, t_{now}, 1 \gg],$$

where $\dot{e}$ is a parameter for a type.

The activity of crying is closely bound to babies in the stage-of-life type, so when the listener hears the sentence "The baby cried" he will understand it in such a way that

[11]This information structure plays a role in our analysis somewhat parallel to, though very different from, the social structure of Sack's analysis.

(1) $T_{baby} : T_{stage\text{-}of\text{-}life}$.

That is to say, this item of information will be available to the listener as he processes the incoming utterance, and will influence the way the input is interpreted.

Since the reader may be familiar to uses of "types" in other disciplines (such as computer science), where they are generally rigid in nature, we should stress that in situation theory, any type will typically be a member of an entire stucture of types, and the applicability of a particular type may well depend upon two or more levels in the of-type hierarchy. For instance, the applicability of the type T_{baby} will be different when it is considered in the light of being in the type $T_{stage\text{-}of\text{-}life}$ as opposed to being in the type T_{family}. In the former case, individuals in the type T_{baby} will be typically and naturally associated with the activity of crying (type T_{crying}); in the latter case they will be typically and naturally associated with having a mother (2-type $T_{mother\text{-}of}$). (In situation-theoretic terms, these associations will be captured by constraints that link types. Those constraints are in general not universals, rather they may depend on, say, individual or cultural factors.) This particular distinction will play a significant role in the analysis that follows.

One immediate question concerns the use of the definite noun phrases 'the baby' and 'the mommy'. Use of the definite article generally entails uniqueness of the referent. In the case of the phrase 'the baby', where, as in the Sacks example, no baby has previously been introduced, one would normally expect this to be part of a more complex descriptive phrase, such as 'the baby of the duchess's maid', or 'the baby on last night's midnight movie'. So just what is it that enables the speaker to open an explanation with the sentence 'The baby cried'? It could be argued that an implicit suggestion for an answer lies in his later discussion of proper openings for 'stories', but this is a part of his article we do not consider here.

For a situation-theoretic analysis, there is no problem here. The situation theorist assumes that all communicative acts involve a *described situation*, that part of the world the act is *about*. Exactly how this described situation is determined varies very much from case to case. For example, the speaker may have witnessed, read about, or even imagined the event she describes. In the Sacks example, the speaker *imagines* a situation in which a baby cried and its mother picked it up. Let s denote that situation.[12]

The situation s will be such that it involves one and only one baby, otherwise the use of the phrase 'the baby' would not be appropriate. In starting a communicative act with the sentence 'The baby cried', the speaker is informing the listener that

[12]It does not affect the mechanics of our analysis whether you think of situations as objects in the speaker and listener's realm — possibly as things they are aware of — or purely as theorist's objects in an abstract ontology adopted to study interaction. All we need to know is that these situations are definite objects available to the theorist as part of a framework for looking at the world. In the case where situations are regarded purely as theorist's abstractions, s will *correspond* to some feature of the interaction—you can think of s as providing us with a *name* for that feature.

she is commencing a description of a situation, s, in which there is exactly one baby, call it b. (Whether or not b is a real individual in the world, or some fictional entity, depends on s. This does not affect the way our analysis proceeds, nor indeed the way people understand the utterance.)

The principal item of information about the described situation that is conveyed by the utterance of the first sentence 'The baby cried' is

$$s \models \ll \text{cries}, b, t_0, 1 \gg$$

where t_0 is the time, prior to the time of utterance, at which the crying took place. In words, in the situation s, the baby b was crying at the time t_0.

Notice that, in the absence of any additional information, the only means available to the listener to identify b is as the referent for the utterance of the phrase 'the baby'. The utterance of this phrase tells the listener two pertinent things about s and b:

(2) $b : T_{baby}$ (i.e. b is of type T_{baby})

where T_{baby} is the type of all babies, and

(3) b is the unique individual of this type in s.

Now let's consider what additional information is conveyed by the utterance of second sentence, 'The mommy picked it up.' Mention of both babies and mommies invokes the family type, T_{family}. This has the following structural components that are relevant to our analysis:

$M(x)$	the property of x being a mother
$B(x)$	the property of x being a baby
$M(x,y)$	the relation of x being the mother of y
T_{mother}	the type of being a mother
T_{baby}	the type of being a baby
$T_{mother\text{-}of}$	the 2-type that relates mothers to their offspring

In the type T_{family}, the type $T_{mother\text{-}of}$ acts as a fundamental one, with the types T_{mother} and T_{baby} being linked to, and potentially derivative on, that type. More precisely, the following structural constraints[13] are salient in the category T_{family}:

$$T_{mother} \Rightarrow \exists \dot{y} T_{mother\text{-}of}$$

$$T_{baby} \Rightarrow \exists \dot{x} T_{mother\text{-}of}$$

where

$$T_{mother} = [\dot{x}, \dot{y} \mid w \models \ll \text{mother-of}, \dot{x}, \dot{y}, t_{now}, 1 \gg].$$

[13]The notion of constraint used here extends that described in [Devlin, 1991].

What do these mean? Well, $T_{mother\text{-}of}$ is a 2-type, the type of all pairs of individuals x, y such that x is the mother of y (at the present time, in the world). The first of the above two constraints says that the type T_{mother} *involves* (or is *linked to*) the type $\exists \dot{y} T_{mother\text{-}of}$. This has the following consequence: in the case where $T_{mother} : T_{family}$ (i.e. T_{mother} is of type T_{family}) and $T_{baby} : T_{family}$, the following implications are salient:

(4) $p : T_{mother} \;\rightarrow\; \exists q\,(p, q : T_{mother\text{-}of})$

(5) $q : T_{baby} \;\rightarrow\; \exists p\,(p, q : T_{mother\text{-}of})$.

These two implications are not constraints. In fact they do not have any formal significance in situation theory. They are purely guides to the reader as to where this is all leading. (4) says that if p is of type T_{mother} (i.e. if p is a mother), then there is an individual q such that the pair p, q is of type $T_{mother\text{-}of}$ (i.e. such that p is the mother of q). The salience of this implication for an agent $\mathcal{A}$ has the consequence that, if $\mathcal{A}$ recognizes that p is a mother then $\mathcal{A}$ will, if possible, look for an individual q of which p is the mother. Analogously for (5).

To continue with our analysis, as in the case of 'the baby', in order for the speaker to make appropriate and informative use of the phrase 'the mommy', the described situation s must contain exactly one individual m who is a mother. In fact we can make a stronger claim: the individual m is the mother of the baby b referred to in the first sentence. For if m were the mother not of b but of some other baby, then the appropriate form of reference would be 'a mother', even in the case were m was the unique mother in s. We can describe the mechanism that produces this interpretation as follows.

Having heard the phrase 'the baby' in the first sentence and 'the mommy' in the second, the following two items of information are salient to the listener:

(6) $m : T_{mother}$

(7) m is the unique individual of this type in s.

In addition, we shall show that the following, third item of information is also salient:

(8) m is the mother of b.

Following the utterance of the first sentence, the listener's cognitive state is such that the type T_{baby} is of type $T_{stage\text{-}of\text{-}life}$. This type has categories that include T_{baby}, T_{child}, $T_{adolescent}$, T_{adult}, all of which have equal ontological status within the stage-of-life type, with none being derivative on any other. But as soon as the phrase 'the mommy' is heard, the combination of 'baby' and 'mommy' switches the emphasis from the type $T_{stage\text{-}of\text{-}life}$ to the type T_{family}, making salient the following propositions:

(9) $T_{baby} : T_{family}$.

(10) $T_{mommy} : T_{family}$.

In the T_{family} category, the various family relationships that bind a family together (and which therefore serve to give this type its status as a type) are more fundamental than the categories they give rise to. In particular, the types T_{baby} and T_{mother} are derivative on the type $T_{mother\text{-}of}$ that relates mothers to their babies.

Now, proposition (9) is the precondition for the salience of implication (5), namely

$$q : T_{baby} \rightarrow \exists p\,(p, q : T_{mother\text{-}of}).$$

Substituting the particular individual b for the variable q, we get

$$b : T_{baby} \rightarrow \exists p\,(p, b : T_{mother\text{-}of}).$$

But by (2), we know that

$$b : T_{baby}.$$

Thus we have the salient information

(11) there is an m such that $m, b : T_{mother\text{-}of}$.

The use of the definite article in the phrase 'the mommy' then makes it natural to take this phrase to refer to the unique m that satisfies (11). Thus the listener naturally takes the phrase 'the mommy' to refer to the baby's mother. This interpretation is reinforced by the completion of the second sentence '...picked it up', since there is an expectation that a mother picks up and comforts her crying baby. This explains how the fact (8) becomes salient to the listener.

It should be noticed that the switch from the salience of one set of constraints to another was caused by the second level of types in the hierarchy. The constraints we were primarily interested in concerned the types T_{mother} and T_{baby}. These types are part of a complex network of inter-relationships (constraints). Just which constraints in this network are salient to the agent is governed by the way the agent encounters the types, that is to say, by the type(s) of those types—for instance, whether T_{baby} is regarded (or encountered) as of type $T_{stage\text{-}of\text{-}life}$ or of type T_{family}. By moving to a second level of typing (i.e. to types of types), we are able to track the way agents may use one set of constraints rather than another, and switch from one set to another. The first level of types allows us to capture the informational connections between two objects; the second level allows us to capture the agent's preference of a particular informational connection. This level of uncertainty is needed or else there could be no negotiation in interaction.

Our analysis thus explicates the information *structure* that the speaker and listener implicitly make use of in order a communicative act to succeed. In particular, it highlights the crucial roles played not only by constraints (the key players in a situation semantic analysis) but also by the internal and hierarchical type-structures. This latter feature is quite new, and takes the analysis a considerable distance from situation semantics. We believe it is a significant tribute to the care

Barwise, Perry, and their colleagues gave to the choice of the ontology for situation theory as a framework to support the study of information and of natural language semantics that it proves to be adequate for a detailed analysis of human interaction such as the one presented here.[14]

AN EXAMPLE FROM INDUSTRY

The fundamental nature of the issues emmbodied in the Sacks example means that the methods we employed in our anaysis have much wider applicability. For instance, in the late 1980s and early 1990s, we analyzed what had gone wrong when a large manufacturer and supplier of mainframe computer systems had tried to automate part of its own information system, namely the data collected in the standard form (the Problem Report Form, or PRF) filled in when an engineer was called out on a repair job.

The PRF was a simple slot-and-filler document on which could be entered various reference numbers to identify the customer and the installed system, the fault as reported by the customer, the date of the report, the date of the engineer's visit, the repair action he took, and any components he replaced.

The PRF was a valuable document, providing the company with an excellent way to track the performance of both their computer systems and their field engineers, as well as the demand for spare parts. In particular, by analyzing the data supplied by the forms, the company could identify and hopefully rectify the weakest components in the design of their systems.

Because of the highly constrained nature of the PRFs, the highly focused nature of the domain — computer fault reporting and repair — and the fact that the important technical information on the forms was all entered by trained computer engineers, the PRFs formed the basis of a highly efficient source of information for all parts of the company. In the early days, when the PRFs were paper documents, experts faced with reading the forms frequently encountered great difficulty understanding exactly what had gone wrong with the customer's system and what the engineer had done to put it right. The PRF was a shared artefact – the focus of interaction between many departments: customer services, spare parts, diagnostics, etc. Information flowed naturally and any uncertainties were cleared up in conversation. When the PRF was computerised, it became an information record in a database, and the information flow between people was interrupted. If a particular PRF had (what company employees referred to as) "good information" in it, it could be easily interpreted and understood well enough to lead to action. If it contained "bad information", it presented a problem.

When an expert system was introduced, the expectation was that it would introduce intelligence into the interrupted information flow, so that the PRF could continue to function as mediated by the expert system. But this did not happen.

[14]Both Barwise and Perry expressed on many occasions a desire to extend their work to look at action and interaction, but they never made such a step.

Things got worse; the information flow was disrupted. Different people (agents) had different perspectives on the information in the PRF. The database representation of this information did not allow for different perspectives, it only encoded what the database designer specified and in the form that the designer specified. There was therefore no flexibility that could allow individual perspectives to be recognized and negotiated, and for people to establish shared understanding.[15]

Applying extensions of the techniques used to analyze the Sacks example, we were able to carry out a detailed analysis of the way social and cultural knowledge affected the information conveyed by the PRFs. This led to a restructuring of the procedures surrounding the completion and use of the documents, resulting in better information flow and improved efficiency in the company.

Furthermore, the additional problem our analysis addressed was to relate the structure of the document to its broader uses in the organisation as a whole. We viewed the PRF as a resource that facilitates (or obstructs, as the case may be) the interaction between different sections of the organisation. In this context, the social significance of the document needs to be understood so that the information flow between different sections may be organised and managed.

An investigation into the uses of the document, as opposed to its structure, brought to light the need to develop a dual perspective — what we called the *document intension* and the *schema of investigation*. The document intension is an "information-structure-skeleton" of the PRF that captures the communicative intent of the various sections of the document, through the use of the constraints that formalize the informational links within the document (essentially its underlying type structure). The schema of investigation traces the information pathways a reader of the document creates in the process of interpretation, schematically presented in Figure 1.

The schema captures formally how the successive application of constraints leads to "perfect" information in the "scene", when everything fits — on the far left of the tree — and also to the "bad PRF" on the far right of the tree. These examples illustrate the strategies that computerized resources capture easily.

However, most of the everyday cases analyzed were not so clear cut. Going from left to right in the tree in Figure 1, if the fault description is clear and the remedial action is not, this would be interpreted as the engineer not knowing his job. Needless to say, no PRF among the hundreds analyzed gave this information explicitly. The most frequent and the most challenging examples were those in the middle of the tree, where the fit had to be established between the fault description, the appropriate remedial action and the resources used in implementing the remedy. This is where most of human interrpretive effort was focused. Sadly, this is also where computerized tools are still grossly inadequate as they are not responsive to the human uses of the information stored in them.

An empirical study of the uses of the PRF in the organization showed that the information contained in the document was needed to support the decisions of cus-

[15]Following Perry and Israel [Israel and Perry, 1990], we can say that a PRF had the information potential that agents could pick up, but the database could not.

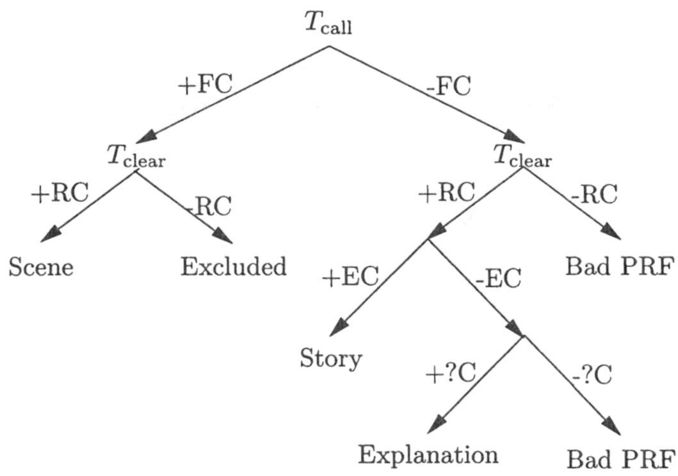

Figure 1. Interpretive grammar

tomer services, fault diagnosis, spare parts, change management, and production, as well as intra-organizational partnerships with the suppliers of various parts of the total system delivered to the customer.

We should stress that, even though both examples presented in this paper, the baby and the fault record, concern particular interactions in particular contexts, the analytic methods used to analyze them, described above, are able to capture the underlying regularities, or uniformities, and hence can be generally applied. This is where the information-based, structural approach can offer advantages over purely descriptive analyses. If our interest were solely the understanding of human action and interaction, that advantage might be of little consequence. It can become significant, however, if we are interested in the design of tools and resources that embody this understanding, and in the organisation of work that recognises the importance of social relationships in everyday practice.

For further details on the PRF example, we refer the reader to our monograph [Devlin and Rosenberg, 1996].

THE UTILITY AND RELEVANCE OF THE INFORMATION STANCE

In our analysis of the Sacks example, we showed how a communicative interaction can be analyzed in terms of information flow, using the framework of situation theory. What makes viewers see a scene the way they do, and why do they choose the precise form of words they use to convey information about that scene? Information may be regarded as (and arguably is, if it is anything at all beyond a manner of speaking) an inert commodity exchanged by the two participants in a linguistic interaction. Hence, adopting the information stance allows us to tease apart the two individuals in a communicative interaction. This allows us to analyze the actions of each with respect to the information transmitted or received, thereby shedding light on some of the intricacies of everyday human–human interaction.

The price that we might pay for this increased level of analytic precision is twofold. First, for some interactions, viewing the interaction as mediated may impede or even skew the analysis. (To go back to an earlier analogy, it would be possible to analyze a wrestling match in terms of some postulated "particles of force" that the two protagonists emit toward one another, but this is unlikely to lead to a better analysis, and in fact will probably obscure the interaction.) Second, it is at least arguable that information simply does not exist — that it is just a way of talking — and that it is more intellectually honest to stick to what is really there.

This last point might be a significant objection if the results of an information-based analysis could be presented only in terms of information. However, when it is possible to adopt the information stance for the purposes of carrying out an analysis, and then present the conclusions without reference to information, as we could with our examination of the Sacks example, that objection surely melts away.

There remains the question as to whether information really does exist. If the matter were to be settled by popular vote, the answer would surely be a resounding "Yes." Indeed, we suspect it would be almost unanimous. This is clearly a significant observation for the relevance of information-based analyses of social phenomena. If information is universally accepted in society, then it is legitimate to analyze social phenomena in terms of information. The results of that analysis (*presented in terms of information*) may then be legitimately presented as conclusions relevant to social science concerns.

In other words, an analysis of a social phenomenon based on information has validity in and of itself. It need not defend itself by an appeal to the information stance. (Although that remains a valid methodological approach.) Given a socially accepted notion of information that appears reasonably stable, an analysis like our study of the Sacks example we presented above turns out to be more than just a "what if" argument, where the intermediate steps are mere ephemera to be discarded once the final conclusion is reached. Rather, each step in the analysis establishes a genuine truth about the world — a truth about the information that flows from one agent to another. Viewed in this way, such an information-based analysis of human action is both valid and genuinely, qualitatively different from other forms of sociological analysis. That difference can be of significance when it comes to applications. A particular strength of the information-based approach, based on a mathematical framework such as situation theory, is that it allows for a formal treatment that can be informed by insights from sociology while at the same time yielding an analysis *that can be applied to the design of information and communication technologies.* This, in fact, was the reason we developed our analytic technique and carried out our analysis of the Sacks example in the first place. The work desribed here represents a genuine, original, and on this occasion highly successful, application of the modern concept of information to the development of understanding of a certain domain of human activity leading to a successful engineering design.

We end on a more speculative note. Today's concept of information assumes — and encourages us to believe — that information has some form of existence beyond its various physical encodings, each of which is often viewed as a container. Futurist commentator John Perry Barlow, co-founder of the Electronic Frontier Foundation is quoted[16] as having said:

> "So far we have placed all of our intellectual protection on the containers and not on the contents. And one of the side effects of digital technology is that it makes those containers irrelevant. Books, CDs, filmstrips — whatever — don't need to exist anymore in order to get ideas out. So whereas we thought we had been in the wine business, suddenly we realized that all along we've been in the bottling business."

The suggestion is that today's digital technologies will completely separate the information from its various representations, which are seen as containers. "Information wants to be free" is a popular rallying cry. But perhaps — and this is definitely where our sympathies lie — social scientist Paul Duguid had it right when he observed that such talk is akin to saying we want to remove the banks and still have the river.

For all that (today's conception of) information has a form of existence, we lean toward the view that what that existence really amounts to is a collective acceptance of the information stance. That is to say, it really is just a way of conceiving of and talking about various aspects of our world.

BIBLIOGRAPHY

[Barwise and Perry, 1983] J. Barwise and J. Perry. *Situations and Attitudes*, Bradford Books, MIT Press, 1983.

[Clark, 1996] H. Clark. *Using Language*, Cambridge University Press, 1996.

[Dennett, 1989] D. Dennett. *The Intentional Stance*, MIT Press, 1989.

[Devlin, 1991] K. Devlin. *Logic and Information*, Cambridge University Press, 1991.

[Devlin and Rosenberg, 1996] K. Devlin and D. Rosenberg. *Language at Work: Analyzing Communication Breakdown in the Workplace to Inform Systems Design*, Stanford University: CSLI Publications and Cambridge University Press, 1996.

[Gumpertz and Hymes, 1972] J. Gumpertz and D. Hymes, eds. *Directions in Sociolinguistics, The Ethnography of Communication*, Holt, Rinehart and Winston Inc., 1972.

[Nunberg, 1996] G. Nunberg, ed. *The Future of the Book*, University of California Press, 1996.

[Israel and Perry, 1990] D. Israel and J. Perry. What is Information?, in *Information, Language and Cognition: Vancouver Studies in Cognitive Science, Vol. I*, University of British Columbia Press, 1990.

[Rosenberg and Hutchinson, 1994] D. Rosenberg and C. Hutchinson, eds. *Design Issues in Computer-Supported Cooperative Work*, London: Springer-Verlag, 1994.

[Sacks, 1972] H. Sacks. On the Analyzability of Stories by Children, in [Gumpertz and Hymes, 1972, pp. 325–345].

[16]In C. W. Beardsley, Uncorking Fine Wine: Copyright Laws, *Mechanical Engineering*, August 1994.

THE PHILOSOPHY OF AI AND THE AI OF PHILOSOPHY

John McCarthy

1 INTRODUCTION

Richmond Thomason [2003] wrote

> The relations between AI and philosophical logic are part of a larger story. It is hard to find a major philosophical theme that doesn't become entangled with issues having to do with reasoning. Implicatures, for instance, have to correspond to inferences that can be carried out by a rational interpreter of discourse. Whatever causality is, causal relations should be inferable in everyday common sense settings. Whatever belief is, it should be possible for rational agents to make plausible inferences about the beliefs of other agents. The goals and standing constraints that inform a rational agent's behavior must permit the formation of reasonable plans.

The relation of AI and philosophy involves many concepts that both subjects include—for example, action, goals, knowledge, belief, and consciousness. However, AI takes what we may call the *designer stance* about these concepts; it asks what kinds of knowledge, belief, consciousness, etc. does a computer system need in order to behave intelligently and how to build them into a computer program. Philosophers have generally taken a more abstract view and asked what are knowledge, etc. The *designer stance* is akin to Daniel Dennett's *design stance*[Dennett, 1978] but not the same. The design stance looks at an existing artifact or organism in terms of what it is designed to do or has evolved to do. The designer stance considers how to design an artifact. This may necessitate giving it knowledge, beliefs, etc., and the ability to plan and execute plans.

Philosophical questions are especially relevant to AI when human-level AI is sought. However, most AI research since the 1970s is not aimed towards human-level AI but at the application of AI theories and techniques to particular problems.

I have to admit dissatisfaction with the lack of ambition displayed by most of my fellow AI researchers. Many useful and interesting programs are written without use of concepts common to AI and philosophy. For example, the language used by

Handbook of the Philosophy of Science. Volume 8: Philosophy of Information
Volume editors: Pieter Adriaans and Johan van Benthem. General editors: Dov M. Gabbay, Paul Thagard and John Woods.

the Deep Blue program that defeated world chess champion Garry Kasparov cannot be used to express "I am a chess program, but consider many more irrelevant moves than a human does." and draw conclusions from it. The designers of the program did not see a need for this capability. Likewise none of the programs that competed in the DARPA Grand Challenge contest to drive a vehicle knew that it was one of 20 competing programs. The DARPA referees prevented the vehicles from seeing each other by making them pause when necessary. A more advanced contest in which one vehicle can pass another might need some awareness of "other minds".

The 1950s AI researchers did think about human-level intelligence. Alan Turing, who pioneered AI, was also the first to emphasize that AI would be realized by computer programs. Now there is more interest in human-level AI and methods to achieve it than in the last 40 years.

[Nilsson, 2005] offers a criterion for telling when for human-level AI has been reached. It is that the system should be teachable to do a wide variety of jobs that humans do—in particular that it should be able to pass the examinations used to select people for these jobs, admitting that passing the exams may be possible without having adequate common sense to do the job. Nilsson is not specific about what kind of teaching is involved, and his criterion is weaker than Lenat's requirement that the system be able to learn from textbooks written for humans. I agree that this is one of the requirements for human-level AI.

[McCarthy, 1996a] also discusses criteria for human-level AI, emphasizing the common sense informatic situation.

Even as the work aimed at human-level AI increases, important methodological differences between AI research and philosophical research are likely to remain. Consider the notion of belief. Philosophers consider belief in general. AI research is likely to continue with systems with very limited beliefs and build up from there. Perhaps these are top-down and bottom-up approaches.

We will discuss several of the concepts common to AI and philosophy in connection with the following example.

A policeman stops a car and says,

> "I'm giving you a ticket for reckless driving. If another car had come over the hill when you passed that BMW, there would have been a head-on collision."

Notice that the example involves a counterfactual conditional "if you had passed ..." with a non-counterfactual consequence "... reckless driving." Less obviously perhaps, a system understanding the sentence must jump into a suitable context and reason within that context, using concepts meaningful in the context. Thus a particular hypothetical head-on collision is in question, not, for example, statistics about how often a head-on collision is fatal.

The philosophy of X, where X is a science, often involves philosophers analyzing the concepts of X and commenting on what concepts are or are not likely to be coherent. AI necessarily shares many concepts with philosophy, e.g. action, consciousness, epistemology (what it is sensible to say about the world), and even free will.

This article treats the philosophy of AI, but section 6 reverses the usual course and analyzes some basic concepts of philosophy from the standpoint of AI. The philosophy of X often involves advice to practitioners of X about what they can and cannot do. Section 6 reverses the usual course and offers advice to philosophers, especially philosophers of mind. One point is that philosophical theories can make sense for us only if they don't preclude human-level artificial systems. Philosophical theories are most useful if they take the *designer stance* and offer suggestions as to what features to put in intelligent systems.

Philosophy of mind studies mind as a phenomenon and studies how thinking, knowledge, and consciousness can be related to the material world. AI is concerned with designing computer programs that think and act. This leads to some different approaches to problems considered in philosophy, and we will argue that it adds new considerations or at least different emphases that philosophers should consider. I take the opportunity of this Handbook to present some ideas and formalisms rather brashly.

Some of the formalisms, e.g. nonmonotonic reasoning and situation calculus, are heavily used in AI systems. Others have not yet been used in computer programs, but I think the problems they address will be important for human-level AI.

2 SOME HISTORICAL REMARKS

Although there were some precursors, serious AI work began in the early 1950s when it became apparent that electronics was advanced enough to do universal computation. Alan Turing recognized in [Turing, 1947] that programming general purpose computers was better than building special purpose machines. This approach depended on AI researchers having access to computers, marginal in the early 50s but nearly universal by the late 1950s.[1]

The 1956 Dartmouth workshop, whose 1955 proposal introduced the term *artificial intelligence* triggered AI as a named field.[2]

My [McCarthy, 1959] triggered work in logical AI, i.e. using mathematical logical languages and reasoning to represent common sense. Progress in logical AI has been continuous, but is still far from human-level.

The Ernst-Newell-Simon *General Problem Solver* (GPS) [Ernst and Newell, 1969] was based on the idea that problem solving could be put in the form of starting with an initial expression and transforming it by a sequence of applications of given rules into a goal expression. Alas, this was an inadequate idea for problem solving in general.

[1] I began thinking about AI in 1948, but my access to computers began in 1955. This converted me to Turing's opinion.

[2] Newell and Simon, who got started first, and who had definite results to present at Dartmouth, used the term *complex information processing* for some years which didn't do justice to their own work.

The first chess programs were written in the 1950s and reached world champion level in the late 90s, through a combination of heuristics and faster computers. Unfortunately, the ideas adequate for champion level chess are inadequate for games like *go* that branch more than chess and which require recognition of parts of a situation.

Marvin Minsky [1963] summarized the ideas available in 1963.

McCarthy and Hayes [1969] got the situation calculus formalism to a large AI audience.

Pat Hayes [1979; 1985] advanced a set of ideas that proved influential in subsequent AI research

David Marr [1982] influenced much work in computer vision with its idea of the 2 1/2 dimensional representation.

The Stanford Artificial Intelligence Laboratory introduced the first robotic arms controlled by programs with input from TV cameras. [Moravec, 1977] described a cart with a TV camera controlled by radio from a time-shared computer.

I will not go much beyond the 1960s in describing AI research in general, because my own interests became too specialized to do the work justice.

3 PHILOSOPHICAL PRESUPPOSITIONS OF AI

That it should be possible to make machines as intelligent as humans involves some philosophical premises, although the possibility is probably accepted by a majority of philosophers. The way we propose to build intelligent machines makes more presuppositions, some of which are likely to be controversial.

This section is somewhat dogmatic, because it doesn't offer detailed arguments for its contentions and doesn't discuss other philosophical points of view except by way of making contrasts.

Our way is called *logical AI*, and involves expressing knowledge in a computer in logical languages and reasoning by logical inference, including nonmonotonic inference. The other main approach to AI involves studying and imitating human neurophysiology. It may also work.

Here are our candidate philosophical presuppositions of logical AI. They are most important for research aimed at human-level AI. There are a lot of them. However, much present AI is too limited in its objectives for it to be important to get the philosophy right.

objective world The world exists independently of humans. The facts of mathematics and physical science are independent of there being people to know them. Intelligent Martians and robots will need to know the same facts as humans.

A robot also needs to believe that the world exists independently of itself and that it cannot learn all about the world. Science tells us that humans evolved in a world which formerly did not contain humans. Given this, it is odd to regard the world as a human construct from sense data. It is even more odd to program a robot to regard the world as its own construct. What the robot believes about the world in general doesn't arise for the limited robots of today, because the languages they are programmed to use can't express assertions about the world in general. This limits what they can learn or can be told—and hence what we can get them to do for us.[3]

In the example, neither the driver nor the policeman will have any problems with the existence of the objective world. Neither should a robot driver or policeman.

correspondence theory of truth A logical robot represents what it *believes* about the world by logical sentences. Some of these beliefs we build in; others come from its observations and still others by induction from its experience. Within the sentences, it uses *terms* to refer to objects in the world.

In every case, we try to design it so that what it will believe about the world is as accurate as possible, though not usually as detailed as possible. Debugging and improving the robot includes detecting false beliefs about the world and changing the way it acquires information to maximize the correspondence between what it believes and the facts of the world.

correspondence theory of reference AI also needs a *correspondence theory of reference* , i.e. that a mental structure can refer to an external object and can be judged by the accuracy of the reference. The terms the robot uses to refer to entities need to correspond to the entities so that the sentences will express facts about these entities. We have in mind both material objects and other entities, e.g. a plan or the electronic structure of the helium atom. The simple case of verification of correspondence of reference is when a robot is asked to pick up block $B3$, and it then picks up that block and not some other block.

As with science, a robot's theories are tested experimentally, but the concepts robots use are hardly ever defined in terms of experiments. Their properties are partially axiomatized, and some axioms relate terms representing concepts to objects in the world via observations.

A robot policeman would need debugging if it thought a car was going 20 mph when it was really going 75 mph. It would also need debugging if its internal visual memory highlighted a cow when it should have highlighted a particular car.

A correspondence theory of reference will necessarily be more elaborate than a theory of truth, because terms refer to objects in the world or to objects in semantic interpretations, whereas sentences refer to truth values. Alas, real

[3]Physics, chemistry, and biology have long been at a level where it more feasible to understand sensation in terms of science than to carry out the project of [Russell, 1914] of constructing science in terms of sensation. The justification of common sense and scientific knowledge is in terms of the whole scientific picture of human sensation and its relation to the world rather than as a construction from sensation.

world theories of reference haven't been much studied. Cognitive scientists and allied philosophers refer to *the symbol grounding problem*, but I'm not sure what they mean.

reality and appearance The important consequence of the correspondence theory is the need to keep in mind the relation between *appearance*, the information coming through the robot's sensors, and *reality*. Only in certain simple cases, e.g. when a program plays chess with typed in moves, does the robot have sufficient access to reality for this distinction to be ignored. A physical robot that played chess by looking at the board and moving pieces would operate on two levels—the abstract level, using (say) algebraic notation for positions and moves, and a concrete level in which a piece on a square has a particular shape, location, and orientation, the latter necessary to recognize an opponent's move and to make its own move on the board. Its vision system would have to compute algebraic representations of positions from TV images.

It is an accident of evolution that unlike bats, we do not have an ultra-sonic sense that would give information about the internal structure of objects.

As common sense and science tell us, the world is three dimensional, and objects usually have complex internal structures. What senses humans and animals have are accidents of evolution. We don't have immediate access to the internal structures of objects or how they are built from atoms and molecules. Our senses and reasoning tell us about objects in the world in complex ways.

Some robots react directly to their inputs without memory or inferences. It is our scientific (i.e. not philosophical) contention that these are inadequate for human-level intelligence, because a robot needs to reason about too many important entities that cannot be fully observed directly.

A robot that reasons about the acquisition of information must itself be aware of these relations. In order that a robot should not always believe what it sees with its own eyes, it must distinguish between appearance and reality.

A robot policeman would also need to be skeptical about whether what it remembered having seen (appearance) corresponded to reality.

third person point of view We ask "How does it (or he) know?", "What does it perceive?" rather than how do I know and what do I perceive. This is compatible with correspondence theories of truth and reference. It applies to how we look at robots, but also to how we want robots to reason about the knowledge of people and other robots.

The interaction between the driver and the policeman involves each reasoning about the other's knowledge.

science Science is substantially correct in what it tells us about the world, and scientific activity is the best way to obtain more knowledge. 20th century corrections to previous scientific knowledge mostly left the old theories as good approximations to reality. Since science separated from philosophy (say at the time of Galileo), scientific theories have been more reliable than philosophy as a source of knowledge.

The policeman typically relies on his radar, although he is unlikely to know much of the science behind it.

mind and brain The human mind is an activity of the human brain. This is a scientific proposition, supported by all the evidence science has discovered so far. However, the dualist intuition of separation between mind and body is related to the fact that it is often important to think about action without acting. Dualist theories may have some use as psychological abstractions. In the case of a programmed robot, the separation between mind and brain (program and computer) can be made quite sharp.

common sense Common sense ways of perceiving the world and common opinion are also mostly correct. When general common sense errs, it can often be corrected by science, and the results of the correction may become part of common sense if they are not too mathematical. Thus common sense has absorbed the notion of inertia. However, its mathematical generalization, the law of conservation of momentum, has made its way into the common sense of only a small fraction of people—even among the people who have taken courses in physics. People who move to asteroids will need to build conservation of momentum and even angular momentum into their intuitions.

From Socrates on, philosophers have found many inadequacies in common sense usage, e.g. common sense notions of the meanings of words. The corrections are often elaborations, making distinctions blurred in common sense usage. Unfortunately, there is no end to possible elaboration of many concepts, and the theories become very complex. However, some of the elaborations seem essential to avoid confusion in some circumstances.

Robots will need both the simplest common sense usages and to be able to tolerate elaborations when required. For this we have proposed three notions—contexts as formal objects [McCarthy, 1993b] and [McCarthy and Buvač, 1997], *elaboration tolerance* [McCarthy, 1999b], and *approximate objects.* [McCarthy, 2000][4]

---

[4]Hilary Putnam [Putnam, 1975] discusses two notions concerning meaning proposed by previous philosophers which he finds inadequate. These are

> (I) That knowing the meaning of a term is just a matter of being in a certain "psychological state" (in the sense of "psychological state" in which states of memory and psychological dispositions are "psychological states"; no one thought that knowing the meaning of a word was a continuous state of consciousness, of course.)
>
> (II) That the meaning of a term (in the sense of "intension") determines its extension

science embedded in common sense Science is embedded in common sense. Galileo taught us that the distance s that a dropped body falls in time t is given by the formula

$$s = \frac{1}{2}gt^2.$$

To use this information, the English or Italian (or their logical equivalent) are just as essential as the formula, and common sense knowledge of the world is required to make the measurements required to use or verify the formula.

common sense expressible in mathematical logic Common sense knowledge and reasoning are expressible as logical formulas and logical reasoning. Some extensions to present mathematical logic are needed.

possibility of AI According to some philosophers' views, artificial intelligence is either a contradiction in terms [Searle, 1984] or intrinsically impossible [Dreyfus, 1992] or [Penrose, 1994]. The methodological basis of these arguments has to be wrong and not just the arguments themselves.

mental qualities treated individually AI has to treat mind in terms of components rather than regarding mind as a unit that necessarily has all the mental features that occur in humans. Thus we design some very simple systems in terms of the beliefs we want them to have and debug them by identifying erroneous beliefs. Its systematic theory allows ascribing minimal beliefs to entities as simple as thermostats, analogously to including 0 and 1 in the number system. Thus a simple thermostat can have as its set of possible beliefs only that the room is too hot or that it is too cold. It does not have to know that it is a thermostat. This led to controversy with philosophers, e.g. John Searle, who think that beliefs can only be ascribed to systems with a large set of mental qualities. [McCarthy, 1979a] treats the thermostat example in detail.

rich ontology Our theories involve many kinds of entity—material objects, situations, properties as objects, contexts, propositions, individual concepts, wishes, intentions. Even when one kind A of entity can be defined in terms of others, we will often prefer to treat A separately, because we may later want to change our ideas of its relation to other entities.

(in the sense that sameness of intension entails sameness of extension).

Suppose Putnam is right in his criticism of the general correctness of (I) and (II). His own ideas are more elaborate.

It may be convenient for a robot to work mostly in contexts within a larger context C_{phil1} in which (I) and (II) (or something even simpler) hold. However, the same robot, if it is to have human level intelligence, must be able to *transcend* C_{phil1} when it has to work in contexts to which Putnam's criticisms of the assumptions of C_{phil1} apply.

It is interesting, but perhaps not necessary for AI at first, to characterize those circumstances in which (I) and (II) are correct.

AI has to consider several related concepts, where many philosophers advocate minimal ontologies. Suppose a man sees a dog. Is seeing a relation between the man and the dog or a relation between the man and an appearance of a dog? Some purport to refute calling seeing a relation between the man and the dog by pointing out that the man may actually see a hologram or picture of the dog. AI needs the relation between the man and the appearance of a dog, the relation between the man and the dog and also the relation between dogs and appearances of them. None need be regarded as most fundamental.

Both the driver and the policeman use enriched ontologies including concepts whose definition in terms of more basic concepts is unknown or even undefined. Thus both have a concept of a car not based on prior knowledge of its parts. The policeman has concepts of and names for offenses for which a ticket is appropriate and those requiring arrest.

natural kinds The entities the robot must refer to often are *rich* with properties the robot cannot know all about. The best example is a *natural kind* like a lemon. A child buying a lemon at a store knows enough properties of the lemons that occur in the stores he frequents to distinguish lemons from other fruits in that particular store. It is a convenience for the child that there isn't a continuum of fruits between lemons and oranges. Distinguishing hills from mountains gives more problems and disagreements. Experts know more properties of lemons than we laymen, but no-one knows all of them. AI systems also have to distinguish between sets of properties that suffice to recognize an object in particular kinds of situation and a general kind.

Curiously, many of the notions studied in philosophy are not natural kinds, e.g. proposition, meaning, necessity. When they are regarded as natural kinds, fruitless arguments about what they really are often take place. AI needs these notions but must be able to work with limited notions of them.

approximate entities Many common sense terms and propositions used successfully in conversation and writing cannot be given agreed-upon if-and-only-if definitions by the participants in a dialog. Examples include "x believes y", which has attracted much philosophical attention but also terms like "location(x)" which have not.

Some people have said that the use of computers requires terms to be defined precisely, but I don't agree. Many approximate entities will have to be considered by computer programs, internally and in communication. However, precision can often be achieved when terms and statements are interpreted in a context appropriate to a particular situation. In human usage, the context itself is not usually specified explicitly, and people understand each other, because the common context is implicit.

Our emphasis on the first class character of approximate entities may be new. It means that we can quantify over approximate entities and also express how an entity is approximate. [McCarthy, 2000] treats approximate entities and approximate theories.

The counterfactual "If another car had come over the hill when you passed ..." is very approximate. It is adequate for communication between the driver and the policeman, but attempts by them to define it more precisely would probably not agree.

There is some overlap between the discussion of approximate entities and philosophical discussions of vagueness. However, our point is the need for approximate entities in AI.

compatibility of determinism and free will A logical robot needs to consider its choices and the consequences of them. Therefore, it must regard itself as having (and indeed has) a kind of *free will* even though it is a deterministic device. In the example, a judge might be offered the excuse that the driver couldn't drop back after he started to pass, because someone was right behind him.

[McCarthy, 2005] formalizes a simple form of deterministic free will. A robot's or human's action sometimes has two stages. The first uses a nondeterministic theory, e.g. *situation calculus*, to compute a set of choices and their consequences and to evaluate the situations that result from performing the actions. The second stage chooses the action whose consequences are regarded as best. The sensation of free will is the situation at the end of the first stage. The choices are calculated, but the action isn't yet decided on or performed. This simple theory should be useful in itself but needs to be elaborated to take into account further aspects of human free will. The need is both philosophical and practical for robot design. One aspect of human free will that is probably unnecessary for robots is weakness of will.

mind-brain distinctions I'm not sure whether this point is philosophical or scientific. The mind corresponds somewhat to software, perhaps with an internal distinction between program and knowledge. Software won't do anything without hardware, but the hardware can be quite simple, e.g. a universal Turing machine or simple stored program computer. Some hardware configurations can run many different programs concurrently, i.e. there can be many minds in the same computer body. Software can also interpret other software.

Confusion about this is the basis of the Searle Chinese room fallacy [Searle, 1984]. The man in the hypothetical Chinese room is interpreting the software of a Chinese personality. Interpreting a program does not require having the knowledge possessed by that program. This would be obvious if people could interpret other personalities at a practical speed, but Chinese room software interpreted by an unaided human might run at 10^{-9} the speed of an actual Chinese.[5]

[5]If Searle would settle for an interaction at the level of Joseph Weizenbaum's [Weizenbaum,

Most AI work does not assume so much philosophy. For example, classifying scenes and other inputs need not assume that there is any reality behind the appearances being classified. However, ignoring reality behind appearance will not lead to human-level AI, and some short term AI goals have also suffered from incorrect, philosophical presumptions, almost always implicit.

Human-level AI also has scientific presuppositions.

4 SCIENTIFIC PRESUPPOSITIONS OF AI

Some of the premises of logical AI are scientific in the sense that they are subject to scientific verification or refutation. This may also be true of some of the premises listed above as philosophical.

innate knowledge The human brain has important innate knowledge, e.g. that the world includes three dimensional objects that usually persist even when not observed. This knowledge was learned by evolution. The existence of innate knowledge was not settled by philosophical analysis of the concept, but is being learned by psychological experiment and theorizing. Acquiring such knowledge by learning from sense data will be quite hard but possible.

Indeed it is worthwhile to build as much knowledge as possible into our robots. The CYC project of Douglas Lenat is an attempt to put a large amount of common sense knowledge into a database.

Identifying human innate knowledge has been the subject of recent psychological research. See [Spelke, 1994] and the discussion in [Pinker, 1997] and the references Pinker gives. In particular, babies and dogs know innately that there are permanent objects and look for them when they go out of sight. We'd better build that into our robots, as well as other innate knowledge psychologists identify. Evolution went to a lot of trouble to acquire knowledge that we needn't require robots to learn from experience. Maybe the childhood preference for natural kind concepts is something robots should have built in.

middle out Humans deal with middle-sized objects and develop our knowledge up and down from the middle. Formal theories of the world must also start from the middle where our experience informs us. Efforts to start from the most basic concepts, e.g. to make a basic ontology, are unlikely to succeed as well as starting in the middle. The ontology must be compatible with the fact that the basic entities in one's initial ontology are not the basic entities in the world. More basic entities, e.g. electrons and quarks, are known less well than the middle entities.

1965], a person could interpret the rules without computer aid—as Weizenbaum recently informed me.

logic level Allen Newell, who did not use logical AI, nevertheless proposed [Newell, 1993] that there was a level of analysis of human rationality that he called the *logic level* at which humans could be regarded as *doing what they thought would achieve their goals.* Many of the systems the Carnegie-Mellon group built, e.g. SOAR, were first designed at the logic level.

universality of intelligence Achieving goals in the world requires that an agent with limited knowledge, computational ability and ability to observe use certain methods. This is independent of whether the agent is human, Martian, or machine. For example, playing chess-like games effectively requires something like alpha-beta pruning.

universal expressiveness of logic This is a proposition analogous to the Turing thesis that Turing machines are computationally universal—anything that can be computed by any machine can be computed by a Turing machine. The *expressiveness thesis* is that anything that can be expressed, can be expressed in first order logic with a suitable collection of functions and predicates.

Some elaboration of the idea is required before it will be as clear as the Turing thesis. First order logic isn't the best way of expressing all that can be expressed any more than Turing machines are the best way of expressing computations. However, with set theory, as axiomatized in first order logic, whatever can be expressed in stronger systems can apparently also be expressed in first order logic.

Gödel's completeness theorem tells us that every sentence p true in all models of a set a of sentences can be deduced. However, nonmonotonic reasoning is needed and used by humans to get consequences true in simple models. Very likely, reflection principles are also needed.

We expect these philosophical and scientific presuppositions to become more important as AI begins to tackle human-level intelligence.

5 COMMON SENSE AND THE COMMON SENSE INFORMATIC SITUATION

The main obstacle to getting computer programs with human-level intelligence is that we don't understand yet how to give them human level common sense. Without common sense, no amount of computer power will give human-level intelligence. Once programs have common sense, improvements in computer power and algorithm design will be directly applicable to making them more intelligent. Understanding common sense is also key to solving many philosophical problems.

The logical AI and knowledge representation communities undertake to study the world and represent common sense knowledge by logical formulas. A competing approach is based on studying the brain and how common sense knowledge is represented in synapses and other neurological structures.

CYC [Lenat, 1995] is a knowledge base with several million common sense facts. Douglas Lenat [Matuszek and Lenat, 2005] has repeatedly emphasized that a key level of common sense will be reached when programs can learn from the Worldwide Web facts about science, history, current affairs, etc. The above cited 2005 paper says

> The original promise of the CYC project—to provide a basis of real world knowledge sufficient to support the sort of learning from language of which humans are capable—has not yet been fulfilled.

Notice the implication that the lack is common sense knowledge rather than the ability to parse English. I agree.

This section is an informal summary of various aspects of common sense. The key phenomenon for both AI and philosophy is what we call the *common sense informatic situation.*

What is common sense?

Common sense is a certain collection of knowledge, reasoning abilities, and perhaps other abilities.

In [McCarthy, 1959] I wrote that the computer programs that had been written up to 1958 lacked common sense. Common sense has proved to be a difficult phenomenon to understand, and the programs of 2005 also lack common sense or have common sense in *bounded informatic situations.* In the 1959 paper, I wrote "We shall therefore say that **a program has common sense if it automatically deduces for itself a sufficiently wide class of immediate consequences of anything it is told and what it already knows.**"

Programs with common sense à la [McCarthy, 1959] are still lacking, and, moreover, the ideas of that paper are not enough. Logical deduction is insufficient, and nonmonotonic reasoning is required. Common sense knowledge is also required.

Here's what I think is a more up-to-date formulation.

A program has common sense if it has sufficient common sense knowledge of the world and suitable inference methods to infer a sufficiently wide class of reasonable consequences of anything it is told and what it already knows. Moreover, many inferences that people consider obvious are not deduced. Some are made by mental simulation and some involve nonmonotonic reasoning.

Requiring some intelligence as part of the idea of common sense gives another formulation.

A program has common sense if it can act effectively in the *common sense informatic situation*, using the available information to achieve its goals.

A program that decides what to do has certain information built in, gets other information from its inputs or observations; still other information is generated by reasoning. Thus it is in a certain *informatic situation.* If the information that has to be used has a common sense character, it will be in what we call the *common sense informatic situation.*

We need to contrast the general *common sense informatic situation* with less general *bounded informatic situations.* The latter are more familiar in science and probably in philosophy.

5.1 Bounded informatic situations

Current (2006) science and technology requires that to write a computer program in some area, construct a database, or even write a formal theory, one has to bound the set of concepts taken into account.

Present formal theories in mathematics and the physical sciences deal with *bounded informatic situations.* A scientist decides informally in advance what phenomena to take into account. For example, much celestial mechanics is done within the Newtonian gravitational theory and does not take into account possible additional effects such as outgassing from a comet or electromagnetic forces exerted by the solar wind. If more phenomena are to be considered, scientists must make new theories—and of course they do.

Likewise present AI formalisms work only in bounded informatic situations. What phenomena to take into account is decided by a person before the formal theory is constructed. With such restrictions, much of the reasoning can be monotonic, but such systems cannot reach human-level ability. For that, the machine will have to decide for itself what information is relevant, and that reasoning will inevitably be partly nonmonotonic.

One example is the simple "blocks world" much studied in AI where the position of a block x is entirely characterized by a sentence $At(x, l)$ or $On(x, y)$, where l is a location or y is another block. The language does not permit saying that one block is partly on another. Moreover, using $On(x, y)$ does not require a previous analysis of the meaning of the word "on" or the concept it represents. Only certain simple axioms are used. This works, because within the context of the kind of simple block stacking program being built, one block is definitely on or not on another, assuming the program never makes the robot put a block in an ambiguous position. Patrick Winston extended the blocks world to allow a block to be supported by two others and discussed structures like arches. See [Winston, 1977].

Another example is the MYCIN [Davis et al. , 1977] expert system in which the ontology (objects considered) includes diseases, symptoms, and drugs, but not patients (there is only one), doctors or events occurring in time. Thus MYCIN cannot be told that the previous patient with the same symptoms died. See [McCarthy, 1983] for more comment on MYCIN.

Systems in a bounded informatic situation are redesigned from the outside when the set of phenomena they take into account is inadequate. However, there is no-one to redesign a human from the outside, so a human has to be able to take new phenomena into account. A human-level AI system needs the same ability to take new phenomena into account.

In general a thinking human is in what we call the *common sense informatic situation*. The known facts are necessarily incomplete. [6]

5.2 The general common sense informatic situation

By the *informatic situation* of an animal, person or computer program, I mean the kinds of information available to it and the reasoning methods available to it. The *common sense informatic situation* is that of a human with ordinary abilities to observe, ordinary innate knowledge, and ordinary ability to reason, especially about the consequences of events that might occur including the consequences of actions it might take. Specialized information, like science and about human institutions such as law, can be learned and embedded in a person's common sense information. In spite of almost 50 years of effort, only modest progress has been made towards making computer systems with human-level common sense abilities. Much more progress has been made with specialized systems in bounded informatic situations.

No-one has a full understanding of what the common sense informatic situation is. I think understanding it is the single biggest problem for AI, and maybe for philosophy and cognitive science. However, it has at least the following features.

beliefs about actions and other events The policeman believes that one car passed another. His beliefs about the effects of events cause him to believe that if another car had come over the hill, there would have been a head-on collision.

elaboration tolerant theories The theory used by the agent is open to new facts and new phenomena. For example, the driver and the policeman could take possible fog into account, or the driver could claim that if another car had been coming he'd have seen the headlights reflected on a barn at the top of the hill. The cop's theory recommended that he reply, "Tell that to the judge."

Another example: A housewife shopping for dinner is at the butcher counter and thinks that her son coming on an airplane at that afternoon likes steak. She decides to check whether the airplane will be in on time. Suddenly a whole different area of common sense knowledge that is not part of the shopping-for-dinner script becomes relevant, i.e. the flight information number of the airline and how to get it if it isn't on her cell phone's telephone list. Section 6 has more on elaboration tolerance.

[6] As discussed in section 4, we live in a world of middle-sized objects which can only be partly observed. Science fiction and scientific and philosophical speculation have often indulged in the *Laplacean fantasy* of super-beings able to predict the future by knowing the positions and velocities of all the particles. That isn't the direction to speculate. More plausible super-beings would be better at using the information that is available to the senses—maybe having more and more sensitive senses, e.g. ultrasound, permitting seeing internal surfaces of objects. Nevertheless, their ability to predict the future and anticipate the consequences of actions they might choose would still be limited by chaotic processes.

incompletely known and incompletely defined entities The objects and other entities under consideration are incompletely known and are not fully characterized by what is known about them. The real cars of the driver and the policeman are incompletely known, and the hypothetical car that might have come over the hill is quite vague. It would not be appropriate for the driver to ask the policeman "What kind of car did you have in mind?" Most of the entities considered are intrinsically not even fully defined. The hypothetical car that might have come over the hill is ill-defined, but so are the actual cars.

nonmonotonic reasoning Elaboration tolerance imposes one requirement on the logic, and this is the ability to do *nonmonotonic reasoning.* The system must reach conclusions that further facts not contradicting the original facts are can alter. For example, when a bird is mentioned, one normally concludes that it can fly. Learning that it is a penguin changes this. There are two major formalisms for doing nonmonotonic reasoning, *circumscription* and *default logic.* Also Prolog programs do nonmonotonic inference when *negation as failure* is used.

Circumscription, [McCarthy, 1980], [McCarthy, 1986], and [Lifschitz, 1993], minimizes the extension of a predicate, keeping the extensions of some others fixed and allowing still others to be varied in achieving the minimum. Circumscription is the logical analog of the calculus of variations in mathematical analysis, but it doesn't so far have as elegant a theory. Here's a basic form of circumscription.

Let a be an axiom with the arguments p (to be minimized), z (which can be varied), and c (which is held constant). Then the circumscription of p, $Circum(a, p, z, c)$ is defined by

$$(1) \quad Circum[a, p, z, c] := a(p, z, c) \wedge (\forall p'\ z')(a(p', z', c) \rightarrow \neg p' < p),$$

where we have the definitions

$$(2) \quad \begin{array}{l} p' < p \equiv p' \leq p \wedge p' \neq p, \\ \text{and} \\ p' \leq p \equiv (\forall x)(p'(x) \rightarrow p(x)). \end{array}$$

Taking into account only some of the phenomena is a nonmonotonic reasoning step. It doesn't matter whether phenomena not taken into account are intentionally left out or if they are unknown to the reasoner.

While nonmonotonic reasoning is essential for both man and machine, it leads to error when an important fact is not taken into account. These are the errors most often noticed. [7]

[7] Here's an extended example from the history of science.

Starting in the middle of the 19th century, Lord Kelvin (William Thomson) undertook to set limits on the age of the earth. He had measurements of the rate of increase of temperature with

[Koons, 2005] contains a good discussion of various kinds of nonmonotonic reasoning.

reasoning in contexts and about contexts In the context of the Sherlock Holmes stories, Holmes is a detective and his mother's maiden name is undefined. In the context of U.S. legal history Holmes is a judge, and his mother's maiden name is Jackson. Bounded theories, usually have a fixed context.

An agent in the common sense informatic situation is often confronted with new contexts. Section 7 is devoted to information in and about contexts as well as relations between information in different contexts.

knowledge of physical objects There is increasing evidence from psychological experiments [Spelke, 1994] that babies have innate knowledge of physical objects and their permanence when they go out of sight. Any common sense system should have this built in. [McCarthy, 1996c], "The well-designed child" discusses what information about the world should be built into a robot.

composition of objects Consider an object composed of parts. It is convenient logically when what we knew about the parts and how they are put together enables us to determine the behavior of the compound object. Indeed this is often true in science and engineering and is often the goal of the search for a scientific theory. . Thus it is quite helpful that the properties of molecules follow from the properties of atoms and their interactions.

The common sense informatic situation is not so convenient logically. The properties of an object are often more readily available than the properties of the parts and their relations.

For example, a baseball has a visible and feelable surface, and we can see and feel the seams and can feel its compliance and its simplest heat transfer properties. We also know, from reading or from seeing a baseball disassembled, something about its innards. However, this knowledge of structure is less usable than the knowledge of the baseball as a whole.

---

depth and of the thermal conductivity of rock. He started with the assumption that the earth was originally molten and computed how long it would have taken for the earth to cool to its present temperature. He first estimated 98 million years and later reduced the estimate to 20–40 million years. This put him into conflict with geologists who already had greater estimates based on counting annual layers in sedimentary rock.

Kelvin's calculations were correct but gave the wrong answer, because no-one until Becquerel's discovery in 1896 knew about radioactive decay, the main source of energy that keeps the earth hot.

Kelvin's reasoning was nonmonotonic. Namely, he assumed that all the sources of energy whose existence could be inferred from his scientific knowledge were all that existed.

Nonmonotonic reasoning is necessary in science as in daily life. There can always be phenomena we don't know about. Indeed there might be another source of energy in the earth besides radioactivity.

Experience tells us that careful nonmonotonic reasoning, taking into account all the sources of information we can find and understand, usually gives good results, but we can never be as certain as we can be of purely mathematical results.

The phenomenon of often knowing more about the whole than about the parts, applies to more than physical objects. It can apply to processes. The phenomenon even existed in mathematics. Euclid's geometry was a powerful logical structure, but the basic concepts were fuzzy.

knowledge of regions in space I don't know how to formulate this precisely nor do I know of comprehensive discussions in the psychological literature, but some such knowledge can be expected to be innate. Evolution has had almost 4 billion years to make it intrinsic. Knowledge of the space on the highway is common to the driver and the policeman in the example.

localization We do not expect events on the moon to influence the physical location of objects on the table. However, we can provide for the possibility that an astronomer looking through a telescope might be so startled by seeing a meteorite collide with the moon that he would fall off his chair and knock an object off the table. Distant causality is a special phenomenon. We take it into account only when we have a specific reason.

knowledge of other actors Babies distinguish faces from other objects very early. Presumably babies have some innate expectations about how other actors may respond to the baby's actions.

self reference In general the informatic situation itself is an object about which facts are known. This human capability is not used in much human reasoning, and very likely animals don't have it.

introspective knowledge This is perhaps a distinctly human characteristic, but some introspective knowledge becomes part of common sense early in childhood, at least by the age of five. By that age, a typical child can remember that it previously thought a box contained candy even when it has learned that it actually contained crayons.

counterfactuals Common sense often involves knowledge of counterfactuals and the ability to infer them from observation and to draw non-counterfactual conclusions from them. In the example, the policeman infers that he should give the driver a ticket from the counterfactual that there would have been a collision if another car had come over the hill. People learn from counterfactual experiences they would rather not have in reality.

bounded informatic situations in contexts Bounded informatic situations have an important relation to the common sense informatic situation. For example, suppose there are some blocks on a table. They are not perfect cubes and they are not precisely aligned. Nevertheless, a simple blocks world theory may be useful for planning building a tower by moving and painting blocks. The bounded theory of the simple blocks world in which the blocks are related only by the $on(x, y, s)$ relation is related to the common sense informatic situation faced by the tower builder. This relation is conveniently expressed using the theory of contexts as objects discussed in section 7 and [McCarthy and Buvač, 1997]. The blocks world theory holds in a subcontext *cblocks* of the common sense theory *c*, and sentences can be *lifted* in either direction between *c* and *cblocks*.

learning A child can learn facts both from experience and from being told. Quite young children can be told about Santa Claus. Unfortunately, no AI systems so far developed (2006 January) can learn facts expressed in natural language on web pages.

Closer to hand, we do not expect objects not touching or connected through intermediate objects to affect each other. Perhaps there is a lot of common sense knowledge of the physical motion of table scale objects and how they affect each other that needs to be expressed as a logical theory.

The difficulties imposed by these requirements are the reason why the goal of Leibniz, Boole and Frege to use logical calculation as the main way of deciding questions in human affairs has not yet been realized. Realizing their goal will require extensions to logic beyond those required to reason in bounded informatic situations. Computer programs operating in the common sense informatic situation also need tools beyond those that have been used so far.

In contrast to the above view, Nagel [Nagel, 1961] treats common sense knowledge as the same kind of knowledge as scientific knowledge, only not systematically tested and justified. This is true of some common sense knowledge, but much common sense knowledge concerns entities that are necessarily ill-defined and knowledge about their relations that is necessarily imprecise.

Shannon's quantitative information theory seems to have little application to the common sense informatic situation. Neither does the Chaitin-Kolmogorov-Solomonoff computational theory. Neither theory concerns what common sense information is.

6 THE AI OF PHILOSOPHY—SOME ADVICE

Van Benthem [1990], tells us that AI is philosophy pursued by other means. That's part of what AI has to do.

AI research attacks problems common to AI and philosophy in a different way. For some philosophical questions, the AI approach is advantageous. In turn AI has already taken advantage of work in analytic philosophy and philosophical logic, and further interactions will help both kinds of endeavor. This section offers reasons why philosophers might be interested in AI approaches to some specific common problems and how AI might benefit from the interaction.

Achieving human-level common sense involves at least partial solutions to many philosophical problems, some of which are long standing in the philosophical, AI, and/or cognitive science literature, and others which have not yet been identified.

Identifying these problems is important for philosophy, for AI, and for cognitive science.

To ascribe certain *beliefs*, *knowledge*, *free will*, *intentions*, *consciousness*, *abilities* or *wants* to a machine or computer program is *legitimate* when such an ascription expresses the same information about the machine that it expresses about a person. It is *useful* when the ascription helps us understand the structure of the machine, its past or future behavior, or how to repair or improve it. It is perhaps never *logically required* even for humans, but expressing reasonably briefly what is actually known about the state of a machine in a particular situation may require ascribing mental qualities or qualities isomorphic to them. Theories of belief, knowledge and wanting can be constructed for machines in a simpler setting than for humans and later applied to humans. Ascription of mental qualities is most straightforward for machines of known structure such as thermostats and computer operating systems, but is *most useful* when applied to entities whose structure is very incompletely known.

While we are quite liberal in ascribing *some* mental qualities even to rather primitive machines, we should be conservative in our criteria for ascribing any *particular* quality. The ascriptions are what [Dennett, 1978] calls taking the *intentional stance.*

Even more important than ascribing mental qualities to existing machines is designing machines to have desired mental qualities.

Here are some features of some AI approaches to common problems of AI and philosophy.

AI starts small. Fortunately, AI research can often make do with small versions of the concepts. These small versions of the concepts and their relations are valid in limited contexts. We discuss three examples here and in section 7, which is about context. These are belief, action in the blocks world, and ownership of purchased objects.

An intelligent temperature control system for a building should be designed to know about the temperatures of particular rooms, the state of various valves, the occupants of rooms, etc. Because the system is not always correct about these facts, we and it should regard them as beliefs. Weather predictions need always be regarded as uncertain, i.e. as beliefs.

It is worthwhile to consider the simplest beliefs first, e.g. those of a thermostat.

A simple thermostat may have just three possible beliefs: the temperature is too cold, okay, or too hot. It behaves according to its current belief, turning the heat on, leaving it as is, or turning it off. It doesn't believe it's a thermostat or believe it believes the room is too cold.

Of course, the behavior of this simple thermostat can be understood without ascribing any beliefs. Beginning a theory of belief with such simple cases has the same advantage as including 1 in the number system. (Ascribing no beliefs to a rock is like including 0.) A temperature control system for a whole building is appropriately ascribed more elaborate beliefs. Ascribing beliefs and other mental qualities is more thoroughly discussed in [McCarthy, 1979a].

A child benefits from knowing that it is one child among others. Likewise, a temperature controller might even benefit from knowing that it is one temperature controller among other such systems. If it learns via the Internet that another system adjusts to snow on the roof, it might modify its program accordingly.

Naive common sense is often right in context. An example is the common sense notion of "x caused y".

There is a context in which "The window was broken by Susan's baseball" is true and "The window was broken, because the building contractor neglected to put a grill in front of it" is not even in the language used by the principal in discussing the punishment of the girl who threw the ball. Such limited contexts are often used and useful. Their relation to more general contexts of causality require study and logical formalization.

theory of action and the frame problem The conditions for an agent achieving goals in the world are very complicated in general, but AI research has developed theories and computer programs of increasing sophistication.

AI has long (since the 1950s anyway) concerned itself with finding sequences of actions that achieve goals. For this AI needs theories of the effects of individual actions, the tree of situations arising from an initial situation, and the effects of sequences of actions. The most used AI formalism for this is the *situation calculus*[8] introduced in [McCarthy and Hayes, 1969]. Its relations to philosophy are discussed in [Thomason, 2003]. There are thorough discussions in [Shanahan, 1997] and [Reiter, 2001], and a new version with *occurrence* axioms as well as the usual *effect axioms* is introduced in [McCarthy, 2002]. Three problems, *the frame problem*, *the qualification problem*, and *the ramification problem* have arisen and are extensively discussed in the AI literature and also in [Thomason, 2003]. The frame problem, also taken up by philosophers, concerns how to avoid stating which *fluents* (aspects of a situation) are unchanged when an action takes place, e.g. avoiding explicitly stating that the color of an object doesn't change when the object is moved.

The basic situation calculus is a non-deterministic (branching) theory of action. AI has also treated deterministic (linear) theories of action. The new formalism of [McCarthy, 2002] permits a treatment [McCarthy, 2005] of a kind of deterministic free will in which a non-deterministic theory serves as part of the deterministic computational mechanism.

AI has considered simple examples that can be subsequently elaborated. The well-known *blocks world* is treated with logical sentences like $On(Block1,$

[8]The event calculus [Mueller, 2006] is an alternative.

$Block2$) or $On(Block1, Block2, S0)$ in which the situation is explicit. Another formalism uses $Value(Location(Block1), S0) = Top(Block2)$. We may also have

$$(3) \quad \begin{array}{l} (\forall s)(\ldots \rightarrow Location(block, Result(Move(block, l), s)) = l) \\ \text{and} \\ (\forall s)(\ldots \rightarrow Color(block, Result(Paint(block, c), s)) = c \end{array}$$

where ... stands for some preconditions for the success of the action. On one hand, such simple action models have been incorporated in programs controlling robot arms that successfully move blocks. On the other hand, the *frame problem* arose in specifying that moving a block didn't change the locations of other blocks or the colors of the blocks. This problem, along with its mates, the qualification problem and the ramification problem, arose in AI research but arise also in studying the effects of action in philosophy.

Note that in the bounded theory of the blocks world as partly described here, there is only one actor, and a block is never partly on one block and partly on another. Elaborations have been made to study these complications, but the methodology of doing the simple cases first has led to good results. Making a full theory of action from scratch is still only a vaguely defined project.

nonmonotonic reasoning Nonmonotonic reasoning is essentially the same topic as defeasible reasoning, long studied in philosophy. What's new since the 1970s is the development of formal systems for nonmonotonic reasoning, e.g. the logic of defaults [Reiter, 1980] and circumscription, [McCarthy, 1980] and [McCarthy, 1986]. There are also computer systems dating from the 1970s that do nonmonotonic reasoning, e.g. Microplanner and Prolog. Nonmonotonic reasoning has been prominent in programs that make plans to achieve goals.

Recent articles in the *Stanford Encyclopedia of Philosophy* have made the connection between AI work in nonmonotonic reasoning and philosophical work on defeasibility. Convenient references are [Thomason, 2003; Koons, 2005], and [Antonelli , 2003].

elaboration tolerance Explicit formalizations of common sense phenomena are almost never complete. There is always more information that can be taken into account. This is independent of whether the phenomena are described in ordinary language or by logical sentences. Theories always have to be elaborated. According to how the theory is written in the first place, the theory may *tolerate* a given elaboration just by adding sentences, which usually requires nonmonotonicity in making inferences from the theory, or the theory may have to be scrapped and a new theory built from scratch. [McCarthy, 1999b] introduces the concept of *elaboration tolerance* and illustrates it with 19 elaborations of the well-known missionaries and cannibals puzzle. The elaborations seem to be straightforward in English but rely on the common sense of the reader. Some of the logical formulations tolerate some of the elaborations just by adding sentences; others don't. One goal is find a logical language in which all the elaborations are additive.

[Lifschitz, 2000] accomplishes 9 of the above-mentioned 19 elaborations in the Causal Calculator of McCain and Turner [McCain and Turner, 1998]. [Shanahan, 1997] has an extensive discussion of elaboration tolerance.

I don't know of discussions of the elaboration tolerance of theories proposed in the philosophical literature.

sufficient complexity usually yields essentially unique interpretations A robot that interacts with the world in a sufficiently complex way gives rise to an essentially unique interpretation of the part of the world with which it interacts. This is an empirical, scientific proposition, but many people, especially philosophers (see [Quine, 1960], [Quine, 1969], [Putnam, 1975], [Dennett, 1971], [Dennett, 1998]), seem to take its negation for granted. There are often many interpretations in the world of short descriptions, but long descriptions almost always admit at most one. As far as I can see, [Quine, 1960] did not discuss the effect of a large context on the indeterminacy of translation—of say *gavagai*.

The most straightforward example is that a simple substitution cipher cryptogram of an English phrase. Thus XYZ could be decrypted as either "cat" or "dog". A simple substitution cryptogram of an English sentence usually has multiple interpretations if the text is less than 21 letters and usually has a unique interpretation if the text is longer than 21 letters. Why 21? It's a measure of the redundancy of English [Shannon and Weaver, 1949]. The redundancy of the sequence of a person's or a robot's interactions with the world is just as real—though clearly much harder to quantify.

approximate objects and theories The idea that entities of philosophical interest are not always well defined can, if you like such attributions, be attributed to Aristotle's

> Our discussion will be adequate if it has as much clearness as the subject matter admits of, for precision is not to be sought for alike in all discussions, any more than in all the products of the crafts. —*Nicomachean Ethics.*

I don't know whether Aristotle pursued the idea further.

I proposed [McCarthy, 2000] that AI requires the formalization of approximate entities that sometimes yields firm logical theories on foundations of semantic quicksand. Thus it is definite that Mount Everest was climbed in 1953 even though it is not definite what rock and ice constitute Mount Everest. A much more approximate concept though still useful is *"The United States wanted in 1990"* applied to "that Iraq would withdraw from Kuwait".

One proposal is to use necessary conditions for a proposition and sufficient conditions but not to strive for conditions that are both necessary and sufficient. These ideas are connected to notions of vagueness that have been discussed by philosophers, but the discussion in the article [Sorensen, 2003] in the Stanford Encyclopedia of Philosophy does not discuss how to formalize essentially vague concepts.

contexts as objects This is an area where, judging from the Stanford Encyclopedia of Philosophy, there is as yet no connection between the rather extensive research in AI that started with [McCarthy, 1993b] and research in philosophy. Since *information in AI* (and in ordinary language) is always presented in a context, section 7 is devoted to a sketch of a theory of contexts as objects.

concepts as objects In natural language, concepts are discussed all the time. Nevertheless, Carnap wrote

> ... it seems that hardly anybody proposes to use different variables for propositions and for truth-values, or different variables for individuals and individual concepts.
> ([Carnap, 1956] , p. 113.

Perhaps Carnap was thinking of [Church, 1951] as the exception. Instead, modal logic is used for expressing certain assertions about propositions, and individual concepts are scarcely formalized at all.

human-level AI will require the ability to express anything humans express in natural language and also to expressions statements about the expressions themselves and their semantics.

[McCarthy, 1979b] proposes distinguishing propositions from truth values and individual concepts from objects in a base domain—and using different variables for them. Here are some examples of the notation. The value of $Mike$ is a person, whereas the value of $MMike$ is a concept—intended to be a concept of that Mike in this case, but that it should be is not a typographical convention. Here are some sentences of a first order language with concepts and objects.

$$\begin{aligned}
&Denot(MMike) = Mike,\\
&Male(Mike),\\
&Denot(MMale(MMike)),\\
&Denot(HHusband(MMary)) = Mike,\\
&Husband(Mary) = Mike,\\
&HHusband(MMary \neq MMike,\\
&(\forall x)(x \neq Husband(Mike)\\
&\rightarrow \neg Exists(HHusband(MMike)).
\end{aligned} \tag{4}$$

The sentence $Denot(MMike) \neq Mike$ might be true under some circumstances.

The distinction between concepts and objects makes it convenient to express some assertions that simpler notations find puzzling. Thus Russell's "I thought your yacht was longer than it is" is treated in [McCarthy, 1979b].

This example and others use functions from objects to concepts of them. Thus we might write $CConcept1(Cicero) = CCicero$. If we also have $Cicero = Tully$, we'll get $CConcept1(Tully) = CCicero$. While we would not ordinarily want $TTully = CCicero$, but since concepts are not characterized by the typography used to write them, this would not be a contradiction.

Some objects have standard concepts, e.g. numbers. We'd like to write $Concept1(3) = 33$, but this conflicts with decimal notation, so it is better to write $Concept1(3) = 3'3$. Consider the true sentences

$$\begin{array}{l} \neg Knew(Kepler, CComposite(NNumber(PPlanets))) \\ \text{and} \\ Knew(Kepler, CComposite(CConcept1(Number(Planets)))). \end{array} \tag{5}$$

The first says that Kepler didn't know the number of planets is composite. The second says that Kepler knew that the number, which happens to be the number of planets, is composite. See also [Maida and Shapiro, 1982] and [Shapiro, 1993] for another AI approach to representing concepts.

These considerations are only a small step in the direction, necessary both for AI and philosophy, of treating concepts as first class objects. [McCarthy, 1997] argues the inadequacy of modal logic for a full treatment of modality. The article incited some vigorous replies.

correspondence theory of reference This is more complicated than the correspondence theory of truth, because the entities to which a term can refer are not just truth values. We recommend that philosophers study the problem of formalizing reference. There isn't even an analog of modal logic for reference.

appearance and reality Science tells us that our limited senses, and indeed any senses we might build into robots, are too limited to observe the world in full detail, i.e. at the atomic level. AI in general, and robotics in particular, must live with this fact and therefore requires a theory of the relations between appearance and reality. This theory must accomodate different levels of detail in both. I haven't got far with this, but [McCarthy, 1999a] gives a small example of the relation between two-dimensional appearance and three-dimensional reality. Realist, especially materialist, philosophers also need to formalize this relationship.

consciousness, especially consciousness of self Humans have a certain amount of ability to observe and reason about their own internal states. For example, I may conclude that I have no way of knowing, short of phoning her, whether my wife is in her office at this moment. Such consciousness of one's internal state is important for achieving goals that do not themselves involve consciousness. [McCarthy, 1996b] discusses what consciousness a robot will need to accomplish the tasks we give it.

7 INFORMATION IN CONTEXTS AND ABOUT CONTEXTS

Information is always transmitted in a context. Indeed a person thinks in a context. For the philosophy of information, information in contexts and the relations among contexts are more important than the Shannon entropy of a text.

This section discusses formalizing contexts as first class objects. The basic relation is $Ist(c, p)$. It asserts that the *proposition* p is true in the *context* c. The most important formulas relate the propositions true in different contexts. Introducing contexts as formal objects will permit axiomatizations in limited contexts to be expanded to *transcend* the original limitations. This seems necessary to provide AI programs using logic with certain capabilities that human fact representation and human reasoning possess. Fully implementing *transcendence* seems to require further extensions to mathematical logic, i.e. beyond the nonmonotonic inference methods first invented in AI and now studied as a new domain of logic.

The expression $Value(c, term)$ giving the value of the expression $term$ in the context c is just as important as $Ist(c, p)$, perhaps more important for applications.

Here are some of the features of a formalized theory of context.

1. There are many kinds of contexts, e.g. the context of Newtonian gravitation and within it the context of the trajectory of a particular spacecraft, the context of a theory formalizing the binary relations $On(x, y)$ and $Above(x, y)$, a situation calculus context with the ternary relations $On(x, y, s)$ and $Above(x, y, s)$, the context of a particular conversation or lecture, the context of a discussion of group theory in French, and the context of the Sherlock Holmes stories.

2. There must be language for expressing the value of a term in a context. For example, we have

$$C0 : Value(Context(ThisArticle), Author) = JohnMcCarthy.$$

3. The theory must provide language for expressing the relations of contexts, e.g. that one context specializes another in time or place, that one context assumes more group theory than another, that one discusses the same subject but in a different language.

4. There must be language for expressing relations between sentences true in related contexts and also for expressing relations between terms in related contexts. When $c1$ is a specialization of $c0$, such rules are called *lifting rules.*

5. Here's an example of a lifting rule associated with databases. Suppose GE (General Electric) sells jet engines to AF (U.S. Air Force) and each organization has a database of jet engines including the price. Assume that the AF context (database) assumes that the price of an engine includes a spare parts kit, whereas the GE context prices them separately. We may have the *lifting formula*

   $$Ist(Outer, Value(AF, Price(engine)) = Value(GE, Price(engine)) \\ +Value(GE, Price(Spare\text{-}Parts\text{-}Kit(engine)))),$$

   expressing in an outer context *Outer* a relation between an expression in the AF context and expressions in the GE context. Others call such formulas *bridging formulas.*

   [McCarthy, 1993b] has an example of lifting a general rule relating predicates $On(x, y)$ and $Above(x, y)$ to a situation with three argument relations $On(x, y, s)$ and $Above(x, y, s)$, in which the third argument s is a situation.

6. We envisage a reasoner that is always in a context. It can *enter* specializations and other modifications of the current context and then reason in it. Afterwards, it can *exit* the inner context, returning to the outer context. In human-level AI systems there will be no outermost context. It will always be possible to *transcend* the outermost context so far named and reason in a new context in which the previous context is an object.

[McCarthy, 1993b] and [McCarthy and Buvač, 1998] present a more detailed theory of formalized contexts. See also [Guha, 1991].

Not included in those papers is the more recent idea that what some AI researchers call "toy theories" may be valid in some contexts, and that a reasoner may do an important part of his thinking in such a limited context.

For example, consider a simple theory of buying and owning. From the point of view of a small child in a store after he has learned that he may not just take something off the shelf, he knows that it is necessary for the parent to buy something in order give it to the child. Call this context *Own*0. The details of buying are unspecified, and this simple notion may last several years. The next level of sophistication involves paying the price of the object. Not only does this notion last longer for the child, but an adult in a grocery store usually operates in this context *Own*1, which admits a straightforward situation calculus axiomatization. Outside of supermarkets, ownership becomes more complicated, e.g. buying a house with a mortgage. Certain of these ownership contexts are understood by the general public and others by lawyer and real estate investors, but no-one has a full theory of ownership.

8 CONCLUSIONS AND REMARKS

Artificial intelligence is based on some philosophical and scientific presuppositions. The simplest forms of AI make fewer presuppositions than AI research aimed at human-level AI. The feature of human-level AI we emphasize is the ability to learn from its experience without being further programmed.

The concreteness of AI research has led to a number of discoveries that are relevant to philosophy, and these are only beginning to be noticed by philosophers. Three of the topics treated in this chapter are formalized nonmonotonic reasoning, formalized contexts, and the need to deal with concepts that have only an approximate meaning in general. Besides what's in this chapter, we particularly recommend [Thomason, 2003].

BIBLIOGRAPHY

[Antonelli , 2003] A. Antonelli. Non-monotonic logic. In E. N. Zalta (Ed.), *The Stanford Encyclopedia of Philosophy*, 2003.

[Carnap, 1956] R. Carnap. *Meaning and Necessity.* University of Chicago Press, 1956.

[Church, 1951] A. Church. The need for abstract entities in semantic analysis. *Proceedings of the American Academy of Arts and Sciences* 80(1):100–112, 1951. Reprinted in *The Structure of Language.* edited by Jerry A. Fodor and Jerrold Katz, Prentice-Hall 1964.

[Davis et al. , 1977] R. Davis, B. Buchanan, and E. Shortliffe. Production rules as a representation for a knowledge-based consultation program. *Artificial Intelligence* 8(1):15–45, 1977.

[Dennett, 1978] D. Dennett. *Brainstorms: Philosophical Essays on Mind and Psychology.* Cambridge: Bradford Books/MIT Press, 1978.

[Dennett, 1998] D. Dennett. *Brainchildren: Essays on Designing Minds.* MIT Press, 1998.

[Dennett, 1971] D. C. Dennett. Intentional systems. *The Journal of Philosophy* 68(4):87–106, 1971.

[Dreyfus, 1992] H. Dreyfus. *What Computers still can't Do.* MIT Press, 1992.

[Ernst and Newell, 1969] G. W. Ernst and A. Newell. *Gps: A CASE Study in Generality and Problem Solving.* New York: Academic Press, 1969.

[Guha, 1991] R. V. Guha. *Contexts: A Formalization and Some Applications.* PhD thesis, Stanford University, 1991. Also published as technical report STAN-CS-91-1399-Thesis, MCC Technical Report Number ACT-CYC-423-91, and available as http://www-formal.stanford.edu/guha/guha.ps.

[Hayes, 1985] P. J. Hayes. The second naive physics manifesto. In H. J.R. and M. R.C. (Eds.), *Formal Theories of the Commonsense World*, 1–36. Ablex, 1985.

[Hayes, 1979] P. J. Hayes. The naive physics manifesto. In D. Michie (Ed.), *Expert systems in the microelectronic age.* Edinburgh, Scotland: Edinburgh University Press, 1979.

[Koons, 2005] R. Koons. Defeasible reasoning. In E. N. Zalta (Ed.), *The Stanford Encyclopedia of Philosophy*, 2005.

[Lenat, 1995] D. B. Lenat. Cyc: A large-scale investment in knowledge infrastructure. *Communications of the ACM* 38(11), 1995.

[Lifschitz, 1993] V. Lifschitz. Circumscription `http://www.cs.utexas.edu/users/vl/mypapers/circumscription.ps`. In *Handbook of Logic in Artificial Intelligence and Logic Programming, Volume 3: Nonmonotonic Reasoning and Uncertain Reasoning.* Oxford University Press, 1993.

[Lifschitz, 2000] V. Lifschitz. Missionaries and cannibals in the causal calculator. In A. G. Cohn, F. Giunchiglia, and B. Selman (Eds.), *KR2000: Principles of Knowledge Representation and Reasoning,Proceedings of the Seventh International conference*, 85–96. Morgan-Kaufman, 2000.

[Maida and Shapiro, 1982] A. S. Maida and S. C. Shapiro. Intensional concepts in propositional semantic networks. *Cognitive Science* 6(4):291–330. Reprinted in R. J. Brachman and H. J. Levesque, eds. Readings in Knowledge Representation, Morgan Kaufmann, Los Altos, CA, 1985, 170-189, 1982.

[Marr, 1982] D. Marr. *Vision.* New York: Freeman, 1982.

[Matuszek and Lenat, 2005] C. M. W. R. K. J. C. D. S. P. S. Matuszek and D. Lenat. Searching for common sense: Populating cyc from the web. In *Proceedings of the Twentieth National Conference on Artificial Intelligence,* Pittsburgh, Pennsylvania, July 2005*, 2005.

[McCain and Turner, 1998] N. McCain and H. Turner. Satisfiability planning with causal theories. In *KR*, 212–223, 1998.

[McCarthy, 1959] J. McCarthy. Programs with Common Sense `http://www-formal.stanford.edu/jmc/mcc59.html`. In *Mechanisation of Thought Processes, Proceedings of the Symposium of the National Physics Laboratory*, 77–84, London, U.K. Her Majesty's Stationery Office, 1959. Reprinted in [McCarthy, 1990].

[McCarthy, 1979a] J. McCarthy. Ascribing mental qualities to machines`http://www-formal.stanford.edu/jmc/ascribing.html`. In M. Ringle (Ed.), *Philosophical Perspectives in Artificial Intelligence*. Harvester Press, 1979. Reprinted in [McCarthy, 1990].

[McCarthy, 1979b] J. McCarthy. First Order Theories of Individual Concepts and Propositions `http://www-formal.stanford.edu/jmc/concepts.html`. In D. Michie (Ed.), *Machine Intelligence*, Vol. 9. Edinburgh: Edinburgh University Press, 1979. Reprinted in [McCarthy, 1990].

[McCarthy, 1980] J. McCarthy. Circumscription—A Form of Non-Monotonic Reasoning `http://www-formal.stanford.edu/jmc/circumscription.html`. *Artificial Intelligence* 13:27–39, 1980. Reprinted in [McCarthy, 1990].

[McCarthy, 1983] J. McCarthy. Some Expert Systems Need Common Sense `http://www-formal.stanford.edu/jmc/someneed.html`. In H. Pagels (Ed.), *Computer Culture: The Scientific, Intellectual and Social Impact of the Computer*, Vol. 426. Annals of the New York Academy of Sciences, 1983.

[McCarthy, 1986] J. McCarthy. Applications of Circumscription to Formalizing Common Sense Knowledge `http://www-formal.stanford.edu/jmc/applications.html`. *Artificial Intelligence* 28:89–116, 1986. Reprinted in [McCarthy, 1990].

[McCarthy, 1990] J. McCarthy. *Formalizing Common Sense: Papers by John McCarthy*. Ablex Publishing Corporation, 1990.

[McCarthy, 1993b] J. McCarthy. Notes on Formalizing Context `http://www-formal.stanford.edu/jmc/context.html`. In *IJCAI93*, 1993.

[McCarthy, 1996a] J. McCarthy. From Here to Human-Level AI `http://www.formal.stanford.edu/jmc/human.html`. In *KR-96*, 640–646, 1996.

[McCarthy, 1996b] J. McCarthy. Making Robots Conscious of their Mental States `http://www-formal.stanford.edu/jmc/consciousness.html`. In S. Muggleton (Ed.), *Machine Intelligence 15*. Oxford University Press, 1996. Appeared in 2000. The web version is improved from that presented at Machine Intelligence 15 in 1995.

[McCarthy, 1996c] J. McCarthy. The well-designed child. `http://www-formal.stanford.edu/jmc/child.html`, 1996

[McCarthy, 1997] J. McCarthy. Modality si! modal logic, no! *Studia Logica* 59(1):29–32, 1997.

[McCarthy, 1999a] J. McCarthy. Appearance and reality `http://www-formal.stanford.edu/jmc/appearance.html`, 1999. *web only for now, and perhaps for the future.* not fully publishable on paper, because it contains an essential imbedded applet.

[McCarthy, 1999b] J. McCarthy. Elaboration tolerance `http://www-formal.stanford.edu/jmc/elaboration.html`, 1999. *web only for now.*

[McCarthy, 2000] J. McCarthy. Approximate objects and approximate theories `http://www.formal.stanford.edu/jmc/approximate.html`. In A. G. Cohn, F. Giunchiglia, and B. Selman (Eds.), *KR2000: Principles of Knowledge Representation and Reasoning,Proceedings of the Seventh International conference*, 519–526. Morgan-Kaufman, 2000.

[McCarthy, 2002] J. McCarthy. Actions and other events in situation calculus`http://www-formal.stanford.edu/jmc/sitcalc.html`. In B. S. A.G. Cohn, F. Giunchiglia (Ed.), *Principles of knowledge representation and reasoning: Proceedings of the eighth international conference (KR2002)*. Morgan-Kaufmann, 2002.

[McCarthy, 2005] J. McCarthy. Simple deterministic free will. See `http://www-formal.stanford.edu/jmc/freewill2.html`, 2005

[McCarthy and Buvač, 1997] J. McCarthy and S. Buvač. Formalizing context (expanded notes). In A. Aliseda, R. v. Glabbeek, and D. Westerståhl (Eds.), *Computing Natural Language*. Center for the Study of Language and Information, Stanford University, 1997.

[McCarthy and Buvač, 1998] J. McCarthy and S. Buvač. Formalizing Context (Expanded Notes). In A. Aliseda, R. v. Glabbeek, and D. Westerståhl (Eds.), *Computing Natural Language*, Vol. 81 of *CSLI Lecture Notes*, 13–50. Center for the Study of Language and Information, Stanford University, 1998.

[McCarthy and Hayes, 1969] J. McCarthy and P. J. Hayes. Some Philosophical Problems from the Standpoint of Artificial Intelligence `http://www-formal.stanford.edu/jmc/mcchay69.html`. In B. Meltzer and D. Michie (Eds.), *Machine Intelligence 4*, 463–502. Edinburgh University Press, 1969. Reprinted in [McCarthy, 1990].

[Minsky, 1963] M. L. Minsky. Steps towards artificial intelligence. In E. A. Feigenbaum and J. Feldman (Eds.), *Computers and Thought*, 406–450. McGraw-Hill, 1963. Originally published in *Proceedings of the Institute of Radio Engineers*, January, 1961 **49:**8–30.

[Moravec, 1977] H. P. Moravec. Towards automatic visual obstacle avoidance. In *IJCAI*, 584, 1977.

[Mueller, 2006] E. T. Mueller. *Common Sense Reasoning.* Morgan Kaufmann, 2006.

[Nagel, 1961] E. Nagel. *The structure of science.* Harcourt, Brace, and World, 1961.

[Newell, 1993] A. Newell. Reflections on the knowledge level. *Artificial Intelligence* 59(1-2):31–38, 1993.

[Nilsson, 2005] N. J. Nilsson. Human-level AI? be serious! *The AI Magazine* 26(4):68–75, 2005.

[Penrose, 1994] R. Penrose. *Shadows of the Mind: A Search for the Missing Science of Consciousness.* Oxford: Oxford University Press, 1994.

[Pinker, 1997] S. Pinker. *How the Mind Works.* Norton, 1997.

[Putnam, 1975] H. Putnam. The meaning of "meaning". In K. Gunderson (Ed.), *Language, Mind and Knowledge*, Vol. VII of *Minnesota Studies in the Philosophy of Science*, 131–193. University of Minnesota Press, 1975.

[Quine, 1969] W. V. O. Quine. Propositional objects. In *Ontological Relativity and other Essays.* Columbia University Press, New York, 1969.

[Quine, 1960] W. V. O. Quine. *Word and Object.* MIT Press, 1960.

[Reiter, 1980] R. Reiter. A Logic for Default Reasoning, *Artificial Intelligence*, 13(1002): 81–132, 1980.

[Reiter, 2001] R. Reiter. *Knowledge in Action.* MIT Press, 2001.

[Russell, 1914] B. Russell. *Our knowledge of the external worrld.* Open Court, 1914.

[Searle, 1984] J. R. Searle. *Minds, Brains, and Science.* Cambridge, Mass.: Harvard University Press, 1984.

[Shanahan, 1997] M. Shanahan. *Solving the Frame Problem, a mathematical investigation of the common sense law of inertia.* MIT Press, 1997.

[Shannon and Weaver, 1949] M. Shannon and W. Weaver. *The Mathematical Theory of Communication.* U. of Illinois Press, 1949.

[Shapiro, 1993] S. C. Shapiro. Belief spaces as sets of propositions. *Journal of Experimental and Theoretical Artificial Intelligence* 5:225–235, 1993. http://www.cse.buffalo.edu/tech-reports/SNeRG-175.ps.

[Sorensen, 2003] R. Sorensen. Vagueness. In E. N. Zalta (Ed.), *The Stanford Encyclopedia of Philosophy*, 2003.

[Spelke, 1994] E. Spelke. Initial knowlege: six suggestions. *Cognition* 50:431–445, 1994.

[Thomason, 2003] R. Thomason. Logic and artificial intelligence. In E. N. Zalta (Ed.), *The Stanford Encyclopedia of Philosophy*, 2003. http://plato.stanford.edu/archives/fall2003/entries/logic-ai/.

[Turing, 1947] A. M. Turing. Lecture to the london mathematical society. In *The Collected Works of A. M. Turing*, Vol. Mechanical Intelligence. North-Holland, 1947. This was apparently the first public introduction of AI, typescript in the King's College archive, the book is 1992.

[van Benthem, 1990] J. van Benthem. Kunstmatige intelligentie: Een voortzetting van de filosofie met andere middelen. *Algemeen Nederlands Tijdschrift voor Wijsbegeerte* 82:83–100, 1990.

[Weizenbaum, 1965] J. Weizenbaum. ELIZA—a computer program for the study of natural language communication between man and machine. *Communications of the Association for Computing Machinery* 9(1):36–45, 1965.

[Winston, 1977] P. H. Winston. *Artificial Intelligence.* Reading, Mass.: Addison Wesley Publishing Co, 1977.

INFORMATION, COMPUTATION, AND COGNITIVE SCIENCE

Margaret A. Boden

Cognitive science views the mind, or mind-brain, as an abstract machine: specifically, as an information-processing machine. In other words, it offers a computational psychology (and anthropology), a computational linguistics, and a computational approach to neuroscience — and, via A-Life, to general biology too. (A-Life, or artificial life, is a close sibling of AI, or artificial intelligence: it uses formal theories and simulations to explore phenomena typical of living things, and to illuminate the nature of life in general.) By the same token, it offers a computational philosophy of mind. More accurately, as we shall see, it offers a range of computational theories in each of these disciplines.

Many non-computational psychologists and neuroscientists use computers as a tool — to handle statistics, for example. What is distinctive about cognitive science is that the computer is not just a tool but a theoretical inspiration. That is, the substantive concepts in the theories of cognitive science are drawn from cybernetics and AI, since they concern abstract matters of information processing and control.

Since the focus of cognitive science is on biological organisms, whether human or non-human, purely technological AI is irrelevant. The paradigm illustration is the "Deep Blue" chess program, which beat the world chess champion Gary Kasparov in 1997. To be sure, the program shared some of its processing with human beings: it employed heuristics such as "Protect your queen", for instance. But the key to its success was the fact that it used dedicated chips enabling it, by processing 200 million positions per second, to consider every legal move for as many as eight steps ahead. Since no human being can do anything remotely like that, Deep Blue's exhaustive look-ahead is of no psychological interest. In general, technological achievements that rely on fundamentally non-human information-processing strategies or computing tricks do not form part of cognitive science.

Cognitive scientists often implement their ideas about brain and behaviour as functioning computer models. So too do other scientists, such as chemists or meteorologists. What is special about models in cognitive science is that the key computational concepts defining the model are substantive terms in the psychological theories concerned. (For instance, whereas models of rainfall are not really wet, models of search in problem-solving really do carry out search.) Such models are invaluable in testing/exploring the implications and coherence of a new theory. But computational theories may be articulated, and scientific questions suggested,

Handbook of the Philosophy of Science. Volume 8: Philosophy of Information
Volume editors: Pieter Adriaans and Johan van Benthem. General editors: Dov M. Gabbay, Paul Thagard and John Woods.

without actually building models. That is especially likely when the phenomena in question are complex and high-level.

Computational theories of hypnosis, for instance, are focussed on the information processing involved in hypnotic thought and action, and on how this differs from normal processing. Zoltan Dienes and Josef Perner [2007] have outlined computational mechanisms whereby hypnosis of varying types can occur. These mechanisms explain why, as many experimenters have reported, some sorts of hypnotic suggestion are easier to communicate than others. The easiest are "motor" examples, such as

> Your arm is becoming so light it is rising in the air.

Next come "challenge" examples, such as

> Your arm is rigid.

Negative cognitive demands (wherein you are ordered to avoid a particular belief or idea) are more difficult still:

> Whenever you count, you'll forget the number four.

And the most difficult of all are positive hallucinations. But even here, there are systematic differences. It is easier to make someone hallucinate as required to the suggestions

> You can taste something sweet

or

> You can hear/feel a mosquito

than to

> You can hear a voice speaking.

Dienes and Perner explain these facts by arguing that the more computational effort goes into performing a task the harder it is to suppress higher-order thoughts of intention. They outline experiments to distinguish between various computational possibilities, and predict how "high" and "low hypnotisable subjects" will differ in tasks where hypnosis is not involved; in addition, they relate their psychological theory to data about the brain. In short, the fact that their theory of mental processing has not been expressed in the form of a computer model does not mean that it is empirically ungrounded or untestable.

The very earliest cognitive science was prompted by information theory, by the cyberneticists' notion of feedback, and by ideas about (analogue) models in the brain [Shannon and Weaver, 1949; Rosenblueth and Wiener, 1950; Craik, 1943]. A particular attraction of information theory, for mid-century experimental psychologists, was its promise that the mind/brain's information-processing capacities/limitations could be not only compared, but measured. For instance, a particular sensory channel could be shown to convey x bits per second. As for the concept of feedback, this — together with the notion of reduction of differences — promised to illuminate some aspects of purposive behaviour [Rosenblueth, Wiener and Bigelow, 1943]. Notoriously, goals and purposes had long caused theoretical, not to say philosophical, headaches for psychology [McDougall, 1911; 1923; Tolman, 1920; 1922; 1932].

Prime examples of psychological work that was part-inspired by information theory included Herbert Simon's [1957] account of decision making in social groups — although this owed much also to his early views on heuristics and satisficing [Simon, 1947]. Information theory was prominent in Donald Broadbent's [1958] studies of perception, communication, and attention (a.k.a. consciousness), including how human beings can interact efficiently with machines. Likewise, it informed George Miller's [1956] account of the limits on short-term memory, and the need for informational recoding by "chunking". And it was a key source for Jerome Bruner's influential experiments on concept learning [Bruner, Goodnow, and Austin, 1956]. All four of these individuals were hugely important for the development of cognitive science.

As for problem-solving, the typical information-theoretic approach was to recall the parlour game Twenty Questions, in which the sensible problem solver aims to halve the number of possibilities at each successive step. This assumes that all possibilities are equally likely. If they are not, then other strategies must be calculated [Attneave, 1959, 5–9]. The core idea, that some rational strategy or other must be used, was picked up by "New Look" psychologists such as Bruner, and was developed also (from rather different roots) in symbolic AI work on heuristic planning [Newell, Shaw, and Simon, 1958; 1959; Newell and Simon, 1961].

(Later, Allen Newell declared that "You Can't Play Twenty Questions with Nature and Win" — by which he meant that the mind's overall computational architecture, its principles of information processing, has to be understood if we are to understand what it does [Newell, 1973]. His account of "production systems" was intended not just as a new form of programming language for AI but as a model of how the mind/brain works [Newell and Simon, 1972]. Other theories of mental architecture are briefly mentioned below.)

Information theory was welcomed by these experimental psychologists, and by some neurophysiologists too, because it offered quantitative measurements and because it dealt with the coding, or transformation, of information as well as its transmission. Psychologists (such as Miller) theorizing about internal representations — chunks, schemas, models ... — could use it to explain why (though as we shall see, not how) these are constructed. In addition, it supported anti-behaviourist centralism without falling into homunculism: positing mysterious 'little men' in the mind/brain.

Strictly speaking, Claude Shannon's information theory was in itself non-psychological. That is, it dealt only with the statistical predictability of messages and message-passing, not with the messages' meaning as such. Notoriously, the core term information was ambiguous. Because it is normally understood as an intentional (that is: meaning-oriented) concept, it was bound to lead psychologists to think about information and communication (another non-intentional term, when used by Shannon) in their everyday, mentalistic, senses. In that sense, after all, someone who is said to have the information that the next train to London leaves at 6 o'clock is deemed to have that thought in their minds, or to be able to conjure up that idea if asked, or at worst to possess a written note which they can interpret in that way.

However, information needed to be supplemented by computation in order to develop what is now known as cognitive science. For information theory alone was not enough to describe the central (mental) processes concerned–which is why it could not explain how chunks, or schemas, are formed. It had provided statistical/formal ideas about machines (initially, telephone systems) which psychologists could apply to humans and other animals. But since computers were not yet available, it had not suggested computer modelling, nor even the general notion of computation that is presupposed in programming. The theoretical focus was on the passage of information, not of control — still less, of step-by-step processing.

That idea was used to explain human thinking in the late 1950s [Newell, Shaw, and Simon, 1958], the explanation resting heavily on a program for problem-solving-by-planning [Newell and Simon, 1961]. A similar theoretical approach was soon recommended by Miller as a way of thinking about all aspects of psychology, in a book called *Plans and the Structure of Behavior* ([Miller, Galanter, and Pribram, 1960]; cf. [Boden, 2006, ch. 6.iv.c]). This brief volume acted as cognitive science's defining manifesto. Besides problem solving, language, and memory, it discussed (albeit very sketchily) instincts, motives, emotion, personality, hypnosis, and psychopathology. The key idea throughout was not information, but programs: not bits, but processes.

If cognitive science's manifesto did not appear until 1960, its seminal paper — Warren McCulloch and Walter Pitts' 'A Logical Calculus of the Ideas Immanent in Nervous Activity' [1943] — had been published many years earlier. Much as information theory was sidelined in the 1960 book, so it had been ignored in the 1943 paper. Programs were not mentioned there either, because the paper was written before the development of digital computers. Instead, it combined early-twentieth century ideas from three distinct disciplines: logic (and logicist philosophy), the mathematical theory of computation, and neurophysiology.

Specifically, the propositional calculus [Russell and Whitehead, 1910] was mapped onto Turing machines [Turing, 1936], and both of these onto 'all-or-none' neurone theory [Sherrington, 1906]. However, McCulloch and Pitts' core notion of TPEs (Temporal Propositional Expressions) represented not timeless logical truths (the focus of the propositional calculus) but temporal steps. These were conceptualized in terms of one or more neurones influencing (exciting or inhibiting) another, but could equally well have been termed computations. Indeed, the paper's definitions of TPEs for computing the basic logical connectives (identity, negation, conjunction, disjunction) were soon adopted by John von Neumann in designing logic-gates for digital computers. This highly abstract paper was seminal in three ways. First, it explicitly declared "the whole of psychology" to be a matter of defining computational systems ("networks") capable of implementing the various psychological phenomena. In other words, it committed itself to what is now recognized as the core aim of cognitive science. Moreover, it spawned both connectionist and logical-symbolic AI. Its emphasis on specific networks of computational units, alias neurones, soon led to early experiments in connectionism. And it later encouraged the development of GOFAI, or Good Old-Fashioned AI [Haugeland, 1985, 112], seen as a way of modelling semantic information: propositions and inferences bearing meaning, or intentional content.

That assumes, of course, that GOFAI symbols can properly be interpreted as having semantic content. Given that McCulloch's logicism was a popular philosophical position at mid-century [Boden, 2006, 4.iii.c], the early AI workers and other cognitive scientists influenced by them easily made this assumption. It would later be stated explicitly as part of the Physical Symbol System hypothesis [Newell, 1980].

However, it was often challenged. Some philosophers objected immediately that the meaning was wholly provided by the human interpreters, including the AI programmers themselves [Mays, 1952]. This objection eventually became notorious, when John Searle [1980] expressed it in his parable of the Chinese Room. Searle's argument has been roundly rebutted, on various grounds, by many people in cognitive science, but it refuses to lie down: the controversy continues [Preston and Bishop, 2002].

That is partly because the notion of intentionality in general is still highly controversial. For example, some philosophers argue that it can be defined in causal/informational terms [Dretske, 1984; 1995] — a position that is implied by the Physical Symbol System hypothesis, too. Others argue that intentionality is evolutionarily based [Millikan, 1984]. If so, then it would follow that meanings might properly be ascribed to A-Life evolutionary robots (see below), though not to designed/programmed systems [Boden, 2001]. In short, the semantic "grounding" of concepts is a contentious matter [Harnad, 1990].

These philosophical controversies are highly relevant to "strong AI", according to which some conceivable computer systems would literally possess meaning, understanding, and intelligence [Searle, 1980]. So they are raised in opposition, for instance, to any work in cognitive science which is based on the Physical Symbol System approach (specifically attacked by Searle in his paper). However, they are rarely relevant to work in 'weak AI". Here, the claim is merely that real psychological processes are sufficiently like (some) computer processes for it to be helpful for psychologists to express their theories in computational, and even programmed, terms. It is that claim, rather than the thesis of strong AI, which motivates cognitive science.

As remarked above, the key concepts of cognitive science are, in a broad sense, computational. "Computational", here, is a shorthand term that covers not only symbolic computation but other types of information processing too. For the concepts in question fall into two camps, cybernetic-dynamical and logical-symbolic.

Some theories in cognitive science are hybrids of both. For example, a model of certain types of clinical apraxia [Norman and Shallice, 1986; Cooper, Shallice and Farringdon, 1995; Cooper *et al.*, 1996] combines connectionism with hierarchical planning as envisaged in GOFAI. Usually, however, they are confined to one side of this intellectual fence. Indeed, the authors often express pungent criticism of, not to say scorn for, the other side.

The history of the field shows changing fashions in these two approaches, each of which has defined increasingly complex and powerful forms of processing over the years. Today, the first is in the ascendant. Connectionist AI, situated robotics, dynamical systems based on differential equations, cellular automata, self-organizing and evolutionary systems, and even decentralized versions of GOFAI (such as work on 'agents" and 'distributed cognition") are all flavour of the month at the beginning of the twenty-first century.

The best-known form of connectionism is PDP (parallel distributed processing), especially those PDP networks which can learn [Rumelhart, McClelland, and the PDP Group, 1986]. These systems are associative memories made up of large numbers of simple, locally interconnected, computational units, whose associations may be positive or negative. Their learning rules are versions of those initially described by the psychophysiologist Donald Hebb [1949]. Representations are stored as a large set of mutually equilibrated connection-weights, distributed across the whole network [Hinton, McClelland and Rumelhart, 1986]. PDP systems can therefore carry out multiple constraint-satisfaction (wherein many, perhaps partly conflicting, constraints are simultaneously satisfied or approximated), and are inherently capable of noise-tolerant pattern recognition. This form of information processing is very difficult to achieve in symbolic AI, not least because the potential errors have to be specifically anticipated; GOFAI programs are notoriously brittle as a result.

Connectionism has provided models of many cognitive and developmental processes. Most of these are based on PDP research. Localist, i.e. non-distributed, connectionism — in which semantic content is represented by only one, or a few, unit/s — is much less popular (although it has recently been robustly defended [Page, 2000]). Besides various series of annual/biennial conference proceedings, collections of psychologically relevant PDP papers abound (e.g. [McClelland, Rumelhart, and the PDP Group, 1986; Ramsey, Stich and Rumelhart, 1991; Holyoak and Barnden, 1993; Elman *et al.*, 1996]).

Some PDP results have had surprising implications for cognitive science as a whole. An early example, and still one of the best-known, was the past-tense learner [Rumelhart and McClelland, 1986]. This network was never provided with any explicit rules for forming the past tense of verbs (such as add -ed to the present-tense form"). Instead, it was trained by being continually presented with pairs of present/past forms (e.g. go/went, run/ran, reach/reached, slip/slipped). In the testing mode, only the present tense would be presented as input, and the network would produce some output in response. At first, all the test-responses would be correct (including the output "went" for the input "go"). Later, there was an apparent regression, as irregular forms were over-regularized (so "go" now elicited "goed"). Eventually, however, the output would always be the correct past-tense form, whether regular (e.g. reached) or irregular (e.g. went).

This system caused an explosion of interest, because it appeared to contradict a hugely influential belief in, not to say a dogma of, classical cognitive science: Noam Chomsky's [1957; 1965] claim that language learning requires the acquisition/development of formal syntactic rules. If Chomsky had applied this claim to language, analogous claims had been made with respect to problem solving by Newell and Simon. Indeed, the GOFAI-based computational approach in general assumed that mental processing involves symbolic rules (including heuristics).

Whether the past-tense learner really did show just what it was claimed to show was, and remains, controversial. Some critics argued that the network did not make exactly the same errors as infants do, and that some of the errors made by children could be explained by GOFAI but not by PDP [Pinker and Prince, 1988]. They added that the system could in principle learn any linguistic regularity, whereas children do not (because natural languages share certain structures, and lack others) and cannot (because some pre-existing bias is needed to enable interesting structure to be picked out). Even non-Chomskians had to admit that if people find it difficult to learn a grammatical structure (mirror-image reversal of word strings, for example), then a PDP simulation should do so too. A further difficulty was raised by GOFAI researchers who argued that connectionism could not allow that the meaning of a sentence depends recursively on the meaning of its component parts [Fodor and Pylyshyn, 1988]. If so, then it could not explain how an infant comes to be able to generate indefinitely many new sentences. Although a PDP model could learn word pairs, it would never be able to learn syntactically structured language.

(Problem solving and planning were seen as similarly out of reach. Indeed, these GOFAI-based critics held that connectionism had no interest for computational psychology. They saw it as addressing the neurological implementation of cognition, rather than cognition as such. Newell, for instance, said that nothing below 100 milliseconds of brain-activity is significant in the study of cognition [Newell, 1980; 1990].)

It turned out later that the psychological data being used on both sides of this late-1980s debate were faulty. For example, psycholinguists had reported, or anyway implied, that–for a while–children always over-regularize all irregular verbs. But they do not: they over-regularize only 5–10% of irregulars, and correct uses co-occur with the incorrect ones. Moreover, psychologists had not reported any irregularizations of regular verbs — which, indeed, a Chomskian would never expect. Yet they do happen. However, both these unexpected phenomena, and many others, fell out 'for free' from PDP networks explored in the early 1990s [Plunkett and Marchman, 1991; 1993].

Even these brief remarks about the past-tense learner illustrate how computer models can lead to theoretical debate, and to empirical advance, in cognitive science. The subtle complexities in the pattern of infants' usage of the past tense, for example, were discovered only as a result of this controversy about the underlying computational mechanisms.

The complaint (above) about recursiveness and componentiality is a reminder of something that is often forgotten. Namely, that many of the strengths of classical symbolic AI have yet to be matched in connectionist and/or situated systems [Boden, 2007, 12.viii-ix and 13.iii.c]. Whether this reflects a point of principle or merely of practice is a hotly contested issue.

For example, Marvin Minsky and Seymour Papert have mounted a fundamental critique of connectionism, first published in 1969 and reiterated twenty years later after the mid-1980s renaissance of interest in that area [Minsky and Papert, 1988]. Their overall charge is that certain types of information processing that are carried out by minds cannot, in principle, be computed by networks based on current ideas about connectionism. Their own (hybrid) theory of mental architecture allows for distributed cognition (they speak of the "society" of mind) and connectionist computation. But it draws as much, or even more, on GOFAI as on neural networks [Minsky, 1985; 2006].

That is no accident, for deductive argument, verbal reasoning, hierarchical planning, and critical self-monitoring are all best modelled in GOFAI terms. Of course, the brain itself is at base a connectionist (parallel-processing) machine. But it does not follow that the virtual machine it uses to do logic, for instance, is not broadly GOFAI in nature. Indeed, some leading connectionists allow that the brain must emulate a (sequential) von Neumann machine in order to accomplish certain logical/linguistic tasks (e.g. [Norman 1986: 541f].). And some of them have tried to model hierarchical structures, such as language or family-trees, in PDP systems (several examples are described in [Hinton, 1990]). So far, however, success has been very limited.

A main attraction of symbolic AI, once the von Neumann computer was available to implement it, was the promise (the logicist assumption) that it could be used to represent specific propositional meanings. Still today, if one wants to model the inferential relations between distinct propositions, one is much better off using a GOFAI approach than a connectionist/dynamical one. Connectionist systems, on the other hand, are better than GOFAI if one wants to model noise-tolerant pattern-recognition [Hinton *et al.*, 1986]. Likewise, a dynamical model of problem-solving [Busemeyer and Townsend, 1993] may simulate how various relevant considerations ebb and flow in conscious reasoning. But it does not explain how those considerations are generated in the first place.

Connectionism is not the only information-processing approach that has been hailed as a much-needed alternative to symbolic AI (and that has failed to match the key strengths of GOFAI): another is situated robotics. Situated robots may be based on "subsumption" architectures [Brooks, 1991a; b], or on dynamical systems [Beer, 1990]. The general principle was foreseen in Simon's description of the apparently-complex pathway of the actually-simple (obstacle-avoiding) ant [Simon, 1962; 1969: ch. 3]. Namely, that the situated system, whether organism or computer, responds directly to environmental cues, in predetermined ways.

In other words, it relies on relatively inflexible sets of inbuilt reflex mechanisms. There is no planning, no deliberation, no choice — and no internal representation. The general motivation is that the behaviour of many animals, such as insects, seems to be like this (a position backed up by extensive neuroscientific and ethological evidence). And a further claim is added: that human behaviour too is mostly, or even entirely, free of such representations.

The latter claim has attracted fierce criticism. David Kirsh [1991], for instance, argues that situationist systems cannot, in principle, compute certain types of information. Specifically, they cannot do those tasks which depend on concept formation. Concepts, he says, are internal representations which enable their possessor to recognize perceptual invariance (as in recognizing many different cats as cats), to reify and combine invariances (in referring predicates to names, for instance, or in drawing inferences), and to reidentify individuals over time. They allow for anticipatory self-control (i.e. planning), and negotiation between (not just scheduling of) potentially conflicting desires. Moreover, they enable one to think counterfactually, and to use the cognitive technology of language to create new abilities and teach them to others. Human adults possess all these information-processing capacities, chimps most of them, dogs some of them, and newborn babies hardly any. Insects, by contrast, do not feature at all on the conceptual radar.

Some models of situated action in animals–crickets, hoverflies, cockroaches, lampreys ... and even frogs [Boden, 2006, 14.vii and 15.vii] — are examples of "neuro-ethology". That is, they take detailed account of the organism's behaviour and neurophysiology, and many have led in turn to further biological investigations (e.g. [Arbib and Cobas, 1991; Arbib and Liaw, 1995; Webb and Scutt, 2000]).

Sometimes, the limits of all possible anatomical arrangements of a certain type have been explored. For instance, consider the patterns of neuromuscular connections that could enable lampreylike creatures — including those which do not actually exist — to swim. Lampreys do not have moveable fins, but swim by rhythmic undulations of the entire body. Computational experiments have shown that there are many network architectures capable of controlling this type of swimming [Ijspeert, Hallam, and Willshaw, 1997]. Moreover, some artificial networks for undulatory swimming, despite being composed of "neurones" closely based on the biological data, were much more efficient than those found in real lampreys. Some had a frequency range five times larger; even when the connections (and their type: excitatory or inhibitory) were fixed to be identical with those in real lampreys, some had frequency ranges three times larger.

That last finding counters the assumption made by some computational psychologists (such as David Marr) that living organisms will in general employ the mathematically optimal solution for a given computational task. This assumption was explicitly used by Marr as a reason for favouring one hypothesis rather than others (e.g. [Marr, 1982]; cf. [Boden, 1986: 63ff., 76]). Evidently, it is mistaken.

It is true, nevertheless, that biological evolution tends on the whole to eliminate inefficiencies. For instance, a number of quick-and-dirty information-processing methods have been evolved in Homo sapiens, whereby a surpisingly wide range of problems are not so much solved by rational thought as dissolved by biologically-inspired guessing [Gigerenzer, 2000]. The reduction of inefficiency is the prime reason why evolutionary computing has become popular in cognitive science (it was used in the lamprey study itself) — and in technological AI as well.

In evolutionary computing, a program, or a specification of a neural network, is randomly mutated by "genetic" algorithms (GAs). A GA causes alterations similar to point mutations and crossovers in biology. There may be many simultaneous mutations within the one program (hence the credit-assignment problem). And the random process of mutation is continually repeated, over successive generations. More accurately, the members of an entire population of 100 or more (initially identical) programs or networks are repeatedly mutated. At each generation, some fitness function is applied — either automatically by the system itself, or interactively by a human being. The fitness function identifies the one or two best (least-worst) members of that generation, which is/are then used as the 'parent/s' in breeding the next generation.

The method was foreseen as a possibility by von Neumann in his writings on cellular automata, and first defined mathematically by John Holland [1962]. Small evolutionary programs were soon written (e.g. [Fogel, Owens, and Walsh 1966]), and key figures in cybernetics-AI–including Minsky, McCulloch, Newell, and John McCarthy — asked how one might evolve program-controlled sensory/motor prostheses for human beings [Fogel and McCulloch, 1970: 271-295]. But with scant results. A major difficulty, in practice, was a version of the "credit-assignment problem" of AI in general: if a program results in satisfactory performance, how can one decide just which aspects of it were (most) responsible? In the case of evolutionary models, how can one identify the mutations that were (most) helpful? Eventually, Holland solved this problem by defining the "bucket-brigade algorithm" [Holland *et al.*, 1986 70–73]. Consequently, and also thanks to 1980s computer power, evolutionary computing began to be pursued in earnest. It is now used to evolve programs, networks, and robot-morphologies (and artworks, too [Whitelaw, 2004]).

After hundreds, or thousands, of generations, the resulting program or network may be successful in achieving the task which the programmers had in mind from the beginning, when they specified the fitness function. But this task is not represented as a goal in the system, as is done in classical AI. (Compare snails, or bees: they do what they do, without knowing/representing what it is that they are managing to achieve.) The final outcome may even be maximally efficient, though in general this is not guaranteed. Moreover, an evolved system may achieve the task that the programmers had in mind in ways which they had never expected, or imagined.

For example, the evolving network may function as the sensorimotor controller of a robot. In that case, the successive generations of the evolving population are simulated: there are not 100 or more real robots scattered across the floor. But after every 500 generations (say), the currently-best network is downloaded into a real robot for testing/confirmation. Even when the task assigned is very simple — such as moving to the centre of the floor and staying there, or navigating from one side of the floor to another — entirely unexpected computational mechanisms may evolve. So sensory "organs" that are not strictly necessary for the task, such as pressure-sensitive whiskers or a second eye, may sometimes happen to lose their connections to the network controller [Cliff, Harvey, and Husbands, 1993], and/or mini-circuits may evolve which act as visual line-orientation detectors comparable with those found in mammals [Harvey, Husbands and Cliff, 1994; Husbands, Harvey and Cliff, 1995].

Evolution is an example of the biologically crucial phenomenon of self-organization, wherein structure spontaneously appears from a less well-ordered base. Just as this description applies to the development of the individual organism, from fertilized egg to embryo and adult, so it applies to the process of evolution itself, wherein new structures and new species emerge over time.

As it happens, one of the researchers responsible for the lamprey models had previously co-authored a seminal study of self-organization in neural networks ([Willshaw and von der Malsburg, 1976]; cf. [Linsker, 1988; 1990]). That study showed that internal structure will emerge spontaneously in an initially random system, given certain very general–and minimal — conditions. Besides suggesting explanations of the development of various information-processing mechanisms (e.g. orientation columns in visual cortex), this undermined simplistic interpretations of the nature/nurture divide [Boden, 2006, chs. 7.vi.g and 14.vi.b]. If a network can organize itself spontaneously, then from the fact that a new-born animal already possesses structure X in its brain, it does not follow that structure X was "innate" in the sense of being specifically coded in the genes.

I said at the outset that cognitive scientists view the mind-brain as an "abstract" information-processing machine. But this is not to say that they endorse the — quintessentially abstract–dogma of strict multiple realizability. This dogma asserts that there are many different ways in which a given computation or information-processing system could be implemented, and that the implementation details are irrelevant in considering the computations concerned.

According to classical functionalism, multiple realizability implies that psychology is autonomous: in other words, biological facts about the brain are irrelevant to it [Putnam, 1960; 1967; Fodor, 1968]. As one computationalist put it: "whether the physical descriptions of the events subsumed by [psychological] generalizations have anything in common is, in an obvious sense, entirely irrelevant to the truth of the generalizations, or to their interestingness, or to their degree of confirmation, or, indeed, to any of their epistemologically important properties" [Fodor, 1974: 14f.]. This doctrine is still used as an argument to counter the objection that metal-and-silicon computers are (physically) very different from neuroprotein, and also as a way of avoiding neuroscientific questions to which, as yet, answers cannot be given.

But, largely due to the advance of neuroscience since multiple realizability was first defined, current cognitive science sometimes attempts a fairly close integration of psychological and neurophysiological data and theories. This is especially evident in neuro-ethological modelling, and in (some) work on connectionism.

The information-processing systems defined by connectionism are broadly inspired by the brain. For example, positive and negative connection-weights echo facilitatory and inhibitory synapses; and Hebbian rules were first defined to describe the results of coactivation among cerebral neurones. Indeed, connectionists typically make much of their neurological roots, when asserting the superiority of their approach over GOFAI. And it is certainly true that connectionist models have sometimes prompted fruitful neuroscientific research (into dyslexia, for example, or pathological action-errors of various kinds).

However, most existing connectionist systems are in fact hugely different from the brain. In general, the component units are computationally far too simple in comparison with real neurones. Moreover, the mathematics that defines the learning rules is usually highly unrealistic. The popular method of back propagation [Rumelhart, Hinton and Williams, 1986a; b], for example, depends on units' being able to transmit information in two directions — which real neurones cannot do.

In recent years, some attempts have been made to model actual neurones more faithfully [O'Reilly and Munakata, 2000] — which is to say, to compromise on the doctrine of multiple realizability. Increasingly, the extent to which the doctrine can be safely followed, or must be specifically challenged, is coming to be seen as an interesting empirical question.

One interesting — and, to many people, counterintuitive — example is the development of "neuromodulatory" information-processing systems called GasNets [Philippides, Husbands and O'Shea, 1998; Philippides *et al.*, 2005]. These are inspired by the discovery of simple chemicals in the brain (such as nitrous oxide) whose diffusion across wide areas alters the computational properties of the individual cells concerned. The size of the diffusion volume matters, and so does the shape of the source — both of which biological facts are simulated in GasNets.

In these models, some nodes scattered across the network can release diffusible 'gases,' which modulate the intrinsic properties of other nodes and connections in various ways, depending on concentration. So one and the same node behaves differently at different times. Given certain gaseous conditions, a particular node will affect another despite there being no direct synaptic link. In other words, it is the interaction between the gas and the electrical connectivities in the system which is crucial. And, since the gas is emitted only on certain occasions, and diffuses and decays at varying rates, this interaction is dynamically complex. So, unlike the usual connectionist system, the pattern of connectivities is not the only important factor determining what types of computation will take place.

In short, multiple realizability should not be used as an excuse for always ignoring what is known about neurophysiology. The neuroscientific facts may even alert us to aspects of computation (e.g. neuromodulation, as in GasNets) which were not formerly suspected. Nevertheless, the theoretical "autonomy" of psychology/computation remains, in the sense that one may–and sometimes, one must — consider what computer scientists would term the virtual machine of the mind/brain, without worrying about the details of its biological implementation.

This is especially true when those details are not yet known. And that, in turn, is most likely to be true when high-level and/or global processes within human personalities are concerned — hypnosis, for example (see above). Where such phenomena are the focus of interest, it may not even be necessary — yet — to worry about implementation in computer models. For if one is considering the computational architecture of the whole mind, many theoretical questions have to be posed, and answered, at a much higher level.

To be sure, Newell's SOAR system is a relatively wide-ranging and inclusive model of the architecture of human cognition, which has been implemented [Rosenbloom, Laird, and Newell, 1993]. So has ACT*, a similarly inclusive model [Anderson, 1983; 1993]. But the prime focus of both SOAR and ACT* is cognition (problem solving, memory ...). Emotion is hardly featured, and personality is ignored.

Two cognitive scientists who have considered such issues at length are Minsky [1985; 2006] and Aaron Sloman [2003, n.d.]. Their accounts are overwhelmingly theoretical, concerned with the general principles of the sorts of computation which might — indeed, must — underlie the rich complexities of adult human minds. One example of these complexities is the emotion of grief. Sloman's analysis of the computational structure of grief makes clear that grief is more than mere feeling [Wright, Sloman, and Beaudoin, 1996]. It involves irrational behaviour driven by obsessional thoughts, continual distraction, depression, anger, and guilt — all of which gradually pass, over many months, as mourning does its work. Just what "work" that is, is explained in terms of the deconstruction and restructuring of fundamental goal-complexes in the bereaved person's mind.

Although both these architectural theories are deeply rooted in practical AI, neither of them is implementable at this stage. Nor will they be (in my opinion) for many, many decades hence. Sloman, however, has provided — and is continually improving — a model of certain aspects of his approach. Namely, his analysis of various types of anxiety: their essential nature (and subtle differences), and the differing psychological/computional conditions in which they arise, and for which they were evolved [Wright and Sloman, 1997].

Emotions in general are seen by Sloman as scheduling mechanisms, by means of which an organism having diverse (potentially conflicting) motives or goals can achieve as many of them as possible. His model of anxiety, then, suggests how varying types of anxiety function so as to shape purposive behaviour in broadly coherent, intelligent, ways. It simulates a nursemaid caring for a dozen babies, each of whom has to be fed, and prevented from falling into a ditch or crawling towards a busy road. Even these few motives, or goals, can conflict. (For instance, while she is feeding a very hungry baby, another may crawl near the ditch.) Further, they can each arise unexpectedly as the result of environmental contingencies. For every simulated baby is an autonomous agent, whose crawling and hungry crying is independent of other babies, and of the nursemaid's actions and motives.

Consequently, the simulated nursemaid's choices about what to do at each moment are computationally complex. They are constrained by her notional embodiment: she only has two hands, so cannot pick up several babies simultaneously; and she cannot be in two places at once. They are guided, too, by the seven different motives she wants to satisfy: feeding, protecting, moving, and rescuing the babies, building a protective fence, patrolling the ditch, and — if no other motive is currently activated — wandering around the nursery.

That is not all, for her decisions are constrained also by the priorities she holds (moving a baby away from the ditch is more urgent than feeding it, even though feeding is just as necessary), and by her assumptions about consequences (falling into the ditch results in a dead baby). Her current perceptions (of the hungriness of each baby, and its location vis-a-vis the ditch and the road) must be taken into account as well. Finally, she must rely on her judgements of urgency — even the hungriest baby can be temporarily ignored, if another is nearing the road — and of danger: sometimes, she must rescue the baby immediately, without stopping to think.

Different emotional processes (different modes of anxiety) are defined within the model to simulate these interacting phenomena. By and large, the babies survive. (In real life, of course, there are other pertinent considerations: personal preferences, for instance, and moral priorities. These are ignored in Sloman's simulation, though not in his background theory. There are many other simplifications too — to the "visual" system, for example; some will soon be overcome, while others are more problematic.)

Such architectural theories cast light on the nature of human freedom [Boden, 2006, ch. 7.i.f-g]. A real nursemaid is free to choose what to do at any time. She can delay feeding one baby so as to finish singing a lullaby to another. She can even ignore the babies entirely for a while, to phone her boyfriend (and if a tragedy results, she can rightly be held responsible). On some occasions, to be sure, she "has no choice": the perceived danger must be averted, now. In Sloman's model, as in real life, the particular type of anxiety that is triggered in such cases preempts any deliberative thinking: the carer just does what needs to be done — "automatically", one might say. However, the sense in which a human being (sometimes) has no choice is fundamentally different from the sense in which an insect (always) has no choice about what to do next.

Only an organism with at least the computational complexity that is implemented in Sloman's simulation, and sketched in his (or Minsky's) architectural theory, is capable of "having no choice" in the human sense (cf. [Dennett, 1984]). In short, freedom does not depend on randomness, or on mysterious spiritual influences: it is an aspect of how our minds work.

Similarly, creativity — also believed by many people to be somehow beyond the reach of science — has been illuminated by computational theories and computer models [Boden, 1990/2004]. For example, distinct types of creativity have been defined, involving the exploration and/or transformation of accepted thinking styles, or unfamiliar combinations of familiar ideas. The latter type, of course, has long been noted by experimental psychologists. What is new is the use of computational ideas in trying to explain just how such combinations can arise [Boden 1990/2004: ch. 6; Hofstadter and FARG, 1995; Fauconnier and Turner, 2002]. Not all computer models that appear to be creative — composing extremely impressive music, for example [Cope, 2001] — are intended as simulations of human creative processes. But even these may suggest some psychologically interesting ideas.

I've argued that the computational approach of cognitive science has already cast light on many areas of psychology, from low-level vision all the way to freedom and creativity, and that it promises more such advances in the future. It has furthered many aspects of anthropology, neuroscience, and biology too [Boden, 2006, chs. 8, 14, and 15 respectively]. But that is not to deny that there are some deeply puzzling problems, alongside the many as-yet-unanswered, though apparently manageable, questions.

These puzzling problems are philosophical, rather than scientific. But all science assumes and implies particular philosophical positions, so this is not a clean divide. Moreover, because of the many disagreements within the philosophy of mind in general, cognitive science has been especially prone to philosophical argument — both from within the field and outside it, and from philosophers and scientists alike.

Two such problems, the nature of intentionality and the possibility of strong AI, were mentioned above. A third is the frame problem, first named by AI scientists [McCarthy and Hayes, 1969] and revisited by countless authors ever afterwards (e.g. [Pylyshyn, 1987; Ford and Hayes, 1992]).

The essence of the frame problem is that the contingencies of real-world events, the complexities of human world-knowledge, and the open texture of words in natural language all militate against cut-and-dried computer programs — and logicist philosophy, too. Whatever obstacles AI modellers anticipate, or whatever features they include in their programmed definitions, something else may turn out to be relevant — and lead the system to fail. Connectionism can counter the frame problem to some extent, avoiding brittleness by means of multiple constraint satisfaction; even so, unanticipated constraints (not represented in the network) may be crucial. GOFAI researchers have developed various types of non-monotonic logic, for use in expert systems and robotics [Boden, 2006, 13.iii.e]. And psycholinguists have tried to corral the notion of relevance [Sperber and Wilson, 1986]. Cognitive science has advanced accordingly. But that is not to say that the frame problem has been fully solved, or ever can be.

Yet another troubling problem is the radical divide between realism and constructivism in philosophy: the Anglo-American (neo-Cartesian) and Continental (neo-Kantian) approaches, respectively. Most cognitive scientists, like scientists in general, adopt the realist view — usually, without even considering the alternative. But some have recently moved in the opposite direction, and their empirical research — in robotics as well as psychology — has been affected accordingly [Varela, Thompson, and Rosch, 1991; Clark, 1997; Harvey, 2005; Wheeler, 2005]. Indeed, one AI scientist has offered an ambitious new definition of "computation" which not only includes intentionality but also seeks to combine the Anglo-American and

Continental viewpoints [Smith, 1996]; however, his ideas (and his writing-style) are highly idiosyncratic, and more readers have been repelled than have been intrigued — still less, convinced.

The realist/constructivist split is arguably the most fundamental dispute in Western philosophy. Among other things, it underlies the huge — and seemingly insuperable — difficulties in explaining how it is possible for phenomenal consciousness to arise in the brain/body. Mind-brain correlations cannot answer this question, and their very existence as normally conceptualized is put into doubt on the constructivist view, wherein to posit the existence of mind or mental representations is itself regarded as illegitimate [Boden, in press: 14.x-xi]. (As an example, consider the Wittgensteinian critique of the references to "mind" and "mind-brain" that are widespread in cognitive science and general neuroscience [Bennett and Hacker, 2003].) The Continental viewpoint also prompted most of Hubert Dreyfus' [1972; 1992] influential criticisms of AI and computational psychology.

My own view is that constructivism is not only anti-scientific but essentially irrational [Boden, 2006, 1.iii.b]. However, there is no knock-down argument for this — a point admitted even by the arch-computationalist Jerry Fodor [1995]. Perhaps there never can be. For sure, a definitive verdict will not be forthcoming tomorrow.

To end on a more positive note: one common objection to cognitive science can be robustly rejected, by recalling the key theme of this chapter. Critics often point out that, over past centuries, the mind/brain has been likened to many different machines. These were always the most up-to-date technology at the time, but were recognized later as inadequate or even grossly misleading analogies for psychology and neuroscience. Computers (so this argument goes) are the current version of this habitual metaphor. They will fall out of favour eventually, like all the others, and cognitive science will expire — probably, to be replaced by theories cast in terms of the next generation of technological gizmos.

If by "computers" these objectors mean today's AI technology, they have a point. After all, the current stock of computational concepts and computing machines has not provided answers to all of cognitive science's questions. (Thirty years ago, and partly because of the more primitive technology then available, the plausible/promising answers were even fewer.) But if they mean computing mechanisms in general, they are mistaken. Much as physicalists do not claim that every aspect of the world can be captured by today's physics, but rather by whatever turns out to be the best theory of physics, so cognitive scientists claim that the mind is to be understood by whatever turns out to be the best theory of computers [Chrisley, 2000]. It remains to be seen just how different from today's notions that theory will be. But, as in physics, much of our current thinking may endure.

In sum, the computer — understood as a generic information-processing machine — is not merely the latest technological metaphor for mind. It is the last, whose implications are being continually enriched.

BIBLIOGRAPHY

[Anderson, 1983] J. R. Anderson. *The Architecture of Cognition*, Cambridge, Mass.: Harvard University Press, 1983.

[Anderson, 1993] J. R. Anderson, ed. *The Rules of the Mind*, Hillsdale, NJ: Lawrence Erlbaum, 1993.

[Arbib and Cobas, 1991] M. A. Arbib and A. Cobas. Schemas for Prey-Catching in Frog and Toad. In J.-A. Meyer and S. W. Wilson (eds.), *From Animals to Animats (Proceedings of the First International Conference on Simulation of Adaptive Behavior)*, Cambridge, Mass.: MIT Press, pp. 142–151, 1991.

[Arbib and Liaw, 1995] M. A. Arbib and J.-S. Liaw. Sensorimotor Transformations in the Worlds of Frogs and Robots, *Artificial Intelligence*, 72: 53-79, 1995.

[Beer, 1990] R. D. Beer. em Intelligence as Adaptive Behavior: An Experiment in Computational Neuroethology, Boston: Academic Press, 1990.

[Bennett and Hacker, 2003] M. R. Bennett and P. M. S. Hacker. *Philosophical Foundations of Neuroscience*, Oxford: Blackwell, 2003.

[Boden, 1988] M. A. Boden. *Computer Models of Mind: Computational Approaches in Theoretical Psychology*, Cambridge: Cambridge University Press, 1988.

[Boden, 1990/2004] M. A. Boden. *The Creative Mind: Myths and Mechanisms*, London: Weidenfeld & Nicolson, 1990. 2nd edn. expanded/revised, London: Routledge, 2004.

[Boden, 2001] M. A. Boden. Life and Cognition. In J. Branquinho (ed.), *The Foundations of Cognitive Science*, Oxford: Oxford University Press, pp. 11–22, 2001.

[Boden, 2006] M. A. Boden. *Mind as Machine: A History of Cognitive Science*, Oxford: Oxford University Press, 2006.

[Broadbent, 1958] D. E. Broadbent. *Perception and Communication*, Oxford: Pergamon Press, 1958.

[Brooks, 1991a] R. A. Brooks. Intelligence Without Representation, *Artificial Intelligence*, 47: 139-159, 1991.

[Brooks, 1991b] R. A. Brooks. Intelligence Without Reason, *Proceedings of the Twelfth International Joint Conference on Artificial Intelligence*, Sydney, 1-27, 1991.

[Bruner *et al.*, 1956] J. S. Bruner, J. Goodnow, and G. Austin. *A Study of Thinking*, (New York: Wiley), 1956

[Busemeyer and Townsend, 1993] J. R. Busemeyer and J. T. Townsend. Decision Field Theory: A Dynamic-Cognitive Approach to Decision Making in an Uncertain Environment, *Psychological Review*, 100: 432-459, 1993.

[Chrisley, 2000] R. L. Chrisley. Transparent Computationalism. In M. Scheutz (ed.), *New Computationalism: Conceptus-Studien 14*, Sankt Augustin: Academia Verlag, pp. 105–121, 2000.

[Chomsky, 1957] A. N. Chomsky. *Syntactic Structures*, 'S-Gravenhage: Mouton, 1957.

[Chomsky, 1965] A. N. Chomsky, *Aspects of the Theory of Syntax*, Cambridge, Mass.: MIT Press, 1965.

[Clark, 1997] A. J. Clark. *Being There: Putting Brain, Body, and World Together Again*, Cambridge, Mass.: MIT Press, 1997.

[Cliff *et al.*, 1993] D. Cliff, I. Harvey, and P. Husbands. Explorations in Evolutionary Robotics, *Adaptive Behavior*, 2: 73-110, 1993.

[Cooper *et al.*, 1996] R. Cooper, J. Fox, J. Farringdon, and T. Shallice. Towards a Systematic Methodology for Cognitive Modelling, *Artificial Intelligence*, 85: 3-44, 1996.

[Cooper *et al.*, 1995] R. Cooper, T. Shallice, and J. Farringdon. Symbolic and Continuous Processes in the Automatic Selection of Actions. In J. Hallam (ed.), *Hybrid Problems, Hybrid Solutions*, Oxford: IOS Press: pp. 27–37, 1995.

[Cope, 2001] D. Cope. *Virtual Music: Computer Synthesis of Musical Style*, Cambridge, Mass.: MIT Press, 2001.

[Craik, 1943] K. J. W. Craik. *The Nature of Explanation*, Cambridge: Cambridge University Press, 1943.

[Dennett, 1984] D. C. Dennett. *Elbow Room: The Varieties of Free Will Worth Wanting*, Cambridge, Mass.: MIT Press, 1984.

[Dienes and Perner, 2007] Z. Dienes and J. Perner. The Cold Control Theory of Hypnosis. To appear in G. Jamieson (ed.), *Hypnosis and Conscious States: The Cognitive Neuroscience Perspective*, pp. 293–314. Oxford: Oxford University Press, 2007.

[Dretske, 1984] F. I. Dretske. *Knowledge and the Flow of Information*, Oxford: Blackwell, 1984.

[Dretske, 1995] F. I. Dretske. *Naturalizing the Mind*, Cambridge, Mass.: MIT Press, 1995.

[Dreyfus, 1972] H. L. Dreyfus. *What Computers Can't Do: A Critique of Artificial Reason*, New York: Harper & Row, 1972.

[Dreyfus, 1992] H. L. Dreyfus. *What Computers Still Can't Do: A Critique of Artificial Reason*, Cambridge, Mass.: MIT Press, 1992.

[Elman *et al.*, 1996] J. L. Elman, E. A. Bates, M. H. Johnson, A. Karmiloff-Smith, D. Parisi, and K. Plunkett. *Rethinking Innateness: A Connectionist Perspective on Development*, Cambridge, Mass.: MIT Press, 1996.

[Fauconnier and Turner, 2002] G. R. Fauconnier and M. Turner. *The Way We Think: Conceptual Blending and the Mind's Hidden Complexities*, New York: Basic Books, 2002.

[Fodor, 1968] J. A. Fodor. *Psychological Explanation: An Introduction to the Philosophy of Psychology*, New York: Random House, 1968.

[Fodor, 1974] J. A. Fodor. Special Sciences, or the Disunity of Science As a Working Hypothesis, *Synthese*, 28: 77-115, 1974.

[Fodor, 1995] J. A. Fodor. Review of John McDowell's Mind and World, *The London Review of Books*, April 20th, 1995. Reprinted in J. A. Fodor, *In Critical Condition: Polemical Essays on Cognitive Science and the Philosophy of Mind*, Cambridge, Mass.: MIT Press, pp. 3–8, 1998.

[Fodor and Pylyshyn, 1988] J. A. Fodor and Z. W. Pylyshyn. Connectionism and Cognitive Architecture: A Critical Analysis, *Cognition*, 28: 3-71, 1988.

[Ford and Hayes, 1992] K. M. Ford and P. J. Hayes, eds. *Reasoning Agents in a Dynamic World: The Frame Problem. Proceedings of the 1st International Workshop on Human and Machine Cognition*, Greenwich, Conn.: JAI Press, 1992.

[Gigerenzer, 2000] G. Gigerenzer. *Adaptive Thinking: Rationality in the Real World*, Oxford: Oxford University Press, 2000.

[Harnad, 1990] S. Harnad. The Symbol Grounding Problem, *Physica D*, 42: 335-346, 1990.

[Harvey, 2005] I. Harvey. Evolution and the Origins of the Rational. In A. J. T. Zilhao, ed., *Cognition, Evolution, and Rationality*, London: Routledge), ch. 6, 2005.

[Harvey *et al.*, 1994] I. Harvey, P. Husbands, and D. Cliff. Seeing the Light: Artificial Evolution, Real Vision. In D. Cliff, P. Husbands, J.-A. Meyer and S. W. Wilson, eds., *From Animals to Animats 3: Proceedings of the Third International Conference on Simulation of Adaptive Behavior*, Cambridge, Mass.: MIT Press, pp. 392–401, 1994.

[Haugeland, 1985] J. Haugeland. *Artificial Intelligence: The Very Idea*, Cambridge, Mass.: MIT Press, 1985.

[Hebb, 1949] D. O. Hebb. *The Organization of Behavior: A Neuropsychological Theory*, New York: Wiley, 1949.

[Hinton, 1990] G. E. Hinton, ed. Connectionist Symbol Processing, Special issue of *Artificial Intelligence*, 46, nos. 1-2, 1990.

[Hinton, *et al.*, 1986] G. E. Hinton, J. L. McClelland, and D. E. Rumelhart. Distributed Representations. In Rumelhart, McClelland and PDP Group 1986: 77-109, 1986.

[Hofstadter and FARG, 1995] D. R. Hofstadter and FARG (The Fluid Analogies Research Group). *Fluid Concepts and Creative Analogies: Computer Models of the Fundamental Mechanisms of Thought*, New York: Basic Books, 1995.

[Holland, 1962] J. H. Holland. Outline for a Logical Theory of Adaptive Systems, *Journal of the Association for Computing Machinery,* 9: 297-314, 1962.

[Holyoak and Barnden, 1993] K. Holyoak and J. Barnden, eds. *Advances in Connectionist and Neural Computation Theory,* 2 vols., Norwood, N.J.: Ablex, 1993.

[Husbands *et al.*, 1995] P. Husbands, I. Harvey, and D. Cliff. Circle in the Round: State Space Attractors for Evolved Sighted Robots, *Journal of Robotics and Autonomous Systems*, 15: 83-106, 1995.

[Ijspeert *et al.*, 1997] A. J. Ijspeert, J. Hallam, and D. Willshaw. Artificial Lampreys: Comparing Naturally and Artificially Evolved Swimming Controllers. In P. Husbands and I. Harvey, eds., *Fourth European Conference on Artificial Life*, Cambridge, Mass.: MIT Press, 256-265, 1997.

[Kirsh, 1991] D. Kirsh. Today the Earwig, Tomorrow Man?, *Artificial Intelligence,* 47: 161-84, 1991.

[Linsker, 1988] R. Linsker. Self-Organization in a Perceptual Network, *Computer,* 21: 105-117, 1988.

[Linsker, 1990] R. Linsker. Perceptual Neural Organization: Some Approaches Based on Network Models and Information Theory, *Annual Review of Neuroscience,* 13: 257-281, 1990.

[McCarthy and Hayes, 1969] J. McCarthy and P. J. Hayes, Some Philosophical Problems from the Standpoint of Artificial Intelligence, In B. Meltzer and D. M. Michie, eds., *Machine Intelligence 4*, Edinburgh: Edinburgh University Press, pp. 463–502, 1969.

[McClelland *et al.*, 1986] J. L. McClelland, D. E. Rumelhart, and the PDP Research Group. *Parallel Distributed Processing: Explorations in the Microstructure of Cognition, Vol. 2, Psychological and Biological Models*, Cambridge, Mass.: MIT Press, 1986.

[McCulloch and Pitts, 1943] W. S. McCulloch and W. H. Pitts. A Logical Calculus of the Ideas Immanent in Nervous Activity, *Bulletin of Mathematical Biophysics,* 5 (1943), 115-133, 1943. Reprinted in [McCulloch, 1965, 19-39].

[McDougall, 1911] W. McDougall. *Body and Mind: A History and a Defense of Animism*, London: Methuen, 1911.

[McDougall, 1923] W. McDougall. *An Outline of Psychology*, London: Methuen, 1923.

[Mays, 1952] W. Mays. Can Machines Think?, *Philosophy,* 27: 148-162, 1952.

[Miller, 1956] G. A. Miller. The Magical Number Seven, Plus or Minus Two: Some Limits on Our Capacity for Processing Information, *Psychological Review,* 63: 81-97, 1956.

[Miller *et al.*, 1960] G. A. Miller, E. Galanter, and K. H. Pribram. *Plans and the Structure of Behavior*, New York: Holt, 1960.

[Millikan, 1984] R. G. Millikan. *Language, Thought, and Other Biological Categories: New Foundations for Realism*, Cambridge, Mass.: MIT Press, 1984.

[Minsky, 1985] M. L. Minsky. *The Society of Mind*, New York: Simon & Schuster, 1985.

[Minsky, 2006] M. L. Minsky. *The Emotion Machine: Commonsense Thinking, ARtificial Intelligence, and the Future of the Human Mind.* New York: Simon and Schuster, 2006.

[Minsky and Papert, 1988] M. L. Minsky and S. A. Papert. Prologue: A View From 1988 and Epilogue: The New Connectionism, in their *Perceptrons: An Introduction to Computational Geometry,* 2nd edn. Cambridge, Mass.: MIT Press, viii–xv & 247–280, 1988.

[Newell, 1973] A. Newell. You Can't Play Twenty Questions with Nature and Win. In W. G. Chase, ed., *Visual Information Processing*, London: Academic Press, 283-308, 1973.

[Newell, 1980] A. Newell. Physical Symbol Systems, *Cognitive Science,* 4: 135-83, 1980.

[Newell, 1990] A. Newell. *Unified Theories of Cognition.* The William James Lectures, 1987 Cambridge, Mass.: Harvard University Press, 1990.

[Newell *et al.*, 1958] A. Newell, J. C. Shaw, and H. A. Simon. Elements of a Theory of Human Problem-Solving, *Psychological Review,* 65: 151-166, 1958.

[Newell and Simon, 1961] A. Newell and H. A. Simon. GPS — A Program that Simulates Human Thought. In H. Billing, ed., *Lernende Automaten*, Munich: Oldenbourg, 109-124, 1961. Reprinted in E. A. Feigenbaum and J. Feldman, eds., *Computers and Thought*, New York: McGraw-Hill, 279-293, 1963.

[Newell and Simon, 1972] A. Newell and H. A. Simon. *Human Problem Solving*, Englewood Cliffs, N.J.: Prentice-Hall, 1972.

[Normann, 1986] D. A. Norman, ed. *Perspectives on Cognitive Science. Papers presented at the first annual meeting of the Cognitive Science Society.* La Jolla, Calif., August 1979. Norwood, N.J.: Ablex, 1986.

[Norman and Shallice, 1986] D. A. Norman and T. Shallice. Attention to Action: Willed and Automatic Control of Behavior. In R. Davidson, G. Schwartz and D. Shapiro, eds., *Consciousness and Self Regulation: Advances in Research and Theory*, Vol. 4, New York: Plenum: 1-18, 1986.

[O'Reilly and Munakata, 2000] R. C. O'Reilly and Y. Munakata. *Computational Explorations in Cognitive Neuroscience: Understanding the Mind by Simulating the Brain*, Cambridge, Mass.: MIT Press, 2000.

[Page, 2000] M. Page. Connectionist Modelling in Psychology: A Localist Manifesto [with peer-commentary], *Behavioral and Brain Sciences,* 23: 443-512, 2000.

[Philippides *et al.*, 1998] A. Philippides, P. Husbands, and M. O'Shea. Neural Signalling — It's a Gas! In L. Niklasson, M. Boden and T. Ziemke, eds. *ICANN98: Proceedings of the 8th International Conference on Artificial Neural Networks*, London: Springer-Verlag, 51-63, 1998.

[Philippides *et al.*, 2005] A. Philippides, P. Husbands, T. Smith, and M. O'Shea. Flexible Couplings: Diffusing Neuromodulators and Adaptive Robotics, *Artificial Life,* 11, 139-160, 2005.

[Plunkett and Marchman, 1991] K. Plunkett and V. Marchman. U-shaped Learning and Frequency Effects in a Multi-Layered Perceptron: Implications for Child Language Acquisiition, *Cognition,* 38: 1-60, 1991.

[Plunkett and Marchman, 1993] K. Plunkett and V. Marchman. From Rote Learning to System Building: Acquiring Verb-Morphology in Children and Connectionist Nets, *Cognition,* 48: 21-69, 1993.

[Preston and Bishop, 2002] J. Preston and M. Bishop, eds. *Views Into the Chinese Room: New Essays on Searle and Artificial Intelligence*, Oxford: Oxford University Press, 2002.

[Putnam, 1960] H. Putnam. Minds and Machines. In S. Hook, ed. , *Dimensions of Mind: A Symposium*, New York: New York University Press, 148-179, 1960.

[Putnam, 1967] H. Putnam. The Nature of Mental States'. First published as 'Psychological Predicates. In W. H. Capitan and D. Merrill, eds., *Art, Mind, and Religion*, Pittsburgh: University of Pittsburgh Press, 37-48, 1967. Reprinted in H. Putnam, *Mind, Language, and Reality: Philosophical Papers*, vol. 2, Cambridge: Cambridge University Press, 429–440, 1975.

[Pylyshyn, 1987] Z. W. Pylyshyn, ed. *The Robot's Dilemma: The Frame Problem in Artificial Intelligence*, Norwood, NJ: Ablex, 1987.

[Ramsey *et al.*, 1991] W. Ramsey, S. Stich, and D. Rumelhart, eds. *Philosophy and Connectionist Thoery*, Hillsdale, N.J.: Lawrence Erlbaum, 1991.

[Rosenbloom *et al.*, 1993] P. S. Rosenbloom, J. E. Laird, and A. Newell, eds. *The SOAR Papers: Research on Integrated Intelligence*, 2 vols, Cambridge, Mass.: MIT Press, 1993.

[Rosenblueth and Weiner, 1950] A. Rosenblueth and N. Wiener. Purposeful and Non-Purposeful Behavior, *Philosophy of Science,* 17: 318-326, 1950.

[Rumelhart *et al.*, 1986a] D. E. Rumelhart, G. E. Hinton, and R. J. Williams. Learning Internal Representations by Error Propagation. In [Rumelhart, McClelland and PDP Group, 1986: 318-362].

[Rumelhart *et al.*, 1986b] D. E. Rumelhart, G. E. Hinton, and R. J. Williams. Learning Representations by Back-Propagating Errors, *Nature,* 323: 533-536, 1986.

[Rumelhart and McClelland, 1986] D. E. Rumelhart and J. L. McClelland. On Learning the Past Tenses of English Verbs. In [McClelland, Rumelhart and PDP Group, 1986, 216-271].

[Rumelhart, McClelland and the PDP Group, 1986] D. E. Rumelhart, J. L. McClelland, and the PDP Research Group. *Parallel Distributed Processing: Explorations in the Microstructure of Cognition*, Vol.1, Foundations, Cambridge, Mass.: MIT Press, 1986.

[Russell and Whitehead, 1910] B. Russell and A. N. Whitehead. *Principia Mathematica,*, vol. I, Cambridge: Cambridge University Press, 1910.

[Searle, 1980] J. R. Searle. Minds, Brains, and Programs, *Behavioral and Brain Sciences,* 3: 417-457, 1980. Includes peer-commentaries, and reply.

[Shannon and Weaver, 1949] C. E. Shannon and W. Weaver. *The Mathematical Theory of Communication*, Urbana: University of Illinois Press, 1949.

[Sherrington, 1906] C. S. Sherrington. *The Integrative Action of the Nervous System.* The Silliman Lectures, Yale, 1904. London: Constable, 1906.

[Simon, 1947] H. A. Simon. *Administrative Behavior: A Study of Decision-Making Processes in Administrative Organization*, New York: Macmillan, 1947.

[Simon, 1957] H. A. Simon. *Models of Man, Social and Rational: Mathematical Essays on Rational Human Behaviour in a Social Setting*, New York: Wiley, 1957.

[Simon, 1962] H. A. Simon. The Architecture of Complexity. In *Proceedings of the American Philosophical Society*, 106, 467-482, 1962.

[Simon, 1969] H. A. Simon. *The Sciences of the Artificial. The Karl Taylor Compton Lectures*, Cambridge, Mass.: MIT Press, 1969.

[Sloman, 2003] A. Sloman. How Many Separately Evolved Emotional Beasties Live Within Us? In R. Trappl, P. Petta and S. Payr, eds., *Emotions in Humans and Artifacts*, Cambridge, Mass.: MIT Press, 29-96, 2003.

[Sloman, undated] A. Sloman. The CogAff group's website: `www.cs.bham.ac.uk/research/cogaff`

[Smith, 1996] B. C. Smith. *On the Origin of Objects*, Cambridge, Mass.: MIT Press, 1996.

[Sperber and Wilson, 1986] D. Sperber and D. Wilson. *Relevance: Communication and Cognition*, Oxford: Blackwell, 1986.

[Tolman, 1920] E. C. Tolman. Instinct and Purpose, *Psychological Review,* 27: 217-233, 1920.

[Tolman, 1922] E. C. Tolman. A New Formula for Behaviorism, *Psychological Review,* 29: 44-53, 1922.

[Tolman, 1932] E. C. Tolman. *Purposive Behavior in Animals and Men*, New York: Appleton-Century-Crofts, 1932.

[Varela *et al.*, 1991] F. J. Varela, E. Thompson, and E. Rosch. *The Embodied Mind: Cognitive Science and Human Experience*, Cambridge, Mass.: MIT Press, 1991.

[Webb and Scutt, 2000] B. Webb and T. Scutt. A Simple Latency-Dependent Spiking-Neuron Model of Cricket Phonotaxis, *Biological Cybernetics,* 82: 247-269, 2000.

[Wheeler, 2005] M. W. Wheeler. *Reconstructing the Cognitive World: The Next Step*, Cambridge, Mass.: MIT Press, 2005.

[Whitelaw, 2004] M. Whitelaw. *Metacreation: Art and Artificial Life*, London: MIT Press, 2004.

[Willshaw and von der Malsburg, 1976] D. J. Willshaw and C. von der Malsburg. How Patterned Neural Connections Can Be Set Up by Self-Organization, *Proceedings of the Royal Society: B,* 194: 431-445, 1976

[Wright and Sloman, 1997] I. P. Wright and A. Sloman. *MINDER1: An Implementation of a Protoemotional Agent Architecture.* Technical Report CSRP-97-1, University of Birmingham, School of Computer Science. (Available from `ftp://ftp.cs.bham.ac.uk/pub/tech-reports/1997/CSRP-97-01.ps.gz`)

[Wright *et al.*, 1996] I. P. Wright, A. Sloman, and L. P. Beaudoin. Towards a Design-Based Analysis of Emotional Episodes, *Philosophy, Psychiatry, and Psychology,* 3: 101-137, 1996.

INFORMATION IN BIOLOGICAL SYSTEMS

John Collier

1 INTRODUCTION

The notion of information has developed in a number of different ways (as discussed in this volume), and many of them have been applied to biology, both usefully and gratuitously, and even misleadingly. These multiple notions of information have not surprisingly led to apparently contradictory claims by authors who have really been talking past each other, although there are also substantive issues at stake. The aim of this chapter is to review some of the ways that notions of information have been used in biology, to disentangle them, and to evaluate their implications and aptness, as well as to point out some of the more widespread confusions.

In particular, I will compare the use of information as a technology of measurement, which does not imply that there is anything present that might be called 'information', with a stronger usage of information in biology that attributes information to biological systems in a non-instrumental way. This distinction between instrumental and substantive uses of information in biological studies often turns on the notion of information used, so it is important in each case to be clear what is at stake. Where there is a choice, I will focus on the substantive use of information in biology. Roughly, substantive use of information uses information in an explanatory way in addition to any representational instruments.[1] I will not discuss what falls under the general heading of *bioinformatics* in this chapter.

It will be impossible to cover all the varied uses of information concepts by biologists, so I will look primarily at cases that seem to be historically significant or else philosophically pivotal (the two often correspond).[2] The central case I will look at is heredity. The association of information with heredity goes back

[1]Sarkar [2000] makes a similar distinction between *heuristic* and *substantive* uses of information, but as an avowed instrumentalist he does not see a clear distinction. In particular, one would assume he sees no special role for explanation in the way that Chomsky [1959], for example, distinguishes between descriptive and explanatory adequacy. I believe that some of Sarkar's obtuseness about the role of information in biological systems is a result of blindness to the distinction, resulting in a failure to consider things relevant to the higher standards required for explanation. Maynard Smith [2000a, 2000b] attributes other problems to Sarkar's misrepresentation of the biology. Between the two problems, there is not much left in Sarkar's objections to information that are not addressed in a more general way in this chapter.

[2]In 1987 I did a search on the last three years of Biological Abstracts (on CD, at the University of Indiana Biology Library). Based on the abstracts, I tried to judge whether the use of information in the paper was required, or was more or less gratuitous. I found as few as seven abstracts that seemed to me to use the information concept in some essential way. The

Handbook of the Philosophy of Science. Volume 8: Philosophy of Information
Volume editors: Pieter Adriaans and Johan van Benthem. General editors: Dov M. Gabbay, Paul Thagard and John Woods.

at least to Weissmann [1904], and was adopted by such disparate biologists as Francis Crick [1958] and Konrad Lorenz [1973]. It is difficult to find well known theoretical biologists who object to the use of information concepts in relation to genetics, and if anything the use of information concepts in biology has increased over the last few decades. Dawkins [1986, p. 112] declared: "If you want to understand life, don't think about vibrant, throbbing gels and oozes, think about information technology." Increasingly, the "throbbing gels and oozes" can themselves be understood as made up of molecular machines that process information [Holzmüller, 1984; Schneider, 1991a; 1991b; 2000; 2006; Darden, 2006]. In order to give a strong grounding in accepted theoretical biology, I will take my lead from the role of information assigned by Maynard Smith and Szathmáry in *The Major Transitions in Evolution* [1995]. They argue that the increase in complexity observed (however roughly) in some lineages results from relatively few major transitions in how genetic information is passed down from generation to generation. As we shall see, things are possibly and probably more complicated than this relatively simple hypothesis, but following it critically will raise some important philosophical issues. Importantly, however, Maynard Smith and Szathmáry use information explanatorily and their views and usages are fairly authoritative; therefore, presumably they pick out authoritative substantive uses of information in biology.

There are a variety of mathematical technologies that can be used for information measurement, but they fall into three general classes [Kolmogorov, 1965]: statistical (e.g., Shannon and Weaver [1949]), combinatorial (a variation on Shannon methods not used directly by Shannon himself), and algorithmic [Chaitin, 1987]. The last has inspired two technologies for information measurement that have been applied to DNA and other biological objects: minimum description length — MDL [Rissanen, 1989] and minimum message length — MML [Wallace and Freeman, 1987]. It is worth noting, however, that the statistical methods are best used on ensembles, whereas the combinatorial and algorithmic methods work best on individuals. This suggests that the latter methods are more appropriate for dealing with information in biological organisms, even though the statistical approach is used so widely that it is often taken to *be* information theory (for more on this, see Winnie [2000]). Despite this, each of the general classes of methods can be used on any particular subject matter with clever adaptation. Thus there is nothing in the mathematical methods themselves that distinguishes the use of information technology in studying the properties of a system from the substantive attribution of information to a system. In particular, the instrumental usefulness of information technologies does not in itself imply the existence of substantive information within the system being studied, at least not without more being said. The instrumental usefulness of information may, for example, reflect epistemic considerations such as how we decide to organize our data. Furthermore,

situation was worse for the entropy concept, which had only two clearly non-gratuitous mentions out of over 200 papers that used it. I would expect that things have not improved, so there is understandable suspicion about the use of these related concepts.

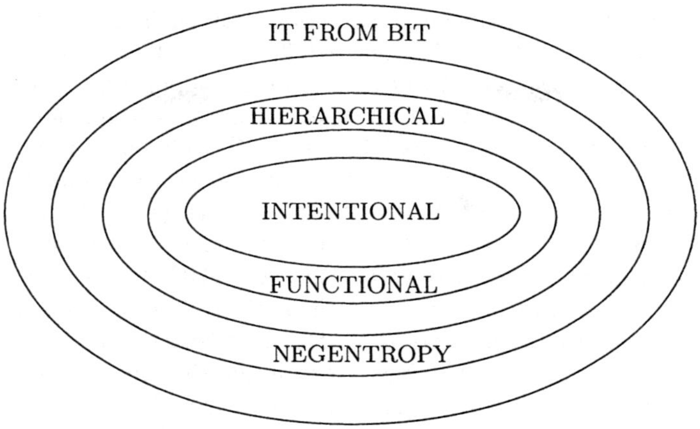

Figure 1. Nesting of major kinds of information

mathematical methods are limited by their nature to the syntactic (formalizable) aspects of information, and are not suited in themselves to dealing with issues of communication, representation, semiosis and meaning, all of which are have an importance in biology that they do not have in, say, physics.[3]

In order to deal with these issues, and evaluate whether or not information in some substantive role has a place in biology, it is useful to give a classification of the ways in which information has been thought to play a substantive role in the sciences. Ignoring many fine distinctions, the basic ways can be placed into an increasingly nested hierarchy: "it from bit", negentropy, hierarchical negentropy, functional information, and meaningful information (see Figure 1). Each inherits the logical and ontological commitments of the containing views, but adds further restrictions.

The most liberal and inclusive view is the "It from bit" view. It has originated independently from so many people that it is pointless to attribute an origin, though it probably goes back to Leibniz' view that the world has a logical structure of perceptions based in the ability to discriminate. The term is due to John Wheeler, and the view has recently been powerfully if controversially championed by Stephen Wolfram [2002]. On this view, any causally (dynamically) grounded distinction makes a difference, thereby ensuring that it is information ([MacKay, 1969], see also [Bateson, 1973]). On this view information is objective, and there is nothing else.

The negentropy view of information is a restriction on the It from bit view.

[3]This is also true of formal methods in general, including the Barwise-Seligman idea of an information channel in distributed systems [Barwise and Seligman, 1997]. Their approach invokes "regularities", which cannot be understood purely formally (accidental regularities do not carry information), but even this informal part of their approach does not imply anything more than non-accidental relations, which can be found in the most basic physics.

Only those Its that are capable of doing work (either directing and using energy or sorting things) count as information. The rest is disorder. This view is due to Schrödinger [1944], though the groundwork was done by Szillard, and the implications were generalized and clarified by Brillouin [1962], though the significance is still controversial [Earman and Norton, 1998; 1999]. The motivation for this view is that work is required for control, and the information in microstates beyond that in macrostates is hidden from view in macroscopic interactions [Collier, 1990b]. Negentropy measures the capacity for control (in bits this is the number of binary discriminations that a system can possibly make).

The next view is a restriction of the negentropic approach to particular levels of a physical hierarchy, so that information is relativized to a cohesive level of an object, such as an organism or a species. The view is due to Brooks and Wiley [1988], Collier [1986; 2003] and Smith [1998].[4] The idea is that not all negentropy is expressed at a given level, and the "Its" available are level relative. This information is a measure of the constraints on the objects within the level; because of their connection to biological and cognitive form, Collier ([1990a], Collier and Hooker, [1999]) calls this *expressed* information *enformation* to distinguish it from other forms of negentropy (for example, statistical information due to nonequilibrium conditions is sometimes called *intropy*).[5] Expressed information is relative to a certain level in a hierarchy [Collier, 2003], an idea that will be clarified below. Expressed information at higher levels is able to control information at lower levels only to a certain degree (sometimes called downward causation), but can control information at its own and higher levels more completely [Collier, 1990b; 2003]. This asymmetry is biologically important.

Restricting further, we have functional information, which is the expressed information that is functional. This sort of information is easily seen as information from outside the system. It has both syntax and semantics (reference), but does not require that the information is information for the system itself.[6] Functional organization is biologically important, of course, but at least one common account of biological functionality tends to suppress the informational aspect. Whether or not we can call information that arises through functionality meaningful has been a subject of some debate. The nature of meaning is the great object of desire for information theory. I will address this issue throughout this chapter, but especially

---

[4]Stan Salthe [1985; 1993] also uses similar technology very widely, but to different effect, as does Robert Ulanowicz [1986; 1997] in ecology. I will not discuss these uses here, as it would take me much further afield from the issues of heredity and function that are my main focus. Wicken [1987] and James Kay and Eric Schneider [1994] specifically avoid this information technology, at least in name, though they have been in close communication with the authors that do explicitly use this levels based information technology. They also use the technology very widely, including ecology.

[5]Collier [1990a] borrowed this term from engineering usage. Lyla Gatlin [1972] called this information stored information, but this name is somewhat misleading, as it does not reflect the dynamical and often active nature that *expressed information* allows, nor its hierarchical nature.

[6]Maturana and Varela [1980] apply this distinction by calling functional organization information *externally*, but not *internally*. I think this distinction is not sustainable [Collier, 2004a; 2004b].

in the final two sections before the conclusion.

Within the scope of meaningful information is intentional information, or cognitive content. At the next level of restriction is social information, though some authors hold that cognitive content depends on language, which is a social activity. I will not discuss these levels further here, which is not to say that either they are unimportant, or are in some sense reducible to the information forms that I do discuss. These forms of information are better discussed in the context of cognition, though the biological roots of cognition are interesting, and connect to various forms of biological information.

INFORMATION AS A TOOL

The use of information theory as a tool in biology is fairly widespread. Biological systems are both complicated and complexly organized, so information theory can be used to calculate or estimate the information content in biological structures from macromolecules to whole organisms, as well as in and between populations. This is not controversial, nor is this sort of application of information theory peculiar to biology. Similarly, communications theory can be used to analyze various biological channels, such as sensory processes, molecular communication, neural communication, intraspecies and interspecies communication and ecological networks in terms of their capacity, connectivity, order, and organization. Algorithmic information theory and its variants, MDL and MML are also useful for the last three, and Charles Bennett's notion of *logical depth* may provide a measure of organization. Another useful tool, at least potentially, is the notion of information channel developed by Barwise and Seligman [1997] for the logic of distributed systems. Biological information channels, whether artifactual or inherent, are nothing if not distributed. Again, there is nothing particularly biological about these applications, and many of them are known from systems theory, electronics and computer science. Some of the applications, however, present interesting issues for the philosophy of biology, especially concerning whether the instrumental use of information is sufficient to explain the use of the idea of information by biologists.

DNA is probably the biological entity most closely associated with information. Maynard Smith and Szathmáry consider *only* hereditary transmission as the basis of information involved in increasing biological complexity, though they do mention major phenotypic changes that opened up new habitats, sensory inputs, and physiological mechanisms for adaptive (and not) radiation without direct change in hereditary channels. They are therefore committed [Maynard Smith and Szathmáry, 1995, p. 8] to a gene-centered approach as outlined in Williams [1966] and explicit in Dawkins [1976]. In light of recent work on developmental systems theory [Griffiths and Grey, 1994; Oyama, 2000; Jablonka and Lamb, 1995] and niche construction ([Odling-Smee *et al.*, 1996], see also [Odling-Smee *et al.*, 2003]), both of which point to non-genetic channels of heredity, it now seems unlikely that the gene-centric view can be sustained. One should not get too excited about his, however, as genes are still very important. Griffiths [2001], Godfrey Smith

[2000] and Sterelny [2000] criticize the gene centred approach offered by Maynard Smith, but seem to carry this over to substantive claims about information in biology, especially with regard to the role of genes, though they vary in the strength of this particular criticism. Griffiths argues for a "parity principle" that seems to lead him to the conclusion that information must be treated the same for all modes of heredity. I see no basis for parity if these means equally important. If it means that all heredity must be treated in terms of information channels, this is unobjectionable. One still has to deal with the issue of coding, however; some heredity is through codings, other heredity is not. Coded heredity is more reliable in both transmission and expression, all other things being equal, and this difference undermines complete parity. I will discuss the importance of coding in more detail below. On the other hand, Maynard Smith's [2000b] response to alternative modes of heredity that these are ultimately grounded in the genes seems to be either wrong or beside the point. All biology is ultimately grounded in chemistry and physics, but that does not mean that there aren't special biological principles. Non-genetic heredity may depend on genetic heredity, but it is much too quick to infer that it can be eliminated in favour of genetic heredity. If there are independent channels of heredity that carry non-genetic information Maynard Smith's 'ultimately grounded' argument fails.

One has to be careful here, though. Sterelny [2000] raises the issue that the regulation of genes depends on the cytoplasm, which is passed down maternally. He suggests that this presents a non-genetic mode of heredity that may contain semantic information. Now it may be true that semantic information is passed on in the cytoplasm, but at least for the regulatory part of the cytoplasm this is best understood not as information, but as part of the information channel that carries genetic information reliably. The cytoplasm is continuous from mother to offspring, and does not reproduce like the genes do. It plays a role in heredity, but not that of carrying information — it provides a continuity of channel for the genetic information, so that its expression is reliable. The same can be said of many environmental conditions: they control the expression of genes to some degree, and hence act as part of the channel for gene expression, affecting what genetic information is expressed. This doesn't rule out the possibility that there may be cellular or environmental information channels that do not carry genetic information. But much of the literature against the primacy of the gene misrepresents the issue of hereditary information by confusing channel conditions with channel content, as Sterelny did. More sophisticated arguments for non-genetic inheritance that make this distinction are required. I will have more to say about channels later.

Despite the likely existence of other channels of inheritance, the most widely assumed view remains that DNA is the predominant if not only focus of biological information transmission. Genetic information is of undeniable importance, so it is worth looking at in more detail. I will look at some alternative channels of inheritance and their significance later. The workings of genes alone have implications for biological information channels, and how best to understand them, and most of the main issues can be understood within the scope of this focus.

DNA is often said to code for proteins, regulation, and various phenotypic properties from chemical networks in the body to social phenomena. None of this is straightforward, not even the mapping of DNA onto proteins.[7] Since many biochemicals (such as, to take an instance, opium) are not proteins, there is no gene that codes for them. This is even more obvious for chemical networks and especially for behavior. Fortunately it is possible to get to the heart of the nature of genetic information without going into details of these complexities, but the technique required for doing so (information channels) perhaps opens up the field for biological information too far. More will be said about this soon. For the time being, focus will be on the channels of genetic heredity in terms of the Weissmannian view of separation of developmental and germ channels, rendered as the "central dogma" of molecular biology. This view is close enough to being true at the molecular level that it is useful.

With the above restrictions, the major transitions in evolution, according to Maynard Smith and Szathmáry are:

Table 1. The Major Transitions (after Maynard Smith and Szathmáry 1995)

1.	Replicating molecules	→	Populations of molecules in compartments
2.	Independent replicators	→	Chromosomes (linked replicators)
3.	RNA as gene and enzyme	→	DNA + protein (genetic code)
4.	Prokaryotes	→	Eukaryotes
5.	Asexual clones	→	Sexual populations
6.	Protists	→	Animals, plants, fungi (cell differentiation)
7.	Solitary individuals	→	Colonies (non-reproductive castes)
8.	Primate societies	→	Human societies

Several things to note about these transitions are a) they occur in only some lineages, so they are not inevitable, but are contingent and confer a fitness advantage only relative to others in their lineage, so post transition organisms don't necessarily replace all those in lineages that have not made the transition in question, b) entities that were capable of independent replication before the transition can replicate only as part of a larger whole after the transition, c) selfish mutations leading to conflict between levels of inclusion are possible in the larger whole — they happen — but there are so many chances for suppressor mutations in the whole that the influence of selfish mutations is rather small, d) the transitions typically involve differentiation and specialization that increase efficiency, and e) the notion of coding appears only in transition 3, but information concepts are applicable from the start.

[7]For discussion of these issues and a general empirically grounded critique of the centrality of the gene see [Keller, 2000]. None of her discussion affects the logic used in this paper.

Maynard Smith and Szathmáry note that a–d are common to all of the transitions, suggesting that there are some general principles at work. Since the early transitions are not part of biological evolution per se, but occur in molecular or chemical evolution, these general principles are not strictly biological. However, they note that point (e) marks the most significant transition — a division of labor that requires coding and translation. The major part of their book, not surprisingly, deals with this transition. I will argue below that transition 3 significantly enhances the role for substantive information by separating through significant dynamical decoupling the roles of the energy and information budgets in prebiotic and living systems, and opens the door for semantic information in biological systems. This distinguishes genetic inheritance from most other forms of inheritance (so-called) enlisted to support the parity thesis.

Arguably, to be alive requires this sort of separation of function and the requisite dynamical decoupling between metabolism and replication ([Brooks and Wiley, 1988; Brooks *et al.*, 1989; Maynard Smith and Szathmáry, 1995], and many others), but nothing incontestable appears to rest on this definition of 'living', since the functional and dynamical separation are a matter of degree. In any case, the definition allows us to distinguish between chemical evolution, in which replication and metabolic processes are not distinct, and biological evolution, in which they are. A useful distinction made by Maynard Smith and Szathmáry is between limited replication and indefinite replication [1995, pp. 41–42]. The former allows only a limited variety of forms dependent on the structure of the replicators, acting as templates. This sort of replication is highly subject to interference from parasitic replicators (ones that replicate at the expense of system organization, but nonetheless use system resources) that limits the size of reliably replicating structures. Limited replication cannot, therefore, support open-ended evolution. Sustained evolution is possible with template reproduction involving complementary base pairing, as with DNA. Is this sort of structure necessary for sustained evolution? If it is, then the dawn of coding is equivalent to the dawn of indefinite replication as well as the distinct decoupling of information transmission and metabolism. This would support the definition of living system given at the start of this paragraph. Unfortunately, as is often the case in biology, the answer is a bit fuzzy: the transition from RNA genes and enzymes to DNA code specializing in information transmission with protein enzymes taking care of the catalysis required for metabolism is not sharp, and the history of the transition is still muddy and incomplete.

The details of transition 3 as they are currently known are given by Maynard Smith and Szathmáry [1995, chapters 5, 6 and 7]. As far as the evolution of the code itself goes, Maynard Smith and Szathmáry describe it using Orgel's phrase: "like coming into focus of an obscure picture." The evolution of the code was gradual, and it came to be more reliable and efficient through selection. Likewise, the replacement of RNA enzymes (ribozymes) by protein enzymes was gradual, and probably evolved in parallel with the evolution of the code. Similar gradual evolution seems to apply to other aspects of transition 3. With the development of protocells (which I will not discuss), longer strands of RNA could be selected at the protocellular level. The fitness advantage comes through the linking of related genes, increasing reliability, but at the expense of some efficiency of replication, since linked genes take longer to reproduce. Other factors were no doubt important. In principle it is possible to gain increased stability through double stranded RNA, but the chemical stability of DNA, produced through the reduction of RNA, gave it a selective advantage (probably appearing first before the evolution of translation and protein enzymes). Its appearance is the final component of transition 3. The result is stable, stored reliably transmitted information that is reliably translatable into proteins. It should be noted that transition 3 has continued in minor detail through the selection of more efficient and reliable components. The transitions are not completely sequential, and the processes making up the transitions are gradual and evolve in parallel.

The general character of the transitions is that they involve greater complexity in how genetic information is translated and transmitted. This increased complexity increases the reliability, speed and/or efficiency of transmission and/or translation, and also opens up new regions of adaptive space that can be occupied. Each of these changes is produced accidentally and is then retained by relative fitness advantages, or so the story goes. The common aspects of the major transitions are shared between biotic and prebiotic evolution, so there is nothing particularly biological about them. Transition 3, however, appears to be a boundary (however fuzzy) between the prebiotic and biotic. If there is something special about biological information, this is where to look.

It from bit, intropic and level intropic substantive views of processes are all found outside of biology, so the relevant level for specifically biological information is the functional level. Instrumentally, the intropic and level intropic views are useful for describing the formation of self-organized structures, their replication and heredity in prebiotic evolution, as Maynard Smith and Szathmáry do in their book. This sort of process occurs in physics and chemistry as well, and continues into biological processes that support functionality. Maturana and Varela [1980] relegate the formation and support of autopoietic structures such as cells to such processes, with functional processes (operational processes) internal to the autopoietic system. They do not apply the notion of information to the internal perspective of autopoietic systems (at least for cells, perhaps not for higher level autopoietic systems), but argue that the concept of information is useful only from an external perspective. Thus, arguably (and it seems to be their considered opinion), information is useful only instrumentally for describing cells, even though they have a robust notion of functionality based in organizational or operational role. Despite their instrumental use of information, many of their followers talk of information internally, using the convenient connection from systems theory between the concept of operationality (function and control) and the concept of information somewhat unquestioningly. Maturana and Varela define autopoiesis in terms of operational closure, indicating that there is no information flow into or out of an autopoietic system, which I believe makes the information concept redundant in discussing the internal operations of the system. This, however, violates the general idea of openness central to most systems theory, and many of their followers have dropped the closure requirement, typically without comment, and allow autopoietic systems to be open to information. This undermines the justification for treating information as useful only for an external description of autopoietic systems, though of course it is still possible to treat information theoretic accounts as only of instrumental value. I have given principled reasons elsewhere [Collier, 2002] for thinking that, contrary to Maturana, even cells are open to information, and that the self-organization process itself requires that the resulting system be open [Collier, 2004a]. Functionality depends on some sort of organizational closure, but it need not be complete and is not complete in biological cases [Collier, 1999b; 2000; 2004b]. These are based in well established principles of open systems theory [Ingarden *et al.*, 1997], so I will not go into more detail here.

The decoupling of energy and information budgets, which is a matter of degree, and increases through the major transitions, permits self-organization within the information system itself [Collier, 1986; 2003; Brooks and Wiley, 1988; Brooks *et al.*, 1989; Layzer 1990]. The degree to which this occurs is presently unclear (it is an empirical matter), but it is a potential source of new organization and information within the information system itself, including within adaptive space itself [Layzer, 1980; Collier, 1998]. This permits "minor transitions", allowing gradual increases in the size of information space. These transitions, like the major transitions, are chance events, but are favoured probabilistically. On the level intropic account they increase both information and entropy (disorder of the lower level) simultaneously, a phenomenon well known in physics [Landsberg, 1984; Layzer, 1990]. Furthermore, as previously mentioned, there are other channels of heredity through the environment by way of niche construction [Odling-Smee *et al.*, 1996; 2003] and developmental systems more generally [Griffiths and Grey, 1994; Oyama 2000]. The interaction of multiple channels of information not only allows the influence of those other channels, but also sets up conditions favourable for further self-organization. Thus Maynard Smith and Szathmáry's focus on DNA is questionable, as there may well be other informationgenic (or *morphogenic*, to use a less awkward word) processes in biology other than DNA selection based on the functionality of its phenotypic expression alone.

Setting this issue aside for the moment, I will focus on the complexities of the instrumental use of information in DNA in order to get clearer what is implied by its use in order to clear the ground for the discussion of the substantive use of information in biology more generally. It is worth noting, however, that the self-organization of information systems and developmental and environmental channels for the inheritance of information can all be understood with the resources of the intropic and level intropic accounts of information, without invocation of a substantive use of functional information tied to genetic information, even though it presupposes the partial decoupling of information and energy (but see below on storage). To understand functional information we need to look more closely at how gene expression and phenotypic selection works.

The route from information stored and transmitted from DNA to the phenotype of an organism is much more complex than the replication of genetic information in reproduction. Replication is fairly well understood, but gene expression, especially in multicellular eukaryotes, is very complex and not very direct. Fortunately, it is possible to avoid the complexities here. James MacLaurin [1998] (see also [Collier, 1999a]) has observed that substantive information has the nice property that if you wiggle something at one end of an information channel the result at the other end will reproduce the aspects of the information transmitted. This means that the complexities of gene expression, such as those discussed by Evelyn Fox Keller [2000], can be ignored in the discussion here, no matter what their scientific interest otherwise.

There seems to be considerable confusion on this issue, with one of the referees of this chapter complaining that issues of gene expression complicate the issue so that one cannot distinguish the genetic component. The nice thing about information, however, is that its effects are carried through a channel without modification in the information to exactly that extent that the information is carried through. Complicating factors do not change the information; they can only reduce the transmitted information. If it is reduced to zero, then there is no genetic information expressed. One has to distinguish here between what happens and how we know it. It might be difficult to distinguish genetic information in the phenotype, but that does not mean that it is not there. Selection of hereditary material depends on its expression in the phenotype, not on whether we can recognize this expression. If we wish to understand the dynamics of evolution, we need to focus on ontological, not epistemological issues. If selection is on a trait, and the trait expresses some genetic information, then the genes are selected. If the genetic (or other hereditary) contribution is not relevant to the trait selection, then there will be no inheritance, so what we have is *not* a case of natural selection. I will have more to say on this shortly.

Marcello Barbieri [2001] describes the "bridge between genes and organism" in two parts. The first part is transcription of DNA into primary transcripts. The second part, with eight steps, is epigenetic. The steps are shwon in Table 1:

Table 2. Epigenesis (after [Barbieri, 2001])

1.	Splicing	→	Messengers
2.	Translation	→	Polypeptides
3.	Folding	→	Proteins
4.	Protein Assembly	→	Organelles
5.	Organelle Assembly	→	Cells
6.	Cell Assembly	→	Tissues
7.	Tissue Assembly	→	Organs
8.	Organ Assembly	→	Organism

These steps correspond to common phenotypic levels. There are various minor steps as well, such as the formation of control networks, membranes and their various inclusions, etc., as well as the back control of earlier processes. For example, opium is not composed of proteins, and its components must be produced under control of the appropriate tissue kind by networks composed of proteins and other substances. Similar observations could be made about behavior and other complex phenotypic states that are not the directly composed of proteins. The minor steps in general do not fit as well into a levels account. For example, networks involving hormones can extend across the whole organism, but their production and action is always local. I am ignoring these sorts of networks in this chapter, since they would need too much space to discuss, and would take me away from the focus on genetic information. Nonetheless, there are interesting issues concerning the extent to which these can be reduced to genetic information, the extent to which they can be analyzed as communication channels, and the way(s) in which they are committed to substantive information accounts. But these interesting issues must be set aside here in favor of the focus on genetic information.[8] If phenotypic changes result from genetic changes, then we have genetic information expressed phenotypically. Knockout techniques are one way to measure such changes, but the relations can be much more complex and subtle. The essential condition is that there is an information channel from the DNA to the relevant trait.

Barwise and Seligman [1997] offer an account of information flow in distributed systems that is widely applicable in biology as well as non-biological systems. The basic idea is of an *infomorphism* based on regular relations between two sets of classifications relating types to tokens. Networks of infomorphisms can be constructed to form information channels that have the intuitively expected properties. However, regularities are not sufficient for the purpose of information channels (accidental regularities do not count for the transmission of information) so the Barwise and Seligman account is already substantive to some degree, despite appearances. The use of types in the formalization suggests that there must be an abstract aspect to infomorphisms, and thus information channels, that is not substantive, but Barwise and Seligman prove that types and tokens are logical duals in the formalization, so there is nothing to stop taking types as substantive. For example, a type can be a filter that sorts out tokens of a particular kind. Realizations of information channels must embody the types, which will be something like a filter. Genetic information is expressed at various levels up to the organism through such channels, and is combined and filtered to produce phenotypic properties both complex and simple, subtle and coarse. The distributed network of channels permits complex interactions to form the information in traits, and there can be (and are) other sources of information that are not genetic (nor

[8]Maynard Smith [2000a] suggests that enzymes should not be understood informationally, but that hormones should. This is in line with the usual notion that hormones signal, but enzymes merely facilitate chemical reactions. Nonetheless there have been some attempts to treat enzyme action biosemiotically. The idea is that they contain a message 'carry out chemical reaction X here'. Given that enzymes are functional, and are not merely chemicals that happen to be there, this idea is not as preposterous as it might at first seem.

environmental, if self-organization occurs within the organism). That such channels exist is beyond doubt; the contributions to phenotypic form from non-genetic sources and the processes involved are less well known. The existence of channels from genes to phenotype will depend on conditions within the organic environment (high acidity, for example, would make expression impossible, blocking the first step — generally the contribution of the internal environment will be more significant). However, despite this non-genetic dependency, the information expressed phenotypically through the cross-level channels is still genetic information. This is sufficient for enough genetic determination of the phenotype to be evolutionarily significant. This is quite aside from the issue of the extent to which the internal environment is itself genetically determined.

The issue of genetic determinism is fraught with confusion. As far as we are concerned with whether heredity or environment (nature or nurture) cause certain traits, we want to know their relative contribution to those traits. However, for evolutionary biology, the important issue is typically not the relative contribution, but whether there is any genetic contribution at all to differences in traits [Wilson, 1988, p. 8]. The idea is that over evolutionary time environmental variation will be statistically insignificant (it will come out in the wash of time), and genetic differences will be all that matters to selection processes. Thus, if something is the product of selection, and thus more adaptive than its alternatives, it will be an adaptation [West-Eberhard, 1992], and presumably will be functional on many standard accounts of biological function [Wright, 1973; Millikan, 1989, Neander, 1991]. If so, this would allow us to bridge the gap between genetic information expression and function, taking us to the next substantive level of information: functional information. I will discuss this further below, but a few more technical details are required first.

The important point here is that the connection between genetic information and function *need not* deal with the complications of gene expression despite numerous recent criticisms of the idea of genetic determination in general, nor need it account for all aspects of phenotypic traits. Even in cases in which there are plausible emergent forms, such as has been claimed for the early development in sea urchins by systems biologists who have studied in full the first 16 stages [Davidson *et al.*, 2002], genetic change can lead to developmental change (different attractors become more likely). Again we see the advantage of the information approach in that it can explain *even across non-reducible levels.*[9]

So far I have kept mostly to the technical application of information theoretic methods to biology. However, in explaining their application, I have shown that at least substantive notions of information are required to make sense of the applications. The substantive notions required so far, though, are not peculiarly biological in any way, until perhaps we come to the issue of adaptation. It is time to look in more detail at the peculiarly biological.

[9]An analysis of the conditions required for this in terms of a formal account of levels and information across levels was given by Collier [2003]. The basic ideas, let alone the details require much too much space to be recounted here.

INFORMATION STORAGE AND PROCESSING

Transition 3 (Table 1), as discussed in the previous section, is the most significant, marking the transition from prebiotic to biotic evolution. At the very least, before transition 3 the processes are questionably biological, while after transition 3 they are clearly biological. Maynard Smith and Szathmáry describe the transition as being to a genetic code from RNA as both gene and enzyme, but I have described this in the previous section as a transition to decoupled information and energy budgets, with specialization for heredity and metabolism.[10] The reason I described it this way in the previous section is that I did not want to presume there, like they do, that the notion of code, which strongly implies a robust sense of information, is required to make sense of the dynamical decoupling. I will set the issue of codes aside until the next section. In this section I will deal only with the implications of the decoupling for the explanatory role of information. Transition 3 also implies the decoupling of genotype and phenotype. Replication after the transition requires storage and transmission of information that is supported by, but largely unaffected by metabolic processes in any way specific to the information stored and transmitted. This information, then, is a candidate for specifically biological information. What is specifically biological about its nature, and why would we call it information?

Perhaps the most obvious reason to call the hereditary processes after transition 3 informational processes is that they involve storage and transmission. Gatlin [1972], for example, places great emphasis on these aspects of genetic information, and gives it no independent characterization, at least not explicitly. However attractive this idea might be, it can't be right, because energy is *also* stored and transmitted in organisms. Almost exclusively, the vehicle for energy transmission and storage in organisms is ATP, but we are not inclined to call ATP information bearing. The reason, I think, is obvious: ATP is not discriminating; information is. A requirement for information to be discriminating is that its embodiment is complex. This is a direct consequence of information theory: the amount of information capacity of a channel can be no greater than the complexity of its embodiment.

While it is theoretically possible for information to be transferred without any net transfer of energy [Bennett, 1973], this can occur only in fully conservative systems, so information transfer will typically also involve energy transfer. Why, then, would we want to refer to information rather than energy in certain biological processes, and especially in the focal case of this article, genetic information transmission? The answer has to do with guidance and control, at least, and possibly function and meaning (semantics) as well. I will deal with guidance and

[10]Godfrey-Smith [2000] argues that coding is not necessary, since proteins could be used to replicate themselves. True, but this would not allow open ended-evolution unless there were some sort of protein code allowing a finite number of proteins to map an open-ended range of possible products. However, Godfrey-Smith's suggestion is consistent with decoupling without coding.

control in this section, and function and meaning in later sections.

Shannon [1949] observed that the notions of information and constraint are interchangeable. The Barwise and Seligman [1997] formalization of the idea of an information channel places the constraints in non-accidental regularities characterized as infomorphisms. These are grounded in classifications that have an embodiment in relations among tokens. However, many purely physical systems can be characterized in the same way[11], so what, if anything, is peculiar to biology? The best answer available is that biological information channels typically show organized complexity [Collier and Hooker, 1999]. They are complex themselves, carry complex information, and are interconnected in complex ways that show considerable logical depth [Bennett, 1985] (see also a similar concept dubbed *sophistication* by Atlan [1972]), indicating organization. In the physical sciences, boundary conditions are typically distinguished from laws governing dynamical transitions, which are regarded as peculiar to typical circumstances. In biology, however, the boundary conditions themselves have considerable regularity, and embody the special laws of the discipline (if any — if not, their closest analogue), or at least the foundation for those laws, such as Mendelian genetics and Fisher's population genetics. This is another aspect of the dynamical decoupling of energy and information in biological systems: the information system is free to form its own regularities, more or less free from any special restrictions of the boundary conditions on the energy budget. In evolutionary time, this has led to the production of more complex informational regularities of the sort described by Maynard Smith and Szathmáry as major transitions, as well as the minor transitions of Brooks and Wiley [1988]. If we were to focus only on the energy budget, most of this organized complexity would be missed. For this reason it is at least inconvenient to reduce biological processes to energy governed ones; whether reduction also misrepresents biological processes requires more investigation.

Inasmuch as the regularities in boundary conditions and their interactions guide changes in the energy of a system, it is natural to refer to them controlling the system. In particular, the genes place boundary conditions on traits, and it is natural to say that they have some control of the traits that are produced epigenetically. Although the traits also have information about the genes, the relation is asymmetrical, since only some of the genetic information is expressed in the traits, and they are at best signs of the genetic information, and the genetic information is not an expression of the traits. The reason for this is that infomorphisms and the logic of distributed systems are not like standard logic: in general one cannot deduce from the knowledge that there is a channel from A to B that changes in the state of B will lead to corresponding changes in the state of A. For example, changing the dials in the control of a nuclear reactor that indicate it is out of control to indicate that it is in control will not put the reactor back into control. At best it will break the channel. However, changing the conditions in the reactor by using the reverse control channel has some hope that the dials will correctly

[11]See [Collier, 1999a] for a general characterization of causal connection that is a case of Barwise and Seligman channels with the classes restricted to dynamical kinds.

indicate the reactor is back in control. In this sense the genes control the traits, but not vice versa.

There are two reasons to reject the idea that genes control traits: 1) control might be regarded as requiring function, but function is not required for explaining gene expression, and 2) control might be regarded as requiring intention, which adds to function some sort of meaningful purpose, but genes are not intentional. This brings us back to the issue of whether the functional kind of substantive information in Figure 1 needs to be invoked in biology, and also raises the issue of intentionality, meaning and semantics. I will address these issues later. Before that I want to address the issue of coding, a notion Maynard Smith and Szathmáry use in their characterization of transition 3, but which I deliberately set aside in this section.

CODES

In the previous section it has been established that the decoupling of metabolism and information required by Transition 3 implies information channels, especially channels from DNA to phenotypic traits. These channels are grounded in classifications grounded in processes that show a regular and organized structure. The regularities are both essential and sufficient for the existence of such channels. Why would Maynard Smith and Szathmáry also require that DNA be a code? It has two major implications for inheritance and variability, required for evolution. Before the discrete character of genes was understood, objections were raised to Darwin's theory of evolution by selection that sexual reproduction would lead to a mixing of genomes, and a tendency to converge towards some intermediate state, which is not what we observe, and not what we need for the origin of species (divergence). The discrete character of genes resolved this problem. It also permits recombination in the reproductive process, and recombination is known to be more effective in creating variable phenotypes than mutations alone. Of course both of these were the result of innovations later than Transition 3 itself, which applies to nonsexual bacteria. Since evolution by selection is not anticipative, these advantages could not have underlain Transition 3, however useful they were in later transitions.

One clear advantage of a code is the reduction of ambiguity in the regularities underlying the information channels involved in gene expression. This leads to an increase in fidelity of reproduction even in nonsexual organisms. The discrete character of changes in the genome is also advantageous even to nonsexual reproducers in that it introduces a degree of modularity into genotypic and resulting phenotypic changes. This modularity presumably makes it easier to change some parts of an organism without changing others. This conservatism is more likely to lead to mutated organisms that are viable. Thus a genetic code has an immediate advantage for even nonsexual organisms, as well as opening up the possibility of later major transitions. It is difficult to see how a more holistic form of heredity could be equally successful, but it is also difficult to rule out the possibility a priori. It is safe to say, however, that if such a mechanism had evolved and become dominant, the evolution of sexual reproduction would not have occurred.

What is a code? Barbieri [2001] argues that codes 1) are rules of correspondence between two independent worlds, 2) give meanings to informational structures and 3) are collective rules that do not depend on individual features of their support structures. The independence in this case is grounded in dynamic decoupling in which at least one of the "worlds" is informational in the sense of the last section. It is not necessary that both "worlds" be informational in this sense, since it is possible for information to be expressed in a non-informational structure or process. Strictly speaking, requirement (3) does not imply discreteness, but discreteness at least greatly enhances the possibility of both collectivity and independence of support structures, as well as the efficiency of the code in the sense of using the same parts in different combinations to express different information. Collier [1986] introduced the notion of a *physical information system* that requires some degree of discreteness, but the advantages, all other things considered, are increased with the degree of discreteness. Such systems satisfy the storage and transmission requirements of the previous section, as well as Barbieri's requirements (1) and (3). The most controversial requirement is the second one, which will be discussed later.

Is the code concept required in biology? Although it is possible to regard hereditary and expression processes entirely in terms of energetic transformations, as in the first section, taking a non-substantive or very weak substantive view of information, genetic information behaves so as to satisfy the requirements of a physical information system, so something more is going on than just transformations of energy. To miss this point is to miss the special character of biological information, not to mention belying the way in which accomplished biologists like Maynard Smith and Szathmáry talk about the systems they study. Although reductionists might argue that such talk is unnecessary, their position is based on a metaphysical view that need not hold, and probably does not hold for gene expression (recall [Davidson *et al.*, 2002]), and possibly for the other forms of biological codes that Barbieri discusses. The reductionist position is thus metaphysically dubious, factually inadequate, and flies in the face of the way experts talk.

Barbieri [2001] points out that the bridge between genes and proteins has one genetic step and at least four levels of epigenetic processes. The first is widely regarded by biologists as a codified assembly, but the epigenetic processes are typically regarded as catalyzed assembly. The difference is between the sort of processes found prior to Transition 3 and those found after the transition. Barbieri argues, however, that the assumption that epigenetic processes are of the older catalyzed form, with no clear distinction between metabolic informational processes, has not been proven. Much of his book takes up an argument that many epigenetic processes are also processes of codified assembly. He provides rich evidence from molecular biology that splicing at step 2, translation at step 3, signal transduction to form organelles, cells and tissues are also codified. Given the advantages of codification described in this section, perhaps this should not be very surprising. It is less easy to accept, however, that the codes require anything more than a syntax (the rules), but Barbieri also argues for a semantics or meaning (condition 2).

INFORMATION AND MEANING

The idea that meaning or semantics is required for fully understanding biological information has been attractive, but also highly controversial. I confess that in my own work so far I have had no need of the idea, but I have taken what I think is a very cautious approach to meaning, and perhaps I have not yet encountered the sorts of problems that require the hypothesis of meaning for their solution. Barbieri trumps his book as a revolutionary manifesto for what he calls "semantic biology". Whether or not he is successful in this, he shows fairly convincingly that the code concept (in its restricted, non-semantic physical information system form) applies far more broadly than is generally accepted.

One thing that is generally agreed is that meaning requires function. A common further requirement is intentionality. Matthen and Levy [1984] (see also [Hershberg and Efroni, 2001; Ahmed and Hashishb, 2003] for more recent views along the same lines, and [Melander, 1993] for the opposite perspective) have argued for intentionality in the immune system, but to the best of my knowledge there are no other arguments for intentionality within biological systems except for the mind. On the other hand, *teleosemantics* [Millikan, 1987; MacDonald and Papineau, 2006] argues for continuity between selection processes and semantic representation, which suggests at least the possibility of non-mental intentionality, but the idea has not been developed beyond the immune system. Barbieri [2001] seems to take it for granted that a code implies a semantics, but this must be non-intentional, if intentionality is peculiar to the mind and perhaps a few specialized systems like the immune system. I will discuss this possibility in the remainder of this chapter, with a discussion first of function, representation and biosemiotics. This discussion will be necessarily cursory, since a complete discussion would be worthy of a book (or several), and almost all of the main aspects are highly controversial.

The standard account of function used in biology is the etiological account [Wright, 1973; Millikan, 1989; Neander, 1991]. According to this account, a trait is functional if it is selected for, meaning that it is an adaptation. On this account, a selected trait will contain information produced by certain genes that are selected along with the trait. Thus, it is sometimes said, the selected genes have information about the environment, as well as about the traits that they express information in. The problem with this account is that meaning is an asymmetrical relationship on most accounts (like control, incidentally). I have already argued that this presents no special problem for the information in traits, but for environmental features the problem is not so easily solved. On the etiological account, it is certainly arguable that since the genes are (indirectly) selected, they are under the control of the environment. Thus the genes don't have the right relation to the environment to have information about it in the same way that they might have information about traits. At best, the genes could be signs of environmental features, not meaningful representations. It all depends on which way the channel goes. One can't have it that both that the genes mean the traits and that they mean the environmental features that selected them. To put it another way, genes are sometimes described as blueprints for the organism, but if this is fair then the etiological account implies that the environment is a blueprint for the genes. Something has gone wrong.[12] Perhaps it is the idea of meaning here, but perhaps it is the etiological view of function.

The etiological account just seems wrong in several very obvious ways. Jeff Foss [1994] has noted that we can typically assign function without knowing anything about etiology (though etiologists will argue that this is often fallible). Alternative accounts of function focus on organizational role [Maturana and Varela, 1980; Rosen, 1991; Cummins, 1983], with selection and resulting adaptation being explained in terms of differences in functionality rather than defining function. On this account the function of genes is heredity and the guidance of ontogeny. Selection acts on the results as a sort of filter, creating a channel guiding the gene pool to greater fitness (see, especially, [Winnie, 2000] for a helpful account). The representational role, if any, remains always in the genes. Perhaps this is a reason to reject the etiological view in favor of the organizational role view of function, but it depends on the cogency of the genes representing. It should be noted that, *mutatis mutandis*, similar problems can be raised for teleosemantics.

Representation typically requires some sort of system of rules that does not depend on their underlying substratum. Physical information systems (or Barbieri codes minus the meaning) have these. What more is required? The usual answer would be *interpretation*. Without interpretation, a representation is useless, nonfunctional. This suggests that we should look for some sort of interpretation in biological systems if we wish to find meaning. On standard accounts of meaning, the interpretation is given by the semantics, which is an abstract relation between symbols and their reference. This will not do for biological representation, however, since the relation has to be embodied in concrete biological processes. In order to correct this deficit, Bend-Olaf Küppers [1990] suggested that we include, along with the syntax (rules) and semantics (reference) of the genes a pragmatics. His view at that time was that the pragmatics was given by selection, but we have seen the problems with *that* view. Küppers has told me since that he has abandoned his earlier view, but the move towards pragmatics is a good one. How do we get a satisfactory biological pragmatics (if we can)?

[12]The symmetry problem has been pointed out in one form or another by Sarkar [2000], Sterelny [2000] and Winnie [2000]. Winnie gives a solution very close to the one I propose. Sterelny and Godfrey-Smith [2000] are also concerned with symmetries between genes and their contexts. I have already argued that the contexts serve the role of channels. Of course there might be regulatory genes that serve a regulation function and code for channel construction. These issues are complex, and need to be untangled, but they do not seem to me to present any special difficulties if we have a suitable account of function that breaks the symmetries.

BIOSEMIOTICS

Biosemiotics is an attempt to apply semiotics to biological systems in order to account for communication from the molecular level through the behavioral and social levels. The dominant approach today is the Copenhagen school (e.g., [Hoffmeyer, 1996]), which takes the semiotics of C.S. Peirce as its starting point. Again, since there are many controversies involved that would take much space to represent, let alone resolve, I will be brief. Peirce believed that pragmatic issues were the basis of meaning, in particular what expectations about the world are attached to a given idea in a way that might guide action. On Peirce's full-fledged view signs are an irreducible whole of three parts, one what we would normally call the symbol, the object (which corresponds roughly to the intensional reference), and the interpretant. This whole is embedded in a system of interpretance that connects to expectations and actions. If these ideas are to be applied to biological systems, the interpretant has to be within the organism, or more accurately, within the relevant biological system. He considers the sunflower, whose flower tends to face the sun very reliably. The direction the flower faces, then, is a good sign of the direction of the sun. However it is not a sign for the sunflower, since there is nothing in the sunflower that makes use of the information in the sign. The effect is a tropism caused by the size of the sunflower, its rapid growth, and the induction of growth inhibiting hormones by sunlight. The direction of the flower itself plays no functional role for the sunflower. Peirce did not know the explanation, but inferred correctly that the direction the flower faces was not a sign for the sunflower. On the other hand, he did not rule out that there could be genuine biological signs.

If we consider DNA as a sign of (at least some aspects of) traits, we need to find an appropriate interpretant within the organism to complete the trinity. As described early in this chapter, genetic information is expressed if differences in the genes make a difference to the traits expressed, no matter how small. This expression is functional on either the organizational role or the etiological accounts. The best candidate for the interpretant in this case is the other coding and catalytic processes involved in epigenesis. *If* this idea can be made out coherently, then there is a good case that DNA contains information about the phenotype of the organism for the organism itself, rather than from merely external view of some anthropomorphizing observer. And this meaning would be about in the semantic sense, with epigenesis providing the pragmatics. John Winnie [2000], though not in the biosemiotic tradition, suggests that parameters whose effects on the components of the system contribute to the likely performance of the system exhausts the "semantic" aspect. In a living system the performance is the contribution to viability of the system, which is subject to selection. This is not very different from Peirce's idea.

SUMMARY AND CONCLUSIONS

The first section showed how information theory can be used descriptively in biology in the case of the genes. This descriptive use is also explanatory to some extent, and invokes substantive information, but in no way that is specifically biological. However, the talk of biologists, and the distinction created by Transition 3, suggests that biological information involves something more than this. It is relatively easy to introduce notions of transmission, control and guidance as substantive, somewhat less easy to convincingly introduce a need for the substantive use of information codes, and much less easy at this time to justify substantive notions of meaning and semantics, though biosemiotics is highly suggestive.

Much of what has been said in this chapter about genetic information applies, *mutatis mutandis*, to other forms of biological information, such as molecular communication, communication in the nervous system, immune system, hormones, pheromones, and behavioral transmission between organisms. There are special aspects of each case, but most of the arguments justifying the use of information concepts in a substantive way carry through to these cases. I hope that the portrait I have given of genetic information is helpful in extending the ideas to other cases.

BIBLIOGRAPHY

[Ahmed and Hashishb, 2006] E. Ahmed and A. H. Hashishb. On modelling the immune system as a complex system. *Theory in Biosciences* **124**: 413–418, 2006.

[Atlan, 1972] H. Atlan. L'organisation biologique et la théorie de l'information. Herman, Paris, 1972.

[Barwise and Seligman, 1997] J. Barwise and J. Seligman. *Information Flow: The Logic of Distributed Systems.* Cambridge: University of Cambridge Press, 1997.

[Barbieri, 2001] M. Barbieri. *The Organic Codes:. The Birth of Semantic Biology.* Acona, Italy: peQuod. Cambridge: Cambridge University Press, reprinted 2002.

[Bateson, 1973] G. Bateson. *Steps to an Ecology of Mind.* London: Paladin, 1973.

[Bennett, 1973] C. H. Bennett. Logical Reversibility of Computation, *IBM Journal of Research and Development* **17**: 525-532, 1973.

[Bennett, 1985] C. H. Bennett. Dissipation, information, computational complexity and the definition of organisation, in D. Pines. Ed. 1985. *Emerging Syntheses In Science. Proceedings of the Founding Workshops of the Santa Fe Institute.* Redwood City, Calif.: Addison West Publishing Company, 1985.

[Brillouin, 1962] L. Brillouin. *Science and Information Theory, 2nd edition.* New York: Academic Press, 1962.

[Brooks, 2000] D. R. Brooks. The nature of the organism: life takes on a life of its own. *Proceedings of the New York Academy of Science.* **901**: 257-265, 2000.

[Brooks, 2001] D. R. Brooks. Evolution in the Information Age: Rediscovering the nature of the organism. *Semiotics, Energy, Evolution and Development*, 2001. `http://www.library.utoronto.ca/see/pages/SEED_Journal.html`

[Brooks, 2002] D. R. Brooks. Taking evolutionary transitions seriously. *Semiotics, Energy, Evolution and Development.* `http://www.library.utoronto.ca/see/pages/SEED_Journal.html`

[Brooks *et al.*, 1989] D. R. Brooks, J. D. Collier, B. A. Maurer, J. D. H. Smith and E. O. Wiley. Entropy and information in biological systems. *Biology and Philosophy* **4**: 407-432, 1989.

[Brooks and McLennan, 1997] D. R. Brooks and D. A. McLennan. Biological signals as material phenomena. *Rev. pensee d'aujord d'hui* 25. pp. 118-127, 1997.

[Brooks and McLennan, 1999] D. R. Brooks and D. A. McLennan. The nature of the organism and the emergence of selection processes and biological signals. In E. Taborsky, ed. *Semiosis, Evolution, Energy: Towards a Reconceptualization of the Sign.* Aachen Shaker Verlag, Bochum Publications in Semiotics New Series. Vol. 3. 185-218, 1999.

[Brooks and Wiley, 1988] D. R. Brooks and E. O. Wiley. *Evolution as Entropy: Toward a Unified Theory of Biology, 2nd edition.* Chicago: University of Chicago Press, 1988.

[Chaitin, 1987] G. J. Chaitin. *Algorithmic Information Theory.* Cambridge: Cambridge University Press, 1987.

[Chomsky, 1959] N. Chomsky. A Review of B. F. Skinner's *Verbal Behavior. Language* **35**: 26-58, 1959.

[Collier, 1990a] J. D. Collier. Intrinsic information. In Philip Hanson (ed) *Information, Language and Cognition: Vancouver Studies in Cognitive Science, Vol. 1.* Oxford: Oxford University Press: 390-409, 1980.

[Collier, 1990b] J. D. Collier. Two faces of Maxwell's demon reveal the nature of irreversibility. *Studies in the History and Philosophy of Science* **21**: 257-268, 1990.

[Collier, 1998] J. D. Collier. Information increase in biological systems: How does adaptation fit? In Gertrudis van der Vijver, Stanley N. Salthe and Manuela Delpos (eds) *Evolutionary Systems.* Dordrecht, Kluwer. pp. 129-140, 1998.

[Collier, 1999a] J. D. Collier. Causation is the transfer of information. Howard Sankey (ed) *Causation, Natural Laws and Explanation.* Dordrecht, Kluwer: 279-331, 1999.

[Collier, 1999b] J. D. Collier.. Autonomy in anticipatory systems: significance for functionality, intentionality and meaning. In: *Computing Anticipatory Systems, CASYS'98 Second International Conference*, edited by D. M. Dubois, American Institute of Physics, Woodbury, New York, AIP Conference Proceedings 465, pp. 7581, 1999.

[Collier, 2000] J. D. Collier. Autonomy and Process Closure as the Basis for Functionality. In *Closure: Emergent Organizations and their Dynamics*, edited by Jerry L.R. Chandler and Gertrudis van de Vijver, Volume 901 of the *Annals of the New York Academy of Science*: 280-291, 2000.

[Collier, 2002] J. D. Collier. What is Autonomy? *Partial Proceedings of CASYS'01: Fifth International Conference on Computing Anticipatory Systems, International Journal of Computing Anticipatory Systems*: **12**: 212-221, published by CHAOS, 2002.

[Collier, 2003] J. D. Collier. 1Hierarchical dynamical information systems with a focus on biology. *Entropy,* **5**: 100-124, 2003.

[Collier, 2004a] J. D. Collier. Self-organisation, individuation and identity, *Revue Internationale de Philosophie*, 2004.

[Collier, 2004b] J. D. Collier. Interactively Open Autonomy Unifies Two Approaches to Function", In: *Computing Anticipatory Systems: CASY'03 - Sixth International Conference*, edited by D. M. Dubois, American Institute of Physics, Melville, New York, AIP Conference Proceedings **718**: 228-235, 2004.

[Collier and Hooker, 1999] J. D. Collier. and C. A. Hooker. Complexly Organised Dynamical Systems. *Open Systems and Information Dynamics,* **6**: 241-302, 1999.

[Crick, 1958] F. Crick. On protein synthesis. *Symposium of the Society of Experimental Biology* **12**: 138-163, 1958.

[Cummins, 1983] R. Cummins. *The Nature of Psychological Explanation.* Cambridge, MA: MIT/Bradford, 1983.

[Darden, 2006] L. Darden. Flow of Information in Molecular Biological Mechanisms. *Biological Theory* **1**: 280-287, 2006.

[Davidson *et al.*, 2002] E. H. Davidson, J. P. Rast, P. Oliveri, A. Ransick, C. Calestani, Chiou-Hwa Yuh, T. Minokawa, G. Amore, V. Hinman, C. Arenas-Mena, O. Otim, C. Titus Brown, C. B. Livi, Pei Yun Lee, R. Revilla, A. G. Rust, Z. Pan, M. J. Schilstra, P. J. C. Clarke, M. I. Arnone, L. Rowen, R. A. Cameron, D. R. McClay, L. Hood, and H. Bolouri. A Genomic Regulatory Network for Development. *Science* **295**: 1669-1678, 2002.

[Dawkins, 1976] R. Dawkins. *The selfish gene.* Oxford: Oxford University Press, 1976.

[Dawkins, 1986] R. Dawkins. *The Blind Watchmaker.* New York: W. W. Norton & Co, 1986.

[Earman and Norton, 1998] J. Earman and J. D. Norton. Exorcist XIV: The Wrath of Maxwell's Demon. Part I. From Maxwell to Szilard. *Studies In History and Philosophy of Science Part B: Studies In History and Philosophy of Modern Physics* **29**: 435-471, 1998.

[Earman and Norton, 1999] J. Earman and J. D. Norton. Exorcist XIV: The Wrath of Maxwell's Demon. Part II. From Szilard to Landauer and Beyond. *Studies In History and Philosophy of Science Part B: Studies In History and Philosophy of Modern Physics.* **30**: 1-40, 1999.

[Foss, 1994] J. Foss. On the evolution of intentionality as seen from the intentional stance. *Inquiry* **37**: 287-310, 1994.

[Gatlin, 1972] L. L. Gatlin. *Information Theory and the Living System*, New York: Columbia University Press, 1972.

[Godfrey-Smith, 2000] P. Godfrey-Smith. Information, Arbitrariness, and Selection: Comments on Maynard Smith. *Philosophy of Science* **67**: 202-207, 2000.

[Griffiths, 2001] P. E. Griffiths. Genetic Information: A Metaphor in Search of a Theory. *Philosophy of Science* **68**: 394-412, 2001.

[Griffiths and Grey, 1994] P. E. Griffiths and R. D. Grey. Developmental Systems Theory and Evolutionary Explanation. *Journal of Philosophy* **91**: 277-304, 1994.

[Hershberg and Efroni, 2001] U. Hershberg and S. Efroni. The immune system and other cognitive systems. *Complexity* **6**: 14-21, 2001.

[Hoffmeyer, 1996] J. Hoffmeyer. *Signs of Meaning in the Universe.* Bloomington: Indiana University Press, 1996.

[Holzmüller, 1984] W. Holzmüller. *Information in Biological Systems: The Role of Macromolecules*, translated by Manfred Hecker. Cambridge: Cambridge University Press, 1984.

[Ingarden *et al.*, 1997] R. S. Ingarden, A. Kossakowski and M. Ohya. *Information Dynamics and Open Systems.* Dordrecht: Kluwer, 1997.

[Jablonka and Lamb, 1995] E. Jablonka and M. J. Lamb. *Epigenetic Inheritance and Evolution: The Lamarckian Dimension.* New York: Oxford University Press, 1995.

[Keller, 2000] E. F. Keller. *The Century of the Gene.* Cambridge, MA: Harvard University Press, 2000.

[Kolmogorov, 1965] A. N. Kolmogorov. Three Approaches to the Quantitative Definition of Information. *Problems of Information Transmission* **1**: 1-7, 1965.

[Küppers, 1990] B.-O. Küppers. *Information and the Origin of Life.* Cambridge: MIT Press, 1990.

[Landsberg, 1984] P. T. Landsberg. Can entropy and 'order' increase together? *Physics Letters* 102A: 171-173, 1984.

[Layzer, 1980] D. Layzer. A macroscopic approach to population genetics. *Journal of Theoretical Biology* **73**: 769788, 1980.

[Layzer, 1990] D. Layzer. *Cosmogenesis: the Growth of Order in the Universe.* New York: Oxford University Press, 1990.

[Lorenz, 1973] K. Z. Lorenz. Analogy as a Source of Knowledge. Nobel Lecture, December 12, 1973.

[Macdonald and Papineau, 2006] G. Macdonald and D. Papineau, eds. *Teleosemantics.* Oxford: Oxford University Press, 2006.

[MacKay, 1969] D. M. MacKay. *Information, Mechanism and Meaning.* Cambridge, MA: MIT Press, 2969.

[MacLaurin, 1998] J. MacLaurin. Reinventing Molecular Weismannian: Information in Evolution. *Biology and Philosophy* **13**: 37-59, 1998.

[Matthen and Levy, 1984] M. Matthen and E. Levy. Teleology, error, and the human immune system. *The Journal of Philosophy* **81**, No. 7: 351-372, 1984.

[Maturana and Varela, 1980] H. R. Maturana and F. J. Varela. *Autopoiesis and Cognition.* Dordrecht: Reidel, 1980.

[Maynard Smith and Szathmáry, 1995] J. Maynard Smith and E. Szathmáry. *The Major Transitions in Evolution.* Oxford: W.H. Freeman Spektrum, 1995.

[Maynard Smith, 2000a] J. Maynard Smith. The Concept of Information in Biology. *Philosophy of Science* **67**: 177-194, 2000.

[Maynard Smith, 2000b] J. Maynard Smith. Reply to Commentaries. *Philosophy of Science* **67**: 214-218, 2000.

[Melander, 1993] P. Melander. How not to explain the errors of the immune system. *Philosophy of Science* **60**: 223-241, 1993.

[Millikan, 1987] R. G. Millikan. *Language, Thought, and Other Biological Categories: New Foundations for Realism.* Cambridge MA: MIT Press, 1987.

[Millikan, 1989] R. G. Millikan. In Defense of Proper Functions. *Philosophy of Science* **56**, 288-302, 1989.

[Neander, 1991] K. Neander. Functions as Selected Effects: The Conceptual Analyst's Defense. *Philosophy of Science*, **58**, 168-184, 1991.

[Odling-Smee *et al.*, 1996] F. J. Odling-Smee, K. N. Laland, and M. W. Feldman. Niche Construction. *Amercian Naturalist* **147**: 641-648, 1996.

[Odling-Smee *et al.*, 2003] F. J. Odling-Smee, K. N. Laland, and M. W. Feldman. *Niche Construction: The Neglected Process in Evolution*. Monographs in Population Biology. 37. Princeton University Press, 2003.

[Oyama, 2000] S. Oyama. *The Ontogeny of Information, 2nd Edition*. Durham, NC: University of North Carolina Press, 2000.

[Rissanen, 1989] J. Rissanen. *Stochastic Complexity in Statistical Inquiry*. Teaneck, NJ: World Scientific, 1989.

[Rosen, 1991] R. Rosen. *Life Itself*. New York: Columbia University Press, 1991.

[Salthe, 1985] S. N. Salthe. *Evolving Hierarchical Systems*. New York: Columbia University Press, 1985.

[Slathe, 1993] S. N. Salthe. *Development and Evolution: Complexity and Change in Biology*. Cambridge, MA: MIT Press, 1993.

[Sarkar, 2000] S. Sarkar. Information in Genetics and Developmental Biology: Comments on Maynard Smith. *Philosophy of Science*, **67**: 208-213, 2000.

[Schneider and Kay, 1994] E. Schneider and J. J. Kay. Life as a Manifestation of the Second Law of Thermodynamics. *Mathematical and Computer Modeling* **19**, No. 6-8, 25-48, 1994.

[Schneider, 1991a] T. D. Schneider. Theory of molecular machines. I. Channel capacity of molecular machines. *Journal of Theoretical Biology* **148**, 83–123, 1991.

[Schneider, 1991b] T. D. Schneider. Theory of molecular machines. II. Energy dissipation from molecular machines. *Journal of Theoretical Biology* **148**, 125–137, 1991.

[Schneider, 2000] T. D. Schneider. Evolution of Biological Information. *Nucleic Acids Research* **28**: 2794-2799, 2000.

[Schneider, 2006] T. D. Schneider. Twenty Years of Delila and Molecular Information Theory, *Biological Theory* **1**: 250-260, 2006.

[Schrödinger, 1944] I. Schrödinger. *What is Life?* Reprinted in *What is Life? And Mind and Matter*. Cambridge: Cambridge University Press, 1944.

[Shannon and Weaver, 1949] C. E. Shannon and W. Weaver. *The Mathematical Theory of Communication*. Urbana: University of Illinois Press, 1949.

[Ulanowicz, 1986] R. E. Ulanowicz. *Growth and Development: Ecosystems Phenomenology*. New York: Springer Verlag, 1986.

[Ulanowicz, 1997] R. E. Ulanowicz. *Ecology, the Ascendant Perspective*. New York: Columbia University Press, 1997.

[Wallace and Freeman, 1987] C. S. Wallace and P. R. Freeman. Estimation and Inference by Compact Coding. *Journal of the Royal Statistical Society, Series B, Methodology* **49**: 240-265, 1987.

[Weber *et al.*, 1988] B. Weber, D. J. Depew, and J. D. Smith, eds. *Information, Entropy and Evolution*. Cambridge, MA: MIT Press, 1988.

[Weber *et al.*, 1989] B. Weber, D. J. Depew, C. Dyke, S, N. Salthe, E. D. Schneider, R. E. Ulanowicz and J. S. Wicken. Evolution in thermodynamic perspective: an ecological approach. *Biology and Philosophy* **4**: 373-406, 1989.

[West-Eberhard, 1992] M. J. West-Eberhard. Adaptation: Current usages. In Keller, E. F. and E. A. Lloyd. Eds. 1992. *Keywords in Evolutionary Biology*. Cambridge, MA: Harvard University Press, 1992.

[Wicken, 1987] J. S. Wicken. *Evolution, Thermodynamics and Information: Extending the Darwinian Paradigm*. New York: Oxford University Press, 1987.

[Williams, 1966] G. C. Williams. *Adaptation and Natural Selection*. Princeton University Press, 1966.

[Wilson, 1988] E. O. Wilson. *On Human Nature*. Cambridge, MA: Harvard University Press, 1988.

[Winnie, 2000] J. A. Winnie. Information and Structure in Molecular Biology: Comments on Maynard Smith. *Philosophy of Science* **67**: 517-526, 2000.

[Wolfram, 2002] S. Wolfram. *A New Kind of Science*. Wolfram Media, 2002.

[Wright, 1973] L. Wright. Functions. *Philosophical Review* **82**: 139-168, 1973.

[Yagil, 1993a] G. Yagil. Complexity analysis of a protein molecule, in J. Demongeot, and V. Capesso (eds) *Mathematics Applied to Biology and Medicine.* Winnipeg: Wuerz Publishing: 305-313, 1993.

[Yagil, 1993b] G. Yagil. On the structural complexity of templated systems, in L. Nadel and D. Stein (eds) 1992 Lectures in Complex Systems. Reading, MA: Addison-Wesley, 1993.

[Yagil, 1995] G. Yagil. Complexity analysis of a self-oganizing vs. a template-directed system, in F. Moran, A. Moreno, J.J. Morleo, and P. Chacón (eds.) *Advances in Artificial Life.* New York: Springer: 179-187, 1995.

INDEX